ENGINEERING
MATHEMATICS
HANDBOOK

OTHER McGRAW-HILL BOOKS OF INTEREST

ENGINEERING MATHEMATICS HANDBOOK

Fourth Edition

Jan J. Tuma

Late Professor of Engineering
Arizona State University

Ronald A. Walsh

Manager: Research & Development
Powercon Corporation, Severn, Maryland

McGraw-Hill

New York San Francisco Washington, D.C. Auckland Bogota
Caracas Lisbon London Madrid Mexico City Milan
Montreal New Delhi San Juan Singapore
Sydney Tokyo Toronto

Library of Congress Cataloging-in-Publication Data

Tuma, Jan J.
   Engineering mathematics handbook / Jan J. Tuma, Ronald A.
Walsh—4th ed.
      p.   cm.
   Includes index.
   ISBN 0-07-065529-4
   1. Engineering mathematics—Handbooks, manuals, etc.
I. Walsh, Ronald A. II. Title.
TA332.T85   1997
515'.15—dc21   97-34589
                                        CIP

McGraw-Hill

A Division of The McGraw·Hill Companies

1 2 3 4 5 6 7 8 9 0 FGR/FGR 9 0 2 1 0 9 8 7

ISBN 0-07-065529-4

*The sponsoring editor for this book was Harold B. Crawford, the editing
supervisor was David E. Fogarty, and the production supervisor was Clare
Stanley. It was set in Baskerville by Bi-Comp, Incorporated.*

Printed and bound by Quebecor/Fairfield.

This book was printed on recycled, acid-free paper
containing a minimum of 50% recycled de-inked fiber.

Contents

Preface

Since the last edition of this handbook was published, there have been many new developments in the scientific and mathematics fields. Mathematics is often referred to as the queen of sciences, and for just cause. Answers to mathematics problems in geometry, trigonometry, calculus, and other branches are precise and exact and have allowed humankind to advance in numerous areas of scientific and engineering study. However, there are domains in which mathematics cannot give us precise answers.

In our modern world, the modern handheld calculators are now assuming some of the roles of high-speed computers in the solution to many difficult mathematics problems. Companies such as Texas Instruments and Hewlett-Packard have recently developed second- and third-generation graphing and equation-solving calculators that can solve an enormous array of difficult problems, once their operations are mastered. Texas Instruments has always used the direct algebraic entry system, which has made these units popular with many university students. Hewlett-Packard used the reverse Polish notation (RPN) system until recently; now its calculators can operate in RPN as well as direct algebraic entry.

Some models of these calculators have as many as 400 preprogrammed engineering equations for direct solution simply by manipulating the variables. These include equations from all branches of science and engineering. The TI-85 and HP-48G will allow you to enter a complex equation with many variables and to then change any variable or variables and have the calculator recalculate the answer to the problem.

In the new sections of this revised 4th edition, you will see examples of some applications of the basic mathematical equations shown throughout this handbook. Sections have been added to the trigonometry chapter and Appendix D. Mollweide and Freudenstein were not forgotten in this fourth edition, and their works are shown because of their extreme importance to mathematics, especially as used in mechanical as well as other branches of engineering. A sample of the use of Newton's method of solving intractable equations is also shown in Appendix D. We will look at electric field problems that were virtually impossible to solve prior to the introduction of the modern high-speed 50- to 200-mHz PCs, wherein 60,000 complex triangles are solved within a time span of only 1 to 1.5 h. This type of computational power is required to solve complex high-voltage electrostatic field problems such as corona discharge phenomena when electrical conductor/insulator systems are designed in the vicinity of a ground plane.

Included in this fourth edition is an extensive conversion table of U.S. Customary System to SI (Système International d'Unités) units. To add to the usefulness of the handbook, I have included a glossary of important mathematical terms in Appendix B. The boxed areas in the book outline important equations and subheadings, making them easier to read and locate. Typical problems are shown with some of the more frequently used equations in Appendix D along with a new section for complex financial uses and engineering economics calculations, with their popularly accepted equations. The simple percentage methods used in many branches of mathematics are also shown.

Some of the examples shown in Appendix D relate to the methods of programming some typical equations in proper form so that they will function on handheld calculators. If the incorrect hierarchy (series) of algebraic insertion, or RPN, is not adhered to, the calculator will give a false or error statement. Improper insertion of a complex equation into a calculator can cause much frustration if bracketing is done incorrectly. Included in this edition is a useful table of factors and prime numbers and an extensive U.S. Customary System–SI measurement conversion section together with proper unit designations. Also included are tables of basic units of measurement for scientific and engineering uses.

The author, editors, and publisher all agree that this edition of this important mathematics reference handbook will prove even more useful than the past editions.

Great credit is due to Prof. Jan J. Tuma for his years of extremely devoted work in compiling the previous editions of this very useful and complete mathematics reference work. The passing of Prof. Jan J. Tuma was a great loss to the scientific and engineering professions.

R. A. Walsh

About the Authors

Jan J. Tuma (deceased) was Professor of Engineering at Arizona State University. He was author of the *Handbook of Structural and Mechanical Matrices and Schaum's Outline of Structural Analysis*, both published by McGraw-Hill.

Ronald A. Walsh is Manager of Research and Development at Powercon Corporation in Severn, Maryland. Earlier, he was an aerospace designer for Martin-Marietta. Mr. Walsh is author of the *Electromechanical Design Handbook*, now in a second edition, and the *McGraw-Hill Machining and Metalworking Handbook*.

1

ALGEBRA

(1) General

Algebra is a systematic investigation of general numbers and their relationships. *General numbers* (real and complex) are letter symbols $(A, B, C, \ldots, a, b, c, \ldots, \alpha, \beta, \gamma, \ldots)$ representing *constant or variable quantities*. The study involves a finite number of *binary operations* (additions, multiplications), governed by algebraic laws and rules of signs.

(2) Algebraic Laws

Commutative law:	Associative law:
$a + b = b + a$	$a + (b + c) = (a + b) + c$
$ab = ba$	$a(bc) = (ab)c$
Distributive law:	**Division law:**
$a(b + c) = ab + ac$	If $ab = 0,$ then $a = 0$ or $b = 0$
$(a + b)c = ac + bc$	

(3) Rules of Signs

Summation	Multiplication	Division
$a + (+b) = a + b$	$(+a)(+b) = +ab$	$\dfrac{+a}{+b} = +\dfrac{a}{b}$
$a + (-b) = a - b$	$(+a)(-b) = -ab$	$\dfrac{+a}{-b} = -\dfrac{a}{b}$
$a - (+b) = a - b$	$(-a)(+b) = -ab$	$\dfrac{-a}{+b} = -\dfrac{a}{b}$
$a - (-b) = a + b$	$(-a)(-b) = +ab$	$\dfrac{-a}{-b} = +\dfrac{a}{b}$

(4) Powers

$a^k = \underbrace{aa \cdots a}_{k \text{ times}}$	$a^0 = 1$	$a^1 = a$

$$a^{-k} = \frac{1}{a^k} \qquad \frac{1}{a^{-k}} = a^k$$

$$a^m a^n = a^{m+n} \qquad (ab)^k = a^k b^k$$

$$\frac{a^m}{a^n} = a^{m-n} \qquad \left(\frac{a}{b}\right)^k = \frac{a^k}{b^k}$$

$$(a^m)^n = +a^{mn} \qquad (a^m)^{-n} = \frac{1}{a^{mn}}$$

$$(\pm a)^{2k} = +a^{2k} \qquad (\pm a)^{2k+1} = \pm a^{2k+1}$$

(5) Roots

$\sqrt[k]{a} = a^{1/k}$	$\sqrt[0]{a} = \infty*$	$\sqrt[1]{a} = a$

$$\sqrt[-k]{a} = \frac{1}{\sqrt[k]{a}} \qquad \frac{1}{\sqrt[-k]{a}} = \sqrt[k]{a}$$

$$\sqrt[m]{a}\sqrt[n]{a} = a^{(m+n)/mn} \qquad \sqrt[k]{ab} = \sqrt[k]{a}\sqrt[k]{b}$$

$$\frac{\sqrt[m]{a}}{\sqrt[n]{a}} = a^{(n-m)/mn} \qquad \sqrt[k]{a/b} = \frac{\sqrt[k]{a}}{\sqrt[k]{b}}$$

$$\sqrt[n]{a^m} = a^{m/n} \qquad \sqrt[-n]{a^m} = \frac{1}{a^{m/n}}$$

$$\sqrt[2k]{+a^{2k}} = \pm a \qquad \sqrt[2k+1]{+a^{2k+1}} = \pm a$$

*If $a > 1$.

(1) Definitions

If $a^x = b$, then x is the logarithm of b to the base a. Logarithms to the *base 10* are called *common,* or *Brigg's,* logarithms and denoted as log x. Logarithms to the *base e* = 2.718 281 828 are called *natural,* or *Napier's,* logarithms and denoted as ln x.

(2) Basic Formulas

$10^\infty = \infty$	$\log \infty = \infty$	$\ln \infty = \infty$	$e^\infty = \infty$
$10^2 = 100$	$\log 100 = 2$	$\ln e^2 = 2$	$e^2 = e^2$
$10^1 = 10$	$\log 10 = 1$	$\ln e = 1$	$e^1 = e$

$10^0 = 1$	$\log 1 = 0$	$\ln 1 = 0$	$e^0 = 1$

$10^{-1} = \frac{1}{10}$	$\log \frac{1}{10} = -1$	$\ln \frac{1}{e} = -1$	$e^{-1} = \frac{1}{e}$
$10^{-2} = \frac{1}{100}$	$\log \frac{1}{100} = -2$	$\ln \frac{1}{e^2} = -2$	$e^{-2} = \frac{1}{e^2}$
$10^{-\infty} = 0$	$\log 0 = -\infty$	$\ln 0 = -\infty$	$e^{-\infty} = 0$

(3) Transformations

$$\log x = \frac{\ln x}{\ln 10} = 0.434\,294\,482 \ldots \ln x$$

$$\ln x = \frac{\log x}{\log e} = 2.302\,585\,093 \ldots \log x$$

(4) Basic Operations

$$\log xy = \log x + \log y$$
$$\log \frac{x}{y} = \log x - \log y$$
$$\log x^k = k \log x$$
$$\log \sqrt[k]{x} = \frac{1}{k} \log x$$
$$\log 10^k = k$$
$$\log e = 0.434\,29 \ldots$$

$$\ln xy = \ln x + \ln y$$
$$\ln \frac{x}{y} = \ln x - \ln y$$
$$\ln x^k = k \ln x$$
$$\ln \sqrt[k]{x} = \frac{1}{k} \ln x$$
$$\ln e^k = k$$
$$\ln 10 = 2.302\,59 \ldots$$

(5) Characteristic of Common Logarithm

A common logarithm consists of an integer, called the *characteristic,* and a decimal, called the *mantissa.* The mantissa is a tabulated value. The characteristic is determined from the following:

$\log 100 = 2.00$	$\log 0.1 = 0.00 - 1 = 9.00 - 10$
$\log 10 = 1.00$	$\log 0.01 = 0.00 - 2 = 8.00 - 10$
$\log 1 = 0.00$	$\log 0.001 = 0.00 - 3 = 7.00 - 10$

$k, n, p = 0, 1, 2, \ldots$ $\qquad$ $\Gamma(\ ) = $ gamma function (Sec. 13.03)

$\overline{B}_k = $ Bernoulli number (Sec. A.03) $\qquad$ $\mathscr{S}_k^{(p)} = $ Stirling number (Sec. A.05)

(1) *n* Factorial $\quad$ (Sec. A.01)

$$n! = n(n-1)(n-2) \cdots (3)(2)(1) = \Gamma(n+1)$$
$$= n(n-1)! \qquad\qquad\qquad = n\Gamma(n)$$
$$= n(n-1)(n-2)! \qquad\qquad = n(n-1)\Gamma(n-1)$$

$$0! = 1 \qquad \text{(by definition)}$$

$$1! = 1, \quad 2! = 2 \cdot 1 = 2, \quad 3! = 3 \cdot 2 \cdot 1 = 6, \ldots$$

(2) Coefficients of τ

$2k-1$	a_{2k-1}
1	+0.422 784 335
3	−0.067 352 301
5	−0.007 385 551
7	−0.001 192 754
9	−0.000 223 155
11	−0.000 044 926
13	−0.000 009 439

(3) *u* Factorial $\quad (-0.5 \le u \le 0.5)\,*$

$$u! = \Gamma(u+1) = e^{\tau}\sqrt{\frac{(1-u)\,u\pi}{(1+u)\sin u\pi}} + \epsilon \qquad |\epsilon| < 5 \times 10^{-10}$$

$$\tau = \sum_{k=1}^{7} a_{2k-1} u^{2k-1} \qquad [\text{for } a_{2k-1} \text{ see (2)}]$$

(4) *v* Factorial $\quad (1 < v \le 4)$

$$v! = (n+u)! = (n+u)(n-1+u) \cdots (1+u)u!$$
$$[\text{for } u! \text{ use (3) or (5)}]$$

(5) Special Values $\quad (|u| < 1)$

x	$x!$
$-u$	$\dfrac{\pi u}{u! \sin \pi u}$
-0.5	$\sqrt{\pi}$
$+0.5$	$0.5\sqrt{\pi}$
$+u$	$\dfrac{\pi u(1-u)}{(1-u)! \sin \pi u}$

(6) *N* Factorial $\quad (4 < N < \infty)\,\dagger$

$$N! = \Gamma(N+1) = e^{\lambda}\left(\frac{N}{e}\right)^N \sqrt{2\pi N} + \epsilon \qquad |\epsilon| < 5 \times 10^{-10}$$

$$\lambda = \frac{\overline{B}_2}{1 \cdot 2}N^{-1} + \frac{\overline{B}_4}{3 \cdot 4}N^{-3} + \frac{\overline{B}_6}{5 \cdot 6}N^{-5} + \frac{\overline{B}_8}{7 \cdot 8}N^{-7} = \frac{N^{-1}}{12} - \frac{N^{-3}}{360} + \frac{N^{-5}}{1,260} - \frac{N^{-7}}{1,680}$$

(7) Double Factorials $\quad$ (Sec. A.01)

$$(2n)!! = 2n(2n-2)(2n-4) \cdots 6 \cdot 4 \cdot 2 = 2^n n! = 2^n \Gamma(n+1) \qquad\qquad 0!! = 1$$

$$(2n-1)!! = (2n-1)(2n-3)(2n-5) \cdots 5 \cdot 3 \cdot 1 = 2^n \Gamma(n + \tfrac{1}{2})/\sqrt{\pi} \qquad 1!! = 1$$

(8) Factorial Polynomials $\quad (x, h = \text{real numbers}, p \ne 1)$

$$X_1^{(p)} = \prod_{k=0}^{p-1}(x-k) = x(x-1)(x-2) \cdots (x-p+1) = \frac{\Gamma(x+1)}{\Gamma(x-p+1)}$$

$$= \sum_{k=1}^{p} \mathscr{S}_k^{(p)} x^k = \mathscr{S}_1^{(p)}x + \mathscr{S}_2^{(p)}x^2 + \mathscr{S}_3^{(p)}x^3 + \cdots + \mathscr{S}_p^{(p)}x^p$$

$\qquad P_1^{(p)} = p!$

$$X_h^{(p)} = \prod_{k=0}^{p-1}(x-kh) = x(x-h)(x-2h) \cdots (x-ph+h) = \frac{h^p \Gamma\left(\frac{x}{h}+1\right)}{\Gamma\left(\frac{x}{h}-p+1\right)}$$

$$= \sum_{k=1}^{p} \mathscr{S}_k^{(p)} h^{p-k} x^k = \mathscr{S}_1^{(p)}h^{p-1}x + \mathscr{S}_2^{(p)}h^{p-2}x^2 + \mathscr{S}_3^{(p)}h^{p-3}x^3 + \cdots + \mathscr{S}_p^{(p)}x^p$$

$\qquad P_2^{(p)} = p!!$

*For $-1 < u < -\frac{1}{2}$ and $\frac{1}{2} < u < 1$ use (3) and (5); see also Sec. A.02.

$\dagger$In nested form: $\lambda = \dfrac{NS}{12}\left(1 - \dfrac{S}{30}\left(1 - \dfrac{2S}{7}\left(1 - \dfrac{3S}{4}\right)\right)\right)$ with $S = N^{-2}$ used as pocket calculator storage.

$$k, n, r = 0, 1, 2, \ldots \qquad a, b = \text{real numbers}$$
$$n! = n \text{ factorial (Sec. 1.03)} \qquad \Lambda\,[\,] = \text{nested sum (Sec. 8.11)}$$

(1) Binomial Theorem $\qquad (b/a = S = \text{pocket calculator storage})$

$$(a+b)^n = a^n + \binom{n}{1}a^{n-1}b + \binom{n}{2}a^{n-2}b^2 + \binom{n}{3}a^{n-3}b^3 + \cdots + b^n = \sum_{k=0}^{n}\binom{n}{k}a^{n-k}b^k$$
$$= a^n\left(1 + \frac{n}{1}S\left(1 + \frac{n-1}{2}S\left(1 + \frac{n-2}{3}S\left(1 + \cdots + \frac{1}{n}S\right)\right)\right)\right) = a^n \overset{n-1}{\underset{k=0}{\Lambda}}\left[1 + \frac{n-k}{1+k}S\right]$$

(2) Binomial Coefficients $\qquad$ (Sec. 18.09)

$$\binom{n}{k} = \frac{n!}{(n-k)!k!} \qquad\qquad \binom{-n}{k} = (-1)^k\binom{n+k-1}{k}$$
$$= \frac{n(n-1)(n-2)\cdots(n-k+1)}{k!} \qquad\qquad = \frac{(-n)(-n-1)\cdots(-n-k+1)}{k!}$$

$\binom{n}{0} = 1$	$\binom{n}{1} = n$	$\binom{n}{k} = \binom{n}{n-k}$	$\binom{2n}{n} = \frac{(2n)!}{(n!)^2} = (-1)^n\binom{-n-1}{n}$
$\binom{n}{n} = 1$	$\binom{n}{n-1} = n$	$\binom{n}{k+1} = \frac{n-k}{k+1}\binom{n}{k}$	$\binom{2n-1}{n} = \frac{n(2n-1)!}{(n!)^2} = (-1)^n\binom{-n}{n}$
$\dfrac{\binom{n}{k}}{\binom{n}{k-1}} = \frac{n+1-k}{k}$	$\dfrac{\binom{n}{k+1}}{\binom{n}{k}} = \frac{n-k}{k+1}$	$\binom{n+1}{k} = \binom{n}{k} + \binom{n}{k-1}$	$\binom{n+1}{k+1} = \binom{n}{k} + \binom{n}{k+1}$

(3) Pascal's Triangle of $\binom{n}{k}$

$(a+b)^0 \to \qquad 1$

$(a+b)^1 \to \qquad 1 \quad 1$

$(a+b)^2 \to \qquad 1 \quad 2 \quad 1$

$(a+b)^3 \to \qquad 1 \quad 3 \quad 3 \quad 1$

$(a+b)^4 \to 1 \quad 4 \quad 6 \quad 4 \quad 1$

(4) Binomial Sums $\qquad$ (Sec. 8.03)

$$\binom{n}{k} + \binom{n-1}{k} + \binom{n-2}{k} + \cdots + \binom{k}{k} = \binom{n+1}{k+1}$$
$$\binom{n}{0} + \binom{n}{1} + \binom{n}{2} + \cdots + \binom{n}{n} = 2^n$$
$$\binom{n}{0} - \binom{n}{1} + \binom{n}{2} - \cdots + (-1)^n\binom{n}{n} = 0$$

(5) Sums of Incomplete Binomial Sequence $\qquad (r \leq n)$

$$\sum_{k=0}^{r}\binom{n}{k} = \overset{r-1}{\underset{k=0}{\Lambda}}\left[1 + \frac{n-k}{1+k}\right] = \left(1 + n\left(1 + \frac{n-1}{2}\left(1 + \frac{n-2}{3}\left(1 + \cdots + \frac{n+1-r}{r}\right)\right)\right)\right)$$

$$\sum_{k=0}^{r}(-1)^k\binom{n}{k} = \overset{r-1}{\underset{k=0}{\Lambda}}\left[1 - \frac{n-k}{1+k}\right] = \left(1 - n\left(1 - \frac{n-1}{2}\left(1 - \frac{n-2}{3}\left(1 - \cdots - \frac{n+1-r}{r}\right)\right)\right)\right)$$

(6) Special Cases $\qquad$ (Sec. 8.12)

$$(1 \pm x)^n = 1 \pm \binom{n}{1}x + \binom{n}{2}x^2 \pm \cdots \qquad \text{If } n = \text{positive integer, the series is finite.}$$

$$\left(1 \pm \frac{1}{n}\right)^n = 1 \pm \binom{n}{1}n^{-1} + \binom{n}{2}n^{-2} \pm \cdots \qquad \text{If } n = \text{negative integer or fraction, the series is}$$
$$\text{infinite.}$$

$$\text{if } n \to \infty, \text{ then } \left(1 + \frac{1}{n}\right)^n = 1 + \frac{1}{1!} + \frac{1}{2!} + \frac{1}{3!} + \cdots = e = 2.718\,281\,828 = \text{Euler's number.}$$

(1) Integral Function

(a) Every integral rational algebraic function of independent variable x,

$$f(x) = a_0 + a_1 x + a_2 x^2 + \cdots + a_n x^n = \sum_{k=0}^{n} a_k x^k$$

can be expressed as a product of real linear factors of the form $c + dx$ and of real irreducible quadratic factors of the form $e + fx + gx^2$.

(b) If two of these functions of the same degree are equal for all values of x, the constant a's of like powers are equal.

$$\sum_{k=0}^{n} a_k x^k = \sum_{k=0}^{n} b_k x^k \qquad\qquad a_0 = b_0, a_1 = b_1, \ldots, a_n = b_n$$

(2) Fractional Function

(a) A rational algebraic function of independent variable x, where $g(x)$ and $h(x)$ are integral rational algebraic functions, is called a *rational algebraic fraction*.

$$f(x) = \frac{g(x)}{h(x)}$$

(b) If the degree of $g(x)$ is less than the degree of $h(x)$, $f(x)$ is called *proper*. If the opposite is true, $f(x)$ is called *improper*.

(c) Every proper, rational algebraic fraction can be resolved into a sum of simpler fractions, whose denominators are of the form $(c + dx)^k$ and $(e + fx + gx^2)^l$, k and l being positive integers.

$$\frac{g(x)}{h(x)} = \sum_{k} \left[\frac{b_{k1}}{(x - x_k)} + \frac{b_{k2}}{(x - x_k)^2} + \cdots + \frac{b_{km}}{(x - x_k)^m} \right]$$

(3) Coefficients

The coefficients b_{kj} are obtained by one of the following methods:

(a) If $m = 1$ (x_k = distinct root), then

$$b_{kj} = \frac{g(x_k)}{f'(x_k)}$$

in which $f'(x_k)$ is the first derivative of $f(x)$ with respect to x evaluated for $x = x_k$.

(b) Multiply both sides of $h(x)$, and use the theorem of Sec. **1.05-1b**.

(c) Multiply both sides by $h(x)$, and differentiate successively. Solve this set of equations for $b_{km}, b_{km-1}, \ldots, b_{k1}$.

The partial fractions corresponding to any pair of complex-conjugate roots $a_k + i\alpha_k$, $a_k - i\alpha_k$ of order m may be combined into

$$c_{kj} \frac{x + d_{kj}}{[(x - a_k)^2 + \alpha_k^2]^j}$$

(1) Quadratic and Biquadratic Equations[1]

$ax^2 + bx + c = 0$	$a \neq 0$

$$x_{1,2} = \frac{-b \pm \sqrt{b^2 - 4ac}}{2a}$$

If a, b, c are real and if

$b^2 - 4ac > 0$ the roots are real and unequal;

$b^2 - 4ac = 0$ the roots are real and equal;

$b^2 - 4ac < 0$ the roots are complex conjugate.

$ax^4 + bx^2 + c = 0$	$a \neq 0$

This reduces by the substitution

$$y = x^2 \qquad \text{to} \qquad y^2 + py + q = 0$$

in which

$$p = b/a \qquad q = c/a$$

$$x_{1,2,3,4} = \pm \sqrt{-\frac{p}{2} \pm \sqrt{\frac{p^2}{4} - q}}$$

(2) Binomial Equations[1]

$x^n - a = 0$	$a \neq 0$

$$x_{1,2,3,\dots,n} = \sqrt[n]{a}\left(\cos\frac{2k\pi}{n} + i\sin\frac{2k\pi}{n}\right)$$

$$\sqrt[n]{i} = \cos\frac{(4k+1)\,\pi}{2n} + i\sin\frac{(4k+1)\,\pi}{2n}$$

$$k = 0, 1, 2, \dots, n-1$$

$x^n + a = 0$	$a \neq 0$

$$x_{1,2,3,\dots,n} = \sqrt[n]{a}\left[\cos\frac{(2k+1)\,\pi}{n} + i\sin\frac{(2k+1)\,\pi}{n}\right]$$

$$\sqrt[n]{-i} = \cos\frac{(4k-1)\,\pi}{2n} + i\sin\frac{(4k-1)\,\pi}{2n}$$

$$k = 0, 1, 2, \dots, n-1$$

(3) Cubic Equations[1]

$ax^3 + bx^2 + cx + d = 0$	$a \neq 0$

This reduces by the substitution

$$x = y - \frac{b}{3a} \qquad \text{to} \qquad y^3 + py + q = 0$$

where

$$p = \frac{1}{3}\left[3\left(\frac{c}{a}\right) - \left(\frac{b}{a}\right)^2\right]$$

$$q = \frac{1}{27}\left[2\left(\frac{b}{a}\right)^3 - 9\left(\frac{b}{a}\right)\left(\frac{c}{a}\right) + 27\left(\frac{d}{a}\right)\right]$$

$$y_1 = u + v \qquad\qquad y_2 = -\frac{u+v}{2} + \frac{u-v}{2}i\sqrt{3}$$

$$y_3 = -\frac{u+v}{2} - \frac{u-v}{2}i\sqrt{3}$$

If $D < 0$, a trigonometric formulation is useful.

$$y_1 = 2\sqrt{\frac{|p|}{3}}\cos\frac{\phi}{3}$$

$$y_2 = -2\sqrt{\frac{|p|}{3}}\cos\frac{\phi+\pi}{3}$$

$$y_3 = -2\sqrt{\frac{|p|}{3}}\cos\frac{\phi-\pi}{3}$$

$$D = \left(\frac{p}{3}\right)^3 + \left(\frac{q}{2}\right)^2$$

$$u = \sqrt[3]{-\tfrac{1}{2}q + \sqrt{D}} = -\sqrt[3]{\tfrac{1}{2}q - \sqrt{D}}$$

$$v = \sqrt[3]{-\tfrac{1}{2}q - \sqrt{D}} = -\sqrt[3]{\tfrac{1}{2}q + \sqrt{D}}$$

If a, b, c, d are real and if

$D > 0$ there are one real and two conjugate complex roots;

$D = 0$ there are three real roots of which at least two are equal;

$D < 0$ there are three real unequal roots.

The value of ϕ is calculated from the expression

$$\phi = \cos^{-1}\frac{-q/2}{\sqrt{|p|^3/27}}$$

[1]For $i = \sqrt{-1}$ refer to Sec. 11.01.

(1) Definition

(a) A determinant of the nth order contains $n \times n$ elements, arranged in n rows and n columns.

$$D = \begin{vmatrix} a_{11} & a_{12} \cdots a_{1n} \\ a_{21} & a_{22} \cdots a_{2n} \\ \cdots\cdots\cdots\cdots \\ a_{n1} & a_{n2} \cdots a_{nn} \end{vmatrix}$$

(b) A minor D_{jk} of the element a_{jk} in the nth-order determinant D is the $(n-1)$st-order determinant obtained from D by deleting the jth row and the kth column.

(2) Evaluation

(a) The cofactor

$$A_{jk} = (-1)^{j+k} D_{jk}$$

(b) The evaluation of the determinant D is accomplished by summing the products of elements of any row or any column into their respective cofactors.

$$D = \sum_{i=1}^{n} a_{ij} A_{ij} = \sum_{j=1}^{n} a_{jk} A_{jk} = \sum_{k=1}^{n} a_{kl} A_{kl} = \cdots$$

(3) Special Cases

(a) Second-order determinant

$$\begin{vmatrix} a_{11} & a_{12} \\ a_{21} & a_{22} \end{vmatrix} = a_{11} a_{22} - a_{21} a_{12}$$

(b) Third-order determinant

$$\begin{vmatrix} a_{11} & a_{12} & a_{13} \\ a_{21} & a_{22} & a_{23} \\ a_{31} & a_{32} & a_{33} \end{vmatrix} = a_{11} A_{11} + a_{21} A_{21} + a_{31} A_{31}$$

where

$$A_{11} = \begin{vmatrix} a_{22} & a_{23} \\ a_{32} & a_{33} \end{vmatrix} \qquad A_{21} = -\begin{vmatrix} a_{12} & a_{13} \\ a_{32} & a_{33} \end{vmatrix} \qquad A_{31} = \begin{vmatrix} a_{12} & a_{13} \\ a_{22} & a_{23} \end{vmatrix}$$

(4) Basic Operations

(a) Equal determinants

Two determinants $|A|$ and $|B|$ are equal if they have the same dimensions and their corresponding elements are equal.

$$a_{jk} = b_{jk}$$

(b) Transpose

The value of a determinant is unchanged if the corresponding rows and columns are interchanged (transpose of determinant).

$$|D| = |D|^T$$

(1) Transformations Inducing no Change

(a) The *value* of a determinant is *unchanged* if the corresponding rows and columns are interchanged.

$$\begin{vmatrix} a_{11} & a_{12} & a_{13} \\ a_{21} & a_{22} & a_{23} \\ a_{31} & a_{32} & a_{33} \end{vmatrix} = \begin{vmatrix} a_{11} & a_{21} & a_{31} \\ a_{12} & a_{22} & a_{32} \\ a_{13} & a_{23} & a_{33} \end{vmatrix} = D$$

(b) If to each element of a row (or column) is *added m times* the corresponding element in another row (or column), the value of the determinant is unchanged.

$$\begin{vmatrix} a_{11} & a_{12} & a_{13} \\ a_{21} & a_{22} & a_{23} \\ a_{31} & a_{32} & a_{33} \end{vmatrix} = \begin{vmatrix} a_{11} & a_{12} & a_{13} \\ a_{21}+ma_{11} & a_{22}+ma_{12} & a_{23}+ma_{13} \\ a_{31} & a_{32} & a_{33} \end{vmatrix}$$

(c) If *two determinants differ* from each other only *in the elements of any one row* (or column), they may be added as follows:

$$\begin{vmatrix} a_{11} & a_{12} & a_{13} \\ a_{21} & a_{22} & a_{23} \\ a_{31} & a_{32} & a_{33} \end{vmatrix} + \begin{vmatrix} b_{11} & b_{12} & b_{13} \\ a_{21} & a_{22} & a_{23} \\ a_{31} & a_{32} & a_{33} \end{vmatrix} = \begin{vmatrix} a_{11}+b_{11} & a_{12}+b_{12} & a_{13}+b_{13} \\ a_{21} & a_{22} & a_{23} \\ a_{31} & a_{32} & a_{33} \end{vmatrix}$$

(2) Transformations Inducing Change

(a) The *sign* of a determinant is *changed* (unchanged) if an odd (even) number of interchanges of any two rows or of any two columns is introduced.

(b) If each element of a row (or column) is multiplied by m, the *new determinant is equal to mD*.

$$\begin{vmatrix} a_{11} & a_{12} & a_{13} \\ ma_{21} & ma_{22} & ma_{23} \\ a_{31} & a_{32} & a_{33} \end{vmatrix} = m\begin{vmatrix} a_{11} & a_{12} & a_{13} \\ a_{21} & a_{22} & a_{23} \\ a_{31} & a_{32} & a_{33} \end{vmatrix} = mD$$

(3) Zero Determinant

(a) If a determinant has *two identical rows* (or columns) or if all the elements of one row (or column) are zero, then the value of the determinant is zero.

(b) If the elements of any row (or column) are *linear combinations* of the corresponding elements of the other rows (or columns), the value of the determinant is zero.

$$\begin{vmatrix} ba_{21}+ca_{31} & ba_{22}+ca_{32} & ba_{23}+ca_{33} \\ a_{21} & a_{22} & a_{23} \\ a_{31} & a_{32} & a_{33} \end{vmatrix} = 0$$

(1) Definition

A *matrix* is a rectangular *array of elements* arranged in rows and columns.

$$[A] = \begin{bmatrix} a_{11} & a_{12} & \cdots & a_{1n} \\ a_{21} & a_{22} & \cdots & a_{2n} \\ \cdots & \cdots & \cdots & \cdots \\ a_{m1} & a_{m2} & \cdots & a_{mn} \end{bmatrix} = [a_{jk}]$$

m = number of rows
n = number of columns
a_{jk} = any element
$m \times n$ = dimension of matrix

(2) Shapes

Rectangular matrix	$m \neq n$	Column matrix	$n = 1$
Square matrix	$m = n$	Row matrix	$m = 1$

(3) Basic Types

(a) Unit matrix

$$\begin{bmatrix} 1 & 0 & 0 \\ 0 & 1 & 0 \\ 0 & 0 & 1 \end{bmatrix}$$

(b) Diagonal matrix

$$\begin{bmatrix} a & 0 & 0 \\ 0 & b & 0 \\ 0 & 0 & c \end{bmatrix}$$

(c) Zero matrix

$$\begin{bmatrix} 0 & 0 & 0 \\ 0 & 0 & 0 \\ 0 & 0 & 0 \end{bmatrix}$$

(d) Symmetrical matrix

$$\begin{bmatrix} a & d & l \\ d & b & f \\ l & f & c \end{bmatrix}$$

(e) Antisymmetrical matrix

$$\begin{bmatrix} 0 & +d & -l \\ -d & 0 & +f \\ +l & -f & 0 \end{bmatrix}$$

(f) Point symmetrical matrix

$$\begin{bmatrix} a & d & c \\ d & b & d \\ c & d & a \end{bmatrix}$$

(4) Basic Operations

(a) Equal matrices

Two matrices $[A]$ and $[B]$ are equal if they have the same dimensions and their corresponding elements are equal.

$$a_{ik} = b_{ik}$$

(b) Sum of matrices

The sum of two or several $m \times n$ matrices is an $m \times n$ matrix, each of whose elements is equal to the sum of the corresponding elements of the initial matrices.

$$a_{ik} + b_{ik} + c_{ik} + \cdots = s_{ik}$$

(c) Scalar-matrix multiplication

The product of a scalar k and an $m \times n$ matrix $[A]$ is an $m \times n$ matrix, each of whose elements is equal to the product of the scalar and the corresponding element of $[A]$.

$$k[A] = [kA]$$

(d) Matrix-matrix multiplication

A product of two rectangular conformable matrices of dimensions $m_1 \times n_1$ and $m_2 \times n_2$ is a rectangular (or square) matrix of dimensions $m_1 \times n_2$ whose elements are equal to the sum of products of the inner elements.

$$\begin{bmatrix} a_{11} & a_{12} & a_{13} \\ a_{21} & a_{22} & a_{23} \end{bmatrix} \begin{bmatrix} b_{11} & b_{12} \\ b_{21} & b_{22} \\ b_{31} & b_{32} \end{bmatrix} = \begin{bmatrix} a_{11}b_{11} + a_{12}b_{21} + a_{13}b_{31} & a_{11}b_{12} + a_{12}b_{22} + a_{13}b_{32} \\ a_{21}b_{11} + a_{22}b_{21} + a_{23}b_{31} & a_{21}b_{12} + a_{22}b_{22} + a_{23}b_{32} \end{bmatrix}$$

Two rectangular matrices are conformable if $n_1 = m_2$.

(1) Definition

The transpose $[A]^T$ of a matrix $[A]$ has each row identical with the corresponding column of $[A]$.

$$[A] = \begin{bmatrix} a \\ b \end{bmatrix} \qquad\qquad [B] = \begin{bmatrix} a & b \\ c & d \end{bmatrix} \qquad\qquad [C] = [a \quad b]$$

$$[A]^T = [a \quad b] \qquad\qquad [B]^T = \begin{bmatrix} a & c \\ b & d \end{bmatrix} \qquad\qquad [C]^T = \begin{bmatrix} a \\ b \end{bmatrix}$$

(2) Special Cases

Transpose of transpose

$$[[A]^T]^T = [A]$$

Transpose of unit matrix

$$[I]^T = [I]$$

Transpose of zero matrix

$$[0]^T = [0]$$

Transpose of diagonal matrix

$$\begin{bmatrix} a & 0 \\ 0 & d \end{bmatrix}^T = \begin{bmatrix} a & 0 \\ 0 & d \end{bmatrix}$$

$$[D]^T = [D]$$

Transpose of symmetrical matrix

$$\begin{bmatrix} a & b \\ b & c \end{bmatrix}^T = \begin{bmatrix} a & b \\ b & c \end{bmatrix}$$

$$[E]^T = [E]$$

Transpose of anti-symmetrical matrix

$$\begin{bmatrix} 0 & b \\ -b & 0 \end{bmatrix}^T = \begin{bmatrix} 0 & -b \\ b & 0 \end{bmatrix}$$

$$[F]^T = -[F]$$

(3) Basic Operations

(a) Transpose of product

The transpose of product of two or more matrices is equal to the product of their transposes in reverse order.

$$\big[[A][B][C][D]\big]^T = [D]^T[C]^T[B]^T[A]^T$$

(b) Matrix-transpose sum and difference

The sum of a square matrix and its transpose is a symmetrical matrix. The difference of a square matrix and its transpose is an antisymmetrical matrix.

$$\underbrace{\begin{bmatrix} a & b \\ c & d \end{bmatrix}}_{[B]} + \underbrace{\begin{bmatrix} a & c \\ b & d \end{bmatrix}}_{[B]^T} = \underbrace{\begin{bmatrix} 2a & b+c \\ c+b & 2d \end{bmatrix}}_{\text{Symmetrical}} \qquad \underbrace{\begin{bmatrix} a & b \\ c & d \end{bmatrix}}_{[B]} - \underbrace{\begin{bmatrix} a & c \\ b & d \end{bmatrix}}_{[B]^T} = \underbrace{\begin{bmatrix} 0 & b-c \\ c-b & 0 \end{bmatrix}}_{\text{Antisymmetrical}}$$

(c) Matrix-transpose product

The product of a square matrix and its transpose, or vice versa, is a symmetrical matrix.

$$\underbrace{\begin{bmatrix} a & b \\ c & d \end{bmatrix}}_{[B]} \underbrace{\begin{bmatrix} a & c \\ b & d \end{bmatrix}}_{[B]^T} = \underbrace{\begin{bmatrix} a^2+b^2 & ac+bd \\ ac+bd & c^2+d^2 \end{bmatrix}}_{\text{Symmetrical}} \qquad \underbrace{\begin{bmatrix} a & c \\ b & d \end{bmatrix}}_{[B]^T} \underbrace{\begin{bmatrix} a & b \\ c & d \end{bmatrix}}_{[B]} = \underbrace{\begin{bmatrix} a^2+c^2 & ab+cd \\ ab+cd & b^2+d^2 \end{bmatrix}}_{\text{Symmetrical}}$$

(d) Matrix resolution

Every unsymmetrical square matrix can be expressed as the sum of a symmetrical matrix and an antisymmetrical matrix.

(1) Definition

The inverse $[A]^{-1}$ of a square matrix $[A]$ is uniquely defined by the conditions

$$[A]^{-1}[A] = [I] = [A][A]^{-1} \quad \text{and} \quad |A| \neq 0$$

If $|A| = 0$, the inverse of $[A]$ does not exist, and $[A]$ is said to be a *singular matrix*. Only nonsingular square matrices have inverses.

$$[A]^{-1} = \begin{bmatrix} a_{11} & a_{12} & \cdots & a_{1n} \\ a_{21} & a_{22} & \cdots & a_{2n} \\ \cdots\cdots\cdots\cdots \\ a_{n1} & a_{n2} & \cdots & a_{nn} \end{bmatrix}^{-1} = \frac{1}{|A|} \begin{bmatrix} A_{11} & A_{21} & \cdots & A_{n1} \\ A_{12} & A_{22} & \cdots & A_{n2} \\ \cdots\cdots\cdots\cdots \\ A_{1n} & A_{2n} & \cdots & A_{nn} \end{bmatrix} = \frac{[A_{jk}]}{|A|}$$

in which $|A|$ = determinant of $[A]$, A_{jk} = cofactor jk of $|A|$, and $[A_{jk}]$ = adjoint matrix of cofactors of $|A|$.

(2) Special Cases

Inverse of inverse	**Inverse of unit matrix**	**Inverse of zero matrix**
$[[A]^{-1}]^{-1} = [A]$	$[I]^{-1} = [I]$	$[0]^{-1} = [0]$
Initial matrix	Unit matrix	Zero matrix
Inverse of diagonal matrix	**Inverse of symmetrical matrix**	**Inverse of antisymmetrical matrix**
$\begin{bmatrix} a & 0 \\ 0 & d \end{bmatrix}^{-1} = \begin{bmatrix} \frac{1}{a} & 0 \\ 0 & \frac{1}{d} \end{bmatrix}$	$\begin{bmatrix} a & b \\ b & c \end{bmatrix}^{-1} = \begin{bmatrix} e & f \\ f & g \end{bmatrix}$	$\begin{bmatrix} 0 & b \\ -b & 0 \end{bmatrix}^{-1} = \begin{bmatrix} 0 & h \\ -h & 0 \end{bmatrix}$
Reciprocal diagonal matrix	Symmetrical matrix	Antisymmetrical matrix

(3) Basic Operations

(a) Inverse of product

The inverse of the product of two or more matrices is equal to the product of their inverses in reverse order.

$$[[A][B][C][D]]^{-1} = [D]^{-1}[C]^{-1}[B]^{-1}[A]^{-1}$$

(b) Normal matrix

A square matrix is said to be normal if it is equal to its transpose. All symmetrical matrices are normal.

$$\text{If } [A] = [A]^T \quad \text{then} \quad [A][A]^T = [A]^T[A] = [A]^2$$

(c) Orthogonal matrix

A square matrix is said to be orthogonal if its transpose is equal to its inverse.

$$\text{If } [A]^T = [A]^{-1} \quad \text{then} \quad [A][A]^T = [A]^T[A] = [I]$$

(1) Basic Laws — Matrices

Commutative law:	**Associative law:**
$A + B = B + A$	$A + (B + C) = (A + B) + C$
$[A + B]^T = [B + A]^T$	$[A + (B + C)]^T = [(A + B) + C]^T$
$[A + B]^{-1} = [B + A]^{-1}$	$[A + (B + C)]^{-1} = [(A + B) + C]^{-1}$
$AB \neq BA$	$A(BC) = (AB)C$
$[AB]^T \neq [BA]^T$	$[A(BC)]^T = [(AB)C]^T$
$[AB]^{-1} \neq [BA]^{-1}$	$[A(BC)]^{-1} = [(AB)C]^{-1}$
Distributive law:	**Division law:**
$A[B + C] = AB + AC$	If $AB = 0$, then A and/or B may or may not
$[A + B]C = AC + BC$	be zero.

(2) Basic Laws—Determinant Matrices $n \times n$ (k = scalar)

$\text{Det}(A^T) = \text{Det}(A)$		$\text{Det}(AB) = \text{Det}(A)\,\text{Det}(B)$
$\text{Det}(A^{-1}) = \dfrac{1}{\text{Det}(A)}$	$\text{Det}(I^{-1}) = 1$	$\text{Det}(AB^{-1}) = \dfrac{\text{Det}(A)}{\text{Det}(B)}$
$\text{Det}(kA) = k^n \, \text{Det}(A)$		$\text{Det}(A^{-1}B^{-1}) = \dfrac{1}{\text{Det}(A)\,\text{Det}(B)}$
$\text{Det}[\text{Adj}(A)] = [\text{Det}(A)]^{n-1}$		$\text{Adj}(AB) = \text{Adj}(B)\,\text{Adj}(A)$

(3) Characteristics

(a) The rank of a matrix is the order of the largest nonzero determinant that can be obtained from the elements of the matrix. The matrix whose order exceeds its rank is singular.

(b) The trace of a matrix is the sum of its diagonal elements.

(4) Relationships of Two Matrices

(a) Equivalence

The square matrix A is equivalent to another square matrix B if there exist nonsingular matrices P and Q such that $A = PBQ$

(b) Congruence

The square matrix A is congruent to another square matrix B if there exists a nonsingular matrix Q such that $A = Q^T B Q$

(c) Similarity

The square matrix A is similar to another square matrix B if there exists a nonsingular matrix Q such that $A = Q^{-1} B Q$

(1) Methods of Solution

(a) System of n simultaneous nonhomogeneous linear equations

$$
\begin{aligned}
a_{11}x_1 + a_{12}x_2 + \cdots + a_{1n}x_n &= b_1 \\
a_{21}x_1 + a_{22}x_2 + \cdots + a_{2n}x_n &= b_2 \\
&\cdots \\
a_{n1}x_1 + a_{n2}x_2 + \cdots + a_{nn}x_n &= b_n
\end{aligned}
$$

has a *unique solution* for the unknowns $x_1, x_2, \ldots, x_n$ if

$$
D = \begin{vmatrix}
a_{11} & a_{12} & \cdots & a_{1n} \\
a_{21} & a_{22} & \cdots & a_{2n} \\
& \cdots \cdots \cdots \\
a_{n1} & a_{n2} & \cdots & a_{nn}
\end{vmatrix} \neq 0
$$

and at least one of the terms $b_1, b_2, \ldots, b_n$ is different from zero.

(b) Determinant solution for the unknowns is

$$
x_1 = \frac{D_1}{D}, \qquad x_2 = \frac{D_2}{D}, \qquad \ldots, \qquad x_n = \frac{D_n}{D}
$$

where the *augmented determinants* are

$$
D_1 = \begin{vmatrix}
b_1 & a_{12} & \cdots & a_{1n} \\
b_2 & a_{22} & \cdots & a_{2n} \\
& \cdots \cdots \cdots \\
b_n & a_{n2} & \cdots & a_{nn}
\end{vmatrix}
\quad
D_2 = \begin{vmatrix}
a_{11} & b_1 & \cdots & a_{1n} \\
a_{21} & b_2 & \cdots & a_{2n} \\
& \cdots \cdots \cdots \\
a_{n1} & b_n & \cdots & a_{nn}
\end{vmatrix}
\quad
D_n = \begin{vmatrix}
a_{11} & a_{12} & \cdots & b_1 \\
a_{21} & a_{22} & \cdots & b_2 \\
& \cdots \cdots \cdots \\
a_{n1} & a_{n2} & \cdots & b_n
\end{vmatrix}
$$

(c) Matrix solution is represented symbolically as

$$
[x] = [A]^{-1}[b]
$$

where $[x]$ is the column matrix of the unknowns, $[A]^{-1}$ is the inverse of $[A]$ (Sec. 1.11), and $[b]$ is the column matrix of the terms b's.

(2) Classification of Solutions

(a) Unique solution. If $D \neq 0$ and $[b] \neq 0$, the system has a unique solution in which some but not all x_j may be zero.

(b) Trivial solution. If $D \neq 0$ and $[b] = 0$, the system has only one solution, $x_1 = x_2 = \cdots = x_n = 0$.

(c) Infinitely many solutions. If $D = 0$ and $[b] = 0$, the system is called homogeneous, and it has infinitely many solutions, one of which is the trivial solution.

(d) No solution. If $D = 0, D_1 = D_2 = \cdots = D_n = 0$, and $[b] \neq 0$, the system has no solution.

(1) Eigenvalues

(a) Initial system

$$
\begin{bmatrix} c_{11} & c_{12} & \cdots & c_{1n} \\ c_{21} & c_{22} & \cdots & c_{2n} \\ \cdots\cdots\cdots\cdots\cdots \\ c_{n1} & c_{n2} & \cdots & c_{nn} \end{bmatrix} \begin{bmatrix} x_1 \\ x_2 \\ \cdots \\ x_n \end{bmatrix} = \lambda \begin{bmatrix} d_{11} & d_{12} & \cdots & d_{1n} \\ d_{21} & d_{22} & \cdots & d_{2n} \\ \cdots\cdots\cdots\cdots\cdots \\ d_{n1} & d_{n2} & \cdots & d_{nn} \end{bmatrix} \begin{bmatrix} x_1 \\ x_2 \\ \cdots \\ x_n \end{bmatrix}
$$

$$\underbrace{}_{[C]} \quad \underbrace{}_{[X]} \qquad \underbrace{}_{[D]} \quad \underbrace{}_{[X]}$$

or simply

$$
\begin{bmatrix} m_{11}-\lambda & m_{12} & \cdots & m_{1n} \\ m_{21} & m_{22}-\lambda & \cdots & m_{2n} \\ \cdots\cdots\cdots\cdots\cdots\cdots\cdots \\ m_{n1} & m_{n2} & \cdots & m_{nn}-\lambda \end{bmatrix} \begin{bmatrix} x_1 \\ x_2 \\ \cdots \\ x_n \end{bmatrix} = \begin{bmatrix} 0 \\ 0 \\ \cdots \\ 0 \end{bmatrix}
$$

$$\underbrace{}_{[K]} \qquad \underbrace{}_{[X]} \quad \underbrace{}_{[0]}$$

where $c_{jk}, d_{jk} =$ given constants. $\lambda, x_j =$ unknowns, and $j = 1, 2, \ldots, n$, $k = 1, 2, \ldots, n$.

(b) Characteristic matrix

$$[K][X] = [0]$$

is called the characteristic matrix equation in which

$$\boxed{[K] = [M] - \lambda[I] = [D]^{-1}[C] - \lambda[I]}$$

is the *characteristic matrix*.

(c) Nontrivial solution of the characteristic matrix equation (Sec. 1.13−2c) exists if and only if det $(K) = 0$, which is a polynomial algebraic equation of nth degree in λ, called the *characteristic equation*. The roots of this equation $\lambda_1, \lambda_2, \ldots, \lambda_n$ are the *eigenvalues* of $[K]$.

(2) Eigenvectors

(a) Definition. Corresponding to each eigenvalue λ_j is a set of values $x_{j1}, x_{j2}, \ldots, x_{jn}$ forming a column matrix $[X_j]$ called the *eigenvector j*.

(b) Orthogonality. If $[M]$ is a symmetrical[1] matrix and $[X_j]$, $[X_k]$ are the eigenvectors corresponding to λ_j, λ_k, respectively, then

$$\boxed{[X_j]^T[M][X_k] = \begin{cases} 0 & \text{if } j \neq k \\ \lambda_j & \text{if } j = k \end{cases}}$$

and the eigenvectors are *orthogonal*.

(c) Normalization. If $[M]$ is the same as in (b) and

$$[Y_j] = [X_j/\sqrt{\lambda_j}] \qquad [Y_k] = [X_k/\sqrt{\lambda_k}]$$

are the *normalized eigenvectors* corresponding to λ_j, λ_k, respectively, then

$$[Y_j]^T[M][Y_k] = \begin{cases} 0 & \text{if } j \neq k \\ 1 & \text{if } j = k \end{cases}$$

and the normalized eigenvectors are also *orthogonal*.

[1]Or hermitian (Sec. 11.05).

(1) Permutations

(a) A permutation is an arrangement of n elements. The number of all possible permutations of n different elements is

$$_nP_n = n(n-1)(n-2)\cdots(3)(2)(1) = n!$$

(b) The number of all different permutations of n elements, among which there are a elements of equal value, is

$$_aP_n = \frac{n(n-1)(n-2)\cdots(3)(2)(1)}{a(a-1)(a-2)\cdots(3)(2)(1)} = \frac{n!}{a!}$$

(c) The number of all different permutations of n elements, among which there are a elements of one equal value and b elements of another equal value, is

$$_{a,b}P_n = \frac{n(n-1)(n-2)\cdots(3)(2)(1)}{a(a-1)(a-2)\cdots(3)(2)(1)b(b-1)(b-2)\cdots(3)(2)(1)} = \frac{n!}{a!b!}$$

(2) Variations

A variation is an arrangement of n elements into a sequence of k terms. The number of all possible variations is

$$_kV_n = \frac{n(n-1)(n-2)\cdots(3)(2)(1)}{(n-k)(n-k-1)(n-k-2)\cdots(3)(2)(1)} = \frac{n!}{(n-k)!} = \binom{n}{k}k!$$

(3) Combinations

A combination is an arrangement (without repetition) of n elements into a sequence of k terms. The number of all possible combinations is

$$_kC_n = \frac{n(n-1)(n-2)\cdots(n-k+2)(n-k+1)}{(n-k)(n-k-1)(n-k-2)\cdots(3)(2)(1)} = \frac{n!}{(n-k)!k!} = \binom{n}{k}$$

(4) Table — Example

Permutations	Elements A, B, C	$n = 3$
	$ABC \quad BCA \quad CAB$ $ACB \quad BAC \quad CBA$	$P_3 = (3)(2)(1) = 6$
Permutations	Elements A, A, C	$n = 3 \qquad a = 2$
	$AAC \quad ACA \quad CAA$	$_2P_3 = \dfrac{(3)(2)(1)}{(2)(1)} = 3$
Variations	Elements A, B, C	$n = 3 \qquad k = 2$
	$AB \quad BC \quad CA$ $BA \quad CB \quad AC$	$_2V_3 = \dfrac{(3)(2)(1)}{1} = 6$
Combinations	Elements A, B, C	$n = 3 \qquad k = 2$
	$AB \quad BC \quad CA$	$_2C_3 = \dfrac{(3)(2)(1)}{(2)(1)} = 3$

2
GEOMETRY

a, b, c = sides	A = area	h = altitude
α, β, γ = angles	R = circumradius	m = median
$2p = a + b + c$	r = inradius	t = bisector

(a) Oblique triangle $(\alpha + \beta + \gamma = 180°)$

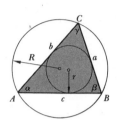

$$A = \sqrt{p(p-a)(p-b)(p-c)} = \frac{abc}{4R} = pr$$

$$A = \frac{ah_a}{2} = \frac{bh_b}{2} = \frac{ch_c}{2} = 2R^2 \sin\alpha \sin\beta \sin\gamma$$

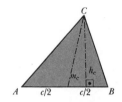

$$A = \frac{ab\sin\gamma}{2} = \frac{bc\sin\alpha}{2} = \frac{ac\sin\beta}{2} = p^2 \tan\frac{\alpha}{2}\tan\frac{\beta}{2}\tan\frac{\gamma}{2}$$

$$A = \frac{a^2\sin\beta\sin\gamma}{2\sin\alpha} = \frac{b^2\sin\alpha\sin\gamma}{2\sin\beta} = \frac{c^2\sin\alpha\sin\beta}{2\sin\gamma} = r^2\cot\frac{\alpha}{2}\cot\frac{\beta}{2}\cot\frac{\gamma}{2}$$

$$h_c = b\sin\alpha = a\sin\beta = \frac{2\sqrt{p(p-a)(p-b)(p-c)}}{c}$$

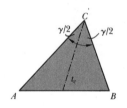

$$R = \frac{c}{2\sin\gamma} = \frac{abc}{4A} = \frac{abc}{4\sqrt{p(p-a)(p-b)(p-c)}}$$

$$r = (p-c)\tan\frac{\gamma}{2} = \frac{2A}{a+b+c} = \frac{2\sqrt{p(p-a)(p-b)(p-c)}}{a+b+c}$$

$$m_c = \sqrt{\frac{a^2}{2} + \frac{b^2}{2} - \frac{c^2}{4}} = \sqrt{b^2 + \left(\frac{c}{2}\right)^2 - bc\cos\alpha}$$

$$t_c = \sqrt{ab\left[1 - \left(\frac{c}{a+b}\right)^2\right]} = \frac{2ab}{a+b}\cos\frac{\gamma}{2}$$

$$a:b:c = \frac{1}{h_a}:\frac{1}{h_b}:\frac{1}{h_c} \qquad h_a:h_b:h_c = \frac{1}{a}:\frac{1}{b}:\frac{1}{c} \qquad \frac{1}{r} = \frac{1}{h_a} + \frac{1}{h_b} + \frac{1}{h_c}$$

(See also Sec. 3.03.)

(b) Right triangle $(\alpha + \beta = 90°)$

$$A = \frac{ab}{2} = \frac{hc}{2}$$

$$h = \frac{ab}{c} \qquad R = \frac{c}{2}$$

$$r = \frac{a+b-c}{2}$$

$$a^2 + b^2 = c^2$$

$$p = \frac{b^2}{c} \qquad q = \frac{a^2}{c}$$

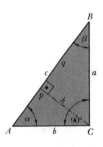

(See also Sec. 3.01.)

(c) Equilateral triangle $(\alpha = \beta = \gamma = 60°)$

$$A = \frac{a^2}{4}\sqrt{3} = \frac{h^2}{3}\sqrt{3}$$

$$h = m = t = \frac{a}{2}\sqrt{3}$$

$$R = \frac{a}{3}\sqrt{3}$$

$$r = \frac{a}{6}\sqrt{3}$$

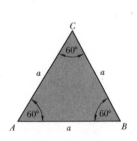

(See also Sec. 3.01.)

α = central angle	a = side	A = area
β = interior angle	n = number of sides	R = circumradius
γ = exterior angle		r = inradius

(a) General polygon

$$\sum_{1}^{n} \beta_j = (n-2)180°$$

$$\sum_{1}^{n} \gamma_j = 360°$$

$$\sum_{1}^{n-2} A_j = A$$

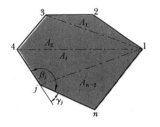

(b) Regular polygon $(n\alpha = 360°, n\beta = (n-2)180°, n\gamma = 360°)$

$$A = \frac{na^2}{4}\cot\frac{\pi}{n} = \frac{nar}{2} = \frac{nR^2}{2}\sin\frac{2\pi}{n}$$

$$R = \frac{a}{2}\csc\frac{\pi}{n} \qquad r = \frac{a}{2}\cot\frac{\pi}{n}$$

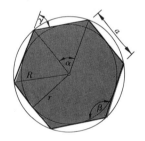

(c) Regular polygon — Table of coefficients

n	$180°/n$	A/a^2	A/R^2	A/r^2	R/a	r/a	R/r
3	60.000	0.433013	1.299038	5.196152	0.577350	0.288675	2.000000
4	45.000	1.000000	2.000000	4.000000	0.707107	0.500000	1.414214
5	36.000	1.720477	2.377642	3.632713	0.850651	0.688191	1.236068
6	30.000	2.598076	2.598076	3.464102	1.000000	0.866025	1.154701
7	25.714*...	3.633914	2.736408	3.371021	1.152383	1.038261	1.109916
8	22.500	4.828427	2.828427	3.313710	1.306563	1.207107	1.082392
9	20.000	6.181825	2.892544	3.275732	1.461902	1.373739	1.064177
10	18.000	7.694208	2.938926	3.249197	1.618034	1.538842	1.051462
12	15.000	11.196154	3.000000	3.215389	1.931852	1.866025	1.035277
15	12.000	17.642362	3.050524	3.188348	2.404867	2.352314	1.022341
16	11.250	20.109363	3.061464	3.182596	2.562917	2.513670	1.019592
20	9.000	31.568769	3.090168	3.167687	3.196228	3.156877	1.012465
24	7.500	45.574519	3.105827	3.159659	3.830649	3.797877	1.008629
32	5.625	81.225378	3.121442	3.151724	5.101151	5.076586	1.004839
48	3.750	183.084812	3.132619	3.146082	7.644910	7.628533	1.002147
64	2.8125*	325.687826	3.136541	3.144114	10.190024	10.177744	1.001206

*$180°/7 = 25.714\,285\,714\cdots°$ (periodic), $180°/64 = 2.8125°$ (finite)

(d) Regular polygon—Area, radius of inscribed and circumscribed circles

Name	Number of sides	Area*	Radius of inscribed circle	Radius of circumscribed circle
Triangle, equilateral................................	3	$0.433\,01l^2$	$0.288\,67l$	$0.577\,35l$
Square..	4	$1.000\,00l^2$	$0.500\,00l$	$0.707\,10l$
Pentagon..	5	$1.720\,48l^2$	$0.688\,19l$	$0.850\,65l$
Hexagon...	6	$2.598\,08l^2$	$0.866\,02l$	$1.000\,0l$
Heptagon..	7	$3.633\,91l^2$	$1.038\,3l$	$1.152\,3l$
Octagon	8	$4.828\,43l^2$	$1.207\,1l$	$1.306\,5l$
Nonagon...	9	$6.181\,82l^2$	$1.373\,7l$	$1.461\,9l$
Decagon..	10	$7.694\,21l^2$	$1.538\,8l$	$1.618\,0l$
Undecagon	11	$9.365\,64l^2$	$1.702\,8l$	$1.774\,7l$
Dodecagon...	12	$11.196\,15l^2$	$1.866\,0l$	$1.931\,8l$

*l = length of one side.

(e) Solutions to compound angles

Given	To find	Equation
α and β	γ	$\cos\gamma = \dfrac{\tan\beta}{\tan\alpha}$
α and β	δ	$\cos\delta = \dfrac{\sin\beta}{\sin\alpha}$
α and γ	β	$\tan\beta = \cos\gamma\,\tan\alpha$
α and γ	δ	$\tan\delta = \cos\alpha\,\tan\gamma$
α and δ	β	$\sin\beta = \sin\alpha\,\cos\delta$
α and δ	γ	$\tan\gamma = \dfrac{\tan\delta}{\cos\alpha}$
β and γ	α	$\tan\alpha = \dfrac{\tan\beta}{\cos\gamma}$
β and γ	δ	$\sin\delta = \cos\beta\,\sin\gamma$
β and δ	α	$\sin\alpha = \dfrac{\sin\beta}{\cos\delta}$
β and δ	γ	$\sin\gamma = \dfrac{\sin\delta}{\cos\beta}$
γ and δ	α	$\cos\alpha = \dfrac{\tan\delta}{\tan\gamma}$
γ and δ	β	$\cos\beta = \dfrac{\sin\delta}{\sin\gamma}$

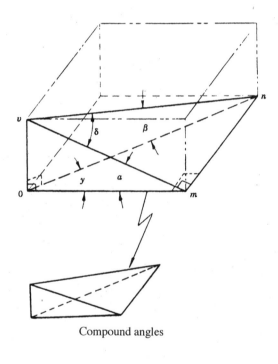

Compound angles

a, b, c, d = sides	$2s = a + b + c + d$	A = area
e, f = diagonals	h = altitude	R = circumradius
$\alpha, \beta, \gamma, \delta$ = angles		r = inradius

(a) Square $\quad (\alpha = \beta = \gamma = \delta = 90°)$

$$e = a\sqrt{2} = 1.4142a \qquad R = \frac{a}{2}\sqrt{2} = 0.7071a$$

$$A = a^2 \qquad\qquad r = \frac{a}{2}$$

(b) Rectangle $\quad (\alpha = \beta = \gamma = \delta = 90°)$

$$e = f = \sqrt{a^2 + b^2} \qquad\qquad R = \frac{e}{2} = \frac{\sqrt{a^2 + b^2}}{2}$$

$$A = ab$$

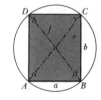

(c) Rhombus $\quad (\alpha + \beta = \gamma + \delta = 180°) \qquad (\alpha = \gamma, \beta = \delta)$

$$e = 2a \cos\frac{\alpha}{2} \qquad f = 2a \sin\frac{\alpha}{2}$$

$$e^2 + f^2 = 4a^2$$

$$h = a \sin\alpha \qquad r = \frac{a}{2}\sin\alpha$$

$$A = ah = a^2 \sin\alpha = \frac{ef}{2}$$

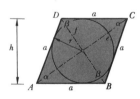

(d) Rhomboid $\quad (\alpha + \beta = \gamma + \delta = 180°) \qquad (\alpha = \gamma, \beta = \delta)$

$$e = \sqrt{a^2 + b^2 - 2ab \cos\beta} \qquad f = \sqrt{a^2 + b^2 - 2ab \cos\alpha}$$

$$e^2 + f^2 = 2(a^2 + b^2)$$

$$h_a = b \sin\alpha \qquad h_b = a \sin\alpha$$

$$A = ah_a = ab \sin\alpha = bh_b$$

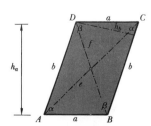

a, b, c, d = sides	$2s = a + b + c + d$	A = area
e, f = diagonals	h = altitude	R = circumradius
$\alpha, \beta, \gamma, \delta$ = angles	$2t = a + b - c + d$	r = inradius

(a) Trapezoid $(\alpha + \delta = \beta + \gamma = 180°)$

$$e = \sqrt{a^2 + b^2 - 2ab \cos \beta}$$

$$f = \sqrt{a^2 + d^2 - 2ad \cos \alpha}$$

$$h = \frac{2}{a-c} \sqrt{t(t-a+c)(t-b)(t-d)}$$

$$A = \frac{(a+c)h}{2}$$

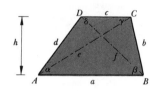

(b) Deltoid $\left(\dfrac{\alpha}{2} + \beta + \dfrac{\gamma}{2} = 180°\right)$

$$e = a \cos \frac{\alpha}{2} + b \cos \frac{\gamma}{2}$$

$$f = 2b \sin \frac{\gamma}{2} = 2a \sin \frac{\alpha}{2}$$

$$A = \frac{ef}{2} = \frac{a^2 \sin \alpha + b^2 \sin \gamma}{2}$$

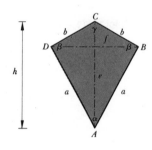

(c) Tangent – Quadrilateral $(a + c = b + d)$

$$A = sr$$

If $\alpha + \gamma = \beta + \delta = 180°$,

$$A = \sqrt{abcd} \qquad r = \frac{\sqrt{abcd}}{s}$$

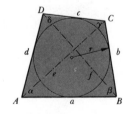

(d) Secant – Quadrilateral $(\alpha + \gamma = \beta + \delta = 180°)$

$$ef = ac + bd = g$$

$$e = \sqrt{\frac{(ad + bc)g}{ab + cd}}$$

$$f = \sqrt{\frac{(ab + cd)g}{ad + bc}} \qquad \sin \omega = \frac{2A}{g}$$

$$A = \sqrt{(s-a)(s-b)(s-c)(s-d)}$$

$$R = \frac{\sqrt{(ab+cd)(ad+bc)g}}{4A}$$

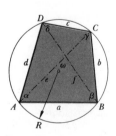

(e) Four-bar linkage

The mathematical relationship of this four-bar linkage was solved by the mathematician Freudenstein in 1953 and is of great importance to mechanical design engineers and engineering mechanics. See the figure:

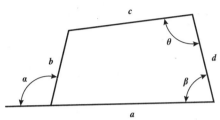

$$L_1 \cos \alpha - L_2 \cos \beta + L_3 = \cos (\alpha - \beta)$$

where

$$L_1 = \frac{a}{d} \qquad L_2 = \frac{a}{b} \qquad \text{and} \qquad L_3 = \frac{b^2 - c^2 + d^2 + a^2}{2bd}$$

Sides: a, b, c, and d; angles: α, β, and θ. See App. D and Index for applications and practical programming uses for the Freudenstein four-bar linkage equation and its variations. Laws of the four-bar linkage and types of mechanism are described here with reference to the accompanying figure.

Grashof's inequality.   The length of the longest link + length of the shortest link < sum of lengths of the two intermediate links.

Types of four-bar mechanisms: (1) If Grashof's inequality is satisfied and b or d is the shortest link, the linkage is a *crank and rocker*; the shortest link is the *crank*, and the opposite link is the *rocker*. (2) If Grashof's inequality is satisfied and the fixed link is the shortest link, the linkage is a *drag linkage*; both cranks can make complete rotations. (3) All other cases: the linkage is a *double-rocker* mechanism (cranks b and d can only oscillate; this will be the case whenever the coupler c is the shortest link).

Grubler's criterion.   The criterion for determining whether any link can rotate 360° is stated by Grubler's criterion (see figure)

$$a + b < c + d$$

C = circumference	S = length of arc	$2l$ = chord
A = area	α = angle	h = altitude
R = radius	D = diameter	v = rise

$$\text{Arc } 1° = \frac{\pi}{180} \qquad\qquad \text{Arc 1 minute} = \frac{\pi}{10,800} \qquad\qquad \text{Arc 1 second} = \frac{\pi}{648,000}$$

$$= 0.017453293 \qquad\qquad\qquad = 0.000290888 \qquad\qquad\qquad = 0.000004848$$

$$\text{radian} \qquad\qquad\qquad\qquad \text{radian} \qquad\qquad\qquad\qquad \text{radian}$$

(a) Circle   **(b) Sector**   **(c) Segment**

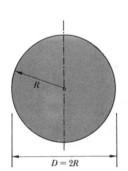 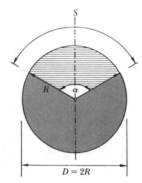 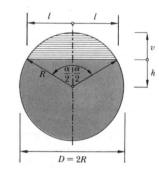

$$C = 2\pi R = \pi D$$

$$A_0 = \pi R^2 = \frac{\pi D^2}{4}$$

$$S = \frac{\pi R \alpha°}{180°} = R\alpha = \frac{D\alpha}{2}$$

$$A = \frac{\pi R^2 \alpha°}{360°} = \frac{R^2 \alpha}{2}$$

$$l = R \sin \frac{\alpha}{2} \quad . \quad h = R \cos \frac{\alpha}{2}$$

$$A = \frac{R^2}{2}\left(\frac{\pi \alpha°}{180°} - \sin \alpha\right)$$

(d) π constants

n	$n\pi$	$\dfrac{1}{n\pi}$	$\dfrac{\pi}{n}$	$\dfrac{n}{\pi}$
1	3.141 592 653 6	0.318 309 886 2	3.141 592 653 6	0.318 309 886 2
2	6.283 185 307 2	0.159 154 943 1	1.570 796 326 8	0.636 619 772 4
3	9.424 777 960 8	0.106 103 295 4	1.047 197 551 2	0.954 929 658 6
4	12.566 370 614 4	0.079 577 471 5	0.785 398 163 4	1.273 239 544 7
5	15.707 963 267 9	0.063 661 977 2	0.628 318 530 7	1.591 549 430 9
6	18.849 555 921 5	0.053 051 647 7	0.523 598 775 6	1.909 859 317 1
7	21.991 148 575 1	0.045 472 840 9	0.448 798 950 5	2.228 169 203 3
8	25.132 741 228 7	0.039 788 735 8	0.392 699 081 7	2.546 479 089 5
9	28.274 338 882 3	0.035 367 765 1	0.349 065 850 4	2.864 788 975 7

γ, δ = peripheral angles	$2l$ = chord length	α, β = central angles

(a) All peripheral angles belonging to the same chord of length $2l$ with vertex on the same side of the circle are equal.

$$\gamma_1 = \gamma_2 = \gamma_3 = \cdots = \gamma \qquad \delta_1 = \delta_2 = \delta_3 = \cdots = \delta$$

(b) Central angle is twice the size of the respective peripheral angle.

$$\alpha = 2\gamma \qquad \beta = 2\delta$$

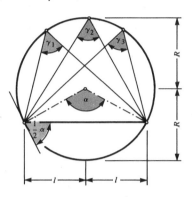
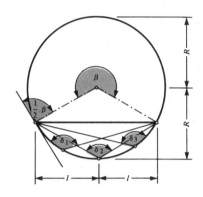

(c) All peripheral angles belonging to the chord of length $2R$ are right angles.

$\gamma_1 = \gamma_2 = \gamma_3 = \cdots = \gamma = \frac{1}{2}\pi$	$\delta_1 = \delta_2 = \delta_3 = \cdots = \delta = \frac{1}{2}\pi$

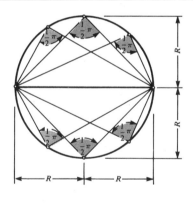
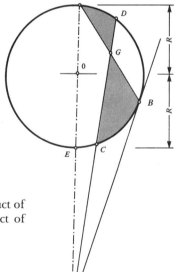

(d) If two chords intersect inside a circle, the product of the segments of one chord equals the product of segments of the other chord.

$$\overline{BG} \cdot \overline{GF} = \overline{CG} \cdot \overline{GD}$$

(e) If a tangent and secant are drawn from a point outside a circle, the tangent length is the mean proportion between the secant length and its external segment.

$$\boxed{\overline{AB} = \sqrt{\overline{AF} \cdot \overline{AE}} = \sqrt{\overline{AD} \cdot \overline{AC}}}$$

(f) Properties of the circle

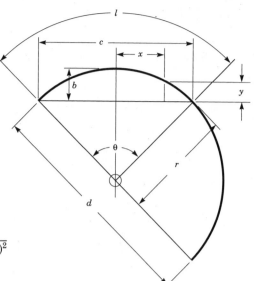

Arc: $l = \dfrac{\pi r \theta}{180}$

Angle: $\theta = \dfrac{(180°)l}{\pi r}$

Radius: $r = \dfrac{4b^2 + c^2}{8b}$ $d = \dfrac{4b^2 + c^2}{4b}$

Chord: $c = 2\sqrt{2br - b^2} = 2r \sin\dfrac{\theta}{2} = d \sin\dfrac{\theta}{2}$

Rise: $b = r - \dfrac{1}{2}\sqrt{4r^2 - c^2} = \dfrac{c}{2}\tan\dfrac{\theta}{4} = 2r \sin^2\dfrac{\theta}{4}$

Rise: $b = r + y - \sqrt{r^2 - x^2}$

where $y = b - r + \sqrt{r^2 - x^2}$ and $x = \sqrt{r^2 - (r + y - b)^2}$

> a = edge $\qquad$ R = circumradius $\qquad$ S = surface
>
> e = diagonal $\qquad$ r = inradius $\qquad$ V = volume
>
> v = altitude $\qquad$ ω = dihedral angle

(a) Tetrahedron $\quad$ **(4 triangles, 6 edges, 4 vertices)** $\quad$ ($\omega = 70° \, 31' \, 44''$)

$$R = \frac{a}{4}\sqrt{6} \qquad r = \frac{a}{12}\sqrt{6} \qquad v = \frac{a}{3}\sqrt{6}$$

$$S = a^2\sqrt{3} \approx 1.7321a^2$$

$$V = \frac{a^3\sqrt{2}}{12} \approx 0.1179a^3$$

(b) Cube $\quad$ **(6 squares, 12 edges, 8 vertices)** $\quad$ ($\omega = 90°$)

$$R = \frac{a}{2}\sqrt{3} \qquad r = \frac{a}{2} \qquad e = a\sqrt{3}$$

$$S = 6a^2$$

$$V = a^3$$

(c) Octahedron $\quad$ **(8 triangles, 12 edges, 6 vertices)** $\quad$ ($\omega = 109° \, 28' \, 16''$)

$$R = \frac{a}{2}\sqrt{2} \qquad r = \frac{a}{6}\sqrt{6}$$

$$S = 2a^2\sqrt{3} \approx 3.4641a^2$$

$$V = \frac{a^3}{3}\sqrt{2} \approx 0.4714a^3$$

(d) Dodecahedron $\quad$ **(12 pentagons, 30 edges, 20 vertices)** $\quad$ ($\omega = 116° \, 33' \, 54''$)

$$R = \frac{a(1 + \sqrt{5})\sqrt{3}}{4} \qquad r = \frac{a}{4}\sqrt{\frac{50 + 22\sqrt{5}}{5}}$$

$$S = 3a^2\sqrt{5(5 + 2\sqrt{5})} \approx 20.6457a^2$$

$$V = \frac{a^3}{4}(15 + 7\sqrt{5}) \approx 7.6631a^3$$

(e) Icosahedron $\quad$ **(20 triangles, 30 edges, 12 vertices)** $\quad$ ($\omega = 138° \, 11' \, 23''$)

$$R = \frac{a}{4}\sqrt{2(5 + \sqrt{5})} \qquad r = \frac{a}{2}\sqrt{\frac{7 + 3\sqrt{5}}{6}}$$

$$S = 5a^2\sqrt{3} \approx 8.6603a^2$$

$$V = \frac{5a^3}{12}(3 + \sqrt{5}) \approx 2.1817a^3$$

(f) Regular polyhedra—Surface and volume

Name	Nature of surface	Surface*	Volume*
Tetrahedron	4 equilateral triangles	$1.732\ 05l^2$	$0.117\ 85l^3$
Hexahedron or cube	6 squares	$6.000\ 00l^2$	$1.000\ 00l^3$
Octahedron	8 equilateral triangles	$3.464\ 10l^2$	$0.471\ 40l^3$
Dodecahedron	12 pentagons	$20.645\ 73l^2$	$7.663\ 12l^3$
Icosahedron	20 equilateral triangles	$8.660\ 25l^2$	$2.181\ 70l^3$

*Surface and volume of regular polyhedra in terms of the length of one edge l.

a, b, c = edges	R = circumradius	A = lateral area
e = diagonal	B = area of base	S = surface
h = lateral edge	v = altitude	V = volume

(a) Rectangular parallelepiped

$$e = \sqrt{a^2 + b^2 + c^2} \qquad R = \frac{\sqrt{a^2 + b^2 + c^2}}{2}$$

$$S = 2(ab + bc + ca)$$

$$V = abc$$

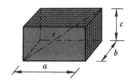

(b) Prism

$2p$ = perimeter of right section

$$A = 2ph$$

$$V = Bv$$

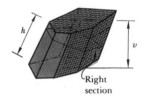

Right section

(c) Right pyramid

$$A = a\sqrt{v^2 + \left(\frac{b}{2}\right)^2} + b\sqrt{v^2 + \left(\frac{a}{2}\right)^2}$$

$$B = ab \qquad\qquad S = A + B$$

$$V = \frac{Bv}{3} \qquad\qquad \text{(valid for any pyramid)}$$

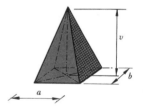

(d) Frustum of right pyramid   (subscript b = bottom, t = top)

$$A = (a_b + a_t)\sqrt{v^2 + \left(\frac{b_b - b_t}{2}\right)^2} + (b_b + b_t)\sqrt{v^2 + \left(\frac{a_b - a_t}{2}\right)^2}$$

$$S = A + a_b b_b + a_t b_t$$

$$V = \frac{v}{3}(B_b + B_t + \sqrt{B_b B_t}) \qquad\qquad \text{(valid for any frustum)}$$

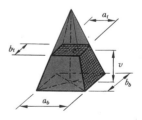

(e) Right wedge

$$A = 2(a + c)\sqrt{v^2 + b^2} + 2b\sqrt{v^2 + (a - c)^2}$$

$$S = A + 4ab$$

$$V = \frac{2bv}{3}(2a + c)$$

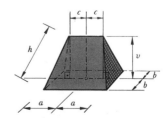

(f) Volume of a wedge

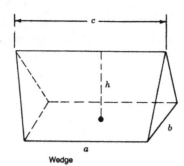

Wedge

$$V = \frac{(2b + c)ah}{6}$$

$\pi = 3.141\,59\cdots$	R = radius	A = lateral area
r = radius	B = area of base	S = surface
h = height	v = altitude	V = volume

(a) Right circular cylinder

$$A = 2\pi R v$$

$$B = \pi R^2 \qquad S = 2\pi R(R+v)$$

$$V = \pi R^2 v$$

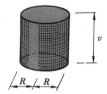

(b) Truncated frustum of right circular cylinder

$$A = \pi R(h_1 + h_2)$$

$$S = \pi R\left[h_1 + h_2 + R + \sqrt{R^2 + \left(\frac{h_2 - h_1}{2}\right)^2}\right]$$

$$V = \pi R^2 \frac{h_1 + h_2}{2}$$

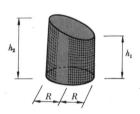

(c) Area and volume of a portion of a cylinder (base edge = diameter)

$$A = 2rh$$

$$V = \frac{2}{3}r^2 h$$

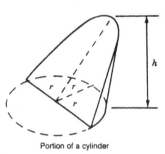

Portion of a cylinder

(d) Area and volume of a portion of a cylinder (special cases)

$$A = \frac{h(ad \pm c \times \text{perimeter of base})}{r \pm c}$$

$$V = \frac{h(\frac{2}{3}a^3 \pm cA)}{r \pm c}$$

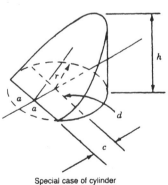

Special case of cylinder

where d = diameter of base circle. Use $+c$ when base area is larger than half the base circle; use $-c$ when base area is smaller than half the base circle.

(e) Hollow right circular cylinder

$$t = R - r \qquad \rho = \frac{R + r}{2}$$

$$A = 2\pi R v$$

$$B = \pi(R^2 - r^2)$$

$$V = \pi v(R^2 - r^2) = 2\pi v t \rho$$

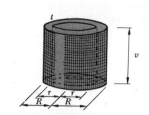

(f) General circular cylinder

$C = $ circumference of right section

$$A = Ch \qquad\qquad B = \pi R^2$$

$$V = Bv$$

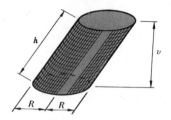

$\pi = 3.141\,59\cdots$	$R =$ radius	$A =$ lateral area
$r =$ radius	$B =$ area of base	$S =$ surface
$h =$ slant height	$v =$ altitude	$V =$ volume

(a) Circular right cone

$$A = \pi R \sqrt{v^2 + R^2} = \pi R h$$

$$B = \pi R^2$$

$$S = \pi R (R + h)$$

$$V = \frac{\pi R^2 v}{3}$$

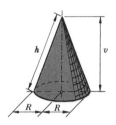

(b) Frustum of right cone   (subscript $b =$ bottom, $t =$ top)

$$A = \pi (R_1 + R_2) \sqrt{v^2 + (R_1 - R_2)^2} = \pi (R_1 + R_2) h$$

$$B_b = \pi R_1{}^2 \qquad\qquad B_t = \pi R_2{}^2$$

$$S = \pi [R_1{}^2 + (R_1 + R_2)h + R_2{}^2]$$

$$V = \frac{\pi v}{3} (R_1{}^2 + R_1 R_2 + R_2{}^2)$$

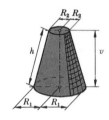

(c) General cone

$$V = \frac{Bv}{3}$$

For the frustum,

$$V = \frac{v_1}{3} (B_b + B_t + \sqrt{B_b B_t})$$

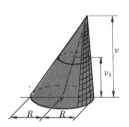

(d) Torus

$$S = 4\pi^2 R r \approx 39.4784 R r$$

$$V = 2\pi^2 R r^2 \approx 19.7392 R r^2$$

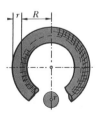

(e) Circular barrel

For circular curvature,

$$V = \tfrac{1}{3}\pi v (2R^2 + r^2)$$

For parabolic curvature,

$$V = \tfrac{1}{15}\pi v (8R^2 + 4Rr + 3r^2)$$

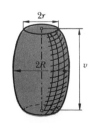

$$\pi = 3.141\,59 \cdots \qquad R = \text{radius} \qquad A = \text{lateral area}$$
$$a, b = \text{radii} \qquad D = \text{diameter} \qquad S = \text{surface}$$
$$v = \text{altitude} \qquad V = \text{volume}$$

(a) Sphere

$$S = 4\pi R^2$$
$$= \pi D^2 = \sqrt[3]{36\pi V^2}$$
$$V = \tfrac{4}{3}\pi R^3$$
$$= \frac{\pi D^3}{6} = \frac{1}{6}\sqrt{\frac{S^3}{\pi}}$$

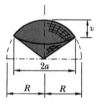

(b) Spherical sector

$$S = \pi R(2v + a)$$
$$V = \frac{2\pi}{3} R^2 v$$

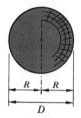

(c) Spherical sector   (one base)

$$a = \sqrt{v(2R - v)}$$
$$A = 2\pi R v$$
$$S = \pi v(4R - v)$$
$$V = \frac{\pi}{3} v^2(3R - v)$$

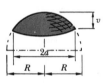

(d) Spherical sector   (two bases)

$$R^2 = a^2 + \left(\frac{a^2 - b^2 - v^2}{2v}\right)^2$$
$$A = 2\pi R v$$
$$S = \pi(2Rv + a^2 + b^2)$$
$$V = \frac{\pi v}{6}(3a^2 + 3b^2 + v^2)$$

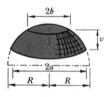

(e) Conical ring

$$S = 2\pi R\left(v + \sqrt{R^2 - \frac{v^2}{4}}\right)$$
$$V = \frac{2\pi}{3} R^2 v$$

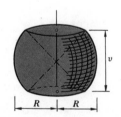

3

TRIGONOMETRY

a, b = legs	c = hypotenuse	A = area
A, B, C = vertices	p, q = segment of c	R = circumradius
α, β, γ = angles	h = height	r = inradius

(1) Relationships

$$a^2 + b^2 = c^2 \qquad \alpha + \beta = 90°$$

$$\sin \alpha = \frac{a}{c} \qquad \sin \beta = \frac{b}{c}$$

$$\cos \alpha = \frac{b}{c} \qquad \cos \beta = \frac{a}{c}$$

$$\tan \alpha = \frac{a}{b} \qquad \tan \beta = \frac{b}{a}$$

$$\cot \alpha = \frac{b}{a} \qquad \cot \beta = \frac{a}{b}$$

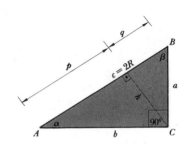

$$\alpha = \sin^{-1}\frac{a}{c} = \cos^{-1}\frac{b}{c} = \tan^{-1}\frac{a}{b} \qquad \beta = \sin^{-1}\frac{b}{c} = \cos^{-1}\frac{a}{c} = \tan^{-1}\frac{b}{a}$$

(2) General Formulas

$$h = a \sin \beta \qquad\qquad p = b \cos \alpha \qquad\qquad q = a \cos \beta$$
$$ = b \sin \alpha \qquad\qquad = h \cot \alpha \qquad\qquad = h \cot \beta$$
$$h = \sqrt{ab \cos \alpha \cos \beta} \qquad c = b \cos \alpha + a \cos \beta$$
$$A = \frac{ab}{2} = \frac{c^2}{4} \sin 2\alpha = \frac{a^2}{2} \cot \alpha = \frac{b^2}{2} \cot \beta \qquad r = \frac{a + b - c}{2} = \frac{c(\sin \alpha + \sin \beta - 1)}{2} \qquad R = \frac{c}{2}$$

(3) Solutions

Known	Solution					
	a	b	c	α	β	A
a, b			$\sqrt{a^2 + b^2}$	$\tan^{-1}\frac{a}{b}$	$\tan^{-1}\frac{b}{a}$	$\frac{ab}{2}$
a, c		$\sqrt{c^2 - a^2}$		$\sin^{-1}\frac{a}{c}$	$\cos^{-1}\frac{a}{c}$	$\frac{a\sqrt{c^2 - a^2}}{2}$
a, α		$a \cot \alpha$	$\frac{a}{\sin \alpha}$		$90° - \alpha$	$\frac{a^2 \cot \alpha}{2}$
b, α	$b \tan \alpha$		$\frac{b}{\cos \alpha}$		$90° - \alpha$	$\frac{b^2 \tan \alpha}{2}$
c, α	$c \sin \alpha$	$c \cos \alpha$			$90° - \alpha$	$\frac{c^2 \sin 2\alpha}{4}$

Note: For definition of $\sin^{-1}(\ )$, $\cos^{-1}(\ )$, $\tan^{-1}(\ )$ see Sec. 6.08.

a, b = legs	c = hypotenuse	$\bar{R}$ = radius of sphere
A, B, C = vertices	h = height	R = circumradius
α, β = angles	A = area	r = inradius

(1) Relationships

In the circle diagram of elements α, c, β, $\bar{a}$, $\bar{b}$, the cosine of any element equals the product of the cotangents of the adjacent elements, and the cosine of any element equals the product of the sines of the opposite elements.

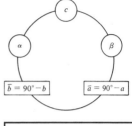

$$\bar{b} = 90° - b \qquad \bar{a} = 90° - a$$

$$
\begin{array}{ll}
\cos \bar{a} = \cot \bar{b} \cot \beta & \cos \bar{a} = \sin \alpha \sin c \\
\cos \bar{b} = \cot \bar{a} \cot \alpha & \cos \bar{b} = \sin \beta \sin c \\
\cos c = \cot \alpha \cot \beta & \cos c = \sin \bar{a} \sin \bar{b} \\
\cos \alpha = \cot \bar{b} \cot c & \cos \alpha = \sin \bar{a} \sin \beta \\
\cos \beta = \cot \bar{a} \cot c & \cos \beta = \sin \bar{b} \sin \alpha
\end{array}
$$

$$+90° < \alpha + \beta < +270°$$
$$-90° < \alpha - \beta < +90°$$

(2) General Formulas $(2p = a + b + c,\ 2\sigma = \alpha + \beta + 90°)$

$$h = \sin^{-1}(\sin \alpha \sin b) = \sin^{-1}(\sin \beta \sin a)$$

$$A = \pi \bar{R}^2 (\alpha + \beta - 90°)/180°$$

$$R = \cot^{-1} \sqrt{\frac{\cos(\sigma - \alpha)\cos(\sigma - \beta)\cos(\sigma - 90°)}{-\cos \sigma}}$$

$$r = \tan^{-1} \sqrt{\frac{\sin(p - a)\sin(p - b)\sin(p - c)}{\sin p}}$$

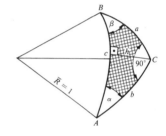

(3) Solutions

Known	Solution				
	a	b	c	α	β
a, b			$\cos^{-1}(\cos a \cos b)$	$\tan^{-1}\dfrac{\tan a}{\sin b}$	$\tan^{-1}\dfrac{\tan b}{\sin a}$
a, c		$\cos^{-1}\dfrac{\cos c}{\cos a}$		$\sin^{-1}\dfrac{\sin a}{\sin c}$	$\cos^{-1}\dfrac{\tan a}{\tan c}$
a, α		$\sin^{-1}\dfrac{\tan a}{\tan \alpha}$	$\sin^{-1}\dfrac{\sin a}{\sin \alpha}$		$\sin^{-1}\dfrac{\cos \alpha}{\cos a}$
b, α	$\tan^{-1}(\sin b \tan \alpha)$		$\tan^{-1}\dfrac{\tan b}{\cos \alpha}$		$\cos^{-1}(\cos b \sin \alpha)$
c, α	$\sin^{-1}(\sin c \sin \alpha)$	$\tan^{-1}(\tan c \cos \alpha)$			$\cot^{-1}(\cos c \tan \alpha)$

Note: For definition of $\sin^{-1}(\ )$, $\cos^{-1}(\ )$, $\tan^{-1}(\ )$, $\cot^{-1}(\ )$ see Sec. 6.08.

a, b, c = sides	h = altitude	A = area
A, B, C = vertices	m = median	R = circumradius
α, β, γ = angles	t = bisector	r = inradius

(1) Basic Laws $(\alpha + \beta + \gamma = 180°)$

(a) Law of sines

$a : b : c = \sin \alpha : \sin \beta : \sin \gamma$

(b) Law of cosines

$a^2 = b^2 + c^2 - 2bc \cos \alpha$

(c) Law of tangents

$(a + b) : (a - b) = \tan \dfrac{\alpha + \beta}{2} : \tan \dfrac{\alpha - \beta}{2}$

(d) Law of projection

$a = b \cos \gamma + c \cos \beta$

(e) Law of angles

$\sin (\alpha + \beta) = \sin \gamma$

$\cos (\alpha + \beta) = -\cos \gamma$

$\tan (\alpha + \beta) = -\tan \gamma$

$\cot (\alpha + \beta) = -\cot \gamma$

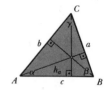

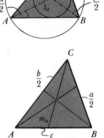

(2) General Formulas $(2p = a + b + c)$

(a) Angles

$$\alpha = 2 \sin^{-1} \sqrt{\frac{(p-b)(p-c)}{bc}}$$

$$= 2 \cos^{-1} \sqrt{\frac{p(p-a)}{bc}}$$

$$= 2 \tan^{-1} \sqrt{\frac{(p-b)(p-c)}{p(p-a)}}$$

(b) Radii

$$R = \frac{a}{2 \sin \alpha}$$

$$r = (p - a) \tan \frac{\alpha}{2}$$

$$r = 4R \sin \frac{\alpha}{2} \sin \frac{\beta}{2} \sin \frac{\gamma}{2}$$

(c) Segments

$$h_a = b \sin \gamma = c \sin \beta$$

$$m_a = \tfrac{1}{2} \sqrt{b^2 + c^2 + 2bc \cos \alpha}$$

$$t_a = \frac{2bc \cos (\alpha/2)}{b + c}$$

(d) Area

$$A = \frac{ah_a}{2} = \frac{ab \sin \gamma}{2}$$

$$= \frac{abc}{4R} = \frac{a^2 \sin \beta \sin \gamma}{2 \sin \alpha}$$

$$= \sqrt{p(p - a)(p - b)(p - c)}$$

Note: Additional formulas are obtained by simultaneous cyclic substitution. For definition of $\sin^{-1}(\ )$, $\cos^{-1}(\ )$, $\tan^{-1}(\ )$ see Sec. 6.08.

Known	Solution* $(2p = a + b + c)$
a, b, c	$\alpha = \cos^{-1}\dfrac{b^2 + c^2 - a^2}{2bc}$ or $\alpha = 2\cos^{-1}\sqrt{\dfrac{p(p-a)}{bc}}$ $\beta = \cos^{-1}\dfrac{a^2 + c^2 - b^2}{2ac}$ or $\beta = 2\cos^{-1}\sqrt{\dfrac{p(p-b)}{ac}}$ $\gamma = 180° - (\alpha + \beta)$ $A = \sqrt{p(p-a)(p-b)(p-c)}$
a, b, α	$\beta = \sin^{-1}\dfrac{b \sin \alpha}{a}$ $\gamma = 180° - (\alpha + \beta)$ $c = \dfrac{a \sin \gamma}{\sin \alpha}$ $A = \dfrac{ab}{2}\sin \gamma$
a, b, γ	$X = 90° - \dfrac{\gamma}{2}$ $Y = \tan^{-1}\left(\dfrac{a-b}{a+b}\cot\dfrac{\gamma}{2}\right)$ $\alpha = X + Y$ $\beta = X - Y$ $c = \sqrt{a^2 + b^2 - 2ab \cos \gamma}$ $A = \dfrac{ab}{2}\sin \gamma$
a, α, β	$\gamma = 180° - (\alpha + \beta)$ $b = \dfrac{a \sin \beta}{\sin \alpha}$ $c = \dfrac{a \sin \gamma}{\sin \alpha} = \dfrac{a \sin (\alpha + \beta)}{\sin \alpha}$ $A = \dfrac{ab}{2}\sin \gamma$
a, β, γ	$\alpha = 180° - (\beta + \gamma)$ $b = \dfrac{a \sin \beta}{\sin \alpha} = \dfrac{a \sin \beta}{\sin (\beta + \gamma)}$ $c = \dfrac{a \sin \gamma}{\sin \alpha} = \dfrac{a \sin \gamma}{\sin (\beta + \gamma)}$ $A = \dfrac{a^2 \sin \beta \sin \gamma}{2 \sin (\beta + \gamma)}$

*$p, X, Y =$ pocket calculator storage.

a, b, c = sides	h = height	$\overline{R}$ = radius of sphere
A, B, C = vertices	ϵ = spherical excess	R = circumradius
α, β, γ = angles	d = spherical defect	r = inradius
$2p = a + b + c$	$2\sigma = \alpha + \beta + \gamma$	A = area

(1) Basic Laws

(a) Law of sines

$$\sin a : \sin b : \sin c = \sin \alpha : \sin \beta : \sin \gamma$$

(b) Law of cosines I

$$\cos a = \cos b \cos c + \sin b \sin c \cos \alpha$$

(c) Law of cosines II

$$\cos \alpha = -\cos \beta \cos \gamma + \sin \beta \sin \gamma \cos a$$

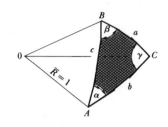

$$0° < a + b + c < +360°$$

$$+180° < \alpha + \beta + \gamma < +540°$$

(2) General Formulas

(a) Delambre's equations

$$\sin \frac{\alpha + \beta}{2} \cos \frac{c}{2} = \cos \frac{a - b}{2} \cos \frac{\gamma}{2}$$

$$\sin \frac{\alpha - \beta}{2} \sin \frac{c}{2} = \sin \frac{a - b}{2} \cos \frac{\gamma}{2}$$

$$\cos \frac{\alpha + \beta}{2} \cos \frac{c}{2} = \cos \frac{a + b}{2} \sin \frac{\gamma}{2}$$

$$\cos \frac{\alpha - \beta}{2} \sin \frac{c}{2} = \sin \frac{a + b}{2} \sin \frac{\gamma}{2}$$

(b) Napier's equations

$$\tan \frac{\alpha + \beta}{2} \cos \frac{a + b}{2} = \cos \frac{a - b}{2} \cot \frac{\gamma}{2}$$

$$\tan \frac{\alpha - \beta}{2} \sin \frac{a + b}{2} = \sin \frac{a - b}{2} \cot \frac{\gamma}{2}$$

$$\tan \frac{a + b}{2} \cos \frac{\alpha + \beta}{2} = \cos \frac{\alpha - \beta}{2} \tan \frac{c}{2}$$

$$\tan \frac{a - b}{2} \sin \frac{\alpha + \beta}{2} = \sin \frac{\alpha - \beta}{2} \tan \frac{c}{2}$$

(c) Circumradius

$$R = \cot^{-1} \sqrt{\frac{\cos (\sigma - \alpha) \cos (\sigma - \beta) \cos (\sigma - \gamma)}{-\cos \sigma}}$$

(d) Inradius

$$r = \tan^{-1} \sqrt{\frac{\sin (p - a) \sin (p - b) \sin (p - c)}{\sin p}}$$

(e) Spherical angle

$$A = \frac{\pi \overline{R}^2 \alpha°}{90°}$$

(f) Spherical triangle

$$A = \frac{\pi \overline{R}^2 \epsilon°}{180°}$$

(g) Angles

$$\alpha = 2 \tan^{-1} \sqrt{\frac{\sin (p - b) \sin (p - c)}{\sin p \sin (p - a)}} \qquad a = 2 \tan^{-1} \sqrt{\frac{-\cos \sigma \cos (\sigma - \alpha)}{\cos (\sigma - \beta) \cos (\sigma - \gamma)}}$$

(h) Spherical excess and defect

$$\epsilon = \alpha + \beta + \gamma - 180° = 4 \tan^{-1} \sqrt{\tan \frac{p}{2} \tan \frac{p - a}{2} \tan \frac{p - b}{2} \tan \frac{p - c}{2}}$$

$$d = 360° - (a + b + c) = 2a - 4 \tan^{-1} \sqrt{\cot \frac{\sigma}{2} \cot \frac{\sigma - \alpha}{2} \tan \frac{\sigma - \beta}{2} \tan \frac{\sigma - \gamma}{2}}$$

Note: Additional formulas are obtained by simultaneous cyclic substitution. For definition of $\sin^{-1}(\ )$, $\cos^{-1}(\ )$, $\tan^{-1}(\ )$, $\cot^{-1}(\ )$ see Sec. 6.08.

Known	Solution* ($\alpha + \beta + \gamma > 180°$)
a, b, c	$p = \dfrac{a+b+c}{2}$ $X = \sqrt{\dfrac{\sin(p-a)\,\sin(p-b)\,\sin(p-c)}{\sin p}}$ $\alpha = 2\tan^{-1}\dfrac{X}{\sin(p-a)} \qquad \beta = 2\tan^{-1}\dfrac{X}{\sin(p-b)} \qquad \gamma = 2\tan^{-1}\dfrac{X}{\sin(p-c)}$
α, β, γ	$\sigma = \dfrac{\alpha+\beta+\gamma}{2}$ $Y = \sqrt{\dfrac{\cos(\sigma-\alpha)\,\cos(\sigma-\beta)\,\cos(\sigma-\gamma)}{-\cos\sigma}}$ $a = 2\tan^{-1}\dfrac{\cos(\sigma-\alpha)}{Y} \qquad b = 2\tan^{-1}\dfrac{\cos(\sigma-\beta)}{Y} \qquad c = 2\tan^{-1}\dfrac{\cos(\sigma-\gamma)}{Y}$
a, b, α	$X = \dfrac{\sin b\,\sin\alpha}{\sin a}$ $Y = \tan\dfrac{a+b}{2}\dfrac{\cos[(\alpha+\beta)/2]}{\cos[(\alpha-\beta)/2]}$ $Z = \tan\dfrac{\alpha+\beta}{2}\dfrac{\cos[(a+b)/2]}{\cos[(a-b)/2]}$ $\beta = \sin^{-1}X \qquad c = 2\tan^{-1}Y \qquad \gamma = 2\cot^{-1}Z$
a, b, γ	$X = \tan^{-1}\dfrac{\cos(\gamma/2)\,\cos[(a-b)/2]}{\sin(\gamma/2)\,\cos[(a+b)/2]}$ $Y = \tan^{-1}\dfrac{\cos(\gamma/2)\,\sin[(a-b)/2]}{\sin(\gamma/2)\,\sin[(a+b)/2]}$ $Z = \dfrac{\cos(\gamma/2)\,\cos[(a-b)/2]}{\sin X}$ $\alpha = X+Y \qquad \beta = X-Y \qquad c = 2\cos^{-1}Z$
a, β, γ	$X = \tan^{-1}\dfrac{\sin(a/2)\,\cos[(\beta-\gamma)/2]}{\cos(a/2)\,\cos[(\beta+\gamma)/2]}$ $Y = \tan^{-1}\dfrac{\sin(a/2)\,\sin[(\beta-\gamma)/2]}{\cos(a/2)\,\sin[(\beta+\gamma)/2]}$ $Z = \dfrac{\cos(a/2)\,\cos[(\beta+\gamma)/2]}{\cos X}$ $b = X+Y \qquad c = X-Y \qquad \alpha = 2\sin^{-1}Z$

*p, σ, X, Y, Z = pocket calculator storage.

a, b, c, d = sides	e, f = diagonals	A = area
A, B, C, D = vertices	$\alpha, \beta, \gamma, \delta$ = angles	ω = central angle $\leq \dfrac{\pi}{2}$

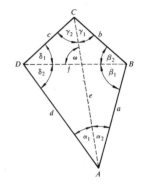

(1) General Formulas $(2s = a + b + c + d)$

$$\alpha = \alpha_1 + \alpha_2 \qquad \beta = \beta_1 + \beta_2$$
$$\gamma = \gamma_1 + \gamma_2 \qquad \delta = \delta_1 + \delta_2$$

$$\alpha + \beta + \gamma + \delta = 360°$$

$$\omega = \cos^{-1}\frac{b^2 + d^2 - a^2 - c^2}{2ef} = \sin^{-1}\frac{2A}{ef}$$

$$A = \frac{ef}{2}\sin\omega = (b^2 + d^2 - a^2 - c^2)\frac{\tan\omega}{4}$$

$$= \sqrt{(s-a)(s-b)(s-c)(s-d) - abcd\,\cos^2\frac{\alpha+\gamma}{2}}$$

(2) Solutions

Known	Solution*		
a, d α, β, δ	$X = 90° - \dfrac{\alpha}{2}$	$Y = \tan^{-1}\left(\dfrac{a-d}{a+d}\cot\dfrac{\alpha}{2}\right)$	$\gamma = 360° - \alpha - \beta - \delta$
	$\beta_1 = X - Y$	$\delta_2 = X + Y$	$f = a\cos(X-Y) + d\cos(X+Y)$
	$b = \dfrac{f\sin(\delta - X - Y)}{\sin\gamma}$		$c = \dfrac{f\sin(\beta - X + Y)}{\sin\gamma}$
a, d $\alpha, \gamma_1, \gamma_2$	$X = \tan^{-1}\dfrac{d\sin\gamma_1}{a\sin(\alpha+\gamma)\sin\gamma_2}$		$Y = \tan^{-1}\dfrac{a\sin\gamma_2}{d\sin(\alpha+\gamma)\sin\gamma_1}$
	$\beta = \cot^{-1}\left[-\dfrac{\cos(\alpha+\gamma-Y)}{\sin(\alpha+\gamma)\cos Y}\right]$		$\delta = \cot^{-1}\left[-\dfrac{\cos(\alpha+\gamma-X)}{\sin(\alpha+\gamma)\cos X}\right]$
	$\alpha_1 = 180° - \delta - \gamma_2$		$\alpha_2 = 180° - \beta - \gamma_1$
	$b = \dfrac{a\sin\alpha_2}{\sin\gamma_1}$		$c = \dfrac{d\sin\alpha_1}{\sin\gamma_2}$
a α_1, α_2 β_1, β_2	$b = \dfrac{a\sin\alpha_2}{\sin(\beta+\alpha_2)}$		$d = \dfrac{a\sin\beta_1}{\sin(\alpha+\beta_1)}$
	$e = \dfrac{a\sin\beta}{\sin(\beta+\alpha_2)}$		$f = \dfrac{a\sin\alpha}{\sin(\alpha+\beta_1)}$
	$c = \sqrt{d^2 + e^2 - 2de\cos\alpha_1} = \sqrt{b^2 + f^2 - 2bf\cos\beta_2}$		

*First four quantities = pocket calculator storage.

3.08 MOLLWEIDE'S EQUATIONS

In the solution of all plane triangles, using trigonometric rules, laws, and procedures, it is wise to always check the calculated answer. The use of the Mollweide equations makes this possible. The Mollweide equations take into account all parts of the plane triangles—right, acute, or obtuse—and allow you to check your calculated answers.

There are two forms of the Mollweide equations, either of which may be used to check the calculated answers to plane triangles:

$$\frac{a + b}{c} = \frac{\cos\left[(A - B)/2\right]}{\sin(C/2)} \quad \text{and} \quad \frac{a - b}{c} = \frac{\sin\left[(A - B)/2\right]}{\cos(C/2)}$$

(1) Procedure

After the triangle is solved, substitute all sides (a, b, and c) and all angles (A, B, and C) into either form of the Mollweide equations and check for a balance between both sides of the equation. A three-decimal-place balance is good, with four, five, or six places indicated for very precise results. The following table shows the angle accuracy obtainable by using the significant figures required in the lengths of the sides as shown in the table.

Required accuracy of the angle	Significant figures required in distances (sides)
10 minutes	3
1 minute	4
10 seconds	5
1 second	6

(2) Proof of the Mollweide Equations

From the Pythagorean theorem, in any right triangle, the square of the hypotenuse is equal to the sum of the squares of the other two sides ($c^2 = a^2 + b^2$), and a triangle with sides in multiples of 3, 4, and 5 will produce an exact right-angle triangle.

We will solve a triangle with sides of 3, 4, and 5 to calculate the internal angles, and then we will substitute all values of sides and angles into the Mollweide equation to check for an equality.

In the accompanying figure of a right triangle, we first calculate the internal angles.

$\tan A = \dfrac{3}{4} = 0.75$

arctan $0.75 = 36.869\ 897\ 65° = $ angle A

$\sin B = \dfrac{4}{5} = 0.80$

arcsin $0.80 = 53.130\ 102\ 35° = $ angle B

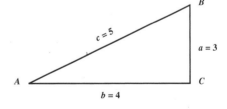

$\angle C = \angle A + \angle B$

$90 = 36.869\ 897\ 65 + 53.130\ 102\ 35$

$90 = 90$

The sum of the internal angles $= 180°$

Substituting the correct sides and computed angles into either of the Mollweide equations will show an equality and prove that the Mollweide equations are valid.

Given:

Side $a = 3$

Side $b = 4$

Side $c = 5$

Angle $A = 36.869\ 897\ 65°$

Angle $B = 53.130\ 102\ 35°$

Angle $C = 90.000\ 000\ 00°$

$$\frac{a - b}{c} = \frac{\sin\left[(A - B)/2\right]}{\cos\left(C/2\right)} \qquad \text{(one form of Mollweide's equation)}$$

$$\frac{3 - 4}{5} = \frac{\sin\left[(36.869\ 897\ 65 - 53.130\ 102\ 35)/2\right]}{\cos\left(90/2\right)}$$

$$\frac{1}{5} = \frac{-0.141\ 421\ 356}{0.707\ 106\ 781} \qquad -0.2 = -0.2 \qquad \text{(equality)}$$

3.09   THE SOLUTION OF TRIANGLES

In right-angled triangles	To solve
Known: Any two sides	Use the Pythagorean theorem to solve for the unknown side; then use the trigonometric functions to solve for the two unknown angles. The third angle is 90°.
Known: Any one side and one angle that is not 90°	Use trigonometric functions to solve for the two unknown sides. The third angle is 180°; two known angles.
Known: Three angles and no sides (*all* triangles)	Cannot be solved because there are an infinite number of triangles which satisfy three known internal angles.
Known: Three sides	Use trigonometric functions to solve for the two unknown angles.

In oblique triangles	To solve
Known: Two sides and any one of two nonincluded angles	Use the law of sines to solve for the second unknown angle. The third angle is 180°; sum of two known angles. Then find the other sides by using the law of sines or the law of tangents.
Known: Two sides and the included angle	Use the law of cosines for one side and the law of sines for the two angles.
Known: Two angles and any one side	Use the law of sines to solve for the other sides or the law of tangents. The third angle is 180°; sum of two known angles.
Known: Three sides	Use the law of cosines to solve for two of the unknown angles. The third angle is 180°; sum of two known angles.
Known: One angle and one side (nonright triangle)	Cannot be solved except under certain conditions. If the triangle is equilateral or isosceles, it may be solved if the known angle is opposite the known side.

(1) Summary of Trigonometric Procedures for Triangles

There are four possible cases in the solution of oblique triangles:

Case 1. Given one side and two angles: a, A, B

Case 2. Given two sides and the angle opposite them: a, b, A or B

Case 3. Given two sides and their included angle: a, b, C

Case 4. Given the three sides: a, b, c

All oblique (nonright-angle) triangles can be solved by use of natural trigonometric functions: the law of sines, the law of cosines, and the angle formula: angle A + angle B + angle C = 180°. This may be done in the following manner:

Case 1. Given a, A, and B, angle C may be found from the angle formula; then sides b and c may be found by using the law of sines twice.

Case 2. Given a, b, and A, angle B may be found by the law of sines, angle C from the angle formula, and side c by the law of sines again.

Case 3. Given a, b, and C, side c may be found by the law of cosines, and angles A and B may be found by the law of sines used twice, or angle A from the law of sines and angle B from the angle formula.

Case 4. Given a, b, and c, the angles may all be found by the law of cosines or angle A may be found from the law of cosines, and angles B and C from the law of sines; or angle A from the law of cosines, angle B from the law of sines, and angle C from the angle formula.

In all cases, the solutions may be checked with the Mollweide equation.

Note: Case 2 is called the *ambiguous* case in which there may be one solution, two solutions, or *no* solution, given a, b, and A.

- If angle $A < 90°$ and $a < b \sin A$, there is *no* solution.
- If angle $A < 90°$ and $a = b \sin A$, there is one solution—a right triangle.
- If angle $A < 90°$ and $b > a > b \sin A$, there are two solutions—oblique triangles.
- If angle $A < 90°$ and $a \geqq b$, there is one solution—an oblique triangle.
- If angle $A < 90°$ and $a \leqq b$, there is *no* solution.
- If angle $A > 90°$ and $a > b$, there is one solution—an oblique triangle.

(2) Signs and Limits of Trigonometric Functions

In a rectangular coordinate system, values for the quadrants are summarized as shown here.

Quadrant II	Y	**Quadrant I**
$(1 - 0) + \sin$		$\sin + (0 - 1)$
$(0 - 1) - \cos$		$\cos + (1 - 0)$
$(\infty - 0) - \tan$		$\tan + (0 - \infty)$
$(0 - \infty) - \cot$		$\cot + (\infty - 0)$
$(\infty - 1) - \sec$		$\sec + (1 - \infty)$
$(1 - \infty) + \csc$		$\csc + (\infty - 1)$
X' ————————————	0	———————— X
Quadrant III		**Quadrant IV**
$(0 - 1) - \sin$		$\sin - (1 - 0)$
$(1 - 0) - \cos$		$\cos + (0 - 1)$
$(0 - \infty) + \tan$		$\tan - (\infty - 0)$
$(\infty - 0) + \cot$		$\cot - (0 - \infty)$
$(1 - \infty) - \sec$		$\sec + (\infty - 1)$
$(\infty - 1) - \csc$	Y'	$\csc - (1 - \infty)$

4

PLANE ANALYTIC GEOMETRY

(1) Systems of Coordinates

(a) Cartesian coordinates

A point P is given by two mutually perpendicular distances x,y (coordinates) measured from two mutually perpendicular axes X,Y (coordinate axes) intersecting at the origin 0.

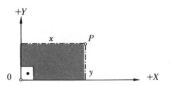

(b) Skew coordinates

A point P is given by the coordinates u,v parallel to two skew axes U,V intersecting at the origin 0.

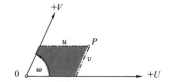

(c) Polar coordinates

A point P is given by two polar coordinates associated with a fixed axis X (polar axis) and a fixed point 0 on this axis (pole). The first coordinate is the radius r, the distance from 0 to P, and the second coordinate is the position angle θ, measured from $+X$ to r.

(2) Relationships

	Cartesian coordinates	Skew coordinates	Polar coordinates
Cartesian coordinates	$x = x$ $y = y$	$x = u + v \cos \omega$ $y = v \sin \omega$	$x = r \cos \theta$ $y = r \sin \theta$
Skew coordinates	$u = x - y \cot \omega$ $v = y \csc \omega$	$u = u$ $v = v$	$u = r\dfrac{\sin (\omega - \theta)}{\sin \omega}$ $v = r\dfrac{\sin \theta}{\sin \omega}$
Polar coordinates	$r = \sqrt{x^2 + y^2}$ $\theta = \tan^{-1}\dfrac{y}{x}$	$r = \sqrt{u^2 + v^2 + 2uv \cos \omega}$ $\theta = \tan^{-1}\dfrac{v \sin \omega}{u + v \cos \omega}$	$r = r$ $\theta = \theta$

(1) Distance of Two Points, Segment in Plane $\overline{P_1 P_2}$

(a) Cartesian coordinates $P_1(x_1, y_1)$; $P_2(x_2, y_2)$

$$d = \sqrt{(x_2 - x_1)^2 + (y_2 - y_1)^2}$$

$$\tan \alpha = \frac{y_2 - y_1}{x_2 - x_1}$$

$$\cos \alpha = \frac{x_2 - x_1}{d}$$

$$\cos \beta = \frac{y_2 - y_1}{d}$$

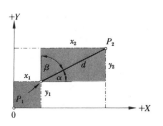

(b) Skew coordinates $P_1(u_1, v_1)$; $P_2(u_2, v_2)$

$$d = \sqrt{(u_2 - u_1)^2 + (v_2 - v_1)^2 + 2(u_2 - u_1)(v_2 - v_1)\cos\omega}$$

$$\tan \alpha = \frac{(v_2 - v_1)\sin\omega}{(u_2 - u_1) + (v_2 - v_1)\cos\omega}$$

$$\cos \alpha = \frac{(u_2 - u_1) + (v_2 - v_1)\cos\omega}{d}$$

$$\cos \gamma = \frac{(v_2 - v_1) + (u_2 - u_1)\cos\omega}{d}$$

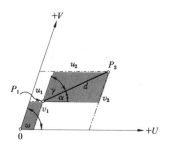

(c) Polar coordinates $P_1(r_1, \theta_1)$; $P_2(r_2, \theta_2)$

$$d = \sqrt{r_1^2 + r_2^2 - 2r_1 r_2 \cos(\theta_1 - \theta_2)}$$

$$\tan \alpha = \frac{r_2 \sin\theta_2 - r_1 \sin\theta_1}{r_2 \cos\theta_2 - r_1 \cos\theta_1}$$

$$\cos \alpha = \frac{r_2 \cos\theta_2 - r_1 \cos\theta_1}{d}$$

$$\cos \beta = \frac{r_2 \sin\theta_2 - r_1 \sin\theta_1}{d}$$

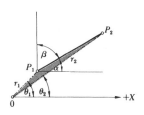

(2) Three Points $P_1(x_1, y_1)$; $P_2(x_2, y_2)$; $P_3(x_3, y_3)$

(a) Area of triangle $(P_1 P_2 P_3)$

$$A = \tfrac{1}{2}\begin{vmatrix} x_1 & y_1 & 1 \\ x_2 & y_2 & 1 \\ x_3 & y_3 & 1 \end{vmatrix}$$

(b) $P_1 P_2 P_3$ on a straight line

$$0 = \begin{vmatrix} x_1 & y_1 & 1 \\ x_2 & y_2 & 1 \\ x_3 & y_3 & 1 \end{vmatrix}$$

(1) Algebraic Transformations

(a) Translation   (b) Rotation   (c) Translation and rotation

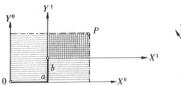 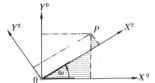 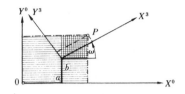

$$x^0 = x^1 + a^0 \qquad\qquad x^0 = x^2 \cos\omega - y^2 \sin\omega \qquad\qquad x^0 = x^3 \cos\omega - y^3 \sin\omega + a^0$$

$$y^0 = y^1 + b^0 \qquad\qquad y^0 = x^2 \sin\omega + y^2 \cos\omega \qquad\qquad y^0 = x^3 \sin\omega + y^3 \cos\omega + b^0$$

$$x^1 = x^0 - a^0 \qquad\qquad x^2 = x^0 \cos\omega + y^0 \sin\omega \qquad\qquad x^3 = (x^0 - a^0)\cos\omega + (y^0 - b^0)\sin\omega$$

$$y^1 = y^0 - b^0 \qquad\qquad y^2 = -x^0 \sin\omega + y^0 \cos\omega \qquad\qquad y^3 = -(x^0 - a^0)\sin\omega + (y^0 - b^0)\cos\omega$$

Note: $0, 1, 2, 3$ are superscripts and not exponents.

(2) Matrix Transformations

(a) Translation   (b) Rotation   (c) Translation and rotation

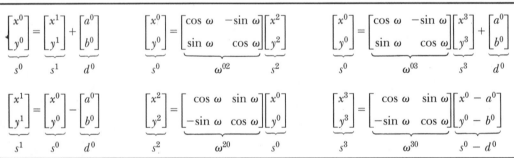

Note: ω matrices are orthogonal, $\det(\omega) = 1$, $\omega^T = \omega^{-1}$.

(3) Transformations in Complex Plane   ($i = \sqrt{-1}$, Sec. 11.01)

(a) Translation

$$(x^0 + iy^0) = (x^1 + iy^1) + (a^0 + ib^0) \qquad\qquad (x^1 + iy^1) = (x^0 + iy^0) - (a^0 + ib^0)$$

(b) Rotation

$$(x^0 + iy^0) = (x^2 + iy^2)\, e^{i\omega} \qquad\qquad (x^2 + iy^2) = (x^0 + iy^0)\, e^{-i\omega}$$

(c) Translation and rotation

$$(x^0 + iy^0) = (x^3 + iy^3)\, e^{i\omega} + (a^0 + ib^0) \qquad\qquad (x^3 + iy^3) = [(x^0 + iy^0) - (a^0 + ib^0)]\, e^{-i\omega}$$

(1) Basic Forms

Direction form

$$y = kx + l$$

Intercept form

$$\frac{x}{a} + \frac{y}{b} = 1$$

Normal form

$$x \cos \beta + y \cos \alpha = n$$

General form

$$Ax + By + C = 0$$

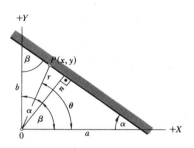

Polar equation

$$r \cos (\theta - \beta) = n$$

(2) Parameters

$A = b$	$\cos \alpha = \pm \dfrac{B}{\sqrt{A^2 + B^2}}$	$\cos \alpha = \pm \dfrac{a}{\sqrt{a^2 + b^2}}$
$B = a$	$\cos \beta = \pm \dfrac{A}{\sqrt{A^2 + B^2}}$	$\cos \beta = \pm \dfrac{b}{\sqrt{a^2 + b^2}}$
$C = -ab$	$n = \pm \dfrac{C}{\sqrt{A^2 + B^2}}$	$n = \pm \dfrac{ab}{\sqrt{a^2 + b^2}}$
$k = \tan \alpha$	$k = -\dfrac{A}{B} \qquad l = -\dfrac{C}{B}$	$k = -\dfrac{b}{a} \qquad l = b$

(3) Two Straight Lines in Plane $(A_1 x + B_1 y + C_1 = 0; \; A_2 x + B_2 y + C_2 = 0)$

$\dfrac{A_1}{B_1} \neq \dfrac{A_2}{B_2}$	Lines intersect at a point	$\dfrac{A_1}{A_2} = \dfrac{B_1}{B_2} = \dfrac{C_1}{C_2}$	Lines coincide
$\dfrac{A_1}{B_1} = -\dfrac{B_2}{A_2}$	Lines are normal	$\dfrac{A_1}{B_1} = \dfrac{A_2}{B_2}$	Lines are parallel

The angle ω between two lines is the clockwise rotation required to transform line 2 into line 1.

$$\tan \omega = \frac{A_1 B_2 - B_1 A_2}{A_1 A_2 + B_1 B_2}$$

(4) Distances $P(x_0, y_0); \; (A_1 x + B_1 y + C_1 = 0; \; A_2 x + B_2 y + C_2 = 0)$

(a) From a line to a point

$$d = \frac{A_1 x_0 + B_1 y_0 + C_1}{\pm \sqrt{A_1^2 + B_1^2}}$$

(b) Between two parallel lines

$$d = \frac{C_2}{\pm \sqrt{A_2^2 + B_2^2}} - \frac{C_1}{\pm \sqrt{A_1^2 + B_1^2}}$$

Note: The sign of the denominator is opposite to the sign of C.

(a) General equation

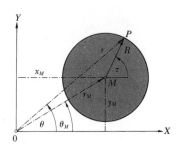

$$Ax^2 + Ay^2 + 2Dx + 2Ey + F = 0$$

Center: $\quad x_M = -\dfrac{D}{A} \qquad y_M = -\dfrac{E}{A}$

Radius: $\quad R = \dfrac{\sqrt{D^2 + E^2 - AF}}{A}$

$$(x - x_M)^2 + (y - y_M)^2 = R^2$$

(b) Parametric equation

$$x = x_M + R \cos \tau \qquad y = y_M + R \sin \tau$$

(c) Polar equation

$$r^2 - 2r_M r \cos (\theta - \theta_M) + r_M{}^2 = R^2$$

(d) Special positions

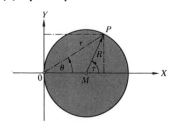

$$x^2 - 2Rx + y^2 = 0$$

$$x = R(1 + \cos \tau)$$

$$y = R \sin \tau$$

$$r = 2R \cos \theta$$

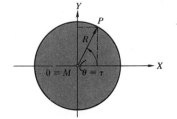

$$\boxed{x^2 + y^2 = R^2}$$

$$x = R \cos \tau$$

$$y = R \sin \tau$$

$$r = R \qquad \theta = \tau$$

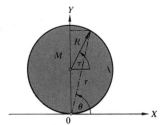

$$x^2 - 2Ry + y^2 = 0$$

$$x = R \cos \tau$$

$$y = R(1 + \sin \tau)$$

$$r = 2R \sin \theta$$

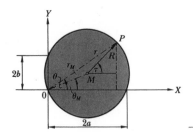

$$x(x - 2a) + y(y - 2b) = 0$$

$$x = a + R \cos \tau$$

$$y = b + R \sin \tau$$

$$r = 2a \cos \theta + 2b \sin \theta$$

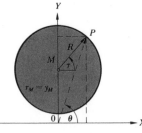

$$x^2 + (y - y_M)^2 = R^2$$

$$x = R \cos \tau$$

$$y = y_M + R \sin \tau$$

$$r^2 - 2rr_M \sin \theta + r_M{}^2 = R^2$$

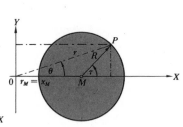

$$(x - x_M)^2 + y^2 = R^2$$

$$x = x_M + R \cos \tau$$

$$y = R \sin \tau$$

$$r^2 - 2rr_M \cos \theta + r_M{}^2 = R^2$$

For definition of x_M, y_M, and R see opposite page.

For definition of k and l see Sec. 4.04.

$$D = \sqrt{R^2 + R^2 k^2 - l^2}$$

$$r_1^2 = x_1^2 + y_1^2$$
$$r_2^2 = x_2^2 + y_2^2$$
$$r_3^2 = x_3^2 + y_3^2$$

(a) Equation of circle given by three points $[P_1(x_1, y_1), P_2(x_2, y_2), P_3(x_3, y_3)]$

$$(x - x_M)^2 + (y - y_M)^2 = R^2$$

$$x_M = \frac{(r_1^2 - r_2^2)(y_2 - y_3) - (y_1 - y_2)(r_2^2 - r_3^2)}{2[(x_1 - x_2)(y_2 - y_3) - (y_1 - y_2)(x_2 - x_3)]}$$

$$y_M = \frac{(r_1^2 - r_2^2)(x_2 - x_3) - (x_1 - x_2)(r_2^2 - r_3^2)}{2[(y_1 - y_2)(x_2 - x_3) - (x_1 - x_2)(y_2 - y_3)]}$$

$$R = \sqrt{(x_1 - x_M)^2 + (y_1 - y_M)^2}$$
$$= \sqrt{(x_2 - x_M)^2 + (y_2 - y_M)^2}$$
$$= \sqrt{(x_3 - x_M)^2 + (y_3 - y_M)^2}$$

(b) Point of intersection of a straight line and a circle.
If $x^2 + y^2 = R^2$ and $y = kx + l$ are, respectively, equations of a circle and a straight line, the coordinates of their points of intersection are

$$x_{1,2} = \frac{-kl \pm \sqrt{D}}{1 + k^2} \qquad y_{1,2} = \frac{l \pm k\sqrt{D}}{1 + k^2}$$

If $D > 0$, the line intersects the circle at two real points.

If $D = 0$, the line is tangent to the circle ($x_1 = x_2$, $y_1 = y_2$).

If $D < 0$, the line does not intersect the circle, and $x_{1,2}$, $y_{1,2}$ are conjugate complex numbers (coordinates of two imaginary points).

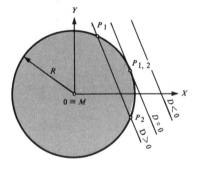

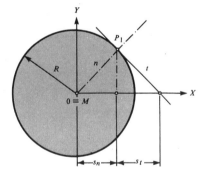

(c) Tangent and normal

Equation of tangent at P_1: $xx_1 + yy_1 = R^2$

Equation of normal at P_1: $yx_1 - xy_1 = R^2$

Length of tangent:

$$t = R\left|\frac{y_1}{x_1}\right|$$

Length of subtangent:

$$s_t = \left|\frac{y_1^2}{x_1}\right|$$

Length of normal:

$$n = R$$

Length of subnormal:

$$s_n = |x_1|$$

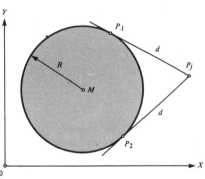

(d) Length of tangent between P_j and the point of contact P_1 or P_2 is

$$d = \sqrt{(x_j - x_M)^2 + (y_j - y_M)^2 - R^2}$$

(a) Notation

$$F = \text{focus} \qquad \overline{PF}_1 + \overline{PF}_2 = 2a \qquad M = \text{center}$$

$$\overline{AB} = 2a \qquad F_1F_2 = 2e \qquad \overline{CD} = 2b$$

Major axis   Minor axis

$$e = \sqrt{a^2 - b^2} \qquad \text{Linear eccentricity}$$

$$\frac{e}{a} = \epsilon < 1 \qquad \text{Numerical eccentricity}$$

$$2p = \frac{2b^2}{a} \qquad 2q = \frac{2a^2}{b}$$

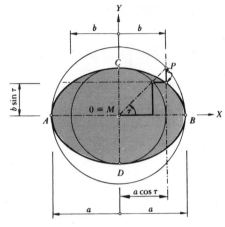

(b) Cartesian equation $(M \equiv 0)$

$$\frac{x^2}{a^2} + \frac{y^2}{b^2} = 1$$

(c) Parametric equation $(M \equiv 0)$

$$x = a \cos \tau \qquad y = b \sin \tau$$

(d) Polar equations

Pole at 0 $\quad r^2 = \dfrac{b^2}{1 - \epsilon^2 \cos^2 \theta}$ $\qquad$ Pole at F_2 $\quad r = \dfrac{p}{1 + \epsilon \cos \theta}$

(e) Normal position

$$Ax^2 + Cy^2 + 2Dx + 2Ey + F = 0$$

Center: $\qquad x_M = -\dfrac{D}{A} \qquad y_M = -\dfrac{E}{C}$

$$a = \sqrt{\frac{CD^2 + AE^2 - ACF}{A^2C}}$$

$$b = \sqrt{\frac{CD^2 + AE^2 - ACF}{AC^2}}$$

$$\frac{(x - x_M)^2}{a^2} + \frac{(y - y_M)^2}{b^2} = 1$$

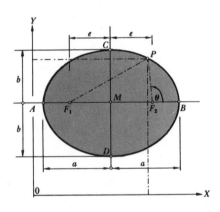

(f) Special positions

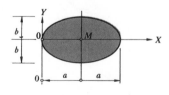

$$y^2 = 2px - \frac{p}{a}x^2$$

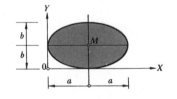

$$\frac{x^2}{a^2} - 2\left(\frac{x}{a} + \frac{y}{b}\right) + \frac{y^2}{b^2} = -1$$

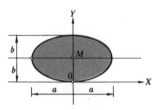

$$x^2 = 2qy - \frac{q}{b}y^2$$

(a) Notation

$$F = \text{focus} \qquad \overline{PF}_1 - \overline{PF}_2 = 2a \qquad M = \text{center}$$

$$\overline{AB} = 2a \qquad \overline{F_1 F_2} = 2e \qquad \overline{CD} = 2b$$

Major axis $\qquad\qquad\qquad\qquad$ Minor axis

$$e = \sqrt{a^2 + b^2} \qquad \text{Linear eccentricity}$$

$$\frac{e}{a} = \epsilon > 1 \qquad \text{Numerical eccentricity}$$

$$2p = \frac{2b^2}{a} \qquad 2q = \frac{2a^2}{b}$$

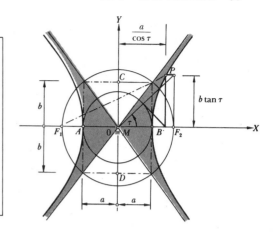

(b) Cartesian equation $\quad (M \equiv 0)$

$$\boxed{\frac{x^2}{a^2} - \frac{y^2}{b^2} = 1}$$

(c) Parametric equation $\quad (M \equiv 0)$

$$x = \frac{a}{\cos \tau} \qquad y = \pm b \tan \tau$$

(d) Polar equations

Pole at $0 \quad r^2 = \dfrac{b^2}{\epsilon^2 \cos^2 \theta - 1} \qquad$ Pole at $F \quad r = \dfrac{p}{1 + \epsilon \cos \theta}$

(e) Normal position

$$Ax^2 - Cy^2 + 2Dx + 2Ey + F = 0$$

Center: $\qquad x_M = -\dfrac{D}{A} \qquad y_M = \dfrac{E}{C}$

$$a = \sqrt{\frac{CD^2 - AE^2 - ACF}{A^2 C}}$$

$$b = \sqrt{\frac{CD^2 - AE^2 - ACF}{AC^2}}$$

$$\frac{(x - x_M)^2}{a^2} - \frac{(y - y_M)^2}{b^2} = 1$$

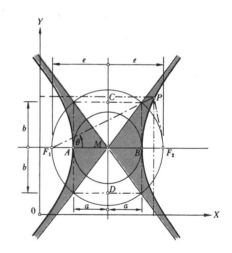

(f) Special positions $\quad$ (Sec. 4.12)

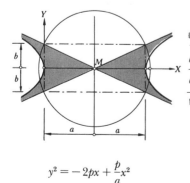

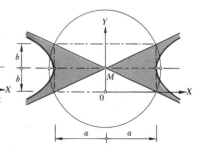

$$y^2 = -2px + \frac{p}{a}x^2 \qquad\qquad \frac{x^2}{a^2} - 2\left(\frac{x}{a} - \frac{y}{b}\right) - \frac{y^2}{b^2} = 1 \qquad\qquad x^2 = -2qy + \frac{2}{b}y^2 + 2a^2$$

(a) Notation

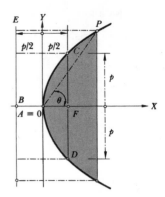

F = focus	$PF = PE$	0 = vertex
$\overline{AF} = \dfrac{p}{2}$	$\overline{BF} = \overline{FC} = p$	$\overline{BA} = \dfrac{p}{2}$

$2p$ = parameter = latus rectum

$\epsilon = 1$ (numerical eccentricity)

(b) Cartesian equation   ($A \equiv 0, X$ **axis**)

$$y^2 = 2px \qquad \begin{cases} p > 0 & \text{Open right} \\ p < 0 & \text{Open left} \end{cases}$$

(c) Parametric equation   ($A \equiv 0, X$ **axis**)

$$x = \frac{p}{2}\tau^2 \qquad y = p\tau$$

(d) Polar equations   (X **axis**)

Pole at 0 $r = 2p \cos \theta\,(1 + \cot^2 \theta)$ Pole at F $r = \dfrac{p}{1 - \cos \theta}$

(e) Normal position   (X **axis**)

$$Cy^2 + 2Dx + 2Ey + F = 0$$

Vertex: $x_A = \dfrac{E^2 - CF}{2CD}$ $y_A = -\dfrac{E}{C}$

$$p = -\frac{D}{C}$$

$$(y - y_A)^2 = 2p(x - x_A)$$

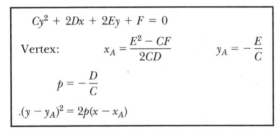

(f) Normal position   (Y **axis**)

$$Ax^2 + 2Dx + 2Ey + F = 0$$

Vertex: $x_A = -\dfrac{D}{A}$ $y_A = \dfrac{D^2 - AF}{2AE}$

$$p = -\frac{E}{A}$$

$$(x - x_A)^2 = 2p(y - y_A)$$

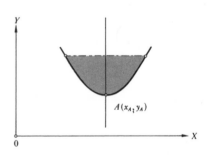

(g) Special cases   (Sec. 4.12)

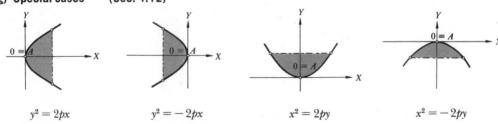

$y^2 = 2px$ $y^2 = -2px$ $x^2 = 2py$ $x^2 = -2py$

5
SPACE ANALYTIC GEOMETRY

(1) Systems of Coordinates

(a) Cartesian coordinates

A point P is given by three mutually perpendicular distances x, y, z (coordinates) measured from three mutually perpendicular YZ, ZX, XY planes, respectively. The lines of intersection of these planes are the coordinate axes X, Y, Z, and the point of their intersection is the origin 0. The right-hand system is shown.

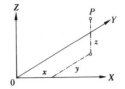

(b) Cylindrical coordinates

A point P is given by its polar coordinates r and θ, in the XY plane, and the cartesian coordinate z.

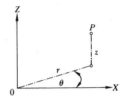

(c) Spherical coordinates

A point P is given by a position angle θ measured from a fixed axis X, a position angle ϕ measured from another fixed axis Z, normal to X, and the position radius ρ, measured from the point of intersection of X and Y, designated as the pole 0.

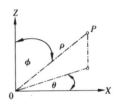

(2) Relationships

	Cartesian	Cylindrical	Spherical
Cartesian	$x = x$ $y = y$ $z = z$	$x = r \cos \theta$ $y = r \sin \theta$ $z = z$	$x = \rho \cos \theta \sin \phi$ $y = \rho \sin \theta \sin \phi$ $z = \rho \cos \phi$
Cylindrical	$r = \sqrt{x^2 + y^2}$ $\theta = \tan^{-1}\dfrac{y}{x}$ $z = z$	$r = r$ $\theta = \theta$ $z = z$	$r = \rho \sin \phi$ $\theta = \theta$ $z = \rho \cos \phi$
Spherical	$\rho = \sqrt{x^2 + y^2 + z^2}$ $\theta = \tan^{-1}\dfrac{y}{x}$ $\phi = \cos^{-1}\dfrac{z}{\sqrt{x^2 + y^2 + z^2}}$	$\rho = \sqrt{r^2 + z^2}$ $\theta = \theta$ $\phi = \cos^{-1}\dfrac{z}{\sqrt{r^2 + z^2}}$	$\rho = \rho$ $\theta = \theta$ $\phi = \phi$

(1) Distance of Two Points, Segment $P_1P_2 = d_{12}$

Segment

$$d_{12} = \sqrt{(x_2 - x_1)^2 + (y_2 - y_1)^2 + (z_2 - z_1)^2}$$

Direction cosines

$$\alpha_{12} = \cos X d_{12} = \frac{x_2 - x_1}{d_{12}} = \frac{d_{12x}}{d_{12}}$$

$$\beta_{12} = \cos Y d_{12} = \frac{y_2 - y_1}{d_{12}} = \frac{d_{12y}}{d_{12}}$$

$$\gamma_{12} = \cos Z d_{12} = \frac{z_2 - z_1}{d_{12}} = \frac{d_{12z}}{d_{12}}$$

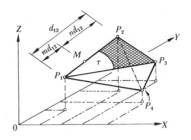

Relationship

$$\alpha^2 + \beta^2 + \gamma^2 = +1$$

(2) Components of Segment d_{12}

$$d_{12x} = d_{12}\alpha_{12} \qquad\qquad d_{12y} = d_{12}\beta_{12} \qquad\qquad d_{12z} = d_{12}\gamma_{12}$$

$$d_{12} = d_{12x}\alpha_{12} + d_{12y}\beta_{12} + d_{12z}\gamma_{12}$$

(3) Coordinates of M Dividing P_1P_2 in Ratio $m : n$

$$x_M = \frac{nx_1 + mx_2}{m + n} \qquad\qquad y_M = \frac{ny_1 + my_2}{m + n} \qquad\qquad z_M = \frac{nz_1 + mz_2}{m + n}$$

(4) Area and Centroid of Triangle with Vertices $P_1P_2P_3$

$$A = \sqrt{A_1^2 + A_2^2 + A_3^2}$$

$$A_1 = \frac{1}{2}\begin{vmatrix} y_1 & z_1 & 1 \\ y_2 & z_2 & 1 \\ y_3 & z_3 & 1 \end{vmatrix} \qquad A_2 = \frac{1}{2}\begin{vmatrix} z_1 & x_1 & 1 \\ z_2 & x_2 & 1 \\ z_3 & x_3 & 1 \end{vmatrix} \qquad A_3 = \frac{1}{2}\begin{vmatrix} x_1 & y_1 & 1 \\ x_2 & y_2 & 1 \\ x_3 & y_3 & 1 \end{vmatrix}$$

$$x_c = \frac{x_1 + x_2 + x_3}{3} \qquad y_c = \frac{y_1 + y_2 + y_3}{3} \qquad z_c = \frac{z_1 + z_2 + z_3}{3}$$

(5) Volume of Tetrahedron with Vertices $P_1 P_2 P_3 P_4$

$$V = \frac{1}{6}\begin{vmatrix} x_1 & y_1 & z_1 & 1 \\ x_2 & y_2 & z_2 & 1 \\ x_3 & y_3 & z_3 & 1 \\ x_4 & y_4 & z_4 & 1 \end{vmatrix}$$

(6) Angle τ between d_{12} and d_{13}

$$\cos \tau = \alpha_{12}\alpha_{13} + \beta_{12}\beta_{13} + \gamma_{12}\gamma_{13}$$

$$= \frac{d_{12x}d_{13x} + d_{12y}d_{13y} + d_{12z}d_{13z}}{d_{12}d_{13}}$$

(1) Basic Equations of a Plane

Direction form

$$z = k_1 x + k_2 y + l$$

Intercept form

$$\frac{x}{a} + \frac{y}{b} + \frac{z}{c} = 1$$

Normal form

$$\alpha x + \beta y + \gamma z = n$$

General form

$$Ax + By + Cz + D = 0$$

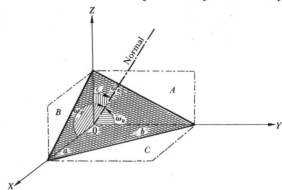

Direction cosines

$$\alpha = \cos \omega_x \qquad \beta = \cos \omega_y \qquad \gamma = \cos \omega_z$$

(2) Relationships

$A = bc$	$\alpha = \dfrac{A}{\pm \sqrt{A^2 + B^2 + C^2}}$	$\alpha = \dfrac{bc}{\pm \sqrt{(ab)^2 + (bc)^2 + (ca)^2}}$
$B = ca$	$\beta = \dfrac{B}{\pm \sqrt{A^2 + B^2 + C^2}}$	$\beta = \dfrac{ca}{\pm \sqrt{(ab)^2 + (bc)^2 + (ca)^2}}$
$C = ab$	$\gamma = \dfrac{C}{\pm \sqrt{A^2 + B^2 + C^2}}$	$\gamma = \dfrac{ab}{\pm \sqrt{(ab)^2 + (bc)^2 + (ca)^2}}$
$D = -abc$	$n = \dfrac{-D}{\pm \sqrt{A^2 + B^2 + C^2}}$	$n = \dfrac{abc}{\pm \sqrt{(ab)^2 + (bc)^2 + (ca)^2}}$
$a = -\dfrac{D}{A}$	$k_1 = -\dfrac{A}{C}$	$k_1 = -\dfrac{c}{a}$
$b = -\dfrac{D}{B}$	$k_2 = -\dfrac{B}{C}$	$k_2 = -\dfrac{c}{b}$
$c = -\dfrac{D}{C}$	$l = -\dfrac{D}{C}$	$l = c$

(3) Plane Passing through a Point P_i in a Given Direction

$$A(x - x_i) + B(y - y_i) + C(z - z_i) = 0$$

(4) Plane Passing through Three Points P_i, P_j, P_k

$$\begin{vmatrix} y_i & z_i & 1 \\ y_j & z_j & 1 \\ y_k & z_k & 1 \end{vmatrix} x + \begin{vmatrix} z_i & x_i & 1 \\ z_j & x_j & 1 \\ z_k & x_k & 1 \end{vmatrix} y + \begin{vmatrix} x_i & y_i & 1 \\ x_j & y_j & 1 \\ x_k & y_k & 1 \end{vmatrix} z = \begin{vmatrix} x_i & y_i & z_i \\ x_j & y_j & z_j \\ x_k & y_k & z_k \end{vmatrix}$$

(5) Distance between the Point P_i and the Plane $Ax + By + Cz + D = 0$

$$d = \frac{Ax_i + By_i + Cz_i + D}{\pm \sqrt{A^2 + B^2 + C^2}}$$

Note: The sign of the denominator is opposite to the sign of D.

(1) Relationships of Two Planes

If two planes are given by their equations as

$$A_1x + B_1y + C_1z + D_1 = 0 \qquad\qquad A_2x + B_2y + C_2z + D_2 = 0$$

then the following are their relationships:

$A_1 : B_1 : C_1 \neq A_2 : B_2 : C_2$	Planes intersect
$A_1 : B_1 : C_1 = A_2 : B_2 : C_2$	Planes are parallel
$A_1A_2 + B_1B_2 + C_1C_2 = 0$	Planes are normal
$A_1 : A_2 = B_1 : B_2 = C_1 : C_2 = D_1 : D_2$	Planes coincide

(2) Angle between the Normals of Two Planes

$$\cos \tau = \frac{A_1A_2 + B_1B_2 + C_1C_2}{\sqrt{A_1^2 + B_1^2 + C_1^2}\,\sqrt{A_2^2 + B_2^2 + C_2^2}}$$

(3) Distance between Two Parallel Planes

$$d = \frac{D_1 - D_2}{\pm\sqrt{A^2 + B^2 + C^2}}$$

$$A_1 = A_2 = A$$
$$B_1 = B_2 = B$$
$$C_1 = C_2 = C$$

(4) Intersection of Two Planes

$$\begin{vmatrix} C_1 & C_2 \\ A_1 & A_2 \end{vmatrix}x + \begin{vmatrix} C_1 & C_2 \\ B_1 & B_2 \end{vmatrix}y + \begin{vmatrix} C_1 & C_2 \\ D_1 & D_2 \end{vmatrix} = 0$$

$$\begin{vmatrix} A_1 & A_2 \\ B_1 & B_2 \end{vmatrix}y + \begin{vmatrix} A_1 & A_2 \\ C_1 & C_2 \end{vmatrix}z + \begin{vmatrix} A_1 & A_2 \\ D_1 & D_2 \end{vmatrix} = 0$$

$$\begin{vmatrix} B_1 & B_2 \\ C_1 & C_2 \end{vmatrix}z + \begin{vmatrix} B_1 & B_2 \\ A_1 & A_2 \end{vmatrix}x + \begin{vmatrix} B_1 & B_2 \\ D_1 & D_2 \end{vmatrix} = 0$$

Each one of these equations represents the projected line of intersection in the respective coordinate plane.

(5) Point of Intersection of Three Planes

$$x = -\frac{\begin{vmatrix} D_1 & B_1 & C_1 \\ D_2 & B_2 & C_2 \\ D_3 & B_3 & C_3 \end{vmatrix}}{\begin{vmatrix} A_1 & B_1 & C_1 \\ A_2 & B_2 & C_2 \\ A_3 & B_3 & C_3 \end{vmatrix}} \qquad y = -\frac{\begin{vmatrix} A_1 & D_1 & C_1 \\ A_2 & D_2 & C_2 \\ A_3 & D_3 & C_3 \end{vmatrix}}{\begin{vmatrix} A_1 & B_1 & C_1 \\ A_2 & B_2 & C_2 \\ A_3 & B_3 & C_3 \end{vmatrix}} \qquad z = -\frac{\begin{vmatrix} A_1 & B_1 & D_1 \\ A_2 & B_2 & D_2 \\ A_3 & B_3 & D_3 \end{vmatrix}}{\begin{vmatrix} A_1 & B_1 & C_1 \\ A_2 & B_2 & C_2 \\ A_3 & B_3 & C_3 \end{vmatrix}}$$

(6) Plane of Symmetry of Two Planes

$$\frac{A_1x + B_1y + C_1z + D_1}{\pm\sqrt{A_1^2 + B_1^2 + C_1^2}} + \frac{A_2x + B_2y + C_2z + D_2}{\pm\sqrt{A_2^2 + B_2^2 + C_2^2}} = 0$$

Note: The sign of the denominator is opposite to the sign of the respective *D*.

(1) General Form

Two linearly independent equations,

$$A_1x + B_1y + C_1z + D_1 = 0 \qquad\qquad A_2x + B_2y + C_2z + D_2 = 0$$

represent a straight line in space. The projections of this line in the coordinate planes and their constants are as follows:

$$\bar{B}x + \bar{A}y + \bar{D}_{xy} = 0 \qquad\qquad \bar{A}z + \bar{C}x + \bar{D}_{zx} = 0 \qquad\qquad \bar{C}y + \bar{B}z + \bar{D}_{yz} = 0$$

$$\bar{A} = \begin{vmatrix} B_1 & C_1 \\ B_2 & C_2 \end{vmatrix} \qquad \bar{D}_{xy} = \begin{vmatrix} C_1 & D_1 \\ C_2 & D_2 \end{vmatrix} = \bar{D}_{yx} \qquad \bar{\alpha} = \frac{A}{\sqrt{\bar{A}^2 + \bar{B}^2 + \bar{C}^2}}$$

$$\bar{B} = \begin{vmatrix} C_1 & A_1 \\ C_2 & A_2 \end{vmatrix} \qquad \bar{D}_{zx} = \begin{vmatrix} A_1 & D_1 \\ A_2 & D_2 \end{vmatrix} = \bar{D}_{xz} \qquad \bar{\beta} = \frac{B}{\sqrt{\bar{A}^2 + \bar{B}^2 + \bar{C}^2}}$$

$$\bar{C} = \begin{vmatrix} A_1 & B_1 \\ A_2 & B_2 \end{vmatrix} \qquad \bar{D}_{yz} = \begin{vmatrix} B_1 & D_1 \\ B_2 & D_2 \end{vmatrix} = \bar{D}_{zy} \qquad \bar{\gamma} = \frac{C}{\sqrt{\bar{A}^2 + \bar{B}^2 + \bar{C}^2}}$$

(2) Direction Form

$y = k_{yx}x + l_{yx}$ $z = k_{zx}x + l_{zx}$	$x = k_{xy}y + l_{xy}$ $z = k_{zy}y + l_{zy}$	$x = k_{xz}z + l_{xz}$ $y = k_{yz}z + l_{yz}$
$k_{yx} = -\dfrac{\bar{B}}{\bar{A}}$ $k_{zx} = -\dfrac{\bar{C}}{\bar{A}}$	$k_{xy} = -\dfrac{\bar{A}}{\bar{B}}$ $k_{zy} = -\dfrac{\bar{C}}{\bar{B}}$	$k_{xz} = -\dfrac{\bar{A}}{\bar{C}}$ $k_{yz} = -\dfrac{\bar{B}}{\bar{C}}$
$l_{yx} = -\dfrac{\bar{D}_{xy}}{\bar{A}}$ $l_{zx} = -\dfrac{\bar{D}_{zx}}{\bar{A}}$	$l_{xy} = -\dfrac{\bar{D}_{yx}}{\bar{B}}$ $l_{zy} = -\dfrac{\bar{D}_{zy}}{\bar{B}}$	$l_{xz} = -\dfrac{\bar{D}_{xz}}{\bar{C}}$ $l_{yz} = -\dfrac{\bar{D}_{yz}}{\bar{C}}$

(3) Straight Line Passing through a Point P_i in a Given Direction

$$\frac{x - x_i}{\bar{\alpha}} = \frac{y - y_i}{\bar{\beta}} = \frac{z - z_i}{\bar{\gamma}}$$

(4) Straight Line Passing through Points P_i and P_j

$$\frac{x - x_i}{x_j - x_i} = \frac{y - y_i}{y_j - y_i} = \frac{z - z_i}{z_j - z_i}$$

(5) Parametric Equation of a Straight Line through a Point P_i

$$x = x_i + \bar{\alpha}t \qquad\qquad y = y_i + \bar{\beta}t \qquad\qquad z = z_i + \bar{\gamma}t$$

(6) Distance of the Point P_k to a Straight Line through a Point P_i

$$d = \sqrt{\begin{vmatrix} \dfrac{x_k - x_i}{\bar{\alpha}} & \dfrac{y_k - y_i}{\bar{\beta}} \end{vmatrix}^2 + \begin{vmatrix} \dfrac{y_k - y_i}{\bar{\beta}} & \dfrac{z_k - z_i}{\bar{\gamma}} \end{vmatrix}^2 + \begin{vmatrix} \dfrac{z_k - z_i}{\bar{\gamma}} & \dfrac{x_k - x_i}{\bar{\alpha}} \end{vmatrix}^2}$$

(1) Straight Line and Plane

If a plane and a straight line are given, respectively, as

$$Ax + By + Cz + D = 0 \qquad \bar{B}x + \bar{A}y + \bar{D}_{xy} = 0 \qquad \bar{A}z + \bar{C}x + \bar{D}_{zx} = 0 \qquad \bar{C}y + \bar{B}z + \bar{D}_{yz} = 0$$

Then the following are their relationships:

$A : B : C \neq \bar{A} : \bar{B} : \bar{C}$	Line and plane intersect
$A : \bar{A} = B : \bar{B} = C : \bar{C}$	Line normal to the plane
$A\bar{A} + B\bar{B} + C\bar{C} = 0$	Line parallel to the plane

Note: The line lies in the plane if they have a common point and $A\bar{A} + B\bar{B} + C\bar{C} = 0$.

(2) Angle between a Straight Line and a Plane

α, β, γ = direction cosines, plane

$\bar{\alpha}, \bar{\beta}, \bar{\gamma}$ = direction cosines, line

$$\sin \tau = \alpha\bar{\alpha} + \beta\bar{\beta} + \gamma\bar{\gamma}$$

(3) Two Straight Lines   (Sec. 5.05 – 3)

$$\frac{x - x_i}{\bar{\alpha}_1} = \frac{y - y_i}{\bar{\beta}_1} = \frac{z - z_i}{\bar{\gamma}_1} \qquad\qquad \frac{x - x_j}{\bar{\alpha}_2} = \frac{y - y_j}{\bar{\beta}_2} = \frac{z - z_j}{\bar{\gamma}_2}$$

(a) The lines are parallel if

$$\bar{\alpha}_1 = \bar{\alpha}_2 \qquad \bar{\beta}_1 = \bar{\beta}_2 \qquad \bar{\gamma}_1 = \bar{\gamma}_2$$

(b) The lines are normal if

$$\bar{\alpha}_1\bar{\alpha}_2 + \bar{\beta}_1\bar{\beta}_2 + \bar{\gamma}_1\bar{\gamma}_2 = 0$$

(c) The lines are coplanar if

$$\begin{vmatrix} x_j - x_i & y_j - y_i & z_j - z_i \\ \bar{\alpha}_1 & \bar{\beta}_1 & \bar{\gamma}_1 \\ \bar{\alpha}_2 & \bar{\beta}_2 & \bar{\gamma}_2 \end{vmatrix} = \Delta = 0$$

(d) The angle between these lines is given by

$$\cos \tau = \bar{\alpha}_1\bar{\alpha}_2 + \bar{\beta}_1\bar{\beta}_2 + \bar{\gamma}_1\bar{\gamma}_2$$

(e) The distance between these lines if $\Delta \neq 0$ is

$$d = \frac{\begin{vmatrix} x_j - x_i & y_j - y_i & z_j - z_i \\ \bar{\alpha}_1 & \bar{\beta}_1 & \bar{\gamma}_1 \\ \bar{\alpha}_2 & \bar{\beta}_2 & \bar{\gamma}_2 \end{vmatrix}}{\sqrt{\begin{vmatrix} \bar{\alpha}_1 & \bar{\alpha}_2 \\ \bar{\beta}_1 & \bar{\beta}_2 \end{vmatrix}^2 + \begin{vmatrix} \bar{\beta}_1 & \bar{\beta}_2 \\ \bar{\gamma}_1 & \bar{\gamma}_2 \end{vmatrix}^2 + \begin{vmatrix} \bar{\gamma}_1 & \bar{\gamma}_2 \\ \bar{\alpha}_1 & \bar{\alpha}_2 \end{vmatrix}^2}}$$

(1) Transformation Matrices

Translation

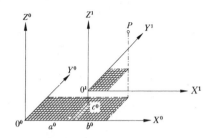

Rotation

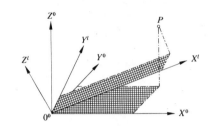

$$\underbrace{\begin{bmatrix} x^0 \\ y^0 \\ z^0 \end{bmatrix}}_{s^0} = \underbrace{\begin{bmatrix} x^1 \\ y^1 \\ z^1 \end{bmatrix}}_{s^1} + \underbrace{\begin{bmatrix} a^0 \\ b^0 \\ c^0 \end{bmatrix}}_{d^0}$$

$$\underbrace{\begin{bmatrix} x^0 \\ y^0 \\ z^0 \end{bmatrix}}_{s^0} = \underbrace{\begin{bmatrix} \alpha_x & \alpha_y & \alpha_z \\ \beta_x & \beta_y & \beta_z \\ \gamma_x & \gamma_y & \gamma_z \end{bmatrix}}_{\omega^{0l}} \underbrace{\begin{bmatrix} x^l \\ y^l \\ z^l \end{bmatrix}}_{s^l}$$

$$\underbrace{\begin{bmatrix} x^1 \\ y^1 \\ z^1 \end{bmatrix}}_{s^1} = \underbrace{\begin{bmatrix} x^0 \\ y^0 \\ z^0 \end{bmatrix}}_{s^0} - \underbrace{\begin{bmatrix} a^0 \\ b^0 \\ c^0 \end{bmatrix}}_{d^0}$$

$$\underbrace{\begin{bmatrix} x^l \\ y^l \\ z^l \end{bmatrix}}_{s^l} = \underbrace{\begin{bmatrix} \alpha_x & \beta_x & \gamma_x \\ \alpha_y & \beta_y & \gamma_y \\ \alpha_z & \beta_z & \gamma_z \end{bmatrix}}_{\omega^{l0}} \underbrace{\begin{bmatrix} x^0 \\ y^0 \\ z^0 \end{bmatrix}}_{s^0}$$

Note: 0, 1, and *l* are superscripts designating the system.

(2) Direction Cosines   (for derivation see Sec. 5.08)

$$\alpha_x = \cos{(x^0 x^l)} = \cos{(x^l x^0)} \qquad \alpha_y = \cos{(x^0 y^l)} = \cos{(y^l x^0)} \qquad \alpha_z = \cos{(x^0 z^l)} = \cos{(z^l x^0)}$$

$$\beta_x = \cos{(y^0 x^l)} = \cos{(x^l y^0)} \qquad \beta_y = \cos{(y^0 y^l)} = \cos{(y^l y^0)} \qquad \beta_z = \cos{(y^0 z^l)} = \cos{(z^l y^0)}$$

$$\gamma_x = \cos{(z^0 x^l)} = \cos{(x^l z^0)} \qquad \gamma_y = \cos{(z^0 y^l)} = \cos{(y^l z^0)} \qquad \gamma_z = \cos{(z^0 z^l)} = \cos{(z^l z^0)}$$

(3) Properties of ω Matrices

$$s^0 = \omega^{0l} s^l$$

$$s^l = \omega^{l0} s^0$$

$$\omega^{0l} \omega^{l0} = I$$

$$\omega^{l0} \omega^{0l} = I$$

$$\omega^{0l} = (\omega^{l0})^T = (\omega^{l0})^{-1}$$

$$\omega^{l0} = (\omega^{0l})^T = (\omega^{0l})^{-1}$$

Note: ω matrices are orthogonal.

(4) Properties of Direction Cosines

$$\alpha_x^2 + \alpha_y^2 + \alpha_z^2 = 1 \qquad\qquad \alpha_x \beta_x + \alpha_y \beta_y + \alpha_z \beta_z = 0$$

$$\beta_x^2 + \beta_y^2 + \beta_z^2 = 1 \qquad\qquad \beta_x \gamma_x + \beta_y \gamma_y + \beta_z \gamma_z = 0$$

$$\gamma_x^2 + \gamma_y^2 + \gamma_z^2 = 1 \qquad\qquad \gamma_x \alpha_x + \gamma_y \alpha_y + \gamma_z \alpha_z = 0$$

Diagonal terms of matrix product $\omega^{0l} \omega^{l0}$

Off-diagonal terms of matrix product $\omega^{0l} \omega^{l0}$

$$\alpha_x \alpha_y + \beta_x \beta_y + \gamma_x \gamma_y = 0 \qquad\qquad \alpha_x^2 + \beta_x^2 + \gamma_x^2 = 1$$

$$\alpha_y \alpha_z + \beta_y \beta_z + \gamma_y \gamma_z = 0 \qquad\qquad \alpha_y^2 + \beta_y^2 + \gamma_y^2 = 1$$

$$\alpha_z \alpha_x + \beta_z \beta_x + \gamma_z \gamma_x = 0 \qquad\qquad \alpha_z^2 + \beta_z^2 + \gamma_z^2 = 1$$

Off-diagonal terms of matrix product $\omega^{l0} \omega^{0l}$

Diagonal terms of matrix product $\omega^{l0} \omega^{0l}$

(1) Successive Rotation

Every space rotation can be resolved into three components, and every component rotation can be computed independently.

Rotation ω^j about Z^0 Rotation ω^k about Y^j Rotation ω^l about X^k

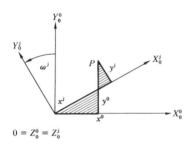
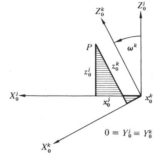
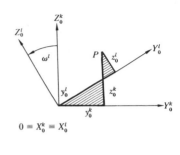

$$[\omega^{oj}] = \begin{bmatrix} \cos\omega^j & -\sin\omega^j & 0 \\ \sin\omega^j & \cos\omega^j & 0 \\ 0 & 0 & 1 \end{bmatrix} \quad [\omega^{jk}] = \begin{bmatrix} \cos\omega^k & 0 & \sin\omega^k \\ 0 & 1 & 0 \\ -\sin\omega^k & 0 & \cos\omega^k \end{bmatrix} \quad [\omega^{kl}] = \begin{bmatrix} 1 & 0 & 0 \\ 0 & \cos\omega^l & -\sin\omega^l \\ 0 & \sin\omega^l & \cos\omega^l \end{bmatrix}$$

$$[\omega^{j0}] = \begin{bmatrix} \cos\omega^j & \sin\omega^j & 0 \\ -\sin\omega^j & \cos\omega^j & 0 \\ 0 & 0 & 1 \end{bmatrix} \quad [\omega^{kj}] = \begin{bmatrix} \cos\omega^k & 0 & -\sin\omega^k \\ 0 & 1 & 0 \\ \sin\omega^k & 0 & \cos\omega^k \end{bmatrix} \quad [\omega^{lk}] = \begin{bmatrix} 1 & 0 & 0 \\ 0 & \cos\omega^l & \sin\omega^l \\ 0 & -\sin\omega^l & \cos\omega^l \end{bmatrix}$$

(2) Successive Matrix Multiplication

The resulting rotational transformation matrix (ω^{0l} or ω^{l0}) is equal to the chain-matrix product of the component matrices, executed in the order of rotation.

$$\omega^{0l} = \omega^{0j}\omega^{jk}\omega^{kl}$$

$$\begin{bmatrix} \alpha_x & \alpha_y & \alpha_z \\ \beta_x & \beta_y & \beta_z \\ \gamma_x & \gamma_y & \gamma_z \end{bmatrix} = \begin{bmatrix} +\cos\omega^j\cos\omega^k & \begin{matrix} -\sin\omega^j\cos\omega^l \\ +\cos\omega^j\sin\omega^k\sin\omega^l \end{matrix} & \begin{matrix} +\sin\omega^j\sin\omega^l \\ +\cos\omega^j\sin\omega^k\cos\omega^l \end{matrix} \\ +\sin\omega^j\cos\omega^k & \begin{matrix} +\cos\omega^j\cos\omega^l \\ +\sin\omega^j\sin\omega^k\sin\omega^l \end{matrix} & \begin{matrix} -\cos\omega^j\sin\omega^l \\ +\sin\omega^j\sin\omega^k\cos\omega^l \end{matrix} \\ -\sin\omega^k & +\cos\omega^k\sin\omega^l & +\cos\omega^k\cos\omega^l \end{bmatrix}$$

$$\omega^{l0} = \omega^{lk}\omega^{kj}\omega^{j0}$$

$$\begin{bmatrix} \alpha_x & \beta_x & \gamma_x \\ \alpha_y & \beta_y & \gamma_y \\ \alpha_z & \beta_z & \gamma_z \end{bmatrix} = \begin{bmatrix} +\cos\omega^j\cos\omega^k & +\sin\omega^j\cos\omega^k & -\sin\omega^k \\ \begin{matrix} -\sin\omega^j\cos\omega^l \\ +\cos\omega^j\sin\omega^k\sin\omega^l \end{matrix} & \begin{matrix} +\cos\omega^j\cos\omega^l \\ +\sin\omega^j\sin\omega^k\sin\omega^l \end{matrix} & +\cos\omega^k\sin\omega^l \\ \begin{matrix} +\sin\omega^j\sin\omega^l \\ +\cos\omega^j\sin\omega^k\cos\omega^l \end{matrix} & \begin{matrix} -\cos\omega^j\sin\omega^l \\ +\sin\omega^j\sin\omega^k\cos\omega^l \end{matrix} & +\cos\omega^k\cos\omega^l \end{bmatrix}$$

$$\eta_x = \cos(x, N)$$
$$\eta_y = \cos(y, N)$$
$$\eta_z = \cos(z, N)$$

$$\Delta_1 = 1 - \cos\theta$$
$$\Delta_2 = \sin\theta$$
$$\Delta_3 = \cos\theta$$

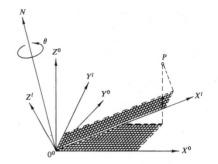

(1) Pure Rotation

The coordinate axes X^0, Y^0, Z^0, are rotated about the fixed axis N of given direction cosines η_x, η_y, η_z through the right-handed angle θ. The relationship of the initial coordinates x^0, y^0, z^0 of a point P to the new coordinates x^l, y^l, z^l of the same point are

$$\underbrace{\begin{bmatrix} x^0 \\ y^0 \\ z^0 \end{bmatrix}}_{s^0} = \underbrace{\begin{bmatrix} \eta_x{}^2\Delta_1 + \Delta_3 & \eta_x\eta_y\Delta_1 + \eta_z\Delta_2 & \eta_x\eta_z\Delta_1 + \eta_y\Delta_2 \\ \eta_y\eta_x\Delta_1 - \eta_z\Delta_2 & \eta_y{}^2\Delta_1 + \Delta_3 & \eta_y\eta_z\Delta_1 + \eta_x\Delta_2 \\ \eta_z\eta_x\Delta_1 + \eta_y\Delta_2 & \eta_z\eta_y\Delta_1 - \eta_x\Delta_2 & \eta_z{}^2\Delta_1 + \Delta_3 \end{bmatrix}}_{\theta^{0l}} \underbrace{\begin{bmatrix} x^l \\ y^l \\ z^l \end{bmatrix}}_{s^l}$$

Inversely, $s^l = \theta^{l0}s^0$
where $\theta^{l0} = (\theta^{0l})^T = (\theta^{0l})^{-1}$

(2) Properties of θ Matrices

The elements of the θ^{0l} matrix are numerically equal to the respective elements of the ω^{0l} matrix in Sec. 5.08−2 such that

$$\theta^{0l} = \omega^{0l} \qquad \theta^{l0} = \omega^{l0}$$

In addition,

$$\theta^{0l} = A\Delta_1 + B\Delta_2 + I\Delta_3 = \omega^{0l}$$

where $I = 3 \times 3$ unit matrix and

$$A = \begin{bmatrix} \eta_x & 0 & 0 \\ 0 & \eta_y & 0 \\ 0 & 0 & \eta_z \end{bmatrix} \begin{bmatrix} 1 & 1 & 1 \\ 1 & 1 & 1 \\ 1 & 1 & 1 \end{bmatrix} \begin{bmatrix} \eta_x & 0 & 0 \\ 0 & \eta_y & 0 \\ 0 & 0 & \eta_z \end{bmatrix} \qquad B = \begin{bmatrix} 0 & \eta_z & -\eta_y \\ -\eta_z & 0 & \eta_x \\ \eta_y & -\eta_x & 0 \end{bmatrix}$$

(3) Inverse Relationship

If the elements of ω^{0l} are known (Sec. 5.08−2), then

$$\cos\theta = \frac{\alpha_x + \beta_y + \gamma_z - 1}{2}$$

$$\eta_x = \sqrt{\frac{\alpha_x - \cos\theta}{1 - \cos\theta}} \qquad \eta_y = \sqrt{\frac{\beta_y - \cos\theta}{1 - \cos\theta}} \qquad \eta_z = \sqrt{\frac{\gamma_z - \cos\theta}{1 - \cos\theta}}$$

6
ELEMENTARY FUNCTIONS

(1) Definitions

Trigonometric functions of an angle ω are defined as follows:

$$\text{sine } \omega = \sin \omega = \frac{y}{R} \qquad\qquad \text{cosecant } \omega = \csc \omega = \frac{R}{y}$$

$$\text{cosine } \omega = \cos \omega = \frac{x}{R} \qquad\qquad \text{secant } \omega = \sec \omega = \frac{R}{x}$$

$$\text{tangent } \omega = \tan \omega = \frac{y}{x} \qquad\qquad \text{cotangent } \omega = \cot \omega = \frac{x}{y}$$

$$\text{versine } \omega = \text{vers } \omega = \frac{R - x}{R} \qquad\qquad \text{coversine } \omega = \text{covers } \omega = \frac{R - y}{R}$$

(2) Angle

The independent variable ω is measured in radians.

$$180° = \pi \qquad\qquad 1° = \frac{\pi}{180} \qquad\qquad 1 \text{ radian} = \frac{180}{\pi}$$

$$= 3.141\,592\,653\,5 \text{ radians} \qquad = 0.017\,453\,292\,5 \text{ radian} \qquad = 57.295\,779\,513\,0°$$

(3) Relationships

$$\sin^2 \omega + \cos^2 \omega = 1$$

$$\tan^2 \omega + 1 = \sec^2 \omega$$

$$\cot^2 \omega + 1 = \csc^2 \omega$$

$$\sin \omega \csc \omega = 1$$

$$\cos \omega \sec \omega = 1$$

$$\tan \omega \cot \omega = 1$$

$$\sin \omega + \text{covers } \omega = +1$$

$$\cos \omega + \text{vers } \omega = +1$$

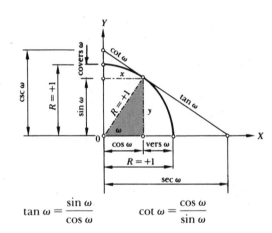

$$\tan \omega = \frac{\sin \omega}{\cos \omega} \qquad\qquad \cot \omega = \frac{\cos \omega}{\sin \omega}$$

(4) Reductions $(k = 0, 1, 2, \ldots)$

	$\pm \alpha$	$\frac{\pi}{2} \pm \alpha$	$2k\pi \pm \alpha$	$(4k + 1)\frac{\pi}{2} \pm \alpha$	$(4k + 2)\frac{\pi}{2} \pm \alpha$	$(4k + 3)\frac{\pi}{2} \pm \alpha$
sin	$\pm \sin \alpha$	$+ \cos \alpha$	$\pm \sin \alpha$	$+ \cos \alpha$	$\mp \sin \alpha$	$- \cos \alpha$
cos	$+ \cos \alpha$	$\mp \sin \alpha$	$+ \cos \alpha$	$\mp \sin \alpha$	$- \cos \alpha$	$\pm \sin \alpha$
tan	$\pm \tan \alpha$	$\mp \cot \alpha$	$\pm \tan \alpha$	$\mp \cot \alpha$	$\pm \tan \alpha$	$\mp \cot \alpha$
cot	$\pm \cot \alpha$	$\mp \tan \alpha$	$\pm \cot \alpha$	$\mp \tan \alpha$	$\pm \cot \alpha$	$\mp \tan \alpha$
sec	$+ \sec \alpha$	$\mp \csc \alpha$	$+ \sec \alpha$	$\mp \csc \alpha$	$- \sec \alpha$	$+ \csc \alpha$
csc	$\pm \csc \alpha$	$+ \sec \alpha$	$\pm \csc \alpha$	$+ \sec \alpha$	$\mp \csc \alpha$	$- \sec \alpha$

(1) Signs in Quadrants

Quadrant	sin	cos	tan	cot	sec	csc	vers	covers
I	+	+	+	+	+	+	+	+
II	+	−	−	−	−	+	+	+
III	−	−	+	+	−	−	+	+
IV	−	+	−	−	+	−	+	+

(2) Graphs

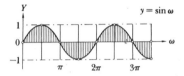

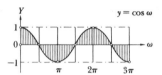

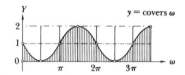

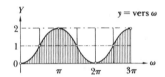

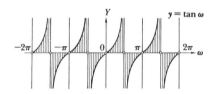

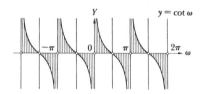

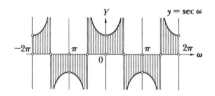

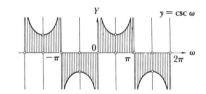

(3) Limit Values

Degrees	Radians	sin ω	cos ω	tan ω	cot ω	sec ω	csc ω	vers ω	covers ω
0°	0	0	+1	0	$\mp\infty$	+1	$\mp\infty$	0	+1
90°	$\dfrac{\pi}{2}$	+1	0	$\pm\infty$	0	$\pm\infty$	+1	+1	0
180°	π	0	−1	0	$\mp\infty$	−1	$\pm\infty$	+2	+1
270°	$\dfrac{3\pi}{2}$	−1	0	$\pm\infty$	0	$\mp\infty$	−1	+1	+2
360°	2π	0	+1	0	$\mp\infty$	+1	$\mp\infty$	0	+1

(1) Functions of Selected Angles

Degrees	Radians	$\sin \alpha$	$\cos \alpha$	$\tan \alpha$	$\cot \alpha$	$\sec \alpha$	$\csc \alpha$	$\text{vers } \alpha$	$\text{covers } \alpha$
0	0	0	$+1$	0	$\mp\infty$	$+1$	$\mp\infty$	0	$+1$
30	$\dfrac{\pi}{6}$	$+\dfrac{1}{2}$	$+\dfrac{\sqrt{3}}{2}$	$+\dfrac{\sqrt{3}}{3}$	$+\sqrt{3}$	$+\dfrac{2\sqrt{3}}{3}$	$+2$	$+1-\dfrac{\sqrt{3}}{2}$	$+\dfrac{1}{2}$
45	$\dfrac{\pi}{4}$	$+\dfrac{\sqrt{2}}{2}$	$+\dfrac{\sqrt{2}}{2}$	$+1$	$+1$	$+\sqrt{2}$	$+\sqrt{2}$	$+1-\dfrac{\sqrt{2}}{2}$	$+1-\dfrac{\sqrt{2}}{2}$
60	$\dfrac{\pi}{3}$	$+\dfrac{\sqrt{3}}{2}$	$+\dfrac{1}{2}$	$+\sqrt{3}$	$+\dfrac{\sqrt{3}}{3}$	$+2$	$+\dfrac{2\sqrt{3}}{3}$	$+\dfrac{1}{2}$	$+1-\dfrac{\sqrt{3}}{2}$
90	$\dfrac{\pi}{2}$	$+1$	0	$\pm\infty$	0	$\pm\infty$	$+1$	$+1$	0
120	$\dfrac{2\pi}{3}$	$+\dfrac{\sqrt{3}}{2}$	$-\dfrac{1}{2}$	$-\sqrt{3}$	$-\dfrac{\sqrt{3}}{3}$	-2	$+\dfrac{2\sqrt{3}}{3}$	$+\dfrac{3}{2}$	$+1-\dfrac{\sqrt{3}}{2}$
135	$\dfrac{3\pi}{4}$	$+\dfrac{\sqrt{2}}{2}$	$-\dfrac{\sqrt{2}}{2}$	-1	-1	$-\sqrt{2}$	$+\sqrt{2}$	$+1+\dfrac{\sqrt{2}}{2}$	$+1-\dfrac{\sqrt{2}}{2}$
150	$\dfrac{5\pi}{6}$	$+\dfrac{1}{2}$	$-\dfrac{\sqrt{3}}{2}$	$-\dfrac{\sqrt{3}}{3}$	$-\sqrt{3}$	$-\dfrac{2\sqrt{3}}{3}$	$+2$	$+1+\dfrac{\sqrt{3}}{2}$	$+\dfrac{1}{2}$
180	π	0	-1	0	$\mp\infty$	-1	$\mp\infty$	$+2$	$+1$
210	$\dfrac{7\pi}{6}$	$-\dfrac{1}{2}$	$-\dfrac{\sqrt{3}}{2}$	$+\dfrac{\sqrt{3}}{3}$	$+\sqrt{3}$	$-\dfrac{2\sqrt{3}}{3}$	-2	$+1+\dfrac{\sqrt{3}}{2}$	$+\dfrac{3}{2}$
225	$\dfrac{5\pi}{4}$	$-\dfrac{\sqrt{2}}{2}$	$-\dfrac{\sqrt{2}}{2}$	$+1$	$+1$	$-\sqrt{2}$	$-\sqrt{2}$	$+1+\dfrac{\sqrt{2}}{2}$	$+1+\dfrac{\sqrt{2}}{2}$
240	$\dfrac{4\pi}{3}$	$-\dfrac{\sqrt{3}}{2}$	$-\dfrac{1}{2}$	$+\sqrt{3}$	$+\dfrac{\sqrt{3}}{3}$	-2	$-\dfrac{2\sqrt{3}}{3}$	$+\dfrac{3}{2}$	$+1+\dfrac{\sqrt{3}}{2}$
270	$\dfrac{3\pi}{2}$	-1	0	$\pm\infty$	0	$\mp\infty$	-1	$+1$	$+2$
300	$\dfrac{5\pi}{3}$	$-\dfrac{\sqrt{3}}{2}$	$+\dfrac{1}{2}$	$-\sqrt{3}$	$-\dfrac{\sqrt{3}}{3}$	$+2$	$-\dfrac{2\sqrt{3}}{3}$	$+\dfrac{1}{2}$	$+1+\dfrac{\sqrt{3}}{2}$
315	$\dfrac{7\pi}{4}$	$-\dfrac{\sqrt{2}}{2}$	$+\dfrac{\sqrt{2}}{2}$	-1	-1	$+\sqrt{2}$	$-\sqrt{2}$	$+1-\dfrac{\sqrt{2}}{2}$	$+1+\dfrac{\sqrt{2}}{2}$
330	$\dfrac{11\pi}{6}$	$-\dfrac{1}{2}$	$+\dfrac{\sqrt{3}}{2}$	$-\dfrac{\sqrt{3}}{3}$	$-\sqrt{3}$	$+\dfrac{2\sqrt{3}}{3}$	-2	$+1-\dfrac{\sqrt{3}}{2}$	$+\dfrac{3}{2}$
360	2π	0	$+1$	0	$\mp\infty$	$+1$	$\mp\infty$	0	$+1$

(2) Functions of Very Small Angles

$$\Delta\omega \leq 0.05 \text{ radian} \qquad \omega > 0.05 \text{ radian} \qquad |\epsilon| \leq 1.25 \times 10^{-3}$$

$$\sin \Delta\omega \cong \Delta\omega \qquad \cos \Delta\omega \cong 1 \qquad \tan \Delta\omega \cong \Delta\omega$$

$$\sin (\omega \pm \Delta\omega) \cong \sin \omega \pm \Delta\omega \cos \omega \qquad \cos (\omega \pm \Delta\omega) \cong \cos \omega \mp \Delta\omega \sin \omega$$

(1) Transformations

	$\sin\omega$	$\cos\omega$	$\tan\omega$	$\cot\omega$	$\sec\omega$	$\csc\omega$
$\sin\omega$	$\sin\omega$	$\pm\sqrt{1-\cos^2\omega}$	$\dfrac{\tan\omega}{\pm\sqrt{1+\tan^2\omega}}$	$\dfrac{1}{\pm\sqrt{1+\cot^2\omega}}$	$\dfrac{\pm\sqrt{\sec^2\omega-1}}{\sec\omega}$	$\dfrac{1}{\csc\omega}$
$\cos\omega$	$\pm\sqrt{1-\sin^2\omega}$	$\cos\omega$	$\dfrac{1}{\pm\sqrt{1+\tan^2\omega}}$	$\dfrac{\cot\omega}{\pm\sqrt{1+\cot^2\omega}}$	$\dfrac{1}{\sec\omega}$	$\dfrac{\pm\sqrt{\csc^2\omega-1}}{-\quad\csc\omega}$
$\tan\omega$	$\dfrac{\sin\omega}{\pm\sqrt{1-\sin^2\omega}}$	$\dfrac{\pm\sqrt{1-\cos^2\omega}}{\cos\omega}$	$\tan\omega$	$\dfrac{1}{\cot\omega}$	$\pm\sqrt{\sec^2\omega-1}$	$\dfrac{1}{\pm\sqrt{\csc^2\omega-1}}$
$\cot\omega$	$\dfrac{\pm\sqrt{1-\sin^2\omega}}{\sin\omega}$	$\dfrac{\cos\omega}{\pm\sqrt{1-\cos^2\omega}}$	$\dfrac{1}{\tan\omega}$	$\cot\omega$	$\dfrac{1}{\pm\sqrt{\sec^2\omega-1}}$	$\pm\sqrt{\csc^2\omega-1}$
$\sec\omega$	$\dfrac{1}{\pm\sqrt{1-\sin^2\omega}}$	$\dfrac{1}{\cos\omega}$	$\pm\sqrt{1+\tan^2\omega}$	$\dfrac{\pm\sqrt{1+\cot^2\omega}}{\cot\omega}$	$\sec\omega$	$\dfrac{\csc\omega}{\pm\sqrt{\csc^2\omega-1}}$
$\csc\omega$	$\dfrac{1}{\sin\omega}$	$\dfrac{1}{\pm\sqrt{1-\cos^2\omega}}$	$\dfrac{\pm\sqrt{1+\tan^2\omega}}{\tan\omega}$	$\pm\sqrt{1+\cot^2\omega}$	$\dfrac{\sec\omega}{\pm\sqrt{\sec^2\omega-1}}$	$\csc\omega$

Note: The sign is governed by the quadrant in which the argument terminates.

(2) Expansions   (n = **integer**)

$$\sin n\omega = n\sin\omega\cos^{n-1}\omega - \binom{n}{3}\sin^3\omega\cos^{n-3}\omega + \binom{n}{5}\sin^5\omega\cos^{n-5}\omega - \cdots$$

$$\cos n\omega = \cos^n\omega - \binom{n}{2}\sin^2\omega\cos^{n-2}\omega + \binom{n}{4}\sin^4\omega\cos^{n-4}\omega - \cdots$$

$$\sin^{2n}\omega = \frac{(-1)^n}{2^{2n-1}}\left[\cos 2n\omega - \binom{2n}{1}\cos(2n-2)\omega + \cdots (-1)^{n-1}\binom{2n}{n-1}\cos 2\omega\right] + \binom{2n}{n}\frac{1}{2^{2n}}$$

$$\sin^{2n-1}\omega = \frac{(-1)^{n-1}}{2^{2n-2}}\left[\sin(2n-1)\omega - \binom{2n-1}{1}\sin(2n-3)\omega + \cdots (-1)^{n-1}\binom{2n-1}{n-1}\sin\omega\right]$$

$$\cos^{2n}\omega = \frac{1}{2^{2n-1}}\left[\cos 2n\omega + \binom{2n}{1}\cos(2n-2)\omega + \cdots + \binom{2n}{n-1}\cos 2\omega\right] + \binom{2n}{n}\frac{1}{2^{2n}}$$

$$\cos^{2n-1}\omega = \frac{1}{2^{2n-2}}\left[\cos(2n-1)\omega + \binom{2n-1}{1}\cos(2n-3)\omega + \cdots + \binom{2n-1}{n-1}\cos\omega\right]$$

(1) Sums of Angles

$$\sin(\alpha \pm \beta) = \sin \alpha \cos \beta \pm \cos \alpha \sin \beta \qquad \cos(\alpha \pm \beta) = \cos \alpha \cos \beta \mp \sin \alpha \sin \beta$$

$$\tan(\alpha \pm \beta) = \frac{\tan \alpha \pm \tan \beta}{1 \mp \tan \alpha \tan \beta} \qquad \cot(\alpha \pm \beta) = \frac{\cot \alpha \cot \beta \mp 1}{\cot \beta \pm \cot \alpha}$$

(2) Sums of Functions

$$\sin \alpha + \sin \beta = 2 \sin \frac{\alpha + \beta}{2} \cos \frac{\alpha - \beta}{2} \qquad \cos \alpha + \cos \beta = 2 \cos \frac{\alpha + \beta}{2} \cos \frac{\alpha - \beta}{2}$$

$$\sin \alpha - \sin \beta = 2 \sin \frac{\alpha - \beta}{2} \cos \frac{\alpha + \beta}{2} \qquad \cos \alpha - \cos \beta = -2 \sin \frac{\alpha - \beta}{2} \sin \frac{\alpha + \beta}{2}$$

$$\tan \alpha \pm \tan \beta = \frac{\sin(\alpha \pm \beta)}{\cos \alpha \cos \beta} \qquad \cot \alpha \pm \cot \beta = \frac{\sin(\beta + \alpha)}{\sin \alpha \sin \beta}$$

$$\sin \alpha + \cos \alpha = \sqrt{2} \sin\left(\frac{\pi}{4} + \alpha\right) \qquad \sin \alpha - \cos \alpha = -\sqrt{2} \cos\left(\frac{\pi}{4} + \alpha\right)$$

$$\tan \alpha + \cot \alpha = 2 \csc 2\alpha \qquad \tan \alpha - \cot \alpha = -2 \cot 2\alpha$$

$$\frac{1 + \tan \alpha}{1 - \tan \alpha} = \tan\left(\frac{\pi}{4} + \alpha\right) \qquad \frac{1 + \cot \alpha}{1 - \cot \alpha} = -\cot\left(\frac{\pi}{4} - \alpha\right)$$

(3) Sums of Sums

$$\sin(\alpha + \beta) + \sin(\alpha - \beta) = 2 \sin \alpha \cos \beta \qquad \cos(\alpha + \beta) + \cos(\alpha - \beta) = 2 \cos \alpha \cos \beta$$

$$\sin(\alpha + \beta) - \sin(\alpha - \beta) = 2 \sin \beta \cos \alpha \qquad \cos(\alpha + \beta) - \cos(\alpha - \beta) = -2 \sin \alpha \sin \beta$$

$$\tan(\alpha + \beta) + \tan(\alpha - \beta) \qquad \cot(\alpha + \beta) + \cot(\alpha - \beta)$$

$$= \frac{\sin 2\alpha}{\cos(\alpha + \beta) \cos(\alpha - \beta)} \qquad = \frac{\sin 2\alpha}{\sin(\alpha + \beta) \sin(\alpha - \beta)}$$

$$\tan(\alpha + \beta) - \tan(\alpha - \beta) \qquad \cot(\alpha + \beta) - \cot(\alpha - \beta)$$

$$= \frac{\sin 2\beta}{\cos(\alpha + \beta) \cos(\alpha - \beta)} \qquad = -\frac{\sin 2\beta}{\sin(\alpha + \beta) \sin(\alpha - \beta)}$$

(1) Multiple Angles

$$\sin 2\alpha = 2 \sin \alpha \cos \alpha$$

$$\cos 2\alpha = 1 - 2 \sin^2 \alpha$$

$$\sin 3\alpha = (\sin \alpha)(3 - 4 \sin^2 \alpha)$$

$$\cos 3\alpha = 4 \cos^3 \alpha - 3 \cos \alpha$$

$$\sin 4\alpha = (4 \sin \alpha \cos \alpha)(1 - 2 \sin^2 \alpha)$$

$$\cos 4\alpha = 8 \cos^4 \alpha - 8 \cos^2 \alpha + 1$$

$$\sin 5\alpha = (\sin \alpha)(5 - 20 \sin^2 \alpha + 16 \sin^4 \alpha)$$

$$\cos 5\alpha = 16 \cos^5 \alpha - 20 \cos^3 \alpha + 5 \cos \alpha$$

$$\tan 2\alpha = \frac{2 \tan \alpha}{1 - \tan^2 \alpha}$$

$$\cot 2\alpha = \frac{\cot^2 \alpha - 1}{2 \cot \alpha}$$

$$\tan 3\alpha = \frac{3 \tan \alpha - \tan^3 \alpha}{1 - 3 \tan^2 \alpha}$$

$$\cot 3\alpha = \frac{3 \cot \alpha - \cot^3 \alpha}{1 - 3 \cot^2 \alpha}$$

$$\tan 4\alpha = \frac{4 \tan \alpha - 4 \tan^3 \alpha}{1 - 6 \tan^2 \alpha + \tan^4 \alpha}$$

$$\cot 4\alpha = \frac{\cot^4 \alpha - 6 \cot^2 \alpha + 1}{4 \cot^3 \alpha - 4 \cot \alpha}$$

$$\tan 5\alpha = \frac{5 \tan \alpha - 10 \tan^3 \alpha + \tan^5 \alpha}{1 - 10 \tan^2 \alpha + 5 \tan^4 \alpha}$$

$$\cot 5\alpha = \frac{5 \cot \alpha - 10 \cot^3 \alpha + \cot^5 \alpha}{1 - 10 \cot^2 \alpha + 5 \cot^4 \alpha}$$

(2) Half Angles $(\alpha < \pi)$

$$\sin \frac{\alpha}{2} = \sqrt{\frac{1 - \cos \alpha}{2}}$$

$$\cos \frac{\alpha}{2} = \sqrt{\frac{1 + \cos \alpha}{2}}$$

$$\tan \frac{\alpha}{2} = \sqrt{\frac{1 - \cos \alpha}{1 + \cos \alpha}} = \frac{\sin \alpha}{1 + \cos \alpha} = \frac{1 - \cos \alpha}{\sin \alpha}$$

$$\cot \frac{\alpha}{2} = \sqrt{\frac{1 + \cos \alpha}{1 - \cos \alpha}} = \frac{1 + \cos \alpha}{\sin \alpha} = \frac{\sin \alpha}{1 - \cos \alpha}$$

(3) Relations $(\alpha < \pi/2)$

$$\sin \alpha = 2 \sin \frac{\alpha}{2} \cos \frac{\alpha}{2}$$

$$\cos \alpha = \cos^2 \frac{\alpha}{2} - \sin^2 \frac{\alpha}{2}$$

$$\tan \alpha = \frac{2 \tan (\alpha/2)}{1 - \tan^2 (\alpha/2)}$$

$$\cot \alpha = \frac{\cot^2 (\alpha/2) - 1}{2 \cot (\alpha/2)}$$

$$= \frac{2 \sin (\alpha/2) \cos (\alpha/2)}{\cos^2 (\alpha/2) - \sin^2 (\alpha/2)}$$

$$= \frac{\cos^2 (\alpha/2) - \sin^2 (\alpha/2)}{2 \sin (\alpha/2) \cos (\alpha/2)}$$

$$\sin \alpha = \sqrt{\frac{1 - \cos 2\alpha}{2}}$$

$$\cos \alpha = \sqrt{\frac{1 + \cos 2\alpha}{2}}$$

$$\tan \alpha = \sqrt{\frac{1 - \cos 2\alpha}{1 + \cos 2\alpha}}$$

$$\cot \alpha = \sqrt{\frac{1 + \cos 2\alpha}{1 - \cos 2\alpha}}$$

$$= \frac{\sin 2\alpha}{1 + \cos 2\alpha} = \frac{1 - \cos 2\alpha}{\sin 2\alpha}$$

$$= \frac{1 + \cos 2\alpha}{\sin 2\alpha} = \frac{\sin 2\alpha}{1 - \cos 2\alpha}$$

(1) Powers

$$\sin^2 \alpha = \tfrac{1}{2}(-\cos 2\alpha + 1)$$

$$\cos^2 \alpha = \tfrac{1}{2}(\cos 2\alpha + 1)$$

$$\sin^3 \alpha = \tfrac{1}{4}(-\sin 3\alpha + 3 \sin \alpha)$$

$$\cos^3 \alpha = \tfrac{1}{4}(\cos 3\alpha + 3 \cos \alpha)$$

$$\sin^4 \alpha = \tfrac{1}{8}(\cos 4\alpha - 4 \cos 2\alpha + 3)$$

$$\cos^4 \alpha = \tfrac{1}{8}(\cos 4\alpha + 4 \cos 2\alpha + 3)$$

$$\sin^5 \alpha = \tfrac{1}{16}(\sin 5\alpha - 5 \sin 3\alpha + 10 \sin \alpha)$$

$$\cos^5 \alpha = \tfrac{1}{16}(\cos 5\alpha + 5 \cos 3\alpha + 10 \cos \alpha)$$

$$\tan^2 \alpha = \frac{1 - \cos 2\alpha}{1 + \cos 2\alpha}$$

$$\cot^2 \alpha = \frac{1 + \cos 2\alpha}{1 - \cos 2\alpha}$$

$$\tan^3 \alpha = \frac{-\sin 3\alpha + 3 \sin \alpha}{\cos 3\alpha + 3 \cos \alpha}$$

$$\cot^3 \alpha = \frac{\cos 3\alpha + 3 \cos \alpha}{-\sin 3\alpha + 3 \sin \alpha}$$

$$\tan^4 \alpha = \frac{\cos 4\alpha - 4 \cos 2\alpha + 3}{\cos 4\alpha + 4 \cos 2\alpha + 3}$$

$$\cot^4 \alpha = \frac{\cos 4\alpha + 4 \cos 2\alpha + 3}{\cos 4\alpha - 4 \cos 2\alpha + 3}$$

$$\tan^5 \alpha = \frac{\sin 5\alpha - 5 \sin 3\alpha + 10 \sin \alpha}{\cos 5\alpha + 5 \cos 3\alpha + 10 \cos \alpha}$$

$$\cot^5 \alpha = \frac{\cos 5\alpha + 5 \cos 3\alpha + 10 \cos \alpha}{\sin 5\alpha - 5 \sin 3\alpha + 10 \sin \alpha}$$

(2) Products

$$\sin \alpha \sin \beta = \tfrac{1}{2}\cos (\alpha - \beta) - \tfrac{1}{2}\cos (\alpha + \beta)$$

$$\tan \alpha \tan \beta = \frac{\cos (\alpha - \beta) - \cos (\alpha + \beta)}{\cos (\alpha - \beta) + \cos (\alpha + \beta)}$$

$$\sin \alpha \cos \beta = \tfrac{1}{2}\sin (\alpha - \beta) + \tfrac{1}{2}\sin (\alpha + \beta)$$

$$\tan \alpha \cot \beta = \frac{\sin (\alpha - \beta) + \sin (\alpha + \beta)}{- \sin (\alpha - \beta) + \sin (\alpha + \beta)}$$

$$\cos \alpha \cos \beta = \tfrac{1}{2}\cos (\alpha - \beta) + \tfrac{1}{2}\cos (\alpha + \beta)$$

$$\cot \alpha \cot \beta = \frac{\cos (\alpha - \beta) + \cos (\alpha + \beta)}{\cos (\alpha - \beta) - \cos (\alpha + \beta)}$$

If $\alpha + \beta + \gamma = \pi$,

$$4 \sin \alpha \sin \beta \sin \gamma = \sin (\alpha + \beta - \gamma) - \sin (\beta + \gamma - \alpha) + \sin (\gamma + \alpha - \beta) - \sin (\alpha + \beta + \gamma)$$

$$4 \sin \alpha \sin \beta \cos \gamma = -\cos (\alpha + \beta - \gamma) + \cos (\beta + \gamma - \alpha) + \cos (\gamma + \alpha - \beta) - \cos (\alpha + \beta + \gamma)$$

$$4 \sin \alpha \cos \beta \cos \gamma = \sin (\alpha + \beta - \gamma) - \sin (\beta + \gamma - \alpha) + \sin (\gamma + \alpha - \beta) - \sin (\alpha + \beta + \gamma)$$

$$4 \cos \alpha \cos \beta \cos \gamma = \cos (\alpha + \beta - \gamma) + \cos (\beta + \gamma - \alpha) + \cos (\gamma + \alpha - \beta) + \cos (\alpha + \beta + \gamma)$$

(1) Definitions

Inverse trigonometric functions are defined as follows:

Trigonometric functions	Inverse trigonometric functions	Principal values
$y = \sin \omega$	$\omega = \sin^{-1} y = \arcsin y$	$-\dfrac{\pi}{2} \leqslant \omega \leqslant +\dfrac{\pi}{2}$
$y = \cos \omega$	$\omega = \cos^{-1} y = \arccos y$	$0 \leqslant \omega \leqslant \pi$
$y = \tan \omega$	$\omega = \tan^{-1} y = \arctan y$	$-\dfrac{\pi}{2} < \omega < +\dfrac{\pi}{2}$
$y = \cot \omega$	$\omega = \cot^{-1} y = \text{arccot } y$	$0 < \omega < \pi$

which means that ω is the arc of an angle of which the trigonometric function is y.

(2) Relationships

$$\sin^{-1} y + \cos^{-1} y = \frac{\pi}{2} \qquad\qquad \tan^{-1} y + \cot^{-1} y = \frac{\pi}{2}$$

$$\sin^{-1}(-y) = -\sin^{-1} y \qquad\qquad \tan^{-1}(-y) = -\tan^{-1} y$$

(3) Graphs

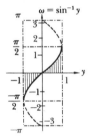  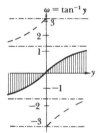 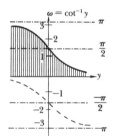

(4) Transformation Table

$y \geq 0$	$\sin^{-1} y$	$\cos^{-1} y$	$\tan^{-1} y$	$\cot^{-1} y$
$\sin^{-1} y$	$\sin^{-1} y$	$\cos^{-1}\sqrt{1 - y^2}$	$\tan^{-1}\dfrac{y}{\sqrt{1 - y^2}}$	$\cot^{-1}\dfrac{\sqrt{1 - y^2}}{y}$
$\cos^{-1} y$	$\sin^{-1}\sqrt{1 - y^2}$	$\cos^{-1} y$	$\tan^{-1}\dfrac{\sqrt{1 - y^2}}{y}$	$\cot^{-1}\dfrac{y}{\sqrt{1 - y^2}}$
$\tan^{-1} y$	$\sin^{-1}\dfrac{y}{\sqrt{1 + y^2}}$	$\cos^{-1}\dfrac{1}{\sqrt{1 + y^2}}$	$\tan^{-1} y$	$\cot^{-1}\dfrac{1}{y}$
$\cot^{-1} y$	$\sin^{-1}\dfrac{1}{\sqrt{1 + y^2}}$	$\cos^{-1}\dfrac{y}{\sqrt{1 + y^2}}$	$\tan^{-1}\dfrac{1}{y}$	$\cot^{-1} y$

(1) Definitions

A hyperbolic function is a combination of e^x and e^{-x} and is introduced as follows:

$$\text{Hyperbolic sine of } x = \sinh x = \frac{e^x - e^{-x}}{2}$$

$$\text{Hyperbolic cosine of } x = \cosh x = \frac{e^x + e^{-x}}{2}$$

$$\text{Hyperbolic tangent of } x = \tanh x = \frac{e^x - e^{-x}}{e^x + e^{-x}}$$

$$\text{Hyperbolic cotangent of } x = \coth x = \frac{e^x + e^{-x}}{e^x - e^{-x}}$$

$$\text{Hyperbolic secant of } x = \operatorname{sech} x = \frac{2}{e^x + e^{-x}}$$

$$\text{Hyperbolic cosecant of } x = \operatorname{csch} x = \frac{2}{e^x - e^{-x}}$$

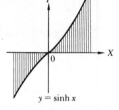

$y = \cosh x$

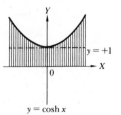

$y = \sinh x$

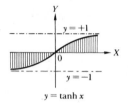

$y = \tanh x$

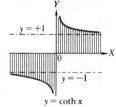

$y = \coth x$

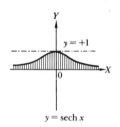

$y = \operatorname{sech} x$

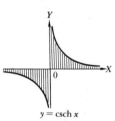

$y = \operatorname{csch} x$

(2) Relationships

$$\cosh^2 x - \sinh^2 x = 1$$

$$\tanh^2 x + \operatorname{sech}^2 x = 1$$

$$\coth^2 x - \operatorname{csch}^2 x = 1$$

$$\tanh x = \frac{\sinh x}{\cosh x}$$

$$\coth x = \frac{\cosh x}{\sinh x}$$

$$\operatorname{sech} x \cosh x = 1$$

$$\operatorname{csch} x \sinh x = 1$$

$$\tanh x \coth x = 1$$

$$\sinh(-x) = -\sinh x$$

$$\operatorname{sech}(-x) = \operatorname{sech} x$$

$$\tanh(-x) = -\tanh x$$

$$\coth(-x) = -\coth x$$

$$\cosh(-x) = \cosh x$$

$$\operatorname{csch}(-x) = -\operatorname{csch} x$$

(3) Limit Values

x	$\sinh x$	$\cosh x$	$\tanh x$	$\coth x$	$\operatorname{sech} x$	$\operatorname{csch} x$
$-\infty$	$-\infty$	$+\infty$	-1	-1	0	0
-1	-1.1752	$+1.5431$	-0.7616	-1.3130	$+0.6480$	-0.8509
0	0	$+1$	0	$\mp\infty$	$+1$	$\mp\infty$
$+1$	$+1.1752$	$+1.5431$	$+0.7616$	$+1.3130$	$+0.6480$	$+0.8509$
$+\infty$	$+\infty$	$+\infty$	$+1$	$+1$	0	0

(1) Transformations

$x > 0$	$\sinh x$	$\cosh x$	$\tanh x$	$\coth x$	$\operatorname{sech} x$	$\operatorname{csch} x$
$\sinh x$	$\sinh x$	$\sqrt{\cosh^2 x - 1}$	$\dfrac{\tanh x}{\sqrt{1 - \tanh^2 x}}$	$\dfrac{1}{\sqrt{\coth^2 x - 1}}$	$\dfrac{\sqrt{1 - \operatorname{sech}^2 x}}{\operatorname{sech} x}$	$\dfrac{1}{\operatorname{csch} x}$
$\cosh x$	$\sqrt{1 + \sinh^2 x}$	$\cosh x$	$\dfrac{1}{\sqrt{1 - \tanh^2 x}}$	$\dfrac{\coth x}{\sqrt{\coth^2 x - 1}}$	$\dfrac{1}{\operatorname{sech} x}$	$\dfrac{\sqrt{1 + \operatorname{csch}^2 x}}{\operatorname{csch} x}$
$\tanh x$	$\dfrac{\sinh x}{\sqrt{1 + \sinh^2 x}}$	$\dfrac{\sqrt{\cosh^2 x - 1}}{\cosh x}$	$\tanh x$	$\dfrac{1}{\coth x}$	$\sqrt{1 - \operatorname{sech}^2 x}$	$\dfrac{1}{\sqrt{1 + \operatorname{csch}^2 x}}$
$\coth x$	$\dfrac{\sqrt{1 + \sinh^2 x}}{\sinh x}$	$\dfrac{\cosh x}{\sqrt{\cosh^2 x - 1}}$	$\dfrac{1}{\tanh x}$	$\coth x$	$\dfrac{1}{\sqrt{1 - \operatorname{sech}^2 x}}$	$\sqrt{1 + \operatorname{csch}^2 x}$
$\operatorname{sech} x$	$\dfrac{1}{\sqrt{1 + \sinh^2 x}}$	$\dfrac{1}{\cosh x}$	$\sqrt{1 - \tanh^2 x}$	$\dfrac{\sqrt{\coth^2 x - 1}}{\coth x}$	$\operatorname{sech} x$	$\dfrac{\operatorname{csch} x}{\sqrt{1 + \operatorname{csch}^2 x}}$
$\operatorname{csch} x$	$\dfrac{1}{\sinh x}$	$\dfrac{1}{\sqrt{\cosh^2 x - 1}}$	$\dfrac{\sqrt{1 - \tanh^2 x}}{\tanh x}$	$\sqrt{\coth^2 x - 1}$	$\dfrac{\operatorname{sech} x}{\sqrt{1 - \operatorname{sech}^2 x}}$	$\operatorname{csch} x$

(2) Expansions　　　(n = **integer**)

$$\sinh nx = n \sinh x \cosh^{n-1} x + \binom{n}{3} \sinh^3 x \cosh^{n-3} x + \binom{n}{5} \sinh^5 x \cosh^{n-5} x + \cdots$$

$$\cosh nx = \cosh^n x + \binom{n}{2} \sinh^2 x \cosh^{n-2} x + \binom{n}{4} \sinh^4 x \cosh^{n-4} x + \cdots$$

$$\sinh^{2n} x = \frac{1}{2^{2n-1}}\left[\cosh 2nx - \binom{2n}{1} \cosh (2n-2)x + \cdots (-1)^{n-1} \binom{2n}{n-1} \cosh 2x \right] + \binom{2n}{n}\frac{1}{2^{2n}}$$

$$\sinh^{2n-1} x = \frac{1}{2^{2n-2}}\left[\sinh (2n-1)x - \binom{2n-1}{1} \sinh (2n-3)x + \cdots (-1)^{n-1}\binom{2n-1}{n-1} \sinh x \right]$$

$$\cosh^{2n} x = \frac{1}{2^{2n-1}}\left[\cosh 2nx + \binom{2n}{1} \cosh (2n-2)x + \cdots + \binom{2n}{n-1} \cosh 2x \right] - \binom{2n}{n}\frac{1}{2^{2n}}$$

$$\cosh^{2n-1} x = \frac{1}{2^{2n-2}}\left[\cosh (2n-1)x + \binom{2n-1}{1} \cosh (2n-3)x + \cdots + \binom{2n-1}{n-1} \cosh x \right]$$

(1) Sums of Angles

$$\sinh(a \pm b) = \sinh a \cosh b \pm \cosh a \sinh b$$

$$\tanh(a \pm b) = \frac{\tanh a \pm \tanh b}{1 \pm \tanh a \tanh b}$$

$$\cosh(a \pm b) = \cosh a \cosh b \pm \sinh a \sinh b$$

$$\coth(a \pm b) = \frac{\coth a \coth b \pm 1}{\coth b \pm \coth a}$$

(2) Sums of Functions

$$\sinh a + \sinh b = 2 \sinh \frac{a+b}{2} \cosh \frac{a-b}{2}$$

$$\tanh a + \tanh b = \frac{\sinh(a+b)}{\cosh a \cosh b}$$

$$\sinh a - \sinh b = 2 \cosh \frac{a+b}{2} \sinh \frac{a-b}{2}$$

$$\tanh a - \tanh b = \frac{\sinh(a-b)}{\cosh a \cosh b}$$

$$\cosh a + \cosh b = 2 \cosh \frac{a+b}{2} \cosh \frac{a-b}{2}$$

$$\coth a + \coth b = \frac{\sinh(a+b)}{\sinh a \sinh b}$$

$$\cosh a - \cosh b = 2 \sinh \frac{a+b}{2} \sinh \frac{a-b}{2}$$

$$\coth a - \coth b = \frac{\sinh(b-a)}{\sinh a \sinh b}$$

$$\sinh a + \cosh a = e^a$$

$$\tanh a + \coth a = 2 \coth 2a$$

$$\sinh a - \cosh a = -e^{-a}$$

$$\tanh a - \coth a = -2 \operatorname{csch} 2a$$

(3) Sums of Sums

$$A = a + b \qquad B = a - b$$

$$\sinh A + \sinh B = 2 \sinh a \cosh b$$

$$\cosh A + \cosh B = 2 \cosh a \cosh b$$

$$\sinh A - \sinh B = 2 \cosh a \sinh b$$

$$\cosh A - \cosh B = 2 \sinh a \sinh b$$

$$\tanh A + \tanh B = \frac{\sinh 2a}{\cosh A \cosh B}$$

$$\coth A + \coth B = \frac{\sinh 2a}{\sinh A \sinh B}$$

$$\tanh A - \tanh B = \frac{\sinh 2b}{\cosh A \cosh B}$$

$$\coth A - \coth B = \frac{-\sinh 2b}{\sinh A \sinh B}$$

(4) Products

$$\sinh a \sinh b = \tfrac{1}{2} \cosh A - \tfrac{1}{2} \cosh B$$

$$\cosh a \cosh b = \tfrac{1}{2} \cosh A + \tfrac{1}{2} \cosh B$$

$$\sinh a \cosh b = \tfrac{1}{2} \sinh A + \tfrac{1}{2} \sinh B$$

$$\tanh a \tanh b = \frac{\tanh a + \tanh b}{\coth a + \coth b}$$

$$\coth a \coth b = \frac{\coth a + \coth b}{\tanh a + \tanh b}$$

(1) Half Angles

$$\sinh \frac{a}{2} = \sqrt{\tfrac{1}{2}(\cosh a - 1)} = \tfrac{1}{2}\sqrt{\cosh a + \sinh a} - \tfrac{1}{2}\sqrt{\cosh a - \sinh a}$$

$$\cosh \frac{a}{2} = \sqrt{\tfrac{1}{2}(\cosh a + 1)} = \tfrac{1}{2}\sqrt{\cosh a + \sinh a} + \tfrac{1}{2}\sqrt{\cosh a - \sinh a}$$

$$\tanh \frac{a}{2} = \frac{\sinh a}{\cosh a + 1} = \frac{\cosh a - 1}{\sinh a} = \sqrt{\frac{\cosh a - 1}{\cosh a + 1}}$$

$$\coth \frac{a}{2} = \frac{\sinh a}{\cosh a - 1} = \frac{\cosh a + 1}{\sinh a} = \sqrt{\frac{\cosh a + 1}{\cosh a - 1}}$$

(2) Multiple Angles

$$\sinh 2a = 2 \sinh a \cosh a \qquad\qquad \cosh 2a = \sinh^2 a + \cosh^2 a$$

$$\tanh 2a = \frac{2 \tanh a}{1 + \tanh^2 a} \qquad\qquad \coth 2a = \frac{1 + \coth^2 a}{2 \coth a}$$

$$\sinh 3a = (\sinh a)(4 \cosh^2 a - 1) \qquad\qquad \cosh 3a = (\cosh a)(4 \sinh^2 a - 1)$$

$$\tanh 3a = \frac{\tanh^3 a + 3 \tanh a}{3 \tanh^2 a + 1} \qquad\qquad \coth 3a = \frac{\coth^3 a + 3 \coth a}{3 \coth^2 a + 1}$$

(3) Powers

$$\sinh^2 a = \tfrac{1}{2}(\cosh 2a - 1) \qquad\qquad \cosh^2 a = \tfrac{1}{2}(\cosh 2a + 1)$$

$$\tanh^2 a = \frac{\cosh 2a - 1}{\cosh 2a + 1} \qquad\qquad \coth^2 a = \frac{\cosh 2a + 1}{\cosh 2a - 1}$$

$$\sinh^3 a = \tfrac{1}{4}(\sinh 3a - 3 \sinh a) \qquad\qquad \cosh^3 a = \tfrac{1}{4}(\cosh 3a + 3 \cosh a)$$

$$\sinh^4 a = \tfrac{1}{8}(\cosh 4a - 4 \cosh 2a + 3) \qquad\qquad \cosh^4 a = \tfrac{1}{8}(\cosh 4a + 4 \cosh 2a + 3)$$

$$\sinh^2 a + \cosh^2 a = \cosh 2a \qquad\qquad (\sinh a \pm \cosh a)^2 = \cosh 2a \pm \sinh 2a$$

$$\sinh^2 a - \cosh^2 a = -1$$

$$\tanh^2 a + \coth^2 a = -8 \frac{\cosh 2a}{\cosh 4a - 1} \qquad\qquad (\tanh a + \coth a)^2 = 4 \frac{\cosh 4a + 1}{\cosh 4a - 1}$$

$$\tanh^2 a - \coth^2 a = 2 \frac{\cosh 4a + 3}{\cosh 4a - 1} \qquad\qquad (\tanh a - \coth a)^2 = \frac{8}{\cosh 4a - 1}$$

$$(\cosh a \pm \sinh a)^n = \cosh na \pm \sinh na$$

(1) Definitions

Inverse hyperbolic functions (area functions) are defined as follows:

Hyperbolic functions	Inverse hyperbolic functions	Principal values
$y = \sinh \omega$	$\omega = \sinh^{-1} y = ar\ \sinh y$	$-\infty \leq y \leq +\infty$
$y = \cosh \omega$	$\omega = \cosh^{-1} y = ar\ \cosh y$	$+1 \leq y \leq +\infty$
$y = \tanh \omega$	$\omega = \tanh^{-1} y = ar\ \tanh y$	$-1 \leq y \leq +1$
$y = \coth \omega$	$\omega = \coth^{-1} y = ar\ \coth y$	$\begin{cases} -1 \leq y \leq -\infty \\ +1 \leq y \leq +\infty \end{cases}$

This means that ω is the area of the hyperbolic segment of which the hyperbolic function is y.

(2) Relationships

$$\sinh^{-1} y = \ln (y + \sqrt{y^2 + 1})$$

$$\tanh^{-1} y = \tfrac{1}{2} \ln \frac{1+y}{1-y} \qquad |y| < 1$$

$$\cosh^{-1} y = \pm \ln (y + \sqrt{y^2 - 1}) \qquad y \geq 1$$

$$\coth^{-1} y = \tfrac{1}{2} \ln \frac{y+1}{y-1} \qquad |y| > 1$$

For other relationships use the logarithmic transformations, for example,

$$\sinh^{-1} y + \cosh^{-1} y = \ln (y + \sqrt{y^2 + 1}) \pm \ln (y + \sqrt{y^2 - 1})$$

(3) Graphs

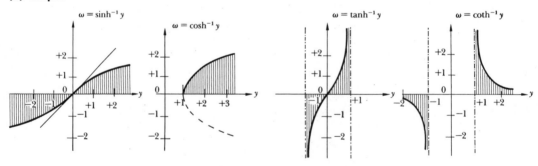

(4) Transformation Table   ($k = +1$ if $y > 0$, or $k = -1$ if $y < 0$)

	$\sinh^{-1} y$	$\cosh^{-1} y$	$\tanh^{-1} y$	$\coth^{-1} y$
$\sinh^{-1} y$	$\sinh^{-1} y$	$k \cosh^{-1} \sqrt{y^2 + 1}$	$\tanh^{-1} \dfrac{y}{\sqrt{y^2 + 1}}$	$\coth^{-1} \dfrac{\sqrt{y^2 + 1}}{y}$
$\cosh^{-1} y$	$k \sinh^{-1} \sqrt{y^2 - 1}$	$\cosh^{-1} y$	$k \tanh^{-1} \dfrac{\sqrt{y^2 - 1}}{y}$	$k \coth^{-1} \dfrac{y}{\sqrt{y^2 - 1}}$
$\tanh^{-1} y$	$\sinh^{-1} \dfrac{y}{\sqrt{1 - y^2}}$	$k \cosh^{-1} \dfrac{1}{\sqrt{1 - y^2}}$	$\tanh^{-1} y$	$\coth^{-1} \dfrac{1}{y}$
$\coth^{-1} y$	$\sinh^{-1} \dfrac{1}{\sqrt{y^2 - 1}}$	$k \cosh^{-1} \dfrac{y}{\sqrt{y^2 - 1}}$	$\tanh^{-1} \dfrac{1}{y}$	$\coth^{-1} y$

7

DIFFERENTIAL CALCULUS

(1) Basic Terms

(a) The quality which changes its value is called a *variable*, and the range of its variation is known as the *interval*.

(b) The quality which remains unchanged is called a *constant*, and its range is a *single number*.

(c) A *function* is a relationship between two or more *variables* and *constants*.

(2) Definitions

(a) A variable y is a *function of another variable x*,

$$y = f(x)$$

if for each value in the range of x there are one or more values in the range of y.

$x =$ independent variable or argument $\qquad y =$ dependent variable

(b) A variable y is a *function of several variables $x_1, x_2, \ldots, x_n$*,

$$y = f(x_1, x_2, \ldots, x_n)$$

if for each set of values in the range of x's there is one or more values in the range of y.

(3) Forms

(a) The three most common forms of representation of a function are *tabular*, *graphical*, and *analytical*.

(b) Analytical representation

Explicit: $y = f(x_1, x_2, \ldots, x_n)$
Implicit: $0 = g(x_1, x_2, \ldots, x_n, y)$
Parametric: $x_1 = h_1(\tau), x_2 = h_2(\tau), \ldots, y = h(\tau)$

(4) Characteristic Properties

(a) Single-valued

A function is single-valued if it has a single value y for any given value of x.

(b) Multivalued

A function is multivalued if it has more than one value of y for any given value of x.

(c) Even and odd

A function is even if $f(-x) = f(x)$ and odd if $f(-x) = -f(x)$.

(d) Periodic

A function is periodic with the period T if $f(x + T) = f(x)$.

(e) Inverse

An inverse of a function $y = f(x)$ is another function $x = g(y)$.

(5) Interval

$a < x < b$	Bounded open interval	(a, b)
$a < x$	Unbounded open interval	(a, ∞)
$x < b$	Unbounded open interval	$(-\infty, b)$
$a \leq x \leq b$	Bounded closed interval	$[a, b]$
$a \leq x$	Unbounded closed interval	$[a, \infty]$
$x \leq b$	Unbounded closed interval	$[-\infty, b]$

(1) Definition of Limit

A variable x is said to have a limit a ($\lim x = a$, or $x \to a$) as x takes on consecutively the values $x_1, x_2, x_3, \ldots, x_n$ if, for every positive number ϵ, however small, the numerical value of

$$|x - a| < \epsilon$$

A function $f(x)$ is said to have a limit b $[\lim_{x \to a} f(x) = b]$ as x takes on consecutively the values $x_1, x_2, x_3, \ldots, x_n$ and approaches a, without assuming the value a, if, for every positive number δ, however small, the numerical value of

$$|f(x) - b| < \epsilon$$

(2) Operations with Limits

$$\lim_{x \to a} [f(x) + g(x)] = \lim_{x \to a} f(x) + \lim_{x \to a} g(x)$$

$$\lim_{x \to a} [f(x)g(x)] = \lim_{x \to a} f(x) \lim_{x \to a} g(x)$$

$$\lim_{x \to a} \frac{f(x)}{g(x)} = \frac{\lim_{x \to a} f(x)}{\lim_{x \to a} g(x)} \qquad \lim_{x \to a} g(x) \neq 0$$

(3) Special Cases

$$\lim_{x \to 0} a^x = 1 \qquad a > 0$$

$$\lim_{x \to 0} (1 + x)^{1/x} = e$$

$$\lim_{x \to 0} \frac{e^x - 1}{x} = 1$$

$$\lim_{x \to \infty} \sqrt[x]{x} = 1$$

$$\lim_{x \to \infty} \left(1 + \frac{y}{x}\right)^x = e^y$$

$$\lim_{x \to \infty} \frac{a^x - 1}{x} = \ln a \qquad a > 0$$

$$\lim_{m \to \infty} \frac{a^m}{m!} = 0$$

$$\lim_{x \to \infty} \left(1 + \frac{1}{x}\right)^x = e$$

$$\lim_{x \to \infty} \frac{x^m}{e^x} = 0$$

$$\lim_{x \to \infty} \frac{(\ln x)^m}{m} = 0$$

$$\lim_{x \to 1} \frac{x - 1}{\ln x} = 1$$

$$\lim_{x \to \infty} \frac{\ln (x + 1)}{x} = 1$$

$$\lim_{x \to 0} \frac{\sin x}{x} = 1$$

$$\lim_{x \to 0} \frac{1 - \cos x}{x} = 0$$

$$\lim_{x \to 0} \frac{\tan x}{x} = 1$$

$$\lim_{m \to \infty} \left(1 + \frac{1}{2} + \frac{1}{3} + \cdots + \frac{1}{m} - \ln m\right) = 0.577\ 215\ 665 \qquad \text{(Euler's constant, Sec. 13.01)}$$

$$\lim_{m \to \infty} \frac{m!}{m^m e^{-m} \sqrt{m}} = \sqrt{2\pi} \qquad \text{(Stirling's formula, Sec. 1.03)}$$

(4) Continuity of Function

(a) A single-valued function is *continuous throughout the neighborhood* of $x = a$ if and only if $\lim_{x \to a} f(x)$ exists and is equal to $f(a)$.

(b) A single-valued function is *continuous in an interval* (a, b) or $[a, b]$ if and only if it is continuous at each point of this interval.

(c) A single-valued function has a *discontinuity of the first kind* at the point $x = a$ if $f(a + 0) \neq f(a - 0)$. The difference of these two values is known as the *jump* (status) of $f(x)$.

(d) A single-valued function is *piecewise-continuous* on a given interval (a, b) or $[a, b]$ if and only if $f(x)$ is continuous throughout this interval except for a finite number of discontinuities of the first kind.

(1) Definitions

(a) First derivative of $y = f(x)$ with respect to x is defined as

$$\boxed{\tan \phi = \frac{dy}{dx} = \lim_{\Delta x \to 0} \frac{\Delta y}{\Delta x} = \lim_{\Delta x \to 0} \frac{f(x + \Delta x) - f(x)}{\Delta x}}$$

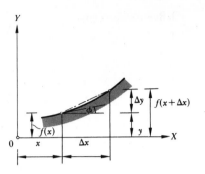

Alternative notations are $f'(x), \dfrac{df(x)}{dx}$, and y'. If $y = f(t)$,

$$\frac{dy}{dt} = \frac{df(t)}{dt} = \dot{y}$$

(b) Second and higher derivatives of the same function are

$$\frac{d^2 y}{dx^2} = \frac{d}{dx}\left(\frac{dy}{dx}\right) = \frac{d}{dx}[f'(x)] = y''$$

$$\frac{d^n y}{dx^n} = \frac{d}{dx}\left(\frac{d^{n-1} y}{dx}\right) = \frac{d}{dx}[f^{(n-1)}(x)] = f^{(n)}(x)$$

(c) First partial derivatives of $y = f(x_1, x_2, \ldots, x_i, x_j, \ldots)$ with respect to one of the independent variables x_i or x_j are

$$\frac{\partial y}{\partial x_i} = \frac{\partial f(\ )}{\partial x_i} = F_i$$

$$\frac{\partial y}{\partial x_j} = \frac{\partial f(\ )}{\partial x_j} = F_j$$

Thus there are as many possible first partial derivatives as there are independent variables.

(d) Second and higher derivatives are

$$F_{jj} = \frac{\partial F_j}{\partial x_j} \qquad F_{ij} = \frac{\partial F_j}{\partial x_i} \qquad F_{ji} = \frac{\partial F_i}{\partial x_j} \qquad F_{ii} = \frac{\partial F_i}{\partial x_i}$$

The same process defines derivatives of any order. When the highest derivatives involved are continuous, the result is independent of the order in which the differentiation is performed.

$$F_{ijj} = F_{jij} = F_{jji} \qquad\qquad F_{ij} = F_{ji}$$

The number of differentiations performed is the *order of the partial derivatives*.

(2) Rules

$$\boxed{y = f(x_1) \qquad x_1 = g(x_2) \qquad x_2 = h(x_3)}$$

$$\frac{dy}{dx_3} = \frac{dy}{dx_1}\frac{dx_1}{dx_2}\frac{dx_2}{dx_3}$$

$$\boxed{y = f(x) \qquad x = g(y) \qquad \frac{dx}{dy} \neq 0}$$

$$\frac{dy}{dx} = \frac{1}{dx/dy} \qquad \frac{d^2 y}{dx^2} = -\frac{d^2 x/dy^2}{(dx/dy)^3}$$

$$\boxed{F(x, y) = 0 \qquad F_y \neq 0}$$

$$\frac{dy}{dx} = -F_x : F_y$$

$$\frac{d^2 y}{dx^2} = -(F_{xx}F_y{}^2 - 2F_x F_{xy}F_y + F_x{}^2 F_{yy}) : F_y{}^3$$

$$\boxed{x = x(t) \qquad y = y(t) \qquad \dot{x}(t) = \frac{dx}{dt} \neq 0 \qquad \dot{y}(t) = \frac{dy}{dt} \neq 0}$$

$$\frac{dy}{dx} = \frac{\dot{y}(t)}{\dot{x}(t)}$$

$$\frac{d^2 y}{dx^2} = \frac{\dot{x}(t)\ddot{y}(t) - \ddot{x}(t)\dot{y}(t)}{[\dot{x}(t)]^3}$$

(1) Rolle's Theorem

If a function $f(x)$ is continuous in the closed interval $[a, b]$ and is differentiable in the open interval (a, b) and if $f(a) = f(b)$, then there is at least one point $(x = c)$ in (a, b) in which

$$\boxed{f'(c) = 0}$$

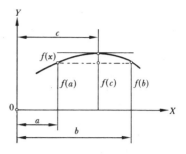

(2) Lagrange's Theorem (first mean-value theorem)

If a function $f(x)$ is continuous in the closed interval $[a, b]$ and is differentiable in the open interval (a, b), then there is at least one point $(x = c)$ in (a, b) in which

$$\boxed{\frac{f(b) - f(a)}{b - a} = f'(c)}$$

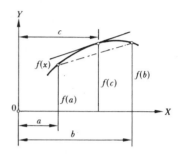

(3) Cauchy's Theorem (second mean-value theorem)

If the functions $f(x)$ and $g(x)$ are continuous in the closed interval $[a, b]$ and are differentiable in the open interval (a, b), if $f(x)$ and $g(x)$ are not simultaneously equal to zero at any point of this open interval, and if $g(a) \neq g(b)$, then there is at least one point $(x = c)$ in (a, b) in which

$$\boxed{\frac{f(b) - f(a)}{g(b) - g(a)} = \frac{f'(c)}{g'(c)}}$$

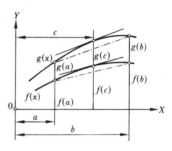

(4) L'Hôpital's Rules

$f(x)$ and $g(x)$ are two continuous functions of x having continuous derivatives at x.

If $f(x)/g(x)$ for $x = a$ is $0/0$ or ∞/∞, then

$$\lim_{x \to a} \frac{f(x)}{g(x)} = \lim_{x \to a} \frac{f'(x)}{g'(x)}$$

If $f(x) - g(x)$ for $x = a$ is $\infty - \infty$, then

$$\lim_{x \to a} [f(x) - g(x)] = \lim_{x \to a} \frac{[1/f(x) - 1/g(x)]'}{[1/f(x)]'[1/g(x)]'}$$

If $f(x)g(x)$ for $x = a$ is $(0)(\infty)$ or $(\infty)(0)$, then

$$\lim_{x \to a} [f(x)g(x)] = \lim_{x \to a} \frac{f'(x)}{[1/g(x)]'}$$

If $f(x)^{g(x)}$ for $x = a$ is 0^0 or ∞^0 or $1^{-\infty}$, then

$$\lim_{x \to a} f(x)^{g(x)} = \lim_{x \to a} e^{g(x)\ln f(x)}$$

u, v, w = differentiable functions of x; u', v', w' = first derivatives of these functions with respect to x; a, b, c, m = constants.

(1) General Formulas

$$(a)' = 0$$

$$(au)' = au'$$

$$(u + v + w + \cdots)' = u' + v' + w' + \cdots$$

$$\left(\frac{uv}{w}\right)' = \left(\frac{uv}{w}\right)\left(\frac{u'}{u} + \frac{v'}{v} - \frac{w'}{w}\right)$$

$$(uv)' = u'v + uv'$$

$$\left(\frac{u}{v}\right)' = \frac{u'v - uv'}{v^2}$$

$$(uvw\cdots)' = (uvw\cdots)\left(\frac{u'}{u} + \frac{v'}{v} + \frac{w'}{w} + \cdots\right)$$

$$(u^v) = vu^{v-1}u' + u^v v' \ln u$$

(2) Algebraic Functions

$$(au^m)' = amu^{m-1}u'$$

$$(a\sqrt{u})' = \frac{a}{2\sqrt{u}}u'$$

$$(a\sqrt[m]{u})' = \frac{a\sqrt[m]{u}}{mu}u'$$

$$\left(\frac{a}{u^m}\right)' = -\frac{am}{u^{m+1}}u'$$

$$\left(\frac{a}{\sqrt{u}}\right)' = -\frac{a}{2\sqrt{u^3}}u'$$

$$\left(\frac{a}{\sqrt[m]{u}}\right)' = -\frac{a}{mu\sqrt[m]{u}}u'$$

(3) Exponential Functions $(e = 2.718\,28\cdots)$

$$(e^{mu})' = me^{mu}u'$$

$$(a^{mu})' = (ma^{mu}\ln a)u'$$

$$(u^{mu})' = mu^{mu}(1 + \ln u)u'$$

$$(e^{-mu})' = -me^{-mu}u'$$

$$(a^{-mu})' = -(m\ln a)a^{-mu}u'$$

$$(u^{-mu})' = -\frac{m(1 + \ln u)}{u^{mu}}u'$$

(4) Logarithmic Functions

$$(\ln au)' = \frac{u'}{u}$$

$$(\log u)' = \frac{u'}{u}\log e$$

$$(\log au)' = \frac{u'}{u}\log e$$

$$\left(\ln\frac{a}{u}\right)' = -\frac{u'}{u}$$

$$\left(\log\frac{1}{u}\right)' = -\frac{u'}{u}\log e$$

$$\left(\log\frac{a}{u}\right)' = -\frac{u'}{u}\log e$$

u = differentiable function of x	u' = first derivative of u with respect to x

(1) Trigonometric Functions

$$(\sin u)' = u' \cos u$$

$$\left(\frac{1}{\sin u}\right)' = -\frac{u' \cos u}{\sin^2 u}$$

$$(\cos u)' = -u' \sin u$$

$$\left(\frac{1}{\cos u}\right)' = \frac{u' \sin u}{\cos^2 u}$$

$$(\tan u)' = \frac{u'}{\cos^2 u}$$

$$\left(\frac{1}{\tan u}\right)' = -\frac{u'}{\sin^2 u}$$

$$(\cot u)' = -\frac{u'}{\sin^2 u}$$

$$\left(\frac{1}{\cot u}\right)' = \frac{u'}{\cos^2 u}$$

$$(\sec u)' = \frac{u' \sin u}{\cos^2 u}$$

$$\left(\frac{1}{\sec u}\right)' = -u' \sin u$$

$$(\csc u)' = -\frac{u' \cos u}{\sin^2 u}$$

$$\left(\frac{1}{\csc u}\right)' = u' \cos u$$

(2) Inverse Trigonometric Functions   (u = **principal value**)

$$(\sin^{-1} u)' = \frac{u'}{\sqrt{1 - u^2}}$$

$$(\cos^{-1} u)' = -\frac{u'}{\sqrt{1 - u^2}}$$

$$(\tan^{-1} u)' = \frac{u'}{1 + u^2}$$

$$(\cot^{-1} u)' = -\frac{u'}{1 + u^2}$$

(3) Hyperbolic Functions

$$(\sinh u)' = u' \cosh u$$

$$(\cosh u)' = u' \sinh u$$

$$(\tanh u)' = \frac{u'}{\cosh^2 u}$$

$$(\coth u)' = -\frac{u'}{\sinh^2 u}$$

$$(\operatorname{sech} u)' = -\frac{u' \sinh u}{\cosh^2 u}$$

$$(\operatorname{csch} u)' = -\frac{u' \cosh u}{\sinh^2 u}$$

(4) Inverse Hyperbolic Functions   (u = **principal value**)

$$(\sinh^{-1} u)' = \frac{u'}{\sqrt{u^2 + 1}}$$

$$(\cosh^{-1} u)' = \frac{u'}{\sqrt{u^2 - 1}}$$

$$(\tanh^{-1} u)' = \frac{u'}{1 - u^2}$$

$$(\coth^{-1} u)' = \frac{u'}{1 - u^2}$$

a, k = constants	m = integer	n = order of derivative

(1) Algebraic Functions

y	$y^{(n)} = \dfrac{d^n y}{dx^n}$	n
$(ax)^m$	$\begin{cases} m(m-1)(m-2)\cdots(m-n+1)\dfrac{y}{x^n} \\ m!\,a^m \\ 0 \end{cases}$	$\begin{matrix} n < m \\ n = m \\ n > m \end{matrix}$
$\sqrt[m]{ax}$	$(-1)^{n-1}(m-1)(2m-1)\cdots[(n-1)m-1]\dfrac{y}{(mx)^n}$	$n \lessgtr m$
$\left(\dfrac{1}{ax}\right)^m$	$(-1)^n m(m+1)(m+2)\cdots(m+n-1)\dfrac{y}{x^n}$	$n \lessgtr m$
$\sqrt[m]{\dfrac{1}{ax}}$	$(-1)^n(m+1)(2m+1)\cdots[(n-1)m+1]\dfrac{y}{(mx)^n}$	$n \lessgtr m$

(2) Exponential and Logarithmic Functions

y	$y^{(n)} = \dfrac{d^n y}{dx^n}$	y	$y^{(n)} = \dfrac{d^n y}{dx^n}$
e^x	e^x	a^x	$(\ln a)^n a^x$
e^{kx}	$k^n e^{kx}$	a^{kx}	$(k \ln a)^n a^{kx}$
$\ln x$	$\dfrac{(-1)^{n-1}(n-1)!}{x^n}$	$\log x$	$\dfrac{(-1)^{n-1}(n-1)!\log e}{x^n}$
$\ln kx$	$\dfrac{(-1)^{n-1}(n-1)!}{x^n}$	$\log kx$	$\dfrac{(-1)^{n-1}(n-1)!\log e}{x^n}$

(3) Trigonometric and Hyperbolic Functions

y	$y^{(n)} = \dfrac{d^n y}{dx^n}$	y	$y^{(n)} = \dfrac{d^n y}{dx^n}$
$\sin x$	$\sin\left(x + \dfrac{n\pi}{2}\right)$	$\cos x$	$\cos\left(x + \dfrac{n\pi}{2}\right)$
$\sin kx$	$k^n \sin\left(kx + \dfrac{n\pi}{2}\right)$	$\cos kx$	$k^n \cos\left(kx + \dfrac{n\pi}{2}\right)$
$\sinh x$	$\begin{cases} \sinh x & (n \text{ even}) \\ \cosh x & (n \text{ odd}) \end{cases}$	$\cosh x$	$\begin{cases} \cosh x & (n \text{ even}) \\ \sinh x & (n \text{ odd}) \end{cases}$
$\sinh kx$	$\begin{cases} k^n \sinh kx & (n \text{ even}) \\ k^n \cosh kx & (n \text{ odd}) \end{cases}$	$\cosh kx$	$\begin{cases} k^n \cosh kx & (n \text{ even}) \\ k^n \sinh kx & (n \text{ odd}) \end{cases}$

k, m, p, r = positive integers　　　　　　　　n = order of derivative
$\bar{B}_r(x)$ = Bernoulli polynomial (Sec. 8.05)　　$\bar{E}_r(x)$ = Euler polynomial (Sec. 8.06)
$X_p^{(r)}$ = factorial polynomial (Sec. 1.03)　　　$\mathscr{S}_k^{(r)}$ = Stirling number (Sec. A.05)

(1) Factorial Functions

y	$y^{(n)} = \dfrac{d^n y}{dx^n}$	n
$\dbinom{x}{r}$	$\begin{cases} \dfrac{n!}{r!x^n}\left[\dbinom{n}{n}\mathscr{S}_n^{(r)}x^n + \dbinom{n+1}{n}\mathscr{S}_{n+1}^{(r)}x^{n+1} + \cdots + \dbinom{r}{n}\mathscr{S}_r^{(r)}x^r\right] \\[2mm] 1 \\[1mm] 0 \end{cases}$	$\begin{matrix} n < r \\[3mm] n = r \\[1mm] n > r \end{matrix}$
$X_p^{(r)}$	$\begin{cases} \dfrac{n!p^r}{x^n}\left[\dbinom{n}{n}\mathscr{S}_n^{(r)}\left(\dfrac{x}{p}\right)^n + \dbinom{n+1}{n}\mathscr{S}_{n+1}^{(r)}\left(\dfrac{x}{p}\right)^{n+1} + \cdots + \dbinom{r}{n}\mathscr{S}_r^{(r)}\left(\dfrac{x}{p}\right)^r\right] \\[2mm] n! \\[1mm] 0 \end{cases}$	$\begin{matrix} n < r \\[3mm] n = r \\[1mm] n > r \end{matrix}$
$x^m X_p^{(r)}$	$\begin{cases} \dfrac{n!p^r x^m}{x^n}\left[\dbinom{m+n}{n}\mathscr{S}_n^{(r)}\left(\dfrac{x}{p}\right)^n + \dbinom{m+n+1}{n}\mathscr{S}_{n+1}^{(r)}\left(\dfrac{x}{p}\right)^{n+1} + \cdots + \dbinom{m+r}{n}\mathscr{S}_r^{(r)}\left(\dfrac{x}{p}\right)^r\right] \\[2mm] \dfrac{(r+m)!}{m!}x^m \\[2mm] \dfrac{(r+m)!}{(r+m-n)!}x^{r+m-n} \\[2mm] n! \\[1mm] 0 \end{cases}$	$\begin{matrix} n < r \\[3mm] n = r \\[3mm] r < n < r+m \\[2mm] n = r+m \\[1mm] n > r+m \end{matrix}$

(2) Bernoulli and Euler Polynomials

y	$y^{(n)} = \dfrac{d^n y}{dx^n}$	y	$y^{(n)} = \dfrac{d^n y}{dx^n}$	n
$\bar{B}_r(x)$	$\begin{cases} n!\dbinom{r}{n}\bar{B}_{r-n}(x) \\[2mm] n! \\[1mm] 0 \end{cases}$	$\bar{E}_r(x)$	$\begin{cases} n!\dbinom{r}{n}\bar{E}_{r-n}(x) \\[2mm] n! \\[1mm] 0 \end{cases}$	$\begin{matrix} n < r \\[3mm] n = r \\[1mm] n > r \end{matrix}$

(3) Product of Two Functions

If u, v are differentiable functions of x and

$$\frac{d^n u}{dx^n} = u^{(n)} \qquad \frac{d^n v}{dx^n} = v^{(n)}$$

are the nth derivatives of u, v, respectively, then

$$\frac{d^n(uv)}{dx^n} = uv^{(n)} + \binom{n}{1}u^{(1)}v^{(n-1)} + \binom{n}{2}u^{(2)}v^{(n-2)} + \cdots + u^{(n)}v$$

a, b = constants > 0 m, k, t = positive integers n = order of derivative

$$F_k = \binom{m}{k}\binom{n}{k}k!\, x^{m-k}b^{n-k} \qquad r = \tfrac{1}{2}(n-1) \qquad s = \tfrac{1}{2}(n-2) \qquad t = \begin{cases} \tfrac{1}{2}m & \text{if } m \leq n \\ \tfrac{1}{2}n & \text{if } m > n \end{cases}$$

$$G_k = \binom{m}{2k}\binom{n}{2k}(2k)!\, x^{m-2k}b^{n-2k} \qquad\qquad H_k = \binom{m}{2k+1}\binom{n}{2k+1}(2k+1)!\, x^{m-2k-1}b^{n-2k-1}$$

y	$y^{(n)} = \dfrac{d^n y}{dx^n}$	
n even or odd $\quad$ $x^m e^{bx}$	$e^{bx}\displaystyle\sum_{k=0}^{2t} F_k$	
$x^m a^{bx}$	$a^{bx}\displaystyle\sum_{k=0}^{2t} (\ln a)^{2t-k}F_k$	
$x^m \ln ax$	$\begin{cases} \dfrac{m!\,\ln ax}{(m-n)!}x^{m-n} - m!(n-1)!x^{m-n}\displaystyle\sum_{k=1}^{n}\dfrac{(-1)^{n-k}}{(m-k)[(n-k)!k!]} & m \geq n \\[3mm] - m!(n-1)!x^{m-n}\displaystyle\sum_{k=1}^{n}\dfrac{(-1)^{n-k}}{(m-k)[(n-k)!k!]} & m < n \end{cases}$	
n even $\quad$ $x^m \sin bx$	$\sin bx\displaystyle\sum_{k=0}^{t}(-1)^kG_k - \cos bx\displaystyle\sum_{k=0}^{s}(-1)^kH_k$	$\binom{m}{m}$
$x^m \cos bx$	$\cos bx\displaystyle\sum_{k=0}^{t}(-1)^kG_k + \sin bx\displaystyle\sum_{k=0}^{s}(-1)^kH_k$	
$x^m \sinh bx$	$\sinh bx\displaystyle\sum_{k=0}^{t}G_k + \cosh bx\displaystyle\sum_{k=0}^{s}H_k$	If $m \leq n$, series terminates with term involving
$x^m \cosh bx$	$\cosh bx\displaystyle\sum_{k=0}^{t}G_k + \sinh bx\displaystyle\sum_{k=0}^{s}H_k$	
n odd $\quad$ $x^m \sin bx$	$-\cos bx\displaystyle\sum_{k=0}^{r}(-1)^kG_k - \sin bx\displaystyle\sum_{k=0}^{r}(-1)^kH_k$	
$x^m \cos bx$	$\sin bx\displaystyle\sum_{k=0}^{r}(-1)^kG_k - \cos bx\displaystyle\sum_{k=0}^{r}(-1)^kH_k$	
$x^m \sinh bx$	$\cosh bx\displaystyle\sum_{k=0}^{r}G_k + \sinh bx\displaystyle\sum_{k=0}^{r}H_k$	
$x^m \cosh bx$	$\sinh bx\displaystyle\sum_{k=0}^{r}G_k + \cosh bx\displaystyle\sum_{k=0}^{r}H_k$	

a, b = constants > 0 m, k, l = positive integers n = order of derivative

$$F_k = \binom{m+k-1}{k}\binom{n}{k}\frac{k!}{(bx)^k} \qquad f^{(k)}(bx) = \frac{d^k f(bx)}{d^k x} \qquad r = \tfrac{1}{2}(n-1) \qquad s = \tfrac{1}{2}(n-2)$$

$$G_k = \binom{m+2k-1}{2k}\binom{n}{2k}\frac{(2k)!}{(bx)^{2k}} \qquad H_k = \binom{m+2k}{2k+1}\binom{n}{2k+1}\frac{(2k+1)!}{(bx)^{2k+1}} \qquad l = \tfrac{1}{2}n$$

y	$y^{(n)} = \dfrac{d^n y}{dx^n}$
$\dfrac{e^{bx}}{x^m}$	$\dfrac{b^n}{x^m} e^{bx} \displaystyle\sum_{k=0}^{n} (-1)^k F_k$
$\dfrac{a^{bx}}{x^m}$	$\dfrac{b^n}{x^m} a^{bx} \displaystyle\sum_{k=0}^{n} (-1)^k (\ln a)^{n-k} F_k$
$\dfrac{\ln ax}{x^m}$	$\dfrac{(m+n-1)!\,\ln ax}{(m-1)!\,x^{m+n}} - \displaystyle\sum_{k=1}^{n} (-1)^{n-k}\binom{n}{k}\binom{m+k-1}{k}\dfrac{k!(n-k-1)!}{x^{m+n}}$
$\dfrac{f(bx)}{x^m}$	$\dfrac{1}{x^m} \displaystyle\sum_{k=0}^{n} (-b)^k F_k f^{(n-k)}(bx)$
$\dfrac{\sin bx}{x^m}$	$-\dfrac{b^n}{x^m}\left[\sin bx \displaystyle\sum_{k=0}^{l} (-1)^k G_k + \cos bx \sum_{k=0}^{s} (-1)^k H_k\right]$
$\dfrac{\cos bx}{x^m}$	$\dfrac{b^n}{x^m}\left[\cos bx \displaystyle\sum_{k=0}^{l} (-1)^k G_k - \sin bx \sum_{k=0}^{s} (-1)^k H_k\right]$
$\dfrac{\sinh bx}{x^m}$	$\dfrac{b^n}{x^m}\left[\sinh bx \displaystyle\sum_{k=0}^{l} G_k - \cosh bx \sum_{k=0}^{s} H_k\right]$
$\dfrac{\cosh bx}{x^m}$	$\dfrac{b^n}{x^m}\left[\cosh bx \displaystyle\sum_{k=0}^{l} G_k - \sinh bx \sum_{k=0}^{s} H_k\right]$
$\dfrac{\sin bx}{x^m}$	$-\dfrac{b^n}{x^m}\left[\cos bx \displaystyle\sum_{k=0}^{r} (-1)^k G_k + \sin bx \sum_{k=0}^{r} (-1)^k H_k\right]$
$\dfrac{\cos bx}{x^m}$	$\dfrac{b^n}{x^m}\left[\sin bx \displaystyle\sum_{k=0}^{r} (-1)^k G_k + \cos bx \sum_{k=0}^{r} (-1)^k H_k\right]$
$\dfrac{\sinh bx}{x^m}$	$\dfrac{b^n}{x^m}\left[\cosh bx \displaystyle\sum_{k=0}^{r} G_k - \sinh bx \sum_{k=0}^{r} H_k\right]$
$\dfrac{\cosh bx}{x^m}$	$\dfrac{b^n}{x^m}\left[\sinh bx \displaystyle\sum_{k=0}^{r} G_k - \cosh bx \sum_{k=0}^{r} H_k\right]$

n even or odd (rows 1–4), *n* even (rows 5–8), *n* odd (rows 9–12)

a, b, c = constants > 0 m, k, t = positive integers n = order of derivative

$$A = \sqrt{a^2 + b^2} \qquad B = \sqrt{a^2 - b^2} \qquad \phi = \tan^{-1}\frac{b}{a} \qquad \psi = \tanh^{-1}\frac{b}{a}$$

$$F_k = \frac{\dbinom{m}{n-k}x^k}{k!} \qquad M_k = \ln bx - k!\sum_{t=1}^{k}\frac{(-1)^t}{t(n-t)!(ax)^t} \qquad r = \begin{cases} m & \text{if } m \le n \\ n & \text{if } n < m \end{cases}$$

y	$y^{(n)} = \dfrac{d^n y}{dx^n}$
$e^{ax}\ln bx$	$a^n e^{ax}\left[\ln bx - n!\sum_{k=1}^{n}\dfrac{(-1)^k}{k(n-k)!(ax)^k}\right]$
$e^{ax}b^{cx}$	$e^{ax}b^{cx}(a + c\ln b)^n$
$e^{ax}\sin bx$	$A^n e^{ax}\sin(bx + n\phi)$
$e^{ax}\cos bx$	$A^n e^{ax}\cos(bx + n\phi)$
$e^{ax}\sinh bx$	$B^n e^{ax}\sinh(bx + n\psi)$
$e^{ax}\cosh bx$	$B^n e^{ax}\cosh(bx + n\psi)$
$x^m e^{ax}\ln bx$	$n!x^{m-n}e^{ax}\sum_{k=0}^{r} a^k F_k M_k$
$x^m e^{ax}b^{cx}$	$n!x^{m-n}e^{ax}b^{cx}\sum_{k=1}^{r}(a + c\ln b)^k F_k$
$x^m e^{ax}\sin bx$	$n!x^{m-n}e^{ax}\sum_{k=0}^{r} A^k \sin(bx + k\phi)F_k$
$x^m e^{ax}\cos bx$	$n!x^{m-n}e^{ax}\sum_{k=0}^{r} A^k \cos(bx + k\phi)F_k$
$x^m e^{ax}\sinh bx$	$n!x^{m-n}e^{ax}\sum_{k=0}^{r} B^k \sinh(bx + k\psi)F_k$
$x^m e^{ax}\cosh bx$	$n!x^{m-n}e^{ax}\sum_{k=0}^{r} B^k \cosh(bx + k\psi)F_k$

n even or odd

a, b, c = constants > 0 m, k, t = positive integers n = order of derivative

$$A = \sqrt{a^2 + b^2} \qquad B = \sqrt{a^2 - b^2} \qquad \phi = \tan^{-1}\frac{b}{a} \qquad \psi = \tanh^{-1}\frac{b}{a}$$

$$F_k = (-1)^{n-k}\binom{m+n-k-1}{n-k}\frac{x^k}{k!} \qquad\qquad M_k = \ln bx - k!\sum_{t=1}^{k}\frac{(-1)^t}{t(n-t)!(ax)^t}$$

y	$y^{(n)} = \dfrac{d^n y}{dx^n}$
$\dfrac{\ln bx}{e^{ax}}$	$\dfrac{(-a)^n}{e^{ax}}\left[\ln bx - n!\sum_{k=0}^{n}\dfrac{1}{k(n-k)!(ax)^k}\right]$
$\dfrac{b^{cx}}{e^{ax}}$	$\dfrac{b^{cx}}{e^{ax}}(c\ln b - a)^n$
$\dfrac{\sin bx}{e^{ax}}$	$(-1)^n\dfrac{A^n}{e^{ax}}\sin(bx - n\phi)$
$\dfrac{\cos bx}{e^{ax}}$	$(-1)^n\dfrac{A^n}{e^{ax}}\cos(bx - n\phi)$
$\dfrac{\sinh bx}{e^{ax}}$	$(-1)^n\dfrac{B^n}{e^{ax}}\sinh(bx - n\psi)$
$\dfrac{\cosh bx}{e^{ax}}$	$(-1)^n\dfrac{B^n}{e^{ax}}\cosh(bx - n\psi)$
$\dfrac{e^{ax}\ln bx}{x^m}$	$n!\dfrac{e^{ax}}{x^{m+n}}\sum_{k=0}^{n}a^k F_k M_k$
$\dfrac{e^{ax}b^{cx}}{x^m}$	$n!\dfrac{e^{ax}b^{cx}}{x^m}\sum_{k=0}^{n}(a + c\ln b)^k F_k$
$\dfrac{e^{ax}\sin bx}{x^m}$	$n!\dfrac{e^{ax}}{x^{m+n}}\sum_{k=0}^{n}A^k\sin(bx + k\phi)F_k$
$\dfrac{e^{ax}\cos bx}{x^m}$	$n!\dfrac{e^{ax}}{x^{m+n}}\sum_{k=0}^{n}A^k\cos(bx + k\phi)F_k$
$\dfrac{e^{ax}\sinh bx}{x^m}$	$n!\dfrac{e^{ax}}{x^{m+n}}\sum_{k=0}^{n}B^k\sinh(bx + k\psi)F_k$
$\dfrac{e^{ax}\cosh bx}{x^m}$	$n!\dfrac{e^{ax}}{x^{m+n}}\sum_{k=0}^{n}B^k\cosh(bx + k\psi)F_k$

n even or odd

$a, b = $ constants $\neq 0$	$k, m = $ positive integers	$n = $ order of derivative
$\alpha = a + b$	$r = \begin{cases} m & \text{if } m \leq n \\ n & \text{if } n < m \end{cases}$	$\beta = a - b$
$\bar{S}_{1k} = e^{\alpha x} - (-1)^k e^{-\alpha x}$		$\bar{S}_{3k} = e^{\beta x} - (-1)^k e^{-\beta x}$
$\bar{S}_{2k} = e^{\alpha x} + (-1)^k e^{-\alpha x}$	$F_k = \binom{n}{m}\binom{m}{n-k}\dfrac{(n-k)!}{2x^{n-k}}$	$\bar{S}_{4k} = e^{\beta x} + (-1)^k e^{-\beta x}$

	y	$y^{(n)} = \dfrac{d^n y}{dx^n}$
n even	$\sin ax \sin bx$	$-\frac{1}{2}(-1)^{n/2}(\alpha^n \cos \alpha x - \beta^n \cos \beta x)$
	$\sin ax \cos bx$	$-\frac{1}{2}(-1)^{n/2}(\alpha^n \sin \alpha x - \beta^n \sin \beta x)$
	$\cos ax \cos bx$	$-\frac{1}{2}(-1)^{n/2}(\alpha^n \cos \alpha x + \beta^n \cos \beta x)$
	$\sinh ax \sinh bx$	$\frac{1}{2}(\alpha^n \cosh \alpha x - \beta^n \cosh \beta x)$
	$\sinh ax \cosh bx$	$\frac{1}{2}(\alpha^n \sinh \alpha x - \beta^n \sinh \beta x)$
	$\cosh ax \cosh bx$	$\frac{1}{2}(\alpha^n \cosh \alpha x + \beta^n \cosh \beta x)$
n even or odd	$x^m \sin ax \sin bx$	$-x^m \displaystyle\sum_{k=0}^{r} F_k\left[\alpha^k \cos\left(\alpha x + \frac{1}{2}k\pi\right) - \beta^k \cos\left(\beta x + \frac{1}{2}k\pi\right)\right]$
	$x^m \sin ax \cos bx$	$-x^m \displaystyle\sum_{k=0}^{r} F_k\left[\alpha^k \sin\left(\alpha x + \frac{1}{2}k\pi\right) - \beta^k \sin\left(\beta x + \frac{1}{2}k\pi\right)\right]$
	$x^m \cos ax \cos bx$	$x^m \displaystyle\sum_{k=0}^{r} F_k\left[\alpha^k \cos\left(\alpha x + \frac{1}{2}k\pi\right) - \beta^k \cos\left(\beta x + \frac{1}{2}k\pi\right)\right]$
	$x^m \sinh ax \sinh bx$	$\frac{1}{2}x^m \displaystyle\sum_{k=0}^{r} F_k(\alpha^k \bar{S}_{2k} - \beta^k \bar{S}_{4k})$
	$x^m \sinh ax \cosh bx$	$\frac{1}{2}x^m \displaystyle\sum_{k=0}^{r} F_k(\alpha^k \bar{S}_{1k} - \beta^k \bar{S}_{3k})$
	$x^m \cosh ax \cosh bx$	$\frac{1}{2}x^m \displaystyle\sum_{k=0}^{r} F_k(\alpha^k \bar{S}_{2k} + \beta^k \bar{S}_{4k})$

a, b, c = constants $\neq 0$ $\qquad\qquad\qquad\qquad$ n = order of derivative

$$\alpha = a + b \qquad A = \sqrt{\alpha^2 + c^2} \qquad B = \sqrt{\beta^2 + c^2} \qquad \beta = a - b$$

$$C = \sqrt{\alpha^2 - c^2} \qquad D = \sqrt{\beta^2 - c^2}$$

$$\phi_A = \tan^{-1}\frac{\alpha}{c} \qquad \phi_B = \tan^{-1}\frac{\beta}{c} \qquad \phi_C = \tanh^{-1}\frac{\alpha}{c} \qquad \phi_D = \tanh^{-1}\frac{\beta}{c}$$

	y	$y^{(n)} = \dfrac{d^n y}{dx^n}$
n odd	$\sin ax \sin bx$	$-\frac{1}{2}(-1)^{(n-1)/2}(\alpha^n \sin \alpha x - \beta^n \sin \beta x)$
	$\sin ax \cos bx$	$-\frac{1}{2}(-1)^{(n-1)/2}(\alpha^n \cos \alpha x - \beta^n \cos \beta x)$
	$\cos ax \cos bx$	$\frac{1}{2}(-1)^{(n-1)/2}(\alpha^n \sin \alpha x + \beta^n \sin \beta x)$
	$\sinh ax \sinh bx$	$\frac{1}{2}(\alpha^n \sinh \alpha x - \beta^n \sinh \beta x)$
	$\sinh ax \cosh bx$	$\frac{1}{2}(\alpha^n \cosh \alpha x - \beta^n \cosh \beta x)$
	$\cosh ax \cosh bx$	$\frac{1}{2}(\alpha^n \sinh \alpha x + \beta^n \sinh \beta x)$
n even or odd	$e^{cx} \sin ax \sin bx$	$-\frac{1}{2}e^{cx}[A^n \cos (\alpha x + n\phi_A) - B^n \cos (\beta x + n\phi_B)]$
	$e^{cx} \sin ax \cos bx$	$-\frac{1}{2}e^{cx}[A^n \sin (\alpha x + n\phi_A) - B^n \sin (\beta x + n\phi_B)]$
	$e^{cx} \cos ax \cos bx$	$\frac{1}{2}e^{cx}[A^n \cos (\alpha x + n\phi_A) - B^n \cos (\beta x + n\phi_B)]$
	$e^{cx} \sinh ax \sinh bx$	$\frac{1}{2}e^{cx}[C^n \cosh (\alpha x + n\phi_C) - D^n \cosh (\alpha x + n\phi_D)]$
	$e^{cx} \sinh ax \cosh bx$	$\frac{1}{2}e^{cx}[C^n \sinh (\alpha x + n\phi_C) - D^n \sinh (\beta x + n\phi_D)]$
	$e^{cx} \cosh ax \cosh bx$	$\frac{1}{2}e^{cx}[C^n \cosh (\alpha x + n\phi_C) + D^n \cosh (\beta x + n\phi_D)]$

$a, b, c, \alpha = $ constants > 0 $\qquad$ $k, m = $ positive integer $\qquad$ $n = $ order of derivative

$$A = \sqrt{a^2 + b^2} \qquad\qquad B = \sqrt{a^2 - c^2} \qquad\qquad \phi_A = \tan^{-1}\frac{b}{a} \qquad\qquad \phi_B = \tan^{-1}\frac{\alpha}{c}$$

$$F_k = \binom{n}{k}\binom{m}{n-k}\frac{(n-k)!}{x^{n-k}} \qquad\qquad\qquad r = \begin{cases} m & \text{if } m \le n \\ n & \text{if } n < m \end{cases}$$

$$S_n = b^n \sin\left(bx + \tfrac{1}{2}n\pi\right) \qquad\qquad\qquad S_{n-k} = b^{n-k}\sin\left[bx + \tfrac{1}{2}(n-k)\pi\right]$$

$$C_n = b^n \cos\left(bx + \tfrac{1}{2}n\pi\right) \qquad\qquad\qquad C_{n-k} = b^{n-k}\cos\left[bx + \tfrac{1}{2}(n-k)\pi\right]$$

	y	$y^{(n)} = \dfrac{d^n y}{dx^n}$	
$n = 2, 6, 10, \cdots$	$\sin ax \sinh ax$	$\lambda_1 \cos ax \cosh ax$	$\lambda_1 = (-1)^{(n-2)/4}2^{n/2}a^n$
	$\sin ax \cosh ax$	$\lambda_1 \cos ax \sinh ax$	
	$\cos ax \sinh ax$	$-\lambda_1 \sin ax \cosh ax$	
	$\cos ax \cosh ax$	$-\lambda_1 \sin ax \sinh ax$	
$n = 4, 8, 12, \cdots$	$\sin ax \sinh ax$	$\lambda_2 \sin ax \sinh ax$	$\lambda_2 = (-1)^{n/4}2^{n/2}a^n$
	$\sin ax \cosh ax$	$\lambda_2 \sin ax \cosh ax$	
	$\cos ax \sinh ax$	$\lambda_2 \cos ax \sinh ax$	
	$\cos ax \cosh ax$	$\lambda_2 \cos ax \cosh ax$	
n even or odd	$e^{\alpha}(a \sin \alpha x \pm b \cos \alpha x)$	$AB^n e^{\alpha} \sin(\alpha x \pm \phi_A + n\phi_B)$	
	$x^m e^{\alpha}(a \sin \alpha x \pm b \cos \alpha x)$	$Ax^m e^{\alpha} \displaystyle\sum_{k=0}^{r} F_k B^k \sin(\alpha x \pm \phi_A + k\phi_B)$	
	$\ln ax \sin bx$	$S_n \ln ax - \displaystyle\sum_{k=1}^{n}\binom{n}{k}(k-1)!\left(-\frac{1}{x}\right)^k S_{n-k}$	
	$\ln ax \cos bx$	$C_n \ln ax - \displaystyle\sum_{k=1}^{n}\binom{n}{k}(k-1)!\left(-\frac{1}{x}\right)^k C_{n-k}$	

a, b, c, α = constants > 0 k, m = positive integers n = order of derivative

$$C = \sqrt{a^2 - b^2} \qquad D = \sqrt{a^2 - c^2} \qquad \phi_C = \tanh^{-1}\frac{b}{a} \qquad \phi_D = \tanh^{-1}\frac{\alpha}{c}$$

$$F_k = \binom{n}{k}\binom{m}{n-k}\frac{(n-k)!}{x^{n-k}} \qquad\qquad r = \begin{cases} m & \text{if } m \le n \\ n & \text{if } n < m \end{cases}$$

$$\bar{S}_n = \tfrac{1}{2}b^n[e^{bx} - (-1)^n e^{-bx}] \qquad\qquad \bar{S}_{n-k} = \tfrac{1}{2}b^{n-k}[e^{bx} - (-1)^{n-k}e^{-bx}]$$

$$\bar{C}_n = \tfrac{1}{2}b^n[e^{bx} + (-1)^n e^{bx}] \qquad\qquad \bar{C}_{n-k} = \tfrac{1}{2}b^{n-k}[e^{bx} + (-1)^{n-k}e^{-bx}]$$

	y	$y^{(n)} = \dfrac{d^n y}{dx^n}$	
$n = 1, 5, 9, \cdots$	$\sin ax \sinh ax$	$\lambda_3(\cosh ax \sin ax + \sinh ax \cos ax)$	$\lambda_3 = (-1)^{(n-1)/4}2^{(n-1)/2}a^n$
	$\sin ax \cosh ax$	$\lambda_3(\cosh ax \cos ax + \sinh ax \sin ax)$	
	$\cos ax \sinh ax$	$-\lambda_3(\sinh ax \sin ax - \cosh ax \cos ax)$	
	$\cos ax \cosh ax$	$-\lambda_3(\sinh ax \cos ax - \cosh ax \sin ax)$	
$n = 3, 7, 11, \cdots$	$\sin ax \sinh ax$	$-\lambda_4(\sinh ax \cos ax - \cosh ax \sin ax)$	$\lambda_4 = (-1)^{(n+1)/4}2^{(n-1)/2}a^n$
	$\sin ax \cosh ax$	$-\lambda_4(\cosh ax \cos ax - \sinh ax \sin ax)$	
	$\cos ax \sinh ax$	$\lambda_4(\cosh ax \cos ax + \sinh ax \sin ax)$	
	$\cos ax \cosh ax$	$\lambda_4(\sinh ax \cos ax + \cosh ax \sin ax)$	
n even or odd	$e^{\alpha}(a \sinh \alpha x \pm b \cosh \alpha x)$	$CD^n e^{\alpha} \sinh (\alpha x \pm \phi_C + n\phi_D)$	
	$x^m e^{\alpha}(a \sinh \alpha x \pm b \cosh \alpha x)$	$Cx^m e^{\alpha} \displaystyle\sum_{k=0}^{r} F_k D^k \sinh (\alpha x \pm \phi_C + k\phi_D)$	
	$\ln ax \sinh bx$	$\bar{S}_n \ln ax - \displaystyle\sum_{k=1}^{n} \binom{n}{k}(k-1)!\left(-\frac{1}{x}\right)^k \bar{S}_{n-k}$	
	$\ln ax \cosh bx$	$\bar{C}_n \ln ax - \displaystyle\sum_{k=1}^{n} \binom{n}{k}(k-1)!\left(-\frac{1}{x}\right)^k \bar{C}_{n-k}$	

$a = \text{constant} > 0$	$k = 1, 2, 3, \ldots$	$n = \text{order of derivative}$
$\alpha = (-1)^{n/2}$		
$\beta = (-1)^{(n+1)/2}$		$S_k = (ka)^n \sin kax$
		$C_k = (ka)^n \cos kax$

y	$y^{(n)} = \dfrac{d^n y}{dx^n}$	
	n even	*n* odd
$\sin^2 ax$	$-\dfrac{\alpha}{2} C_2$	$-\dfrac{\beta}{2} S_2$
$\sin^3 ax$	$-\dfrac{\alpha}{4}(S_3 - 3S_1)$	$-\dfrac{\beta}{4}(C_3 - 3C_1)$
$\sin^4 ax$	$\dfrac{\alpha}{8}(C_4 - 4C_2)$	$\dfrac{\beta}{8}(S_4 - 4S_2)$
$\sin^5 ax$	$\dfrac{\alpha}{16}(S_5 - 5S_3 + 10S_1)$	$\dfrac{\beta}{16}(C_5 - 5C_3 + 10C_1)$
$\sin^6 ax$	$-\dfrac{\alpha}{32}(C_6 - 6C_4 + 15C_2)$	$-\dfrac{\beta}{32}(S_6 - 6S_4 + 15S_2)$
$\sin^7 ax$	$-\dfrac{\alpha}{64}(S_7 - 7S_5 + 21S_3 - 35S_1)$	$-\dfrac{\beta}{64}(C_7 - 7C_5 + 21C_3 - 35C_1)$
$\cos^2 ax$	$\dfrac{\alpha}{2} C_2$	$\dfrac{\beta}{2} S_2$
$\cos^3 ax$	$\dfrac{\alpha}{4}(C_3 + 3C_1)$	$\dfrac{\beta}{4}(S_3 + 3S_1)$
$\cos^4 ax$	$\dfrac{\alpha}{8}(C_4 + 4C_2)$	$\dfrac{\beta}{8}(S_4 + 4S_2)$
$\cos^5 ax$	$\dfrac{\alpha}{16}(C_5 + 5C_3 + 10C_1)$	$\dfrac{\beta}{16}(S_5 + 5S_3 + 10S_1)$
$\cos^6 ax$	$\dfrac{\alpha}{32}(C_6 + 6C_4 + 15C_2)$	$\dfrac{\beta}{32}(S_6 + 6S_4 + 15S_2)$
$\cos^7 ax$	$\dfrac{\alpha}{64}(C_7 + 7C_5 + 21C_3 + 35C_1)$	$\dfrac{\beta}{64}(S_7 + 7S_5 + 21S_3 + 35S_1)$

Note: For higher derivatives of powers not shown above ($k > 7$), use the expansions of Sec. 6.04–2 and the general derivative formulas of Sec. 7.07–3.

$a = \text{constant} > 0$	$k = 1, 2, 3, \ldots$	$n = \text{order of derivative}$
$\bar{S}_k = (ka)^n \sinh kax$		$\bar{C}_k = (ka)^n \cosh kax$

y	$y^{(n)} = \dfrac{d^n y}{dx^n}$	
	n even	n odd
$\sinh^2 ax$	$\dfrac{1}{2}\bar{C}_2$	$\dfrac{1}{2}\bar{S}_2$
$\sinh^3 ax$	$\dfrac{1}{4}(\bar{S}_3 - 3\bar{S}_1)$	$\dfrac{1}{4}(\bar{C}_3 - 3\bar{C}_1)$
$\sinh^4 ax$	$\dfrac{1}{8}(\bar{C}_4 - 4\bar{C}_2)$	$\dfrac{1}{8}(\bar{S}_4 - 4\bar{S}_2)$
$\sinh^5 ax$	$\dfrac{1}{16}(\bar{S}_5 - 5\bar{S}_3 + 10\bar{S}_1)$	$\dfrac{1}{16}(\bar{C}_5 - 5\bar{C}_3 + 10\bar{C}_1)$
$\sinh^6 ax$	$\dfrac{1}{32}(\bar{C}_6 - 6\bar{C}_4 + 15\bar{C}_2)$	$\dfrac{1}{32}(\bar{S}_6 - 6\bar{S}_4 + 15\bar{S}_2)$
$\sinh^7 ax$	$\dfrac{1}{64}(\bar{S}_7 - 7\bar{S}_5 + 21\bar{S}_3 - 35\bar{S}_1)$	$\dfrac{1}{64}(\bar{C}_7 - 7\bar{C}_5 + 21\bar{C}_3 - 35\bar{C}_1)$
$\cosh^2 ax$	$\dfrac{1}{2}\bar{C}_2$	$\dfrac{1}{2}\bar{S}_2$
$\cosh^3 ax$	$\dfrac{1}{4}(\bar{C}_3 + 3\bar{C}_1)$	$\dfrac{1}{4}(\bar{S}_3 + 3\bar{S}_1)$
$\cosh^4 ax$	$\dfrac{1}{8}(\bar{C}_4 + 4\bar{C}_2)$	$\dfrac{1}{8}(\bar{S}_4 + 4\bar{S}_2)$
$\cosh^5 ax$	$\dfrac{1}{16}(\bar{C}_5 + 5\bar{C}_3 + 10\bar{C}_1)$	$\dfrac{1}{16}(\bar{S}_5 + 5\bar{S}_3 + 10\bar{S}_1)$
$\cosh^6 ax$	$\dfrac{1}{32}(\bar{C}_6 + 6\bar{C}_4 + 15\bar{C}_2)$	$\dfrac{1}{32}(\bar{S}_6 + 6\bar{S}_4 + 15\bar{S}_2)$
$\cosh^7 ax$	$\dfrac{1}{64}(\bar{C}_7 + 7\bar{C}_5 + 21\bar{C}_3 + 35\bar{C}_1)$	$\dfrac{1}{64}(\bar{S}_7 + 7\bar{S}_5 + 21\bar{S}_3 + 35\bar{S}_1)$

Note: For higher derivatives of powers not shown above ($k > 7$), use the expansions of Sec. 6.10–2 and the general derivative formulas of Sec. 7.07–3.

(1) Differential

The first differential of $y = f(x)$ is

$$\boxed{dy = \frac{dy}{dx} dx = df(x) = f'(x)\, dx = y'\, dx}$$

The higher differentials are obtained by successive differentiations of the first differential.

$$d^2y = f''(x)\, dx^2 \qquad\qquad d^ny = f^{(n)}(x)\, dx^n$$

If $x = x(t)$ and $y = y(t)$, then

$$dx = \frac{\partial[x(t)]}{\partial t} dt = \dot{x}\, dt \qquad\qquad\qquad dy = \frac{\partial[y(t)]}{\partial t} dt = \dot{y}\, dt$$

(2) Total Differential of $z = f(x, y),\, x = x(t),\, y = y(t)$

$$dz = \frac{\partial z}{\partial x} dx + \frac{\partial z}{\partial y} \partial y \qquad\qquad\qquad dz = \frac{\partial z}{\partial x}\frac{\partial x}{\partial t} dt + \frac{\partial z}{\partial y}\frac{\partial y}{\partial t} dt$$

(3) Exact Differential

In order that

$$dF[x, y] = A(x, y)\, dx + B(x, y)\, dy$$

$$dF[x, y, z] = A(x, y, z)\, dx + B(x, y, z)\, dy + C(x, y, z)\, dz$$

be the exact differentials, the following (necessary and sufficient) conditions must be satisfied, respectively:

$$\frac{\partial A}{\partial y} = \frac{\partial B}{\partial x} \qquad\qquad \frac{\partial B}{\partial z} = \frac{\partial C}{\partial y} \qquad\qquad \frac{\partial C}{\partial x} = \frac{\partial A}{\partial z}$$

(4) Jacobian Determinant

If $x = x(t),\, y = y(t)$, and $z = z(t)$,

$$A(x, y, z, t) = 0 \qquad\qquad B(x, y, z, t) = 0 \qquad\qquad C(x, y, z, t) = 0$$

and if the jacobian determinant

$$J\begin{bmatrix} A, B, C \\ x, y, z \end{bmatrix} = \begin{vmatrix} \dfrac{\partial A}{\partial x} & \dfrac{\partial A}{\partial y} & \dfrac{\partial A}{\partial z} \\[2mm] \dfrac{\partial B}{\partial x} & \dfrac{\partial B}{\partial y} & \dfrac{\partial B}{\partial y} \\[2mm] \dfrac{\partial C}{\partial x} & \dfrac{\partial C}{\partial y} & \dfrac{\partial C}{\partial z} \end{vmatrix} = \frac{\partial(A, B, C)}{\partial(x, y, z)} \neq 0$$

then the derivatives and the differentials of A, B, C are given by

$$\begin{bmatrix} \dfrac{\partial A}{\partial t} \\[3mm] \dfrac{\partial B}{\partial t} \\[3mm] \dfrac{\partial C}{\partial t} \end{bmatrix} = \begin{bmatrix} \dfrac{\partial A}{\partial x} & \dfrac{\partial A}{\partial y} & \dfrac{\partial A}{\partial z} \\[2mm] \dfrac{\partial B}{\partial x} & \dfrac{\partial B}{\partial y} & \dfrac{\partial B}{\partial z} \\[2mm] \dfrac{\partial C}{\partial x} & \dfrac{\partial C}{\partial y} & \dfrac{\partial C}{\partial z} \end{bmatrix} \begin{bmatrix} \dfrac{\partial x}{\partial t} \\[3mm] \dfrac{\partial y}{\partial t} \\[3mm] \dfrac{\partial z}{\partial t} \end{bmatrix}$$

The higher derivatives and differentials can be found in the same manner.

8
SEQUENCES AND
SERIES

(1) Series of Constant Terms

(a) A sequence is a set of numbers u_1, u_2, u_3, ... arranged in a prescribed order and formed according to a definite rule. Each member of the sequence is called a *term*, and the sequence is defined by the number of terms as *finite* or *infinite*.

(b) An infinite series

$$\sum_{i=1}^{\infty} u_i = u_1 + u_2 + u_3 + \cdots$$

is the sum of an infinite sequence.
If

$$s_n = \sum_{i=1}^{n} u_i \qquad \text{and} \qquad \lim_{n \to \infty} s_n = S$$

exist, the series is called *convergent*, and S is the sum. The infinite series which does not converge is *divergent*.

(c) An absolutely convergent series is a series whose absolute terms form a convergent series.

$$\sum_{i=1}^{\infty} |u_i| = |S|$$

(d) A double series

$$\sum_{i=0}^{\infty} \sum_{x=0}^{\infty} a_{ix} = \sum_{i=0}^{\infty} \left(\sum_{x=0}^{\infty} a_{ix} \right) = \sum_{x=0}^{\infty} \left(\sum_{i=0}^{\infty} a_{ix} \right)$$

converges to the limit D if

$$\lim_{\substack{m \to \infty \\ n \to \infty}} \sum_{i=0}^{m} \left(\sum_{x=0}^{n} a_{ix} \right) = D$$

(e) A product series

$$\sum_{i=0}^{\infty} \sum_{x=0}^{\infty} a_i b_x = \left(\sum_{i=0}^{\infty} a_i \right) \left(\sum_{x=0}^{\infty} b_x \right)$$

is an *absolutely convergent series* if

$$\sum_{i=0}^{\infty} a_i \qquad \text{and} \qquad \sum_{x=0}^{\infty} b_x$$

are *absolutely convergent series*.

(2) Series of Functions

(a) A power series in $x - a$ is of the form

$$S(x) = \sum_{i=0}^{\infty} c_i (x - a)^i = c_0 + c_1(x - a) + c_2(x - a)^2 + c_3(x - a)^3 \cdots$$

where a and c_0, c_1, c_2, c_3, ... are *constants*. For every power series there exists a value $b \geq 0$, such that $S(x)$ is absolutely convergent for all $|x| < b$ and divergent for all $|x| > b$. Then b is the *radius of convergence*, and the totality $-b < x < b$ is the *interval of convergence*.

(b) A function series in x is of the form

$$F(x) = \sum_{i=0}^{\infty} a_i f_i(x) = a_0 f_0(x) + a_1 f_1(x) + a_2 f_2(x) + \cdots$$

where a_0, a_1, a_2, ... are constants. The series converges to $F(x)$ if the sequence of partial sums converges to $F(x)$.

(1) Tests of Convergence

The following tests are available for the analysis of convergence and divergence of the series

$$\sum_{n=1}^{\infty} u_n = u_1 + u_2 + u_3 + \cdots$$

in which *each term is a constant.*

(a) Comparison test

$$\boxed{\text{If } |u_n| < a|v_n|}$$

where a is a constant independent of n and v_n is the nth term of another series which is known to be absolutely convergent, the series consisting of u terms is also absolutely convergent.

$$\boxed{\text{If } |u_n| > a|v_n|}$$

and the series consisting of v terms is known to be absolutely divergent, then the series consisting of u terms is also divergent.

(b) Cauchy's test (nth root test)

$$\boxed{\text{If } \lim_{n\to\infty} \sqrt[n]{|u_n|} = L}$$

then the series is absolutely convergent for $L < 1$ and divergent for $L > 1$, and the test fails for $L = 1$.

(c) D'Alambert's test (ratio test)

$$\boxed{\text{If } \lim_{n\to\infty} \left|\frac{u_{n+1}}{u_n}\right| = L}$$

then the series is absolutely convergent for $L < 1$ and divergent for $L > 1$, and the test fails for $L = 1$.

(d) Raabe's test

$$\boxed{\text{If } \lim_{n\to\infty} n \left(1 - \left|\frac{u_{n+1}}{u_n}\right|\right) = L}$$

then the series is absolutely convergent for $L > 1$ and divergent for $L < 1$, and the test fails for $L = 1$.

(e) Integral test

If each term of the u series is a function of its suffix n and if the function $f(x)$ which represents $f(n)$ when $x = n$ is continuous and monotonic, the given series is convergent if

$$\boxed{\int_n^{\infty} f(x)\, dx = \lim_{M\to\infty} \int_n^M f(x)\, dx}$$

is convergent, and it is divergent if this integral is divergent.

(2) Operations with Absolutely Convergent Series

(a) The terms of an absolutely convergent series can be rearranged in any order, and the new series will converge to the same sum.

(b) The sum, difference, and product of two or more absolutely convergent series is absolutely convergent.

$$a, b, c = \text{signed numbers} \qquad n = \text{number of terms} \qquad \alpha = (-1)^{n+1}$$
$$A_k, B_k, C_k = k\text{th term of series} \qquad k, m = \text{positive integers} \qquad \beta = (-1)^{k+1}$$

(1) Arithmetic Series $\qquad A_k = a + (k-1)b \qquad B_k = \beta A_k$

$$\sum_{k=1}^{n} A_k = a + (a+b) + (a+2b) + \cdots + [a + (n-1)b] = \frac{n}{2}(A_1 + A_n) = \frac{n}{2}[2a + (n-1)b] \qquad n < \infty$$

$$\sum_{k=1}^{n} B_k = a - (a+b) + (a+2b) - \cdots + \alpha[a + (n-1)b] = \frac{1+\alpha}{2}[a + b(n-\tfrac{1}{2})] - \frac{nb}{2} \qquad n < \infty$$

(2) Geometric Series $\qquad A_k = ab^{k-1} \qquad C_k = a^{m+1-k}b^{k-1}$

$$\sum_{k=1}^{n} A_k = a + ab + ab^2 + \cdots + ab^{n-1} = \frac{bA_n - A_1}{b-1} = \frac{a(b^n - 1)}{b-1} \qquad b \neq 1 \qquad n < \infty$$

$$\sum_{k=1}^{\infty} A_k = a + ab + ab^2 + \cdots = \frac{a}{1-b} \qquad -1 < b < 1 \qquad n = \infty$$

$$\sum_{k=1}^{m+1} C_k = a^m + a^{m-1}b + a^{m-2}b^2 + \cdots + b^m = \frac{a^n - b^n}{a-b} \qquad a \neq b \qquad n = m+1 < \infty$$

$$\sum_{k=1}^{\infty} C_k = a^m + a^{m-1}b + a^{m-2}b^2 + \cdots = \frac{a^{m+1}}{a-b} \qquad -1 < b < 1 \qquad a > b \qquad n = \infty$$

(3) Arithmogeometric Series $\qquad |c| \neq 1 \qquad A_k = [a + (k-1)b]c^{k-1} \qquad B_k = \beta A_k$

$$\sum_{k=1}^{n} A_k = a + (a+b)c + (a+2b)c^2 + \cdots + [a + (n-1)b]c^{n-1} = \frac{nbc^n}{c-1} - \frac{[a - c(a-b)](c^n - 1)}{(c-1)^2} \qquad n < \infty$$

$$\sum_{k=1}^{\infty} A_k = a + (a+b)c + (a+2b)c^2 + \cdots = \frac{a + (b-a)c}{(1-c)^2} \qquad -1 < c < 1 \qquad n = \infty$$

$$\sum_{k=1}^{n} B_k = a - (a+b)c + (a+2b)c^2 - \cdots \alpha[a + (n-1)n]c^{n-1} = \frac{\alpha nbc^n}{1+c} + \frac{[a + c(a-b)](\alpha c^n + 1)}{(1+c)^2} \qquad n < \infty$$

$$\sum_{k=1}^{\infty} B_k = a - (a+b)c + (a+2b)c^2 - \cdots = \frac{a - (b-a)c}{(1+c)^2} \qquad -1 < c < 1 \qquad n = \infty$$

(4) Series of Binomial Coefficients $\qquad$ **(Secs. 1.04 and 8.11)** $\qquad n = 1, 2, 3, \ldots \qquad n < \infty$

$$\binom{2n}{2} + \binom{2n}{4} + \binom{2n}{6} + \cdots + \binom{2n}{2n} = 2^{2n-1} - 1 \qquad \binom{2n-1}{1} + \binom{2n-1}{3} + \binom{2n-1}{5} + \cdots + \binom{2n-1}{2n-1} = 2^{2n-2}$$

$$\binom{n}{1} + 2\binom{n}{2} + 3\binom{n}{3} + \cdots + n\binom{n}{n} = n2^{n-1} \qquad \binom{n}{1} + 3\binom{n}{2} + 5\binom{n}{3} + \cdots + (2n-1)\binom{n}{n} = (n-1)2^n + 1$$

$$\binom{n}{1}^2 + \binom{n}{2}^2 + \binom{n}{3}^2 + \cdots + \binom{n}{n}^2 = \binom{2n}{n} - 1 \qquad \binom{n}{1}^2 + 2\binom{n}{2}^2 + 3\binom{n}{3}^2 + \cdots + n\binom{n}{n}^2 = \frac{n}{2}\binom{2n}{n}$$

(5) Harmonic Series $\qquad \psi(n) = \text{Digamma function (Sec. 13.04)} \qquad n < \infty$

$$\sum_{k=1}^{n} \frac{1}{k} = C + \psi(n) \qquad \sum_{k=1}^{n} \frac{1}{2k-1} = \tfrac{1}{2}[C + \psi(n - \tfrac{1}{2})] + \ln 2 \qquad C = 0.577\,215\,665$$

a, b = signed numbers	n = number of terms $\\$ k, m = positive integers	$\alpha = (-1)^{n+1}$ $\\$ $\beta = (-1)^{k+1}$

(1) Series of Algebraic Terms

	$n < \infty$	$a > 0$	$n = \infty$	$0 < a < 1$
	$\displaystyle\sum_{k=1}^{n} ka^k = \frac{na^{n+1}}{a-1} - \frac{a(a^n-1)}{(a-1)^2}$		$\displaystyle\sum_{k=1}^{\infty} ka^k = \frac{a}{(1-a)^2}$	
	$\displaystyle\sum_{k=1}^{n} \beta ka^k = \frac{\alpha na^{n+1}}{1+a} + \frac{a(1+\alpha a^n)}{(1+a)^2}$		$\displaystyle\sum_{k=1}^{\infty} \beta ka^k = \frac{a}{(1+a)^2}$	
	$\displaystyle\sum_{k=1}^{n} (2k-1)a^k = \frac{2na^{n+1}}{a-1} - \frac{a(a+1)(a^n-1)}{(a-1)^2}$		$\displaystyle\sum_{k=1}^{\infty} (2k-1)a^k = \frac{a(1+a)}{(1-a)^2}$	
	$\displaystyle\sum_{k=1}^{n} \beta(2k-1)a^k = \frac{2\alpha na^{n+1}}{1+a} + \frac{a(a-1)(1+\alpha a^n)}{(1+a)^2}$		$\displaystyle\sum_{k=1}^{\infty} \beta(2k-1)a^k = \frac{a(1-a)}{(1+a)^2}$	
	$\displaystyle\sum_{k=1}^{n} \frac{k}{b^k} = \frac{b(b^n-1)-n(b-1)}{b^n(b-1)^2}$	$b > 0$	$\displaystyle\sum_{k=1}^{\infty} \frac{k}{b^k} = \frac{b}{(b-1)^2}$	$b > 1$
	$\displaystyle\sum_{k=1}^{n} \frac{\beta k}{b^k} = \frac{b(1+\alpha b^n)+\alpha n(1+b)}{b^n(1+b)^2}$		$\displaystyle\sum_{k=1}^{\infty} \frac{\beta k}{b^k} = \frac{b}{(b+1)^2}$	
	$\displaystyle\sum_{k=1}^{n} \frac{2k-1}{b^k} = \frac{(b+1)(b^n-1)-2n(b-1)}{b^n(b-1)^2}$		$\displaystyle\sum_{k=1}^{\infty} \frac{2k-1}{b^k} = \frac{b+1}{(b-1)^2}$	
	$\displaystyle\sum_{k=1}^{n} \frac{\beta(2k-1)}{b^k} = \frac{(b-1)(1+\alpha b^n)+2n(1+\alpha b)}{b^n(1+b)^2}$		$\displaystyle\sum_{k=1}^{\infty} \frac{\beta(2k-1)}{b^k} = \frac{b+1}{(b+1)^2}$	

(2) Series of Trigonometric Terms

	$n < \infty$	$a > 0$	$n < \infty$	$a > 0$
	$\displaystyle\sum_{k=1}^{n} \sin ka = \frac{\sin(na/2)\,\sin[(n+1)a/2]}{\sin(a/2)}$		$\displaystyle\sum_{k=1}^{n} \sin(2k-1)a = \frac{\sin^2 na}{\sin a}$	
	$\displaystyle\sum_{k=1}^{n} \beta \sin ka = \frac{\sin a + \alpha \sin na + \alpha \sin(n+1)a}{2(1+\cos a)}$		$\displaystyle\sum_{k=1}^{n} \beta \sin(2k-1)a = \frac{\alpha \sin 2na}{2\cos a}$	
	$\displaystyle\sum_{k=1}^{n} \cos ka = \frac{\sin(na/2)\,\cos[(n+1)a/2]}{\sin(a/2)}$		$\displaystyle\sum_{k=1}^{n} \cos(2k-1)a = \frac{\sin na \cos na}{\sin a}$	
	$\displaystyle\sum_{k=1}^{n} \beta \cos ka = \frac{1+\cos a + \alpha \cos na + \alpha \cos(n+1)a}{2(1+\cos a)}$		$\displaystyle\sum_{k=1}^{n} \beta \cos(2k-1)a = \frac{\alpha \cos 2na + 1}{2\cos a}$	
	$\displaystyle\sum_{k=1}^{n} k \sin ka = \frac{\sin(n+1)a}{4\sin^2(a/2)} - \frac{(n+1)\cos[(2n+1)a/2]}{2\sin(a/2)}$		$\displaystyle\sum_{k=1}^{n} \sin^2 ka = \frac{n}{2} - \frac{\sin na \cos(n+1)a}{2\sin a}$	
	$\displaystyle\sum_{k=1}^{n} k \cos ka = \frac{\cos(n+1)a-1}{4\sin^2(a/2)} + \frac{(n+1)\sin[(2n+1)a/2]}{2\sin(a/2)}$		$\displaystyle\sum_{k=1}^{n} \cos^2 ka = \frac{n}{2} + \frac{\sin na \cos(n+1)a}{2\sin a}$	

$\bar{B}_m$ = Bernoulli number
B_m = auxiliary Bernoulli number
$\bar{B}_m(x)$ = Bernoulli polynomial

$Z(\ )$ = zeta function (Sec. A.06)
m = positive integer
$\alpha = (-1)^{m+1}$

(1) Bernoulli Numbers   (Sec. A.03-2)

(a) Generating function

$$\frac{x}{e^x - 1} = \sum_{m=0}^{\infty} \bar{B}_m \frac{x^m}{m!} = \frac{\bar{B}_0}{0!} + \frac{\bar{B}_1 x}{1!} + \frac{\bar{B}_2 x^2}{2!} + \frac{\bar{B}_3 x^3}{3!} + \cdots \qquad |x| < 2\pi$$

where

$\bar{B}_0 = 1$	$\bar{B}_2 = \frac{1}{6}$	$\bar{B}_4 = -\frac{1}{30}$	$\bar{B}_6 = \frac{1}{42}$	$\bar{B}_8 = -\frac{1}{30}$	$\bar{B}_{10} = \frac{5}{66}$ $\cdots$
$\bar{B}_1 = -\frac{1}{2}$	$\bar{B}_3 = 0$	$\bar{B}_5 = 0$	$\bar{B}_7 = 0$	$\bar{B}_9 = 0$	$\bar{B}_{11} = 0$ $\cdots$

are *Bernoulli numbers* of order $m = 0, 1, 2, 3, \ldots$.

(b) Auxiliary generating function

$$2 - \frac{x}{2} \cot \frac{x}{2} = \sum_{m=0}^{\infty} B_m \frac{x^{2m}}{(2m)!} = \frac{B_0}{0!} + \frac{B_1 x^2}{2!} + \frac{B_2 x^4}{4!} + \frac{B_3 x^6}{6!} + \cdots \qquad |x| < \pi$$

where

$B_0 = 1$	$B_1 = \frac{1}{6}$	$B_2 = \frac{1}{30}$	$B_3 = \frac{1}{42}$	$B_4 = \frac{1}{30}$	$B_5 = \frac{5}{66}$ $\cdots$

are *auxiliary Bernoulli numbers* of order $m = 0, 1, 2, 3, \ldots$.

(c) Series representation. For $m = 1, 2, 3, \ldots,$

$$B_m = \alpha \bar{B}_{2m} = 2 \frac{(2m)!}{(2\pi)^{2m}} \left(\frac{1}{1^{2m}} + \frac{1}{2^{2m}} + \frac{1}{3^{2m}} + \frac{1}{4^{2m}} + \cdots \right) = 2 \frac{(2m)!}{(2\pi)^{2m}} Z(2m)$$

(2) Bernoulli Polynomials   (Sec. A.03-1)

(a) Definition. The Bernoulli polynomial $\bar{B}_m(x)$ of order $m = 0, 1, 2, 3, \ldots$ is defined as

$$\bar{B}_m(x) = x^m \bar{B}_0 + \binom{m}{1} x^{m-1} \bar{B}_1 + \binom{m}{2} x^{m-2} \bar{B}_2 + \cdots + \binom{m}{m} \bar{B}_m$$

where $\bar{B}_0, \bar{B}_1, \bar{B}_2, \ldots, \bar{B}_m$ are Bernoulli numbers defined in (1a) above.

(b) First six polynomials

$\bar{B}_0(x) = 1$	$\bar{B}_3(x) = x^3 - \frac{3}{2}x^2 + \frac{1}{2}x$
$\bar{B}_1(x) = x - \frac{1}{2}$	$\bar{B}_4(x) = x^4 - 2x^3 + x^2 - \frac{1}{30}$
$\bar{B}_2(x) = x^2 - x + \frac{1}{6}$	$\bar{B}_5(x) = x^5 - \frac{5}{2}x^4 + \frac{5}{3}x^3 - \frac{1}{6}x$

(c) Properties   ($m > 0$)

$$\bar{B}_{2m}(0) = \bar{B}_{2m} = \alpha B_m \qquad \bar{B}_{2m+1}(0) = \bar{B}_{2m+1} = 0 \qquad \frac{d\bar{B}_m(x)}{dx} = m\bar{B}_{m-1}(x)$$

$\bar{E}_m$ = Euler number
E_m = auxiliary Euler number
$\bar{E}_m(x)$ = Euler polynomial

$\bar{Z}(\ )$ = complementary zeta function (Sec. A.06)
m = positive integer
$\alpha = (-1)^{m+1}$

(1) Euler Numbers   (Sec. A.04−2)

(a) Generating function

$$\frac{2\sqrt{e^x}}{e^x+1} = \sum_{m=0}^{\infty} \frac{\bar{E}_m}{2^m}\frac{x^m}{m!} = \frac{\bar{E}_0}{0!} + \frac{\bar{E}_1 x}{2(1!)} + \frac{\bar{E}_2 x^2}{4(2!)} + \frac{\bar{E}_3 x^3}{8(3!)} + \cdots \qquad |x| < \pi$$

where

$\bar{E}_0 = 1$	$\bar{E}_2 = -1$	$\bar{E}_4 = 5$	$\bar{E}_6 = -61$	$\bar{E}_8 = 1{,}385$	$\bar{E}_{10} = -50{,}521$	$\cdots$
$\bar{E}_1 = 0$	$\bar{E}_3 = 0$	$\bar{E}_5 = 0$	$\bar{E}_7 = 0$	$\bar{E}_9 = 0$	$\bar{E}_{11} = 0$	$\cdots$

are *Euler numbers* of order $m = 0, 1, 2, 3, \ldots$.

(b) Auxiliary generating function

$$\sec x = \sum_{m=0}^{\infty} E_m \frac{x^{2m}}{(2m)!} = \frac{E_0}{0!} + \frac{E_1 x^2}{2!} + \frac{E_2 x^4}{4!} + \frac{E_3 x^6}{6!} + \cdots \qquad |x| < \pi$$

where

$E_0 = 1$	$E_1 = 1$	$E_2 = 5$	$E_3 = 61$	$E_4 = 1{,}385$	$E_5 = 50{,}521$	$\cdots$

are *auxiliary Euler numbers* of order $m = 0, 1, 2, 3, \ldots$.

(c) Series representation. For $m = 1, 2, 3, \ldots$,

$$E_m = -\alpha\bar{E}_{2m} = 2\left(\frac{2}{\pi}\right)^{2m+1}(2m)!\left(\frac{1}{1^{2m+1}} - \frac{1}{3^{2m+1}} + \frac{1}{5^{2m+1}} - \cdots\right) = 2\left(\frac{2}{\pi}\right)^{2m+1}(2m)!\bar{Z}(2m+1)$$

(2) Euler Polynomials   (Sec. A.04−1)

(a) Definition. The Euler polynomial $\bar{E}_m(x)$ of order $m = 0, 1, 2, 3, \ldots$ is defined as

$$\bar{E}_m(x) = \frac{2}{m+1}\left[(1-2)\binom{m+1}{1}x^m\bar{B}_1 + (1-2^2)\binom{m+1}{2}x^{m-1}\bar{B}_2 + \cdots + (1-2^{m+1})\binom{m+1}{m+1}\bar{B}_{m+1}\right]$$

where $\bar{B}_1, \bar{B}_2, \bar{B}_3, \ldots, \bar{B}_{m+1}$ are Bernoulli numbers defined in Sec. 8.05−1a.

(b) First six polynomials

$\bar{E}_0(x) = 1$	$\bar{E}_3(x) = x^3 - \frac{3}{2}x^2 + \frac{1}{4}$
$\bar{E}_1(x) = x - \frac{1}{2}$	$\bar{E}_4(x) = x^4 - 2x^3 + x$
$\bar{E}_2(x) = x^2 - x$	$\bar{E}_5(x) = x^5 - \frac{5}{2}x^4 + \frac{5}{2}x^2 - \frac{1}{2}$

(c) Properties   ($m > 0$)

$$\bar{E}_{2m}(\tfrac{1}{2}) = 2^{-2m}\bar{E}_{2m} = -2^{-2m}\alpha E_m \qquad \bar{E}_{2m+1}(\tfrac{1}{2}) = \bar{E}_{2m+1} = 0 \qquad \frac{d\bar{E}_m(x)}{dx} = m\bar{E}_{m-1}(x)$$

$$\bar{B}_{m+1}(\;) = \text{Bernoulli polynomial (Sec. 8.05)} \qquad n = \text{number of terms}$$

$$\bar{E}_m(\;) = \text{Euler polynomial (Sec. 8.06)} \qquad k, m = \text{positive integers}$$

$$\mathscr{P}_{m,n},\, \mathscr{P}^*_{m,n},\, \overline{\mathscr{P}}_{m,n},\, \overline{\mathscr{P}}^*_{m,n} = \text{sums} \qquad \alpha = (-1)^{n+1} \qquad \beta = (-1)^{k+1}$$

(1) Monotonic Series $\quad (n < \infty)$

(a) General cases

$$\sum_{k=1}^{n} k^m = 1^m + 2^m + 3^m + \cdots + n^m = \frac{\bar{B}_{m+1}(n+1) - \bar{B}_{m+1}(0)}{m+1} = \mathscr{P}_{m,n}$$

$$\sum_{k=1}^{n} (2k)^m = 2^m + 4^m + 6^m + \cdots + (2n)^m = 2^m \mathscr{P}_{m,n}$$

$$\sum_{k=1}^{n} (2k-1)^m = 1^m + 3^m + 5^m + \cdots + (2n-1)^m = 2^m \sum_{k=0}^{m} \left[\binom{m}{k} \frac{\mathscr{P}_{m-k,n}}{2^k} \right] = \mathscr{P}^*_{m,n}$$

(b) Particular cases

m	$\mathscr{P}_{m,n}$	$\mathscr{P}^*_{m,n}$
1	$\dfrac{n(n+1)}{2}$	n^2
2	$\dfrac{n(n+1)(2n+1)}{6}$	$\dfrac{n(4n^2-1)}{3}$
3	$\left[\dfrac{n(n+1)}{2}\right]^2$	$n^2(2n^2-1)$
4	$\dfrac{n^5}{5} + \dfrac{n^4}{2} + \dfrac{n^3}{3} - \dfrac{n}{30}$	$\dfrac{(2n)^5}{10} - \dfrac{(2n)^3}{3} + \dfrac{7(2n)}{30}$

(2) Alternating Series $\quad (n < \infty)$

(a) General cases

$$\sum_{k=1}^{n} \beta k^m = 1^m - 2^m + 3^m - \cdots \alpha n^m = \frac{\bar{E}_m(n+1) - \alpha \bar{E}_m(0)}{2} = \overline{\mathscr{P}}_{m,n}$$

$$\sum_{k=1}^{n} \beta(2k)^m = 2^m - 4^m + 6^m - \cdots \alpha(2n)^m = 2^m \overline{\mathscr{P}}_{m,n}$$

$$\sum_{k=1}^{n} \beta(2k-1)^m = 1^m - 3^m + 5^m - \cdots \alpha(2n-1)^m = 2^m \sum_{k=0}^{m} \left[\beta\binom{m}{k} \frac{\overline{\mathscr{P}}_{m-k,n}}{2^k} \right] = \overline{\mathscr{P}}^*_{m,n}$$

(b) Particular cases

m	$\overline{\mathscr{P}}_{m,n}$	$\overline{\mathscr{P}}^*_{m,n}$
1	$\dfrac{\alpha n + (1+\alpha)/2}{2}$	αn
2	$\dfrac{\alpha n(n+1)}{2}$	$\dfrac{4\alpha n^2 - (\alpha+1)}{2}$
3	$\dfrac{\alpha n^2(2n+3) - (\alpha+1)/2}{4}$	$\alpha n(4n^2-3)$
4	$\dfrac{\alpha n(n^3 + 2n^2 - 1)}{2}$	$\dfrac{8\alpha n^2(2n^2-3) + 5(\alpha+1)}{2}$

$Z(\ )$ = zeta function (Sec. A.06)

$\bar{Z}(\ )$ = complementary zeta function (Sec. A.06)

$\mathscr{D}_{m,n}, \mathscr{D}^*_{m,n}, \overline{\mathscr{D}}_{m,n}, \overline{\mathscr{D}}^*_{m,n}$ = sums

n = number of terms

k, m = positive integers

$\beta = (-1)^{k+1}$

(1) Monotonic Series $(n = \infty)$

(a) General cases

$$\sum_{k=1}^{\infty} \frac{1}{k^m} = \frac{1}{1^m} + \frac{1}{2^m} + \frac{1}{3^m} + \cdots = Z(m) = \mathscr{D}_{m,n}$$

$$\sum_{k=1}^{\infty} \frac{1}{(2k)^m} = \frac{1}{2^m} + \frac{1}{4^m} + \frac{1}{6^m} + \cdots = \frac{1}{2^m} Z(m) = \frac{1}{2^m} \mathscr{D}_{m,n}$$

$$\sum_{k=1}^{\infty} \frac{1}{(2k-1)^m} = \frac{1}{1^m} + \frac{1}{3^m} + \frac{1}{5^m} + \cdots = \left(1 - \frac{1}{2^m}\right) Z(m) = \mathscr{D}^*_{m,n}$$

(b) Particular cases   (Sec. A.06)

m	$\mathscr{D}_{m,n}$	$\mathscr{D}^*_{m,n}$
1	∞	∞
2	$\dfrac{\pi^2}{6} = 1.644\,934\,066\,848\,226$	$\dfrac{\pi^2}{8} = 1.233\,700\,550\,136\,170$
3	$1.202\,056\,903\,159\,594$	$1.051\,799\,790\,264\,644$
4	$\dfrac{\pi^4}{90} = 1.082\,323\,233\,711\,114$	$\dfrac{\pi^4}{96} = 1.014\,678\,031\,604\,192$

(2) Alternating Series $(n = \infty)$

(a) General cases

$$\sum_{k=1}^{\infty} \frac{\beta}{k^m} = \frac{1}{1^m} - \frac{1}{2^m} + \frac{1}{3^m} - \cdots = \left(1 - \frac{2}{2^m}\right) Z(m) = \overline{\mathscr{D}}_{m,n}$$

$$\sum_{k=1}^{\infty} \frac{\beta}{(2k)^m} = \frac{1}{2^m} - \frac{1}{4^m} + \frac{1}{6^m} - \cdots = \frac{1}{2^m}\left(1 - \frac{2}{2^m}\right) Z(m) = \frac{1}{2^m} \overline{\mathscr{D}}_{m,n}$$

$$\sum_{k=1}^{\infty} \frac{\beta}{(2k-1)^m} = \frac{1}{1^m} - \frac{1}{3^m} + \frac{1}{5^m} - \cdots = \overline{Z}(m) = \overline{\mathscr{D}}^*_{m,n}$$

(b) Particular cases   (Sec. A.06)

m	$\overline{\mathscr{D}}_{m,n}$	$\overline{\mathscr{D}}^*_{m,n}$
1	$\ln 2 = 0.693\,147\,180\,559\,945$	$\dfrac{\pi}{4} = 0.785\,398\,163\,397\,448$
2	$\dfrac{\pi^2}{12} = 0.822\,467\,033\,424\,113$	$0.915\,965\,594\,177\,219$
3	$0.901\,542\,677\,369\,696$	$\dfrac{\pi^3}{32} = 0.968\,946\,146\,259\,369$
4	$\dfrac{7\pi^4}{720} = 0.947\,032\,829\,497\,246$	$0.988\,944\,551\,741\,105$

(1) Tests of Convergence

The following tests are available for the analysis of the convergence of the series

$$F(x) = \sum_{n=1}^{\infty} f_n(x) = f_1(x) + f_2(x) + f_3(x) + \cdots$$

in which each term is a function.

(a) Cauchy's test for uniform convergence

A series of real (or complex) functions converges uniformly on $F(x)$ in $[a, b]$ if for every real number $\epsilon > 0$ there exists a real number $N > 0$, independent of x in $[a, b]$, such that

$$|F(x) - f_n(x)| < \epsilon \qquad \text{for all } n > N$$

This is a necessary and sufficient condition for uniform convergence for a function series.

(b) Weierstrass's test for uniform and absolute convergence

A series of real (or complex) functions converges uniformly and absolutely on every $F(x)$ in $[a, b]$ if

$$|f_n(x)| \leq M_n \qquad \text{for all } n$$

and $M_1 + M_2 + M_3 + \cdots$ is a convergent comparison series of real positive terms. Since this test establishes the absolute (as well as the uniform) convergence, it is applicable only to series which converge absolutely. It must be noted that a function series may converge uniformly but not absolutely, and vice versa.

(c) Dirichlet's test for uniform convergence

If
$$\sum_{n=1}^{\infty} a_n = a_1 + a_2 + a_3 + \cdots$$
is a monotonic decreasing sequence of real numbers

then the infinite function series
$$\sum_{n=1}^{\infty} a_n f_n(x) = a_1 f_1(x) + a_2 f_2(x) + a_3 f_3(x) + \cdots$$

converges uniformly on a set $G(x)$ of values of x if the infinite series

$$\sum_{n=1}^{\infty} f_n(x) = f_1(x) + f_2(x) + f_3(x) + \cdots$$

converges uniformly on the same set $G(x)$ of values of x.

(2) Properties of Uniformly Convergent Function Series

(a) Theorem of continuity

If any term of a uniformly convergent function series is a continuous function of x in $[a, b]$, then the sum of the series is also a continuous function of x in $[a, b]$.

(b) Theorem of differentiability of a function series

A uniformly convergent series in (a, b) can be differentiated term by term in (a, b). If each term of the differentiated series is continuous and the differentiated series is uniformly convergent in (a, b), then it will converge to the derivative of the function it represents in (a, b).

(c) Theorem of integrability of a function series

A uniformly convergent series in $[a, b]$ can be integrated term by term in $[a, b]$, and the integrated series will converge uniformly to the integral of the function it represents in $[a, b]$.

(1) Interval of Convergence (ratio test)

The power series in the real (or complex) variable x

$$S(x) = \sum_{n=0}^{\infty} a_n x^n = a_0 + a_1 x + a_2 x^2 + \cdots$$

where the coefficients $a_0, a_1, a_2, \ldots$ are real or complex numbers, independent of x,

is *convergent* if $\qquad \lim_{n \to \infty} \left| \dfrac{a_{n+1} x^{n+1}}{a_n x^n} \right| = \lim_{n \to \infty} \left| \dfrac{a_{n+1}}{a_n} \right| |x| = r|x| < 1$

and is *divergent* if $\qquad \lim_{n \to \infty} \left| \dfrac{a_{n+1} x^{n+1}}{a_n x^n} \right| = \lim_{n \to \infty} \left| \dfrac{a_{n+1}}{a_n} \right| |x| = r|x| > 1$

The interval of convergence is then

$$r|x| < 1 \qquad \text{or} \qquad -\frac{1}{r} < x < \frac{1}{r}$$

and it is symmetrical about the origin of x. The series is convergent in this interval and diverges outside this interval. It may or may not converge at the end points of the interval.

(2) Uniform and Absolute Convergence

The power series which converges in the interval

$$\alpha < x < \beta$$

converges absolutely and uniformly for every value of x within this interval. Since a uniformly convergent series represents a continuous function, a *uniformly convergent series defines a continuous function within the interval of convergence*.

(3) Operations with Power Series

(a) Uniqueness theorem

If two power series

$$S(x) = \sum_{n=0}^{\infty} a_n x^n \qquad \text{and} \qquad S(x) = \sum_{n=0}^{\infty} b_n x^n$$

converge to the same sum $S(x)$ for all real values of x, then

$$a_0 = b_0, a_1 = b_1, a_2 = b_2, \ldots$$

(b) Summation theorem

Two power series can be added or subtracted term by term for each value of x common to their interval of convergence.

(c) Product theorem

Two power series can be multiplied term by term for each value of x common to their interval of convergence. Thus

$$\left(\sum_{m=0}^{\infty} a_m x^m \right) \left(\sum_{n=0}^{\infty} b_n x^n \right) = \sum_{m=0}^{\infty} \sum_{n=0}^{\infty} a_m b_n x^{m+n}$$

(d) Theorem of differentiability and integrability

A power series can be differentiated and integrated term by term in any closed interval if and only if this interval lies entirely within the interval of uniform convergence of the power series.

a, b = signed numbers	k, n = positive integers
$\Lambda[\ ]$ = nested sum	$n + 1$ = number of terms

(1) Series of Constant Terms

(a) Geometric series   (Sec. 8.03)

$$\sum_{k=0}^{n} \bar{\beta}ab^k = a \pm ab + ab^2 \pm ab^3 + \cdots + \bar{\alpha}ab^n$$

$$= a(1 \pm b(1 \pm b(1 \pm b(1 \pm \cdots \pm b)))) = a \bigwedge_{k=1}^{n} \left[1 \pm \frac{kb}{k} \right]$$

$$\bar{\alpha} = (\pm 1)^n$$
$$\bar{\beta} = (\pm 1)^k$$

(b) Series of factorials   (Sec. 1.03)

$$\sum_{k=0}^{n} \bar{\beta}k! = 0! \pm 1! + 2! \pm 3! + \cdots + \bar{\alpha}n!$$
$$= \quad (1 \pm 1(1 \pm 2(1 \pm 3(1 \pm \cdots \pm n)))) = \bigwedge_{k=1}^{n} [1 \pm k]$$

$$\sum_{k=0}^{n} \bar{\beta}(2k+1)! = 1! \pm 3! + 5! \pm 7! + \cdots + \bar{\alpha}(2n+1)!$$
$$= \quad (1 \pm 2 \cdot 3(1 \pm 4 \cdot 5(1 \pm 6 \cdot 7(1 \pm \cdots \pm 2n(2n+1))))) = \bigwedge_{k=1}^{n} [1 \pm 2k(2k+1)]$$

(c) Series of double factorials   (Sec. 1.03)

$$\sum_{k=0}^{n} \bar{\beta}(2k)!! = 0!! \pm 2!! + 4!! \pm 6!! + \cdots + \bar{\alpha}(2n)!!$$
$$= \quad (1 \pm 2(1 \pm 4(1 \pm 6(1 \pm \cdots \pm 2n)))) = \bigwedge_{k=1}^{n} [1 \pm 2k]$$

$$\sum_{k=0}^{n} \bar{\beta}(2k+1)!! = 1!! \pm 3!! + 5!! \pm 7!! + \cdots + \bar{\alpha}(2n+1)!!$$
$$= \quad (1 \pm 3(1 \pm 5(1 \pm 7(1 \pm \cdots \pm (2n+1))))) = \bigwedge_{k=1}^{n} [1 \pm (2k+1)]$$

(d) Series of binomial coefficients   (Sec. 1.04)

$$\sum_{k=0}^{n} \bar{\beta}\binom{n}{k} = \binom{n}{0} \pm \binom{n}{1} + \binom{n}{2} \pm \binom{n}{3} + \cdots + \bar{\alpha}\binom{n}{n}$$

$$= \quad \left(1 \pm \frac{n}{1}\left(1 \pm \frac{n-1}{2}\left(1 \pm \frac{n-2}{3}\left(1 \pm \cdots \pm \frac{1}{n}\right)\right)\right)\right) = \bigwedge_{k=1}^{n} \left[1 \pm \frac{n+1-k}{k}\right]$$

(2) Power Series

(a) Basic form

$$b_k = \frac{a_k}{a_{k-1}}$$

$$\sum_{k=0}^{n} \bar{\beta}a_k x^k = a_0 \pm a_1 x + a_2 x^2 \pm a_3 x^3 + \cdots + \bar{\alpha}a_n x^n$$
$$= a_0(1 \pm b_1 x(1 \pm b_2 x(1 \pm b_3 x(1 \pm \cdots \pm b_n x)))) = a_0 \bigwedge_{k=1}^{n} [1 \pm b_k x]$$

(b) First derivative

$$\frac{d}{dx}\left[\sum_{k=0}^{n} \bar{\beta}a_k x^k\right] = \sum_{k=1}^{n} \bar{\beta}k a_k x^{k-1} = \pm a_1 + 2a_2 x \pm 3a_3 x^2 + \cdots + \bar{\alpha}n a_n x^{n-1}$$

$$= \pm a_1\left(1 \pm \frac{2b_2 x}{1}\left(1 \pm \frac{3b_3}{2}x\left(1 \pm \cdots \pm \frac{nb_n}{n-1}x\right)\right)\right) = \pm a_1 \bigwedge_{k=2}^{n}\left[1 \pm \frac{kb_k}{(k-1)}x\right]$$

(c) Indefinite integral   (C = constant of integration)

$$\int\left[\sum_{k=0}^{n} \bar{\beta}a_k x^k\right]dx = \sum_{k=0}^{n} \bar{\beta}\frac{a_k x^{k+1}}{k+1} = a_0 x + \frac{a_1}{2}x^2 + \frac{a_2}{3}x^3 + \frac{a_3}{4}x^4 + \cdots + \bar{\alpha}\frac{a_n}{n+1}x^{n+1} + C$$

$$= a_0 x\left(1 \pm \frac{b_1}{2}x\left(1 \pm \frac{2b_2}{3}x\left(1 \pm \frac{3b_3}{4}x\left(1 \pm \cdots \pm \frac{nb_n}{n+1}x\right)\right)\right)\right) + C = a_0 x \bigwedge_{k=1}^{n}\left[1 \pm \frac{kb_k}{k+1}x\right] + C$$

n = signed number	k, p, q = positive integers
N = nested	$\bigwedge [\ ] = N - \text{sum (Sec. 8.11)}$

(1) Basic Cases

(a) Symbolic form　　(Sec. 1.04)

$$(1 \pm x)^n = 1 \pm \binom{n}{1}x + \binom{n}{2}x^2 \pm \binom{n}{3}x^3 + \cdots \begin{cases} n = 0, 1, 2, \ldots, x \geqq 0 & \text{Finite series} \\ \\ n \neq 0, 1, 2, \ldots, |x| < 1 & \text{Infinite convergent series} \\ \\ n \neq 0, 1, 2, \ldots, |x| > 1 & \text{Infinite divergent series} \end{cases}$$

(b) Standard form　　$(x^2 < 1, n \neq 0, 1, 2, \ldots)$

$$(1 \pm x)^n = 1 \pm \frac{n}{1!}x + \frac{n(n-1)}{2!}x^2 \pm \frac{n(n-1)(n-2)}{3!}x^3 + \cdots = \sum_{k=0}^{\infty} (\pm 1)^k \binom{n}{k}x^k$$

(c) Nested form　　$(x^2 < 1, n \neq 0, 1, 2, \ldots)$

$$(1 \pm x)^n = \left(1 \pm \frac{n}{1}x\left(1 \pm \frac{n-1}{2}x\left(1 \pm \frac{n-2}{3}x(1 \pm \cdots)\right)\right)\right) = \bigwedge_{k=1}^{\infty}\left[1 \pm \frac{n+k-1}{k}x\right]$$

(2) Special Cases in Nested Form　　$(x^2 < 1)$

$$A = \frac{x^*}{q}$$

n	N-series	N-sum
-1	$(1 \mp x(1 \mp x(1 \mp x(1 \mp x(1 \mp \cdots)))))$	$\bigwedge_{k=1}^{\infty}\left[1 \mp \frac{k}{k}x\right]$
-2	$\left(1 \mp \frac{2}{1}x\left(1 \mp \frac{3}{2}x\left(1 \mp \frac{4}{3}x\left(1 \mp \frac{5}{4}x(1 \mp \cdots)\right)\right)\right)\right)$	$\bigwedge_{k=1}^{\infty}\left[1 \mp \frac{k+1}{k}x\right]$
$-p$	$\left(1 \mp \frac{p}{1}x\left(1 \mp \frac{p+1}{2}x\left(1 \mp \frac{p+2}{3}x\left(1 \mp \frac{p+3}{4}x(1 \mp \cdots)\right)\right)\right)\right)$	$\bigwedge_{k=1}^{\infty}\left[1 \mp \frac{p+k-1}{k}x\right]$
$\dfrac{1}{2}$	$1 \pm \frac{x}{2}\left(1 \mp \frac{1}{2}\left(\frac{x}{2}\right)\left(1 \mp \frac{3}{3}\left(\frac{x}{2}\right)\left(1 \mp \frac{5}{4}\left(\frac{x}{2}\right)(1 \mp \cdots)\right)\right)\right)$	$1 \pm \frac{x}{2}\bigwedge_{k=1}^{\infty}\left[1 \mp \frac{2k-1}{k+1}\frac{x}{2}\right]$
$\dfrac{1}{3}$	$1 \pm \frac{x}{3}\left(1 \mp \frac{2}{2}\left(\frac{x}{3}\right)\left(1 \mp \frac{5}{3}\left(\frac{x}{3}\right)\left(1 \mp \frac{8}{4}\left(\frac{x}{3}\right)(1 \mp \cdots)\right)\right)\right)$	$1 \pm \frac{x}{3}\bigwedge_{k=1}^{\infty}\left[1 \mp \frac{3k-1}{k+1}\frac{x}{3}\right]$
$\dfrac{1}{q}$	$1 \pm A\left(1 \mp \frac{q-1}{2}A\left(1 \mp \frac{2q-1}{3}A\left(1 \mp \frac{3q-1}{4}A(1 \mp \cdots)\right)\right)\right)$	$1 \pm A\bigwedge_{k=1}^{\infty}\left[1 \mp \frac{kq-1}{k+1}A\right]$
$-\dfrac{1}{2}$	$1 \mp \frac{x}{2}\left(1 \mp \frac{3}{2}\left(\frac{x}{2}\right)\left(1 \mp \frac{5}{3}\left(\frac{x}{2}\right)\left(1 \mp \frac{7}{4}\left(\frac{x}{2}\right)(1 \mp \cdots)\right)\right)\right)$	$1 \mp \frac{x}{2}\bigwedge_{k=1}^{\infty}\left[1 \mp \frac{2k+1}{k+1}\frac{x}{2}\right]$
$-\dfrac{1}{3}$	$1 \mp \frac{x}{3}\left(1 \mp \frac{4}{2}\left(\frac{x}{3}\right)\left(1 \mp \frac{7}{3}\left(\frac{x}{3}\right)\left(1 \mp \frac{10}{4}\left(\frac{x}{3}\right)(1 \mp \cdots)\right)\right)\right)$	$1 \mp \frac{x}{3}\bigwedge_{k=1}^{\infty}\left[1 \mp \frac{3k+1}{k+1}\frac{x}{3}\right]$
$-\dfrac{1}{q}$	$1 \mp A\left(1 \mp \frac{q+1}{2}A\left(1 \mp \frac{2q+1}{3}A\left(1 \mp \frac{3q+1}{4}A(1 \mp \cdots)\right)\right)\right)$	$1 \mp A\bigwedge_{k=1}^{\infty}\left[1 \mp \frac{kq+1}{k+1}A\right]$
$\dfrac{p}{q}$	$1 \pm pA\left(1 \mp \frac{q-p}{2}A\left(1 \mp \frac{2q-p}{3}A\left(1 \mp \frac{3q-p}{4}A(1 \mp \cdots)\right)\right)\right)$	$1 \pm pA\bigwedge_{k=1}^{\infty}\left[1 \mp \frac{kq-p}{k+1}A\right]$
$-\dfrac{p}{q}$	$1 \mp pA\left(1 \mp \frac{q+p}{2}A\left(1 \mp \frac{2q+p}{3}A\left(1 \mp \frac{3q+p}{4}A(1 \mp \cdots)\right)\right)\right)$	$1 \mp pA\bigwedge_{k=1}^{\infty}\left[1 \mp \frac{kq+p}{k+1}A\right]$

*A = pocket calculator storage.

(1) Single Variable

(a) MacLaurin's series at $x = 0$

If a function $f(x)$ is continuous and single-valued and has all derivatives on an interval including $x = 0$, then

$$f(x) = f(0) + \frac{f'(0)}{1!}x + \frac{f''(0)}{2!}x^2 + \cdots + \frac{f^{(n)}(0)}{n!}x^n + R_n$$

in which $R_n = \dfrac{f^{(n+1)}(\theta x)}{(n+1)!}x^{n+1}$ $0 < \theta < 1$

This series represents $f(x)$ for those values of x for which $R_n \to 0$ as $n \to \infty$.

(b) Taylor's series at $x = a$

If a function $f(x)$ is continuous and single-valued and has all derivatives on an interval including $x = a$, then

$$f(x) = f(a) + \frac{f'(a)}{1!}(x - a) + \frac{f''(a)}{2!}(x - a)^2 + \cdots + \frac{f^{(n)}(a)}{n!}(x - a)^n + R_n$$

in which $R_n = \dfrac{f^{(n+1)}(\theta x)}{(n+1)!}(x - a)^{n+1}$ $a < \theta x < x$

The series represents $f(x)$ for those values of x for which $R_n \to 0$ as $n \to \infty$.

(c) Modified Taylor's series at $x = a + h$

If a function $f(x)$ is continuous and single-valued and has all derivatives on an interval including $x = a + h$, then

$$f(a + h) = f(a) + \frac{f'(a)}{1!}h + \frac{f''(a)}{2!}h^2 + \cdots + \frac{f^{(n)}(a)}{n!}h^n + R_n$$

in which $R_n = \dfrac{f^{(n+1)}(a + \theta h)}{(n+1)!}h^{n+1}$ $a < a + \theta h < a + h$

(2) Two Variables

Taylor's series for a function of two variables is

$$f(x + a, y + b) = f(x, y) + \frac{1}{1!}D_1[f(x, y)] + \frac{1}{2!}D_2[f(x, y)] + \cdots + \frac{1}{n!}D_n[f(x, y)] + R_n$$

in which $D_n = \left(a\dfrac{\partial}{\partial x} + b\dfrac{\partial}{\partial y}\right)^n$ and $R_n = \dfrac{1}{(n+1)!}D_{n+1}[f(x + \theta_1 a, y + \theta_2 b)]$

or at $x = 0, y = 0$, $0 < \theta_1 < 1, 0 < \theta_2 < 1$

$$f(x, y) = f(0, 0) + \frac{1}{1!}D_1[f(0, 0)] + \frac{1}{2!}D_2[f(0, 0)] + \cdots + \frac{1}{n!}D_n[f(0, 0)] + R_n$$

in which $D_n = \left(x\dfrac{\partial}{\partial x} + y\dfrac{\partial}{\partial y}\right)^n$ and $R_n = \dfrac{1}{(n+1)!}D_{n+1}[f(\theta_1 x, \theta_2 y)]$

$$0 < \theta_1 < 1, 0 < \theta_2 < 1$$

m = real signed number	j, k, r = positive integers

(1) Basic Operations

$a_0 = 1, b_0 = 1$

(a) Sum of two series

$$\sum_{k=0}^{\infty} a_k x^k \pm \sum_{k=0}^{\infty} b_k x^k = \sum_{k=0}^{\infty} (a_k \pm b_k) x^k$$

(b) Product and quotient of two series

$\left(\sum_{k=0}^{\infty} a_k x^k \right)\left(\sum_{k=0}^{\infty} b_k x^k \right) = \sum_{k=0}^{\infty} A_k x^k$	$A_0 = 1, A_1 = a_1 + b_1, \cdots, A_k = b_k + \sum_{j=1}^{k} a_j b_{k-j}$
$\left(\sum_{k=0}^{\infty} a_k x^k \right) : \left(\sum_{k=0}^{\infty} b_k x^k \right) = \sum_{k=0}^{\infty} B_k x^k$	$B_0 = 1, B_1 = a_1 - b_1, \cdots, B_k = a_k - \sum_{j=1}^{k} b_j B_{k-j}$

(2) Powers and Roots of a Series

$a_0 = 1, \omega_j = jm - k + j$

(a) General case

$\left(\sum_{k=0}^{\infty} a_k x^k \right)^m = \sum_{k=0}^{\infty} C_{m,k} x^k$	$C_{m,0} = 1, C_{m,1} = ma_1 C_{m,0}, \cdots, C_{m,k} = \frac{1}{k}\left[\sum_{j=1}^{k} \omega_j a_j C_{m,k-j} \right]$

(b) Particular cases

m	ω_j	m	ω_j	m	ω_j	m	ω_j
2	$3j - k$	-1	$-k$	$\frac{1}{2}$	$3j/2 - k$	$-\frac{1}{2}$	$j/2 - k$
3	$4j - k$	-2	$-k - j$	$\frac{1}{3}$	$4j/3 - k$	$-\frac{1}{3}$	$2j/3 - k$
4	$5j - k$	-3	$-k - 2j$	$\frac{1}{4}$	$5j/4 - k$	$-\frac{1}{4}$	$3j/4 - k$

(3) Special Operations

$a_0 = 0, b_0 = 0$

(a) Substitution. If $y = a_1 x + a_2 x^2 + a_3 x^3 + \cdots$, then

$$\sum_{r=0}^{\infty} b_r y^r = D_1 x + D_2 x^2 + D_3 x^3 + \cdots = \sum_{r=1}^{\infty} D_r x^r$$

$$D_1 = b_1 C_{1,1}, \ D_2 = b_1 C_{1,2} + b_2 C_{2,2}, \ D_3 = b_1 C_{1,3} + b_2 C_{2,3} + b_3 C_{3,3} \cdots$$

where $C_{r,k}$ is $C_{m,k}$ given in (2a) above, $b_1, b_2, \ldots, b_r$ are known values, and $C_{r,r} = a_1^r$.

(b) Reversion. If $y = x - a_2 x^2 - a_3 x^3 - a_4 x^4 - a_5 x^5 - a_6 x^6 - a_7 x^7 - \cdots$, then

$$x = y + R_2 y^2 + R_3 y^3 + R_4 y^4 + R_5 y^5 + R_6 y^6 + R_7 y^7 + \cdots = y + \sum_{k=2}^{\infty} R_k y^k$$

$$R_2 = a_2 \qquad\qquad R_3 = 2a_2^2 + a_3$$

$$R_4 = 5a_2^3 + 5a_2 a_3 + a_4 \qquad R_5 = 14a_2^4 + 21a_2^2 a_3 + 3a_3^2 + a_5$$

$$R_6 = 42a_2^5 + 84a_2^2 a_3 + 28(a_2^2 a_4 + a_2 a_3^2) + 7(a_2 a_3 + a_3 a_4) + a_6$$

$$R_7 = 132a_2^6 + 330a_2^4 a_3 + 60(3a_2^2 a_3^2 + 2a_2^3 a_4) + 12(6a_2 a_3 a_4 + 3a_2^2 a_5 + a_3^3) + 4(2a_2 a_6 + 2a_3 a_5 + a_4^2) + a_7 \cdots$$

B_k = auxiliary Bernoulli number (Sec. 8.05) E_k = auxiliary Euler number (Sec. 8.06)

$$\alpha_k = \frac{3^{2k+1} - 3}{(2k + 1)!} \qquad \beta_k = \frac{4^{2k} - 4}{(2k + 2)!} \qquad \gamma_k = \frac{3^{2k} + 3}{(2k)!} \qquad \delta_k = \frac{4^{2k} + 4^{k+1}}{(2k)!}$$

(1) Trigonometric Functions[1]

$f(x)$	Standard series	Nested series	Interval
$\sin x$	$\displaystyle\sum_{k=1}^{\infty} (-1)^{k+1} \frac{x^{2k-1}}{(2k-1)!}$	$\displaystyle x \bigwedge_{k=1}^{\infty} \left[1 - \frac{x^2}{2k(2k+1)}\right]$	
$\sin^2 x$	$\displaystyle\frac{1}{2} \sum_{k=1}^{\infty} (-1)^{k+1} \frac{(2x)^{2k}}{(2k)!}$	$\displaystyle x^2 \bigwedge_{k=1}^{\infty} \left[1 - \frac{(2x)^2}{(2k+1)(2k+2)}\right]$	
$\sin^3 x$	$\displaystyle\frac{1}{4} \sum_{k=1}^{\infty} (-1)^{k+1} \alpha_k x^{2k+1}$	$\displaystyle x^3 \bigwedge_{k=1}^{\infty} \left[1 - \frac{\alpha_{k+1}}{\alpha_k} x^2\right]$	$-\infty < x < \infty$
$\sin^4 x$	$\displaystyle\frac{1}{8} \sum_{k=1}^{\infty} (-1)^{k+1} \beta_k x^{2k+2}$	$\displaystyle\frac{x^3}{16} \bigwedge_{k=1}^{\infty} \left[1 - \frac{\beta_{k+1}}{\beta_k} x^2\right]$	
$\cos x$	$\displaystyle\sum_{k=0}^{\infty} (-1)^{k} \frac{x^{2k}}{(2k)!}$	$\displaystyle \bigwedge_{k=1}^{\infty} \left[1 - \frac{x^2}{(2k-1)(2k)}\right]$	
$\cos^2 x$	$\displaystyle 1 + \frac{1}{2} \sum_{k=0}^{\infty} (-1)^{k} \frac{(2x)^{2k}}{(2k)!}$	$\displaystyle 1 - x^2 \bigwedge_{k=1}^{\infty} \left[1 - \frac{(2x)^2}{(2k+1)(2k+2)}\right]$	
$\cos^3 x$	$\displaystyle\frac{1}{4} \sum_{k=0}^{\infty} (-1)^{k} \gamma_k x^{2k}$	$\displaystyle\frac{1}{4} x^2 \bigwedge_{k=1}^{\infty} \left[1 - \frac{\gamma_{k+1}}{\gamma_k} x^2\right]$	$-\infty < x < \infty$
$\cos^4 x$	$\displaystyle 1 + \frac{1}{8} \sum_{k=0}^{\infty} (-1)^{k} \delta_k x^{2k}$	$\displaystyle 1 - 2x^2 \bigwedge_{k=1}^{\infty} \left[1 - \frac{\delta_{k+1}}{\delta_k} x^2\right]$	
$\tan x$	$\displaystyle\frac{1}{x} \sum_{k=1}^{\infty} a_k (2x)^{2k}$	$\displaystyle x \bigwedge_{k=1}^{\infty} \left[1 + \frac{a_{k+1}}{a_k} (2x)^2\right]$	$|x| < \dfrac{\pi}{2}$
$\cot x$	$\displaystyle\frac{1}{x} - \frac{1}{x} \sum_{k=1}^{\infty} b_k (2x)^{2k}$	$\displaystyle\frac{1}{x} - \frac{x}{3} \bigwedge_{k=1}^{\infty} \left[1 + \frac{b_{k+1}}{b_k} (2x)^2\right]$	$0 < |x| < \pi$
$\sec x$	$\displaystyle 1 + \sum_{k=1}^{\infty} c_k x^{2k}$	$\displaystyle 1 + \frac{x^2}{2} \bigwedge_{k=1}^{\infty} \left[1 + \frac{c_{k+1}}{c_k} x^2\right]$	$|x| < \dfrac{\pi}{2}$
$\csc x$	$\displaystyle\frac{1}{x} + \frac{1}{x} \sum_{k=1}^{\infty} d_k x^{2k}$	$\displaystyle\frac{1}{x} + \frac{x}{6} \bigwedge_{k=1}^{\infty} \left[1 + \frac{d_{k+1}}{d_k} x^2\right]$	$0 < |x| < \pi$

(2) Factors a_k, b_k, c_k, d_k for $k = 1, 2, \ldots, 5$*

k	$a_k = \dfrac{4^k - 1}{(2k)!} B_k$		$b_k = \dfrac{1}{(2k)!} B_k$		$c_k = \dfrac{1}{(2k)!} E_k$		$d_k = \dfrac{4^k - 2}{(2k)!} B_k$	
1	2.500 000 000	(−01)	8.333 333 333	(−02)	5.000 000 000	(−01)	1.666 666 667	(−01)
2	2.083 333 333	(−02)	1.388 888 889	(−03)	2.083 333 333	(−01)	1.944 444 444	(−02)
3	2.083 333 333	(−03)	3.306 878 307	(−05)	8.472 222 222	(−02)	2.050 264 550	(−03)
4	2.108 134 921	(−04)	8.267 195 767	(−07)	3.435 019 841	(−02)	2.099 867 725	(−04)
5	2.135 692 240	(−05)	2.087 675 699	(−08)	1.392 223 325	(−02)	2.133 604 564	(−05)

[1]For $\Sigma(\ )$ and $\Lambda[\ ]$ refer to Secs. 8.01 and 8.11, respectively.
*For $k = 6, 7, \ldots, 10$ see opposite page.

B_k = auxiliary Bernoulli number (Sec. 8.05) E_k = auxiliary Euler number (Sec. 8.06)

$$\alpha_k = \frac{3^{2k+1} - 3}{(2k + 1)!} \qquad \beta_k = \frac{4^{2k} - 4}{(2k + 2)!} \qquad \gamma_k = \frac{3^{2k} + 3}{(2k)!} \qquad \delta_k = \frac{4^{2k} + 4^{k+1}}{(2k)!}$$

(1) Hyperbolic Functions[1]

$f(x)$	Standard series	Nested series	Interval
$\sinh x$	$\displaystyle\sum_{k=1}^{\infty} \frac{x^{2k-1}}{(2k-1)!}$	$\displaystyle x \bigwedge_{k=1}^{\infty} \left[1 + \frac{x^2}{2k(2k+1)}\right]$	
$\sinh^2 x$	$\displaystyle\frac{1}{2}\sum_{k=1}^{\infty} \frac{(2x)^{2k}}{(2k)!}$	$\displaystyle x^2 \bigwedge_{k=1}^{\infty} \left[1 + \frac{(2x)^2}{(2k+1)(2k+2)}\right]$	
$\sinh^3 x$	$\displaystyle\frac{1}{4}\sum_{k=1}^{\infty} \alpha_k x^{2k+1}$	$\displaystyle x^3 \bigwedge_{k=1}^{\infty} \left[1 + \frac{\alpha_{k+1}}{\alpha_k} x^2\right]$	$-\infty < x < \infty$
$\sinh^4 x$	$\displaystyle\frac{1}{8}\sum_{k=1}^{\infty} \beta_k x^{2k+2}$	$\displaystyle\frac{x^4}{16} \bigwedge_{k=1}^{\infty} \left[1 + \frac{\beta_{k+1}}{\beta_k} x^2\right]$	
$\cosh x$	$\displaystyle\sum_{k=0}^{\infty} \frac{x^{2k}}{(2k)!}$	$\displaystyle\bigwedge_{k=1}^{\infty} \left[1 + \frac{x^2}{(2k-1)(2k)}\right]$	
$\cosh^2 x$	$\displaystyle 1 + \frac{1}{2}\sum_{k=0}^{\infty} \frac{(2x)^{2k}}{(2k)!}$	$\displaystyle 1 + x^2 \bigwedge_{k=1}^{\infty} \left[1 + \frac{(2x)^2}{(2k+1)(2k+2)}\right]$	
$\cosh^3 x$	$\displaystyle\frac{1}{4}\sum_{k=0}^{\infty} \gamma_k x^{2k}$	$\displaystyle\frac{3}{4}x^2 \bigwedge_{k=1}^{\infty} \left[1 + \frac{\gamma_{k+1}}{\gamma_k} x^2\right]$	$-\infty < x < \infty$
$\cosh^4 x$	$\displaystyle 1 + \frac{1}{8}\sum_{k=0}^{\infty} \delta_k x^{2k}$	$\displaystyle 1 + 2x^2 \bigwedge_{k=1}^{\infty} \left[1 + \frac{\delta_{k+1}}{\delta_k} x^2\right]$	
$\tanh x$	$\displaystyle\frac{1}{x}\sum_{k=1}^{\infty} (-1)^{k+1} a_k (2x)^{2k}$	$\displaystyle x \bigwedge_{k=1}^{\infty} \left[1 - \frac{a_{k+1}}{a_k}(2x)^2\right]$	$\|x\| < \dfrac{\pi}{2}$
$\coth x$	$\displaystyle\frac{1}{x} + \frac{1}{x}\sum_{k=1}^{\infty} (-1)^{k+1} b_k (2x)^{2k}$	$\displaystyle\frac{1}{x} + \frac{x}{3} \bigwedge_{k=1}^{\infty} \left[1 - \frac{b_{k+1}}{b_k}(2x)^2\right]$	$0 < \|x\| < \pi$
$\operatorname{sech} x$	$\displaystyle 1 + \sum_{k=1}^{\infty} (-1)^k c_k x^{2k}$	$\displaystyle 1 - \frac{x^2}{2} \bigwedge_{k=1}^{\infty} \left[1 - \frac{c_{k+1}}{c_k} x^2\right]$	$\|x\| < \dfrac{\pi}{2}$
$\operatorname{csch} x$	$\displaystyle\frac{1}{x} - \frac{1}{x}\sum_{k=1}^{\infty} (-1)^{k+1} d_k x^{2k}$	$\displaystyle\frac{1}{x} - \frac{x}{6} \bigwedge_{k=1}^{\infty} \left[1 - \frac{d_{k+1}}{d_k} x^2\right]$	$0 < \|x\| < \pi$

(2) Factors a_k, b_k, c_k, d_k for $k = 6, 7, \ldots, 10$*

k	$a_k = \dfrac{4^k - 1}{(2k)!} B_k$	$b_k = \dfrac{1}{(2k)!} B_k$	$c_k = \dfrac{1}{(2k)!} E_k$	$d_k = \dfrac{4^k - 2}{(2k)!} B_k$
6	2.163 875 862 (−06)	5.284 190 139 (−10)	5.642 496 810 (−03)	2.163 347 443 (−06)
7	2.192 460 960 (−07)	1.338 253 653 (−11)	2.286 819 095 (−03)	2.192 327 134 (−07)
8	2.221 426 982 (−08)	3.389 680 296 (−13)	5.268 129 274 (−04)	2.221 393 085 (−08)
9	2.250 776 066 (−09)	8.586 062 056 (−15)	2.756 231 338 (−04)	2.250 767 480 (−09)
10	2.280 512 946 (−10)	2.174 868 699 (−16)	1.522 343 222 (−04)	2.280 510 771 (−10)

[1]For $\Sigma(\ )$ and $\Lambda[\ \ ]$ refer to Secs. 8.01 and 8.11, respectively.
*For $k = 1, 2, \ldots, 5$ see opposite page.

a = positive constant	$e = 2.718\ 281\ 828\ \cdots$ (Sec. 104)

(1) Exponential Functions[1]

$f(x)$	Standard series	Nested series	Interval
e	$\displaystyle\sum_{k=0}^{\infty}\frac{1}{k!}$	$\displaystyle\bigwedge_{k=1}^{\infty}\left[1+\frac{1}{k}\right]$	
e^x	$\displaystyle\sum_{k=0}^{\infty}\frac{x^k}{k!}$	$\displaystyle\bigwedge_{k=1}^{\infty}\left[1+\frac{x}{k}\right]$	$-\infty < x < \infty$
e^{-x}	$\displaystyle\sum_{k=0}^{\infty}\frac{(-x)^k}{k!}$	$\displaystyle\bigwedge_{k=1}^{\infty}\left[1-\frac{x}{k}\right]$	
a	$\displaystyle\sum_{k=0}^{\infty}\frac{(\ln a)^k}{k!}$	$\displaystyle\bigwedge_{k=1}^{\infty}\left[1+\frac{\ln a}{k}\right]$	
a^x	$\displaystyle\sum_{k=0}^{\infty}\frac{(x\ln a)^k}{k!}$	$\displaystyle\bigwedge_{k=1}^{\infty}\left[1+\frac{x\ln a}{k}\right]$	$-\infty < x < \infty$
a^{-x}	$\displaystyle\sum_{k=0}^{\infty}\frac{(-x\ln a)^k}{k!}$	$\displaystyle\bigwedge_{k=1}^{\infty}\left[1-\frac{x\ln a}{k}\right]$	

(2) Logarithmic Functions[1]

$f(x)$	Standard series	Nested series	Interval
$\ln x$	$2\left(\dfrac{x-1}{x+1}\right)\displaystyle\sum_{k=0}^{\infty}\frac{1}{2k+1}\left(\frac{x-1}{x+1}\right)^{2k}$	$2\left(\dfrac{x-1}{x+1}\right)\displaystyle\bigwedge_{k=1}^{\infty}\left[1+\frac{2k-1}{2k+1}\left(\frac{x-1}{x+1}\right)^2\right]$	$0 < x < \infty$
	$(x-1)\displaystyle\sum_{k=0}^{\infty}\frac{(1-x)^k}{k+1}$	$(x-1)\displaystyle\bigwedge_{k=1}^{\infty}\left[1-\frac{k(x-1)}{k+1}\right]$	$0 < x < 2$
	$\dfrac{x-1}{x}\displaystyle\sum_{k=0}^{\infty}\frac{1}{k+1}\left(\frac{x-1}{x}\right)^{k}$	$\dfrac{x-1}{x}\displaystyle\bigwedge_{k=1}^{\infty}\left[1+\frac{k(x-1)}{(k+1)x}\right]$	$\dfrac{1}{2}\leq x < \infty$
$\ln(x+1)$	$x\displaystyle\sum_{k=0}^{\infty}\frac{(-x)^k}{k+1}$	$x\displaystyle\bigwedge_{k=1}^{\infty}\left[1-\frac{kx}{k+1}\right]$	$-1 < x \leq 1$
$\ln(x-1)$	$-x\displaystyle\sum_{k=0}^{\infty}\frac{x^k}{k+1}$	$-x\displaystyle\bigwedge_{k=1}^{\infty}\left[1+\frac{kx}{k+1}\right]$	$-1 \leq x < 1$
$\ln\dfrac{x+1}{x-1}$	$\dfrac{2}{x}\displaystyle\sum_{k=0}^{\infty}\frac{1}{2k+1}\left(\frac{1}{x}\right)^{2k}=2\coth^{-1}x$	$\dfrac{2}{x}\displaystyle\bigwedge_{k=1}^{\infty}\left[1+\frac{2k-1}{2k+1}\left(\frac{1}{x}\right)^2\right]$	$|x| > 1$
$\ln\dfrac{1+x}{1-x}$	$2x\displaystyle\sum_{k=0}^{\infty}\frac{x^{2k}}{2k+1}=2\tanh^{-1}x$	$2x\displaystyle\bigwedge_{k=1}^{\infty}\left[1+\frac{(2k-1)x^2}{2k+1}\right]$	$|x| < 1$

[1]For $\Sigma(\ )$ and $\Lambda[\ ]$ refer to Secs. 8.01 and 8.11, respectively.

( )!! = double factorial (Sec. 1.03)	$\dbinom{2k}{k} = \dfrac{(2k)!}{(k!)^2} = 2^k\dfrac{(2k-1)!!}{k!}$ (Sec. 1.04)

(1) Inverse Trigonometric Functions[1]

$f(x)$	Standard series	Nested series	Interval
$\sin^{-1}x$	$x\displaystyle\sum_{k=0}^{\infty}\frac{1}{2k+1}\binom{2k}{k}\left(\frac{x}{2}\right)^{2k}$	$x\displaystyle\bigwedge_{k=1}^{\infty}\left[1+\frac{(2k-1)^2x^2}{2k(2k+1)}\right]$	$|x|<1$
$\cos^{-1}x$	$\dfrac{\pi}{2}-\sin^{-1}x$ (for $\sin^{-1}x$ use the series above)		
$\tan^{-1}x$	$x\displaystyle\sum_{k=0}^{\infty}(-1)^k\frac{x^{2k}}{2k+1}$	$x\displaystyle\bigwedge_{k=1}^{\infty}\left[1-\frac{2k-1}{2k+1}x^2\right]$	$|x|<1$
	$\dfrac{\pi}{2}-\dfrac{1}{x}\displaystyle\sum_{k=0}^{\infty}\frac{(-1)^k}{(2k+1)x^{2k}}$	$\dfrac{\pi}{2}-\dfrac{1}{x}\displaystyle\bigwedge_{k=1}^{\infty}\left[1-\frac{2k-1}{(2k+1)x^2}\right]$	$|x|\geq1$
$\cot^{-1}x$	$\dfrac{\pi}{2}-\tan^{-1}x$ (for $\tan^{-1}x$ use the respective series above)		
$\sec^{-1}x$	$\dfrac{\pi}{2}-\dfrac{1}{x}\displaystyle\sum_{k=0}^{\infty}\frac{1}{2k+1}\binom{2k}{k}\left(\frac{1}{2x}\right)^{2k}$	$\dfrac{\pi}{2}-\dfrac{1}{x}\displaystyle\bigwedge_{k=1}^{\infty}\left[1+\frac{(2k-1)^2}{2k(2k+1)x^2}\right]$	$|x|>1$
$\csc^{-1}x$	$\dfrac{1}{x}\displaystyle\sum_{k=0}^{\infty}\frac{1}{2k+1}\binom{2k}{k}\left(\frac{1}{2x}\right)^{2k}$	$\dfrac{1}{x}\displaystyle\bigwedge_{k=1}^{\infty}\left[1+\frac{(2k-1)^2}{2k(2k+1)x^2}\right]$	

(2) Inverse Hyperbolic Functions[1]

$f(x)$	Standard series	Nested series	Interval
$\sinh^{-1}x$	$x\displaystyle\sum_{k=0}^{\infty}\frac{(-1)^k}{2k+1}\binom{2k}{k}\left(\frac{x}{2}\right)^{2k}$	$x\displaystyle\bigwedge_{k=1}^{\infty}\left[1-\frac{(2k-1)^2x^2}{2k(2k+1)}\right]$	$|x|<1$
	$\ln 2x-\displaystyle\sum_{k=1}^{\infty}\frac{(-1)^k}{2k}\binom{2k}{k}\left(\frac{1}{2x}\right)^{2k}$	$\ln 2x+\dfrac{1}{4x^2}\displaystyle\bigwedge_{k=1}^{\infty}\left[1-\frac{k(2k+1)}{2(k+1)^2x^2}\right]$	$x\geq1$
$\cosh^{-1}x$	$\ln 2x-\displaystyle\sum_{k=1}^{\infty}\frac{1}{2k}\binom{2k}{k}\left(\frac{1}{2x}\right)^{2k}$	$\ln 2x-\dfrac{1}{4x^2}\displaystyle\bigwedge_{k=1}^{\infty}\left[1+\frac{k(2k+1)}{2(k+1)^2x^2}\right]$	$x\geq1$
$\tanh^{-1}x$	$x\displaystyle\sum_{k=0}^{\infty}\frac{x^{2k}}{2k+1}$	$x\displaystyle\bigwedge_{k=1}^{\infty}\left[1+\frac{2k-1}{2k+1}x^2\right]$	$|x|<1$
$\coth^{-1}x$	$\dfrac{1}{x}\displaystyle\sum_{k=0}^{\infty}\frac{1}{(2k+1)x^{2k}}$	$\dfrac{1}{x}\displaystyle\bigwedge_{k=1}^{\infty}\left[1+\frac{2k-1}{(2k+1)x^2}\right]$	$|x|>1$
$\text{sech}^{-1}x$	$\ln\dfrac{2}{x}-\displaystyle\sum_{k=1}^{\infty}\frac{1}{2k}\binom{2k}{k}\left(\frac{x}{2}\right)^{2k}$	$\ln\dfrac{2}{x}-\dfrac{x^2}{4}\displaystyle\bigwedge_{k=1}^{\infty}\left[1+\frac{k(2k+1)x^2}{2(k+1)^2}\right]$	$0<x<1$
$\text{csch}^{-1}x$	$\ln\dfrac{2}{x}-\displaystyle\sum_{k=1}^{\infty}\frac{(-1)^k}{2k}\binom{2k}{k}\left(\frac{x}{2}\right)^{2k}$	$\ln\dfrac{2}{x}+\dfrac{x^2}{4}\displaystyle\bigwedge_{k=1}^{\infty}\left[1-\frac{k(2k+1)x^2}{2(k+1)^2}\right]$	$0<x<1$

[1]For $\Sigma(\ )$ and $\bigwedge[\ ]$ refer to Secs. 8.01 and 8.11, respectively.

$$a, b = \text{constants} > 0 \qquad\qquad \omega = \tan^{-1}\frac{b}{a} \qquad\qquad c = \sqrt{a^2 + b^2}$$

(1) Trigonometric and Hyperbolic Functions with Algebraic Argument[1]

$f(x)$	Standard series	Interval
$\cos(a + bx)$	$\cos a \sum\limits_{k=0}^{\infty} (-1)^k \dfrac{(bx)^{2k}}{(2k)!} - \sin a \sum\limits_{k=0}^{\infty} (-1)^k \dfrac{(bx)^{2k+1}}{(2k+1)!}$	
$\sin(a + bx)$	$\sin a \sum\limits_{k=0}^{\infty} (-1)^k \dfrac{(bx)^{2k}}{(2k)!} + \cos a \sum\limits_{k=0}^{\infty} (-1)^k \dfrac{(bx)^{2k+1}}{(2k+1)!}$	$|a + bx| < \infty$
$\cosh(a + bx)$	$\cosh a \sum\limits_{k=0}^{\infty} \dfrac{(bx)^{2k}}{(2k)!} + \sinh a \sum\limits_{k=0}^{\infty} \dfrac{(bx)^{2k+1}}{(2k+1)!}$	
$\sinh(a + bx)$	$\sinh a \sum\limits_{k=0}^{\infty} \dfrac{(bx)^{2k}}{(2k)!} + \cosh a \sum\limits_{k=0}^{\infty} \dfrac{(bx)^{2k+1}}{(2k+1)!}$	

(2) Winkler's Functions of the First Kind[1]

$$e = 2.718\ 281\ 828 \cdots \text{ (Sec. 1.04)}$$

$f(x)$	Standard series	$f(x)$	Standard series	Interval
$e^{ax}\cos bx$	$\sum\limits_{k=0}^{\infty} \dfrac{(cx)^k \cos k\omega}{k!}$	$e^{ax}\cos ax$	$\sum\limits_{k=0}^{\infty} M_{1,k}(ax)^k$	
$e^{ax}\sin bx$	$\sum\limits_{k=0}^{\infty} \dfrac{(cx)^k \sin k\omega}{k!}$	$e^{ax}\sin ax$	$\sum\limits_{k=0}^{\infty} M_{2,k}(ax)^k$	$|ax| < \infty$
$e^{ax}(\cos bx + \sin bx)$	$\sum\limits_{k=0}^{\infty} \dfrac{\sqrt{2}(cx)^k \sin(\frac{1}{4}\pi + k\omega)}{k!}$	$e^{ax}(\cos ax + \sin ax)$	$\sum\limits_{k=0}^{\infty} M_{3,k}(ax)^k$	
$e^{ax}(\cos bx - \sin bx)$	$\sum\limits_{k=0}^{\infty} \dfrac{\sqrt{2}(cx)^k \cos(\frac{1}{4}\pi + k\omega)}{k!}$	$e^{ax}(\cos ax - \sin ax)$	$\sum\limits_{k=0}^{\infty} M_{1,k}(ax)^k$	

(3) Factors $M_{1,k}, M_{2,k}, M_{3,k}, M_{4,k}$ for $k = 0, 1, 2, \ldots, 12$

k	$M_{1,k}$ $\dfrac{(\sqrt{2})^k \cos\frac{1}{4}k\pi}{k!}$	$M_{2,k}$ $\dfrac{(\sqrt{2})^k \sin\frac{1}{4}k\pi}{k!}$	$M_{3,k}$ $\dfrac{(\sqrt{2})^{k+1}}{k!}\sin\left[\dfrac{(k+1)\pi}{4}\right]$	$M_{1,k}$ $\dfrac{(\sqrt{2})^{k+1}}{k!}\cos\left[\dfrac{(k+1)\pi}{4}\right]$
0	1	0	1	1
1	1/1!	1/1!	2/1!	0
2	0	2/2!	2/2!	− 2/2!
3	− 2/3!	2/3!	0	− 4/3!
4	− 4/4!	0	− 4/4!	− 4/4!
5	− 4/5!	− 4/5!	− 8/5!	0
6	0	− 8/6!	− 8/6!	8/6!
7	8/7!	− 8/7!	0	16/7!
8	16/8!	0	16/8!	16/8!
9	16/9!	16/9!	32/9!	0
10	0	32/10!	32/10!	−32/10!
11	−32/11!	32/11!	0	−64/11!
12	−64/12!	0	−64/12!	−64/12!

[1]For $\Sigma(\ )$ refer to Sec. 8.01.

$$a, b = \text{constants} > 0 \qquad\qquad \omega = \tan^{-1}\frac{b}{a} \qquad\qquad c = \sqrt{a^2 + b^2}$$

(1) Trigonometric and Hyperbolic Functions with Algebraic Argument[1]

$f(x)$	Standard series	Interval
$\cos (a - bx)$	$\cos a \displaystyle\sum_{k=0}^{\infty} (-1)^k \frac{(bx)^{2k}}{(2k)!} + \sin a \sum_{k=0}^{\infty} (-1)^k \frac{(bx)^{2k+1}}{(2k + 1)!}$	
$\sin (a - bx)$	$\sin a \displaystyle\sum_{k=0}^{\infty} (-1)^k \frac{(bx)^{2k}}{(2k)!} - \cos a \sum_{k=0}^{\infty} (-1)^k \frac{(bx)^{2k+1}}{(2k + 1)!}$	
$\cosh (a - bx)$	$\cosh a \displaystyle\sum_{k=0}^{\infty} \frac{(bx)^{2k}}{(2k)!} - \sinh a \sum_{k=0}^{\infty} \frac{(bx)^{2k+1}}{(2k + 1)!}$	$\|a + bx\| < \infty$
$\sinh (a - bx)$	$\sinh a \displaystyle\sum_{k=0}^{\infty} \frac{(bx)^{2k}}{(2k)!} - \cosh a \sum_{k=0}^{\infty} \frac{(bx)^{2k+1}}{(2k + 1)!}$	

(2) Winkler's Functions of the Second Kind[1] $e = 2.718\ 281\ 828 \cdots$ (Sec. 1.04)

$f(x)$	Standard series	$f(x)$	Standard series	Interval
$e^{-ax} \cos bx$	$\displaystyle\sum_{k=0}^{\infty} \frac{(-cx)^k \cos k\omega}{k!}$	$e^{-ax} \cos ax$	$\displaystyle\sum_{k=0}^{\infty} N_{1,k}(ax)^k$	
$e^{-ax} \sin bx$	$-\displaystyle\sum_{k=0}^{\infty} \frac{(-cx)^k \sin k\omega}{k!}$	$e^{-ax} \sin ax$	$\displaystyle\sum_{k=0}^{\infty} N_{2,k}(ax)^k$	
$e^{-ax} (\cos bx + \sin bx)$	$-\displaystyle\sum_{k=0}^{\infty} \frac{\sqrt{2}(-cx)^k \cos (\frac{1}{4}\pi + k\omega)}{k!}$	$e^{-ax} (\cos ax + \sin ax)$	$\displaystyle\sum_{k=0}^{\infty} N_{3,k}(ax)^k$	$\|ax\| < \infty$
$e^{-ax} (\cos bx - \sin bx)$	$\displaystyle\sum_{k=0}^{\infty} \frac{\sqrt{2}(-cx)^k \sin (\frac{1}{4}\pi + k\omega)}{k!}$	$e^{-ax} (\cos ax - \sin ax)$	$\displaystyle\sum_{k=0}^{\infty} N_{4,k}(ax)^k$	

(3) Factors $N_{1,k}, N_{2,k}, N_{3,k}, N_{4,k}$ for $k = 0, 1, 2, \ldots, 12$

k	$N_{1,k}$ $\dfrac{(-\sqrt{2})^k \cos \frac{1}{4}k\pi}{k!}$	$N_{2,k}$ $-\dfrac{(-\sqrt{2})^k \sin \frac{1}{4}k\pi}{k!}$	$N_{3,k}$ $-\dfrac{(-\sqrt{2})^{k+1}}{k!} \cos \left[\dfrac{(k + 1)\pi}{4}\right]$	$N_{4,k}$ $-\dfrac{(\sqrt{2})^{k+1}}{k!} \cos \left[\dfrac{(k + 1)\pi}{4}\right]$
0	1	0	1	1
1	$-1/1!$	$1/1!$	0	$-2/1!$
2	0	$-2/2!$	$-2/2!$	$2/2!$
3	$2/3!$	$2/3!$	$4/3!$	0
4	$-4/4!$	0	$-4/4!$	$-4/4$
5	$4/5!$	$-4/5!$	0	$8/5!$
6	0	$8/6!$	$8/6!$	$-8/6!$
7	$-8/7!$	$-8/7!$	$-16/7!$	0
8	$16/8!$	0	$16/8!$	$16/8!$
9	$-16/9!$	$16/9!$	0	$-32/9!$
10	0	$-32/10!$	$-32/10!$	$32/10!$
11	$32/11!$	$32/11!$	$64/11!$	0
12	$-64/12!$	0	$-64/12!$	$-64/12!$

[1]For $\Sigma(\ \ )$ refer to Sec. 8.01.

$a = \text{constant} > 0$

$B_k = \text{auxiliary Bernoulli number}$ (Sec. 8.05)

(1) Rayleigh's Functions[1]

$f(x)$	Standard series	Nested series	Interval
$\frac{1}{2}(\cosh ax + \cos ax)$	$\sum_{k=0}^{\infty} \frac{(ax)^{4k}}{(4k)!}$	$\bigwedge_{k=1}^{\infty} \left[1 + \frac{(4k-4)!(ax)^4}{(4k)!} \right]$	
$\frac{1}{2}(\sinh ax + \sin ax)$	$\sum_{k=0}^{\infty} \frac{(ax)^{4k+1}}{(4k+1)!}$	$ax \bigwedge_{k=1}^{\infty} \left[1 + \frac{(4k-3)!(ax)^4}{(4k+1)!} \right]$	$\infty < ax < \infty$
$\frac{1}{2}(\cosh ax - \cos ax)$	$\sum_{k=0}^{\infty} \frac{(ax)^{4k+2}}{(4k+2)!}$	$\frac{(ax)^2}{2} \bigwedge_{k=1}^{\infty} \left[1 + \frac{(4k-2)!(ax)^4}{(4k+2)!} \right]$	
$\frac{1}{2}(\sinh ax - \sin ax)$	$\sum_{k=0}^{\infty} \frac{(ax)^{4k+3}}{(4k+3)!}$	$\frac{(ax)^3}{6} \bigwedge_{k=1}^{\infty} \left[1 + \frac{(4k-1)!(ax)^4}{(4k+3)!} \right]$	
$\frac{1}{2}(\tanh ax + \tan ax)$	$\frac{1}{ax} \sum_{k=1}^{\infty} \mathscr{A}_k (2ax)^{4k-2}$	$ax \bigwedge_{k=1}^{\infty} \left[1 + \frac{\mathscr{A}_{k+1}}{\mathscr{A}_k} (2ax)^4 \right]$	$|ax| < \frac{\pi}{2}$
$\frac{1}{2}(\tanh ax - \tan ax)$	$-\frac{1}{ax} \sum_{k=1}^{\infty} \mathscr{B}_k (2ax)^{4k}$	$-\frac{ax}{3} \bigwedge_{k=1}^{\infty} \left[1 + \frac{\mathscr{B}_{k+1}}{\mathscr{B}_k} (2ax)^4 \right]$	

(2) Logarithmic Functions with Trigonometric Argument[1]   ln ( ) = natural logarithm (Sec. 1.02)

$f(x)$	Standard series	Nested series	Interval
$\ln (\cos ax)$	$-\sum_{k=1}^{\infty} \mathscr{C}_k (2ax)^{2k}$	$-\frac{1}{2}(ax)^2 \bigwedge_{k=1}^{\infty} \left[1 + \frac{\mathscr{C}_{k+1}}{\mathscr{C}_k} (2ax)^2 \right]$	$|ax| < \frac{\pi}{2}$
$\ln (\sin ax) - \ln (ax)$	$-\sum_{k=1}^{\infty} \mathscr{D}_k (2ax)^{2k}$	$-\frac{1}{6}(ax)^2 \bigwedge_{k=1}^{\infty} \left[1 + \frac{\mathscr{D}_{k+1}}{\mathscr{D}_k} (2ax)^2 \right]$	$0 < |ax| < \infty$
$\ln (\tan ax)$	$\ln (\sin ax) - \ln (\cos ax)$ (for evaluations use the series above)		$0 < |ax| < \frac{\pi}{2}$

(3) Factors $\mathscr{A}_k$, $\mathscr{B}_k$, $\mathscr{C}_k$, $\mathscr{D}_k$ for $k = 1, 2, \ldots, 10$

k	$\mathscr{A}_k$ $\dfrac{4^{2k-1}-1}{(4k-2)!} B_{2k-1}$	$\mathscr{B}_k$ $\dfrac{4^k-1}{(4k)!} B_{2k}$	$\mathscr{C}_k$ $\dfrac{4^k-1}{(2k)(2k)!} B_k$	$\mathscr{D}_k$ $\dfrac{1}{(2k)(2k)!} B_k$
1	2.500 000 000 (−01)	2.083 333 333 (−02)	1.250 000 000 (−01)	4.166 666 667 (−02)
2	2.083 333 333 (−03)	2.108 134 921 (−04)	5.208 333 333 (−03)	3.472 222 222 (−04)
3	2.135 692 240 (−05)	2.163 875 862 (−06)	4.722 222 222 (−04)	5.511 463 845 (−06)
4	2.192 460 960 (−07)	2.221 426 982 (−08)	2.635 168 651 (−05)	1.033 399 471 (−07)
5	2.250 776 066 (−09)	2.280 512 946 (−10)	2.135 692 240 (−06)	2.087 675 699 (−09)
6	2.310 629 816 (−11)	2.341 241 724 (−12)	1.803 229 885 (−07)	4.403 491 783 (−11)
7	2.372 101 712 (−13)	2.043 441 525 (−14)	1.566 043 543 (−08)	9.558 954 664 (−13)
8	2.435 195 511 (−16)	2.467 368 808 (−16)	1.388 391 864 (−09)	2.118 550 185 (−14)
9	2.499 967 242 (−17)	2.532 996 436 (−18)	1.250 431 148 (−10)	4.770 034 476 (−16)
10	2.566 461 197 (−19)	2.600 369 646 (−20)	1.402 564 730 (−11)	1.087 434 350 (−17)

[1]For $\Sigma(\ )$ and $\Lambda[\ \ ]$ refer to Secs. 8.01 and 8.11, respectively.

$$a = \text{constant} > 0 \qquad\qquad T_r = \left[\frac{d^r \tan ax}{dx^r}\right]_{ax=0}$$

(1) Krylov's Functions[1]

$f(x)$	Standard series	Nested series	Interval
$\cosh ax \cos ax$	$\displaystyle\sum_{k=0}^{\infty} (-4)^k \frac{(ax)^{4k}}{(4k)!}$	$\displaystyle\bigwedge_{k=1}^{\infty}\left[1 - \frac{4(4k-4)!}{(4k)!}(ax)^4\right]$	
$\frac{1}{2}(\cosh ax \sin ax$ $+ \sinh ax \cos ax)$	$\displaystyle\sum_{k=0}^{\infty} (-4)^k \frac{(ax)^{4k+1}}{(4k+1)!}$	$\displaystyle\bigwedge_{k=1}^{\infty}\left[1 - \frac{4(4k-3)!}{(4k+1)!}(ax)^4\right]$	
$\frac{1}{2}\sinh ax \sin ax$	$\displaystyle\sum_{k=0}^{\infty} (-4)^k \frac{(ax)^{4k+2}}{(4k+2)!}$	$\displaystyle\bigwedge_{k=1}^{\infty}\left[1 - \frac{4(4k-2)!}{(4k+2)!}(ax)^4\right]$	$-\infty < ax < \infty$
$\frac{1}{4}(\cosh ax \sin ax$ $- \sinh ax \cos ax)$	$\displaystyle\sum_{k=0}^{\infty} (-4)^k \frac{(ax)^{4k+3}}{(4k+3)!}$	$\displaystyle\bigwedge_{k=1}^{\infty}\left[1 - \frac{4(4k-1)!}{(4k+3)!}(ax)^4\right]$	
$\cosh ax \sin ax$	$\displaystyle\frac{x}{\sqrt{2}}\sum_{k=0}^{\infty} \alpha_k \frac{(ax)^{2k}}{(2k+1)!}$	$\displaystyle\frac{x}{\sqrt{2}}\bigwedge_{k=1}^{\infty}\left[1 + \frac{\alpha_{k+1}(ax)^2}{\alpha_k 2k(2k+1)}\right]$	
$\sinh ax \cos ax$	$\displaystyle\frac{x}{\sqrt{2}}\sum_{k=0}^{\infty} \beta_k \frac{(ax)^{2k}}{(2k+1)!}$	$\displaystyle\frac{x}{\sqrt{2}}\bigwedge_{k=1}^{\infty}\left[1 + \frac{\beta_{k+1}(ax)^2}{\beta_k 2k(2k+1)}\right]$	

(2) Exponential Functions with Trigonometric Argument[1]

$e = 2.718\ 281\ 828 \ldots$ (Sec. 1.04)

$f(x)$	Standard series	Nested series	Interval
$e^{\cos ax}$	$\displaystyle e\sum_{k=0}^{\infty} \mathcal{E}_k(ax)^{2k}$	$\displaystyle e\bigwedge_{k=1}^{\infty}\left[1 + \frac{\mathcal{E}_k}{\mathcal{E}_{k-1}}(ax)^2\right]$	$-\infty < ax < \infty$
$e^{\sin ax}$	$\displaystyle \sum_{k=0}^{\infty} \mathcal{F}_k(ax)^k$	Not applicable	
$e^{\tan ax}$	$\displaystyle \sum_{k=0}^{\infty} \mathcal{G}_k(ax)^k$	$\displaystyle\bigwedge_{k=1}^{\infty}\left[1 + \frac{\mathcal{G}_k}{\mathcal{G}_{k-1}}(ax)\right]$	$\|ax\| < \frac{\pi}{2}$

(3) Factors α_k, β_k, $\mathcal{E}_k$, $\mathcal{F}_k$, $\mathcal{G}_k$, for $k = 0, 1, 2, \ldots, 9$

k	α_k	β_k	$\mathcal{E}_k$ $\displaystyle\sum_{r=1}^{k} \frac{\mathcal{E}_{k-r}\cos r\pi}{2k(2k-2r)!(2r-1)!}$	$\mathcal{F}_k$ $\displaystyle\sum_{r=1}^{k} \frac{\mathcal{F}_{k-r}\sin \frac{1}{2}r\pi}{k(k-r)!(r-1)!}$	$\mathcal{G}_k$ $\displaystyle\sum_{r=1}^{k} \frac{\mathcal{G}_{k-r}T_r}{k(k-r)!(r-1)!}$
0	1	1	1.000 000 000 (+00)	1.000 000 000 (+00)	1.000 000 000 (+00)
1	1	−1	−5.000 000 000 (−01)	1.000 000 000 (+00)	1.000 000 000 (+00)
2	−1	−1	1.666 666 667 (−01)	5.000 000 000 (−01)	5.000 000 000 (−01)
3	−1	1	−4.305 555 556 (−02)	0.000 000 000 (+00)	5.000 000 000 (−01)
4	−1	−1	9.399 801 587 (−03)	−1.250 000 000 (−01)	3.750 000 000 (−01)
5	1	−1	−1.806 657 848 (−03)	−6.666 666 667 (−02)	3.333 333 333 (−01)
6	1	1	3.138 799 536 (−04)	−4.166 666 667 (−03)	2.458 333 333 (−01)
7	1	−1	−5.016 685 851 (−05)	1.111 111 111 (−03)	1.902 777 778 (−01)
8	−1	−1	7.470 220 791 (−06)	5.381 944 444 (−03)	1.373 263 889 (−01)
9	−1	1	−1.046 251 198 (−06)	1.763 668 430 (−04)	1.132 082 231 (−01)

[1]For $\Sigma(\ )$ and $\Lambda[\ \ ]$ refer to Secs. 8.01 and 8.11, respectively.

(1) Finite Products

$$1 + x^{2n} = \left(x^2 + 2x\cos\frac{\pi}{2n} + 1\right)\left(x^2 + 2x\cos\frac{3\pi}{2n} + 1\right)\cdots\left(x^2 + 2x\cos\frac{(2n-1)\pi}{2n} + 1\right)$$

$$1 + x^{2n+1} = (1+x)\left[\left(x^2 - 2x\cos\frac{\pi}{2n+1} + 1\right)\left(x^2 - 2x\cos\frac{3\pi}{2n+1} + 1\right)\cdots\left(x^2 - 2x\cos\frac{(2n-1)\pi}{2n+1} + 1\right)\right]$$

$$\sin 2nx = n\sin 2x\left[(1 - a_1\sin^2 x)(1 - a_2\sin^2 x)(1 - a_3\sin^2 x)\cdots(1 - a_{n-1}\sin^2 x)\right]$$

$$\sin(2n+1)x = n\sin x\left[(1 - b_1\sin^2 x)(1 - b_2\sin^2 x)(1 - b_3\sin^2 x)\cdots(1 - b_n\sin^2 x)\right]$$

$$a_r = \frac{1}{\sin^2(r\pi/2n)} \qquad b_r = \frac{1}{\sin^2[r\pi/(2n+1)]} \qquad r = 1, 2, 3, \ldots$$

$$\cos 2nx = (1 - c_1\sin^2 x)(1 - c_3\sin^2 x)(1 - c_5\sin^2 x)\cdots(1 - c_{2n-1}\sin^2 x)$$

$$\cos(2n+1)x = \cos x\left[(1 - d_1\sin^2 x)(1 - d_3\sin^2 x)(1 - d_5\sin^2 x)\cdots(1 - d_{2n-1}\sin^2 x)\right]$$

$$c_r = \frac{1}{\sin^2(r\pi/4n)} \qquad d_r = \frac{1}{\sin^2[r\pi/(4n+2)]} \qquad r = 1, 3, 5, \ldots$$

(2) Infinite Products

$$\sin nx = nx\left[1 - \left(\frac{nx}{\pi}\right)^2\right]\left[1 - \left(\frac{nx}{2\pi}\right)^2\right]$$
$$\times\left[1 - \left(\frac{nx}{3\pi}\right)^2\right]\cdots$$

$$\sinh nx = nx\left[1 + \left(\frac{nx}{\pi}\right)^2\right]\left[1 + \left(\frac{nx}{2\pi}\right)^2\right]$$
$$\times\left[1 + \left(\frac{nx}{3\pi}\right)^2\right]\cdots$$

$$\cos nx = \left[1 - \left(\frac{2nx}{\pi}\right)^2\right]\left[1 - \left(\frac{2nx}{3\pi}\right)^2\right]$$
$$\times\left[1 - \left(\frac{2nx}{5\pi}\right)^2\right]\cdots$$

$$\cosh nx = \left[1 + \left(\frac{2nx}{\pi}\right)^2\right]\left[1 + \left(\frac{2nx}{3\pi}\right)^2\right]$$
$$\times\left[1 + \left(\frac{2nx}{5\pi}\right)^2\right]\cdots$$

$$\sin(x+y) = \left[\left(1 + \frac{y}{x}\right)\left(1 + \frac{y}{\pi + x}\right)\left(1 - \frac{y}{\pi - x}\right)\left(1 + \frac{y}{2\pi + x}\right)\left(1 - \frac{y}{2\pi - x}\right)\cdots\right]\sin x$$

$$\sin(x-y) = \left[\left(1 - \frac{y}{x}\right)\left(1 + \frac{y}{\pi - x}\right)\left(1 - \frac{y}{\pi + x}\right)\left(1 + \frac{y}{2\pi - x}\right)\left(1 - \frac{y}{2\pi + x}\right)\cdots\right]\sin x$$

$$\cos(x+y) = \left[\left(1 + \frac{2y}{\pi + 2x}\right)\left(1 - \frac{2y}{\pi - 2x}\right)\left(1 + \frac{2y}{3\pi + 2x}\right)\left(1 - \frac{2y}{3\pi - 2x}\right)\cdots\right]\cos x$$

$$\cos(x-y) = \left[\left(1 + \frac{2y}{\pi - 2x}\right)\left(1 - \frac{2y}{\pi + 2x}\right)\left(1 + \frac{2y}{3\pi - 2x}\right)\left(1 - \frac{2y}{3\pi + 2x}\right)\cdots\right]\cos x$$

$$\frac{\pi}{2} = \left(\frac{2}{1}\right)\left(\frac{2}{3}\right)\left(\frac{4}{3}\right)\left(\frac{4}{5}\right)\left(\frac{6}{5}\right)\left(\frac{6}{7}\right)\cdots$$

$$\frac{\sin x}{x} = \cos\frac{x}{2}\cos\frac{x}{4}\cos\frac{x}{8}\cos\frac{x}{16}\cdots$$

9
INTEGRAL
CALCULUS

(1) Definitions

$F(x)$ is an *indefinite integral* (antiderivative) of $f(x)$ if

$$\frac{dF(x)}{dx} = f(x)$$

Since the derivative of $F(x) + C$ is also equal to $f(x)$, all integrals of $f(x)$ are included in the expression

$$\int f(x)\, dx = F(x) + C$$

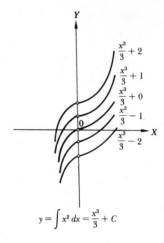

$$y = \int x^2\, dx = \frac{x^3}{3} + C$$

in which $f(x)$ is called the *integrand* and C is an *arbitrary constant*. Because of the indeterminacy of C, there is an infinite number of $F(x) + C$, differing by their relative position to X axis only. The adjacent graph illustrates the meaning of C for a given function.

(2) Integration Methods

(a) Antiderivative method

If $f(x)$ is a derivative of a known function, the integral is this function plus the constant of integration.

$$\int \frac{dF(x)}{dx}\, dx = F(x) + C$$

(b) Integration by parts

If the integrand can be expressed as a product of two functions

$$f(x) = u(x)v'(x)$$

then

$$\int u(x)v'(x)\, dx = u(x)v(x) - \int u'(x)v(x)\, dx$$

in which the integral of the right side may be known or can be calculated by one or more repetitions of the same.

(c) Substitution method

The introduction of a new variable

$$x = \phi(t) \qquad dx = \phi'(t)\, dt$$

yields

$$\int f(x)\, dx = \int f(\phi(t))\phi'(t)\, dt + C$$

The integral of this transformed function may be known, or can be calculated by other methods. The most common substitutions are listed in Secs. 9.07 and 9.08. For particular cases refer to Chap. 19.

(d) Integration by series

If the integrand can be expressed as a uniformly convergent series of powers of x (within its interval of convergence) and if the result of integration of this series term by term is also a uniformly convergent series, the sum of this series is also the value of the integral.

(1) Definitions

If $f(x)$ is continuous in the closed interval $[a, b]$ and this interval is divided into n equal parts by the points $a, x_1, x_2, \ldots, x_{n-1}, b$ such that $\Delta x = (b - a)/n$, then the *definite integral* of $f(x)$ with respect to x, between the limits $x = a$ to $x = b$, is

$$\int_a^b f(x)\, dx = \lim_{n \to \infty} \sum_1^n f(X_i)\, \Delta x = \left[\int f(x)\, dx\right]_a^b = \Big[F(x)\Big]_a^b = F(b) - F(a)$$

where $F(x)$ is a function, the derivative of which with respect to x is $f(x)$.

The numbers a and b are called, respectively, the *lower* and *upper limits of integration*, and $[a, b]$ is called the *range of integration*.

Geometrically, the definite integral of $f(x)$ with respect to x, between limits $x = a$ to $x = b$, is the *area* bounded by $f(x)$, the X axis, and the verticals through the end points of a and b.

(2) Rules of Limits

$$\int_a^b = -\int_b^a \qquad\qquad \int_a^b + \int_b^c = \int_a^c \qquad\qquad \int_a^c - \int_b^c = \int_a^b \qquad\qquad \int_a^a = 0$$

(3) Fundamental Theorems

$$\int_a^b dx = b - a \qquad\qquad\qquad \int_a^b \lambda f(x)\, dx = \lambda \int_a^b f(x)\, dx$$

$$\int_a^b (f(x) + g(x))\, dx = \int_a^b f(x)\, dx + \int_a^b g(x)\, dx$$

$$\int_a^b f(x) \frac{dg(x)}{dx}\, dx = \Big[f(x)g(x)\Big]_a^b - \int_a^b \frac{df(x)}{dx} g(x)\, dx$$

$$\int_a^x f(t)\, dt = F(x) - F(a) \qquad\qquad\qquad \int_a^{\phi(x)} f(t)\, dt = F[\phi(x)] - F(a)$$

$$\frac{d}{dx} \int_a^x f(t)\, dt = f(x) \qquad\qquad\qquad \frac{d}{dx} \int_a^{\phi(x)} f(t)\, dt = F[\phi(x)] \frac{d\phi(x)}{dx}$$

$$\frac{\partial}{\partial \alpha} \int_{\phi_1(\alpha)}^{\phi_2(\alpha)} f(x, \alpha)\, dx = \int_{\phi_1(\alpha)}^{\phi_2(\alpha)} \frac{\partial f(x, \alpha)}{\partial \alpha}\, dx + f(\phi_2(\alpha), \alpha) \frac{\partial \phi_2(\alpha)}{\partial \alpha} - f(\phi_1(\alpha), \alpha) \frac{\partial \phi_1(\alpha)}{\partial \alpha}$$

For additional relationships and particular cases refer to Chap. 20.

(4) Mean Values

$$\text{Arithmetic mean value} = \frac{\int_a^b f(x)\, dx}{b - a} \qquad\qquad \text{Quadratic mean value} = \sqrt{\frac{\int_a^b f(x)^2\, dx}{b - a}}$$

(1) Notation

> u = differentiable function of x $u', u'', \ldots, u^{(n)}$ = successive derivatives of u
>
> $U_1, U_2, \ldots, U_n$ = successive integrals of u a, b, m = constants
>
> The constant of integration, C, is omitted.

(2) Relationships

$$\int (0)\, dx = \text{constant} \qquad\qquad\qquad \int (a)\, dx = ax$$

$$\int f'(x)\, dx = f(x) \qquad\qquad\qquad \int f(a)f'(x)\, dx = f(a)f(x)$$

$$\int f(x)\, dx = xf(0) + \frac{x^2}{2!}f'(0) + \frac{x^3}{3!}f''(0) + \cdots \qquad \int f(x)\, dx = xf(x) - \frac{x^2}{2!}f'(x) + \frac{x^3}{3!}f''(x) - \cdots$$

$$\int f(0)\, dx = xf(0) = f(0)e^{\ln x} \qquad\qquad \int f(x)\, dx = xf(x) - \int xf'(x)\, dx$$

$$\int uu'\, dx = \frac{u^2}{2} \qquad\qquad\qquad \int \frac{u'}{u}\, dx = \ln u$$

$$\int u^m u'\, dx = \frac{u^{m+1}}{m+1} \qquad\qquad \int \frac{u'}{u^m}\, dx = -\frac{1}{(m-1)u^{m-1}} \qquad m \neq 1$$

$$\int (a+bu)^m u'\, dx = \frac{(a+bu)^{m+1}}{b(m+1)} \qquad\qquad m \neq -1,\, b \neq 0$$

$$\int \frac{u'\, dx}{a+bu} = \frac{\ln(a+bu)}{b} \qquad\qquad b \neq 0$$

$$\int \frac{u'\, dx}{(a+bu)^m} = -\frac{1}{b(m-1)(a+bu)^{m-1}} \qquad\qquad m \neq 1,\, b \neq 0$$

$$\int \frac{u'\, dx}{\sqrt{a+bu}} = \frac{2\sqrt{a+bu}}{b} \qquad\qquad b \neq 0$$

$$\int \frac{u'\, dx}{\sqrt{(a+u)(b+u)}} = 2\ln\left(\sqrt{a+u} + \sqrt{b+u}\right)$$

$$\int ux\, dx = xU_1 - U_2$$

$$\int ux^2\, dx = x^2 U_1 - 2xU_2 + 2U_3$$

$$\int ux^m\, dx = x^m U_1 - mx^{m-1}U_2 + m(m-1)x^{m-2}U_3 - m(m-1)(m-2)x^{m-3}U_4 + \cdots$$

(1) Notation

$$u, v = \text{differentiable functions of } x$$

$$\left.\begin{array}{l} u', u'', \ldots, u^{(n)} \\ v', v'', \ldots, v^{(n)} \end{array}\right\} = \text{successive derivatives of } u \text{ and } v, \text{ respectively}$$

$$\left.\begin{array}{l} U_1, U_2, \ldots, U_n \\ V_1, V_2, \ldots, V_n \end{array}\right\} = \text{successive integrals of } u \text{ and } v, \text{ respectively}$$

$$a, b, m = \text{constants}$$

The constant of integration, C, is omitted.

(2) Relationships

$$\int (u + v) \, dx = \int u \, dx + \int v \, dx$$

$$\int (u + v)^m \, dx = x(u + v) - \frac{x^2}{2!}(u' + v') + \frac{x^3}{3!}(u'' + v'') - \cdots$$

$$= \int u(u + v)^{m-1} \, dx + \int v(u + v)^{m-1} \, dx$$

$$\int uv \, dx = U_1 v - \underbrace{U_2 v' + U_3 v'' - U_4 v''' + \cdots}$$

$$= U_1 v - \int U_1 v' \, dx$$

$$= uV_1 - \underbrace{u'V_2 + u''V_3 - u'''V_4 + \cdots}$$

$$= uV_1 - \int u'V_1 \, dx$$

$$\int uv' \, dx = uv - \int u'v \, dx \qquad\qquad \int \frac{uv'}{v^2} \, dx = -\frac{u}{v} + \int \frac{u'}{v} \, dx$$

$$\int u'v \, dx = uv - \int uv' \, dx \qquad\qquad \int \frac{u'v - uv'}{v^2} \, dx = \frac{u}{v}$$

$$\int (u'v + uv') \, dx = uv \qquad\qquad\quad \int \frac{u'v - uv'}{uv} \, dx = \ln \frac{u}{v}$$

$$\int \frac{u'v - uv'}{(u + v)^2} \, dx = -\frac{v}{u + v} \qquad\qquad \int \frac{u'v - uv'}{u^2 + v^2} \, dx = \tan^{-1} \frac{u}{v}$$

$$\int \frac{u'v - uv'}{(u - v)^2} \, dx = -\frac{v}{u - v} \qquad\qquad \int \frac{u'v - uv'}{u^2 - v^2} \, dx = \tfrac{1}{2} \ln \frac{u + v}{u - v}$$

$$\int u^{(n+1)} v \, dx = u^{(n)} v - u^{(n-1)} v' + u^{(n-2)} v'' - \cdots (-1)^{n+1} \int uv^{(n+1)} \, dx$$

If $u'' = a^2 u$ and $v'' = b^2 v$, then   If $u'' = a^2 u$ and $v'' = -b^2 v$, then

$$\int uv \, dx = \frac{u'v - uv'}{a^2 - b^2} \qquad\qquad\qquad \int uv \, dx = \frac{u'v - uv'}{a^2 + b^2}$$

(Ref. 9.05)

In the following integral formulas, $u = f(x)$, a, m = constants. The constant of integration, C, is omitted.

$$\int u^m \, du = \frac{u^{m+1}}{m+1} \qquad m \neq -1 \qquad\qquad \int \frac{du}{u} = \ln |u| \qquad u \neq 0$$

$$\int \frac{1}{u^m} \, du = \frac{u^{1-m}}{1-m} \qquad m \neq 1 \qquad\qquad \int \sqrt[n]{u^m} \, du = \frac{nu\sqrt[n]{u^m}}{m+n} \qquad m \neq -n$$

$$\int \frac{du}{a^2 + u^2} = \frac{1}{a} \tan^{-1} \frac{u}{a} = -\frac{1}{a} \cot^{-1} \frac{u}{a} \qquad a \neq 0$$

$$\int \frac{du}{a^2 - u^2} = \frac{1}{a} \tanh^{-1} \frac{u}{a} = \frac{1}{2a} \ln \frac{a+u}{a-u} \qquad u^2 < a^2$$

$$\int \frac{du}{u^2 - a^2} = \frac{1}{a} \coth^{-1} \frac{u}{a} = \frac{1}{2a} \ln \frac{u+a}{u-a} \qquad u^2 > a^2$$

$$\int \frac{du}{\sqrt{a^2 + u^2}} = \sinh^{-1} \frac{u}{a} = \ln (u + \sqrt{u^2 + a^2})$$

$$\int \frac{du}{\sqrt{a^2 - u^2}} = \sin^{-1} \frac{u}{a} = -\cos^{-1} \frac{u}{a}$$

$$\int \frac{du}{\sqrt{u^2 - a^2}} = \cosh^{-1} \frac{u}{a} = \ln (u + \sqrt{u^2 - a^2})$$

$$\int \frac{du}{u\sqrt{a^2 + u^2}} = -\frac{1}{a} \operatorname{csch}^{-1} \frac{u}{a} = -\frac{1}{a} \sinh^{-1} \frac{a}{u} = -\frac{1}{a} \ln \frac{a + \sqrt{a^2 + u^2}}{u}$$

$$\int \frac{du}{u\sqrt{a^2 - u^2}} = -\frac{1}{a} \operatorname{sech}^{-1} \frac{u}{a} = -\frac{1}{a} \cosh^{-1} \frac{a}{u} = -\frac{1}{a} \ln \frac{a + \sqrt{a^2 - u^2}}{u}$$

$$\int \frac{du}{u\sqrt{u^2 - a^2}} = \frac{1}{a} \sec^{-1} \frac{u}{a} = \cos^{-1} \frac{a}{u}$$

$$\int \sqrt{a^2 + u^2} \, du = \frac{u}{2} \sqrt{a^2 + u^2} + \frac{a^2}{2} \sinh^{-1} \frac{u}{a} = \frac{u}{2} \sqrt{a^2 + u^2} + \frac{a^2}{2} \ln (u + \sqrt{a^2 + u^2})$$

$$\int \sqrt{a^2 - u^2} \, du = \frac{u}{2} \sqrt{a^2 - u^2} + \frac{a^2}{2} \sin^{-1} \frac{u}{a} = \frac{u}{2} \sqrt{a^2 - u^2} - \frac{a^2}{2} \cos^{-1} \frac{u}{a}$$

$$\int \sqrt{u^2 - a^2} \, du = \frac{u}{2} \sqrt{u^2 - a^2} - \frac{a^2}{2} \cosh^{-1} \frac{u}{a} = \frac{u}{2} \sqrt{u^2 - a^2} - \frac{a^2}{2} \ln (u + \sqrt{u^2 - a^2})$$

$$\int \frac{u \, du}{u^4 + a^4} = \frac{1}{2a^2} \tan^{-1} \frac{u^2}{a^2}$$

$$\int \frac{u \, du}{u^4 - a^4} = \frac{1}{4a^2} \ln \frac{u^2 - a^2}{u^2 + a^2}$$

[1]For additional cases refer to Secs. 19.02 through 19.40.

In the following integral formulas, $u = f(x)$ and $a, b, m = $ constants. The constant of integration, C, is omitted.

$$\int a^u \, du = \frac{a^u}{\ln a} \qquad\qquad \int \frac{du}{a^u} = -\frac{1}{a^u \ln a}$$

$$\int e^u \, du = e^u \qquad\qquad \int \ln u \, du = u \ln u - u$$

$$\int \sin au \, du = \frac{-1}{a} \cos au \qquad\qquad \int \sinh au \, du = \frac{1}{a} \cosh au$$

$$\int \cos au \, du = \frac{1}{a} \sin au \qquad\qquad \int \cosh au \, du = \frac{1}{a} \sinh au$$

$$\int \tan au \, du = \frac{-1}{a} \ln \cos au \qquad\qquad \int \tanh au \, du = \frac{1}{a} \ln \cosh au$$

$$\int \cot au \, du = \frac{1}{a} \ln \sin au \qquad\qquad \int \coth au \, du = \frac{1}{a} \ln \sinh au$$

$$\int \sec au \, du = \frac{1}{2a} \ln \frac{1 + \sin au}{1 - \sin au} \qquad\qquad \int \operatorname{sech} au \, du = \frac{1}{a} \tan^{-1} \sinh au$$

$$\int \csc au \, du = \frac{-1}{2a} \ln \frac{1 + \cos au}{1 - \cos au} \qquad\qquad \int \operatorname{csch} au \, du = \frac{1}{a} \ln \tanh \frac{au}{2}$$

$$\int \frac{du}{\sin au} = \frac{-1}{2a} \ln \frac{1 + \cos au}{1 - \cos au} \qquad\qquad \int \frac{du}{\sinh au} = \frac{1}{a} \ln \tanh \frac{au}{2}$$

$$\int \frac{du}{\cos au} = \frac{1}{2a} \ln \frac{1 + \sin au}{1 - \sin au} \qquad\qquad \int \frac{du}{\cosh au} = \frac{1}{a} \tan^{-1} \sinh au$$

$$\int \frac{du}{\tan au} = \frac{1}{a} \ln \sin au \qquad\qquad \int \frac{du}{\tanh au} = \frac{1}{a} \ln \sinh au$$

$$\int \frac{du}{\cot au} = \frac{-1}{a} \ln \cos au \qquad\qquad \int \frac{du}{\coth au} = \frac{1}{a} \ln \cosh au$$

$$\int \frac{du}{\sec au} = \frac{1}{a} \sin au \qquad\qquad \int \frac{du}{\operatorname{sech} au} = \frac{1}{a} \sinh au$$

$$\int \frac{du}{\csc au} = \frac{-1}{a} \cos au \qquad\qquad \int \frac{du}{\operatorname{csch} au} = \frac{1}{a} \cosh au$$

$$\int e^{au} \sin bu \, du = \frac{e^{au}}{a^2 + b^2} (a \sin bu - b \cos bu)$$

$$\int e^{au} \cos bu \, du = \frac{e^{au}}{a^2 + b^2} (a \cos bu + b \sin bu)$$

---

[1]For additional cases refer to Secs. 19.41 through 19.86.

Integral	Substitution
$\displaystyle\int \frac{dx}{(x-a)^m}$	$\dfrac{1}{x-a}=t \qquad dx=-\dfrac{dt}{t^2}$
$\displaystyle\int \frac{f'(x)}{f(x)}\,dx$	$f(x)=t \qquad f'(x)\,dx=dt$
$\displaystyle\int x^m(a+bx^n)^p\,dx$	If $p=\alpha/\beta$ is a fraction and $(m+1)/n$ is an integer, use $$t=\sqrt[\beta]{a+bx^n}$$ If $p=\alpha/\beta$ is a fraction and $(m+1)/n+p$ is an integer, use $$t=\sqrt[\beta]{\dfrac{a+bx^n}{x^n}}$$ If p is an integer, use the binomial expansion.
$\displaystyle\int f(x^2)\,dx$	$x^2=t \qquad 2x\,dx=dt$
$\displaystyle\int f\left[x;\left(\dfrac{ax+b}{cx+d}\right)^m\right]dx$	$\left(\dfrac{ax+b}{cx+d}\right)^m=t^m$
$\displaystyle\int f\left[x;\left(\dfrac{ax+b}{cx+d}\right)^m;\left(\dfrac{ax+b}{cx+d}\right)^n;\dots\right]dx$	$\left(\dfrac{ax+b}{cx+d}\right)^r=t^r$ r is the least common multiple of m and n.
$\displaystyle\int f(x;\sqrt{ax^2+bx+c})\,dx$	If $a>0$, use $$x=\frac{t^2-c}{b-2t\sqrt{a}}$$ $$dx=2\frac{-t^2\sqrt{a}+bt-c\sqrt{a}}{(b-2t\sqrt{a})^2}\,dt$$ If $c>0$, use $$x=\frac{2t\sqrt{c}-b}{a-t^2}$$ $$dx=\frac{2a\sqrt{c}-2bt+2t^2\sqrt{c}}{(a-t^2)^2}\,dt$$ If $ax^2+bx+c=a(x-\alpha)(x-\beta)$, use $$x=\frac{t^2\alpha-a\beta}{t^2-a} \qquad dx=\frac{2at(\beta-\alpha)}{(t^2-a)^2}\,dt$$
$\displaystyle\int f(x;\sqrt[m]{ax+b})\,dx$	$\sqrt[m]{ax+b}=t \qquad dx=\dfrac{mt^{m-1}\,dt}{a}$

Integral and substitution	Transformations	
$\int f(x;\ \sqrt{a^2-x^2})\ dx$	$\sin t = \dfrac{x}{a}$	$\cos t = \dfrac{\sqrt{a^2-x^2}}{a}$
$x = a\sin t \qquad dx = a\cos t\ dt$	$\tan t = \dfrac{x}{\sqrt{a^2-x^2}}$	$\cot t = \dfrac{\sqrt{a^2-x^2}}{x}$
$\int f(x;\ \sqrt{a^2-x^2})\ dx$	$\sinh t = \dfrac{x}{\sqrt{a^2-x^2}}$	$\cosh t = \dfrac{a}{\sqrt{a^2-x^2}}$
$x = a\tanh t \qquad dx = \dfrac{a\ dt}{\cosh^2 t}$	$\tanh t = \dfrac{x}{a}$	$\coth t = \dfrac{a}{x}$
$\int f(x;\ \sqrt{a^2+x^2})\ dx$	$\sin t = \dfrac{x}{\sqrt{a^2+x^2}}$	$\cos t = \dfrac{a}{\sqrt{a^2+x^2}}$
$x = a\tan t \qquad dx = \dfrac{a\ dt}{\cos^2 t}$	$\tan t = \dfrac{x}{a}$	$\cot t = \dfrac{a}{x}$
$\int f(x;\ \sqrt{a^2+x^2})\ dx$	$\sinh t = \dfrac{x}{a}$	$\cosh t = \dfrac{\sqrt{a^2+x^2}}{a}$
$x = a\sinh t \qquad dx = a\cosh t\ dt$	$\tanh t = \dfrac{x}{\sqrt{a^2+x^2}}$	$\coth t = \dfrac{\sqrt{a^2+x^2}}{x}$
$\int f(x;\ \sqrt{x^2-a^2})\ dx$	$\sin t = \dfrac{\sqrt{x^2-a^2}}{x}$	$\cos t = \dfrac{a}{x}$
$x = \dfrac{a}{\cos t} \qquad dx = \dfrac{a\sin t\ dt}{\cos^2 t}$	$\tan t = \dfrac{\sqrt{x^2-a^2}}{a}$	$\cot t = \dfrac{a}{\sqrt{x^2-a^2}}$
$\int f(x;\ \sqrt{x^2-a^2})\ dx$	$\sinh t = \dfrac{\sqrt{x^2-a^2}}{a}$	$\cosh t = \dfrac{x}{a}$
$x = a\cosh t \qquad dx = a\sinh t\ dt$	$\tanh t = \dfrac{\sqrt{x^2-a^2}}{x}$	$\coth t = \dfrac{x}{\sqrt{x^2-a^2}}$
$\int f(\sin x;\ \cos x;\ \tan x;\ \cot x)\ dx$	$\sin x = \dfrac{2t}{1+t^2}$	$\cos x = \dfrac{1-t^2}{1+t^2}$
$\tan\dfrac{x}{2} = t \qquad dx = \dfrac{2dt}{1+t^2}$	$\tan x = \dfrac{2t}{1-t^2}$	$\cot x = \dfrac{1-t^2}{2t}$
$\int f(\sinh x;\ \cosh x;\ \tanh x;\ \coth x)\ dx$	$\sinh x = \dfrac{2t}{1-t^2}$	$\cosh x = \dfrac{1+t^2}{1-t^2}$
$\tanh\dfrac{x}{2} = t \qquad dx = \dfrac{2dt}{1-t^2}$	$\tanh x = \dfrac{2t}{1+t^2}$	$\coth x = \dfrac{1+t^2}{2t}$
$\int f(e^x)\ dx$	$\int f(a^x)\ dx$	
$e^x = t \qquad dx = \dfrac{dt}{t}$	$a^x = t \qquad dx = \dfrac{dt}{t\ln t}$	

The calculation of a double integral is performed by successive evaluation of two definite integrals.

(1) Cartesian Coordinates

If P_1, P_2 and Q_1, Q_2 are points on a contour enclosing area A, selected so that they identify extreme coordinates x and y, respectively, then

$$A = \iint\limits_A f(x, y)\, dx\, dy = \int_a^b \left[\int_{f_1(x)}^{f_2(x)} f(x, y)\, dy \right] dx$$

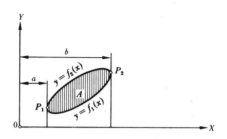

where the boundary of A consists of two continuous curves $y = f_1(x)$ and $y = f_2(x)$. The boundary of A is met by a line parallel to the Y axis in at most two points, and $x = a$, $x = b$ are the extreme values of x on A; or

$$A = \iint\limits_A f(x, y)\, dx\, dy = \int_c^d \left[\int_{g_1(y)}^{g_2(y)} f(x, y)\, dx \right] dy$$

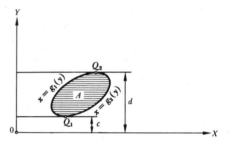

where the boundary of A consists of two continuous curves $x = g_1(y)$ and $x = g_2(y)$. The boundary of A is met by a line parallel to the X axis in at most two points, and $y = c$, $y = d$ are the extreme values of y on A.

If neither of these conditions is satisfied, the area A must be divided in two or more portions. Then

$$A = \iint\limits_A f(x, y)\, dx\, dy$$

$$= \iint\limits_{A_1} f(x, y)\, dx\, dy + \iint\limits_{A_2} f(x, y)\, dx\, dy$$

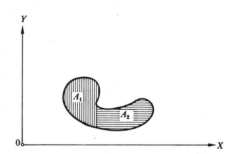

(2) Polar Coordinates

If S_1, S_2 are points on a contour enclosing area A, selected so that they identify extreme polar coordinates, then

$$A = \iint\limits_A f(r, \theta)\, r\, dr\, d\theta = \int_\alpha^\beta \left[\int_{f_1(\theta)}^{f_2(\theta)} f(r, \theta)\, r\, dr \right] d\theta$$

where the boundary of A consists of two continuous curves $r = f_1(\theta)$ and $r = f_2(\theta)$. The boundary of A is met by two tangents from 0 in at most two points, and $\theta = \alpha$, $\theta = \beta$, are the extreme values of θ on A.

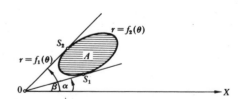

(3) Interpretation

If $f(x, y)$ has the same sign over A, the double integral may be interpreted as the *volume of a vertical cylinder* bounded below by the region A projected in the XY plane and above by the surface $z = f(x, y)$.

a, b, c = constants > 0 k, m, p, r = positive integers

$\binom{x}{r}$ = binomial coefficient (Sec. 1.04) $X_p^{(r)}$ = factorial polynomial (Sec. 1.03)

$\mathscr{S}_k^{(r)}$ = Stirling number (Sec. A.05) $\bar{B}_m(x)$ = Bernoulli polynomial (Sec. 8.05)

$\alpha = m/p$ $(\ )!$ = factorial (Sec. 1.03) $\bar{E}_m(x)$ = Euler polynomial (Sec. 8.06)

$f(x)$	$\int_0^x \int_0^x f(x)\,dx\,dx$	$f(x)$	$\int_0^x \int_0^x f(x)\,dx\,dx$
$\dfrac{1}{x}$	$x(\ln x - 1)$	$\sqrt{\dfrac{1}{x}}$	$\dfrac{4x\sqrt{x}}{3}$
x^m	$\dfrac{x^{m+2}}{(m+1)(m+2)}$	$\dfrac{1}{x^m}$	$\dfrac{x^2}{(m-1)(m-2)x^m}$
x^c	$\dfrac{x^{c+2}}{(c+1)(c+2)}$	$\dfrac{1}{x^c}$	$\dfrac{x^2}{(c-1)(c-2)x^c}$
$\sqrt[p]{x^m}$	$\dfrac{x^2\sqrt[p]{x^m}}{(\alpha+1)(\alpha+2)}$	$\sqrt[p]{\dfrac{1}{x^m}}$	$\dfrac{x^2}{(\alpha-1)(\alpha-2)}\sqrt[p]{\dfrac{1}{x^m}}$
e^{ax}	$\dfrac{1}{a^2}(e^{ax} - ax - 1)$	b^{ax}	$\dfrac{b^{ax} - ax\ln b - 1}{(a\ln a)^2}$
e^{-ax}	$\dfrac{1}{a^2}(e^{-ax} + ax - 1)$	b^{-ax}	$\dfrac{b^{-ax} + ax\ln b - 1}{(a\ln b)^2}$
$\ln ax$	$\dfrac{(2\ln ax - 3)x^2}{4}$	$\log ax$	$\dfrac{(2\ln ax - 3)x^2}{4\ln 10}$
$\sin ax$	$\dfrac{1}{a^2}(ax - \sin ax)$	$\sinh ax$	$\dfrac{1}{a^2}(\sinh ax - ax)$
$\cos ax$	$\dfrac{1}{a^2}(1 - \cos ax)$	$\cosh ax$	$\dfrac{1}{a^2}(\cosh ax - 1)$
$\binom{x}{r}$	$\dfrac{1}{r!}\sum_{k=1}^{r}\dfrac{x^{k+2}}{(k+1)(k+2)}\mathscr{S}_k^{(r)}$	$X_p^{(r)}$	$p^r\sum_{k=1}^{r}\dfrac{x^{k+2}}{(k+1)(k+2)}\mathscr{S}_k^{(r)}$
$x^m\binom{x}{r}$	$\dfrac{1}{r!}\sum_{k=1}^{r}\dfrac{x^{m+k+2}}{(k+1)(k+2)}\mathscr{S}_k^{(r)}$	$x^m X_p^{(r)}$	$p^r\sum_{k=1}^{r}\dfrac{x^{m+k+2}}{(k+1)(k+2)}\mathscr{S}_k^{(r)}$
$\bar{B}_m(x)$	$\dfrac{\bar{B}_{m+2}(x) - \bar{B}_{m+2}(0)}{(m+1)(m+2)} - \dfrac{x\bar{B}_{m+1}(0)}{m+1}$	$\bar{E}_m(x)$	$\dfrac{\bar{E}_{m+2}(x) - \bar{E}_{m+2}(0)}{(m+1)(m+2)} - \dfrac{x\bar{E}_{m+1}(0)}{m+1}$

The calculation of a triple integral is performed by successive evaluation of three definite integrals.

(1) Cartesian Coordinates

$$V = \iiint_V f(x, y, z)\, dV = \int_a^b \int_{y_1(x)}^{y_2(x)} \int_{z_1(x,y)}^{z_2(x,y)} f(x, y, z)\, dz\, dy\, dx$$

(2) Cylindrical Coordinates

$$V = \iiint_V f(r, \theta, z)\, dV = \int_\alpha^\beta \int_{r_1(\theta)}^{r_2(\theta)} \int_{z_1(r,\theta)}^{z_2(r,\theta)} f(r, \theta, z)\, r\, dz\, d\theta\, dr$$

(3) Spherical Coordinates

$$V = \iiint_V f(r, \phi, \theta)\, dV = \int_\alpha^\beta \int_{\phi_1(\theta)}^{\phi_2(\theta)} \int_{r_1(\phi,\theta)}^{r_2(\phi,\theta)} f(r, \phi, \theta)\, r^2 \sin \phi\, dr\, d\phi\, d\theta$$

(4) Interpretation

If $f(x, y, z) = 1$, the triple integral may be interpreted as the *volume enclosed by the region V*.

(5) Curvilinear Coordinates

If $x = x(u, v, w)$, $y = y(u, v, w)$, $z = z(u, v, w)$, and $f(x, y, z) = g(u, v, w)$, then

$$V = \iiint_V f(x, y, z)\, dV = \int_{u_1}^{u_2} \int_{v_1(u)}^{v_2(u)} \int_{w_1(u,v)}^{w_2(u,v)} g(u, v, w) \underbrace{\frac{\partial(x, y, z)}{\partial(u, v, w)}}_{J}\, du\, dv\, dw$$

in which

$$J = \frac{\partial(x, y, z)}{\partial(u, v, w)} = \begin{vmatrix} \dfrac{\partial x}{\partial u} & \dfrac{\partial y}{\partial u} & \dfrac{\partial z}{\partial u} \\[2mm] \dfrac{\partial x}{\partial v} & \dfrac{\partial y}{\partial v} & \dfrac{\partial z}{\partial v} \\[2mm] \dfrac{\partial x}{\partial w} & \dfrac{\partial y}{\partial w} & \dfrac{\partial z}{\partial w} \end{vmatrix}$$

In the case of a double integral,

$$J = \frac{\partial(x, y)}{\partial(u, v)} = \begin{vmatrix} \dfrac{\partial x}{\partial u} & \dfrac{\partial y}{\partial u} \\[2mm] \dfrac{\partial x}{\partial v} & \dfrac{\partial y}{\partial v} \end{vmatrix}$$

Note: The order of integration is arbitrary; thus a *double integral* can be evaluated in *two ways*, and a *triple integral* can be evaluated in *six ways*.

a, b, c = constants > 0 k, m, p, r = positive integers

$\dbinom{x}{r}$ = binomial coefficient (Sec. 1.04) $X_p^{(r)}$ = factorial polynomial (Sec. 1.03)

$\mathscr{S}_k^{(r)}$ = Stirling number (Sec. A.05) $\bar{B}_m(x)$ = Bernoulli polynomial (Sec. 8.05)

$\alpha = m/p$ $\bar{E}_m(x)$ = Euler polynomial (Sec. 8.06)

$f(x)$	$\int_0^x \int_0^x \int_0^x f(x)\,(dx)^3$	$f(x)$	$\int_0^x \int_0^x \int_0^x f(x)\,(dx)^3$
$\dfrac{1}{x}$	$\dfrac{(2 \ln ax - 3)x^2}{4}$	$\sqrt{\dfrac{1}{x}}$	$\dfrac{8x^2\sqrt{x}}{15}$
x^m	$\dfrac{x^{m+3}}{(m+1)(m+2)(m+3)}$	$\dfrac{1}{x^m}$	$-\dfrac{x^3}{(m-1)(m-2)(m-3)x^m}$
x^c	$\dfrac{x^{c+3}}{(c+1)(c+2)(c+3)}$	$\dfrac{1}{x^c}$	$-\dfrac{x^3}{(c-1)(c-2)(c-3)x^c}$
$\sqrt[p]{x^m}$	$\dfrac{x^3\sqrt[p]{x^m}}{(\alpha+1)(\alpha+2)(\alpha+3)}$	$\sqrt[p]{\dfrac{1}{x^m}}$	$-\dfrac{x^3}{(\alpha-1)(\alpha-2)(\alpha-3)}\sqrt[p]{\dfrac{1}{x^m}}$
e^{ax}	$\dfrac{1}{a^3}\left[e^{ax} - \dfrac{1}{2}(ax)^2 - ax - 1\right]$	b^{ax}	$\dfrac{b^{ax} - \frac{1}{2}(ax \ln b)^2 - ax \ln b - 1}{(a \ln b)^3}$
e^{-ax}	$-\dfrac{1}{a^3}\left[e^{-ax} - \dfrac{1}{2}(ax)^2 + ax - 1\right]$	b^{-ax}	$-\dfrac{b^{-ax} - \frac{1}{2}(ax \ln b)^2 + ax \ln b - 1}{(a \ln b)^3}$
$\ln ax$	$\dfrac{(\ln ax - 11)x^3}{6}$	$\log ax$	$\dfrac{(\ln ax - 11)x^3}{6 \ln 10}$
$\sin ax$	$\dfrac{1}{a^3}\left[\cos ax + \dfrac{1}{2}(ax)^2 - 1\right]$	$\sinh ax$	$\dfrac{1}{a^3}\left[\cosh ax - \dfrac{1}{2}(ax)^2 - 1\right]$
$\cos ax$	$\dfrac{1}{a^3}(ax - \sin ax)$	$\cosh ax$	$\dfrac{1}{a^3}(\sinh ax - ax)$
$\dbinom{x}{r}$	$\dfrac{1}{r!}\displaystyle\sum_{k=1}^{r}\dfrac{x^{k+3}}{(k+1)(k+2)(k+3)}\mathscr{S}_k^{(r)}$	$X_p^{(r)}$	$p^r\displaystyle\sum_{k=1}^{r}\dfrac{x^{k+3}}{(k+1)(k+2)(k+3)}\mathscr{S}_k^{(r)}$
$x^m\dbinom{x}{r}$	$\dfrac{1}{r!}\displaystyle\sum_{k=1}^{r}\dfrac{x^{m+k+3}}{(k+1)(k+2)(k+3)}\mathscr{S}_k^{(r)}$	$x^m X_p^{(r)}$	$p^r\displaystyle\sum_{k=1}^{r}\dfrac{x^{m+k+3}}{(k+1)(k+2)(k+3)}\mathscr{S}_k^{(r)}$
$\bar{B}_m(x)$	$\dfrac{\bar{B}_{m+3}(x) - \bar{B}_{m+3}(0)}{(m+1)(m+2)(m+3)}$ $-\displaystyle\sum_{k=1}^{2}\dfrac{m!\,x^{3-k}\bar{B}_{m+k}(0)}{(m+k)!(3-k)!}$	$\bar{E}_m(x)$	$\dfrac{\bar{E}_{m+3}(x) - \bar{E}_{m+3}(0)}{(m+1)(m+2)(m+3)}$ $-\displaystyle\sum_{k=1}^{2}\dfrac{m!\,x^{3-k}\bar{E}_{m+k}(0)}{(m+k)!(3-k)!}$

a, b, c = constants > 0 $\qquad$ k, m, p = positive integers $\qquad$ n = order of integration

( )! = factorial (Sec. 1.03) $\qquad\qquad$ $\Gamma(\ )$ = gamma function (Sec. 13.03)

$$F_{k-1} = \frac{(bx)^{k-1}}{(k-1)!} \qquad\qquad F_{2k} = \frac{(bx)^{2k}}{(2k)!} \qquad\qquad F_{2k-1} = \frac{(bx)^{2k-1}}{(2k-1)!}$$

$f(x)$	$\int_0^x \int_0^x \cdots \int_0^x f(x)\,(dx)^n$	
x^m	$\dfrac{m!}{(m+n)!} x^{m+n}$	
x^c	$\dfrac{\Gamma(c+1)}{\Gamma(c+n+1)} x^{c+n}$	
$\sqrt[p]{x^m}$	$\dfrac{x^n}{(\alpha+1)(\alpha+2)(\alpha+3)\cdots(\alpha+n)} \sqrt[p]{x^m}$	$\left(\alpha = \dfrac{m}{p}\right)$
e^{ax}	$\dfrac{1}{a^n}\left[e^{ax} - \displaystyle\sum_{k=0}^{n-1} \dfrac{(ax)^k}{k!}\right]$	
e^{-ax}	$\dfrac{(-1)^n}{a^n}\left[e^{-ax} - \displaystyle\sum_{k=0}^{n-1} \dfrac{(-ax)^k}{k!}\right]$	
$\sin bx$	$\dfrac{\sin(bx-n\phi)}{b^n} + x^n \displaystyle\sum_{k=1}^{n} \dfrac{\sin k\phi}{(n-k)!(bx)^k}$	$\phi = \tfrac{1}{2}\pi$
$\cos bx$	$\dfrac{\cos(bx-n\phi)}{b^n} - x^n \displaystyle\sum_{k=1}^{n} \dfrac{\cos k\phi}{(n-k)!(bx)^k}$	
$\sinh bx$	$\begin{cases} \dfrac{\sinh bx}{b^n} - \dfrac{1}{b^n}\displaystyle\sum_{k=1}^{n/2} F_{2k-1} = G_1 & (n \text{ even}) \\[3ex] \dfrac{\cosh bx}{b^n} - \dfrac{1}{b^n}\displaystyle\sum_{k=0}^{(n-1)/2} F_{2k} = G_2 & (n \text{ odd}) \end{cases}$	G_1, G_2, G_3, G_4 = equivalents used on the opposite page
$\cosh bx$	$\begin{cases} \dfrac{\cosh bx}{b^n} - \dfrac{1}{b^n}\displaystyle\sum_{k=1}^{n/2} F_{2k} = G_3 & (n \text{ even}) \\[3ex] \dfrac{\sinh bx}{b^n} - \dfrac{1}{b^n}\displaystyle\sum_{k=0}^{(n-1)/2} F_{2k-1} = G_4 & (n \text{ odd}) \end{cases}$	
$\sin^2 bx$	$-\dfrac{\sin[2bx-(n-1)\phi]}{4b^n} + \dfrac{\frac{1}{2}x^n}{n!} - \dfrac{1}{4b^n}\displaystyle\sum_{k=1}^{n} \dfrac{F_{k-1}}{k}\sin[(k-1)\phi]$	$\phi = \tfrac{1}{2}\pi$
$\cos^2 bx$	$\dfrac{\sin[2bx-(n-1)\phi]}{4b^n} + \dfrac{\frac{1}{2}x^n}{n!} + \dfrac{1}{4b^n}\displaystyle\sum_{k=1}^{n} \dfrac{F_{k-1}}{k}\sin[(k-1)\phi]$	

For definition of $a, b, c, k, m, n, p, (\ )!, \Gamma(\ )$, see opposite page.

$$M_n = \frac{p^n}{(m + p)(m + 2p)(m + 3p) \cdots (m + np)}$$

$$M_k = \frac{p^k}{(m + p)(m + 2p)(m + 3p) \cdots (m + kp)} \qquad\qquad N_k = \frac{(\pm bx)^{k-1}}{k!} \sin\left[2a - \tfrac{1}{2}(n - 1)\pi\right]$$

$f(x)$	$\displaystyle\int_0^x \int_0^x \cdots \int_0^x f(x)\,(dx)^n$	
$(a \pm bx^p)^m$	$\displaystyle a^m x^n \sum_{k=0}^{n} \frac{(pk)!}{(n + pk)!} \binom{m}{k} \left(\frac{\pm bx^p}{a}\right)^k$	
$(a \pm bx^c)^m$	$\displaystyle a^m x^n \sum_{k=0}^{n} \frac{\Gamma(ck + 1)}{\Gamma(n + ck + 1)} \binom{m}{k} \left(\frac{\pm bx^c}{a}\right)^k$	
$\sqrt[p]{(a \pm bx)^m}$	$\displaystyle M_n \sqrt[p]{(a \pm bx)^m} - x^n \sqrt[p]{a^m} \sum_{k=1}^{n} \frac{M_k}{(n - k)!} \left(\frac{a}{x}\right)^k$	
b^{ax}	$\displaystyle \frac{1}{(a \ln b)^n}\left[b^{ax} - \sum_{k=0}^{n-1} \frac{(ax \ln b)^k}{k!} \right]$	
b^{-ax}	$\displaystyle \frac{1}{(-a \ln b)^n}\left[b^{-ax} - \sum_{k=0}^{n-1} \frac{(-ax \ln b)^k}{k!} \right]$	
$\sin(a \pm bx)$	$\displaystyle \frac{\sin(a \pm bx - n\phi)}{(\pm b)^n} - x^n \sum_{k=1}^{n} \frac{\sin(a - k\phi)}{(n - k)!(\pm bx)^k}$	$\phi = \tfrac{1}{2}\pi$
$\cos(a \pm bx)$	$\displaystyle \frac{\cos(a \pm bx - n\phi)}{(\pm b)^n} - x^n \sum_{k=1}^{n} \frac{\cos(a - k\phi)}{(n - k)!(\pm bx)^k}$	
$\sinh(a \pm bx)$	$\begin{cases} G_3 \sinh a \pm G_1 \cosh a & (n \text{ even}) \\[2em] G_4 \sinh a \pm G_2 \cosh a & (n \text{ odd}) \end{cases}$	$G_1, G_2, G_3, G_4 =$ equivalents defined on the opposite page
$\cosh(a \pm bx)$	$\begin{cases} G_3 \cosh a \pm G_1 \sinh a & (n \text{ even}) \\[2em] G_4 \cosh a \pm G_2 \sinh a & (n \text{ odd}) \end{cases}$	
$\sin^2(a \pm bx)$	$\displaystyle -\frac{\sin[2(a \pm bx) - (n - 1)\phi]}{4(\pm b)^n} - \frac{1}{4(\pm b)^n} \sum_{k=1}^{n} N_k + \frac{\tfrac{1}{2}x^n}{n!}$	$\phi = \tfrac{1}{2}\pi$
$\cos^2(a \pm bx)$	$\displaystyle \frac{\sin[2(a \pm bx) - (n - 1)\phi]}{4(\pm b)^n} + \frac{1}{4(\pm b)^n} \sum_{k=1}^{n} N_k + \frac{\tfrac{1}{2}x^n}{n!}$	

a, b = constants > 0 n = order of integral

k, m, p, r = positive integers $X_p^{(r)}$ = factorial polynomial (Sec. 1.03)

$\binom{x}{r}$ = binomial coefficients (Sec. 1.04) $\bar{B}_m(x)$ = Bernoulli polynomial (Sec. 8.05)

$\mathscr{S}_k^{(r)}$ = Stirling number (Sec. A.05) $\bar{E}_m(x)$ = Euler polynomial (Sec. 8.06)

$M_k = \dfrac{\sin\left(\frac{1}{2}k\pi\right)}{(n-k)!x^k}\left[\dfrac{1}{(a+b)^k} - \dfrac{1}{(a-b)^k}\right]$ $N_k = \dfrac{\cos\left(\frac{1}{2}k\pi\right)}{(n-k)!x^k}\left[\dfrac{1}{(a+b)^k} - \dfrac{1}{(a-b)^k}\right]$

$f(x)$	$\int_0^x \int_0^x \int_0^x \cdots \int_0^x f(x)\,(dx)^n$	
$\binom{x}{r}$	$\dfrac{1}{r!}\sum\limits_{k=1}^{r} \dfrac{x^{k+n}}{(k+1)(k+2)(k+3)\cdots(k+n)}\mathscr{S}_k^{(r)}$	
$X_p^{(r)}$	$p^r\sum\limits_{k=1}^{r} \dfrac{x^{k+n}}{(k+1)(k+2)(k+3)\cdots(k+n)}\mathscr{S}_k^{(r)}$	
$x^m X_p^{(r)}$	$p^r\sum\limits_{k=1}^{r} \dfrac{x^{k+m+n}}{(k+1)(k+2)(k+3)\cdots(k+n)}\mathscr{S}_k^{(r)}$	
$\bar{B}_m(x)$	$\dfrac{\bar{B}_{m+n}(x) - \bar{B}_{m+n}(0)}{(m+1)(m+2)(m+3)\cdots(m+n)} - \sum\limits_{k=1}^{n-1}\dfrac{m!x^{n-k}\bar{B}_{m+k}(0)}{(m+k)!(n-k)!}$	
$\bar{E}_m(x)$	$\dfrac{\bar{E}_{m+n}(x) - \bar{E}_{m+n}(0)}{(m+1)(m+2)(m+3)\cdots(m+n)} - \sum\limits_{k=1}^{n-1}\dfrac{m!x^{n-k}\bar{E}_{m+k}(0)}{(m+k)!(n-k)!}$	
$e^{ax}\sin bx$	$\dfrac{e^{ax}}{R^n}\sin(bx-n\phi) + x^n\sum\limits_{k=1}^{n}\dfrac{\sin k\phi}{(n-k)!(Rx)^k}$	$\phi = \tan^{-1}\dfrac{b}{a}$
$e^{-ax}\sin bx$	$(-1)^n\dfrac{e^{-ax}}{R^n}\sin(bx+n\phi) - x^n\sum\limits_{k=1}^{n}\dfrac{\sin k\phi}{(n-k)!(-Rx)^k}$	
$e^{ax}\cos bx$	$\dfrac{e^{ax}}{R^n}\cos(bx-n\phi) - x^n\sum\limits_{k=1}^{n}\dfrac{\cos k\phi}{(n-k)!(Rx)^k}$	$R = \sqrt{a^2+b^2}$
$e^{-ax}\cos bx$	$(-1)^n\dfrac{e^{-ax}}{R^n}\cos(bx+n\phi) - x^n\sum\limits_{k=1}^{n}\dfrac{\cos k\phi}{(n-k)!(-Rx)^k}$	
$\sin ax \sin bx$	$-\dfrac{\cos\left[(a+b)x - \frac{1}{2}n\pi\right]}{2(a+b)^n} + \dfrac{\cos\left[(a-b)x - \frac{1}{2}n\pi\right]}{2(a-b)^n} + \dfrac{x^n}{2}\sum\limits_{k=1}^{n} N_k$	
$\sin ax \cos bx$	$\dfrac{\sin\left[(a+b)x - \frac{1}{2}n\pi\right]}{2(a+b)^n} + \dfrac{\sin\left[(a-b)x - \frac{1}{2}n\pi\right]}{2(a-b)^n} + \dfrac{x^n}{2}\sum\limits_{k=1}^{n} M_k$	
$\cos ax \cos bx$	$\dfrac{\cos\left[(a+b)x - \frac{1}{2}n\pi\right]}{2(a+b)^n} + \dfrac{\cos\left[(a-b)x - \frac{1}{2}n\pi\right]}{2(a-b)^n} - \dfrac{x^n}{2}\sum\limits_{k=1}^{n} N_k$	

10
VECTOR ANALYSIS

(1) Definitions

(a) Scalar is a quantity defined by magnitude (signed number) only and designated by ordinary letters such as $a, b, c, \ldots, r, \ldots, A, B, C, \ldots, R, \ldots, \alpha, \beta, \gamma, \ldots$. Examples of scalars are length, time, temperature, and mass.

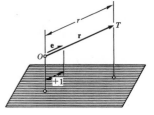

(b) Vector is a quantity defined by magnitude (scalar) and direction (line of action and sense) and designated by boldface letters such as $\mathbf{a}, \mathbf{b}, \mathbf{c}, \ldots, \mathbf{r}, \ldots, \mathbf{A}, \mathbf{B}, \mathbf{C}, \ldots, \mathbf{R}, \ldots$. Examples of vectors are force, moment, displacement, velocity, and acceleration.

(c) Graphical representation of a vector $\mathbf{r}$ is a directed segment given by its initial point O (origin of vector) and its end point T (terminus of vector). The magnitude r of the vector $\mathbf{r}$ is the length of this segment.

$$r = |\mathbf{r}| = \overline{OT}$$

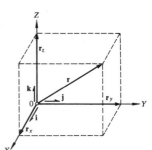

(d) Unit vector $\mathbf{e}$ is a vector of unit magnitude. Any vector $\mathbf{r}$ can be represented analytically as the product of its magnitude r and its unit vector $\mathbf{e}$.

$$\mathbf{r} = r\mathbf{e} \qquad \text{or} \qquad \mathbf{e} = \frac{\mathbf{r}}{r}$$

(2) Components and Magnitudes

(a) Resolution A vector $\mathbf{r}$ may be resolved into any number of components. In the right-handed cartesian coordinate system, $\mathbf{r}$ is resolved into three mutually perpendicular components, each parallel to the respective coordinate axis.

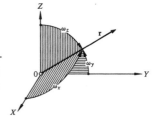

$$\boxed{\mathbf{r} = \mathbf{r}_x + \mathbf{r}_y + \mathbf{r}_z = r_x\mathbf{i} + r_y\mathbf{j} + r_z\mathbf{k} = r\mathbf{e}}$$

where $\mathbf{r}_x, \mathbf{r}_y, \mathbf{r}_z$ are the vector components, r_x, r_y, r_z are their magnitudes, $\mathbf{i}, \mathbf{j}, \mathbf{k}$ are the unit vectors in the X, Y, Z axes, respectively, r is the magnitude of $\mathbf{r}$ and $\mathbf{e}$ is its unit vector.

(b) Unit vectors $\mathbf{i}, \mathbf{j}, \mathbf{k}$, and $\mathbf{e}$ are inversely defined as

$$\boxed{\mathbf{i} = \frac{\mathbf{r}_x}{r_x} \qquad \mathbf{j} = \frac{\mathbf{r}_y}{r_y} \qquad \mathbf{k} = \frac{\mathbf{r}_z}{r_x} \qquad \mathbf{e} = \frac{r_x}{r}\mathbf{i} + \frac{r_y}{r}\mathbf{j} + \frac{r_z}{k}\mathbf{k}}$$

(c) Magnitude r is given by the magnitude of components r_x, r_y, r_z as

$$r = |\mathbf{r}| = \sqrt{r_x^2 + r_y^2 + r_z^2} = r_x\frac{\mathbf{i}}{\mathbf{e}} + r_y\frac{\mathbf{j}}{\mathbf{e}} + r_z\frac{\mathbf{k}}{\mathbf{e}} = \alpha r_x + \beta r_y + \gamma r_z$$

(d) Direction cosines α, β, γ of $\mathbf{r}$ defined as the cosines of the angles $\omega_x, \omega_y, \omega_z$ measured to $\mathbf{r}$ from the positive X, Y, Z axes, respectively, are

$$\boxed{\cos \omega_x = \frac{r_x}{r} = \frac{\mathbf{i}}{\mathbf{e}} = \alpha \qquad \cos \omega_y = \frac{r_y}{r} = \frac{\mathbf{j}}{\mathbf{e}} = \beta \qquad \cos \omega_x = \frac{r_x}{r} = \frac{\mathbf{k}}{\mathbf{e}} = \gamma}$$

and their relations are

$$\alpha^2 + \beta^2 + \gamma^2 = 1 \qquad \text{and} \qquad \mathbf{e} = \alpha\mathbf{i} + \beta\mathbf{j} + \gamma\mathbf{k}$$

(1) Vector Addition and Subtraction

(a) Sum of vectors a *and* **b** is a vector **c** formed by placing the initial point of **b** on the terminal point of **a** and joining the initial point of **a** to the terminal point of **b**.

$$a + b = c$$

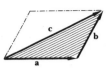

(b) Difference of vectors a *and* **b** is a vector **d** formed by placing the initial point of **b** on the initial point of **a** and joining the terminal point of **b** with the terminal point of **a**.

$$a - b = d$$

(c) Difference of vectors b *and* **a** is a vector **e** formed by placing the initial point of **b** on the initial point of **a** and joining the terminal point of **a** with terminal point of **b**.

$$b - a = e$$

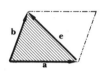

(2) Scalar – Vector Laws   (**a, b** = vectors; m, n = scalars)

$m\mathbf{a} = \mathbf{a}m$	Commutative law	$(m + n)\mathbf{a} = m\mathbf{a} + n\mathbf{a}$	Distributive law
$m(n\mathbf{a}) = (mn)\mathbf{a}$	Associative law	$m(\mathbf{a} + \mathbf{b}) = m\mathbf{a} + m\mathbf{b}$	Distributive law

(3) Vector Summation Laws

(a) Commutative law

$$\mathbf{a} + \mathbf{b} + \mathbf{c} = \mathbf{b} + \mathbf{c} + \mathbf{a} = \mathbf{c} + \mathbf{a} + \mathbf{b} = \mathbf{f}$$

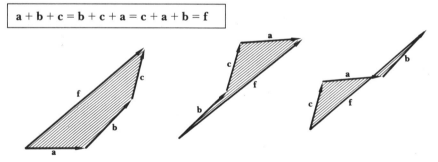

(b) Associative law

$$(\mathbf{a} + \mathbf{b}) + \mathbf{c} = \mathbf{a} + (\mathbf{b} + \mathbf{c}) = \mathbf{f}$$

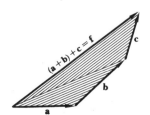

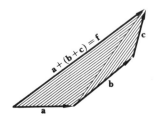

(1) Scalar Product

(a) Dot product (sclara product, inner product) of two vectors **a** and **b** denoted as **a · b** or (**ab**) is defined as the product of their magnitudes a, b and the cosine of the angle ω between them. *The result is a scalar.*

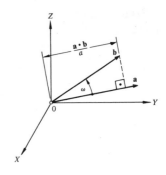

$$\mathbf{a} \cdot \mathbf{b} = ab \cos \omega = a_x b_x + a_y b_y + a_z b_z$$

$$\omega = \cos^{-1} \frac{\mathbf{a} \cdot \mathbf{b}}{ab} \qquad 0 < \omega < \pi$$

(b) Product laws (m, n = scalars)

Commutative: $\mathbf{a} \cdot \mathbf{b} = \mathbf{b} \cdot \mathbf{a}$
Associative: $m\mathbf{a} \cdot n\mathbf{b} = mn\mathbf{a} \cdot \mathbf{b}$
Distributive: $\mathbf{a} \cdot (\mathbf{b} + \mathbf{c}) = \mathbf{a} \cdot \mathbf{b} + \mathbf{a} \cdot \mathbf{c}$

(c) Special cases ($\mathbf{a} \neq 0, \mathbf{b} \neq 0$)

Two vectors **a** and **b** are *normal* ($\omega = 90°$) if

$\mathbf{a} \cdot \mathbf{b} = 0$

From this,

$\mathbf{i} \cdot \mathbf{j} = \mathbf{j} \cdot \mathbf{k} = \mathbf{k} \cdot \mathbf{i} = 0$

Two vectors **a** and **b** are *parallel* ($\omega = 0°$) if

$\mathbf{a} \cdot \mathbf{b} = ab$

From this,

$\mathbf{i} \cdot \mathbf{i} = \mathbf{j} \cdot \mathbf{j} = \mathbf{k} \cdot \mathbf{k} = 1$

(2) Vector Product

(a) Cross product (vector product) of two vectors **a** and **b** denoted as **a × b** or [**ab**] is defined as the product of their magnitudes, the sine of the angle ω between them, and the unit vector **n** normal to their plane. *The result is a vector.*

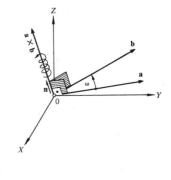

$$\mathbf{a} \times \mathbf{b} = ab \sin \omega \, \mathbf{n} = \begin{vmatrix} \mathbf{i} & \mathbf{j} & \mathbf{k} \\ a_x & a_y & a_z \\ b_x & b_y & b_z \end{vmatrix} = \begin{bmatrix} 0 & -a_z & a_y \\ a_x & 0 & -a_x \\ -a_y & a_x & 0 \end{bmatrix} \begin{bmatrix} b_x \\ b_y \\ b_z \end{bmatrix}$$

$$\omega = \sin^{-1} \frac{|\mathbf{a} \times \mathbf{b}|}{ab} \qquad 0 < \omega < \pi$$

(b) Product laws (m, n = scalars)

Noncommutative: $\mathbf{a} \times \mathbf{b} = -\mathbf{b} \times \mathbf{a}$
Associative: $(m\mathbf{a}) \times (n\mathbf{b}) = mn(\mathbf{a} \times \mathbf{b})$
Distributive: $\mathbf{a} \times (\mathbf{b} + \mathbf{c}) = \mathbf{a} \times \mathbf{b} + \mathbf{a} \times \mathbf{c}$

(c) Special cases ($\mathbf{a} \neq 0, \mathbf{b} \neq 0$)

Two vectors **a** and **b** are *normal* ($\omega = 90°$) if

$\mathbf{a} \times \mathbf{b} = ab\mathbf{n}$

From this

$\mathbf{i} \times \mathbf{j} = \mathbf{k} \qquad \mathbf{j} \times \mathbf{k} = \mathbf{i} \qquad \mathbf{k} \times \mathbf{i} = \mathbf{j}$

Two vectors **a** and **b** are *parallel* ($\omega = 0°$) if

$\mathbf{a} \times \mathbf{b} = 0$

From this,

$\mathbf{i} \times \mathbf{i} = \mathbf{j} \times \mathbf{j} = \mathbf{k} \times \mathbf{k} = 0$

(1) Triple Products

(a) Scalar triple product (result a scalar)

$$\mathbf{a} \cdot (\mathbf{b} \times \mathbf{c}) = \mathbf{b} \cdot (\mathbf{c} \times \mathbf{a}) = \mathbf{c} \cdot (\mathbf{a} \times \mathbf{b}) = (\mathbf{a} \times \mathbf{b}) \cdot \mathbf{c} = (\mathbf{b} \times \mathbf{c}) \cdot \mathbf{a} = (\mathbf{c} \times \mathbf{a}) \cdot \mathbf{b} = [\mathbf{abc}]$$

$$[\mathbf{abc}] = \begin{vmatrix} a_x & a_y & a_z \\ b_x & b_y & b_z \\ c_x & c_y & c_z \end{vmatrix} = \begin{vmatrix} a_x & b_x & c_x \\ a_y & b_y & c_y \\ a_z & b_z & c_z \end{vmatrix} = [a_x \ a_y \ a_z] \begin{bmatrix} 0 & -b_z & b_y \\ b_z & 0 & -b_x \\ -b_y & b_x & 0 \end{bmatrix} \begin{bmatrix} c_x \\ c_y \\ c_z \end{bmatrix}$$

(b) Vector triple product (result a vector)

$$\mathbf{a} \times (\mathbf{b} \times \mathbf{c}) = (\mathbf{a} \cdot \mathbf{c})\mathbf{b} - (\mathbf{a} \cdot \mathbf{b})\mathbf{c} = \begin{vmatrix} \mathbf{b} & \mathbf{c} \\ \mathbf{a} \cdot \mathbf{b} & \mathbf{a} \cdot \mathbf{c} \end{vmatrix} = \begin{bmatrix} 0 & -a_z & a_y \\ a_x & 0 & -a_x \\ -a_y & a_x & 0 \end{bmatrix} \begin{bmatrix} 0 & -b_z & b_y \\ b_z & 0 & -b_x \\ -b_y & b_x & 0 \end{bmatrix} \begin{bmatrix} c_x \\ c_y \\ c_z \end{bmatrix}$$

$$(\mathbf{a} \times \mathbf{b}) \times \mathbf{c} = (\mathbf{a} \cdot \mathbf{c})\mathbf{b} - (\mathbf{b} \cdot \mathbf{c})\mathbf{a} = \begin{vmatrix} \mathbf{b} & \mathbf{a} \\ \mathbf{c} \cdot \mathbf{b} & \mathbf{c} \cdot \mathbf{a} \end{vmatrix} = \begin{bmatrix} 0 & -c_z & c_y \\ c_x & 0 & -c_x \\ -c_y & c_x & 0 \end{bmatrix} \begin{bmatrix} 0 & -b_z & b_y \\ b_z & 0 & -b_x \\ -b_y & b_x & 0 \end{bmatrix} \begin{bmatrix} a_x \\ a_y \\ a_z \end{bmatrix}$$

(2) Special Products

(a) Double scalar product (result a scalar)

$$(\mathbf{a} \times \mathbf{b}) \cdot (\mathbf{c} \times \mathbf{d}) = \begin{vmatrix} \mathbf{a} \cdot \mathbf{c} & \mathbf{b} \cdot \mathbf{c} \\ \mathbf{a} \cdot \mathbf{d} & \mathbf{b} \cdot \mathbf{d} \end{vmatrix} = [b_x \ b_y \ b_z] \begin{bmatrix} 0 & a_z & -a_y \\ -a_z & 0 & a_x \\ a_y & -a_y & 0 \end{bmatrix} \begin{bmatrix} 0 & -c_z & c_y \\ c_z & 0 & -c_x \\ -c_y & c_x & 0 \end{bmatrix} \begin{bmatrix} d_x \\ d_y \\ d_z \end{bmatrix}$$

$$(\mathbf{a} \times \mathbf{b})^2 = \begin{vmatrix} \mathbf{a}^2 & \mathbf{a} \cdot \mathbf{b} \\ \mathbf{a} \cdot \mathbf{b} & \mathbf{b}^2 \end{vmatrix} = [b_x \ b_y \ b_z] \begin{bmatrix} 0 & a_z & -a_y \\ -a_z & 0 & a_x \\ a_y & -a_x & 0 \end{bmatrix} \begin{bmatrix} 0 & -a_z & a_y \\ a_z & 0 & -a_x \\ -a_y & a_x & 0 \end{bmatrix} \begin{bmatrix} b_x \\ b_y \\ b_z \end{bmatrix}$$

(b) Double vector product (result a vector)

$$(\mathbf{a} \times \mathbf{b}) \times (\mathbf{c} \times \mathbf{d}) = \begin{vmatrix} \mathbf{i} & \mathbf{j} & \mathbf{k} \\ \begin{vmatrix} a_y & a_z \\ b_y & b_z \end{vmatrix} & \begin{vmatrix} a_z & a_x \\ b_z & b_x \end{vmatrix} & \begin{vmatrix} a_x & a_y \\ b_x & b_y \end{vmatrix} \\ \begin{vmatrix} c_y & c_z \\ d_y & d_z \end{vmatrix} & \begin{vmatrix} c_z & c_x \\ d_z & d_x \end{vmatrix} & \begin{vmatrix} c_x & c_y \\ d_x & d_y \end{vmatrix} \end{vmatrix} = -\mathbf{a}[\mathbf{bcd}] + \mathbf{b}[\mathbf{cda}] = \mathbf{c}[\mathbf{dab}] - \mathbf{d}[\mathbf{abc}]$$

(c) Double scalar triple product (result a scalar)

$$[\mathbf{a} \ \mathbf{b} \ \mathbf{c}][\mathbf{d} \ \mathbf{e} \ \mathbf{f}] = \begin{vmatrix} \mathbf{a} \cdot \mathbf{d} & \mathbf{a} \cdot \mathbf{e} & \mathbf{a} \cdot \mathbf{f} \\ \mathbf{b} \cdot \mathbf{d} & \mathbf{b} \cdot \mathbf{e} & \mathbf{b} \cdot \mathbf{f} \\ \mathbf{c} \cdot \mathbf{d} & \mathbf{c} \cdot \mathbf{e} & \mathbf{c} \cdot \mathbf{f} \end{vmatrix} = \begin{vmatrix} \mathbf{a} \\ \mathbf{b} \\ \mathbf{c} \end{vmatrix} [\mathbf{d} \ \mathbf{e} \ \mathbf{f}]$$

(d) Products of orthogonal vectors

If $\mathbf{a} \times (\mathbf{b} \times \mathbf{c}) = \mathbf{b} \times (\mathbf{c} \times \mathbf{a}) = \mathbf{c} \times (\mathbf{a} \times \mathbf{b})$ then $\mathbf{a}, \mathbf{b}, \mathbf{c}$ are *orthogonal*.
If $\mathbf{a} \times \mathbf{b} = \mathbf{c}$ $\mathbf{b} \times \mathbf{c} = \mathbf{a}$ $\mathbf{c} \times \mathbf{a} = \mathbf{b}$ then $\mathbf{a}, \mathbf{b}, \mathbf{c}$ are *orthogonal unit vectors*, and $\mathbf{a} \times (\mathbf{b} \times \mathbf{c}) = \mathbf{b} \times (\mathbf{c} \times \mathbf{a}) = \mathbf{c} \times (\mathbf{a} \times \mathbf{b}) = 0$

(1) Ordinary Derivatives

(a) Definitions of limit, continuity, and vector function are formally identical with the definitions of scalar functions.

If $\mathbf{r} = \mathbf{r}(t) = x(t)\mathbf{i} + y(t)\mathbf{j} + z(t)\mathbf{k}$

$\qquad$ = vector function of scalar variable t

then

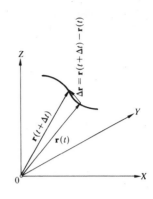

$$\frac{d\mathbf{r}}{dt} = \dot{\mathbf{r}} = \frac{dx(t)}{dt}\mathbf{i} + \frac{dy(t)}{dt}\mathbf{j} + \frac{dz(t)}{dt}\mathbf{k} = \dot{x}\mathbf{i} + \dot{y}\mathbf{j} + \dot{z}\mathbf{k}$$

$$\frac{d\dot{\mathbf{r}}}{dt} = \ddot{\mathbf{r}} = \frac{d\dot{x}(t)}{dt}\mathbf{i} + \frac{d\dot{y}(t)}{dt}\mathbf{j} + \frac{d\dot{z}(t)}{dt}\mathbf{k} = \ddot{x}\mathbf{i} + \ddot{y}\mathbf{j} + \ddot{z}\mathbf{k}$$

(b) Basic formulas $\qquad [\alpha = \alpha(t) = \text{scalar function of } t]$

$$\frac{d}{dt}(\mathbf{r}_1 + \mathbf{r}_2) = \frac{d\mathbf{r}_1}{dt} + \frac{d\mathbf{r}_2}{dt} \qquad\qquad \frac{d}{dt}(\mathbf{r}_1 \cdot \mathbf{r}_2) = \mathbf{r}_1 \cdot \frac{d\mathbf{r}_2}{dt} + \frac{d\mathbf{r}_1}{dt} \cdot \mathbf{r}_2$$

$$\frac{d}{dt}(\alpha\mathbf{r}) = \alpha\frac{d\mathbf{r}}{dt} + \frac{d\alpha}{dt}\mathbf{r} \qquad\qquad \frac{d}{dt}(\mathbf{r}_1 \times \mathbf{r}_2) = \mathbf{r}_1 \times \frac{d\mathbf{r}_2}{dt} + \frac{d\mathbf{r}_1}{dt} \times \mathbf{r}_2$$

$$\frac{d}{dt}(\mathbf{r}_1 \cdot \mathbf{r}_2 \times \mathbf{r}_3) = \mathbf{r}_1 \cdot \mathbf{r}_2 \times \frac{d\mathbf{r}_3}{dt} + \mathbf{r}_1 \cdot \frac{d\mathbf{r}_2}{dt} \times \mathbf{r}_3 + \frac{d\mathbf{r}_1}{dt} \cdot \mathbf{r}_2 \times \mathbf{r}_3$$

$$\frac{d}{dt}[\mathbf{r}_1 \times (\mathbf{r}_2 \times \mathbf{r}_3)] = \mathbf{r}_1 \times \left(\mathbf{r}_2 \times \frac{d\mathbf{r}_3}{dt}\right) + \mathbf{r}_1 \times \left(\frac{d\mathbf{r}_2}{dt} \times \mathbf{r}_3\right) + \frac{d\mathbf{r}_1}{dt} \times (\mathbf{r}_2 \times \mathbf{r}_3)$$

(2) Partial Derivatives

If $\mathbf{r}(x, y, z)$ is a *vector function of several scalar variables* x, y, z, then partial derivatives are obtained by differentiating the magnitude of each component as a scalar function.

$$\frac{\partial\mathbf{r}}{\partial x} = \frac{\partial\mathbf{r}(x, y, z)}{\partial x} \qquad \frac{\partial\mathbf{r}}{\partial y} = \frac{\partial\mathbf{r}(x, y, z)}{\partial y} \qquad \frac{\partial\mathbf{r}}{\partial z} = \frac{\partial\mathbf{r}(x, y, z)}{\partial z}$$

$$\frac{\partial^2\mathbf{r}}{\partial x^2} = \frac{\partial}{\partial x}\left(\frac{\partial\mathbf{r}}{\partial x}\right) \qquad \frac{\partial^2\mathbf{r}}{\partial y^2} = \frac{\partial}{\partial y}\left(\frac{\partial\mathbf{r}}{\partial y}\right) \qquad \frac{\partial^2\mathbf{r}}{\partial z^2} = \frac{\partial}{\partial z}\left(\frac{\partial\mathbf{r}}{\partial z}\right)$$

$$\frac{\partial^2\mathbf{r}}{\partial x\partial y} = \frac{\partial}{\partial x}\left(\frac{\partial\mathbf{r}}{\partial y}\right) \qquad \frac{\partial^2\mathbf{r}}{\partial y\partial z} = \frac{\partial}{\partial y}\left(\frac{\partial\mathbf{r}}{\partial z}\right) \qquad \frac{\partial^2\mathbf{r}}{\partial z\partial x} = \frac{\partial}{\partial z}\left(\frac{\partial\mathbf{r}}{\partial x}\right)$$

$$\frac{\partial(\mathbf{r}_1 \cdot \mathbf{r}_2)}{\partial x} = \mathbf{r}_1 \cdot \frac{\partial\mathbf{r}_2}{\partial x} + \frac{\partial\mathbf{r}_1}{\partial x} \cdot \mathbf{r}_2$$

$$\frac{\partial^2(\mathbf{r}_1 \cdot \mathbf{r}_2)}{\partial x\partial y} = \mathbf{r}_1 \cdot \frac{\partial^2\mathbf{r}_2}{\partial x\partial y} + \frac{\partial\mathbf{r}_1}{\partial x} \cdot \frac{\partial\mathbf{r}_2}{\partial y} + \frac{\partial\mathbf{r}_1}{\partial y} \cdot \frac{\partial\mathbf{r}_2}{\partial x} + \frac{\partial^2\mathbf{r}_1}{\partial x\partial y} \cdot \mathbf{r}_2$$

$$\frac{\partial(\mathbf{r}_1 \times \mathbf{r}_2)}{\partial x} = \mathbf{r}_1 \times \frac{\partial\mathbf{r}_2}{\partial x} + \frac{\partial\mathbf{r}_1}{\partial x} \times \mathbf{r}_2$$

$$\frac{\partial^2(\mathbf{r}_1 \times \mathbf{r}_2)}{\partial x\partial y} = \mathbf{r}_1 \times \frac{\partial^2\mathbf{r}_2}{\partial x\partial y} + \frac{\partial\mathbf{r}_1}{\partial x} \times \frac{\partial\mathbf{r}_2}{\partial y} + \frac{\partial\mathbf{r}_1}{\partial y} \times \frac{\partial\mathbf{r}_2}{\partial x} + \frac{\partial^2\mathbf{r}_1}{\partial x\partial y} \times \mathbf{r}_2$$

(3) Partial Differentials

$$d(\mathbf{r}_1 \cdot \mathbf{r}_2) = \mathbf{r}_1 \cdot d\mathbf{r}_2 + d\mathbf{r}_1 \cdot \mathbf{r}_2$$

$$d(\mathbf{r}_1 \times \mathbf{r}_2) = \mathbf{r}_1 \times d\mathbf{r}_2 + d\mathbf{r}_1 \times \mathbf{r}_2$$

(4) Total Differentials

$$d\mathbf{r} = \frac{\partial\mathbf{r}}{\partial x}dx + \frac{\partial\mathbf{r}}{\partial y}dy + \frac{d\mathbf{r}}{\partial z}dz$$

$$d\mathbf{r} = \left(\frac{\partial\mathbf{r}}{\partial x}\frac{\partial x}{\partial t} + \frac{\partial\mathbf{r}}{\partial y}\frac{\partial y}{\partial t} + \frac{\partial\mathbf{r}}{\partial z}\frac{\partial z}{\partial t}\right)dt$$

(1) Indefinite Integral

(a) Definition

The *indefinite integral of a vector function* $\mathbf{f}(t)$ of a single scalar variable t is the anti-derivative of that vector function.

$$\mathbf{f}(t) = \dot{\mathbf{F}}(t) = \frac{d\mathbf{F}(t)}{dt}$$

$$\int \mathbf{f}(t)\, dt = \int \dot{\mathbf{F}}(t)\, dt = \mathbf{F}(t) + \mathbf{c}$$

$\mathbf{c} =$ vector constant of integration

(b) General formulas

$\mathbf{E} =$ constant vector
$m =$ constant scalar
$t =$ variable scalar
$\mathbf{F}(t) =$ vector function of t
$\dot{\mathbf{F}}(t) =$ first derivative of $\mathbf{F}(t)$ with respect to t
$\ddot{\mathbf{F}}(t) =$ second derivative of $\mathbf{F}(t)$ with respect to t

$$\int m\dot{\mathbf{F}}(t)\, dt = m\mathbf{F}(t) + \mathbf{c}$$

$$\int \mathbf{E}\cdot\dot{\mathbf{F}}(t)\, dt = \mathbf{E}\cdot\mathbf{F}(t) + \mathbf{c}$$

$$\int \mathbf{E}\times\dot{\mathbf{F}}(t)\, dt = \mathbf{E}\times\mathbf{F}(t) + \mathbf{c}$$

$$\int \mathbf{F}(t)\cdot\dot{\mathbf{F}}(t)\, dt = \frac{\mathbf{F}^2(t)}{2} + \mathbf{c}$$

$$\int \mathbf{F}(t)\times\ddot{\mathbf{F}}(t)\, dt = \mathbf{F}(t)\times\dot{\mathbf{F}}(t) + \mathbf{c}$$

$$\int\int \ddot{\mathbf{F}}(t)\, dt\, dt = \mathbf{F}(t) + \mathbf{c}_1 t + \mathbf{c}_2$$

(2) Definite Integral

(a) Definition

The *definite integral of a vector function* $\mathbf{f}(t)$ of a single scalar variable t is the limit of a sum between the lower and the upper limit.

$$\int_{t=a}^{t=b} \mathbf{f}(t)\, dt = \int_{t=a}^{t=b} \dot{\mathbf{F}}(t)\, dt = \mathbf{F}(b) - \mathbf{F}(a)$$

$a, b =$ scalar constants

(b) Interchange of limits

$$\int_{t=a}^{t=b} \mathbf{f}(t)\, dt = -\int_{t=b}^{t=a} \mathbf{f}(t)\, dt = \mathbf{F}(b) - \mathbf{F}(a)$$

(c) Decomposition of limits

$$\int_{t=a}^{t=b} \mathbf{f}(t)\, dt = \int_{t=a}^{t=c} \mathbf{f}(t)\, dt + \int_{t=c}^{t=b} \mathbf{f}(t)\, dt$$

$$= \mathbf{F}(b) - \mathbf{F}(a)$$

(d) Sum of several functions

$$\int_{t=a}^{t=b} [\mathbf{f}_1(t) + \mathbf{f}_2(t) + \mathbf{f}_3(t)]\, dt = \int_{t=a}^{t=b} \mathbf{f}_1(t)\, dt + \int_{t=a}^{t=b} \mathbf{f}_2(t)\, dt + \int_{t=a}^{t=b} \mathbf{f}_3(t)\, dt$$

(e) Constant factors $\mathbf{E}, m$

$$\int_{t=a}^{t=b} \mathbf{E} \times \mathbf{f}(t)\, dt = \mathbf{E} \times \int_{t=a}^{t=b} \mathbf{f}(t)\, dt \qquad \int_{t=a}^{t=b} m\mathbf{f}(t) = m \int_{t=a}^{t=b} \mathbf{f}(t)\, dt$$

$$\int_{t=a}^{t=b} \mathbf{E} \cdot \mathbf{f}(t)\, dt = \mathbf{E} \cdot \int_{t=a}^{t=b} \mathbf{f}(t)\, dt$$

(1) Line Integral

(a) Definition

The *line integral of a vector* **F** over a path C is defined as the integral of the scalar product of **F** and the elemental path $d\mathbf{r}$ along C. **r** defines a curve which has continuous derivatives, and **F** is continuous along C.

$$\int_a^b \mathbf{F} \cdot d\mathbf{r} = \int_a^b F_x \, dx + \int_a^b F_y \, dy + \int_a^b F_z \, dz$$

(b) Theorems

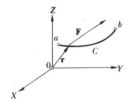

If $\mathbf{F} = \nabla\phi = \dfrac{\partial\phi}{\partial x}\mathbf{i} + \dfrac{\partial\phi}{\partial y}\mathbf{j} + \dfrac{\partial\phi}{\partial z}\mathbf{k}$ and C is *an open curve*, then

$$\int_a^b \nabla\phi \cdot d\mathbf{r} \text{ is } \textit{independent of the path} \text{ between } a \text{ and } b.$$

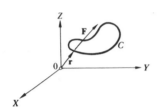

If $\mathbf{F} = \nabla\phi = \dfrac{\partial\phi}{\partial x}\mathbf{i} + \dfrac{\partial\phi}{\partial y}\mathbf{j} + \dfrac{\partial\phi}{\partial z}\mathbf{k}$ and C is *a closed curve*, then

$$\oint \nabla\phi \cdot d\mathbf{r} \text{ is } \textit{equal to zero}.$$

(2) Surface Integral

(a) A normal surface vector is a vector whose length is equal to the area bounded by a closed curve C and whose direction is perpendicular to the plane of the area.

$$d\mathbf{S} = ds\,\mathbf{n}$$

(b) Flow of a scalar field

$$\mathbf{P} = \int\int_S \phi \, d\mathbf{S} = \mathbf{i} \int\int_{\Sigma yz} \phi \, dy \, dz + \mathbf{j} \int\int_{\Sigma xz} \phi \, dx \, dz + \mathbf{k} \int\int_{\Sigma xy} \phi \, dx \, dy$$

(c) Scalar flow of a vector field

$$Q = \int\int_S \mathbf{F} \cdot d\mathbf{S} = \int\int_{\Sigma yz} F_x \, dy \, dz + \int\int_{\Sigma xz} F_y \, dx \, dz + \int\int_{\Sigma xy} F_z \, dx \, dy$$

(d) Vector flow of a vector field

$$\mathbf{R} = \int\int_S \mathbf{F} \times d\mathbf{S} = \int\int_{\Sigma yz} (F_z\mathbf{j} - F_y\mathbf{k}) \, dy \, dz + \int\int_{\Sigma xz} (F_x\mathbf{k} - F_z\mathbf{i}) \, dx \, dz + \int\int_{\Sigma xy} (F_y\mathbf{i} - F_x\mathbf{j}) \, dx \, dy$$

(e) Closed surface integral notation

$$\mathbf{P} = \oint_S \phi \, d\mathbf{S} \qquad Q = \oint_S \mathbf{F} \cdot d\mathbf{S} \qquad \mathbf{R} = \oint_S \mathbf{F} \times d\mathbf{S}$$

Note: In integrals **P**, Q, **R** each integral is taken over projection of the surface on the respective coordinate plane.

(1) Gauss' Theorem

A *scalar flow of a vector function* **F** through a closed surface S is equal to the integral of $\nabla \cdot \mathbf{F}$ over the volume V bounded by S.

$$\underbrace{\oint_S \mathbf{F} \cdot d\mathbf{S}}_{\substack{\text{Closed surface} \\ \text{integral}}} = \underbrace{\int_V \nabla \cdot \mathbf{F} \, dV}_{\substack{\text{Closed volume} \\ \text{integral}}}$$

In cartesian coordinates

$$\oint_S \mathbf{F} \cdot d\mathbf{S} = \iint_{\Sigma yz} F_x \, dy \, dz + \iint_{\Sigma zx} F_y \, dz \, dx + \iint_{\Sigma xy} F_z \, dx \, dy = \iiint_{\Sigma xyz} \left(\frac{\partial F_x}{\partial x} + \frac{\partial F_y}{\partial y} + \frac{\partial F_z}{\partial z} \right) dx \, dy \, dz$$

Thus a *closed volume integral* can be reduced to a *closed surface integral*.

(2) Stokes' Theorem

A *circulation of a vector function* **F** about a closed path C is equal to a vector flow of the same vector function over an arbitrary surface bounded by C.

$$\underbrace{\oint_C \mathbf{F} \cdot d\mathbf{r}}_{\substack{\text{Closed path} \\ \text{integral}}} = \underbrace{\int_S (\nabla \times \mathbf{F}) \cdot d\mathbf{S}}_{\substack{\text{Surface} \\ \text{integral}}}$$

In *cartesian coordinates*

$$\oint_C \mathbf{F} \cdot d\mathbf{r} = \int_C (F_x \, dx + F_y \, dy + F_z \, dz)$$
$$= \iint_{\Sigma yz} \left(\frac{\partial F_z}{\partial y} - \frac{\partial F_y}{\partial z} \right) dy \, dz + \iint_{\Sigma zx} \left(\frac{\partial F_x}{\partial z} - \frac{\partial F_z}{\partial x} \right) dz \, dx + \iint_{\Sigma xy} \left(\frac{\partial F_y}{\partial x} - \frac{\partial F_x}{\partial y} \right) dx \, dy$$

Thus a *surface integral* can be reduced to a *line integral*.

(3) Green's Theorem

If in Gauss' theorem $\mathbf{F} = \alpha \nabla \beta$, where α, β are scalar functions of x, y, z, then

$$\underbrace{\oint_S (\alpha \nabla \beta) \cdot d\mathbf{S}}_{\substack{\text{Closed surface} \\ \text{integral}}} = \underbrace{\int_V \nabla_* \cdot (\alpha \nabla \beta) \, dV}_{\substack{\text{Closed volume} \\ \text{integral}}} = \underbrace{\int_V (\alpha \nabla^2 \beta) + \nabla \alpha \cdot \nabla \beta) \, dV}_{\substack{\text{From operator's} \\ \text{formula}}}$$

If $\alpha = +1$,

$$\oint_S \nabla \beta \cdot d\mathbf{S} = \int_V \nabla^2 \beta \, dV$$

and in *cartesian coordinates*

$$\iint_{\Sigma yz} \frac{\partial \beta}{\partial x} \, dy \, dz + \iint_{\Sigma zx} \frac{\partial \beta}{\partial y} \, dz \, dx + \iint_{\Sigma xy} \frac{\partial \beta}{\partial z} \, dx \, dy = \iiint_{\Sigma xyz} \left(\frac{\partial^2 \beta}{\partial x^2} + \frac{\partial^2 \beta}{\partial y^2} + \frac{\partial^2 \beta}{\partial z^2} \right) dx \, dy \, dz$$

(1) Basic Equations

(a) Coordinates

$$
\begin{array}{ccc}
a = \sqrt{x^2 + y^2} & x = a\cos\theta & \dot{x} = \dfrac{dx}{dt} \\[2mm]
\theta = \tan^{-1}\dfrac{y}{x} & y = a\sin\theta & \dot{y} = \dfrac{dy}{dt} \\[2mm]
h = z & z = h & \dot{z} = \dfrac{dz}{dt}
\end{array}
$$

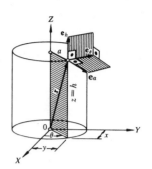

where $a = a(t)$, $\theta = \theta(t)$, $h = h(t)$ are the time-dependent *cylindrical coordinates* and $x = x(t)$, $y = y(t)$, $z = z(t)$ are their time-dependent cartesian counterparts.

(b) Time derivatives

$$
\dot{a} = \frac{da}{dt} = \frac{x\dot{x} + y\dot{y}}{a} \qquad \dot{\theta} = \frac{d\theta}{dt} = \frac{x\dot{y} - y\dot{x}}{a^2} \qquad \dot{h} = \frac{dh}{dt} = \dot{z}
$$

(c) Position vector and its time derivatives

$$
\mathbf{r} = a\mathbf{e}_a + h\mathbf{e}_h
$$

$$
\dot{\mathbf{r}} = \frac{d\mathbf{r}}{dt} = \dot{a}\mathbf{e}_a + a\dot{\theta}\mathbf{e}_\theta + \dot{h}\mathbf{e}_h
$$

$$
\ddot{\mathbf{r}} = \frac{d^2\mathbf{r}}{dt^2} = (\ddot{a} - a\dot{\theta}^2)\mathbf{e}_a + (a\ddot{\theta} + 2\dot{a}\dot{\theta})\mathbf{e}_\theta + \ddot{h}\mathbf{e}_h
$$

where $\dot{a}, \dot{\theta}, \dot{h}$ are derived in (b), $\ddot{a} = d^2a/dt^2$, $\ddot{\theta} = d^2\theta/dt^2$, $\ddot{h} = d^2h/dt^2$, and $\mathbf{e}_a, \mathbf{e}_\theta, \mathbf{e}_h$ are the respective unit vectors.

(2) Matrix Transformations

$$
C_\theta = \cos\theta \qquad S_\theta = \sin\theta \qquad \dot{C}_\theta = -\dot{\theta}\sin\theta \qquad \dot{S}_\theta = \dot{\theta}\cos\theta
$$

(a) Time derivatives of coordinates

$$
\begin{bmatrix} \dot{a} \\ a\dot{\theta} \\ \dot{h} \end{bmatrix} =
\begin{bmatrix} C_\theta & S_\theta & 0 \\ -S_\theta & C_\theta & 0 \\ 0 & 0 & 1 \end{bmatrix}
\begin{bmatrix} \dot{x} \\ \dot{y} \\ \dot{z} \end{bmatrix}
\qquad
\begin{bmatrix} \dot{x} \\ \dot{y} \\ \dot{z} \end{bmatrix} =
\begin{bmatrix} C_\theta & -S_\theta & 0 \\ S_\theta & C_\theta & 0 \\ 0 & 0 & 1 \end{bmatrix}
\begin{bmatrix} \dot{a} \\ a\dot{\theta} \\ \dot{h} \end{bmatrix}
$$

(b) Transformations of unit vectors

$$
\begin{bmatrix} \mathbf{e}_a \\ \mathbf{e}_\theta \\ \mathbf{e}_h \end{bmatrix} =
\begin{bmatrix} C_\theta & S_\theta & 0 \\ -S_\theta & C_\theta & 0 \\ 0 & 0 & 1 \end{bmatrix}
\begin{bmatrix} \mathbf{i} \\ \mathbf{j} \\ \mathbf{k} \end{bmatrix}
\qquad
\begin{bmatrix} \mathbf{i} \\ \mathbf{j} \\ \mathbf{k} \end{bmatrix} =
\begin{bmatrix} C_\theta & -S_\theta & 0 \\ S_\theta & C_\theta & 0 \\ 0 & 0 & 1 \end{bmatrix}
\begin{bmatrix} \mathbf{e}_a \\ \mathbf{e}_\theta \\ \mathbf{e}_h \end{bmatrix}
$$

(c) Time derivatives of unit vectors

$$
\begin{bmatrix} \dot{\mathbf{e}}_a \\ \dot{\mathbf{e}}_\theta \\ \dot{\mathbf{e}}_h \end{bmatrix} =
\begin{bmatrix} 0 & \dot{\theta} & 0 \\ -\dot{\theta} & 0 & 0 \\ 0 & 0 & 0 \end{bmatrix}
\begin{bmatrix} \mathbf{e}_a \\ \mathbf{e}_\theta \\ \mathbf{e}_h \end{bmatrix} =
\begin{bmatrix} \dot{C}_\theta & \dot{S}_\theta & 0 \\ -\dot{S}_\theta & \dot{C}_\theta & 0 \\ 0 & 0 & 0 \end{bmatrix}
\begin{bmatrix} \mathbf{i} \\ \mathbf{j} \\ \mathbf{k} \end{bmatrix}
$$

(1) Basic Equations

(a) Coordinates

$$b = \sqrt{x^2 + y^2 + z^2} \qquad x = b \sin\phi \cos\theta \qquad \dot{x} = \frac{dx}{dt}$$

$$\phi = \cos^{-1}\frac{z}{\sqrt{x^2 + y^2}} \qquad y = b \sin\phi \sin\theta \qquad \dot{y} = \frac{dy}{dt}$$

$$\theta = \tan^{-1}\frac{y}{x} \qquad z = b \cos\phi \qquad \dot{z} = \frac{dz}{dt}$$

where $b = b(t)$, $\phi = \phi(t)$, $\theta = \theta(t)$ are the time-dependent *spherical coordinates* and $x = x(t)$, $y = y(t)$, $z = z(t)$ are their time-dependent cartesian counterparts.

(b) Time derivatives $(a = \sqrt{x^2 + y^2})$

$$\dot{b} = \frac{db}{dt} = \frac{x\dot{x} + y\dot{y} + z\dot{z}}{b} \qquad \dot{\phi} = \frac{d\phi}{dt} = \frac{z(x\dot{x} - y\dot{y}) - \dot{z}a^2}{ab^2} \qquad \dot{\theta} = \frac{d\theta}{dt} = \frac{\dot{x}y - y\dot{x}}{a^2}$$

(c) Position vector and its time derivatives

$$\mathbf{r} = b\mathbf{e}_b \qquad\qquad\qquad \dot{\mathbf{r}} = \frac{d\mathbf{r}}{dt} = \dot{b}\mathbf{e}_b + b\dot{\phi}\mathbf{e}_\phi + b\dot{\theta}\sin\phi\,\mathbf{e}_\theta$$

$$\ddot{\mathbf{r}} = \frac{d^2\mathbf{r}}{dt^2} = (\ddot{b} - b\dot{\phi}^2 - b\dot{\theta}^2\sin^2\phi)\mathbf{e}_b + (2\dot{b}\dot{\phi} + b\ddot{\phi} - b\dot{\theta}^2\sin\phi\cos\phi)\mathbf{e}_\phi$$
$$+ (2\dot{b}\dot{\theta}\sin\phi + b\ddot{\theta}\sin\phi + 2b\dot{\theta}\dot{\phi}\cos\phi)\mathbf{e}_\theta$$

where the single and double overdot and $\mathbf{e}_b$, $\mathbf{e}_\phi$, $\mathbf{e}_\theta$ have meanings similar to those in Sec. 10.07.

(2) Matrix Transformations

$$C_\phi = \cos\phi \qquad \dot{C}_\phi = -\dot{\phi}\sin\phi \qquad C_\theta = \cos\theta \qquad \dot{C}_\theta = -\dot{\theta}\sin\theta$$
$$S_\phi = \sin\phi \qquad \dot{S}_\phi = \dot{\phi}\cos\phi \qquad S_\theta = \sin\theta \qquad \dot{S}_\theta = \dot{\theta}\cos\theta$$

(a) Time derivatives of coordinates

$$\begin{bmatrix} \dot{b} \\ b\dot{\phi} \\ b\dot{\theta} \end{bmatrix} = \begin{bmatrix} S_\phi C_\theta & S_\phi S_\theta & C_\phi \\ C_\phi C_\theta & C_\phi S_\theta & -S_\phi \\ -S_\phi S_\theta & S_\phi C_\theta & 0 \end{bmatrix} \begin{bmatrix} \dot{x} \\ \dot{y} \\ \dot{z} \end{bmatrix} \qquad \begin{bmatrix} \dot{x} \\ \dot{y} \\ \dot{z} \end{bmatrix} = \begin{bmatrix} S_\phi C_\theta & C_\phi C_\theta & -S_\phi S_\theta \\ S_\phi S_\theta & C_\phi S_\theta & S_\phi C_\theta \\ C_\phi & -S_\phi & 0 \end{bmatrix} \begin{bmatrix} \dot{b} \\ b\dot{\phi} \\ b\dot{\theta} \end{bmatrix}$$

(b) Transformations of unit vectors

$$\begin{bmatrix} \mathbf{e}_b \\ \mathbf{e}_\phi \\ \mathbf{e}_\theta \end{bmatrix} = \begin{bmatrix} S_\phi C_\theta & S_\phi S_\theta & C_\phi \\ C_\phi C_\theta & C_\phi S_\theta & -S_\phi \\ -S_\theta & C_\theta & 0 \end{bmatrix} \begin{bmatrix} \mathbf{i} \\ \mathbf{j} \\ \mathbf{k} \end{bmatrix} \qquad \begin{bmatrix} \mathbf{i} \\ \mathbf{j} \\ \mathbf{k} \end{bmatrix} = \begin{bmatrix} S_\phi C_\theta & C_\phi C_\theta & -S_\theta \\ S_\phi S_\theta & C_\phi S_\theta & C_\theta \\ C_\phi & -S_\phi & 0 \end{bmatrix} \begin{bmatrix} \mathbf{e}_b \\ \mathbf{e}_\phi \\ \mathbf{e}_\theta \end{bmatrix}$$

(c) Time derivatives of unit vectors $(\lambda_1 = \dot{\theta}\cos\phi,\ \lambda_2 = \dot{\theta}\sin\phi)$

$$\begin{bmatrix} \dot{\mathbf{e}}_b \\ \dot{\mathbf{e}}_\phi \\ \dot{\mathbf{e}}_\theta \end{bmatrix} = \begin{bmatrix} 0 & \dot{\phi} & \lambda_2 \\ -\dot{\phi} & 0 & \lambda_1 \\ -\lambda_2 & -\lambda_1 & 0 \end{bmatrix} \begin{bmatrix} \mathbf{e}_b \\ \mathbf{e}_\phi \\ \mathbf{e}_\theta \end{bmatrix} = \begin{bmatrix} S_\phi \dot{C}_\theta + \dot{S}_\phi C_\theta & S_\phi \dot{S}_\theta + \dot{S}_\phi S_\theta & \dot{C}_\phi \\ C_\phi \dot{C}_\theta + \dot{C}_\phi C_\theta & C_\phi \dot{S}_\theta + \dot{C}_\phi S_\theta & -\dot{S}_\phi \\ -\dot{S}_\theta & \dot{C}_\theta & 0 \end{bmatrix} \begin{bmatrix} \mathbf{i} \\ \mathbf{j} \\ \mathbf{k} \end{bmatrix}$$

(1) Differential Elements

The *curvilinear coordinates* u_1, u_2. u_3 are said to be *orthogonal* if the coordinate curves are mutually perpendicular at every point. The differential element of length $ds_m = h_m \, du_m$ $(m = 1, 2, 3)$ is defined by the differential element of the respective coordinate and the corresponding *scaling factor*.

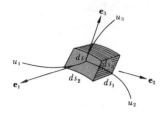

(2) Vector Function

The vector function $\mathbf{r} = r_1\mathbf{e}_1 + r_2\mathbf{e}_2 + r_3\mathbf{e}_3$ is determined by *scalar functions* r_m and the respective *unit vector* $\mathbf{e}_m$, tangent to u_m.

$$\mathbf{e}_1 = \frac{1}{h_1}\frac{\partial \mathbf{r}}{\partial u_1} \qquad \mathbf{e}_2 = \frac{1}{h_2}\frac{\partial \mathbf{r}}{\partial u_2} \qquad \mathbf{e}_3 = \frac{1}{h_3}\frac{\partial \mathbf{r}}{\partial u_3}$$

(3) Table of h_m and ds_m

	x	y	z	a	θ	z	b	θ	ϕ
ds_m	dx	dy	dz	da	$a\,d\theta$	dz	db	$b \sin\phi \, d\theta$	$b\,d\phi$
h_m	1	1	1	1	a	1	1	$b \sin\phi$	b

(4) Differential Operators — General Formulas

(a) Gradient of scalar function $f =$ Grad f

$$\nabla f = \frac{1}{h_1}\frac{\partial f}{\partial u_1}\mathbf{e}_1 + \frac{1}{h_2}\frac{\partial f}{\partial u_2}\mathbf{e}_2 + \frac{1}{h_3}\frac{\partial f}{\partial u_3}\mathbf{e}_3$$

(b) Divergence of vector function $\mathbf{V} =$ Div $\mathbf{V}$

$$\nabla \cdot \mathbf{V} = \frac{1}{h_1 h_2 h_3}\left[\frac{\partial}{\partial u_1}(V_1 h_2 h_3) + \frac{\partial}{\partial u_2}(h_1 V_2 h_3) + \frac{\partial}{\partial u_3}(h_1 h_2 V_3)\right]$$

(c) Curl of vector function $\mathbf{V} =$ Curl $\mathbf{V} =$ Rot $\mathbf{V}$

$$\nabla \times \mathbf{V} = \frac{1}{h_1 h_2 h_3} \begin{vmatrix} h_1\mathbf{e}_1 & \dfrac{\partial}{\partial u_1} & h_1 V_1 \\[2ex] h_2\mathbf{e}_2 & \dfrac{\partial}{\partial u_2} & h_2 V_2 \\[2ex] h_3\mathbf{e}_3 & \dfrac{\partial}{\partial u_3} & h_3 V_3 \end{vmatrix}$$

(d) Laplacian of scalar function f

$$\nabla^2 f = \frac{1}{h_1 h_2 h_3}\left[\frac{\partial}{\partial u_1}\left(\frac{h_2 h_3}{h_1}\frac{\partial f}{\partial u_1}\right) + \frac{\partial}{\partial u_2}\left(\frac{h_1 h_3}{h_2}\frac{\partial f}{\partial u_2}\right) + \frac{\partial}{\partial u_3}\left(\frac{h_1 h_2}{h_3}\frac{\partial f}{\partial u_3}\right)\right]$$

(e) General operator's formulas

$$\nabla(f_1 + f_2) = \nabla f_1 + \nabla f_2 \qquad\qquad \nabla \cdot (f\mathbf{V}) = f(\nabla \cdot \mathbf{V}) + (\nabla f) \cdot \mathbf{V}$$

$$\nabla \cdot (\mathbf{V}_1 + \mathbf{V}_2) = \nabla \cdot \mathbf{V}_1 + \nabla \cdot \mathbf{V}_2 \qquad\qquad \nabla \times (f\mathbf{V}) = f(\nabla \times \mathbf{V}) + (\nabla f) \times \mathbf{V}$$

$$\nabla \times (\mathbf{V}_1 + \mathbf{V}_2) = \nabla \times \mathbf{V}_1 + \nabla \times \mathbf{V}_2 \qquad\qquad \nabla \cdot (\nabla f) = \nabla^2 f$$

(1) Cartesian Operators   (Sec. 10.05)

(a) Gradient

$$\nabla f = \frac{\partial f}{\partial x}\mathbf{i} + \frac{\partial f}{\partial y}\mathbf{j} + \frac{\partial f}{\partial z}\mathbf{k}$$

(b) Divergence

$$\nabla \cdot \mathbf{V} = \frac{\partial V_x}{\partial x} + \frac{\partial V_y}{\partial y} + \frac{\partial V_z}{\partial z}$$

(c) Curl

$$\nabla \times \mathbf{V} = \begin{vmatrix} \mathbf{i} & \frac{\partial}{\partial x} & V_x \\ \mathbf{j} & \frac{\partial}{\partial y} & V_y \\ \mathbf{k} & \frac{\partial}{\partial z} & V_z \end{vmatrix}$$

(d) Laplacian

$$\nabla^2 f = \frac{\partial^2 f}{\partial x^2} + \frac{\partial^2 f}{\partial y^2} + \frac{\partial^2 f}{\partial z^2}$$

(2) Cylindrical Operators   (Sec. 10.09)

(a) Gradient

$$\nabla f = \frac{\partial f}{\partial a}\mathbf{e}_a + \frac{1}{a}\frac{\partial f}{\partial \theta}\mathbf{e}_\theta + \frac{\partial f}{\partial z}\mathbf{e}_z$$

(b) Divergence

$$\nabla \cdot \mathbf{V} = \frac{1}{a}\left[\frac{\partial(aV_a)}{\partial a} + \frac{\partial V_\theta}{\partial \theta} + a\frac{\partial V_z}{\partial z}\right]$$

(c) Curl

$$\nabla \times \mathbf{V} = \frac{1}{a}\begin{vmatrix} \mathbf{e}_a & \frac{\partial}{\partial a} & V_a \\ a\mathbf{e}_\theta & \frac{\partial}{\partial \theta} & aV_\theta \\ \mathbf{e}_z & \frac{\partial}{\partial z} & V_z \end{vmatrix}$$

(d) Laplacian

$$\nabla^2 f = \frac{1}{a}\left[\frac{\partial}{\partial a}\left(a\frac{\partial f}{\partial a}\right) + \frac{1}{a}\frac{\partial}{\partial \theta}\left(\frac{\partial f}{\partial \theta}\right) + \frac{\partial}{\partial z}\left(a\frac{\partial f}{\partial z}\right)\right]$$

(3) Spherical Operators   (Sec. 10.10)

(a) Gradient

$$\nabla f = \frac{\partial f}{\partial b}\mathbf{e}_b + \frac{1}{b \sin \phi}\frac{\partial f}{\partial \theta}\mathbf{e}_\theta + \frac{1}{b}\frac{\partial f}{\partial \phi}\mathbf{e}_\phi$$

(b) Divergence

$$\nabla \cdot \mathbf{V} = \frac{1}{b^2 \sin \phi}\left[\frac{\partial}{\partial b}(b^2 \sin \phi V_b) + \frac{\partial}{\partial \theta}(bV_\theta) + \frac{\partial}{\partial \phi}(b \sin \phi V_\phi)\right]$$

(c) Curl

$$\nabla \times \mathbf{V} = \frac{1}{b^2 \sin \phi}\begin{vmatrix} \mathbf{e}_b & \frac{\partial}{\partial b} & V_b \\ b \sin \phi \mathbf{e}_\theta & \frac{\partial}{\partial \theta} & b \sin \phi V_\theta \\ b\mathbf{e}_\phi & \frac{\partial}{\partial \phi} & bV_\phi \end{vmatrix}$$

(d) Laplacian

$$\nabla^2 f = \frac{1}{b^2 \sin \phi}\left[\frac{\partial}{\partial b}\left(b^2 \sin \phi \frac{\partial f}{\partial b}\right) + \frac{\partial}{\partial \theta}\left(\frac{1}{\sin \phi}\frac{\partial f}{\partial \theta}\right) + \frac{\partial}{\partial \phi}\left(\sin \phi \frac{\partial f}{\partial \phi}\right)\right]$$

(1) System of Orthogonal Curves

A system of circles through two points A, B, and another system of circles orthogonal to the first system shown in the adjacent figure define the circular coordinate system and generate the following coordinate surfaces.

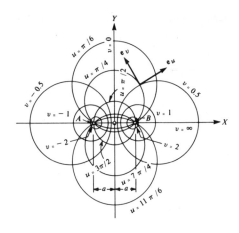

(2) Bipolar Cylindrical Surface

(a) Coordinate surface is generated by the translation of the orthogonal circles defined in (1) above along the Z axis. The equations of this surface area

$$x + (y - a \cot u)^2 = a^2 \csc^2 u$$
$$(x - a \coth v)^2 + y^2 = a^2 \operatorname{csch}^2 u \qquad z = z$$

(b) Coordinates   (Sec. 10.11-1)

$$r = a (\cosh v - \cos u)^{-1}$$
$$x = r \sinh v \qquad y = r \sin u \qquad z = z$$

(c) Scaling factors   (Sec. 10.11-2)

$$h_1{}^2 = h_2{}^2 = r^2$$
$$h_3{}^2 = 1$$

(d) Differential operators   (Sec. 10.11-4)

$$\nabla f = \frac{1}{r}\left(\frac{\partial f}{\partial u}\mathbf{e}_u + \frac{\partial f}{\partial v}\mathbf{e}_v\right) + \frac{\partial f}{\partial z}e_z$$

$$\nabla \cdot \mathbf{V} = \frac{1}{r^2}\left[\frac{\partial}{\partial u}(rV_u) + \frac{\partial}{\partial v}(rV_v) + \frac{\partial}{\partial z}(r^2 V_z)\right]$$

$$\nabla^2 f = \frac{1}{r^2}\left(\frac{\partial^2 f}{\partial u^2} + \frac{\partial^2 f}{\partial v^2}\right) + \frac{\partial^2 f}{\partial z^2}$$

$$\nabla \times \mathbf{V} = \frac{1}{r^2}
\begin{vmatrix}
r\mathbf{e}_u & \dfrac{\partial}{\partial u} & rV_u \\[2mm]
r\mathbf{e}_v & \dfrac{\partial}{\partial v} & rV_v \\[2mm]
\mathbf{e}_z & \dfrac{\partial}{\partial z} & V_z
\end{vmatrix}$$

where f is a scalar function, $\mathbf{V}$ is a vector function, and $\mathbf{e}_u$, $\mathbf{e}_v$, $\mathbf{e}_z$ are the unit vectors of the circular cylinder.

(3) Toroidal Surface

(a) Coordinate surface is generated by the rotation of the orthogonal circles defined in (1) above about the Y axis, which is designated the Z axis. The equations of this surface are

$$(x^2 + y^2) + (z - a \cot v)^2 = a^2 \csc^2 v \qquad (\sqrt{x^2 + y^2} - a \coth u)^2 + z^2 = a^2 \operatorname{scsh}^2 u \qquad y = x \tan \phi$$

(b) Coordinates   (Sec. 10.11-1)

$$x = \frac{a \sinh v \cos \phi}{\cosh v - \cos u} \qquad y = \frac{a \sinh v \sin \phi}{\cosh v - \cos u}$$

$$z = \frac{a \sin u}{\cosh v - \cos u}$$

(c) Scaling factors   (Sec. 10.11-2)

$$h_1{}^2 = h_2{}^2 = r^2 = \frac{a^2}{(\cosh v - \cos u)^2}$$

$$h_3{}^2 = r^2 \sinh^2 v \qquad g = \sinh v$$

(d) Differential operators   (Sec. 10.11-4)

$$\nabla f = \frac{1}{r}\left(\frac{\partial f}{\partial u}\mathbf{e}_u + \frac{\partial f}{\partial v}\mathbf{e}_v\right) + \frac{1}{gr}\frac{\partial f}{\partial \phi}\mathbf{e}_\phi$$

$$\nabla \cdot \mathbf{V} = \frac{1}{gr^3}\left[\frac{\partial}{\partial u}(gr^2 V_u) + \frac{\partial}{\partial v}(gr^2 V_v) + \frac{\partial}{\partial \phi}(r^2 V_\phi)\right]$$

$$\nabla^2 f = \frac{1}{gr^3}\left[\frac{\partial}{\partial u}\left(gr\frac{\partial f}{\partial u}\right) + \frac{\partial}{\partial v}\left(gr\frac{\partial f}{\partial v}\right) + \frac{\partial}{\partial \phi}\left(\frac{r}{g}\frac{\partial f}{\partial \phi}\right)\right]$$

$$\nabla \times \mathbf{V} = \frac{1}{g}
\begin{vmatrix}
\mathbf{e}_u & \dfrac{\partial}{\partial u} & V_u \\[2mm]
\mathbf{e}_v & \dfrac{\partial}{\partial v} & V_v \\[2mm]
g\mathbf{e}_\phi & \dfrac{\partial}{\partial \phi} & gV_\phi
\end{vmatrix}$$

where f is a scalar function, $\mathbf{V}$ is a vector function, and $\mathbf{e}_u$, $\mathbf{e}_v$, $\mathbf{e}_\phi$ are the unit vectors of the circular toroid.

(1) System of Orthogonal Curves

An orthogonal system of confocal parabolas with focus F shown in the adjacent figure defines the orthogonal parabolic coordinate system and generates the following coordinate surfaces.

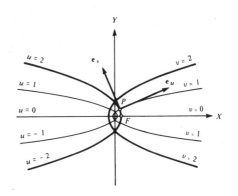

(2) Parabolic Cylindrical Surface

(a) Coordinate surface is generated by the translation of the orthogonal curves defined in (1) above along the Z axis. The equations of this surface are

$$\frac{x^2}{u^2} = -2y + u^2 \qquad \frac{x^2}{v^2} = 2y + v^2 \qquad z = z$$

(b) Coordinates   **(Sec. 10.11-1)**

$x = \frac{1}{2}(u^2 - v^2)$

$y = uv \qquad\qquad z = z$

(c) Scaling factors   **(Sec. 10.11-2)**

$h_1^2 = h_2^2 = h^2 = u^2 + v^2$

$h_3^2 = 1$

(d) Differential operators   **(Sec. 10.11-4)**

$$\nabla f = \frac{1}{h}\left(\frac{\partial f}{\partial u}\mathbf{e}_u + \frac{\partial f}{\partial v}\mathbf{e}_v\right) + \frac{\partial f}{\partial z}\mathbf{e}_z$$

$$\nabla \cdot \mathbf{V} = \frac{1}{h^2}\left[\frac{\partial}{\partial u}(hV_u) + \frac{\partial}{\partial v}(hV_v) + \frac{\partial}{\partial z}(h^2 V_z)\right]$$

$$\nabla^2 f = \frac{1}{h^2}\left(\frac{\partial^2 f}{\partial u^2} + \frac{\partial^2 f}{\partial v^2}\right) + \frac{\partial^2 f}{\partial z^2}$$

$$\nabla \times \mathbf{V} = \frac{1}{h^2}\begin{vmatrix} h\mathbf{e}_u & \dfrac{\partial}{\partial u} & hV_u \\[6pt] h\mathbf{e}_v & \dfrac{\partial}{\partial v} & hV_v \\[6pt] \mathbf{e}_z & \dfrac{\partial}{\partial z} & V_z \end{vmatrix}$$

where f is a scalar function, $\mathbf{V}$ is a vector function, and $\mathbf{e}_u$, $\mathbf{e}_v$, $\mathbf{e}_z$ are the unit vectors of the parabolic cylinder.

(3) Paraboloidal Surface

(a) Coordinate surface is generated by the rotation of the orthogonal curves defined in (1) above about the X axis, which is designated the Z axis. The equations of this surface are

$$\frac{x^2 + y^2}{2} = -2z + u^2 \qquad\qquad \frac{x^2 + y^2}{v^2} = 2z + v^2 \qquad\qquad z = x \tan \phi$$

(b) Coordinates   **(Sec. 10.11-1)**

$x = uv \cos\phi \qquad\qquad y = uv \sin\phi$

$z = \frac{1}{2}(u^2 - v^2)$

(c) Scaling factors   **(Sec. 10.11-2)**

$h_1^2 = h_2^2 = h^2 = u^2 + v^2 \qquad\qquad h_3^2 = u^2 v^2$

$g = uv/h$

(d) Differential operators   **(Sec. 10.11-4)**

$$\nabla f = \frac{1}{h}\left(\frac{\partial f}{\partial u}\mathbf{e}_u + \frac{\partial f}{\partial v}\mathbf{e}_v\right) + \frac{1}{uv}\frac{\partial f}{\partial \phi}\mathbf{e}_\phi$$

$$\nabla \cdot \mathbf{V} = \frac{1}{h^2 uv}\left[\frac{\partial}{\partial u}(huvV_u) + \frac{\partial}{\partial v}(huvV_v)\right.$$

$$\left. + \frac{\partial}{\partial \phi}(h^2 V_\phi)\right]$$

$$\nabla^2 f = \frac{1}{h^2 u}\frac{\partial}{\partial u}\left(u\frac{\partial f}{\partial u}\right) + \frac{1}{h^2 v}\frac{\partial}{\partial v}\left(v\frac{\partial f}{\partial v}\right) + \frac{1}{u^2 v^2}\frac{\partial^2 f}{\partial \phi^2}$$

$$\nabla \times \mathbf{V} = \frac{1}{uv}\begin{vmatrix} \mathbf{e}_u & \dfrac{\partial}{\partial u} & V_u \\[6pt] \mathbf{e}_v & \dfrac{\partial}{\partial v} & V_v \\[6pt] g\mathbf{e}_\phi & \dfrac{\partial}{\partial \phi} & gV_\phi \end{vmatrix}$$

where f is a scalar function, $\mathbf{V}$ is a vector function, and $\mathbf{e}_u$, $\mathbf{e}_v$, $\mathbf{e}_\phi$ are the unit vectors of the rotational paraboloid.

(1) System of Orthogonal Curves

An orthogonal system of confocal ellipses and hyperbolas with foci F_1, F_2 shown in the adjacent figure defines the orthogonal elliptic coordinate system and generates the following coordinate surfaces.

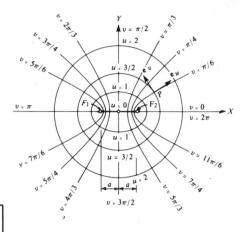

(2) Elliptic Cylindrical Coordinate Surface

The surface is generated by the translation of the orthogonal curves defined in (1) above along the Z axis. The coordinates, scaling factors, and the Laplacian of scalar function f (Sec. 10.11) are

$$x = a \cosh u \cos v \qquad y = a \sinh u \sin v \qquad z = z$$

$$h_1^2 = h_2^2 = a^2(\sinh^2 u + \sin^2 v) \qquad h_3^2 = 1$$

$$\nabla^2 f = \frac{1}{a^2(\sinh^2 u + \sin^2 v)}\left(\frac{\partial^2 f}{\partial u^2} + \frac{\partial^2 f}{\partial v^2}\right) + \frac{\partial^2 f}{\partial x^2}$$

(3) Prolate Spheroidal Coordinate Surface

The surface is generated by the rotation of the orthogonal curves defined in (1) above about the X axis, which is designated the Z axis. The coordinates, scaling factors, and the Laplacian of scalar function f (Sec. 10.11) are

$$x = a \sinh \xi \sin \eta \cos \phi \qquad y = a \sinh \xi \sin \eta \sin \phi \qquad z = a \cosh \xi \cos \eta$$

$$h_1^2 = h_2^2 = a^2(\sinh^2 \xi + \sin^2 \eta) \qquad h_3^2 = a^2 \sinh^2 \xi \sin^2 \eta$$

$$\nabla^2 f = \frac{1}{a^2(\sinh^2 \xi + \sin^2 \eta) \sinh \xi} \frac{\partial}{\partial \xi}\left(\sinh \xi \frac{\partial f}{\partial \xi}\right)$$

$$+ \frac{1}{a^2(\sinh^2 \xi + \sin^2 \eta) \sin \eta} \frac{\partial}{\partial \eta}\left(\sin \eta \frac{\partial f}{\partial \eta}\right) + \frac{1}{a^2 \sin^2 \xi \sin^2 \eta} \frac{\partial^2 f}{\partial \phi^2}$$

(4) Oblate Spheroidal Coordinate Surface

The surface is generated by the rotation of the orthogonal curves defined in (1) above about the Y axis, which is designated the Z axis. The coordinates, scaling factors, and the Laplacian of scalar function f (Sec. 10.11) are

$$x = a \cosh \xi \cos \eta \cos \phi \qquad y = a \cosh \xi \cos \eta \sin \phi \qquad z = a \sinh \xi \sin \eta$$

$$h_1^2 = h_2^2 = a^2(\sinh^2 \xi + \sin^2 \eta) \qquad h_3^2 = a^2 \cosh^2 \xi \cos^2 \eta$$

$$\nabla^2 f = \frac{1}{a^2(\sinh^2 \xi + \sin^2 \eta) \cosh \xi} \frac{\partial}{\partial \xi}\left(\cosh \xi \frac{\partial f}{\partial \xi}\right)$$

$$+ \frac{1}{a^2(\sinh^2 \xi + \sin^2 \eta) \cos \eta} \frac{\partial}{\partial \eta}\left(\cos \eta \frac{\partial f}{\partial \eta}\right) + \frac{1}{a^2 \cosh^2 \xi \cos^2 \eta} \frac{\partial^2 f}{\partial \phi^2}$$

11
FUNCTIONS OF A COMPLEX VARIABLE

$$p = a + bi \qquad r = \sqrt{a^2 + b^2} \qquad \phi = \tan^{-1}\frac{b}{a} \qquad a, b, c, d = \text{real numbers}$$

(1) Algebraic Forms

(a) Imaginary number. The second root of a negative number,

$$\sqrt{-b^2} = b\sqrt{-1} = bi$$

is called the *imaginary number*. The basis of imaginary numbers is the *imaginary unit i*.

(b) Complex number consists of a real part and an imaginary part,

$$p = a + bi \qquad q = a - bi$$

where q is the *conjugate* of p.

$$\sqrt{-1} = i \qquad i^{4k+1} = i$$
$$i^2 = -1 \qquad i^{4k+2} = -1$$
$$i^3 = -i \qquad i^{4k+3} = -i$$
$$i^4 = 1 \qquad i^{4k+4} = 1$$
$$k = 0, 1, 2, \ldots$$

(c) Basic operations

$p = a + bi \qquad q = a - bi$	$p = a + bi \qquad q = c + di$

$$p + q = 2a \qquad p - q = 2bi \qquad\qquad p \pm q = (a \pm c) + (b \pm d)i$$

$$pq = a^2 + b^2 \qquad\qquad\qquad pq = (ac - bd) + (ad + bc)i$$

$$\frac{p}{q} = \frac{a^2 + 2abi - b^2}{a^2 + b^2} \qquad\qquad \frac{p}{q} = \frac{(ac + bd) + (bc - ad)i}{c^2 + d^2}$$

(d) Complex surds

$$\sqrt{\pm bi} = \sqrt{\frac{b}{2}} \pm i\sqrt{\frac{b}{2}} = \frac{\sqrt{2b}}{2}(1 \pm i) \qquad \sqrt{a + bi} \pm \sqrt{a - bi} = \sqrt{2(a \pm \sqrt{a^2 + b^2})}$$

$$\sqrt{\pm i} = \sqrt{\frac{1}{2}} \pm i\sqrt{\frac{1}{2}} = \frac{\sqrt{2}}{2}(1 \pm i) \qquad \sqrt{a \pm bi} = \sqrt{\frac{\sqrt{a^2 + b^2} + a}{2}} \pm i\sqrt{\frac{\sqrt{a^2 + b^2} - a}{2}}$$

(2) Transcendent Forms

(a) Complex number $p = a + bi$ can be represented as a point in the *complex plane* (Argand or Gauss plane).

$$p = a + bi = re^{i\phi} = r(\cos\phi + i\sin\phi) \qquad \phi = \tan^{-1}\frac{b}{a}$$

where $e = 2.718\,281\,828\ldots$ (Sec. 1.04).

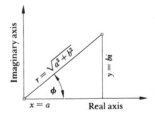

(b) Basic operations $(m, n = 1, 2, 3, \ldots)\ (k = 0, 1, 2, \ldots, n-1)$

$p_1 = a_1 + b_1 i = r_1(\cos\phi_1 + i\sin\phi_1)$	$p_2 = a_2 + b_2 i = r_2(\cos\phi_2 + i\sin\phi_2)$

$$p_1 p_2 = r_1 r_2[\cos(\phi_1 + \phi_2) + i\sin(\phi_1 + \phi_2)] \qquad p_1 : p_2 = (r_1 : r_2)[\cos(\phi_1 - \phi_2) + i\sin(\phi_1 - \phi_2)]$$

$$p^{m/n} = r^{m/n}(\cos\phi + i\sin\phi)^{m/n} = r^{m/n}\left[\cos\frac{m}{n}(\phi + 2k\pi) + i\sin\frac{m}{n}(\phi + 2k\pi)\right]$$

$$\sqrt[n]{1} = \cos\frac{2k\pi}{n} + i\sin\frac{2k\pi}{n} \qquad\qquad \sqrt[n]{-1} = \cos\frac{(2k+1)\pi}{n} + i\sin\frac{(2k+1)\pi}{n}$$

$$z = x + iy \qquad r = \sqrt{x^2 + y^2} \qquad \phi = \tan^{-1}\frac{y}{x} \qquad x, y = \text{real independent variables}$$

(1) Function of Complex Variable

(a) Complex variable. If x and y are two independent real variables, then

$$z = x + iy \qquad w = f(z)$$

are called the *complex variable z* and the *function of complex variable w*, respectively.

(b) Function of complex variable is said to be *analytic* (regular, holomorphic) at a point z_0, if its derivative with respect to z exists at that point and at every point in some neighborhood of that point.

(c) Rules of operations with analytic functions are formally identical with those of functions of real variables, including their representation by power series.

(2) Exponential Functions

(a) Basic relations $(m, n = 1, 2, 3, \ldots)$ $(k = 0, \pm 1, \pm 2, \ldots)$

$$z^{m/n} = r^{m/n}\left(\cos\frac{m\phi}{n} + i\sin\frac{m\phi}{n}\right) = r^{m/n}e^{im\phi/n}$$

$$e^{ix} = 1 + \frac{ix}{1!} + \frac{(ix)^2}{2!} + \frac{(ix)^3}{3!} + \cdots + \frac{(ix)^n}{n!} + \cdots = \cos x + i\sin x$$

$$e^{-ix} = 1 - \frac{ix}{1!} + \frac{(ix)^2}{2!} - \frac{(ix)^3}{3!} + \cdots \pm \frac{(ix)^n}{n!} + \cdots = \cos x - i\sin x$$

$$e^{2k\pi i} = 1$$
$$e^{(2k+1)\pi i} = -1$$
$$e^{\phi+2k\pi i} = e^{\phi}$$
$$e^{\phi+(2k+1)\pi i} = -e^{\phi}$$
$$i^i = (0.20788\ldots)e^{2k\pi}$$

(b) Derivative and integral

$$\frac{d(e^{(a+bi)x})}{dx} = (a + bi)e^{(a+bi)x} = (a + bi)e^{ax}(\cos bx + i\sin bx)$$

$$\int e^{(a+bi)x}\,dx = \frac{e^{(a+bi)x}}{a+bi} = \frac{e^{ax}}{a^2+b^2}[(a\cos bx + b\sin bx) + i(a\sin bx - b\cos bx)]$$

(3) Logarithmic Functions

(a) Basic relations $(k = 0, \pm 1, \pm 2, \ldots)$

$$\ln z = \ln re^{i(\phi+2k\pi)} = \ln r + i(\phi + 2k\pi)$$

$$\ln a = \ln a + 2k\pi i \qquad\qquad \ln bi = \ln b + i\left(\frac{\pi}{2} + 2k\pi\right)$$

$$\ln (-a) = \ln a + i(2k+1)\pi \qquad \ln (-bi) = \ln b + i\left(\frac{3\pi}{2} + 2k\pi\right)$$

$$\ln 1 = 2k\pi i$$
$$\ln (-1) = (2k+1)\pi i$$
$$\ln i = (4k+1)\frac{\pi i}{2}$$
$$\ln (-i) = (4k+3)\frac{\pi i}{2}$$

(b) Derivative and Integral

$$\frac{d[\ln (a+bi)x]}{dx} = \frac{d[\ln (a+bi)]}{dx} + \frac{d(\ln x)}{dx} = \frac{1}{x}$$

$$\int [\ln (a+bi)x]\,dx = x[\ln (a+bi)x - 1] = x\left(\ln \sqrt{a^2+b^2} - 1 + i\tan^{-1}\frac{b}{a} + \ln x\right)$$

$z = x + iy$	a, b = real numbers	$k = 0, 1, 2, \ldots$ $\qquad n = 1, 2, 3, \ldots$
$p = a + bi$	$A = \frac{1}{2}\sqrt{(1+a)^2 + b^2}$	$B = \frac{1}{2}\sqrt{(1-a)^2 + b^2}$

(1) Trigonometric Functions

(a) Basic relations

$\sin ix = \sin (ix \pm 2k\pi) = \quad i \sinh x$	$\sin x = \sin (x \pm 2k\pi) = -i \sinh ix$
$\cos ix = \cos (ix \pm 2k\pi) = \quad \cosh x$	$\cos x = \cos (x \pm 2k\pi) = \quad \cosh ix$
$\tan ix = \tan (ix \pm k\pi) = \quad i \tanh x$	$\tan x = \tan (x \pm k\pi) = -i \tanh ix$
$\cot ix = \cot (ix \pm k\pi) = -i \coth x$	$\cot x = \cot (x \pm k\pi) = \quad i \coth ix$
$\sec ix = \sec (ix \pm 2k\pi) = \quad \text{sech} x$	$\sec x = \sec (x \pm 2k\pi) = \quad \text{sech} ix$
$\csc ix = \csc (ix \pm 2k\pi) = -i \text{csch} x$	$\csc x = \csc (x \pm 2k\pi) = \quad i \text{csch} ix$

(b) Complex variable argument

$$\sin z = \sin (z \pm 2k\pi) = -i \sinh iz = \sin x \cosh y + i \cos x \sinh y$$

$$\cos z = \cos (z \pm 2k\pi) = \quad \cosh iz = \cos x \cosh y - i \sin x \sinh y$$

$$\tan z = \tan (z \pm k\pi) = -i \tanh iz = \frac{\sin 2x + i \sinh 2y}{\cos 2x + \cosh 2y}$$

(c) Identities $\quad (k = 0, 1, 2, \ldots, n-1)$

$$\sin x = \frac{e^{ix} - e^{-ix}}{2i}$$

$$(\cos x \pm i \sin x)^n = \cos nx \pm i \sin nx$$

$$\cos x = \frac{e^{ix} + e^{-ix}}{2}$$

$$\sqrt[n]{\cos x \pm i \sin x} = \cos \frac{x + 2k\pi}{n} \pm i \sin \frac{x + 2k\pi}{n}$$

(2) Inverse Trigonometric Functions

(a) Basic relations

$\sin^{-1} ix = \quad i \sinh^{-1} x$	$\cos^{-1} x = \pm i \cosh^{-1} x$	$\tan^{-1} ix = \quad i \tanh^{-1} x$
$\csc^{-1} ix = -i \cosh^{-1} x$	$\sec^{-1} x = \pm i \, \text{sech}^{-1} x$	$\cot^{-1} ix = -i \coth^{-1} x$

(b) Complex variable argument

$$\sin^{-1} z = -\cos^{-1} z + \frac{\pi}{2}(1 \pm 4k) = -i \sinh^{-1} iz = -i \ln (iz \pm \sqrt{1 - z^2}) \pm 2k\pi$$

$$\cos^{-1} z = -\sin^{-1} z + \frac{\pi}{2}(1 \pm 4k) = -i \cosh^{-1} z = -i \ln (iz \pm \sqrt{z^2 - 1}) \pm 2k\pi$$

$$\tan^{-1} z = -\cot^{-1} z + \frac{\pi}{2}(1 \pm 2k) = -i \tanh^{-1} iz = -i \ln \sqrt{\frac{1 + iz}{1 - iz}} \pm k\pi$$

(c) Complex number argument[1]

$\sin^{-1} p = (-1)^k[\sin^{-1}(A - B) + i\theta \cosh^{-1}(A + B)] \pm k\pi$	
$\cos^{-1} p = \pm[\cos^{-1}(A - B) - i\theta \cosh^{-1}(A + B) \pm 2k\pi]$	$\theta = \begin{cases} +1 & \text{if } b \geq 0 \\ \\ -1 & \text{if } b < 0 \end{cases}$
$\tan^{-1} p = \frac{i}{2}\left[(1 \pm 2k)\pi - \tan^{-1}\frac{1 + b}{a} - \tan^{-1}\frac{1 - b}{a}\right] + i\theta \ln \sqrt{\frac{C}{D}}$	

[1]For C, D see opposite page.

$z = x + iy$	$a, b = $ real numbers	$k = 0, 1, 2, \ldots$	$n = 1, 2, 3, \ldots$
$p = a + bi$	$C = \frac{1}{2}\sqrt{(1 + b)^2 + a^2}$	$D = \frac{1}{2}\sqrt{(1 - b)^2 + a^2}$	

(1) Hyperbolic Functions

(a) Basic relations

$\sinh ix = \sinh (ix \pm 2k\pi i) = \ i \sin x$	$\sinh x = \sinh (x \pm 2k\pi i) = -i \sin ix$
$\cosh ix = \cosh (ix \pm 2k\pi i) = \ \cos x$	$\cosh x = \cosh (x \pm 2k\pi i) = \ \cos ix$
$\tanh ix = \tanh (ix \pm k\pi i) \ = \ i \tan x$	$\tanh x = \tanh (x \pm k\pi i) \ = -i \tan ix$
$\coth ix = \coth (ix \pm k\pi i) \ = -i \cot x$	$\coth x = \coth (x \pm k\pi i) \ = \ i \cot ix$
$\operatorname{sech} ix = \operatorname{sech} (ix \pm 2k\pi i) = \ \sec x$	$\operatorname{sech} x = \operatorname{sech} (x \pm 2k\pi i) = \ \sec ix$
$\operatorname{csch} ix = \operatorname{csch} (ix \pm 2k\pi i) = -i \csc x$	$\operatorname{csch} x = \operatorname{csch} (x \pm 2k\pi i) = \ i \csc ix$

(b) Complex variable argument

$$\sinh z = \sinh (z \pm 2k\pi i) = -i \sin iz = \sinh x \cos y + i \cosh x \sin y$$

$$\cosh z = \cosh (z \pm 2k\pi i) = \ \cos iz = \cosh x \cos y + i \sinh x \sin y$$

$$\tanh z = \tanh (z \pm k\pi i) \ = -i \tan iz = \frac{\sinh 2x + i \sin 2y}{\cosh 2x + \cos 2y}$$

(c) Identities　　$(k = 0, 1, 2, \ldots, n - 1)$

$$\sinh ix = \frac{e^{ix} - e^{-ix}}{2}$$

$$(\cosh x \pm \sinh x)^n = \cosh nx \pm \sinh nx$$

$$\cosh ix = \frac{e^{ix} + e^{-ix}}{2}$$

$$\sqrt[n]{\cosh x \pm \sinh x} = \cosh \frac{x + 2k\pi i}{n} \pm \sinh \frac{x + 2k\pi i}{n}$$

(2) Inverse Hyperbolic Functions

(a) Basic relations

$\sinh^{-1} ix = \ i \sin^{-1} x$	$\cosh^{-1} x = \pm i \cos^{-1} x$	$\tanh^{-1} ix = \ i \tan^{-1} x$
$\operatorname{csch}^{-1} ix = -i \csc^{-1} x$	$\operatorname{sech}^{-1} x = \pm i \sec^{-1} x$	$\coth^{-1} ix = -i \cot^{-1} x$

(b) Complex variable argument

$$\sinh^{-1} z = -i \sin^{-1} iz = \ln (\sqrt{z^2 + 1} + z) \pm 2k\pi i = -\ln (\sqrt{z^2 + 1} - z) \pm 2k\pi i$$

$$\cosh^{-1} z = \ i \cos^{-1} z = \ln (z + \sqrt{z^2 - 1}) \pm 2k\pi i = -\ln (z - \sqrt{z^2 - 1}) \pm 2k\pi i$$

$$\tanh^{-1} z = -i \tan^{-1} iz = \ln \sqrt{\frac{1 + z}{1 - z}} \pm k\pi i$$

(c) Complex number argument[1]

$\sinh^{-1} p = (-1)^k [\cosh^{-1} (C + D) + i\theta \sin^{-1} (C - D)] \pm k\pi i$	$$\theta = \begin{cases} +1 & \text{if } b \geq 0 \\ \\ -1 & \text{if } b < 0 \end{cases}$$
$\cosh^{-1} p = \pm [\cosh^{-1} (A + B) + i\theta \cos^{-1} (A - B) \pm k\pi i]$	
$\tanh^{-1} p = \dfrac{i}{2}\left[(1 \pm 2k)\pi - \tan^{-1}\dfrac{1 + a}{b} - \tan^{-1}\dfrac{1 - a}{b} \right] + \ln \sqrt{\dfrac{A}{B}}$	

[1]For A, B see opposite page.

$[A], [B]$ = real matrices of order $m \times m$ $[I]$ = unit matrix of order $m \times m$

$[\ ]^T$ = transpose (Sec. 1.10) $[\ ]^{-1}$ = inverse (Sec. 1.11)

(a) Complex matrix $[M]$, its conjugate $[\overline{M}]$, and its associate $[M*]$ are, respectively,

$$[M] = [A] + i[B] = [A + iB] \qquad [\overline{M}] = [A] - i[B] = [A - iB]$$
$$[M*] = [\overline{M}]^T = [A]^T - i[B]^T = [A - iB]^T$$

(b) Hermitian matrix is the complex generalization of the *normal matrix* (Sec. 1.11).

If $[M] = [M*]$ then $[M][M*] = [M*][M] = [M]^2$

(c) Unitary matrix is the complex generalization of the *orthogonal matrix* (Sec. 1.11).

If $[M*] = [M]^{-1}$ then $[M][M*] = [M*][M] = [I]$

(d) Involutory matrix is the complex generalization of the *orthonormal matrix* (Sec. 1.11).

If $[M] = [M*] = [M]^{-1}$ then $[M][M] = [I]$

(2) Operations

(a) Summation and resolution

$[M] + [\overline{M}] = 2[A]$ $[M] - [\overline{M}] = 2i[B]$ $[M] = \frac{1}{2}[M + M*] + \frac{1}{2}[M - M*]$

(b) Product of two hermitian matrices is a hermitian matrix, product of two unitary matrices is a unitary matrix, and product of two involutory matrices is an involutory matrix.

(3) Classification

Real matrices $B = 0$			Complex matrices $B \neq 0$		
Normal	$A = A^T$		**Hermitian**	$M = M*$	
Antinormal	$A = -A^T$		**Antihermitian**	$M = -M*$	
Orthogonal		$A^T = A^{-1}$	**Unitary**		$M* = M^{-1}$
Antiorthogonal		$A^T = -A^{-1}$	**Antiunitary**		$M* = -M^{-1}$
Orthonormal	$A = A^T = A^{-1}$		**Involutory**	$M = M* = M^{-1}$	
Antiorthonormal	$A = -A^T = -A^{-1}$		**Anti-involutory**	$M = -M* = -M^{-1}$	

12
FOURIER SERIES

(1) Basic Case

(a) Definition

Any *single-valued function* $f(\theta)$ that is *continuous* except for a *finite number of discontinuities* in an interval $-\pi < \theta < +\pi$, and has a finite number of maxima and minima in this interval may be represented by a *convergent Fourier series*.

$$f(\theta) = \frac{a_0}{2} + a_1 \cos \theta + a_2 \cos 2\theta + a_3 \cos 3\theta + \cdots + b_1 \sin \theta + b_2 \sin 2\theta + b_3 \sin 3\theta + \cdots$$

$$= \frac{a_0}{2} + \sum_{n=1}^{\infty} (a_n \cos n\theta + b_n \sin n\theta)$$

If $f(\theta)$ is a *periodic function* of θ with *period* 2π,

$$a_n = \frac{1}{\pi} \int_{-\pi}^{+\pi} f(\theta) \cos n\theta \, d\theta \qquad n = 0, 1, 2, \ldots \qquad b_n = \frac{1}{\pi} \int_{-\pi}^{+\pi} f(\theta) \sin n\theta \, d\theta \qquad n = 1, 2, \ldots$$

(b) Phase angles α and β

The cosine and sine terms in the Fourier series may be combined in a single cosine or sine series with phase angles α or β, respectively.

$$f(\theta) = \frac{A_0}{2} + \sum_{n=1}^{\infty} A_n \cos(n\theta + \alpha_n)$$

$$A_n = \sqrt{a_n^2 + b_n^2}$$

$$\alpha_n = \tan^{-1}\left(-\frac{b_n}{a_n}\right)$$

$$f(\theta) = \frac{B_0}{2} + \sum_{n=1}^{\infty} B_n \sin(n\theta + \beta_n)$$

$$B_n = \sqrt{a_n^2 + b_n^2}$$

$$\beta_n = \tan^{-1}\frac{a_n}{b_n}$$

(2) Special Cases

(a) Change in variable $\left(\theta = \frac{\pi x}{l} \text{ and } -l < x < +l\right)$

$$f(x) = \frac{\bar{a}_0}{2} + \sum_{n=1}^{\infty} \left(\bar{a}_n \cos \frac{n\pi x}{l} + \bar{b}_n \sin \frac{n\pi x}{l}\right)$$

$$\bar{a}_n = \frac{1}{l} \int_{-l}^{+l} f(x) \cos \frac{n\pi x}{l} \, dx$$

$$\bar{b}_n = \frac{1}{l} \int_{-l}^{+l} f(x) \sin \frac{n\pi x}{l} \, dx$$

(b) Change in variable $\left(\theta = \frac{2\pi t}{T} \text{ and } -\frac{T}{2} < t < +\frac{T}{2}\right)$

$$f(t) = \frac{a_0^*}{2} + \sum_{n=1}^{\infty} \left(a_n^* \cos \frac{2n\pi t}{T} + b_n^* \sin \frac{2n\pi t}{T}\right)$$

$$a_n^* = \frac{2}{T} \int_{-T/2}^{+T/2} f(t) \cos \frac{2n\pi t}{T} \, dt$$

$$b_n^* = \frac{2}{T} \int_{-T/2}^{+T/2} f(t) \sin \frac{2n\pi t}{T} \, dt$$

(1) Change in Limits

In the development of Fourier series the *limits of integral may be changed* (shifting of interval) as shown.

	$-2\pi < \theta < 0$	$\phi < \theta < \phi + 2\pi$	$0 < \theta < 2\pi$
a_n	$\dfrac{1}{\pi}\displaystyle\int_{-2\pi}^{0} f(\theta)\cos n\theta\, d\theta$	$\dfrac{1}{\pi}\displaystyle\int_{\phi}^{\phi+2\pi} f(\theta)\cos n\theta\, d\theta$	$\dfrac{1}{\pi}\displaystyle\int_{0}^{2\pi} f(\theta)\cos n\theta\, d\theta$
b_n	$\dfrac{1}{\pi}\displaystyle\int_{-2\pi}^{0} f(\theta)\sin n\theta\, d\theta$	$\dfrac{1}{\pi}\displaystyle\int_{\phi}^{\phi+2\pi} f(\theta)\sin n\theta\, d\theta$	$\dfrac{1}{\pi}\displaystyle\int_{0}^{2\pi} f(\theta)\sin n\theta\, d\theta$
	$-2l < x < 0$	$a < x < a + 2l$	$0 < x < 2l$
$\overline{a}_n$	$\dfrac{1}{l}\displaystyle\int_{-2l}^{0} f(x)\cos\dfrac{n\pi x}{l}\, dx$	$\dfrac{1}{l}\displaystyle\int_{a}^{a+2l} f(x)\cos\dfrac{n\pi x}{l}\, dx$	$\dfrac{1}{l}\displaystyle\int_{0}^{2l} f(x)\cos\dfrac{n\pi x}{l}\, dx$
$\overline{b}_n$	$\dfrac{1}{l}\displaystyle\int_{-2l}^{0} f(x)\sin\dfrac{n\pi x}{l}\, dx$	$\dfrac{1}{l}\displaystyle\int_{a}^{a+2l} f(x)\sin\dfrac{n\pi x}{l}\, dx$	$\dfrac{1}{l}\displaystyle\int_{0}^{2l} f(x)\cos\dfrac{n\pi x}{l}\, dx$
	$-T < t < 0$	$C < t < C + T$	$0 < t < T$
a_n^*	$\dfrac{2}{T}\displaystyle\int_{-T}^{0} f(t)\cos\dfrac{2n\pi t}{T}\, dt$	$\dfrac{2}{T}\displaystyle\int_{C}^{C+T} f(t)\cos\dfrac{2n\pi t}{T}\, dt$	$\dfrac{2}{T}\displaystyle\int_{0}^{T} f(t)\cos\dfrac{2n\pi t}{T}\, dt$
b_n^*	$\dfrac{2}{T}\displaystyle\int_{-T}^{0} f(t)\sin\dfrac{2n\pi t}{T}\, dt$	$\dfrac{2}{T}\displaystyle\int_{C}^{C+T} f(t)\sin\dfrac{2n\pi t}{T}\, dt$	$\dfrac{2}{T}\displaystyle\int_{0}^{T} f(t)\sin\dfrac{2n\pi t}{T}\, dt$

(2) Identities

In the Fourier series expansion the following identities are useful:

	n	n even	n odd	$\dfrac{n}{2}$ odd	$\dfrac{n}{2}$ even
$\sin n\pi$	0	0	0	0	0
$\cos n\pi$	$(-1)^n$	$+1$	-1	$+1$	$+1$
$\sin\dfrac{n\pi}{2}$		0	$(-1)^{n-1/2}$	0	0
$\cos\dfrac{n\pi}{2}$		$(-1)^{n/2}$	0	-1	$+1$

If all derivatives are finite and the series is convergent, then

$$\frac{f'(2\pi) - f'(0)}{n^2} - \frac{f'''(2\pi) - f'''(0)}{n^4} + \cdots = \int_{0}^{2\pi} f(\theta)\cos n\theta\, d\theta$$

$$1 - \frac{1}{3} + \frac{1}{5} - \frac{1}{7} + \cdots = \frac{\pi}{4} \qquad\qquad 1 + \frac{1}{2^2} + \frac{1}{3^2} + \frac{1}{4^2} + \cdots = \frac{\pi^2}{6}$$

$$1 - \frac{1}{3^3} + \frac{1}{5^3} - \frac{1}{7^3} + \cdots = \frac{\pi^3}{32} \qquad\qquad 1 - \frac{1}{2^2} + \frac{1}{3^2} - \frac{1}{4^2} + \cdots = \frac{\pi^2}{12}$$

$$1 + \frac{1}{3^4} + \frac{1}{5^4} + \frac{1}{7^4} + \cdots = \frac{\pi^4}{96} \qquad\qquad 1 + \frac{1}{2^4} + \frac{1}{3^4} + \frac{1}{4^4} + \cdots = \frac{\pi^4}{90}$$

(1) Closed Form $(s = \text{constant}; n = 1, 2, \ldots)$

$$\sum_{n=1}^{\infty} s^n \sin nx = \frac{s \sin x}{1 - 2s \cos x + s^2} \qquad s^2 < 1$$

$$\sum_{n=0}^{\infty} s^n \cos nx = \frac{1 - s \cos x}{1 - 2s \cos x + s^2} \qquad s^2 < 1$$

$$\sum_{n=1}^{\infty} \frac{s^n}{n} \sin nx = \tan^{-1} \frac{s \sin x}{1 - s \cos x} \qquad s^2 \leqslant 1$$

$$\sum_{n=1}^{\infty} \frac{s^n}{n} \cos nx = \ln \frac{1}{\sqrt{1 - 2s \cos x + s^2}} \qquad s^2 \leqslant 1$$

$$\sum_{n=1}^{\infty} \frac{\sin nx}{n} = \frac{\pi - x}{2} \qquad\qquad \sum_{n=1}^{\infty} \frac{\cos nx}{n} = \frac{1}{2} \ln \frac{1}{2(1 - \cos x)}$$

$$\sum_{n=1}^{\infty} \frac{\sin nx}{n^3} = \frac{\pi^2 x}{6} - \frac{\pi x^2}{4} + \frac{x^3}{12} \quad 0 < x < 2\pi \qquad \sum_{n=1}^{\infty} \frac{\cos nx}{n^2} = \frac{\pi^2}{6} - \frac{\pi x}{2} + \frac{x^2}{4} \quad 0 < x < 2\pi$$

$$\sum_{n=1}^{\infty} \frac{\sin nx}{n^5} = \frac{\pi^4 x}{90} - \frac{\pi^2 x^3}{36} + \frac{\pi x^4}{48} - \frac{x^5}{240} \qquad \sum_{n=1}^{\infty} \frac{\cos nx}{n^4} = \frac{\pi^4}{90} - \frac{\pi^2 x^2}{12} + \frac{\pi x^3}{12} - \frac{x^4}{48}$$

(2) Complex Form

Since the *exponential* and *trigonometric functions* are *connected* by

$$\cos \theta = \frac{e^{i\theta} + e^{-i\theta}}{2} \qquad\qquad \sin \theta = \frac{e^{i\theta} - e^{-i\theta}}{2i}$$

with $\omega_n = \frac{n\pi}{l}$ and $n = 0, \pm 1, \pm 2, \ldots$, then

$$f(x) = \frac{1}{2} \left(C_0 + \sum_{n=1}^{\infty} C_n e^{i\omega_n x} + \sum_{n=1}^{\infty} D_n e^{-i\omega_n x} \right)$$

in which

$$C_n = \frac{1}{l} \int_{-l}^{l} f(x) e^{-i\omega_n x} \, dx \qquad\qquad D_n = \frac{1}{l} \int_{-l}^{l} f(x) e^{i\omega_n x} \, dx$$

and

$$C_n = \bar{a}_n - i\bar{b}_n \qquad\qquad D_n = \bar{a}_n + i\bar{b}_n$$

or in simpler form

$$f(x) = \frac{1}{2} \sum_{n=-\infty}^{+\infty} C_n e^{-i\omega_n x}$$

The set of coefficients $\{C_n\}$ is called the *spectrum* of $f(x)$.

(1) Even Functions

$$\bar{a}_n = \frac{2}{l} \int_0^l f(x) \cos \frac{n\pi x}{l} \, dx$$

$$\bar{b}_n = 0 \qquad n = 0, 1, 2, 3$$

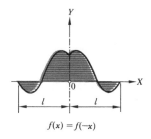

$$f(x) = f(-x)$$

$$\bar{a}_{2n} = \frac{2}{l} \int_0^l f(x) \cos \frac{2n\pi x}{l} \, dx$$

$$\bar{b}_{2n} = \frac{2}{l} \int_0^l f(x) \sin \frac{2n\pi x}{l} \, dx$$

$$\bar{a}_{2n+1} = 0 \qquad \bar{b}_{2n+1} = 0 \qquad n = 0, 1, 2, \ldots$$

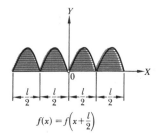

$$f(x) = f\left(x + \frac{l}{2}\right)$$

(2) Odd Functions

$$\bar{b}_n = \frac{2}{l} \int_0^l f(x) \sin \frac{n\pi x}{l} \, dx$$

$$\bar{a}_n = 0 \qquad n = 1, 2, 3, \ldots$$

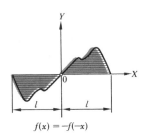

$$f(x) = -f(-x)$$

$$\bar{a}_{2n+1} = \frac{2}{l} \int_0^l f(x) \cos \frac{(2n+1)\pi x}{l} \, dx$$

$$\bar{b}_{2n+1} = \frac{2}{l} \int_0^l f(x) \sin \frac{(2n+1)\pi x}{l} \, dx$$

$$\bar{a}_{2n} = 0 \qquad \bar{b}_{2n} = 0 \qquad n = 0, 1, 2, \ldots$$

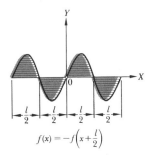

$$f(x) = -f\left(x + \frac{l}{2}\right)$$

$$\bar{b}_{2n+1} = \frac{4}{l} \int_0^{l/2} f(x) \sin \frac{(2n+1)\pi x}{l} \, dx$$

$$\bar{a}_{2n} = 0 \qquad \bar{b}_n = 0 \qquad n = 0, 1, 2, \ldots$$

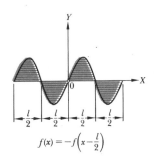

$$f(x) = -f\left(x - \frac{l}{2}\right)$$

$$f(x) = \frac{A_0}{2} + \sum_{n=1}^{\infty} \left(A_n \cos \frac{n\pi x}{L} + B_n \sin \frac{n\pi x}{L} \right) \qquad \alpha = n\pi a$$

(1)

$$\frac{A_0}{2} = p$$

$$A_n = 0$$

$$B_n = 0$$

(2)

$$\frac{A_0}{2} = 0$$

$$A_n = 0$$

$$B_n = \frac{4p}{n\pi}$$

$n = 1, 3, 5, \ldots$

(3)

$$A_0 = pa$$

$$A_n = \frac{2p \sin \alpha}{n\pi}$$

$$B_n = 0$$

$n = 1, 2, 3, \ldots$

(4)

$$\frac{A_0}{2} = 0$$

$$A_n = 0$$

$$B_n = \frac{2p(1 - \cos \alpha)}{n\pi}$$

$n = 1, 3, 5, \ldots$

(5)

Case 1 − Case 3

(6)

$$\frac{\text{Case 3 + Case 4}}{2}$$

[1]In Secs. 12.05 through 12.13 Fourier coefficients $\bar{a}_0, \bar{a}_n, \bar{b}_n$ introduced in Sec. 12.01 (2a) are denoted as A_0, A_n, B_n, respectively, to eliminate the potential conflict with the position factors a and b.

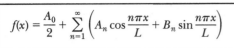

$$f(x) = \frac{A_0}{2} + \sum_{n=1}^{\infty}\left(A_n \cos\frac{n\pi x}{L} + B_n \sin\frac{n\pi x}{L}\right)$$

$\alpha = n\pi a, \beta = n\pi b$

---

(7)

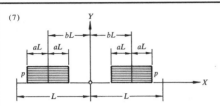

$\dfrac{A_0}{2} = 2pa$ $n = 1, 2, 3, \ldots$

$A_n = \dfrac{4p \sin\alpha \cos\beta}{n\pi}$

$B_n = 0$

---

(8)

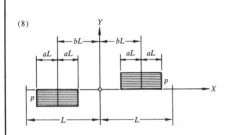

$\dfrac{A_0}{2} = 0$ $n = 1, 2, 3, \ldots$

$A_n = 0$

$B_n = \dfrac{4p \sin\alpha \sin\beta}{n\pi}$

---

(9)

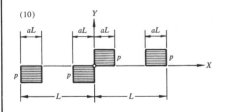

$\dfrac{A_0}{2} = 2pa$ $n = 2, 4, 6, \ldots$

$A_n = \dfrac{4p \sin\alpha}{n\pi}$

$B_n = 0$

---

(10)

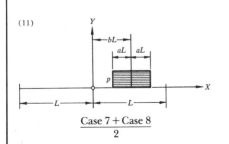

$\dfrac{A_0}{2} = 0$ $n = 1, 3, 5, \ldots$

$A_n = 0$

$B_n = \dfrac{4p(1 - \cos\alpha)}{n\pi}$

---

(11)

$$\frac{\text{Case 7} + \text{Case 8}}{2}$$

(12)

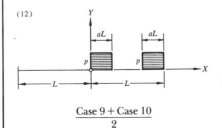

$$\frac{\text{Case 9} + \text{Case 10}}{2}$$

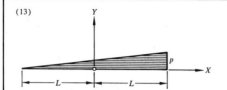

$$f(x) = \frac{A_0}{2} + \sum_{n=1}^{\infty} \left(A_n \cos\frac{n\pi x}{L} + B_n \sin\frac{n\pi x}{L} \right)$$

$$i = \sqrt{-1}$$

(13)

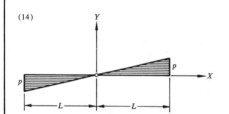

$$\frac{A_0}{2} = \frac{p}{2}$$

$$A_n = 0$$

$$B_n = -\frac{p i^{2n}}{n\pi}$$

$n = 1, 2, 3, \dots$

(14)

$$\frac{A_0}{2} = 0$$

$$A_n = 0$$

$$B_n = -\frac{2p i^{2n}}{n\pi}$$

$n = 1, 2, 3, \dots$

(15)

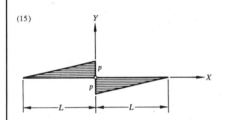

$$\frac{A_0}{2} = 0$$

$$A_n = 0$$

$$B_n = -\frac{2p}{n\pi}$$

$n = 1, 2, 3, \dots$

(16)

$$\frac{A_0}{2} = \frac{p}{2}$$

$$A_n = 0$$

$$B_n = -\frac{p}{n\pi}$$

$n = 1, 2, 3, \dots$

(17)

$$\text{Case 13} - \frac{\text{Case 25}}{2}$$

(18)

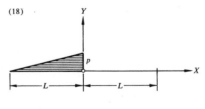

Case 16 − Case 17

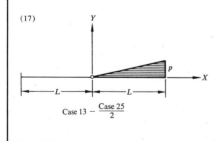

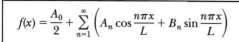

$f(x) = \dfrac{A_0}{2} + \displaystyle\sum_{n=1}^{\infty} \left(A_n \cos \dfrac{n\pi x}{L} + B_n \sin \dfrac{n\pi x}{L} \right)$	$i = \sqrt{-1}$

(19) 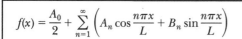	$\dfrac{A_0}{2} = \dfrac{p}{2}$ $\qquad$ $n = 1, 2, 3, \ldots$ $A_n = 0$ $B_n = \dfrac{pi^{2n}}{n\pi}$

(20) 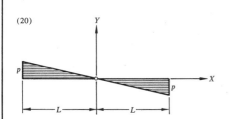	$\dfrac{A_0}{2} = 0$ $\qquad$ $n = 1, 2, 3, \ldots$ $A_n = 0$ $B_n = \dfrac{2pi^{2n}}{n\pi}$

(21)	$\dfrac{A_0}{2} = 0$ $\qquad$ $n = 1, 2, 3, \ldots$ $A_n = 0$ $B_n = \dfrac{2p}{n\pi}$

(22) 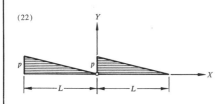	$\dfrac{A_0}{2} = \dfrac{p}{2}$ $\qquad$ $n = 1, 2, 3, \ldots$ $A_n = 0$ $B_n = \dfrac{p}{n\pi}$

(23) 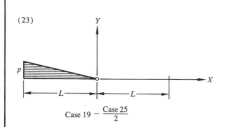 Case 19 $-\ \dfrac{\text{Case 25}}{2}$	(24) 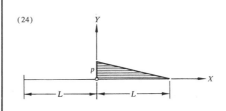 Case 22 $-$ Case 23

$$f(x) = \frac{A_0}{2} + \sum_{n=1}^{\infty} \left(A_n \cos \frac{n\pi x}{L} + B_n \sin \frac{n\pi x}{L} \right) \qquad\qquad \alpha = n\pi a$$

(25) 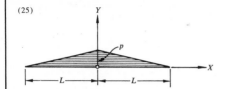	$\dfrac{A_0}{2} = \dfrac{p}{2}$ $\qquad n = 1, 3, 5, \ldots$ $A_n = \dfrac{4p}{n^2\pi^2}$ $B_n = 0$
(26) 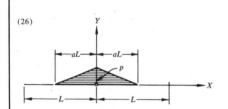	$\dfrac{A_0}{2} = \dfrac{pa}{2}$ $\qquad n = 1, 2, 3, \ldots$ $A_n = \dfrac{2p(1 - \cos\alpha)}{n^2\pi^2 a}$ $B_n = 0$
(27) 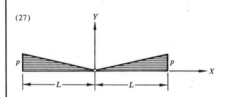	$\dfrac{A_0}{2} = \dfrac{p}{2}$ $\qquad n = 1, 3, 5, \ldots$ $A_n = -\dfrac{4p}{n^2\pi^2}$ $B_n = 0$
(28) 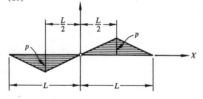	$\dfrac{A_0}{2} = 0$ $\qquad n = 1, 3, 5 \ldots$ $A_n = 0$ $\qquad i = \sqrt{-1}$ $B_n = \dfrac{8p}{n^2\pi^2}\; i^{n-1}$

(29) 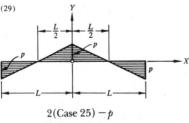 $2(\text{Case } 25) - p$	(30) 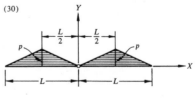 $2(\text{Case } 25 - \text{Case } 26)$

$f(x) = \dfrac{A_0}{2} + \displaystyle\sum_{n=1}^{\infty}\left(A_n\cos\dfrac{n\pi x}{L} + B_n\sin\dfrac{n\pi x}{L}\right)$	$\alpha = n\pi a,\ \beta = n\pi b$

(31)

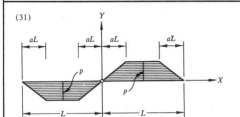

$$\frac{A_0}{2} = 0 \qquad\qquad n = 1, 3, 5, \ldots$$

$$A_n = 0$$

$$B_n = \frac{4p\sin\alpha}{n\pi\alpha}$$

(32)

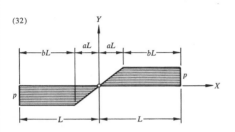

$$\frac{A_0}{2} = 0 \qquad\qquad n = 1, 2, 3, \ldots$$

$$A_n = 0 \qquad\qquad i = \sqrt{-1}$$

$$B_n = -\frac{2pi^{2n}}{n\pi}\left(1 + \frac{\sin\beta}{\alpha}\right)$$

(33)

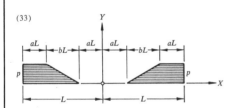

$$\frac{A_0}{2} = \frac{p}{2} \qquad\qquad n = 1, 3, 5, \ldots$$

$$A_n = -\frac{4p\cos\alpha}{n\pi\beta}$$

$$B_n = 0$$

(34)

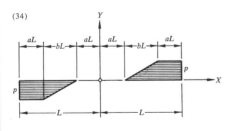

$$\frac{A_0}{2} = 0 \qquad\qquad n = 1, 2, 3, \ldots$$

$$A_n = 0 \qquad\qquad i = \sqrt{-1}$$

$$B_n = -\frac{2pi^{2n}}{n\pi}\left[1 + \frac{1 + (-1)^n}{\beta}\sin\alpha\right]$$

(35)

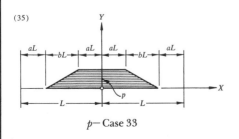

p– Case 33

(36)

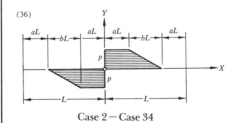

Case 2 – Case 34

$$f(x) = \frac{A_0}{2} + \sum_{n=1}^{\infty}\left(A_n \cos\frac{n\pi x}{L} + B_n \sin\frac{n\pi x}{L}\right)$$

$i = \sqrt{-1}$

(37)

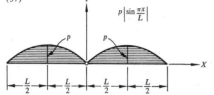

$p\left|\sin\frac{\pi x}{L}\right|$

$$\frac{A_0}{2} = \frac{2p}{\pi}$$

$$A_n = \frac{-4p}{(n-1)(n+1)\pi}$$

$$B_n = 0$$

$n = 2, 4, 6, \ldots$

(38)

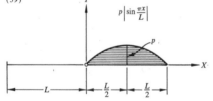

$p\left|\cos\frac{\pi x}{L}\right|$

$$\frac{A_0}{2} = \frac{2p}{\pi}$$

$$A_n = \frac{-4pi^n}{(n-1)(n+1)\pi}$$

$$B_n = 0$$

$n = 2, 4, 6, \ldots$

(39)

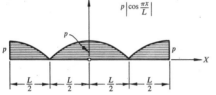

$p\left|\sin\frac{\pi x}{L}\right|$

$$\frac{A_0}{2} = \frac{p}{\pi} + \frac{p}{2}\sin\frac{\pi x}{L}$$

$$A_n = -\frac{2p}{(n-1)(n+1)\pi}$$

$$B_n = 0$$

$n = 2, 4, 6, \ldots$

(40)

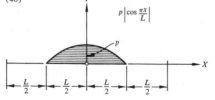

$p\left|\cos\frac{\pi x}{L}\right|$

$$\frac{A_0}{2} = \frac{p}{\pi} + \frac{p}{2}\cos\frac{\pi x}{L}$$

$$A_n = \frac{-2pi^n}{(n-1)(n+1)\pi}$$

$$B_n = 0$$

$n = 2, 4, 6, \ldots$

(41)

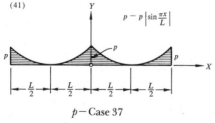

$p - p\left|\sin\frac{\pi x}{L}\right|$

p — Case 37

(42)

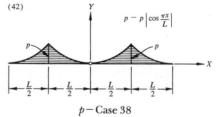

$p - p\left|\cos\frac{\pi x}{L}\right|$

p — Case 38

$$f(x) = \frac{A_0}{2} + \sum_{n=1}^{\infty} \left(A_n \cos \frac{n\pi x}{L} + B_n \sin \frac{n\pi x}{L} \right)$$

$$i = \sqrt{-1}$$

---

(43)

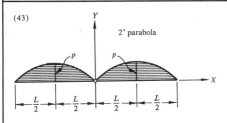

2° parabola

$$\frac{A_0}{2} = \frac{2p}{3}$$

$$A_n = -\frac{4p}{n^2\pi^2}$$

$$B_n = 0$$

$$n = 1, 2, 3, \ldots$$

---

(44)

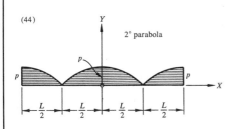

2° parabola

$$\frac{A_0}{2} = \frac{2p}{3}$$

$$A_n = -\frac{4pi^{2n}}{n^2\pi^2}$$

$$B_n = 0$$

$$n = 1, 2, 3, \ldots$$

---

(45)

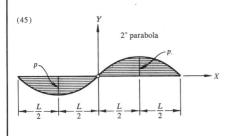

2° parabola

$$\frac{A_0}{2} = 0$$

$$A_n = 0$$

$$B_n = \frac{32p}{n^3\pi^3}$$

$$n = 1, 3, 5, \ldots$$

---

(46)

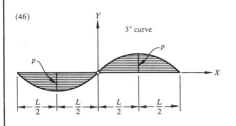

3° curve

$$f(x) = \frac{8px(L-x)(L+x)}{3L^3}$$

$$\frac{A_0}{2} = 0 \qquad A_n = 0$$

$$B_n = \frac{32pi^{2n}}{n^3\pi^3}$$

$$n = 1, 2, 3, \ldots$$

---

(47)

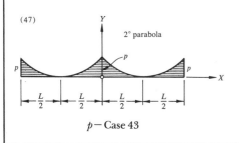

2° parabola

$$p - \text{Case } 43$$

(48)

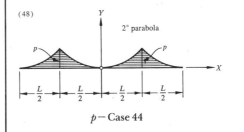

2° parabola

$$p - \text{Case } 44$$

$$f(x) = \frac{A_0}{2} + \sum_{n=1}^{\infty} \left(A_n \cos \frac{n\pi x}{L} + B_n \sin \frac{n\pi x}{L} \right)$$

$$\alpha = n\pi a$$

---

(49)

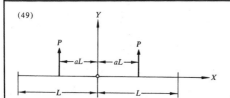

$$\frac{A_0}{2} = \frac{P}{L}$$

$$A_n = \frac{2P \cos \alpha}{L}$$

$$B_n = 0$$

$n = 1, 2, 3, \ldots$

---

(50)

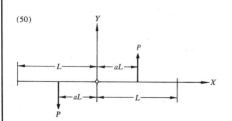

$$\frac{A_0}{2} = 0$$

$$A_n = 0$$

$$B_n = \frac{2P \sin \alpha}{L}$$

$n = 1, 2, 3, \ldots$

---

(51)

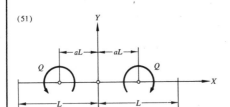

$$\frac{A_0}{2} = 0$$

$$A_n = \frac{2n\pi Q \sin \alpha}{L^2}$$

$$B_n = 0$$

$n = 1, 2, 3, \ldots$

---

(52)

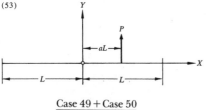

$$\frac{A_0}{2} = 0$$

$$A_n = 0$$

$$B_n = \frac{2n\pi Q \cos \alpha}{L^2}$$

$n = 1, 2, 3, \ldots$

---

(53)

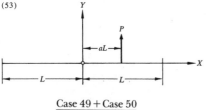

$$\frac{\text{Case } 49 + \text{Case } 50}{2}$$

(54)

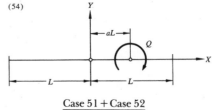

$$\frac{\text{Case } 51 + \text{Case } 52}{2}$$

13

HIGHER
TRANSCENDENT
FUNCTIONS

(1) Definition

Integrals which cannot be evaluated as finite combinations of elementary functions are called *integral functions*. The most typical functions in this group evaluated by *series expansion* (Sec. 9.01) are given below.

(2) Integral–Sine, –Cosine, and –Exponential Functions

$$\text{Si}(x) = \int_0^x \frac{\sin x}{x}\,dx = x - \frac{1}{3}\frac{x^3}{3!} + \frac{1}{5}\frac{x^5}{5!} - \frac{1}{7}\frac{x^7}{7!} + \cdots$$

$$\text{Si}(\infty) = \frac{\pi}{2}$$

$$\text{Ci}(x) = \int_{+\infty}^x \frac{\cos x}{x}\,dx = C + \ln x - \frac{1}{2}\frac{x^2}{2!} + \frac{1}{4}\frac{x^4}{4!} + \cdots$$

$$\text{Ci}(\infty) = 0$$

$$\text{Ei}(x) = \int_{+\infty}^x \frac{e^{-x}}{x}\,dx = C + \ln x - x + \frac{1}{2}\frac{x^2}{2!} - \frac{1}{3}\frac{x^3}{3!} + \cdots$$

$$\text{Ei}(\infty) = 0$$

$$\bar{\text{Ei}}(x) = \int_{-\infty}^x \frac{e^x}{x}\,dx = C + \ln x + x + \frac{1}{2}\frac{x^2}{2!} + \frac{1}{3}\frac{x^3}{3!} + \cdots$$

$$\bar{\text{Ei}}(\infty) = \infty$$

$$C = \int_{+\infty}^0 e^{-x} \ln x\,dx = 0.577\,215\,665 = \text{Euler's constant} \qquad \text{(Sec. 7.02)}$$

(3) Fresnel Integrals

$$\int_0^x \frac{\sin x}{\sqrt{x}}\,dx = 2\sqrt{x}\left(\frac{1}{3}\frac{x}{1!} - \frac{1}{7}\frac{x^3}{3!} + \frac{1}{11}\frac{x^5}{5!} - \cdots\right)$$

$$\int_0^{+\infty} \frac{\sin x}{\sqrt{x}}\,dx = \sqrt{\frac{\pi}{2}}$$

$$\int_0^x \frac{\cos x}{\sqrt{x}}\,dx = 2\sqrt{x}\left(1 - \frac{1}{5}\frac{x^2}{2!} + \frac{1}{9}\frac{x^4}{4!} - \frac{1}{13}\frac{x^6}{6!} + \cdots\right)$$

$$\int_0^{+\infty} \frac{\cos x}{\sqrt{x}}\,dx = \sqrt{\frac{\pi}{2}}$$

$$S(x) = \sqrt{\frac{2}{\pi}} \int_0^x \sin x^2\,dx$$

$$S(-x) = -S(x)$$

$$S(0) = 0$$

$$= \sqrt{\frac{2}{\pi}}\left(\frac{1}{1!}\frac{x^3}{3} - \frac{1}{3!}\frac{x^7}{7} + \frac{1}{5!}\frac{x^{11}}{11} - \cdots\right)$$

$$S(\infty) = \tfrac{1}{2}$$

$$C(x) = \sqrt{\frac{2}{\pi}} \int_0^x \cos x^2\,dx$$

$$C(-x) = -C(x)$$

$$C(0) = 0$$

$$= \sqrt{\frac{2}{\pi}}\left(\frac{1}{0!}\frac{x}{1} - \frac{1}{2!}\frac{x^5}{5} + \frac{1}{4!}\frac{x^9}{9} - \cdots\right)$$

$$C(\infty) = \tfrac{1}{2}$$

(4) Error Function

$$\text{erf}(x) = \frac{2}{\sqrt{\pi}} \int_0^x e^{-x^2}\,dx$$

$$\text{erf}(-x) = -\text{erf}(x)$$

$$\text{erf}(0) = 0$$

$$= \frac{2}{\sqrt{\pi}}\left(\frac{1}{0!}\frac{x}{1} - \frac{1}{1!}\frac{x^3}{3} + \frac{1}{2!}\frac{x^5}{5} - \frac{1}{3!}\frac{x^7}{7} + \cdots\right)$$

$$\text{erf}(\infty) = 1$$

(a) Sine integral

$$\text{Si}(x) = \int_0^x \frac{\sin x}{x}\,dx$$

x	0	1	2	3	4	5	6	7	8	9	x
0.	0.0000	0.0999	0.1996	0.2985	0.3965	0.4931	0.5881	0.6812	0.7721	0.8605	0.
1.	0.9461	1.0287	1.1080	1.1840	1.2562	1.3247	1.3892	1.4496	1.5058	1.5578	1.
2.	1.6054	1.6487	1.6876	1.7222	1.7525	1.7785	1.8004	1.8182	1.8321	1.8422	2.
3.	1.8487	1.8517	1.8514	1.8481	1.8419	1.8331	1.8219	1.8086	1.7934	1.7765	3.
4.	1.7582	1.7387	1.7184	1.6973	1.6758	1.6541	1.6325	1.6110	1.5900	1.5696	4.

(b) Cosine integral

$$\text{Ci}(x) = \int_{+\infty}^x \frac{\cos x}{x}\,dx$$

x	0	1	2	3	4	5	6	7	8	9	x
0.	$-\infty$	-1.7279	-1.0422	-0.6492	-0.3788	-0.1778	-0.0223	$+0.1051$	$+0.1983$	$+0.2761$	0.
1.	$+0.3374$	$+0.3849$	$+0.4205$	$+0.4457$	$+0.4620$	$+0.4704$	$+0.4717$	$+0.4670$	$+0.4568$	$+0.4419$	1.
2.	$+0.4230$	$+0.4005$	$+0.3751$	$+0.3472$	$+0.3173$	$+0.2859$	$+0.2533$	$+0.2201$	$+0.1865$	$+0.1529$	2.
3.	$+0.1196$	$+0.0870$	$+0.0553$	$+0.0247$	-0.0045	-0.0321	-0.0580	-0.0819	-0.1038	-0.1235	3.
4.	-0.1410	-0.1562	-0.1690	-0.1795	-0.1877	-0.1935	-0.1970	-0.1984	-0.1976	-0.1948	4.

(c) Exponential integral

$$\text{Ei}(x) = \int_{+\infty}^x \frac{e^{-x}}{x}\,dx$$

x	0	1	2	3	4	5	6	7	8	9	x
0.	$-\infty$	-1.8229	-1.2227	-0.9057	-0.7024	-0.5598	-0.4544	-0.3738	-0.3106	-0.2602	0.
1.	-0.2194	-0.1860	-0.1584	-0.1355	-0.1162	-0.1000	-0.0863	-0.0747	-0.0647	-0.0562	1.
2.	-0.0489	-0.0426	-0.0372	-0.0325	-0.0284	-0.0249	-0.0219	-0.0192	-0.0169	-0.0148	2.

(d) Error integral

$$\text{erf}(x) = \frac{2}{\sqrt{\pi}} \int_0^x e^{-x^2}\,dx$$

x	0	1	2	3	4	5	6	7	8	9	x
0.	0.0000	0.1125	0.2227	0.3286	0.4284	0.5205	0.6039	0.6778	0.7421	0.7969	0.
1.	0.8427	0.8802	0.9103	0.9340	0.9523	0.9661	0.9764	0.9838	0.9891	0.9928	1.
2.	0.9953	0.9970	0.9981	0.9989	0.9994	0.9996	0.9998	0.9999	0.9999	1.0000	2.

x = signed real number ($\|x\| \leq \infty$)	$m, n, r = 0, 1, 2, \ldots$
u = signed real number ($\|u\| \leq 1$)	$n!$ = n factorial (Sec. 1.03)

(1) Gamma Function Γ

(a) Definition

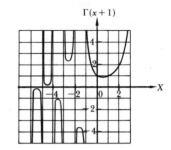

$\Gamma(x+1)$

$$\Gamma(x) = \int_0^\infty t^{x-1} e^{-t}\, dt \qquad x > 0$$

$$\Gamma(x) = \lim_{n \to \infty} \frac{n^x n!}{x(x+1)(x+2)\cdots(x+n)}$$

(b) Functional equations   (Sec. 1.03)

$$\Gamma(x+r) = (x+r-1)(x+r-2)\cdots(x+1)\Gamma(x+1) \qquad \Gamma(n+1) = n(n-1)(n-2)\cdots 2 \cdot 1 = n!$$

$$\Gamma(x-r) = \frac{\Gamma(x+1)}{(x-r)(x-r+1)(x-r+2)\cdots(x-1)x} \qquad \begin{aligned}\Gamma(n) &= (n-1)(n-2)(n-3)\cdots 2 \cdot 1\\ &= (n-1)!\end{aligned}$$

(c) Reflection equations   (Sec. A.02)

$$\Gamma(u)\Gamma(-u) = \frac{-\pi}{u \sin u\pi} \qquad\qquad \Gamma(u)\Gamma(1-u) = \frac{\pi}{\sin u\pi}$$

$$\Gamma(1+u)\Gamma(-u) = \frac{-\pi}{\sin u\pi} \qquad\qquad \Gamma(1+u)\Gamma(1-u) = \frac{u\pi}{\sin u\pi}$$

(d) Special values   (Sec. A.01)

$$\Gamma(-n) = \infty \qquad\qquad \Gamma(0) = \infty \qquad \Gamma(1) = 1 \qquad\qquad \Gamma(2) = 1$$

$$\Gamma(n+\tfrac{1}{2}) = \frac{(2n)!\sqrt{\pi}}{n!2^{2n}} \qquad \Gamma(\tfrac{1}{2}) = \sqrt{\pi} \qquad \Gamma(\tfrac{3}{2}) = \frac{\sqrt{\pi}}{2} \qquad \Gamma(-n+\tfrac{1}{2}) = \frac{(-1)^n n! 2^{2n}\sqrt{\pi}}{(2n)!}$$

(2) Pi Function – Π

(a) Definition

$$\Pi(x) = \Gamma(x+1) = \int_0^\infty t^x e^{-t}\, dt \qquad\qquad x > 0$$

(b) Recursion formulas

$$\Pi(n) = n(n-1)(n-2)\cdots 1 = n! \qquad \Pi(n) = \frac{\Pi(n+1)}{n+1} \qquad \Pi(n) = n\Pi(n-1)$$

(3) Beta Function – B

(a) Definition

$$\mathrm{B}(x, y) = \int_0^1 t^{x-1}(1-t)^{y-1}\, dt = \frac{\Gamma(x)\Gamma(y)}{\Gamma(x+y)} \qquad\qquad x > 0, y > 0$$

(b) Relations

$$\mathrm{B}(m, n) = \mathrm{B}(n, m) \qquad\qquad \mathrm{B}(m, n) = \frac{(m-1)!(n-1)!}{(m+n-1)!}$$

$C = 0.577\,215\,665 = $ Euler's constant (Sec. 7.02) $\qquad k, m, n, r = 0, 1, 2, \ldots$

$\Gamma(\ ) = $ gamma function (Sec. 13.03) $\qquad\qquad$ Z$(\ ) = $ zeta function (Sec. A.06)

(1) Digamma Function ψ

(a) Definition

$$\psi(x) = \frac{d}{dx}\left\{\ln[\Gamma(x+1)]\right\} = \frac{\dfrac{d\,\Gamma(x+1)}{dx}}{\Gamma(x+1)}$$

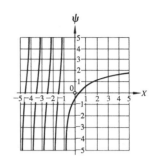

$$\psi(x) = \lim_{n \to \infty}\left(\ln n - \sum_{k=1}^{n}\frac{1}{x+k}\right) = \sum_{k=1}^{\infty}\left(\frac{1}{k} - \frac{1}{x+k}\right) - C$$

(b) Functional equations   (Sec. 8.03)

$$\psi(x+r) = \psi(x) + \sum_{k=1}^{r}\frac{1}{x+k} \qquad\qquad \psi(n) = \sum_{k=1}^{n}\frac{1}{k} - C$$

$$\psi(x-r) = \psi(x) - \sum_{k=1}^{r}\frac{1}{x+1-k} \qquad \psi(n-\tfrac{1}{2}) = 2\sum_{k=1}^{n}\frac{1}{2k-1} - \ln 4 - C$$

(c) Reflection equations

$$\psi(u) = 1 + \frac{1}{2u} - \frac{1}{(1+u)(1-u)} - \frac{\pi}{2\tan u\pi} - C - \sum_{k=1}^{8} b_{2k}u^{2k} - (|\varepsilon| \le 5 \times 10^{-11}) \qquad |u| \le 0.5$$

$b_2 = 0.202\,056\,903$	$b_4 = 0.036\,927\,755$	$b_6 = 0.008\,349\,277$	$b_8 = 0.002\,008\,393$
$b_{10} = 0.000\,494\,189$	$b_{12} = 0.000\,122\,713$	$b_{14} = 0.000\,030\,588$	$b_{16} = 0.000\,007\,637$

(d) Special values

$$\psi(-n) = \pm\infty \qquad \psi(0) = -C \qquad \psi(1) = 1 - C \qquad \psi(n) = 1 + \frac{1}{2} + \frac{1}{3} + \cdots + \frac{1}{n} - C$$

$$\psi(N) = \ln N + \frac{1}{2N} - \frac{N^{-2}}{12}\left(1 - \frac{N^{-2}}{10}\left(1 - \frac{10N^{-2}}{21}\left(1 - \frac{21N^{-2}}{20}\right)\right)\right) - (|\varepsilon| \le 5 \times 10^{-11}) \qquad N \ge 5$$

(2) Polygamma Function $\psi^{(m)}$

(a) Definition

$$\psi^{(m)}(x) = \frac{d^m}{dx^m}[\psi(x)] = \frac{d^{m+1}}{dx^{m+1}}\left\{\ln[\Gamma(x+1)]\right\} = (-1)^{m+1}m!\sum_{k=1}^{\infty}\frac{1}{(x+k)^{m+1}}$$

(b) Integral argument

$$\psi^{(1)}(n) = Z(2) - \sum_{k=1}^{n}\frac{1}{k^2} \qquad\qquad \psi^{(m)}(n) = (-1)^{m+1}m!\left[Z(m+1) - \sum_{k=1}^{n}\frac{1}{k^{m+1}}\right]$$

(c) Reflection equation

$$\psi^{(m)}(r \pm u) = \psi^{(m)}(\pm u) \mp (-1)^{m+1}m!\sum_{k=1}^{r}\frac{1}{(k \pm u)^{m+1}} \qquad |u| \le 0.5$$

(1) Elliptic Integrals, Normal Form, Formulas

$$F(k, x) = \int_0^x \frac{dx}{\sqrt{(1 - x^2)(1 - k^2 x^2)}} = F(k, \phi) = \int_0^\phi \frac{d\phi}{\sqrt{1 - k^2 \sin^2 \phi}}$$

$$E(k, x) = \int_0^x \sqrt{\frac{1 - k^2 x^2}{1 - x^2}} \, dx = E(k, \phi) = \int_0^\phi \sqrt{1 - k^2 \sin^2 \phi} \, d\phi$$

where $k = \sin \omega =$ modulus (given constant) in the interval $0 \leqslant k \leqslant +1$, $x =$ independent variable in interval $-1 \leqslant x \leqslant +1$.

(2) Elliptic Integrals, Complete Form, Formulas

$$F\left(k, \frac{\pi}{2}\right) = K = \frac{\pi}{2}\left\{1 + \left(\frac{1}{2}\right)^2 k^2 + \left[\frac{(1)(3)}{(2)(4)}\right]^2 k^4 + \left[\frac{(1)(3)(5)}{(2)(4)(6)}\right]^2 k^6 + \cdots\right\}$$

$$E\left(k, \frac{\pi}{2}\right) = E = \frac{\pi}{2}\left\{1 - \left(\frac{1}{2}\right)^2 \frac{k^2}{1} - \left[\frac{(1)(3)}{(2)(4)}\right]^2 \frac{k^4}{3} - \left[\frac{(1)(3)(5)}{(2)(4)(6)}\right]^2 \frac{k^6}{5} - \cdots\right\}$$

(3) Elliptic Integrals, Degenerated Form

$$F(0, x) = \sin^{-1} x \qquad K = \frac{\pi}{2} \qquad\qquad F(1, x) = \tanh^{-1} x \qquad K = \infty$$

$$E(0, x) = \sin^{-1} x \qquad E = \frac{\pi}{2} \qquad\qquad E(1, x) = x \qquad\qquad E = 1$$

13.06 ELLIPTIC FUNCTIONS

(1) Definition

If $u = F(k, \phi)$, the inverse function is designated as $\phi = \text{am } u$ and is called the *elliptic function of Jacobi.*

$$x = \text{sn } u = \sin(\text{am } u) \qquad\qquad \text{sn}^2 u + \text{cn}^2 u = 1$$

$$\sqrt{1 - x^2} = \text{cn } u = \cos(\text{am } u) \qquad\qquad \text{dn}^2 u + k^2 \text{sn}^2 u = 1$$

$$\sqrt{1 - k^2 x^2} = \text{dn } u = \sqrt{1 - k^2 \text{sn}^2 u} \qquad \text{dn}^2 u - k^2 \text{cn}^2 u = 1 - k^2$$

$$\text{sn } u = u - \frac{(1 + k^2) u^3}{3!} + \frac{(1 + 14k^2 + k^4) u^5}{5!} - \cdots$$

$$\text{cn } u = 1 - \frac{u^2}{2!} + \frac{(1 + 4k^2) u^4}{4!} - \frac{(1 + 44k^2 + 16k^4) u^6}{6!} + \cdots$$

$$\text{dn } u = 1 - \frac{k^2 u^2}{2!} + \frac{k^2(4 + k^2) u^4}{4!} - \frac{k^2(16 + 44k^2 + k^4) u^6}{6!} + \cdots$$

$$\text{sn } (k = 0) = \sin u \qquad \text{cn } (k = 0) = \cos u \qquad \text{dn } (k = 0) = 1$$

$$\text{sn } (k = 1) = \tanh u \qquad \text{cn } (k = 1) = \frac{1}{\cosh u} \qquad \text{dn } (k = 1) = \frac{1}{\cosh u}$$

(2) Derivatives

$$\frac{d}{dx}(\text{sn } u) = \text{cn } u \text{ dn } u$$

$$\frac{d}{dx}(\text{cn } u) = -\text{sn } u \text{ dn } u$$

$$\frac{d}{dx}(\text{dn } u) = -k^2 \text{sn } u \text{ cn } u$$

(3) Integrals

$$\int \text{sn } u \, du = \frac{1}{k} \ln(\text{dn } u - k \text{ cn } u)$$

$$\int \text{cn } u \, du = \frac{1}{k} \cos^{-1}(\text{dn } u)$$

$$\int \text{dn } u \, du = \sin^{-1}(\text{sn } u)$$

(a) First kind

$$F(k, \phi) = \int_0^\phi \frac{d\phi}{\sqrt{1 - k^2 \sin^2 \phi}} \qquad k = \sin \omega$$

ω / ϕ	0°	10°	20°	30°	40°	50°	60°	70°	80°	90°	ϕ / ω
k	0	0.1737	0.3420	0.5000	0.6428	0.7660	0.8660	0.9397	0.9848	1.0000	k
0°	0.0000	0.0000	0.0000	0.0000	0.0000	0.0000	0.0000	0.0000	0.0000	0.0000	0°
10°	0.1745	0.1746	0.1746	0.1748	0.1749	0.1751	0.1752	0.1753	0.1754	0.1754	10°
20°	0.3491	0.3493	0.3499	0.3508	0.3520	0.3533	0.3545	0.3555	0.3561	0.3564	20°
30°	0.5236	0.5243	0.5263	0.5294	0.5334	0.5379	0.5422	0.5459	0.5484	0.5493	30°
40°	0.6981	0.6997	0.7043	0.7117	0.7213	0.7323	0.7436	0.7535	0.7604	0.7629	40°
50°	0.8727	0.8756	0.8842	0.8982	0.9173	0.9401	0.9647	0.9876	1.0044	1.0107	50°
60°	1.0472	1.0519	1.0660	1.0896	1.1226	1.1643	1.2126	1.2619	1.3014	1.3170	60°
70°	1.2217	1.2286	1.2495	1.2853	1.3372	1.4068	1.4944	1.5959	1.6918	1.7354	70°
80°	1.3963	1.4057	1.4344	1.4846	1.5597	1.6660	1.8125	2.0119	2.2653	2.4363	80°
90°	1.5708	1.5828	1.6200	1.6858	1.7868	1.9356	2.1565	2.5046	3.1534	∞	90°

(b) Second kind

$$E(k, \phi) = \int_0^\phi \sqrt{1 - k^2 \sin^2 \phi}\, d\phi \qquad k = \sin \omega$$

ω / ϕ	0°	10°	20°	30°	40°	50°	60°	70°	80°	90°	ϕ / ω
k	0	0.1737	0.3420	0.5000	0.6428	0.7660	0.8660	0.9397	0.9848	1.0000	k
0°	0.0000	0.0000	0.0000	0.0000	0.0000	0.0000	0.0000	0.0000	0.0000	0.0000	0°
10°	0.1745	0.1745	0.1744	0.1743	0.1742	0.1740	0.1739	0.1738	0.1737	0.1736	10°
20°	0.3491	0.3489	0.3483	0.3473	0.3462	0.3450	0.3438	0.3429	0.3422	0.3420	20°
30°	0.5236	0.5229	0.5209	0.5179	0.5141	0.5100	0.5061	0.5029	0.5007	0.5000	30°
40°	0.6981	0.6966	0.6921	0.6851	0.6763	0.6667	0.6575	0.6497	0.6446	0.6428	40°
50°	0.8727	0.8698	0.8614	0.8483	0.8317	0.8134	0.7954	0.7801	0.7697	0.7660	50°
60°	1.0472	1.0426	1.0290	1.0076	0.9801	0.9493	0.9184	0.8914	0.8728	0.8660	60°
70°	1.2217	1.2149	1.1949	1.1632	1.1221	1.0750	1.0266	0.9830	0.9514	0.9397	70°
80°	1.3963	1.3870	1.3597	1.3161	1.2590	1.1926	1.1225	1.0565	1.0054	0.9848	80°
90°	1.5708	1.5589	1.5238	1.4675	1.3931	1.3055	1.2111	1.1184	1.0401	1.0000	90°

13.08 OTHER ELLIPTIC INTEGRALS, NORMAL FORM

$$\lambda = \sqrt{1 - k^2 \sin^2 \phi} \qquad k = \sin \omega$$

$$\kappa = \frac{F - E}{k^2} \qquad l = \cos \omega$$

$$\int_0^\phi \frac{\sin^2 \phi \, d\phi}{\lambda} = \kappa \qquad\qquad \int_0^\phi \frac{\cos^2 \phi \, d\phi}{\lambda} = F - \kappa$$

$$\int_0^\phi \frac{\sin^2 \phi \, d\phi}{\lambda^2} = \frac{(F - \kappa)\lambda - \sin \phi \cos \phi}{l^2 \lambda} \qquad\qquad \int_0^\phi \frac{\cos^2 \phi \, d\phi}{\lambda^2} = \frac{\kappa\lambda + \sin \phi \cos \phi}{\lambda}$$

ω	$F\left(k,\frac{\pi}{2}\right)$	$E\left(k,\frac{\pi}{2}\right)$	ω	$F\left(k,\frac{\pi}{2}\right)$	$E\left(k,\frac{\pi}{2}\right)$	ω	$F\left(k,\frac{\pi}{2}\right)$	$E\left(k,\frac{\pi}{2}\right)$
0°	1.5708	1.5708	50°	1.9356	1.3055	82°0′	3.3699	1.0278
1°	1.5709	1.5707	51°	1.9539	1.2963	82°12′	3.3946	1.0267
2°	1.5713	1.5703	52°	1.9729	1.2870	82°24′	3.4199	1.0256
3°	1.5719	1.5697	53°	1.9927	1.2776	82°36′	3.4460	1.0245
4°	1.5727	1.5689	54°	2.0133	1.2682	82°48′	3.4728	1.0234
5°	1.5738	1.5678	55°	2.0347	1.2587	83°0′	3.5004	1.0223
6°	1.5751	1.5665	56°	2.0571	1.2492	83°12′	3.5288	1.0213
7°	1.5767	1.5650	57°	2.0804	1.2397	83°24′	3.5581	1.0202
8°	1.5785	1.5632	58°	2.1047	1.2301	83°36′	3.5884	1.0192
9°	1.5805	1.5611	59°	2.1300	1.2206	83°48′	3.6196	1.0182
10°	1.5828	1.5589	60°	2.1565	1.2111	84°0′	3.6519	1.0172
11°	1.5854	1.5564	61°	2.1842	1.2015	84°12′	3.6853	1.0163
12°	1.5882	1.5537	62°	2.2132	1.1921	84°24′	3.7198	1.0153
13°	1.5913	1.5507	63°	2.2435	1.1826	84°36′	3.7557	1.0144
14°	1.5946	1.5476	64°	2.2754	1.1732	84°48′	3.7930	1.0135
15°	1.5981	1.5442	65°	2.3088	1.1638	85°0′	3.8317	1.0127
16°	1.6020	1.5405	66°	2.3439	1.1546	85°12′	3.8721	1.0118
17°	1.6061	1.5367	67°	2.3809	1.1454	85°24′	3.9142	1.0110
18°	1.6105	1.5326	68°	2.4198	1.1362	85°36′	3.9583	1.0102
19°	1.8151	1.5283	69°	2.4610	1.1273	85°48′	4.0044	1.0094
20°	1.6200	1.5238	70°0′	2.5046	1.1184	86°0′	4.0528	1.0087
21°	1.6252	1.5191	70°30′	2.5273	1.1140	86°12′	4.1037	1.0079
22°	1.6307	1.5142	71°0′	2.5507	1.1096	86°24′	4.1574	1.0072
23°	1.6365	1.5090	71°30′	2.5749	1.1053	86°36′	4.2142	1.0065
24°	1.6426	1.5037	72°0′	2.5998	1.1011	86°48′	4.2746	1.0059
25°	1.6490	1.4981	72°30′	2.6256	1.0968	87°0′	4.3387	1.0053
26°	1.6557	1.4924	73°0′	2.6521	1.0927	87°12′	4.4073	1.0047
27°	1.6627	1.4864	73°30′	2.6796	1.0885	87°24′	4.4812	1.0041
28°	1.6701	1.4803	74°0′	2.7081	1.0844	87°36′	4.5609	1.0036
29°	1.6777	1.4740	74°30′	2.7375	1.0804	87°48′	4.6477	1.0031
30°	1.6858	1.4675	75°0′	2.7681	1.0764	88°0′	4.7427	1.0026
31°	1.6941	1.4608	75°30′	2.7998	1.0725	88°12′	4.8479	1.0022
32°	1.7028	1.4539	76°0′	2.8327	1.0686	88°24′	4.9654	1.0017
33°	1.7119	1.4469	76°30′	2.8669	1.0648	88°36′	5.0988	1.0014
34°	1.7214	1.4397	77°0′	2.9026	1.0611	88°48′	5.2327	1.0010
35°	1.7313	1.4323	77°30′	2.9397	1.0574	89°0′	5.4349	1.0008
36°	1.7415	1.4248	78°0′	2.9786	1.0538	89°6′	5.5402	1.0006
37°	1.7522	1.4171	78°30′	3.0192	1.0502	89°12′	5.6579	1.0005
38°	1.7633	1.4092	79°0′	3.0617	1.0468	89°18′	5.7914	1.0005
39°	1.7748	1.4013	79°30′	3.1064	1.0434	89°24′	5.9455	1.0003
40°	1.7868	1.3931	80°0′	3.1534	1.0401	89°30′	6.1278	1.0002
41°	1.7992	1.3849	80°12′	3.1729	1.0388	89°36′	6.3509	1.0001
42°	1.8122	1.3765	80°24′	3.1928	1.0375	89°42′	6.6385	1.0001
43°	1.8256	1.3680	80°36′	3.2132	1.0363	89°48′	7.0440	1.0000
44°	1.8396	1.3594	80°48′	3.2340	1.0350	89°54′	7.7371	1.0000
45°	1.8541	1.3506	81°0′	3.2553	1.0338	90°0′	∞	1.0000
46°	1.8692	1.3418	81°12′	3.2771	1.0326			
47°	1.8848	1.3329	81°24′	3.2995	1.0313			
48°	1.9011	1.3238	81°36′	3.3223	1.0302			
49°	1.9180	1.3147	81°48′	3.3458	1.0290			

(Ref. 13.04, p. 85)

14

ORDINARY DIFFERENTIAL EQUATIONS

(1) Definitions

A differential equation is an *algebraic* or *transcendent equality* involving differentials or derivatives. The *order* of a differential equation is the order of the highest derivative. The *degree* of a differential equation is the algebraic degree of the highest derivative. The *number of independent variables* defines the differential equation as an *ordinary differential equation* (single independent variable) or a *partial differential equation* (two or more independent variables). A *homogeneous* differential equation has all terms of the same degree in the independent variable and its derivatives. If one or more terms do not involve the independent variable, the differential equation is *nonhomogeneous*. A homogeneous equation obtained from a nonhomogeneous equation by setting the nonhomogeneous terms equal to zero is the *reduced equation*. A *linear differential equation* consists of linear terms. A linear term is one which is of the first degree in the independent variables and their derivatives.

(2) Solution

An ordinary differential equation of order n given in general form as

$$F[x, y(x), y'(x), y''(x), \ldots, y^{(n)}(x)] = 0$$

has a *general solution* (general integral)

$$y = y(x, C_1, C_2, \ldots, C_n)$$

where $C_1, C_2, \ldots, C_n$ are *arbitrary* and *independent constants*. A general solution of a reduced differential equation is the *complementary solution* (complementary function). A *particular solution* is a special case of the general solution with definite values assigned to the constants. A *singular solution* is a solution which cannot be obtained from the general solution by specifying the values of the arbitrary constants.

14.02 SPECIAL FIRST-ORDER DIFFERENTIAL EQUATIONS

(1) Direct Separation

Equation

$$y' = f(x)$$

$$y' = g(y)$$

$$y' = f(x)g(y)$$

$$y' = \frac{f(x)}{g(y)}$$

$$y' = \frac{g(y)}{f(x)}$$

Solution

$$y = \int f(x) \, dx + C$$

$$x = \int \frac{dy}{g(y)} + C$$

$$\int \frac{dy}{g(y)} = \int f(x) \, dx + C$$

$$\int g(y) \, dy = \int f(x) \, dx + C$$

$$\int \frac{dy}{g(y)} = \int \frac{dx}{f(x)} + C$$

(2) Separation by Substitution

Equation, substitution

$$y f(x, y) \, dx + x g(x, y) \, dy = 0$$

$$y = rx$$

$$y f(x, y) \, dx + x g(x, y) \, dy = 0$$

$$y = \frac{r}{x}$$

Solution

$$\int \frac{dx}{x} = \int \frac{g(r) \, dr}{f(r) + r g(r)} + C$$

$$\int \frac{dx}{x} = \int \frac{f(r) \, dr}{r[g(r) - f(r)]} + C$$

(1) Direct Solution　　　$(a = \text{constant})$

Equation	**Solution**
$y'' = a$	$y = \frac{1}{2} a x^2 + C_1 x + C_2$
$y'' = f(x)$	$y = \iint f(x)\, dx\, dx + C_1 x + C_2$

(2) Substitution　　　$(y' = \psi)$

Equation

$y'' = f(y)$

$y'' = f(y')$

$y'' = f(x, y')$

$y'' = f(y, y')$

Solution

$$\int \psi\, d\psi = \int f(y)\, dy + C$$

$$x = \pm \int \frac{dy}{\sqrt{2 \int f(y)\, dy + C_1}} + C_2$$

$$\int \frac{d\psi}{f(\psi)} = \int dx + C$$

$$x = \int \frac{d\psi}{f(\psi)} + C_1 \qquad\qquad y = \int \frac{\psi\, d\psi}{f(\psi)} + C_2$$

$$\psi' = f(x, \psi)$$

$$\psi = f(x, C_1) \qquad\qquad y = \int f(x, C_1)\, dx + C_2$$

$$\psi \frac{d\psi}{dy} = f(y, \psi)$$

$$\psi = f(y, C_1) \qquad\qquad x = \int \frac{dy}{f(y, C_1)} + C_2$$

14.04　nth-ORDER DIFFERENTIAL EQUATION, SPECIAL CASE

Equation

$$\boxed{\frac{d^{(n)}y}{dx^n} = f(x)}$$

Solution

$$y = \frac{1}{(n-1)!} \int_0^x f(\tau)\,(x - \tau)^{n-1}\, d\tau + g(x)$$

$$g(x) = C_0 + C_1 x + C_2 x^2 + \cdots + C_{n-1} x^{n-1}$$

14.05　EXACT DIFFERENTIAL EQUATION

If M and N are functions of (x, y) and $\partial M / \partial y = \partial N / \partial x$, then $M\, dx + N\, dy = 0$ is an exact differential equation, the solution of which is

$$\int M\, dx + \int \left(N - \int \frac{\partial M}{\partial y}\, dx \right) dy + C = 0 \qquad \text{or} \qquad \int N\, dy + \int \left(M - \int \frac{\partial N}{\partial x}\, dy \right) dx + C = 0$$

If the condition $\partial M / \partial y = \partial N / \partial x$ is not satisfied, there exists a function $\psi(x, y) = \psi$ (*integrating factor*) such that

$$\frac{\partial (\psi M)}{\partial y} = \frac{\partial (\psi N)}{\partial x}$$

(1) Variable Coefficients $\boxed{y' + P(x)y = Q(x)}$

Condition   **Solution**

$Q(x) = 0$

$$y = C \exp\left[-\int P(x)\,dx\right]$$

$Q(x) \neq 0$

$$y = \exp\left[-\int P(x)\,dx\right]$$
$$\times \left\{\int Q(x) \exp\left[\int P(x)\,dx\right]dx + C\right\}$$

(2) Constant Coefficients $(A, B, \alpha, \beta = \text{constants})$ $\boxed{y' + By = Q(x)}$

Condition   **Solution**

$Q(x) = 0$
$$y = Ce^{-Bx}$$

$Q(x) = A$
$$y = Ce^{-Bx} + \frac{A}{B}$$

$Q(x) = Ax$
$$y = Ce^{-Bx} + \frac{A}{B}\left(x - \frac{1}{B}\right)$$

$Q(x) = Ax^2$
$$y = Ce^{-Bx} + \frac{A}{B}\left(x^2 - \frac{2x}{B} + \frac{2}{B^2}\right)$$

$Q(x) = Af(x)$
$$y = Ce^{-Bx} + \frac{A}{B}\left[f(x) - \frac{f'(x)}{B} + \frac{f''(x)}{B^2} - \cdots\right]$$

$Q(x) = Ae^{\alpha x}$
$$y = Ce^{-Bx} + \frac{A}{\alpha + B}e^{\alpha x}$$

$Q(x) = A \sin \beta x$
$$y = Ce^{-Bx} + \frac{A(B \sin \beta x - \beta \cos \beta x)}{\beta^2 + B^2}$$

$Q(x) = A \cos \beta x$
$$y = Ce^{-Bx} + \frac{A(B \cos \beta x + \beta \sin \beta x)}{\beta^2 + B^2}$$

$Q(x) = Ae^{\alpha x} \sin \beta x$
$$y = Ce^{-Bx} + \frac{A[(\alpha + B) \sin \beta x - \beta \cos \beta x]}{\beta^2 + (\alpha + B)^2}$$

$Q(x) = Ae^{\alpha x} \cos \beta x$
$$y = Ce^{-Bx} + \frac{A[\beta \sin \beta x + (\alpha + B) \cos \beta x]}{\beta^2 + (\alpha + B)^2}$$

(3) Bernoulli's Equation $\boxed{y' + P(x)y = Q(x)y^n}$

Substitution:
$$y = z^{1/(1-n)} \qquad y' = \frac{1}{1-n}z^{n/(1-n)}z'$$

Reduced equation:
$$z' + (1-n)P(x)z = (1-n)Q(x)$$

Solution:
$$y^{1-n} = \exp\left[(n-1)\int P(x)\,dx\right]\left\{(1-n)\int Q(x)\exp\left[(1-n)\int P(x)\,dx\right]dx + C\right\}$$

(1) Standard Form

A linear differential equation of order n with constant coefficients is given as

$$y^{(n)} + a_1 y^{(n-1)} + a_2 y^{(n-2)} + \cdots + a_n y = f(x)$$

where $f(x)$ is an *arbitrary function* of x.

(2) General Solution

For such an equation, the general solution is

$$y = y_C + y_P$$

where $y_C = $ *complementary function* and $y_P = $ *particular solution*.

(3) Complementary Function

By the substitution $y = e^{\lambda x}$ the reduced differential equation transforms into

$$f(\lambda) e^{\lambda x} = (\lambda^n + a_1 \lambda^{n-1} + a_2 \lambda^{n-2} + \cdots + a_n) e^{\lambda x} = 0$$

where $f(\lambda) = 0$ is the *characteristic equation*, the roots of which take one of the forms given below (or their combinations) and yield the *coefficients of the complementary functions* given below.

Roots	Complementary function
Real, distinct $\lambda_1 \neq \lambda_2 \neq \cdots \neq \lambda_{n-1} \neq \lambda_n$	$y_C = C_1 e^{\lambda_1 x} + C_2 e^{\lambda_2 x} + \cdots + C_n e^{\lambda_n x}$
Real, repeated $\lambda_1 = \lambda_2 = \cdots = \lambda_{n-1} = \lambda_n$	$y_C = (C_1 + C_2 x + C_3 x + \cdots + C_n x^{n-1}) e^{\lambda x}$
Complex, distinct $\lambda_1 = \alpha + \beta i \qquad \lambda_2 = \alpha - \beta i$ $\cdots\cdots\cdots\cdots\cdots\cdots\cdots\cdots$ $\lambda_{n-1} = \gamma + \delta i \qquad \lambda_n = \gamma - \delta i$	$y_C = e^{\alpha x}(C_1 \cos \beta x + C_2 \sin \beta x)$ $\quad + \cdots + e^{\gamma x}(C_{n-1} \cos \delta x + C_n \sin \delta x)$
Complex, repeated $\lambda_1 = \lambda_3 = \cdots \lambda_{n-1} = \alpha + \beta i$ $\lambda_2 = \lambda_4 = \cdots \lambda_n = \alpha - \beta i$	$y_C = e^{\alpha x}(C_1 + C_3 x + \cdots + C_{n-1} x^{n/2-1}) \cos \beta x$ $\quad + e^{\alpha x}(C_2 + C_4 x + \cdots + C_n x^{n/2}) \sin \beta x$

(4) Particular Solution

$$y_P = e^{\lambda_n x} \int e^{(\lambda_{n-1} - \lambda_n)x} \int e^{(\lambda_{n-2} - \lambda_{n-1})x} \cdots \int e^{(\lambda_1 - \lambda_2)x} \int e^{-\lambda_1 x} f(x) \, (dx)^n$$

$$y'' + ay = f(x) \qquad \lambda^2 + a = 0 \qquad \boxed{a \neq 0}$$

(1) Complementary Solution (a = real number)

$$
\begin{array}{ll}
a < 0 & a > 0 \\
\lambda_{1,2} = \pm\sqrt{-a} = \pm\alpha & \lambda_{1,2} = \pm\sqrt{-a} = \pm i\alpha \\
y_C = A \cosh \alpha x + B \sinh \alpha x & y_C = A \cos \alpha x + B \sin \alpha x
\end{array}
$$

(2) Particular Solution (c, b, ω = real numbers)

$$y_P = \frac{1}{a}\left[f(x) - \frac{f''(x)}{a} + \frac{f^{1v}(x)}{a^2} - \frac{f^{v1}(x)}{a^3} + \cdots \right]$$

$$\boxed{\begin{array}{l} u = a + b^2 - \omega^2 \\ v = 2b\omega \end{array}}$$

$f(x)$	y_P	$f(x)$	y_P
c	$\dfrac{c}{a}$	cx^2	$\dfrac{c}{a}\left(x^2 - \dfrac{2}{a}\right)$
cx	$\dfrac{cx}{a}$	ce^{bx}	$\dfrac{ce^{bx}}{a+b^2}$
$c \cos \omega x$	$\dfrac{c \cos \omega x}{a - \omega^2}$	$ce^{bx} \cos \omega x$	$c\dfrac{u \cos \omega x + v \sin \omega x}{u^2 + v^2} e^{bx}$
$c \sin \omega x$	$\dfrac{c \sin \omega x}{a - \omega^2}$	$ce^{bx} \sin \omega x$	$c\dfrac{u \sin \omega x - v \cos \omega x}{u^2 + v^2} e^{bx}$

14.09 THIRD-ORDER DIFFERENTIAL EQUATION

$$y''' + py'' + qy' + ry = f(x) \qquad \lambda^3 + p\lambda^2 + q\lambda + r = 0 \qquad \boxed{p, q, r \neq 0}$$

(1) Complementary Solution (p, q, r = real numbers, D given in Sec. 1.05)

$$
\begin{array}{lll}
D < 0 & \lambda_1 = \alpha, \lambda_2 = \beta, \lambda_3 = \gamma & y_C = Ae^{\alpha x} + Be^{\beta x} + Ce^{\gamma x} \\
D = 0 & \lambda_1 = \lambda_2 = \alpha, \lambda_3 = \gamma & y_C = (A + Bx)e^{\alpha x} + Ce^{\gamma x} \\
D = 0 & \lambda_1 = \lambda_2 = \lambda_3 = \alpha & y_C = (A + Bx + Cx^2)e^{\alpha x} \\
D > 0 & \lambda_1, \lambda_2 = \alpha \pm i\beta, \lambda_3 = \gamma & y_C = (A \cos \beta x + B \sin \beta x)e^{\alpha x} + Ce^{\gamma x}
\end{array}
$$

(2) Particular Solution (b, c, ω = real numbers)

$$y_P = \sum_{k=0}^{\infty} s_k f^{(k)}(x) \qquad s_0 = \frac{1}{r} \qquad s_1 = -\frac{q}{r^2} \qquad s_2 = -\frac{pr - q^2}{r^3}$$

$$s_k = -\frac{1}{r}(qs_{k-1} + ps_{k-2} + s_{k-3})$$

$$\boxed{\begin{array}{l} u = r + qb + pb^2 + b^3 - (p + 3b)\omega^2 \\ v = (q + 2pb + 3b^2)\omega - \omega^3 \end{array}}$$

$f(x)$	y_P	$f(x)$	y_P
c	$\dfrac{c}{r}$	cx^2	$\dfrac{c}{r}\left[x^2 - \dfrac{2q}{r}x + \dfrac{2(q^2 - pr)}{r^2}\right]$
cx	$\dfrac{c}{r}\left(x - \dfrac{q}{r}\right)$	ce^{bx}	$\dfrac{ce^{bx}}{b^3 + pb^2 + qb + r}$
$c \cos \omega x$	$c\dfrac{u \cos \omega x + v \sin \omega x^*}{u^2 + v^2}$	$ce^{bx} \cos \omega x$	$c\dfrac{u \cos \omega x + v \sin \omega x}{u^2 + v^2} e^{bx}$
$c \sin \omega x$	$c\dfrac{u \sin \omega x - v \cos \omega x^*}{u^2 + v^2}$	$ce^{bx} \sin \omega x$	$c\dfrac{u \sin \omega x - v \cos \omega x}{u^2 + v^2} e^{bx}$

*$*b = 0$ in u and v.

$$y'' + py' + qy = f(x) \qquad \lambda^2 + p\lambda + q = 0 \qquad p, q \neq 0$$

(1) Complementary Solution   (p, q = real numbers, $D = p^2 - 4q$)

$D > 0$	$D = 0$	$D < 0$
$\lambda_{1,2} = \dfrac{-p \pm \sqrt{D}}{2} = \alpha, \beta$	$\lambda_{1,2} = -\dfrac{p}{2} = \lambda$	$\lambda_{1,2} = \dfrac{-p \pm \sqrt{D}}{2} = \alpha \pm i\beta$
$y_C = A e^{\alpha x} + B e^{\beta x}$	$y_C = (A + Bx) e^{\lambda x}$	$y_C = (A \cos \beta x + B \sin \beta x) e^{\alpha x}$

(2) Particular solution   (b, c, ω = real numbers)

$$y_P = \frac{1}{q}\left\{ f(x) + \sum_{k=1}^{\infty} \left(\frac{-1}{q}\right)^k \left[(p-q)^k + q^k \right] f^{(k)}(x) \right\}$$

$$\boxed{\begin{aligned} u &= q + bp + b^2 - \omega^2 \\ v &= (2b + p)\omega \end{aligned}}$$

$f(x)$	y_P	$f(x)$	y_P
c	$\dfrac{c}{q}$	cx^2	$\dfrac{c}{q}\left[x^2 - \dfrac{2px}{q} + \dfrac{2(p^2 - q)}{q^2} \right]$
cx	$\dfrac{c}{q}\left(x - \dfrac{p}{q} \right)$	ce^{bx}	$\dfrac{ce^{bx}}{q + pb + b^2}$
$c \cos \omega x$	$c\,\dfrac{u \cos \omega x + v \sin \omega x}{u^2 + v^2}$*	$ce^{bx} \cos \omega x$	$c\,\dfrac{u \cos \omega x + v \sin \omega x}{u^2 + v^2} e^{bx}$
$c \sin \omega x$	$c\,\dfrac{u \sin \omega x - v \cos \omega x}{u^2 + v^2}$*	$ce^{bx} \sin \omega x$	$c\,\dfrac{u \sin \omega x - v \cos \omega x}{u^2 + v^2} e^{bx}$

*$b = 0$ in u and v.

14.11   FOURTH-ORDER DIFFERENTIAL EQUATION

$$y^{iv} + ay = f(x) \qquad \lambda^4 + a = 0 \qquad a \neq 0$$

(1) Complementary Solution   (a = real number)

$a > 0$ $\lambda_1 = (1+i)\alpha$ $\lambda_2 = (1-i)\alpha$ $\lambda_3 = -(1+i)\alpha$ $\lambda_4 = -(1-i)\alpha$

$\alpha = \sqrt[4]{\dfrac{a}{4}}$ $y_C = e^{\alpha x}(A \cos \alpha x + B \sin \alpha x) + e^{-\alpha x}(C \cos \alpha x + D \sin \alpha x)$

$a < 0$ $\lambda_1 = \beta$ $\lambda_2 = -\beta$ $\lambda_3 = i\beta$ $\lambda_4 = -i\beta$

$\beta = \sqrt[4]{-a}$ $y_C = A \cosh \beta x + B \sinh \beta x + C \cos \beta x + D \sin \beta x$

(2) Particular Solution   (b, c, ω = real numbers)

$$y_P = \frac{1}{a}\left[f(x) - \frac{f^{iv}(x)}{a} + \frac{f^{viii}(x)}{a^2} - \frac{f^{xxii}(x)}{a^3} + \cdots \right]$$

$$\boxed{\begin{aligned} u &= a + (b^2 - \omega^2)^2 - 4\omega^2 b^2 \\ v &= 4b(b^2 - \omega^2)\omega \end{aligned}}$$

$f(x)$	y_P	$f(x)$	y_P
c	$\dfrac{c}{a}$	cx^2	$\dfrac{cx^2}{a}$
cx	$\dfrac{cx}{a}$	ce^{bx}	$\dfrac{ce^{bx}}{a + b^4}$
$c \cos \omega x$	$\dfrac{c \cos \omega}{a + \omega^4}$	$ce^{bx} \cos \omega x$	$c\,\dfrac{u \cos \omega x + v \sin \omega x}{u^2 + v^2} e^{bx}$
$c \sin \omega x$	$\dfrac{c \sin \omega x}{a + \omega^4}$	$ce^{bx} \sin \omega x$	$c\,\dfrac{u \sin \omega x - v \cos \omega x}{u^2 + v^2} e^{bx}$

$$y^{iv} + py'' + qy = f(x) \qquad \lambda^4 + p\lambda^2 + q = 0 \qquad \boxed{p, q \neq 0}$$

(1) Complementary Solution   (p, q = real numbers, $D = p^2 - 4q$)

Conditions		Solution
$p > 0$	$D > 0$	$\lambda_{1,2} = \pm i \sqrt{\tfrac{1}{2}(p + \sqrt{D})} = \pm i\alpha \qquad \lambda_{3,4} = \pm i \sqrt{\tfrac{1}{2}(p - \sqrt{D})} = \pm i\beta$ $y_C = A \cos \alpha x + B \sin \alpha x + C \cos \beta x + D \sin \beta x$
	$D = 0$	$\lambda_1 = \lambda_2 = i\sqrt{\tfrac{p}{2}} = i\lambda \qquad\qquad \lambda_3 = \lambda_4 = -i\sqrt{\tfrac{p}{2}} = -i\lambda$ $y_C = (A + Bx)\cos \lambda x + (C + Dx)\sin \lambda x$
	$D < 0$	$\lambda_{1,2} = \pm i\sqrt{\tfrac{p}{2} + \tfrac{i}{2}\sqrt{-D}} = \pm \tfrac{i}{\sqrt{2}}\left(\sqrt{\sqrt{q} + \tfrac{p}{2}} + i\sqrt{\sqrt{q} - \tfrac{p}{2}}\right) = \pm(\bar\alpha i - \bar\beta)$ $\lambda_{3,4} = \pm i\sqrt{\tfrac{p}{2} - \tfrac{i}{2}\sqrt{-D}} = \pm \tfrac{i}{\sqrt{2}}\left(\sqrt{\sqrt{q} + \tfrac{p}{2}} - i\sqrt{\sqrt{q} - \tfrac{p}{2}}\right) = \pm(\bar\alpha i + \bar\beta)$ $y_C = \cos \bar\alpha x(A \cosh \bar\beta x + B \sinh \bar\beta x) + \sin \bar\alpha x(C \cosh \bar\beta x + D \sinh \bar\beta x)$
$p < 0$	$D > 0$	$\lambda_{1,2} = \pm \sqrt{\tfrac{1}{2}(-p + \sqrt{D})} = \pm\alpha \qquad \lambda_{3,4} = \pm \sqrt{\tfrac{1}{2}(-p - \sqrt{D})} = \pm\beta$ $y_C = A \cosh \alpha x + B \sinh \alpha x + C \cosh \beta x + D \sinh \beta x$
	$D = 0$	$\lambda_1 = \lambda_2 = \sqrt{-\tfrac{p}{2}} = \lambda \qquad\qquad \lambda_3 = \lambda_4 = -\sqrt{-\tfrac{p}{2}} = -\lambda$ $y_C = (A + Bx)\cosh \lambda x + (C + Dx)\sinh \lambda x$
	$D < 0$	$\lambda_{1,2} = \pm \sqrt{-\tfrac{p}{2} + \tfrac{i}{2}\sqrt{-D}} = \pm \tfrac{1}{\sqrt{2}}\left(\sqrt{\sqrt{q} - \tfrac{p}{2}} + i\sqrt{\sqrt{q} + \tfrac{p}{2}}\right) = \pm(\bar\alpha + i\bar\beta)$ $\lambda_{3,4} = \pm \sqrt{-\tfrac{p}{2} - \tfrac{i}{2}\sqrt{-D}} = \pm \tfrac{1}{\sqrt{2}}\left(\sqrt{\sqrt{q} - \tfrac{p}{2}} - i\sqrt{\sqrt{q} + \tfrac{p}{2}}\right) = \pm(\bar\alpha - i\bar\beta)$ $y_C = \cosh \bar\alpha x(A \cos \bar\beta x + B \sin \bar\beta x) + \sinh \bar\alpha x(C \cos \bar\beta x + D \sin \bar\beta x)$

(2) Particular Solution   (b, c, ω = real numbers)

$$y_P = \sum_{k=0}^{\infty} s_{2k} f^{(2k)}(x) \qquad f^{(0)} = f(x) \qquad f^{(2k)}(x) = d^{2k} f(x)/dx^{2k}$$

$$s_0 = \frac{1}{q} \qquad s_2 = -\frac{p}{q^2} \qquad s_4 = \frac{p^2 - q}{q^3} \qquad s_6 = -\frac{p^3 - 2pq}{q^4} \qquad \boxed{s_{2k} = -\frac{1}{q}(p s_{2k-2} + s_{2k-4})}$$

$f(x)$	y_P	$f(x)$	y_P
c	$\dfrac{c}{q}$	cx^2	$\dfrac{c}{q}\left(x^2 - \dfrac{2p}{q}\right)$
cx	$\dfrac{cx}{q}$	ce^{bx}	$\dfrac{ce^{bx}}{q + pb^2 + b^4}$
$c \cos \omega x$	$c\dfrac{\cos \omega x}{q - p\omega^2 + \omega^4}$	$c \sin \omega x$	$c\dfrac{\sin \omega x}{q - p\omega^2 + \omega^4}$

(1) Standard Form

Euler's differential equation of order n is given as

$$a_0 x^n y^{(n)} + a_1 x^{n-1} y^{(n-1)} + a_2 x^{n-2} y^{(n-2)} + \cdots + a_n y = f(x)$$

where $f(x)$ is an arbitrary function of x.

(2) Complementary Solution

By the substitution $y = x^\lambda$ the reduced differential equation transforms into

$$f(\lambda) x^\lambda = \left[a_0 \frac{\lambda!}{(\lambda - n)!} + a_1 \frac{\lambda!}{(\lambda - n + 1)!} + a_2 \frac{\lambda!}{(\lambda - n + 2)!} + \cdots + a_n \right] x^\lambda = 0$$

where $f(\lambda) = 0$ is the *characteristic equation*, the roots of which take one of the forms (or their combinations) given in Sec. 14.07(3).

(3) Particular Solution

In general, there is no *general method of finding a particular solution* of this differential equation. A method which sometimes yields a solution is to assume a series

$$y_P \doteq \frac{1}{a_n} [f(x) + A_1 f'(x) + A_2 f''(x) + \cdots].$$

where $A_1, A_2, \ldots$ are functions given by the following conditions:

$$A_1 = -x\bar{a}_{n-1}$$

$$A_2 = -x^2 \bar{a}_{n-2} - x\bar{a}_{n-1} A_1$$

$$A_3 = -x^3 \bar{a}_{n-3} - x^2 \bar{a}_{n-2} A_1 - x a_{n-1} A_2$$

$$\cdots \cdots \cdots \cdots \cdots \cdots \cdots \cdots \cdots \cdots \cdots$$

$$A_n = -x^n \bar{a}_0 - x^{n-1} \bar{a}_1 A_1 - x^{n-2} \bar{a}_2 A_2 - x^{n-3} \bar{a}_3 A_3 - \cdots$$

$$A_{n+1} = 0 - x^n \bar{a}_0 A_1 - x^{n-1} \bar{a}_1 A_2 - x^{n-2} \bar{a}_2 A_3 - \cdots$$

$$\cdots \cdots \cdots \cdots \cdots \cdots \cdots \cdots \cdots \cdots \qquad \bar{a}_j = \frac{a_j}{a_n}$$

14.14 SECOND-ORDER EULER'S DIFFERENTIAL EQUATION

$$x^2 y'' + bxy' + cy = f(x) \qquad \lambda^2 + p\lambda + q = 0$$

(1) Complementary Solution $\quad (D = p^2 - 4q)$

$D > 0$	$D = 0$	$D < 0$
$\lambda_{1,2} = \dfrac{-p \pm \sqrt{D}}{2} = \alpha, \beta$	$\lambda_{1,2} = -\dfrac{p}{2} = \lambda$	$\lambda_{1,2} = \dfrac{-p \pm \sqrt{D}}{2} = \alpha \pm i\beta$
$y_C = A x^\alpha + B x^\beta$	$y_C = x^\lambda (A \ln|x| + B)$	$y_C = x^\alpha [A \cos(\beta \ln|x|) + B \sin(\beta \ln|x|)]$

(2) Particular Solution

The method of function series [Sec. 14.12(3)] frequently yields a particular solution.

(1) Concept

Some differential equations, particularly homogeneous linear equations with variable coefficients, can be solved by assuming a solution in the form of an *infinite power series* such as

$$y = b_0 + b_1 x + b_2 x^2 + \cdots + b_r x^r + \cdots = \sum_{r=0}^{\infty} b_r x^r$$

the first and higher derivatives of which are

$$y' = \sum_{r=0}^{\infty} r b_r x^{r-1} \qquad\qquad y'' = \sum_{r=0}^{\infty} r(r-1) b_r x^{r-2}, \ldots$$

After these expressions have been substituted in the given differential equation, the terms in like power of x are combined, and the coefficients of each power of x are set equal to zero. The system of equations thus obtained yields the values of the constants of the series of which n (corresponding to the order) must remain unknown (arbitrary) as the constants of integration.

(2) Transformation

Frequently it serves to an advantage to transform the given differential equation

$$y'' + a(x)y' + b(x)y = 0$$

by the substitution of

$$y = y(x) = u(x) \exp\left[-\tfrac{1}{2}\int a(x)\,dx\right] \qquad \text{to} \qquad u'' + \left[b(x) - \frac{a(x)'}{2} - \left(\frac{a(x)}{2}\right)^2\right]u = 0$$

(3) Orthogonal Polynomials

A *set of polynomials* $p_n(x)$ $(n = 0, 1, 2, \ldots)$ of degree n in x is orthogonal in the interval (a, b) with respect to the weight function $w(x)$ if

$$\int_a^b w(x)p_m(x)p_n(x)\,dx = \begin{cases} 0 & \text{for } m \neq n \\ a_n & \text{for } m = n \end{cases} \qquad m, n = 0, 1, 2, \ldots$$

Under certain conditions, this relationship admits the representation of a function in the form

$$f(x) = \sum_{n=0}^{\infty} C_n p_n(x) \qquad \text{with} \qquad C_n = \frac{1}{a_n}\int_a^b f(x)w(x)p_n(x)\,dx$$

(4) Classical Orthogonal Polynomials

Classical orthogonal polynomials of particular interest are designated by the names of their discoverers. The *Legendre* (Sec. 14.17), *Chebyshev* (Sec. 14.18), *Laguerre* (Sec. 14.19), and *Hermite polynomials* (Sec. 14.20) have *two typical properties*:

(a) $p_n(x)$ *satisfies the differential equation*

$$a(x)y'' + b(x)y' + cy = 0$$

where $a(x), b(x)$ are independent of n, c, and n, c are independent of x.

(b) The polynomial *can be represented by the generalized Rodrigues formula* as

$$p_n(x) = \frac{1}{k_n w(x)} \frac{d^n}{dx^n}\{w(x)[\phi(x)]^n\}$$

where $\phi(x)$ is a polynomial of the first or second degree.

(1) Gauss' Differential Equation   (α, β, γ = constants)

$$x(1-x)y'' - [(\alpha+\beta+1)x - \gamma]y' - \alpha\beta y = 0$$

(2) Solution   ($y = A_1 y_1 + A_2 y_2$)

$$y_1 = F(\alpha, \beta, \gamma, x)$$

$$= 1 + \frac{\alpha\beta}{\gamma}\frac{x}{1!} + \frac{\alpha(\alpha+1)\beta(\beta+1)}{\gamma(\gamma+1)}\frac{x^2}{2!} + \frac{\alpha(\alpha+1)(\alpha+2)\beta(\beta+1)(\beta+2)}{\gamma(\gamma+1)(\gamma+2)}\frac{x^3}{3!} + \cdots$$

$$y_2 = x^{1-\gamma}F(\alpha-\gamma+1, \beta-\gamma+1, 2-\gamma, x)$$

$$= x^{1-\gamma}\left[1 + \frac{(\alpha-\gamma+1)(\beta-\gamma+1)}{(2-\gamma)}\frac{x}{1!}\right.$$

$$\left. + \frac{(\alpha-\gamma+1)(\alpha-\gamma+2)(\beta-\gamma+1)(\beta-\gamma+2)}{(2-\gamma)(3-\gamma)}\frac{x^2}{2!} + \cdots\right] \quad |x| < 1, \gamma \neq 0, 1, 2, \ldots$$

$$y_1 = F(\alpha, \beta, \alpha+\beta-\gamma+1, 1-x)$$

$$y_2 = (1-x)^{\gamma-\alpha-\beta}F(\gamma-\beta, \gamma-\alpha, \gamma-\alpha-\beta+1, 1-x) \quad |x-1| < 1, \alpha+\beta-\gamma \neq 0, 1, 2, \ldots$$

$$y_1 = x^{-\alpha}F(\alpha, \alpha-\gamma+1, \alpha-\beta+1, x^{-1})$$

$$y_2 = x^{-\beta}F(\beta, \beta-\gamma+1, \beta-\alpha+1, x^{-1}) \quad |x| > 1, \alpha-\beta \neq 0, 1, 2, \ldots$$

14.17   CONFLUENT HYPERGEOMETRIC DIFFERENTIAL EQUATION

(1) Kummer's Differential Equation   (β, γ = constants)

$$xy'' + (\gamma - x)y' - \beta y = 0$$

(2) Solution   ($y = A_1 y_1 + A_2 y_2$)

$$y_1 = F(\beta, \gamma, x)$$

$$= 1 + \frac{\beta}{\gamma}\frac{x}{1!} + \frac{\beta(\beta+1)}{\gamma(\gamma+1)}\frac{x^2}{2!} + \cdots$$

$$y_2 = x^{1-\gamma}F(\beta-\gamma+1, 2-\gamma, x)$$

$$= x^{1-\gamma}\left[1 + \frac{\beta-\gamma+1}{2-\gamma}\frac{x}{1!} + \frac{(\beta-\gamma+1)(\beta-\gamma+2)}{(2-\gamma)(3-\gamma)}\frac{x^2}{2!} + \cdots\right] \quad \gamma \neq 0, 1, 2, \ldots$$

$\bar{P}_n(x), \bar{Q}_n(x)$ = Legendre functions

$P_n(x), Q_n(x)$ = Legendre polynomials

$\bigwedge [\,]$ = nested sum (Sec. 8.11)

$(\,)!!$ = double factorial (Sec. 1.03)

(1) Differential Equation

$$(1 - x^2)y'' - 2xy' + n(n + 1)y = 0$$

$$-\infty < n < +\infty$$

Interval: $[-1, +1]$
Weight: $w(x) = 1$

(2) Solution for All n: $y = A_1 \bar{P}_n(x) + A_2 \bar{Q}_n(x)$

$$\bar{P}_n(x) = \bigwedge_{k=0}^{\infty} \left[1 - \frac{(n-2k)(n+2k+1)}{(2k+1)(2k+2)} S \right]$$

$$= \left(1 - \frac{n(n+1)}{1 \cdot 2} S \left(1 - \frac{(n-2)(n+3)}{3 \cdot 4} S \left(1 - \frac{(n-4)(n+5)}{5 \cdot 6} S \left(1 - \cdots \right) \right) \right) \right)$$

$$\bar{Q}_n(x) = x \bigwedge_{k=0}^{\infty} \left[1 - \frac{(n-2k-1)(n+2k+2)}{(2k+2)(2k+3)} S \right]$$

$$= x \left(1 - \frac{(n-1)(n+2)}{2 \cdot 3} S \left(1 - \frac{(n-3)(n+4)}{4 \cdot 5} S \left(1 - \frac{(n-5)(n+6)}{6 \cdot 7} S \left(1 - \cdots \right) \right) \right) \right)$$

$S = x^2$

(3) Solution for $n = 0, 1, 2, \ldots$: $y = A_1 P_n(x) + A_2 Q_n(x)$

$$P_n(x) = \frac{(2n-1)!!}{n!} x^n \bigwedge_{k=0}^{t} \left[1 - \frac{(n-2k)(n-2k-1)}{(2k+2)(2n-2k-1)} R \right]$$

$$= \frac{(2n-1)!!}{n!} x^n \left(1 - \frac{n(n-1)}{2(2n-1)} R \left(1 - \frac{(n-2)(n-3)}{4(2n-3)} R \left(1 - \frac{(n-4)(n-5)}{6(2n-5)} R \left(1 - \cdots \right) \right) \right) \right)$$

$$Q_n(x) = \left(\ln \sqrt{\frac{x+1}{x-1}} \right) P_n(x) - \sum_{k=1}^{n} \frac{P_{k-1}(x) P_{n-k}(x)}{k}$$

$R = x^{-2}$

where $P(x)$ series terminates with $t = n/2$ for even n and with $t = (n-1)/2$ for odd n.

(4) Special Values

$$P_n(0) = \begin{cases} (-1)^{n/2} \dfrac{(n-1)!!}{n!!} & n \text{ even} \\ 0 & n \text{ odd} \end{cases}$$

$$Q_n(0) = \begin{cases} 0 & n \text{ even} \\ (-1)^{(n+1)/2} \dfrac{(n-1)!!}{n!!} & n \text{ odd} \end{cases}$$

$$P_n(1) = 1 \qquad P_n(-1) = (-1)^n$$

$$Q_n(1) = \infty \qquad Q_n(-1) = \infty(-1)^{n+1}$$

(5) Relations   (Sec. A.14)

$$P_{n+1}(x) = \frac{2n+1}{n+1} x P_n(x) - \frac{n}{n+1} P_{n-1}(x)$$

$$Q_{n+1}(x) = \frac{2n+1}{n+1} x Q_n(x) - \frac{n}{n+1} Q_{n-1}(x)$$

$$\int_{-1}^{+1} P_m(x) P_n(x) \, dx = \begin{cases} 0 & m \neq n \\ \dfrac{2}{2n+1} & m = n \end{cases}$$

$$\int_{-1}^{+1} Q_m(x) Q_n(x) \, dx = \begin{cases} 0 & m \neq n \\ \dfrac{2}{2n+1} & m = n \end{cases}$$

(1) Graphs of $P_n(x)$

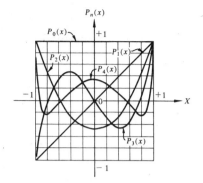

(2) Graphs of $Q_n(x)$

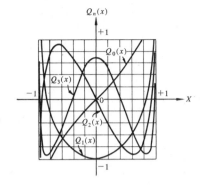

(3) Table of $P_n(x)$

$$P_n(-x) = (-1)^n P_n(x) \qquad |x| \leq 1$$

n \ x	0.0	0.2	0.4	0.6	0.8
0	+1.000 000 0	+1.000 000 0	+1.000 000 0	+1.000 000 0	+1.000 000 0
1	0.000 000 0	+0.200 000 0	+0.400 000 0	+0.600 000 0	+0.800 000 0
2	−0.500 000 0	−0.440 000 0	−0.260 000 0	+0.040 000 0	+0.460 000 0
3	0.000 000 0	−0.280 000 0	−0.440 000 0	−0.360 000 0	+0.080 000 0
4	+0.375 000 0	+0.232 000 0	+0.113 000 0	−0.408 000 0	−0.233 000 0
5	0.000 000 0	+0.307 520 0	+0.270 640 0	−0.152 640 0	−0.399 520 0
6	−0.312 500 0	−0.080 576 0	+0.292 636 0	+0.172 096 0	−0.391 796 0
7	0.000 000 0	−0.293 516 8	−0.014 590 4	+0.322 598 4	−0.239 651 2
8	+0.273 437 5	−0.039 564 8	−0.266 999 3	+0.212 339 2	−0.016 655 3
9	0.000 000 0	+0.245 957 1	−0.188 763 6	−0.046 103 0	+0.187 855 3
10	−0.246 093 8	+0.129 072 0	+0.096 835 1	−0.243 662 7	+0.300 529 8
11	0.000 000 0	−0.174 346 4	+0.245 553 1	−0.237 192 7	+0.288 213 4
12	+0.225 585 9	−0.185 136 9	+0.099 488 2	−0.049 414 1	+0.166 441 6

(4) Table of $Q_n(x)$

$$Q_n(-x) = (-1)^{n+1} Q_n(x) \qquad |x| \leq 1$$

n \ x	0.0	0.2	0.4	0.6	0.8
0	0.000 000 0	+0.202 732 6	+0.423 648 9	+0.693 147 2	+1.098 612 3
1	−1.000 000 0	−0.959 953 5	−0.830 540 4	−0.584 111 7	−0.121 110 2
2	0.000 000 0	−0.389 202 3	−0.710 148 7	−0.872 274 1	−0.694 638 3
3	+0.666 666 7	+0.509 901 6	+0.080 261 1	−0.482 866 3	−0.845 444 4
4	0.000 000 0	+0.470 367 3	+0.588 794 3	+0.147 195 0	−0.662 643 4
5	−0.533 333 3	−0.238 589 1	+0.359 723 0	+0.545 264 7	−0.277 851 0
6	0.000 000 0	−0.479 455 4	−0.226 865 1	+0.477 127 8	+0.144 688 8
7	+0.457 142 9	+0.026 421 5	−0.476 862 4	+0.064 286 9	+0.453 123 1
8	0.000 000 0	+0.429 431 5	−0.159 139 8	−0.345 164 0	+0.553 081 9
9	−0.406 349 2	+0.138 743 9	+0.303 638 7	−0.448 329 9	+0.432 993 1
10	0.000 000 0	−0.333 765 6	+0.373 991 2	−0.200 448 5	+0.160 375 2
11	+0.369 408 4	−0.253 568 6	+0.009 580 9	+0.177 967 6	−0.148 694 4
12	0.000 000 0	+0.208 750 5	−0.335 479 9	+0.388 407 2	−0.375 007 2

$\bar{T}_n(x), \bar{U}_n(x)$ = Chebyshev functions

$T_n(x), U_n(x)$ = Chebyshev polynomials

$\Lambda[\ ]$ = nested sum (Sec. 8.11)

(1) Differential Equation

$$(1-x^2)y'' - xy' + n^2 y = 0 \qquad -\infty < n < +\infty$$

Interval: $[-1, +1]$

Weight: $w(x) = \dfrac{1}{\sqrt{1-x^2}}$

(2) Solution for All n: $y = A_1 \bar{T}_n(x) + A_2 \bar{U}_n(x)$

$$\bar{T}_n(x) = \bigwedge_{k=0}^{\infty}\left[1 - \frac{n^2 - (2k)^2}{(2k+1)(2k+2)}S\right]$$

$S = x^2$

$$= \left(1 - \frac{n^2}{1\cdot 2}S\left(1 - \frac{n^2 - 2^2}{3\cdot 4}S\left(1 - \frac{n^2 - 4^2}{5\cdot 6}S(1 - \cdots)\right)\right)\right)$$

$$\bar{U}_n(x) = x\bigwedge_{k=0}^{\infty}\left[1 - \frac{n^2 - (2k+1)^2}{(2k+2)(2k+3)}S\right]$$

$$= x\left(1 - \frac{n^2 - 1}{2\cdot 3}S\left(1 - \frac{n^2 - 3^2}{4\cdot 5}S\left(1 - \frac{n^2 - 5^2}{6\cdot 7}S(1 - \cdots)\right)\right)\right)$$

(3) Solution for $n = 0, 1, 2, \ldots$: $y = A_1 T_n(x) + A_2 U_n(x)$

$$T_n(x) = x^n\bigwedge_{k=0}^{t}\left[1 - \frac{(n-2k)(n-2k-1)}{(2k+1)(2k+2)}R\right]$$

$R = \dfrac{1-x^2}{x^2}$

$$= x^n\left(1 - \frac{n(n-1)}{1\cdot 2}R\left(1 - \frac{(n-2)(n-3)}{3\cdot 4}R\left(1 - \frac{(n-4)(n-5)}{5\cdot 6}R(1 - \cdots)\right)\right)\right)$$

$$U_n(x) = \frac{1}{n}\sqrt{1-x^2}\frac{dT_n(x)}{dx} \qquad U_0(x) = \sin^{-1}x$$

where $T_n(x)$ series terminates with $t = n/2$ for even n and with $t = (n-1)/2$ for odd n.

(4) Special Values

$$T_n(0) = \begin{cases}(-1)^{n/2} & n \text{ even}\\ 0 & n \text{ odd}\end{cases} \qquad U_n(0) = \begin{cases}0 & n \text{ even}\\ (-1)^{n+1} & n \text{ odd}\end{cases}$$

$$T_n(1) = 1 \qquad T_n(-1) = (-1)^n \qquad U_n(1) = 0 \qquad U_n(-1) = 0$$

(5) Relations (Sec. A.14)

$$T_{n+1}(x) = 2xT_n(x) - T_{n-1}(x) \qquad\qquad U_{n+1}(x) = 2xU_n(x) - U_{n-1}(x)$$

$$\int_{-1}^{+1}\frac{T_m(x)T_n(x)}{\sqrt{1-x^2}}\,dx = \begin{cases}0 & m \neq n\\ \frac{\pi}{2} & m = n \neq 0\\ \pi & m = n = 0\end{cases} \qquad \int_{-1}^{+1}\frac{U_m(x)U_n(x)}{\sqrt{1-x^2}}\,dx = \begin{cases}0 & m \neq n\\ \frac{\pi}{2} & m = n \neq 0\\ 0 & m = n = 0\end{cases}$$

(1) Graphs of $T_n(x)$

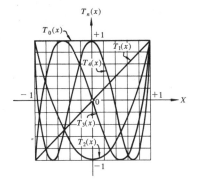

(2) Graphs of $U_n(x)$

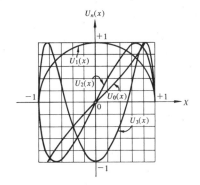

(3) Table of $T_n(x)$

$$T_n(-x) = (-1)^n T_n(x) \qquad |x| \le 1$$

n \ x	0.0	0.2	0.4	0.6	0.8
0	+1.000 000 0	+1.000 000 0	+1.000 000 0	+1.000 000 0	+1.000 000 0
1	0.000 000 0	+0.200 000 0	+0.400 000 0	+0.600 000 0	+0.800 000 0
2	−1.000 000 0	−0.920 000 0	−0.680 000 0	−0.280 000 0	+0.280 000 0
3	0.000 000 0	−0.568 000 0	−0.944 000 0	−0.936 000 0	−0.352 000 0
4	+1.000 000 0	+0.692 800 0	−0.075 200 0	−0.843 200 0	−0.843 200 0
5	0.000 000 0	+0.845 120 0	+0.883 840 0	−0.075 840 0	−0.997 120 0
6	−1.000 000 0	−0.354 752 0	+0.782 272 0	+0.752 192 0	−0.752 192 0
7	0.000 000 0	−0.987 020 8	−0.258 022 4	+0.978 470 4	−0.206 387 2
8	+1.000 000 0	−0.040 056 3	−0.988 689 9	+0.421 972 5	+0.421 972 5
9	0.000 000 0	+0.970 998 3	−0.532 929 5	−0.472 103 4	+0.881 543 2
10	−1.000 000 0	+0.428 455 6	+0.562 346 3	−0.988 496 6	+0.988 496 6
11	0.000 000 0	−0.799 616 0	+0.982 806 6	−0.714 092 5	+0.700 051 4
12	+1.000 000 0	−0.748 302 4	+0.223 899 0	+0.131 585 6	+0.131 585 6

(4) Table of $U_n(x)$

$$U_n(-x) = (-1)^{n+1} U_n(x) \qquad |x| \le 1$$

n \ x	0.0	0.2	0.4	0.6	0.8
0	0.000 000 0	+0.201 357 9	+0.411 516 8	+0.643 501 1	+0.927 295 2
1	+1.000 000 0	+0.979 795 9	+0.916 515 1	+0.800 000 0	+0.600 000 0
2	0.000 000 0	+0.391 918 4	+0.733 212 1	+0.960 000 0	+0.960 000 0
3	−1.000 000 0	−0.823 028 6	−0.329 945 5	+0.352 000 0	+0.936 000 0
4	0.000 000 0	−0.721 129 8	−0.997 168 5	−0.537 600 0	+0.537 600 0
5	+1.000 000 0	+0.534 576 7	−0.467 789 3	−0.997 120 0	−0.075 840 0
6	0.000 000 0	+0.934 960 5	+0.622 937 1	−0.658 944 0	−0.658 944 0
7	−1.000 000 0	−0.160 592 5	+0.966 138 9	+0.206 387 2	−0.978 470 4
8	0.000 000 0	−0.999 197 5	+0.149 740 6	+0.906 608 6	−0.906 608 6
9	+1.000 000 0	−0.239 086 5	−0.846 159 4	+0.881 543 2	−0.472 103 4
10	0.000 000 0	+0.903 562 9	−0.826 668 1	+0.151 243 7	+0.151 243 1
11	−1.000 000 0	+0.600 511 7	+0.184 824 9	−0.700 051 4	+0.714 092 4
12	0.000 000 0	−0.663 358 2	+0.974 528 0	−0.991 304 8	+0.991 304 7

$\bar{L}_n(x), \bar{L}_n^*(x)$ = Laguerre functions	$L_n^*(x)$ = Laguerre auxiliary function
$L_n(x)$ = Laguerre polynomial	$\Lambda\,[\,]$ = nested sum (Sec. 8.11)

(1) Differential Equation

$xy'' + (1-x)y' + ny = 0$	$-\infty < n < +\infty$

Interval: $[0, \infty]$
Weight: $W(x) = e^{-x}$

(2) Solution for All n: $y = A_1 \bar{L}_n(x) + A_2 \bar{L}_n^*(x)$

$$\bar{L}_n(x) = (-x)^n \overset{\infty}{\underset{k=0}{\Lambda}} \left[1 - \frac{(n-k)^2}{(k+1)x} \right]$$

$$= (-x)^n \left(1 - \frac{n^2}{x} \left(1 - \frac{(n-1)^2}{2x} \left(1 - \frac{(n-2)^2}{3x} \left(1 - \cdots \right) \right) \right) \right)$$

$$\bar{L}_n^*(x) = \frac{e^x}{x^{n+1}} \overset{\infty}{\underset{k=0}{\Lambda}} \left[1 + \frac{(n+k+1)^2}{(k+1)x} \right]$$

$$= \frac{e^x}{x^{n+1}} \left(1 + \frac{(n+1)^2}{x} \left(1 + \frac{(n+2)^2}{2x} \left(1 + \frac{(n+3)^3}{3x} \left(1 + \cdots \right) \right) \right) \right)$$

(3) Solution for $n = 0, 1, 2, \ldots$: $y = A_1 L_n(x) + A_2 L_n^*(x)$

$$L_n(x) = n! \overset{n}{\underset{k=0}{\Lambda}} \left[1 - \frac{(n-k)x}{(k+1)^2} \right]$$

$$= n! \left(1 - \frac{nx}{1 \cdot 1} \left(1 - \frac{(n-1)x}{2 \cdot 2} \left(1 - \frac{(n-2)x}{3 \cdot 3} \left(1 - \cdots \right) \right) \right) \right)$$

$$L_n^*(x) = L_n(x) \ln x + \sum_{k=1}^{\infty} C_{n,k} x$$

where $C_{n,1} = n+1$ and $C_{n,k} = \frac{1}{k^2} \left[(k-1) C_{n,k-1} + \frac{n+1}{k!} \binom{-n}{k-1} \right]$.

(4) Special Values and Relations   **Sec. A.14**

$$L_{n+1}(x) = (2n+1-x) L_n(x) - n^2 L_{n-1}(x) \qquad L_n(0) = n! \qquad L_n^*(0) = \infty$$

$$\int_0^\infty e^{-x} L_m(x) L_n(x)\, dx = \begin{cases} 0 & m \neq n \\ (n!)^2 & m = n \end{cases} \qquad \int_0^\infty x^p e^{-x} L_n(x)\, dx = \begin{cases} 0 & p < n \\ (-1)^n (n!)^2 & p = n \end{cases}$$

(5) Table[1] of $L_n(x)/n!$ $\boxed{0 \leq x \leq \infty}$

n \ x	0.5	1.0	3.0	5.0
0	+ 1.00000 00000	+ 1.00000 00000	+ 1.00000 00000	+ 1.00000 00000
1	+ 0.50000 00000	0.00000 00000	− 2.00000 00000	− 4.00000 00000
2	+ 0.12500 00000	− 0.50000 00000	− 0.50000 00000	+ 3.50000 00000
3	− 0.14583 33333	− 0.66666 66667	+ 1.00000 00000	+ 2.66666 66667
4	− 0.33072 91667	− 0.62500 00000	+ 1.37500 00000	− 1.29166 66667
5	− 0.44557 29167	− 0.46666 66667	+ 0.85000 00000	− 3.16666 66667
6	− 0.50414 49653	− 0.25694 44444	− 0.01250 00000	− 2.09027 77778
7	− 0.51833 92237	− 0.04047 61905	− 0.74642 85714	+ 0.32539 68254
8	− 0.49836 29984	+ 0.15399 30556	− 1.10870 53571	+ 2.23573 90873
9	− 0.45291 95204	+ 0.30974 42681	− 1.06116 07143	+ 2.69174 38272
10	− 0.38937 44141	+ 0.41894 59325	− 0.70002 23214	+ 1.75627 61795
11	− 0.31390 72988	+ 0.48013 41791	− 0.18079 95130	+ 0.10754 36909
12	− 0.23164 96389	+ 0.49621 22235	+ 0.34035 46063	− 1.44860 42948

[1](Ref. 14.01, p. 800.)

$\bar{H}_n(x)$, $\bar{H}_n^*(x)$ = Hermite functions Λ [ ] = nested sum (Sec. 8.11)

$H_n(x)$, $H_n^*(x)$ = Hermite polynomial   ( )!! = double factorial (Sec. 1.03)

(1) Differential Equation

$$y'' - 2xy' + 2ny = 0 \qquad -\infty < n < +\infty$$

Interval: $[-\infty, \infty]$
Weight: $w(x) = e^{-x^2}$

(2) Solution for All n: $y = A_1\bar{H}_n(x) + A_2\bar{H}_n^*(x)$

$$\bar{H}_n(x) = \overset{\infty}{\underset{k=0}{\Lambda}}\left[1 - \frac{2(n-2k)}{(1+2k)(2+2k)}S\right] = \left(1 - \frac{2n}{1\cdot 2}S\left(1 - \frac{2(n-2)}{3\cdot 4}S\left(1 - \frac{2(n-4)}{5\cdot 6}S\left(1 - \cdots\right)\right)\right)\right)$$

$$\bar{H}_n^*(x) = x\overset{\infty}{\underset{k=0}{\Lambda}}\left[1 - \frac{2(n-1-2k)}{(2+2k)(3+3k)}S\right]$$

$$S = x^2$$

$$= x\left(1 - \frac{2(n-1)}{2\cdot 3}S\left(1 - \frac{2(n-3)}{4\cdot 5}S\left(1 - \frac{2(n-5)}{6\cdot 7}S\left(1 - \cdots\right)\right)\right)\right)$$

(3) Solution for $n = 0, 1, 2, \ldots$: $y = A_1 H_n(x) + A_2 H_n^*(x)$

$$H_n(x) = (2x)^n\overset{n}{\underset{k=0}{\Lambda}}\left[1 - \frac{(n-k)(n-k-1)}{k+1}R\right] = H_n^*(x)$$

$$R = \frac{1}{4x^2}$$

$$= (2x)^n\left(1 - \frac{n(n-1)}{1}R\left(1 - \frac{(n-2)(n-3)}{2}R\left(1 - \frac{(n-4)(n-5)}{3}R\left(1 - \cdots\right)\right)\right)\right)$$

where $A_1 = \begin{cases} i^n\dfrac{n!C}{(n/2)!} & n\text{ even} \\ 0 & n\text{ odd} \end{cases}$ $A_2 = \begin{cases} 0 & n\text{ even} \\ 2i^{n-1}\dfrac{n!C}{[(n-1)/2]!} & n\text{ odd} \end{cases}$

and C = arbitrary constant.

(4) Special Values and Relations   Sec. A.14

$$H_{n+1}(x) = 2xH_n(x) - 2n H_{n-1}(x) \qquad \frac{d}{dx}[H_{n+1}(x)] = 2(n+1)H_n(x)$$

$$H_n(0) = \begin{cases} (\sqrt{-2})^n(n-1)!! & n\text{ even} \\ 0 & n\text{ odd} \end{cases} \qquad \int_{-\infty}^{+\infty}e^{-x^2}H_m(x)H_n(x)\,dx = \begin{cases} 0 & m \neq n \\ 2^n n!\sqrt{\pi} & m = n \end{cases}$$

(5) Table[1] of $H_n(x)$

$$H_n(-x) = (-1)^n H_n(x) \qquad |x| \leq \infty$$

n \ x	0.5		1.0		3.0		5.0	
0	+1.00000		+1.00000		+1.00000 00		1.00000 00000	
1	+1.00000		+2.00000		+6.00000 00		1.00000 00000	(+01)
2	−1.00000		+2.00000		+3.40000 00	(+01)	9.80000 00000	(+01)
3	−5.00000		−4.00000		+1.80000 00	(+02)	9.40000 00000	(+02)
4	+1.00000		−2.00000	(+01)	+8.76000 00	(+02)	8.81200 00000	(+03)
5	+4.10000	(+01)	−8.00000	(+00)	+3.81600 00	(+03)	8.06000 00000	(+04)
6	+3.10000	(+01)	+1.84000	(+02)	+1.41360 00	(+04)	7.17880 00000	(+05)
7	−4.61000	(+02)	+4.64000	(+02)	+3.90240 00	(+04)	6.21160 00000	(+06)
8	−8.95000	(+02)	−1.64800	(+03)	+3.62400 00	(+04)	5.20656 80000	(+07)
9	+6.48100	(+03)	−1.07200	(+04)	−4.06944 00	(+05)	4.21271 20000	(+08)
10	+2.25910	(+04)	+8.22400	(+03)	−3.09398 40	(+06)	3.27552 97600	(+09)
11	−1.07029	(+05)	+2.30848	(+05)	−1.04250 24	(+07)	2.43298 73600	(+10)
12	−6.04031	(+05)	+2.80768	(+05)	+5.51750 40	(+06)	1.71237 08128	(+11)

[1](Ref. 14.01, p. 802.)

(1) Differential Equation

$$x^2y'' + xy' + (x^2 - n^2)y = 0$$

(2) Solution

$$y = A_1 J_n(x) + A_2 J_{-n}(x) \qquad n \neq 0, 1, 2, \ldots \qquad\qquad y = A_1 J_n(x) + A_2 Y_n(x) \qquad \text{all } n$$

(3) Bessel Functions of the First Kind of Order n

$$J_n(x) = \sum_{k=0}^{\infty} \frac{(-1)^k (x/2)^{2k+n}}{k!\,\Gamma(k+1+n)}$$

$$= \frac{x^n}{2^n \Gamma(1+n)}\left[1 - \frac{x^2}{2(2+2n)} + \frac{x^4}{(2)(4)(2+2n)(4+2n)} - \cdots \right]$$

$$J_{-n}(x) = \sum_{k=0}^{\infty} \frac{(-1)^k (x/2)^{2k-n}}{k!\,\Gamma(k+1-n)}$$

$$= \frac{x^{-n}}{2^{-n} \Gamma(1-n)}\left[1 - \frac{x^2}{2(2-2n)} + \frac{x^4}{(2)(4)(2-2n)(4-2n)} - \cdots \right]$$

$$= (-1)^n J_n(x) \qquad n = 0, 1, 2, \ldots$$

$$J_0(x) = 1 - \frac{(x/2)^2}{(1!)^2} + \frac{(x/2)^4}{(2!)^2} - \frac{(x/2)^6}{(3!)^2} + \cdots$$

$$J_1(x) = \frac{x}{2}\left[1 - \frac{(x/2)^2}{2(1!)^2} + \frac{(x/2)^4}{3(2!)^2} - \frac{(x/2)^6}{4(3!)^2} + \cdots \right] = -\frac{d}{dx}[J_0(x)]$$

(4) Bessel Functions of the Second Kind of Order n

$$Y_n(x) = \begin{cases} \dfrac{J_n(x)\cos n\pi - J_{-n}(x)}{\sin n\pi} & n \neq 0, 1, 2, \ldots \\[3mm] \lim\limits_{p \to n} \dfrac{J_p(x)\cos p\pi - J_{-p}(x)}{\sin p\pi} & n = 0, 1, 2, \ldots \end{cases}$$

$$Y_{-n}(x) = (-1)^n Y_n(x) \qquad n = 0, 1, 2, \ldots$$

$$Y_0(x) = \frac{2}{\pi}\left[\left(\ln\frac{x}{2} + C\right)J_0(x) + \frac{2}{1}J_2(x) - \frac{2}{2}J_4(x) + \frac{2}{3}J_6(x) - \cdots \right]$$

$$Y_1(x) = \frac{2}{\pi}\left[\left(\ln\frac{x}{2} + C\right)J_1(x) - \frac{1}{x} - \frac{1}{2}J_1(x) + \frac{9}{4}J_3(x) - \cdots \right] = -\frac{d}{dx}[Y_0(x)]$$

where $\qquad C = 0.577\,215\,665 \qquad\qquad$ (Sec. 7.02)

(1) Recurrence Relations

$$J_{n+1}(x) = \frac{2n}{x} J_n(x) - J_{n-1}(x)$$

$$= \frac{n}{x} J_n(x) - \frac{d}{dx}[J_n(x)]$$

$$= -x^n \frac{d}{dx}[x^{-n} J_n(x)]$$

$$J_{n-1}(x) = \frac{2n}{x} J_n(x) - J_{n+1}(x)$$

$$= \frac{n}{x} J_n(x) + \frac{d}{dx}[J_n(x)]$$

$$= x^{-n} \frac{d}{dx}[x^n J_n(x)]$$

$$\frac{d}{dx}[J_n(x)] = \tfrac{1}{2}[J_{n-1}(x) - J_{n+1}(x)]$$

$$\int x^{n+1} J_n(x)\, dx = x^{n+1} J_{n+1}(x)$$

$$Y_{n+1}(x) = \frac{2n}{x} Y_n(x) - Y_{n-1}(x)$$

$$= \frac{n}{x} Y_n(x) - \frac{d}{dx}[Y_n(x)]$$

$$= -x^n \frac{d}{dx}[x^{-n} Y_n(x)]$$

$$Y_{n-1}(x) = \frac{2n}{x} Y_n(x) - Y_{n+1}(x)$$

$$= \frac{n}{x} Y_n(x) + \frac{d}{dx}[Y_n(x)]$$

$$= x^{-n} \frac{d}{dx}[x^n Y_n(x)]$$

$$\frac{d}{dx}[Y_n(x)] = \tfrac{1}{2}[Y_{n-1}(x) - Y_{n+1}(x)]$$

$$\int x^{n+1} Y_n(x)\, dx = x^{n+1} Y_{n+1}(x)$$

(2) Half-Odd Integers   (See also Sec. 14.28–3)

$$J_{1/2}(x) = \sqrt{\frac{2}{\pi x}} \sin x$$

$$J_{3/2}(x) = \sqrt{\frac{2}{\pi x}} \left(\frac{\sin x}{x} - \cos x \right)$$

$$J_{5/2}(x) = \sqrt{\frac{2}{\pi x}} \left[\left(\frac{3}{x^2} - 1 \right) \sin x - \frac{3}{x} \cos x \right]$$

$$J_{-1/2}(x) = \sqrt{\frac{2}{\pi x}} \cos x$$

$$J_{-3/2}(x) = -\sqrt{\frac{2}{\pi x}} \left(\frac{\cos x}{x} + \sin x \right)$$

$$J_{-5/2}(x) = \sqrt{\frac{2}{\pi x}} \left[\left(\frac{3}{x^2} - 1 \right) \cos x + \frac{3}{x} \sin x \right]$$

$$Y_{1/2}(x) = -\sqrt{\frac{2}{\pi x}} \cos x$$

$$Y_{3/2}(x) = -\sqrt{\frac{2}{\pi x}} \left(\frac{\cos x}{x} + \sin x \right)$$

$$Y_{5/2}(x) = -\sqrt{\frac{2}{\pi x}} \left[\left(\frac{3}{x^2} - 1 \right) \cos x + \frac{3}{x} \sin x \right]$$

$$Y_{-1/2}(x) = \sqrt{\frac{2}{\pi x}} \sin x$$

$$Y_{-3/2}(x) = -\sqrt{\frac{2}{\pi x}} \left(\frac{\sin x}{x} + \cos x \right)$$

$$Y_{-5/2}(x) = \sqrt{\frac{2}{\pi x}} \left[\left(\frac{3}{x^2} - 1 \right) \sin x - \frac{3}{x} \cos x \right]$$

(3) Hankel Functions of Order n

$$H_n^{(1)}(x) = J_n(x) + i Y_n(x)$$

$$J_n(x) = \frac{H_n^{(1)}(x) + H_n^{(2)}(x)}{2}$$

$$H_n^{(2)}(x) = J_n(x) - i Y_n(x)$$

$$Y_n(x) = \frac{H_n^{(1)}(x) - H_n^{(2)}(x)}{2i}$$

(1) Asymptotic Approximation

$$J_n(x) \approx \sqrt{\frac{2}{\pi x}} \cos\left(x - \frac{n\pi}{2} - \frac{\pi}{4}\right) \qquad x > 25$$

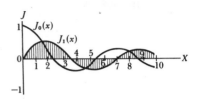

(2) Numerical Values $\boxed{J_0(x)}$ $x = 0 - 10$

x	0	1	2	3	4	5	6	7	8	9	x
0.	1.0000	0.9975	0.9900	0.9776	0.9604	0.9385	0.9120	0.8812	0.8463	0.8075	0.
1.	0.7652	0.7196	0.6711	0.6201	0.5669	0.5118	0.4554	0.3980	0.3400	0.2818	1.
2.	0.2239	0.1667	0.1104	0.0555	0.0025	−0.0484	−0.0968	−0.1424	−0.1850	−0.2243	2.
3.	−0.2601	−0.2921	−0.3202	−0.3443	−0.3643	−0.3801	−0.3918	−0.3992	−0.4026	−0.4018	3.
4.	−0.3971	−0.3887	−0.3766	−0.3610	−0.3423	−0.3205	−0.2961	−0.2693	−0.2404	−0.2097	4.
5.	−0.1776	−0.1443	−0.1103	−0.0758	−0.0412	−0.0068	0.0270	0.0599	0.0917	0.1220	5.
6.	0.1506	0.1773	0.2017	0.2238	0.2433	0.2601	0.2740	0.2851	0.2931	0.2981	6.
7.	0.3001	0.2991	0.2951	0.2882	0.2786	0.2663	0.2516	0.2346	0.2154	0.1944	7.
8.	0.1717	0.1475	0.1222	0.0960	0.0692	0.0419	0.0146	−0.0125	−0.0392	−0.0653	8.
9.	−0.0903	−0.1142	−0.1367	−0.1577	−0.1768	−0.1939	−0.2090	−0.2218	−0.2323	−0.2403	9.
10.	−0.2459										

(3) Numerical Values $\boxed{J_1(x)}$ $x = 0 - 10$

x	0	1	2	3	4	5	6	7	8	9	x
0.	0.0000	0.0499	0.0995	0.1483	0.1960	0.2423	0.2867	0.3290	0.3688	0.4059	0.
1.	0.4401	0.4709	0.4983	0.5220	0.5419	0.5579	0.5699	0.5778	0.5815	0.5812	1.
2.	0.5767	0.5683	0.5560	0.5399	0.5202	0.4971	0.4708	0.4416	0.4097	0.3754	2.
3.	0.3391	0.3009	0.2613	0.2207	0.1792	0.1374	0.0955	0.0538	0.0128	−0.0272	3.
4.	−0.0660	−0.1033	−0.1386	−0.1719	−0.2028	−0.2311	−0.2566	−0.2791	−0.2985	−0.3147	4.
5.	−0.3276	−0.3371	−0.3432	−0.3460	−0.3453	−0.3414	−0.3343	−0.3241	−0.3110	−0.2951	5.
6.	−0.2767	−0.2559	−0.2329	−0.2081	−0.1816	−0.1538	−0.1250	−0.0953	−0.0652	−0.0349	6.
7.	−0.0047	0.0252	0.0543	0.0826	0.1096	0.1352	0.1592	0.1813	0.2014	0.2192	7.
8.	0.2346	0.2476	0.2580	0.2657	0.2708	0.2731	0.2728	0.2697	0.2641	0.2559	8.
9.	0.2453	0.2324	0.2174	0.2004	0.1816	0.1613	0.1395	0.1166	0.0928	0.0684	9.
10.	0.0435										

(4) Asymptotic Series for Large x

$$J_n(x) \approx \sqrt{\frac{2}{\pi x}}\left[\cos\psi\left(1 - \frac{\alpha_1\alpha_3}{2!} + \frac{\alpha_1\alpha_3\alpha_5\alpha_7}{4!} - \cdots\right) - \sin\psi\left(\frac{\alpha_1}{1!} - \frac{\alpha_1\alpha_3\alpha_5}{3!} + \cdots\right)\right]$$

$$\psi = x - \frac{n\pi}{2} - \frac{\pi}{4} \qquad \alpha_k = \frac{4n^2 - k^2}{8x} \qquad k = 1, 2, \ldots \qquad x > 15$$

(1) Asymptotic Approximation

$$Y_n(x) \approx \sqrt{\frac{2}{\pi x}} \sin\left(x - \frac{n\pi}{2} - \frac{\pi}{4}\right) \qquad x > 25$$

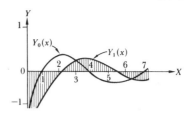

(2) Numerical Values

$$\boxed{Y_0(x)} \qquad\qquad x = 0 - 10$$

x	0	1	2	3	4	5	6	7	8	9	x
0.	$-\infty$	-1.5342	-1.0811	-0.8073	-0.6060	-0.4445	-0.3085	-0.1907	-0.0868	0.0056	0.
1.	0.0883	0.1622	0.2281	0.2865	0.3379	0.3824	0.4204	0.4520	0.4774	0.4968	1.
2.	0.5104	0.5183	0.5208	0.5181	0.5104	0.4981	0.4813	0.4605	0.4359	0.4079	2.
3.	0.3769	0.3431	0.3071	0.2691	0.2296	0.1890	0.1477	0.1061	0.0645	0.0234	3.
4.	-0.0169	-0.0561	-0.0938	-0.1296	-0.1633	-0.1947	-0.2235	-0.2494	-0.2723	-0.2921	4.
5.	-0.3085	-0.3216	-0.3313	-0.3374	-0.3402	-0.3395	-0.3354	-0.3282	-0.3177	-0.3044	5.
6.	-0.2882	-0.2694	-0.2483	-0.2251	-0.1999	-0.1732	-0.1452	-0.1162	-0.0864	-0.0563	6.
7.	-0.0259	0.0042	0.0339	0.0628	0.0907	0.1173	0.1424	0.1658	0.1872	0.2065	7.
8.	0.2235	0.2381	0.2501	0.2595	0.2662	0.2702	0.2715	0.2700	0.2659	0.2592	8.
9.	0.2499	0.2383	0.2245	0.2086	0.1907	0.1712	0.1502	0.1279	0.1045	0.0804	9.
10.	0.0557										

(3) Numerical Values

$$\boxed{Y_1(x)} \qquad\qquad x = 0 - 10$$

x	0	1	2	3	4	5	6	7	8	9	x
0.	$-\infty$	-6.4590	-3.3238	-2.2931	-1.7809	-1.4715	-1.2604	-1.1032	-0.9781	-0.8731	0.
1.	-0.7812	-0.6981	-0.6211	-0.5485	-0.4791	-0.4123	-0.3476	-0.2847	-0.2237	-0.1644	1.
2.	-0.1070	-0.0517	0.0015	0.0523	0.1005	0.1459	0.1884	0.2276	0.2635	0.2959	2.
3.	0.3247	0.3496	0.3707	0.3879	0.4010	0.4102	0.4154	0.4167	0.4141	0.4078	3.
4.	0.3979	0.3846	0.3680	0.3484	0.3260	0.3010	0.2737	0.2445	0.2136	0.1812	4.
5.	0.1479	0.1137	0.0792	0.0445	0.0101	-0.0238	-0.0568	-0.0887	-0.1192	-0.1481	5.
6.	-0.1750	-0.1998	-0.2223	-0.2422	-0.2596	-0.2741	-0.2857	-0.2945	-0.3002	-0.3029	6.
7.	-0.3027	-0.2995	-0.2934	-0.2846	-0.2731	-0.2591	-0.2428	-0.2243	-0.2039	-0.1817	7.
8.	-0.1581	-0.1331	-0.1072	-0.0806	-0.0535	-0.0262	0.0011	0.0280	0.0544	0.0799	8.
9.	0.1043	0.1275	0.1491	0.1691	0.1871	0.2032	0.2171	0.2287	0.2379	0.2447	9.
10.	0.2490										

(4) Asymptotic Series for Large x

$$Y_n(x) \approx \sqrt{\frac{2}{\pi x}}\left[\sin\psi\left(1 - \frac{\alpha_1\alpha_3}{2!} + \frac{\alpha_1\alpha_3\alpha_5\alpha_7}{4!} - \cdots\right) + \cos\psi\left(\frac{\alpha_1}{1!} - \frac{\alpha_1\alpha_3\alpha_5}{3!} + \cdots\right)\right]$$

$$\psi = x - \frac{n\pi}{2} - \frac{\pi}{4} \qquad \alpha_k = \frac{4n^2 - k^2}{8x} \qquad k = 1, 2, \ldots \qquad x > 15$$

(1) Differential Equation

$$x^2 y'' + xy' - (x^2 + n^2)y = 0$$

(2) Solution

$$y = A_1 I_n(x) + A_2 I_{-n}(x) \qquad n \neq 0, 1, 2, \ldots \qquad y = A_1 I_n(x) + A_2 K_n(x) \qquad \text{all } n$$

(3) Modified Bessel Functions of the First Kind of Order n

$$I_n(x) = i^{-n} J_n(ix) = \sum_{k=0}^{\infty} \frac{(x/2)^{2k+n}}{k!\,\Gamma(k+1+n)}$$

$$= \frac{x^n}{2^n \Gamma(1+n)}\left[1 + \frac{x^2}{2(2+2n)} + \frac{x^4}{(2)(4)(2+2n)(4+2n)} + \cdots\right]$$

$$I_{-n}(x) = i^n J_{-n}(ix) = \sum_{k=0}^{\infty} \frac{(x/2)^{2k-n}}{k!\,\Gamma(k+1-n)}$$

$$= \frac{x^{-n}}{2^{-n}\Gamma(1-n)}\left[1 + \frac{x^2}{2(2-2n)} + \frac{x^4}{(2)(4)(2-2n)(4-2n)} + \cdots\right]$$

$$I_{-n}(x) = I_n(x) \qquad n = 0, 1, 2, \ldots$$

$$I_0(x) = 1 + \frac{(x/2)^2}{(1!)^2} + \frac{(x/2)^4}{(2!)^2} + \frac{(x/2)^6}{(3!)^2} + \cdots$$

$$I_1(x) = \frac{x}{2}\left[1 + \frac{(x/2)^2}{2(1!)^2} + \frac{(x/2)^4}{3(2!)^2} + \frac{(x/2)^6}{4(3!)^2} + \cdots\right] = \frac{d}{dx}[I_0(x)]$$

(4) Modified Bessel Functions of the Second Kind of Order n

$$K_n(x) = \begin{cases} \dfrac{\pi}{2}\dfrac{I_{-n}(x) - I_n(x)}{\sin n\pi} & n \neq 0, 1, 2, \ldots \\[3mm] \lim\limits_{p \to n}\dfrac{\pi}{2}\dfrac{I_{-p}(x) - I_p(x)}{\sin p\pi} & n = 0, 1, 2, \ldots \end{cases}$$

$$K_{-n}(x) = K_n(x) \qquad n = 0, 1, 2, \ldots$$

$$K_0(x) = -\left[\left(\ln\frac{x}{2} + C\right)I_0(x) - \frac{2}{1}I_2(x) - \frac{2}{2}I_4(x) - \frac{2}{3}I_6(x) - \cdots\right]$$

$$-K_1(x) = -\left[\left(\ln\frac{x}{2} + C\right)I_1(x) - \frac{1}{x} - \frac{1}{2}I_1(x) - \frac{9}{4}I_3(x) - \cdots\right] = \frac{d}{dx}[K_0(x)]$$

where $\qquad C = 0.577\,215\,665 \qquad$ (Sec. 7.02)

(1) Recurrence Relations

$$I_{n+1}(x) = -\frac{2n}{x}I_n(x) + I_{n-1}(x)$$

$$= -\frac{n}{x}I_n(x) + \frac{d}{dx}[I_n(x)]$$

$$= x^n\frac{d}{dx}[x^{-n}I_n(x)]$$

$$I_{n-1}(x) = \frac{2n}{x}I_n(x) + I_{n+1}(x)$$

$$= \frac{n}{x}I_n(x) + \frac{d}{dx}[I_n(x)]$$

$$= x^{-n}\frac{d}{dx}[x^n I_n(x)]$$

$$\frac{d}{dx}[I_n(x)] = \tfrac{1}{2}[I_{n+1}(x) + I_{n-1}(x)]$$

$$\int x^{n+1}I_n(x)\,dx = x^{n+1}I_{n+1}(x)$$

$$K_{n+1}(x) = \frac{2n}{x}K_n(x) + K_{n-1}(x)$$

$$= \frac{n}{x}K_n(x) - \frac{d}{dx}[K_n(x)]$$

$$= -x^n\frac{d}{dx}[x^{-n}K_n(x)]$$

$$K_{n-1}(x) = -\frac{2n}{x}K_n(x) + K_{n+1}(x)$$

$$= -\frac{n}{x}K_n(x) - \frac{d}{dx}[K_n(x)]$$

$$= -x^{-n}\frac{d}{dx}[x^n K_n(x)]$$

$$\frac{d}{dx}[K_n(x)] = -\tfrac{1}{2}[K_{n+1}(x) + K_{n-1}(x)]$$

$$\int x^{n+1}K_n(x)\,dx = -x^{n+1}K_{n+1}(x)$$

(2) Half-Odd Integers   (See also Sec. 14.29-3)

$$I_{1/2}(x) = \sqrt{\frac{2}{\pi x}}\sinh x$$

$$I_{3/2}(x) = -\sqrt{\frac{2}{\pi x}}\left(\frac{\sinh x}{x} - \cosh x\right)$$

$$I_{5/2}(x) = \sqrt{\frac{2}{\pi x}}\left[\left(\frac{3}{x^2} + 1\right)\sinh x - \frac{3}{x}\cosh x\right]$$

$$I_{-1/2}(x) = \sqrt{\frac{2}{\pi x}}\cosh x$$

$$I_{-3/2}(x) = -\sqrt{\frac{2}{\pi x}}\left(\frac{\cosh x}{x} - \sinh x\right)$$

$$I_{-5/2}(x) = \sqrt{\frac{2}{\pi x}}\left[\left(\frac{3}{x^2} + 1\right)\cosh x - \frac{3}{x}\sinh x\right]$$

$$K_{1/2}(x) = e^{-x}\sqrt{\frac{\pi}{2x}}$$

$$K_{3/2}(x) = e^{-x}\sqrt{\frac{\pi}{2x}}\left(\frac{1}{x} + 1\right)$$

$$K_{5/2}(x) = e^{-x}\sqrt{\frac{\pi}{2x}}\left(\frac{2}{x^2} + \frac{2}{x} + 1\right)$$

$$K_{-1/2}(x) = e^{-x}\sqrt{\frac{\pi}{2x}}$$

$$K_{-3/2}(x) = e^{-x}\sqrt{\frac{\pi}{2x}}\left(\frac{1}{x} + 1\right)$$

$$K_{-5/2}(x) = e^{-x}\sqrt{\frac{\pi}{2x}}\left(\frac{2}{x^2} + \frac{2}{x} + 1\right)$$

(3) Transformations $(n = p \pm 1/2,\ p = 0, 1, 2, \ldots)$

$$J_{p+1/2}(x) = (-1)^p Y_{-p-1/2}(x) = (-1)^p\sqrt{\frac{2}{\pi x}}x^{p+1}\left(\frac{1}{x}\frac{d}{dx}\right)^p\frac{\sin x}{x}$$

$$Y_{p+1/2}(x) = -(-1)^p J_{-p-1/2}(x) = -(-1)^p\sqrt{\frac{2}{\pi x}}x^{p+1}\left(\frac{1}{x}\frac{d}{dx}\right)^p\frac{\cos x}{x}$$

$$I_{p+1/2}(x) = \sqrt{\frac{2}{\pi x}}x^{p+1}\left(\frac{1}{x}\frac{d}{dx}\right)^p\frac{\sinh x}{x}$$

$$K_{p+1/2}(x) = K_{-p-1/2}(x) = (-1)^p\sqrt{\frac{\pi}{2x}}x^{p+1}\left(\frac{1}{x}\frac{d}{dx}\right)^p\frac{e^{-x}}{x}$$

(1) Asymptotic Approximation

$$I_n(x) \approx \sqrt{\frac{1}{2\pi x}}\, e^x \qquad x > 25$$

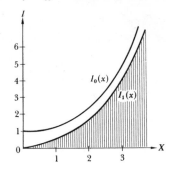

(2) Numerical Values $I_0(x)$ $x = 0 - 10$

x	0	1	2	3	4	5	6	7	8	9	x
0.	1.000	1.003	1.010	1.023	1.040	1.063	1.092	1.126	1.167	1.213	0.
1.	1.266	1.326	1.394	1.469	1.553	1.647	1.750	1.864	1.990	2.128	1.
2.	2.280	2.446	2.629	2.830	3.049	3.290	3.553	3.842	4.157	4.503	2.
3.	4.881	5.294	5.747	6.243	6.785	7.378	8.028	8.739	9.517	10.37	3.
4.	11.30	12.32	13.44	14.67	16.01	17.48	19.09	20.86	22.79	24.91	4.
5.	27.24	29.79	32.58	35.65	39.01	42.69	46.74	51.17	56.04	61.38	5.
6.	67.23	73.66	80.72	88.46	96.96	106.3	116.5	127.8	140.1	153.7	6.
7.	168.6	185.0	202.9	222.7	244.3	268.2	294.3	323.1	354.7	389.4	7.
8.	427.6	469.5	515.6	566.3	621.9	683.2	750.5	824.4	905.8	995.2	8.
9.	1,094	1,202	1,321	1,451	1,595	1,753	1,927	2,119	2,329	2,561	9.
10.	2,816										

(3) Numerical Values $I_1(x)$ $x = 0 - 10$

x	0	1	2	3	4	5	6	7	8	9	x
0.	0.0000	0.0501	0.1005	0.1517	0.2040	0.2579	0.3137	0.3719	0.4329	0.4971	0.
1.	0.5652	0.6375	0.7147	0.7973	0.8861	0.9817	1.085	1.196	1.317	1.448	1.
2.	1.591	1.745	1.914	2.098	2.298	2.517	2.755	3.016	3.301	3.613	2.
3.	3.953	4.326	4.734	5.181	5.670	6.206	6.793	7.436	8.140	8.913	3.
4.	9.759	10.69	11.71	12.82	14.05	15.39	16.86	18.48	20.25	22.20	4.
5.	24.34	26.68	29.25	32.08	35.18	38.59	42.33	46.44	50.95	55.90	5.
6.	61.34	67.32	73.89	81.10	89.03	97.74	107.3	117.8	129.4	142.1	6.
7.	156.0	171.4	188.3	206.8	227.2	249.6	274.2	301.3	381.1	363.9	7.
8.	399.9	439.5	483.0	531.0	583.7	641.6	705.4	775.5	852.7	937.5	8.
9.	1,031	1,134	1,247	1,371	1,508	1,658	1,824	2,006	2,207	2,428	9.
10.	2,671										

(4) Asymptotic Series for Large x

$$I_n(x) \approx \frac{e^x}{\sqrt{2\pi x}}\left(1 - \frac{\alpha_1}{1!} + \frac{\alpha_1 \alpha_3}{2!} - \cdots\right)$$

$$\alpha_k = \frac{4n^2 - k^2}{8x} \qquad k = 1, 2, \ldots \qquad x > 15$$

(1) Asymptotic Approximation

$$K_n(x) \approx \sqrt{\frac{\pi}{2x}}e^{-x} \qquad x > 25$$

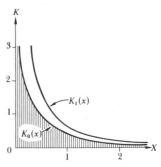

(2) Numerical Values $\boxed{K_0(x)}$ $x = 0 - 10$

x	0	1	2	3	4	5	6	7	8	9	x
0.	∞	2.4271	1.7527	1.3725	1.1145	0.9244	0.7775	0.6605	0.5653	0.4867	0.
1.	0.4210	0.3656	0.3185	0.2782	0.2437	0.2138	0.1880	0.1655	0.1459	0.1288	1.
2.	0.1139	0.1008	0.08927	0.07914	0.07022	0.06235	0.05540	0.04926	0.04382	0.03901	2.
3.	0.03474	0.03095	0.02759	0.02461	0.02196	0.01960	0.01750	0.01563	0.01397	0.01248	3.
4.	0.01116	$0.0^2 9980$	$0.0^2 8927$	$0.0^2 7988$	$0.0^2 7149$	$0.0^2 6400$	$0.0^2 5730$	$0.0^2 5132$	$0.0^2 4597$	$0.0^2 4119$	4.
5.	$0.0^2 3691$	$0.0^2 3308$	$0.0^2 2966$	$0.0^2 2659$	$0.0^2 2385$	$0.0^2 2139$	$0.0^2 1918$	$0.0^2 1721$	$0.0^2 1544$	$0.0^2 1386$	5.
6.	$0.0^2 1244$	$0.0^2 1117$	$0.0^2 1003$	$0.0^3 9001$	$0.0^3 8083$	$0.0^3 7259$	$0.0^3 6520$	$0.0^3 5857$	$0.0^3 5262$	$0.0^3 4728$	6.
7.	$0.0^3 4248$	$0.0^3 3817$	$0.0^3 3431$	$0.0^3 3084$	$0.0^3 2772$	$0.0^3 2492$	$0.0^3 2240$	$0.0^3 2014$	$0.0^3 1811$	$0.0^3 1629$	7.
8.	$0.0^3 1465$	$0.0^3 1317$	$0.0^3 1185$	$0.0^3 1066$	$0.0^4 9588$	$0.0^4 8626$	$0.0^4 7761$	$0.0^4 6983$	$0.0^4 6283$	$0.0^4 5654$	8.
9.	$0.0^4 5088$	$0.0^4 4579$	$0.0^4 4121$	$0.0^4 3710$	$0.0^4 3339$	$0.0^4 3006$	$0.0^4 2706$	$0.0^4 2436$	$0.0^4 2193$	$0.0^4 1975$	9.
10.	$0.0^4 1778$										

(3) Numerical Values $\boxed{K_1(x)}$ $x = 0 - 10$

x	0	1	2	3	4	5	6	7	8	9	x
0.	∞	9.8538	4.7760	3.0560	2.1844	1.6564	1.3028	1.0503	0.8618	0.7165	0.
1.	0.6019	0.5098	0.4346	0.3725	0.3208	0.2774	0.2406	0.2094	0.1826	0.1597	1.
2.	0.1399	0.1227	0.1079	0.09498	0.08372	0.07389	0.06528	0.05774	0.05111	0.04529	2.
3.	0.04016	0.03563	0.03164	0.02812	0.02500	0.02224	0.01979	0.01763	0.01571	0.01400	3.
4.	0.01248	0.01114	$0.0^2 9938$	$0.0^2 8872$	$0.0^2 7923$	$0.0^2 7078$	$0.0^2 6325$	$0.0^2 5654$	$0.0^2 5055$	$0.0^2 4521$	4.
5.	$0.0^2 4045$	$0.0^2 3619$	$0.0^2 3239$	$0.0^2 2900$	$0.0^2 2597$	$0.0^2 2326$	$0.0^2 2083$	$0.0^2 1866$	$0.0^2 1673$	$0.0^2 1499$	5.
6.	$0.0^2 1344$	$0.0^2 1205$	$0.0^2 1081$	$0.0^3 9691$	$0.0^3 8693$	$0.0^3 7799$	$0.0^3 6998$	$0.0^3 6280$	$0.0^3 5636$	$0.0^3 5059$	6.
7.	$0.0^3 4542$	$0.0^3 4078$	$0.0^3 3662$	$0.0^3 3288$	$0.0^3 2953$	$0.0^3 2653$	$0.0^3 2383$	$0.0^3 2141$	$0.0^3 1924$	$0.0^3 1729$	7.
8.	$0.0^3 1554$	$0.0^3 1396$	$0.0^3 1255$	$0.0^3 1128$	$0.0^3 1014$	$0.0^4 9120$	$0.0^4 8200$	$0.0^4 7374$	$0.0^4 6631$	$0.0^4 5964$	8.
9.	$0.0^4 5364$	$0.0^4 4825$	$0.0^4 4340$	$0.0^4 3904$	$0.0^4 3512$	$0.0^4 3160$	$0.0^4 2843$	$0.0^4 2559$	$0.0^4 2302$	$0.0^4 2072$	9.
10.	$0.0^4 1865$										

(4) Asymptotic Series for Large x

$$K_n(x) \approx \sqrt{\frac{\pi}{2x}}e^{-x}\left(1 + \frac{\alpha_1}{1!} + \frac{\alpha_1 \alpha_3}{2!} + \cdots\right)$$

$$\alpha_k = \frac{4n^2 - k^2}{8x} \qquad k = 1, 2, \ldots \qquad x > 15$$

(1) Differential Equation

$$x^2 y'' + xy' \pm ia^2 x^2 y = 0$$

(2) Solution

$$y = A_1 [\text{Ber }(ax) \mp i\,\text{Bei }(ax)] + A_2 [\text{Ker }(ax) \mp i\,\text{Kei }(ax)]$$

(3) Ber, Bei **Functions**

$$\text{Ber }(ax) = 1 - \frac{(ax/2)^4}{(2!)^2} + \frac{(ax/2)^8}{(4!)^2} - \cdots$$

$$\text{Bei }(ax) = \frac{(ax/2)^2}{(1!)^2} - \frac{(ax/2)^6}{(3!)^2} + \frac{(ax/2)^{10}}{(5!)^2} - \cdots$$

(4) Ker, Kei **Functions**

$$\text{Ker }(ax) = -\left(\ln\frac{ax}{2} + C\right)\text{Ber }(ax) + \frac{\pi}{4}\text{Bei }(ax) - \lambda_2 + \lambda_4 - \cdots$$

$$\text{Kei }(ax) = -\left(\ln\frac{ax}{2} + C\right)\text{Bei }(ax) - \frac{\pi}{4}\text{Ber }(ax) + \lambda_1 - \lambda_3 - \cdots$$

where $\qquad C = 0.577\,215\,665 \qquad \lambda_k = \frac{(ax/2)^{2k}}{(k!)^2}\left(1 + \frac{1}{2} + \frac{1}{3} + \cdots + \frac{1}{k}\right)$ (Sec. 13.04)

(5) Ber′, Bei′ **Functions**

$$\text{Ber}'(ax) = -\frac{2a(ax/2)^3}{(2!)^2} + \frac{4a(ax/2)^7}{(4!)^2} - \cdots = \frac{d}{dx}[\text{Ber }(ax)]$$

$$\text{Bei}'(ax) = \frac{a(ax/2)}{(1!)^2} - \frac{3a(ax/2)^5}{(3!)^2} + \cdots = \frac{d}{dx}[\text{Bei }(ax)]$$

(6) Ker′, Kei′ **Functions**

$$\text{Ker}'(ax) = -\left(\ln\frac{ax}{2} + C\right)\text{Ber}'(ax) - \frac{1}{x}\text{Ber }(ax) + \frac{\pi}{4}\text{Bei}'(ax) - \lambda_2' + \lambda_4' - \cdots = \frac{d}{dx}[\text{Ker }(ax)]$$

$$\text{Kei}'(ax) = -\left(\ln\frac{ax}{2} + C\right)\text{Bei}'(ax) - \frac{1}{x}\text{Bei }(ax) - \frac{\pi}{4}\text{Ber}'(ax) + \lambda_1' - \lambda_3' + \cdots = \frac{d}{dx}[\text{Kei }(ax)]$$

where $\qquad \lambda_k' = \frac{d\lambda_k}{dx}$

(7) Relations

$$\text{Ber }(x) + i\,\text{Bei }(x) = J_0(xi\sqrt{i}) = I_0(x\sqrt{i})$$

$$\text{Ker }(x) + i\,\text{Kei }(x) = K_0(x\sqrt{i})$$

$$\int x\,\text{Ber }(x)\,dx = x\,\text{Bei}'(x) \qquad\qquad \int x\,\text{Bei }(x)\,dx = -x\,\text{Ber}'(x)$$

$$\int x\,\text{Ker }(x)\,dx = x\,\text{Kei}'(x) \qquad\qquad \int x\,\text{Kei }(x)\,dx = -x\,\text{Ker}'(x)$$

(1) Differential Equation

$$x^2 y'' + xy' \pm (ix^2 + n^2) y = 0$$

(2) Solution

$$y = A_1 [\text{Ber}_n (x) \mp i \text{Bei}_n (x)] + A_2 [\text{Ker}_n (x) \mp i \text{Kei}_n (x)]$$

(3) Ber$_n$, Bei$_n$ Functions

$$\text{Ber}_n (x) = \sum_{k=0}^{\infty} \frac{(-1)^{n+k} (x/2)^{n+2k}}{k!(n+k)!} \cos \frac{(n+2k)\pi}{4}$$

$$\text{Bei}_n (x) = \sum_{k=0}^{\infty} \frac{(-1)^{n+k+1} (x/2)^{n+2k}}{k!(n+k)!} \sin \frac{(n+2k)\pi}{4} \qquad n = 0, 1, 2, \ldots \qquad k = 0, 1, 2, \ldots$$

(4) Ker$_n$, Kei$_n$ Functions

$$\text{Ker}_n (x) = -\left(\ln \frac{x}{2} + C\right) \text{Ber}_n (x) + \frac{\pi}{4} \text{Bei}_n (x) + \frac{1}{2} \sum_{k=0}^{n-1} \frac{(n-k-1)!(x/2)^{2k-n}}{k!} \cos \frac{(3n+2k)\pi}{4}$$

$$+ \frac{1}{2} \sum_{k=0}^{\infty} \frac{(x/2)^{2k+n}}{k!(n+k)!} (\tau_k + \tau_{k+n}) \cos \frac{(3n+2k)\pi}{4}$$

$$\text{Kei}_n (x) = -\left(\ln \frac{x}{2} + C\right) \text{Bei}_n (x) - \frac{\pi}{4} \text{Ber}_n (x) - \frac{1}{2} \sum_{k=0}^{n-1} \frac{(n-k-1)!(x/2)^{2k-n}}{k!} \sin \frac{(3n+2k)\pi}{4}$$

$$+ \frac{1}{2} \sum_{k=0}^{\infty} \frac{(x/2)^{2k+n} (\tau_k + \tau_{k+n})}{k!(n+k)!} \sin \frac{(3n+2k)\pi}{4}$$

where $\qquad \tau_k = 1 + \dfrac{1}{2} + \dfrac{1}{3} + \cdots + \dfrac{1}{k} \qquad \tau_{k+n} = 1 + \dfrac{1}{2} + \dfrac{1}{3} + \cdots + \dfrac{1}{k+n}$

$$C = 0.577\,215\,665 \qquad k, n = 0, 1, 2, \ldots \qquad \text{(Sec. 13.04)}$$

(5) Relations

$$\text{Ber}_n (x) + i \text{Bei}_n (x) = J_n(xi\sqrt{i}) = i^n I_n(x\sqrt{i})$$

$$\text{Ber}_n (x) + \text{Bei}_n (x) = -\frac{x}{n\sqrt{2}} [\text{Bei}_{n-1} (x) + \text{Bei}_{n+1} (x)]$$

$$\text{Ber}_n (x) - \text{Bei}_n (x) = -\frac{x}{n\sqrt{2}} [\text{Ber}_{n-1} (x) + \text{Ber}_{n+1} (x)]$$

$$\text{Ker}_n (x) + i \text{Kei}_n (x) = i^{-n} K_n(x\sqrt{i})$$

$$\text{Ker}_n (x) + \text{Kei}_n (x) = -\frac{x}{n\sqrt{2}} [\text{Kei}_{n-1} (x) + \text{Kei}_{n+1} (x)]$$

$$\text{Ker}_n (x) - \text{Kei}_n (x) = -\frac{x}{n\sqrt{2}} [\text{Ker}_{n-1} (x) + \text{Kei}_{n+1}(x)]$$

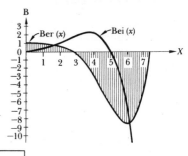

(1) Asymptotic Approximation

$$\text{Ber}(x) \approx \frac{e^{x/\sqrt{2}}}{\sqrt{2\pi x}} \cos\left(\frac{x}{\sqrt{2}} - \frac{\pi}{8}\right)$$

$$\text{Bei}(x) \approx \frac{e^{x/\sqrt{2}}}{\sqrt{2\pi x}} \sin\left(\frac{x}{\sqrt{2}} - \frac{\pi}{8}\right)$$

$$x > 25$$

(2) Numerical Values   Ber(*x*) $x = 0 - 10$

x	0	1	2	3	4	5	6	7	8	9	x
0.	1.0000	1.0000	1.0000	0.9999	0.9996	0.9990	0.9980	0.9962	0.9936	0.9890	0.
1.	0.9844	0.9771	0.9676	0.9554	0.9401	0.9211	0.8979	0.8700	0.8367	0.7975	1.
2.	0.7517	0.6987	0.6377	0.5680	0.4890	0.4000	0.3001	0.1887	0.0651	−0.0714	2.
3.	−0.2214	−0.3855	−0.5644	−0.7584	−0.9680	−1.1936	−1.4353	−1.6933	−1.9674	−2.2576	3.
4.	−2.5634	−2.8843	−3.3295	−3.5679	−3.9283	−4.2991	−4.6784	−5.0639	−5.4531	−5.8429	4.
5.	−6.2301	−6.6107	−6.9803	−7.3344	−7.6674	−7.9736	−8.2466	−8.4794	−8.6644	−8.7937	5.
6.	−8.8583	−8.8491	−8.7561	−8.5688	−8.2762	−7.8669	−7.3287	−6.6492	−5.8155	−4.8146	6.
7.	−3.6329	−2.2571	−0.6737	1.1308	3.1695	5.4550	7.9994	10.814	13.909	17.293	7.
8.	20.974	24.957	29.245	33.840	38.738	43.936	49.423	55.187	61.210	67.469	8.
9.	73.936	80.576	87.350	94.208	101.10	107.95	114.70	121.26	127.54	133.43	9.
10.	138.84										

(3) Numerical Values   Bei(*x*) $x = 0 - 10$

x	0	1	2	3	4	5	6	7	8	9	x
0.	0.0000	$0.0^2 2500$	0.01000	0.02250	0.04000	0.06249	0.08998	0.1224	0.1599	0.2023	0.
1.	0.2496	0.3017	0.3587	0.4204	0.4867	0.5576	0.6327	0.7120	0.7953	0.8821	1.
2.	0.9723	1.0654	1.1610	1.2585	1.3575	1.4572	1.5569	1.6557	1.7529	1.8472	2.
3.	1.9376	2.0228	2.1016	2.1723	2.2334	2.2832	2.3199	2.3413	2.3454	2.3300	3.
4.	2.2927	2.2309	2.1422	2.0236	1.8726	1.6860	1.4610	1.1946	0.8837	0.5251	4.
5.	0.1160	−0.3467	−0.8658	−1.4443	−2.0845	−2.7890	−3.5597	−4.3986	−5.3068	−6.2854	5.
6.	−7.3347	−8.4545	−9.6437	−10.901	−12.223	−13.607	−15.047	−16.538	−18.074	−19.644	6.
7.	−21.239	−22.848	−24.456	−26.049	−27.609	−29.116	−30.548	−31.882	−33.092	−34.147	7.
8.	−35.017	−35.667	−36.061	−36.159	−35.920	−35.298	−34.246	−32.714	−30.651	−28.003	8.
9.	−24.713	−20.724	−15.976	−10.412	−3.9693	3.4106	11.787	21.218	31.758	43.459	9.
10.	56.371										

(4) Asymptotic Series for Large *x*

$$\text{Ber}(x) \approx \frac{e^{x/\sqrt{2}}}{\sqrt{2\pi x}}\{\cos\phi[1 + (1^2)\beta_1 + (1^2)(3^2)\beta_2 + \cdots] + \sin\phi[(1^2)\gamma_1 + (1^2)(3^2)\gamma_2 + \cdots]\}$$

$$\text{Bei}(x) \approx \frac{e^{x/\sqrt{2}}}{\sqrt{2\pi x}}\{\sin\phi[1 + (1^2)\beta_1 + (1^2)(3^2)\beta_2 + \cdots] - \cos\phi[(1^2)\gamma_1 + (1^2)(3^2)\gamma_2 + \cdots]\}$$

$$\phi = \frac{x}{\sqrt{2}} - \frac{\pi}{8} \qquad \beta_k = \frac{\cos(k\pi/4)}{k!(8x)^k} \qquad \gamma_k = \frac{\sin(k\pi/4)}{k!(8x)^k} \qquad k = 1, 2, \ldots \qquad x > 15$$

(1) Asymptotic Approximation

$$\text{Ker}\,(x) \approx \sqrt{\frac{\pi}{2x}}\,e^{-x/\sqrt{2}}\cos\left(\frac{x}{\sqrt{2}}+\frac{\pi}{8}\right)$$

$$\text{Kei}\,(x) \approx -\sqrt{\frac{\pi}{2x}}\,e^{-x/\sqrt{2}}\sin\left(\frac{x}{\sqrt{2}}+\frac{\pi}{8}\right)$$

$x > 25$

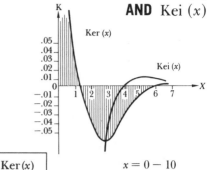

(2) Numerical Values | Ker (x) | $x = 0 - 10$

x	0	1	2	3	4	5	6	7	8	9	x
0.	∞	2.4205	1.7331	1.3372	1.0626	0.8559	0.6931	0.5614	0.4529	0.3625	0.
1.	0.2867	0.2228	0.1689	0.1235	0.08513	0.05293	0.02603	$0.0^2 3691$	-0.01470	-0.02966	1.
2.	-0.04166	-0.05111	-0.05834	-0.06367	-0.06737	-0.06969	-0.07083	-0.07097	-0.07030	-0.06894	2.
3.	-0.06703	-0.06468	-0.06198	-0.05903	-0.05590	-0.05264	-0.04932	-0.04597	-0.04265	-0.03937	3.
4.	-0.03618	-0.03308	-0.03011	-0.02726	-0.02456	-0.02200	-0.01960	-0.01734	-0.01525	-0.01330	4.
5.	-0.01151	$-0.0^2 9865$	$-0.0^2 8359$	$-0.0^2 6989$	$-0.0^2 5749$	$-0.0^2 4632$	$-0.0^2 3632$	$-0.0^2 2740$	$-0.0^2 1952$	$-0.0^2 1258$	5.
6.	$-0.0^3 6530$	$-0.0^3 1295$	$-0.0^3 3191$	$-0.0^3 6991$	$-0.0^2 1017$	$-0.0^2 1278$	$-0.0^2 1488$	$-0.0^2 1653$	$-0.0^2 1777$	$-0.0^2 1866$	6.
7.	$0.0^2 1922$	$0.0^2 1951$	$0.0^2 1956$	$0.0^2 1940$	$0.0^2 1907$	$0.0^2 1860$	$0.0^2 1800$	$0.0^2 1731$	$0.0^2 1655$	$0.0^2 1572$	7.
8.	$0.0^2 1486$	$0.0^2 1397$	$0.0^2 1306$	$0.0^2 1216$	$0.0^2 1126$	$0.0^2 1037$	$0.0^3 9511$	$0.0^3 8675$	$0.0^3 7871$	$0.0^3 7102$	8.
9.	$0.0^3 6372$	$0.0^3 5681$	$0.0^3 5030$	$0.0^3 4422$	$0.0^3 3855$	$0.0^3 3330$	$0.0^3 2846$	$0.0^3 2402$	$0.0^3 1996$	$0.0^3 1628$	9.
10.	$0.0^3 1295$										

(3) Numerical Values | Kei (x) | $x = 0 - 10$

x	0	1	2	3	4	5	6	7	8	9	x
0.	-0.7854	-0.7769	-0.7581	-0.7331	-0.7038	-0.6716	-0.6374	-0.6022	-0.5664	-0.5305	0.
1.	-0.4950	-0.4601	-0.4262	-0.3933	-0.3617	-0.3314	-0.3026	-0.2752	-0.2494	-0.2251	1.
2.	-0.2024	-0.1812	-0.1614	-0.1431	-0.1262	-0.1107	-0.09644	-0.08342	-0.07157	-0.06083	2.
3.	-0.05112	-0.04240	-0.03458	-0.02762	-0.02145	-0.01600	-0.01123	$0.0^2 7077$	$-0.0^3 3487$	$0.0^3 4108$	3.
4.	$0.0^2 2198$	$0.0^2 4386$	$0.0^2 6194$	$0.0^2 7661$	$0.0^2 8826$	$0.0^2 9721$	0.01038	0.01083	0.01110	0.01121	4.
5.	0.01119	0.01105	0.01082	0.01051	0.01014	$0.0^2 9716$	$0.0^2 9255$	$0.0^2 8766$	$0.0^2 8258$	$0.0^2 7739$	5.
6.	$0.0^2 7216$	$0.0^2 6696$	$0.0^2 6183$	$0.0^2 5681$	$0.0^2 5194$	$0.0^2 4724$	$0.0^2 4274$	$0.0^2 3846$	$0.0^2 3440$	$0.0^2 3058$	6.
7.	$0.0^2 2700$	$0.0^2 2366$	$0.0^2 2057$	$0.0^2 1770$	$0.0^2 1507$	$0.0^2 1267$	$0.0^2 1048$	$0.0^3 8498$	$0.0^3 6714$	$0.0^3 5117$	7.
8.	$0.0^3 3696$	$0.0^3 2440$	$0.0^3 1339$	$0.0^4 3809$	$-0.0^4 4449$	$-0.0^3 1149$	$-0.0^3 1742$	$-0.0^3 2233$	$-0.0^3 2632$	$-0.0^3 2949$	8.
9.	$-0.0^3 3192$	$-0.0^3 3368$	$-0.0^3 3486$	$-0.0^3 3552$	$-0.0^3 3574$	$-0.0^3 3557$	$-0.0^3 3508$	$-0.0^3 3430$	$-0.0^3 3329$	$-0.0^3 3210$	9.
10.	$-0.0^3 3075$										

(4) Asymptotic Series for Large x

$$\text{Ker}\,(x) \approx \sqrt{\frac{\pi}{2x}}\,e^{-x/\sqrt{2}}\{\cos\eta\,[1-(1^2)\beta_1+(1^2)(3^2)\beta_2-\cdots]+\sin\eta\,[(1^2)\gamma_1-(1^2)(3^2)\gamma_2+\cdots]\}$$

$$\text{Kei}\,(x) \approx \sqrt{\frac{\pi}{2x}}\,e^{-x/\sqrt{2}}\{-\cos\eta\,[(1^2)\gamma_1-(1^2)(3^3)\gamma_2+\cdots]-\sin\eta\,[1-(1^2)\beta_1+(1^2)(3^2)\beta_2-\cdots]\}$$

$$\eta=\frac{x}{\sqrt{2}}+\frac{\pi}{8} \qquad \beta_k=\frac{\cos\,(k\pi/4)}{k!\,(8x)^k} \qquad \gamma_k=\frac{\sin\,(k\pi/4)}{k!\,(8x)^k} \qquad k=1,2,\ldots \qquad x>15$$

$$J_n(x + y) = \sum_{k=-\infty}^{+\infty} J_k(x) J_{n-k}(y) \qquad n = 0, \pm 1, \pm 2, \ldots$$

$$1 = J_0(x) + 2J_2(x) + \cdots + 2J_{2n}(x) + \cdots$$

$$x = 2[J_1(x) + 3J_3(x) + \cdots + (2n + 1) J_{2n+1}(x) + \cdots]$$

$$x^2 = 8[J_2(x) + 4J_4(x) + \cdots + n^2 J_{2n}(x) + \cdots]$$

$$\sin x = 2[J_1(x) - J_3(x) + J_5(x) - \cdots]$$

$$\cos x = J_0(x) - 2[J_2(x) - J_4(x) + \cdots]$$

$$\sinh x = 2[I_1(x) + I_3(x) + I_5(x) + \cdots]$$

$$\cosh x = I_0(x) + 2[I_2(x) + I_4(x) + \cdots]$$

$$\sin (x \sin \omega) = 2[J_1(x) \sin \omega + J_3(x) \sin 3\omega + \cdots]$$

$$\sin (x \cos \omega) = 2[J_1(x) \cos \omega - J_3(x) \cos 3\omega - \cdots]$$

$$\cos (x \sin \omega) = J_0(x) + 2[J_2(x) \cos 2\omega + J_4(x) \cos 4\omega + \cdots]$$

$$\cos (x \cos \omega) = J_0(x) - 2[J_2(x) \cos 2\omega - J_4(x) \cos 4\omega + \cdots]$$

14.37 DEFINITE INTEGRALS INVOLVING BESSEL FUNCTIONS

$$\int_0^\infty J_n(\alpha x) \, dx = \frac{1}{\alpha} \qquad n > -1 \qquad\qquad \int_0^\infty J_n(\alpha x) \frac{dx}{x} = \frac{1}{n} \qquad n = 1, 2, \ldots$$

$$\int_0^\infty e^{-\alpha x} J_0(\beta x) \, dx = \frac{1}{\sqrt{\alpha^2 + \beta^2}} \qquad\qquad \int_0^\infty e^{-\alpha x} J_1(\beta x) \, dx = \frac{1}{\beta} \left(1 - \frac{\alpha}{\sqrt{\alpha^2 + \beta^2}} \right)$$

$$\int_0^\infty J_n(\alpha x) \sin \beta x \, dx = \begin{cases} \dfrac{\sin (n \sin^{-1} (\beta/\alpha))}{\sqrt{\alpha^2 - \beta^2}} & 0 < \beta < \alpha \\[3mm] \dfrac{\alpha^n \cos (n\pi/2)}{\sqrt{b^2 - \alpha^2}(\beta + \sqrt{\beta^2 - \alpha^2})^n} & 0 < \alpha < \beta \end{cases} \Bigg\} \; n > -2$$

$$\int_0^\infty J_n(\alpha x) \cos \beta x \, dx = \begin{cases} \dfrac{\cos [n \cos^{-1} (\beta/\alpha)]}{\sqrt{\alpha^2 - \beta^2}} & 0 < \beta < \alpha \\[3mm] \dfrac{-\alpha^n \sin (n\pi/2)}{\sqrt{\beta^2 - \alpha^2}(\beta + \sqrt{\beta^2 - \alpha^2})^n} & 0 < \alpha < \beta \end{cases} \Bigg\} \; n > -1$$

$$\int_0^\infty \frac{J_m(x) J_n(x)}{x} \, dx = \begin{cases} \dfrac{2}{(m^2 - n^2)\pi} \sin \dfrac{(m - n)\pi}{2} & m \neq n \\[3mm] \dfrac{1}{2m} & m = n \end{cases} \Bigg\} \; m + n > 0$$

15
PARTIAL
DIFFERENTIAL
EQUATIONS

(1) Definition

A partial differential equation of order n is *a functional equation* involving at least one nth partial derivative of the unknown function $\Phi(x_1, x_2, \ldots, x_m)$ of two or more variables $x_1, x_2, \ldots, x_m$.

(2) Solution

A partial differential equation of order n given in general form as

$$F\left(x_1, x_2, \ldots, x_m, \Phi; \frac{\partial \Phi}{\partial x_1}, \frac{\partial \Phi}{\partial x_2}, \ldots, \frac{\partial \Phi}{\partial x_m}; \frac{\partial^2 \Phi}{\partial x_1^2}, \frac{\partial^2 \Phi}{\partial x_2^2}, \ldots, \frac{\partial^2 \Phi}{\partial x_m^2}; \cdots \right) = 0$$

has a *general solution* (general integral), which involves *arbitrary functions*. A *particular solution* (particular integral) is a special case of the general solution with *specific functions* substituted for the arbitrary functions. These specific functions are generated by the given conditions (boundary conditions, initial conditions). Many partial differential equations admit *singular solutions* (singular integrals) unrelated to the general integral.

(3) Separation of Variables

Although the separation-of-variables method is not universally applicable, it offers a convenient and simple solution of many partial differential equations in engineering and the physical sciences. The underlying idea is to *separate* the given partial differential equation into *a set of ordinary differential equations*, the solution of which is known.

The *assumed solution* is

$$\Phi(x_1, x_2, \ldots, x_m) = \Phi_1(x_1)\Phi_2(x_2) \cdots \Phi_m(x_m)$$

the linear combination of which is the general solution. This procedure is applicable when the substitution of the assumed solution in the given partial differential equation transforms this equation into

$$\psi_1\left(x_1; \frac{d\Phi_1}{\partial x_1}; \ldots\right) + \psi_2\left(x_2, x_3, \ldots, x_m; \frac{d\Phi_2}{dx_2}; \ldots\right) = 0$$

A setting $\psi_1 = C$ and $\psi_2 = -C$ breaks the equation $\psi_1 + \psi_2 = 0$ into an ordinary differential equation and a partial differential equation, the second of which (if necessary) can be the subject of a repeated separation process.

(4) Classification

Partial differential equations are classified according to the type of conditions imposed by the problem. Linear differential equations of the second order in the two variables given as

$$Au_{xx} + Bu_{xy} + Cu_{yy} + Du_x + Eu_y + Fu + G = 0$$

where $u_{xx}, u_{xy}, \ldots$ are the partial derivatives with respect to x and/or y and $A, B, \ldots$, are functions of x and y, are classified as *elliptic, hyperbolic, or parabolic* according to whether

$$\Delta = \begin{vmatrix} A & B \\ B & C \end{vmatrix}$$

is *positive, negative, or zero*. The concept of these definitions can be extended for differential equations with more than two variables.

a, b, c = given constants $u = u(x, y)$ $P = P(x, y, u)$	$u_x = \dfrac{du}{dx}$ $Q = Q(x, y, u)$	A, B, C = integration constants $u_y = \dfrac{du}{dy}$ $R = R(x, y, u)$

(1) Quasilinear Equation

For

$$P(x, y, u)\frac{du}{dx} + Q(x, y, u)\frac{du}{dy} = R(x, y, z)$$

the *characteristic equation* is

$$dx : dy : du = P : Q : R \qquad \text{or} \qquad \frac{dx}{P} = \frac{dy}{Q} = \frac{du}{R}$$

from which

$$f(x, y, u) = A \qquad g(x, y, u) = B$$

and the *general solution* is

$$F[f(x, y, u) ; g(x, y, u)] = 0$$

where $F[\ ]$ is an *arbitrary function*.

(2) Special Linear Equations

Differential equation	Solution
$a\,u_x \pm b\,u_y = 0$	$u = f(ay \mp bx)$
$y\,u_x - x\,u_y = 0$	$u = f(x^2 + y^2)$
$y\,u_x + x\,u_y = u$	$u = (x + y)f(x^2 - y^2)$
$xu\,u_y + yu\,u_x = cx$	$u^2 = 2cy + f(x^2 + y^2)$
$a\,u_x + b\,u_y = c$	$F\left[\left(\dfrac{x}{a} - \dfrac{u}{c}\right); \left(\dfrac{y}{b} - \dfrac{u}{c}\right)\right] = 0$
$x\,u_x + y\,u_y = u$	$F\left(\dfrac{x}{u}; \dfrac{y}{u}\right) = 0$
$(x - a)\,u_x + (y - b)\,u_y = u - c$	$F\left(\dfrac{x - a}{u - c}; \dfrac{y - b}{u - c}\right) = 0$
$yu\,u_x + xu\,u_y = xy$	$F[(x^2 - y^2) ; (x^2 - u^2)] = 0$

(3) Special Nonlinear Equations

Differential equation	Solution	
$u_x\,u_y = c$	$u = Ax + By + C$	$B = \dfrac{c}{A}$
$u_x\,u_y = xy$	$u = Ax^2 + By^2 + C$	$B = \dfrac{1}{4A}$
$u_x\,u_y = u_x + u_y$	$u = Ax + By + C$	$B = \dfrac{A}{A - 1}$
$u_x\,u_y = x\,u_x + y\,u_y$	$u = Ax^2 + xy + By^2 + C$	$B = \dfrac{1}{4A}$

A, B, C = integration constants $\qquad$ α, β, γ = separation constants

$J(\ ), Y(\ )$ = Bessel functions (Sec. 14.24)

(1) Two-dimensional Cases

(a) Rectangular coordinates

$$\frac{\partial^2\Phi}{\partial x^2} + \frac{\partial^2\Phi}{\partial y^2} = 0$$

Solution:

$$\Phi_\alpha = (A_1 e^{i\alpha x} + A_2 e^{-i\alpha x})(B_1 e^{\alpha y} + B_2 e^{-\alpha y}) \qquad \alpha \neq 0$$

$$\Phi_0 = (A_1 + A_2 x)(B_1 + B_2 y) \qquad \alpha = 0$$

(b) Polar coordinates

$$\frac{\partial^2\Phi}{\partial r^2} + \frac{1}{r}\frac{\partial\Phi}{\partial r} + \frac{1}{r^2}\frac{\partial^2\Phi}{\partial\phi^2} = 0$$

Solution:

$$\Phi_\alpha = (A_1 r^\alpha + A_2 r^{-\alpha})(B_1 e^{\alpha i\phi} + B_2 e^{-\alpha i\phi}) \qquad \alpha \neq 0$$

$$\Phi_0 = (A_1 + A_2 \ln r)(B_1 + B_2 \phi) \qquad \alpha = 0$$

(2) Three-dimensional Cases

(a) Rectangular coordinates

$$\frac{\partial^2\Phi}{\partial x^2} + \frac{\partial^2\Phi}{\partial y^2} + \frac{\partial^2\Phi}{\partial z^2} = 0$$

Solution: $(\alpha^2 + \beta^2 + \gamma^2 = 0)$

$$\phi_{\alpha\beta\gamma} = (A_1 e^{\alpha x} + A_2 e^{-\alpha x})(B_1 e^{\beta y} + B_2 e^{-\beta y})(C_1 e^{\gamma z} + C_2 e^{-\gamma z}) \qquad \alpha \neq 0, \beta \neq 0, \gamma \neq 0$$

$$\phi_{000} = (A_1 + A_2 x)(B_1 + B_2 y)(C_1 + C_2 z) \qquad \alpha = 0, \beta = 0, \gamma = 0$$

(b) Cylindrical coordinates

$$\frac{\partial^2\Phi}{\partial r^2} + \frac{1}{r}\frac{\partial\Phi}{\partial r} + \frac{1}{r^2}\frac{\partial^2\Phi}{\partial\phi^2} + \frac{\partial^2\Phi}{\partial z^2} = 0$$

Solution:

$$\phi_{\alpha\beta} = [A_1 J_\alpha(i\beta r) + A_2 Y_\alpha(i\beta r)](B_1 e^{i\alpha\phi} + B_2 e^{-i\alpha\phi})(C_1 e^{i\beta z} + C_2 e^{-i\beta z}) \qquad \alpha \neq 0, \beta \neq 0$$

$$\phi_{00} = (A_1 + A_2 \ln r)(B_1 + B_2 \phi)(C_1 + C_2 z) \qquad \alpha = 0, \beta = 0$$

$$\phi_{\alpha 0} = (A_1 + A_2 \ln r)(B_1 e^{i\alpha\phi} + B_2 e^{-i\alpha\phi})(C_1 + C_2 z) \qquad \alpha \neq 0, \beta = 0$$

$$\phi_{0\beta} = J_0(i\beta r)(B_1 + B_2 \phi)(C_1 e^{i\beta z} + C_2 e^{-i\beta z}) \qquad \alpha = 0, \beta \neq 0$$

A, B, C = integration constants

$J(\ ), Y(\ )$ = Bessel functions (Sec. 14.24)

α, β, γ = separation constants

k = given constant

(1) Two-dimensional Cases

(a) Rectangular coordinates

$$\frac{\partial^2\Phi}{\partial x^2} + \frac{\partial^2\Phi}{\partial y^2} + k^2\Phi = 0$$

Solution: $(\alpha^2 + \beta^2 = k^2)$

$$\Phi_{\alpha\beta} = (A_1 e^{i\alpha x} + A_2 e^{-i\alpha x})(B_1 e^{i\beta y} + B_2 e^{-i\beta y}) \qquad \alpha \neq 0$$

$$\Phi_{0k} = (A_1 + A_2 x)(B_1 e^{iky} + B_2 e^{-iky}) \qquad \alpha = 0$$

(b) Polar coordinates

$$\frac{\partial^2\Phi}{\partial r^2} + \frac{1}{r}\frac{\partial\Phi}{\partial r} + \frac{1}{r^2}\frac{\partial^2\Phi}{\partial \phi^2} + k^2\Phi = 0$$

Solution:

$$\Phi_{\alpha k} = [A_1 J_\alpha(ikr) + A_2 Y_\alpha(ikr)](B_1 e^{i\alpha\phi} + B_2 e^{-i\alpha\phi}) \qquad \alpha \neq 0$$

$$\Phi_{0k} = J_0(ikr)(B_1 + B_2\phi) \qquad \alpha = 0$$

(2) Three-dimensional Cases

(a) Rectangular coordinates

$$\frac{\partial^2\Phi}{\partial x^2} + \frac{\partial^2\Phi}{\partial y^2} + \frac{\partial^2\Phi}{\partial z^2} + k^2\Phi = 0$$

Solution: $(\alpha^2 + \beta^2 + \gamma^2 = k^2)$

$$\Phi_{\alpha\beta\gamma} = (A_1 e^{i\alpha x} + A_2 e^{-i\alpha x})(B_1 e^{i\beta y} + B_2 e^{-i\beta y})(C_1 e^{i\gamma z} + C_2 e^{-i\gamma z}) \qquad \alpha \neq 0, \beta \neq 0, \gamma \neq 0$$

$$\Phi_{00k} = (A_1 + A_2 x)(B_1 + B_2 y)(C_1 e^{ikz} + C_2 e^{-ikz}) \qquad \alpha = 0, \beta = 0$$

(b) Cylindrical coordinates

$$\frac{\partial^2\Phi}{\partial r^2} + \frac{1}{r}\frac{\partial\Phi}{\partial r} + \frac{1}{r^2}\frac{\partial^2\Phi}{\partial \phi^2} + \frac{\partial^2\Phi}{\partial z^2} + k^2\Phi = 0$$

Solution:

$$\Phi_{\alpha\beta k} = [A_1 J_\alpha(i\gamma r) + A_2 Y_\alpha(i\gamma r)](B_1 e^{i\alpha\phi} + B_2 e^{-i\alpha\phi})(C_1 e^{i\beta z} + C_2 e^{-i\beta z}) \qquad \alpha \neq 0, \beta \neq 0$$

$$\Phi_{00k} = J_0(ikr)(B_1 + B_2\phi)(C_1 + C_2 z) \qquad \alpha = 0, \beta = 0$$

$$\gamma = \sqrt{k^2 - \beta^2}$$

A, B, C = integration constants	$\alpha, \beta, \gamma, \lambda, \omega$ = separation constants
c = given constant	$J(\ ), Y(\ )$ = Bessel functions (Sec. 14.24)

(1) Rectangular Coordinates

(a) One-dimensional case

$$\frac{\partial^2 \Phi}{\partial x^2} - \frac{1}{c^2}\frac{\partial \Phi}{\partial t} = 0$$

Solution: $(\alpha = \lambda, \omega = c^2 \lambda^2)$

$$\Phi_\alpha = (A_1 e^{i\alpha x} + A_2 e^{-i\alpha x}) e^{-\omega t} \qquad \alpha \neq 0$$

$$\Phi_0 = A_1 + A_2 x \qquad \alpha = 0$$

(b) Two-dimensional case

$$\frac{\partial^2 \Phi}{\partial x^2} + \frac{\partial^2 \Phi}{\partial y^2} - \frac{1}{c^2}\frac{\partial \Phi}{\partial t} = 0$$

Solution: $(\alpha^2 + \beta^2 = \lambda^2, \omega = c^2 \lambda^2)$

$$\Phi_{\alpha\beta} = (A_1 e^{i\alpha x} + A_2 e^{-i\alpha x})(B_1 e^{i\beta y} + B_2 e^{-i\beta y}) e^{-\omega t} \qquad \alpha \neq 0, \beta \neq 0$$

$$\Phi_{0\lambda} = (A_1 + A_2 x)(B_1 e^{i\lambda y} + B_2 e^{-i\lambda y}) e^{-\omega t} \qquad \alpha = 0, \beta = \lambda$$

(c) Three-dimensional case

$$\frac{\partial^2 \Phi}{\partial x^2} + \frac{\partial^2 \Phi}{\partial y^2} + \frac{\partial^2 \Phi}{\partial z^2} - \frac{1}{c^2}\frac{\partial \Phi}{\partial t} = 0$$

Solution: $(\alpha^2 + \beta^2 + \gamma^2 = \lambda^2, \omega = c^2 \lambda^2)$

$$\Phi_{\alpha\beta\gamma} = (A_1 e^{i\alpha x} + A_2 e^{-i\alpha x})(B_1 e^{i\beta y} + B_2 e^{-i\beta y})(C_1 e^{i\gamma z} + C_2 e^{-i\gamma z}) e^{-\omega t} \qquad \alpha \neq 0, \beta \neq 0, \gamma \neq 0$$

$$\Phi_{00\lambda} = (A_1 + A_2 x)(B_1 + B_2 y)(C_1 e^{i\lambda z} + C_2 e^{-i\lambda z}) e^{-\omega t} \qquad \alpha = 0, \beta = 0, \gamma = \lambda$$

(2) Cylindrical Coordinates[1]

(a) Two-dimensional case

$$\frac{\partial^2 \Phi}{\partial r^2} + \frac{1}{r} + \frac{\partial \Phi}{\partial r} + \frac{\partial^2 \Phi}{\partial \phi^2} - \frac{1}{c^2}\frac{\partial \Phi}{\partial t} = 0$$

Solution: $(\alpha = 1, 2, 3, \ldots)(\omega = c^2 \lambda^2)$

$$\Phi_{\alpha\beta} = [A_1 J_\alpha(i\lambda r) + A_2 Y_\alpha(i\lambda r)](B_1 e^{i\alpha\phi} + B_2 e^{-i\alpha\phi}) e^{-\omega t} \qquad \alpha \neq 0, \lambda \neq 0$$

$$\Phi_{0\lambda} = J_0(i\lambda r)(B_1 + B_2 \phi) e^{-\omega t} \qquad \alpha = 0, \lambda \neq 0$$

(b) Three-dimensional case

$$\frac{\partial^2 \Phi}{\partial r^2} + \frac{1}{r}\frac{\partial \Phi}{\partial r} + \frac{1}{r^2}\frac{\partial^2 \Phi}{\partial \phi^2} + \frac{\partial^2 \Phi}{\partial z^2} - \frac{1}{c^2}\frac{\partial \Phi}{\partial t} = 0$$

Solution: $(\alpha = 1, 2, 3, \ldots)(\omega = c^2 \lambda^2)$

$$\Phi_{\alpha\beta\gamma} = [A_1 J_\alpha(i\gamma r) + A_2 Y_\alpha(i\gamma r)](B_1 e^{i\alpha\phi} + B_2 e^{-i\alpha\phi})(C_1 e^{i\beta z} + C_2 e^{-i\beta z}) e^{-\omega t}$$

$$\alpha \neq 0, \beta \neq 0, \gamma = \sqrt{\lambda^2 - \beta^2}$$

$$\Phi_{00\lambda} = J_0(i\lambda r)(B_1 + B_2 \phi)(C_1 + C_2 z) e^{-\omega t} \qquad \alpha = 0, \beta = 0, \gamma = \lambda$$

[1] In all axially symmetrical cases $\alpha = 0$.

A, B, C, D = integration constants	$\alpha, \beta, \gamma, \lambda, \omega$ = separation constants
c = given constant	$J(\ ), Y(\ )$ = Bessel functions (Sec. 14.24)

(1) Rectangular Coordinates

(a) One-dimensional case

$$\frac{\partial^2 \Phi}{\partial x^2} - \frac{1}{c^2}\frac{\partial^2 \Phi}{\partial t^2} = 0$$

Solution: $(\alpha = \lambda, \omega = c\lambda)$

$$\Phi_\alpha = (A_1 e^{i\alpha x} + A_2 e^{-i\alpha x})(B_1 e^{i\omega t} + B_2 e^{-i\omega t}) \qquad \alpha \neq 0$$

$$\Phi_0 = A_1 + A_2 x \qquad \alpha = 0$$

(b) Two-dimensional case

$$\frac{\partial^2 \Phi}{\partial x^2} + \frac{\partial^2 \Phi}{\partial y^2} - \frac{1}{c^2}\frac{\partial^2 \Phi}{\partial t^2} = 0$$

Solution: $(\alpha^2 + \beta^2 = \lambda^2, \omega = c\lambda)$

$$\Phi_{\alpha\beta} = (A_1 e^{i\alpha x} + A_2 e^{-i\alpha x})(B_1 e^{i\beta y} + B_2 e^{-i\beta y})(C_1 e^{i\omega t} + C_2 e^{-i\omega t}) \qquad \alpha \neq 0, \beta \neq 0$$

$$\Phi_{0\lambda} = (A_1 + A_2 x)(B_1 e^{i\lambda y} + B_2 e^{-i\lambda y})(C_1 e^{i\omega t} + C_2 e^{-i\omega t}) \qquad \alpha = 0, \beta = \lambda$$

(c) Three-dimensional case

$$\frac{\partial^2 \Phi}{\partial x^2} + \frac{\partial^2 \Phi}{\partial y^2} + \frac{\partial^2 \Phi}{\partial z^2} - \frac{1}{c^2}\frac{\partial^2 \Phi}{\partial t^2} = 0$$

Solution: $(\alpha^2 + \beta^2 + \gamma^2 = \lambda^2, \omega = c\lambda)$

$$\Phi_{\alpha\beta\gamma} = (A_1 e^{i\alpha x} + A_2 e^{-i\alpha x})(B_1 e^{i\beta y} + B_2 e^{-i\beta y})(C_1 e^{i\gamma z} + C_2 e^{-i\gamma z})(D_1 e^{i\omega t} + D_2 e^{-i\omega t})$$

$$\alpha \neq 0, \beta \neq 0, \gamma \neq 0$$

$$\Phi_{00\lambda} = (A_1 + A_2 x)(B_1 + B_2 y)(C_1 e^{i\lambda z} + C_2 e^{-i\lambda z})(D_1 e^{i\omega t} + D_2 e^{-i\omega t}) \qquad \alpha = 0, \beta = 0, \gamma = \lambda$$

(2) Cylindrical Coordinates[1]

(a) Two-dimensional case

$$\frac{\partial^2 \Phi}{\partial r^2} + \frac{1}{r}\frac{\partial \Phi}{\partial r} + \frac{1}{r^2}\frac{\partial^2 \Phi}{\partial \phi^2} - \frac{1}{c^2}\frac{\partial^2 \Phi}{\partial t^2} = 0$$

Solution: $(\alpha = 1, 2, 3, \ldots)(\omega = c\lambda)$

$$\Phi_{\alpha\beta} = [A_1 J_\alpha(i\lambda r) + A_2 Y_\alpha(i\lambda r)](B_1 e^{i\alpha\phi} + B_2 e^{-i\alpha\phi})(C_1 e^{i\omega t} + C_2 e^{-i\omega t}) \qquad \alpha \neq 0, \lambda \neq 0$$

$$\Phi_{0\lambda} = J_0(i\lambda r)(B_1 + B_2 \phi)(C_1 e^{i\omega t} + C_2 e^{-i\omega t}) \qquad \alpha = 0, \lambda \neq 0$$

(b) Three-dimensional case

$$\frac{\partial^2 \Phi}{\partial r^2} + \frac{1}{r}\frac{\partial \Phi}{\partial r} + \frac{1}{r^2}\frac{\partial^2 \Phi}{\partial \phi^2} + \frac{\partial^2 \Phi}{\partial z^2} - \frac{1}{c^2}\frac{\partial^2 \Phi}{\partial t^2} = 0$$

Solution: $(\alpha = 1, 2, 3, \ldots)(\omega = c\lambda)$

$$\Phi_{\alpha\beta\gamma} = [A_1 J_\alpha(i\gamma r) + A_2 Y(i\gamma r)](B_1 e^{i\alpha\phi} + B_2 e^{-i\alpha\phi})(C_1 e^{i\beta z} + C_2 e^{-i\beta z})(D_1 e^{i\omega t} + D_2 e^{-i\omega t})$$

$$\alpha \neq 0, \beta \neq 0, \gamma = \sqrt{\lambda^2 - \beta^2}$$

$$\Phi_{00\lambda} = J_0(i\lambda r)(B_1 + B_2 \phi)(C_1 + C_2 z)(D_1 e^{i\omega t} + D_2 e^{-i\omega t}) \qquad \alpha = 0, \beta = 0, \gamma = \lambda$$

[1]In all axially symmetrical cases $\alpha = 0$.

$u = u(x, t)$ = longitudinal displacement	$v = v(x, t)$ = transverse displacement
a, b, l = given constants	$A, B, C, D, \ldots$ = integration constants
$f(x, t)$ = forcing function	ω = angular frequency (eigenvalue)
$X(x), Y(x)$ = eigenvectors (Sec. 1.14)	$Q(t), T(t)$ = time functions

(1) Longitudinal Vibration

(a) General solution

$$a\frac{\partial^2 u}{\partial x^2} - b\frac{\partial^2 u}{\partial t^2} = -f(x, t) \qquad \lambda_k = \omega_k\sqrt{\frac{b}{a}} \qquad k = 1, 2, 3, \ldots$$

$$u(x, t) = \sum_{k=1}^{\infty} X_k(x) T_k(t)$$

$$X_k(x) = A_k \cos \lambda_k x + B_k \frac{\sin \lambda_k x}{\lambda_k} \qquad Q_k(t) = \int_0^l f(x, t) X_k(x)\, dx$$

$$T_k(t) = C_k \frac{\cos \omega_k t}{\Delta_k} + D_k \frac{\sin \omega_k t}{\Delta_k \omega_k} + \frac{b}{\Delta_k \omega_k} \int_0^t Q_k(\tau) \sin \omega_k(t - \tau)\, d\tau$$

(b) Integration constants

$$A_k = X_k(+0) \qquad B_k = X_k{}'(+0) \qquad \Delta_k = \int_0^l X_k{}^2(x)\, dx$$

$$C_k = \int_0^l u(x, 0) X_k(x)\, dx \qquad\qquad D_k = \int_0^l \dot{u}(x, 0) X_k(x)\, dx$$

$$\text{where } X_k'(+0) = \frac{dX(x)}{dx}\bigg|_{x=0}, \quad \dot{u}(x, 0) = \frac{du(x, t)}{dt}\bigg|_{t=0}$$

(2) Transverse Vibration

(a) General solution

$$a\frac{\partial^4 v}{\partial x^4} + b\frac{\partial^2 v}{\partial t^2} = f(x, t) \qquad \lambda_k = \omega_k \sqrt[4]{\frac{b\omega_k{}^2}{a}} \qquad k = 1, 2, 3, \ldots$$

$$v(x, t) = \sum_{k=1}^{\infty} Y_k(x) T_k(t)$$

$$Y_k(x) = A_k \Psi_1 + B_k \Psi_2 + C_k \Psi_3 + D_k \Psi_4 \qquad Q_k(t) = \int_0^l f(x, t) Y_k(x)\, dx$$

$$\Psi_1 = \frac{\cosh \lambda_k x + \cos \lambda_k x}{2} \qquad\qquad \Psi_2 = \frac{\sinh \lambda_k x + \sin \lambda_k x}{2\lambda_k}$$

$$\Psi_3 = \frac{\cosh \lambda_k x - \cos \lambda_k x}{2\lambda_k{}^2} \qquad\qquad \Psi_4 = \frac{\sinh \lambda_k x - \sin \lambda_k x}{2\lambda_k{}^3}$$

$$T_k(t) = E_k \frac{\cos \omega_k t}{\Delta_k} + F_k \frac{\sin \omega_k t}{\Delta_k \omega_k} + \frac{b}{\Delta_k \omega_k} \int_0^t Q_k(\tau) \sin \omega_k(t - \tau)\, d\tau$$

(b) Integration constants

$$A_k = Y_k(+0) \qquad B_k = Y_k'(+0) \qquad C_k = Y''(+0) \qquad D_k = Y'''(+0)$$

$$\Delta_k = \int_0^l Y_k{}^2(x)\, dx \qquad E_k = \int_0^l v(x, 0) Y_k(x)\, dx \qquad F_k = \int_0^l \dot{v}(x, 0) Y_k(x)\, dx$$

$$\text{where } Y'(+0) = \frac{dY(x)}{dx}\bigg|_{x=0}, \quad Y''(+0) = \frac{d^2 Y(x)}{dx^2}\bigg|_{x=0}, \quad \ldots, \quad \dot{v}(x, 0) = \frac{dv(x, t)}{dt}\bigg|_{t=0}.$$

16
LAPLACE
TRANSFORMS

(1) Definition

If $F(t)$ is a *piecewise-continuous function* of the real variable $t \geqslant 0$, then the *Laplace transform* of $F(t)$ is

$$\mathcal{L}\{F(t)\} = \int_0^\infty e^{-st} F(t)\, dt = f(s)$$

where s = real or complex and $e = 2.71828 \cdots$.

(2) Basic Relationships $(a, b = \text{real numbers}, n = 1, 2, 3, \ldots)$

$$\mathcal{L}(a) = \frac{a}{s} \qquad \mathcal{L}(t) = \frac{1}{s^2} \qquad \mathcal{L}(t^n) = \frac{n!}{s^{n+1}}$$

$$\mathcal{L}\{aF(t)\} = af(s) \qquad\qquad \mathcal{L}\{tF(t)\} = -f'(s)$$

$$\mathcal{L}\{F(at)\} = \frac{1}{a}f\left(\frac{s}{a}\right) \qquad\qquad \mathcal{L}\left\{F\left(\frac{t}{a}\right)\right\} = af(as)$$

$$\mathcal{L}\left\{\frac{1}{t}F(t)\right\} = \int_s^\infty f(s)\, ds \qquad\qquad \mathcal{L}\{t^n F(t)\} = (-1)^n f^{(n)}(s)$$

$$\mathcal{L}\{e^{at}F(t)\} = f(s - a) \qquad\qquad \mathcal{L}\{F(t - a)\} = e^{-as}f(s)$$

$$\mathcal{L}\{aF_1(t) + bF_2(t)\} = \mathcal{L}\{aF_1(t)\} + \mathcal{L}\{bF_2(t)\} = af_1(s) + bf_2(s)$$

(3) Derivatives

$$\mathcal{L}\{F'(t)\} = sf(s) - F(+0)$$

$$\mathcal{L}\{F''(t)\} = s^2 f(s) - sF(+0) - F'(+0)$$

$$\mathcal{L}\{F^{(n)}(t)\} = s^n f(s) - s^{n-1}F(+0) - s^{n-2}F'(+0) - \cdots - F^{(n-1)}(+0)$$

(4) Integrals $[u = u(t), v = v(t)]$

$$\mathcal{L}\left\{\int_0^t F(u)\, du\right\} = \frac{f(s)}{s} \qquad\qquad \mathcal{L}\left\{\int_0^t \int_0^u F(v)\, dv\, du\right\} = \frac{f(s)}{s^2}$$

$$\mathcal{L}\left\{\int_0^t \int_0^t \cdots \int_0^t F(t)\,(dt)^n\right\} = \mathcal{L}\left\{\int_0^t \frac{(t - \tau)^{n-1}}{(n - 1)!} F(\tau)\, d\tau\right\} = \frac{f(s)}{s^n}$$

$$\mathcal{L}\left\{\int_0^t F_1(t - \tau)F_2(\tau)\, d\tau\right\} = \mathcal{L}\left\{\int_0^t F_1(\tau)F_2(t - \tau)\, d\tau\right\} = f_1(s) \cdot f_2(s)$$

where $f_1(s) = \mathcal{L}\{F_1(t)\}$ and $f_2(s) = \mathcal{L}\{F_2(t)\}$.

(5) Periodic Functions $[G(T + t) = G(t), H(T + t) = -H(t)]$

$$\mathcal{L}\{G(T + t)\} = \frac{\int_0^T e^{-st}G(t)\, dt}{1 - e^{-sT}} \qquad\qquad \mathcal{L}\{H(T + t)\} = \frac{\int_0^T e^{-st}H(t)\, dt}{1 - e^{+sT}}$$

(1) Definition

The *inverse Laplace transform* of $f(s)$ defined in Sec. 16.01-1 is

$$\mathcal{L}^{-1}\{f(s)\} = \frac{1}{2\pi i} \int_{c-i\infty}^{c+i\infty} e^{st} f(s)\, ds = F(t)$$

where c is chosen so that all the singular points of $f(s)$ lie to the left of the line $\mathrm{Re}(s) = c$ in the complex s plane. Not every function $f(s)$ has an inverse Laplace transform.

(2) Basic Relationships   (a, b = real numbers, $n = 1, 2, 3, \ldots$)

$$\mathcal{L}^{-1}\left\{\frac{a}{s}\right\} = a \qquad\qquad \mathcal{L}^{-1}\left\{\frac{1}{s^2}\right\} = t \qquad\qquad \mathcal{L}^{-1}\left\{\frac{1}{s^n}\right\} = \frac{t^{n-1}}{(n-1)!}$$

$$\mathcal{L}^{-1}\{af(s+b)\} = ae^{-bt}F(t) \qquad\qquad \mathcal{L}^{-1}\{af(s-b)\} = ae^{bt}F(t)$$

$$\mathcal{L}^{-1}\{f(as)\} = \frac{1}{a}F\left(\frac{t}{a}\right) \qquad\qquad \mathcal{L}^{-1}\left\{f\left(\frac{s}{a}\right)\right\} = aF(at)$$

$$\mathcal{L}^{-1}\{sf(s)\} = \frac{dF(t)}{dt} \qquad \text{if } F(0) = 0$$

$$\mathcal{L}^{-1}\{s^n f(s)\} = \frac{d^n F(t)}{dt^n} \qquad \text{if } F(0) = F'(0) = \cdots = F^{n-1}(0) = 0$$

$$\mathcal{L}^{-1}\left\{\frac{f(s)}{s}\right\} = \int_0^t F(t)\, dt \qquad\qquad \mathcal{L}^{-1}\left\{\frac{f(s)}{s^n}\right\} = \int_0^t \frac{(t-\tau)^{n-\ell}}{(n-1)!} F(\tau)\, d\tau$$

$$\mathcal{L}^{-1}\left\{\sum_{k=1}^{r} a_k f_k(s)\right\} = a_1 F_1(t) + a_2 F_2(t) + \cdots + a_r F_r(t)$$

(3) Derivatives

$$\mathcal{L}^{-1}\{f'(s)\} = -tF(t) \qquad\qquad \mathcal{L}^{-1}\{f^{(n)}(s)\} = (-t)^n F(t)$$

(4) Convolutions

$$\mathcal{L}^{-1}\{f_1(s) \bullet f_2(s)\} = \begin{cases} \displaystyle\int_0^t F_1(t-\tau)F_2(\tau)\, d\tau \\[2em] \displaystyle\int_0^t F_1(\tau)F_2(t-\tau)\, d\tau \end{cases}$$

$$\mathcal{L}^{-1}\{f_1(s) \bullet f_2(s) \bullet f_3(s)\} = \begin{cases} \displaystyle\int_0^t F_1(t-\tau)F_2(\tau)F_3(\tau)\, d\tau \\[2em] \displaystyle\int_0^t F_1(\tau)F_2(t-\tau)F_3(\tau)\, d\tau \\[2em] \displaystyle\int_0^t F_1(\tau)F_2(\tau)F_3(t-\tau)\, d\tau \end{cases}$$

where $\mathcal{L}^{-1}\{f_1(s)\} = F_1(t)$, $\mathcal{L}^{-1}\{f_2(s)\} = F_2(t)$, $\mathcal{L}^{-1}\{f_3(s)\} = F_3(t)$.

(5) Tables of Laplace Transforms

Three sets of Laplace transform tables are given in this chapter. Sections 16.03–16.28 show the inverses of algebraic, irrational algebraic, transcendental, and higher transcendental Laplace transforms. Sections 16.29–16.34 display the Laplace transforms of piecewise functions defined by graphs, and, finally Secs. 16.35–16.49 tabulate the Laplace transform solutions of the most frequently encountered differential equations.

| $\dfrac{1}{s}$ | $\dfrac{s^m}{s^n(s-a)}$ | $\dfrac{s^m}{s^n(s-a)^2}$ |

| $a = $ constant $\neq 0$ | $\alpha = at$ | $m, n = $ positive integers $\neq 0$ |

$f(s)$	$F(t)$
$\dfrac{1}{s}$	1
$\dfrac{1}{s^2}$	t
$\dfrac{1}{s^n}$	$\dfrac{t^{n-1}}{(n-1)!}$
$\dfrac{1}{s-a}$	e^α
$\dfrac{1}{s(s-a)}$	$\dfrac{1}{a}(e^\alpha - 1)$
$\dfrac{1}{s^2(s-a)}$	$\dfrac{1}{a^2}(e^\alpha - \alpha - 1)$
$\dfrac{1}{s^3(s-a)}$	$\dfrac{1}{a^3}\left((e^\alpha - \dfrac{\alpha^2}{2} - \alpha - 1\right)$
$\dfrac{1}{s^4(s-a)}$	$\dfrac{1}{a^4}\left((e^\alpha - \dfrac{\alpha^3}{6} - \dfrac{\alpha^2}{2} - \alpha - 1\right)$
$\dfrac{1}{s^5(s-a)}$	$\dfrac{1}{a^5}\left((e^\alpha - \dfrac{\alpha^4}{24} - \dfrac{\alpha^3}{6} - \dfrac{\alpha^2}{2} - \alpha - 1\right)$
$\dfrac{s}{(s-a)^2}$	$e^\alpha(\alpha + 1)$
$\dfrac{1}{(s-a)^2}$	te^α
$\dfrac{1}{s(s-a)^2}$	$\dfrac{1}{a^2}[e^\alpha(\alpha - 1) + 1]$
$\dfrac{1}{s^2(s-a)^2}$	$\dfrac{1}{a^3}[e^\alpha(\alpha - 2) + \alpha + 2]$
$\dfrac{1}{s^3(s-a)^2}$	$\dfrac{1}{a^4}\left[e^\alpha(\alpha - 3) + \dfrac{\alpha^2}{2} + 2\alpha + 3\right]$
$\dfrac{1}{s^4(s-a)^2}$	$\dfrac{1}{a^5}\left[e^\alpha(\alpha - 4) + \dfrac{\alpha^3}{6} + \alpha^2 + 3\alpha + 4\right]$

$\dfrac{s^m}{s^n(s-a)^3}$	$\dfrac{s^m}{s^n(s-a)^4}$	$\dfrac{1}{s^n(s-a)^p}$

$a = \text{constant} \neq 0$	$\alpha = at$	$m, n, p = \text{positive integer} \neq 0$

$f(s)$	$F(t)$
$\dfrac{s^2}{(s-a)^3}$	$\dfrac{1}{2}(2 + 4\alpha + \alpha^2)e^\alpha$
$\dfrac{s}{(s-a)^3}$	$\dfrac{1}{2a}(2\alpha + \alpha^2)e^\alpha$
$\dfrac{1}{(s-a)^3}$	$\dfrac{t^2}{2}e^\alpha$
$\dfrac{1}{s(s-a)^3}$	$\dfrac{1}{a^3}\left[\left(\dfrac{\alpha^2}{2} - \alpha + 1\right)e^\alpha - 1\right]$
$\dfrac{1}{s^2(s-a)^3}$	$\dfrac{1}{a^4}\left[\left(\dfrac{\alpha^2}{2} - 2\alpha + 3\right)e^\alpha - \alpha - 3\right]$
$\dfrac{1}{s^3(s-a)^3}$	$\dfrac{1}{a^5}\left[\left(\dfrac{\alpha^2}{2} - 3\alpha + 6\right)e^\alpha - \dfrac{\alpha^2}{2} - 3\alpha - 6\right]$
$\dfrac{s^3}{(s-a)^4}$	$\dfrac{1}{6}(6 + 18\alpha + 9\alpha^2 + \alpha^3)e^\alpha$
$\dfrac{s^2}{(s-a)^4}$	$\dfrac{1}{6a}(6\alpha + 6\alpha^2 + \alpha^3)e^\alpha$
$\dfrac{s}{(s-a)^4}$	$\dfrac{1}{6a^2}(3\alpha^2 + \alpha^3)e^\alpha$
$\dfrac{1}{(s-a)^4}$	$\dfrac{t^3}{6}e^\alpha$
$\dfrac{1}{s(s-a)^4}$	$\dfrac{1}{a^4}\left[\left(\dfrac{\alpha^3}{6} - \dfrac{\alpha^2}{2} + \alpha - 1\right)e^\alpha + 1\right]$
$\dfrac{1}{s^2(s-a)^4}$	$\dfrac{1}{a^5}\left[\left(\dfrac{\alpha^3}{6} + \alpha^2 + 3\alpha - 4\right)e^\alpha + \alpha + 4\right]$
$\dfrac{1}{(s-a)^p}$	$\dfrac{t^{p-1}}{(p-1)!}e^\alpha$
$\dfrac{1}{s(s-a)^p}$	$\dfrac{1}{a^p}\left[\left(\sum\limits_{k=1}^{p}(-1)^{k+1}\dfrac{\alpha^{n-k}}{(n-k)!}\right)e^\alpha + (-1)^p\right]$
$\dfrac{1}{s^2(s-a)^p}$	$\dfrac{1}{a^{p+1}}\left[\left(\sum\limits_{k=1}^{p}k(-1)^{k+1}\dfrac{\alpha^{n-k}}{(n-k)!}\right)e^\alpha + (-1)^p(\alpha + p)\right]$

$\dfrac{s^m}{s^n(s-a)(s-b)}$	$\dfrac{s^m}{s^n(s-a)(s-b)^2}$	$\dfrac{s^m}{s^n(s-a)^2(s-b)^2}$

a, b = constants $\neq 0$	$\alpha = at$		$\beta = bt$
m, n = positive integers $\neq 0$	$A = t = -\dfrac{2}{a-b}$	$B = t + \dfrac{2}{a-b}$	$C = \dfrac{1}{a-b}$

$f(s)$	$F(t)$
$\dfrac{s}{(s-a)(s-b)}$	$C(ae^\alpha - be^\beta)$
$\dfrac{1}{(s-a)(s-b)}$	$C(e^\alpha - e^\beta)$
$\dfrac{1}{s(s-a)(s-b)}$	$C\left[\dfrac{1}{a}(e^\alpha - 1) - \dfrac{1}{b}(e^\beta - 1)\right]$
$\dfrac{1}{s^2(s-a)(s-b)}$	$C\left[\dfrac{1}{a^2}(e^\alpha - \alpha - 1) - \dfrac{1}{b^2}(e^\beta - \beta - 1)\right]$
$\dfrac{1}{s^3(s-a)(s-b)}$	$C\left[\dfrac{1}{a^3}\left(e^\alpha - \dfrac{\alpha^2}{2} - \alpha - 1\right) - \dfrac{1}{b^3}\left(e^\beta - \dfrac{\beta^2}{2} - \beta - 1\right)\right]$
$\dfrac{s^2}{(s-a)(s-b)^2}$	$C^2[a^2e^\alpha - b^2e^\beta] - C(\beta + 2)e^\beta$
$\dfrac{s}{(s-a)(s-b)^2}$	$C^2(ae^\alpha - be^\beta) - C(\beta + 1)e^\beta$
$\dfrac{1}{(s-a)(s-b)^2}$	$C^2(e^\alpha - e^\beta) - Cte^\beta$
$\dfrac{1}{s(s-a)(s-b)^2}$	$C^2\left[\dfrac{1}{a}(e^\alpha - 1) - \dfrac{1}{b}(e^\beta - 1)\right] - C\dfrac{1 + (\beta - 1)e^\beta}{b^2}$
$\dfrac{1}{s^2(s-a)(s-b)^2}$	$C^2\left[\dfrac{1}{a^2}(e^\alpha - \alpha - 1) - \dfrac{1}{b^2}(e^\beta - \beta - 1)\right] - C\dfrac{2 + \beta - (2 - b)e^\beta}{b^3}$
$\dfrac{s^2}{(s-a)^2(s-b)^2}$	$C^2[(2 + aA)ae^\alpha + (2 + bB)bC^\beta]$
$\dfrac{s}{(s-a)^2(s-b)^2}$	$C^2[(1 + aA)e^\alpha + (1 + bB)e^\beta]$
$\dfrac{1}{(s-a)^2(s-b)^2}$	$C^2(Ae^\alpha + Be^\beta)$
$\dfrac{1}{s(s-a)^2(s-b)^2}$	$C^2\left[(aA - 1)\dfrac{e^\alpha}{a^2} + (bB - 1)\dfrac{e^\beta}{b^2}\right] + \dfrac{1}{(ab)^2}$
$\dfrac{1}{s^2(s-a)^2(s-b)^2}$	$C^2\left[(aA - 2)\dfrac{e^\alpha}{a^3} + (bB - 2)\dfrac{e^\beta}{b^3}\right] + \dfrac{t}{(ab)^2} + \dfrac{2(a + b)}{(ab)^3}$

$$\frac{s^m}{s^n(s-a)(s-b)^p(s-c)^q}$$

a, b, c = constants $\neq 0$	$\alpha = at$	$\beta = bt$	$\gamma = ct$

$$A_1 = \frac{1}{(a-b)(a-c)} \qquad B_1 = \frac{1}{(b-a)(b-c)} \qquad C_1 = \frac{1}{(c-a)(c-b)}$$

$$A_2 = \frac{1}{(a-b)(a-c)^2} \qquad B_2 = \frac{1}{(b-a)(b-c)^2} \qquad C_2 = \frac{1}{(c-a)^2(c-b)^2}$$

$$A_3 = \frac{1}{(a-b)^2(a-c)^2} \qquad B_3 = \frac{3b-2a-c}{(b-a)^2(b-c)^3} \qquad C_3 = \frac{3c-2a-b}{(c-a)^2(c-b)^3}$$

$f(s)$	$F(t)$
$\dfrac{s^2}{(s-a)(s-b)(s-c)}$	$a^2 A_1 e^\alpha + b^2 B_1 e^\beta + c^2 C_1 e^\gamma$
$\dfrac{s}{(s-a)(s-b)(s-c)}$	$a A_1 e^\alpha + b B_1 e^\beta + c C_1 e^\gamma$
$\dfrac{1}{(s-a)(s-b)(s-c)}$	$A_1 e^\alpha + B_1 e^\beta + C_1 e^\gamma$
$\dfrac{1}{s(s-a)(s-b)(s-c)}$	$\dfrac{A_1}{a}(e^\alpha - 1) + \dfrac{B_1}{b}(e^\beta - 1) + \dfrac{C_1}{c}(e^\gamma - 1)$
$\dfrac{1}{s^2(s-a)(s-b)(s-c)}$	$\dfrac{A_1}{a^2}(\alpha^\alpha - \alpha - 1) + \dfrac{B_1}{b^2}(e^\beta - \beta - 1) + \dfrac{C_1}{c^2}(e^\gamma - \gamma - 1)$
$\dfrac{1}{s^3(s-a)(s-b)(s-c)}$	$\dfrac{A_1}{a^3}(e^\alpha - \tfrac{1}{2}\alpha^2 - \alpha - 1) + \dfrac{B_1}{b^3}(e^\beta - \tfrac{1}{2}\beta^2 - \beta - 1) + \dfrac{C_1}{c^3}(e^\gamma - \tfrac{1}{2}\gamma^2 - \gamma - 1)$
$\dfrac{s}{(s-a)(s-b)(s-c)^2}$	$a A_2 e^\alpha + b B_2 e^\beta + c C_2 e^\gamma$
$\dfrac{1}{(s-a)(s-b)(s-c)^2}$	$A_2 e^\alpha + B_2 e^\beta + (C_1 t + C_2)e^\gamma$
$\dfrac{1}{s(s-a)(s-b)(s-c)^2}$	$\dfrac{A_2}{a}(e^\alpha - 1) + \dfrac{B_2}{b}(e^\beta - 1) + \left(\dfrac{C_1}{c^2} + \dfrac{C_2}{c}\right)(e^\gamma - 1) + \dfrac{C_1}{c^2}\gamma$
$\dfrac{1}{s^2(s-a)(s-b)(s-c)^2}$	$\dfrac{A_2}{a^2}(e^\alpha - \alpha - 1) + \dfrac{B_2}{b^2}(e^\beta - \beta - 1) + \left(\dfrac{C_1}{c^2} + \dfrac{C_2}{c}\right)(e^\gamma - \gamma - 1) + \dfrac{C_1}{c^2}\left(\dfrac{\gamma^2}{2} + \gamma\right)$
$\dfrac{1}{(s-a)(s-b)^2(s-c)^2}$	$A_3 e^\alpha + (B_2 t + B_3)e^\beta + (C_2 t + C_3)e^\gamma$
$\dfrac{1}{s(s-a)(s-b)^2(s-c)^2}$	$\dfrac{A_3}{a}(e^\alpha - 1) + \left(\dfrac{B_2}{b^2} + \dfrac{B_3}{b}\right)(e^\beta - 1) + \left(\dfrac{C_2}{c^2} + \dfrac{C_3}{c}\right)(e^\gamma - 1) + \dfrac{B_2}{b}\beta + \dfrac{C_2}{c}\gamma$

$$\frac{s^m}{s^n(s^2 + a^2)^p}$$

| $a = $ constant $\neq 0$ | $m, n, p = $ positive integers $\neq 0$ | $\alpha = at$ |

$f(s)$	$F(t)$
$\dfrac{s}{s^2 + a^2}$	$\cos \alpha$
$\dfrac{1}{s^2 + a^2}$	$\dfrac{1}{a} \sin \alpha$
$\dfrac{1}{s(s^2 + a^2)}$	$\dfrac{1}{a^2}(1 - \cos \alpha)$
$\dfrac{1}{s^3(s^2 + a^2)}$	$\dfrac{1}{a^3}(\alpha - \sin \alpha)$
$\dfrac{1}{s^2(s^2 + a^2)}$	$\dfrac{1}{a^4}\left(-1 + \dfrac{\alpha^2}{2} + \cos \alpha\right)$
$\dfrac{s^2}{(s^2 + a^2)^2}$	$\dfrac{1}{2a}(\sin \alpha + \alpha \cos \alpha)$
$\dfrac{s}{(s^2 + a^2)^2}$	$\dfrac{\alpha}{2a^2} \sin \alpha$
$\dfrac{1}{(s^2 + a^2)^2}$	$\dfrac{1}{2a^3}(\sin \alpha - \alpha \cos \alpha)$
$\dfrac{1}{s(s^2 + a^2)^2}$	$\dfrac{1}{2a^4}(2 - 2\cos \alpha - \alpha \sin \alpha)$
$\dfrac{1}{s^2(s^2 + a^2)^2}$	$\dfrac{1}{2a^5}(2\alpha + \alpha \cos \alpha - 3 \sin \alpha)$
$\dfrac{s^2}{(s^2 + a^2)^3}$	$\dfrac{1}{8a^3}[(1 + \alpha^2)\sin \alpha - \alpha \cos \alpha]$
$\dfrac{s}{(s^2 + a^2)^3}$	$\dfrac{\alpha}{8a^4}(\sin \alpha - \alpha \cos \alpha)$
$\dfrac{1}{(s^2 + a^2)^3}$	$\dfrac{1}{8a^5}[(3 - \alpha^2)\sin \alpha - 3\alpha \cos \alpha]$
$\dfrac{1}{s(s^2 + a^2)^3}$	$\dfrac{1}{8a^6}[(\alpha^2 - 8)\cos \alpha - 5\alpha \sin \alpha + 8]$
$\dfrac{1}{s^2(s^2 + a^2)^3}$	$\dfrac{1}{8a^7}[(\alpha^2 - 5)\sin \alpha + 7\alpha \cos \alpha + 8\alpha]$

$$\frac{s^m}{s^n(s^2 - a^2)^p}$$

$a = $ constant $\neq 0$	$m, n, p = $ positive integers $\neq 0$	$\alpha = at$

$f(s)$	$F(t)$
$\dfrac{s}{s^2 - a^2}$	$\cosh \alpha$
$\dfrac{1}{s^2 - a^2}$	$\dfrac{1}{a} \sinh \alpha$
$\dfrac{1}{s(s^2 - a^2)}$	$\dfrac{1}{a^2} (\cosh \alpha - 1)$
$\dfrac{1}{s^3(s^2 - a^2)}$	$\dfrac{1}{a^3} (\sinh \alpha - \alpha)$
$\dfrac{1}{s^2(s^2 - a^2)}$	$\dfrac{1}{a^4} \left(\cosh \alpha - \dfrac{\alpha^2}{2} - 1 \right)$
$\dfrac{s^2}{(s^2 - a^2)^2}$	$\dfrac{1}{2a} (\sinh \alpha + \alpha \cosh \alpha)$
$\dfrac{s}{(s^2 - a^2)^2}$	$\dfrac{\alpha}{2a^2} \sinh \alpha$
$\dfrac{1}{(s^2 - a^2)^2}$	$\dfrac{1}{2a^3} (\alpha \cosh \alpha - \sinh \alpha)$
$\dfrac{1}{s(s^2 - a^2)^2}$	$\dfrac{1}{2a^4} (2 - 2 \cosh \alpha + \alpha \sin \alpha)$
$\dfrac{1}{s^2(s^2 - a^2)^2}$	$\dfrac{1}{2a^5} (2\alpha + \alpha \cosh \alpha - 3 \sinh \alpha)$
$\dfrac{s^2}{(s^2 - a^2)^3}$	$\dfrac{1}{8a^3} [\alpha \cosh \alpha - (1 - \alpha^2) \sinh \alpha]$
$\dfrac{s}{(s^2 - a^2)^3}$	$\dfrac{\alpha}{8a^4} (\alpha \cosh \alpha - \sinh \alpha)$
$\dfrac{1}{(s^2 - a^2)^3}$	$\dfrac{1}{8a^5} [(3 + \alpha^2) \sinh \alpha - 3\alpha \cosh \alpha]$
$\dfrac{1}{s(s^2 - a^2)^3}$	$\dfrac{1}{8a^6} [(8 + \alpha^2) \cosh \alpha - 5\alpha \sinh \alpha - 8]$
$\dfrac{1}{s^2(s^2 - a^2)^3}$	$\dfrac{1}{8a^7} [(5 + \alpha^2) \sinh \alpha - 7\alpha \cosh \alpha - 8\alpha]$

$\dfrac{s^m}{s''(s-a)(s^2+b^2)}$	$\dfrac{s^m}{s''(s-a)^2(s^2+b^2)}$

$a, b = \text{constants} \neq 0$ $\qquad R = \sqrt{a^2+b^2}$ $\qquad\qquad \alpha = at$ $\qquad\qquad \beta = bt$

$$A_1 = \frac{t}{R^2} - \frac{2a}{R^4} \qquad\qquad B_1 = \frac{a^2-b^2}{bR^4} \qquad\qquad C_1 = \frac{2}{R^4} + \frac{1}{4R^2}$$

$$A_2 = A_1 - \frac{1}{aR^2} \qquad\qquad B_2 = \frac{2a}{R^4} \qquad\qquad C_2 = C_1 + \frac{1}{aR^2} + \frac{B_1}{b}$$

$$A_3 = A_2 - \frac{1}{aR^2} \qquad\qquad\qquad\qquad\qquad C_3 = C_2 + \frac{1}{aR^2} + \frac{B_2}{b}$$

$f(s)$	$F(t)$
$\dfrac{s^2}{(s-a)(s^2+b^2)}$	$\dfrac{1}{R^2}(a^2 e^\alpha + ab\sin\beta + b^2\cos\beta)$
$\dfrac{s}{(s-a)(s^2+b^2)}$	$\dfrac{1}{R^2}(ae^\alpha - a\cos\beta + b\sin\beta)$
$\dfrac{1}{(s-a)(s^2+b^2)}$	$\dfrac{1}{R^2}\left(e^\alpha - \dfrac{a\sin\beta + b\cos\beta}{b}\right)$
$\dfrac{1}{s(s-a)(s^2+b^2)}$	$\dfrac{1}{R^2}\left(\dfrac{e^\alpha}{a} + \dfrac{a\cos\beta - b\sin\beta}{b^2}\right) - \dfrac{1}{ab^2}$
$\dfrac{1}{s^2(s-a)(s^2+b^2)}$	$\dfrac{1}{R^2}\left(\dfrac{e^\alpha}{a^2} + \dfrac{a\sin\beta + b\cos\beta}{b^3}\right) - \dfrac{1+\alpha}{a^2b^2}$
$\dfrac{1}{s^3(s-a)(s^2+b^2)}$	$\dfrac{1}{R^2}\left(\dfrac{e^\alpha}{a^3} - \dfrac{a\cos\beta - b\sin\beta}{b^4}\right) - \dfrac{\alpha+\alpha^2}{a^3b^2} + \dfrac{a^2-b^2}{a^3b^4}$
$\dfrac{s^3}{(s-a)^2(s^2+b^2)}$	$A_1 a^3 e^\alpha - b^3 B_1\cos\beta + b^3 B_2\sin\beta + \dfrac{1+a+a^2}{R^2}e^\alpha$
$\dfrac{s^2}{(s-a)^2(s^2+b^2)}$	$A_1 a^2 e^\alpha - b^2 B_1\sin\beta - b^2 B_2\cos\beta + \dfrac{1+a}{R^2}e^\alpha$
$\dfrac{s}{(s-a)^2(s^2+b^2)}$	$A_1 a e^\alpha + b B_1\cos\beta - b B_2\sin\beta + \dfrac{e^\alpha}{R^2}$
$\dfrac{1}{(s-a)^2(s^2+b^2)}$	$A_1 e^\alpha + B_1\cos\beta + B_2\cos\beta$
$\dfrac{1}{s(s-a)^2(s^2+b^2)}$	$A_2\dfrac{e^\alpha}{a} - \dfrac{B_1}{b}\cos\beta + \dfrac{B_2}{b}\sin\beta + C_1$
$\dfrac{1}{s^2(s-a)^2(s^2+b^2)}$	$A_3\dfrac{e^\alpha}{a^2} - \dfrac{B_1}{b^2}\sin\beta - \dfrac{B_2}{b^2}\cos\beta + C_2$

$$\frac{s^m}{s''(s-a)(s-b)(s^2+c^2)} \qquad \frac{s^m}{s''(s-a)(s-b)^2(s^2+c^2)}$$

$a, b, c = \text{constants} \neq 0 \qquad \alpha = at \qquad \beta = bt \qquad \gamma = ct$

$$A_1 = \frac{1}{(a-b)(a^2+c^2)} \qquad B_1 = \frac{1}{(a-b)(b^2+c^2)} \qquad C_1 = \frac{1}{c\sqrt{(c^2-ab)^2+c^2(a+b)^2}}$$

$$A_2 = \frac{1}{(a-b)^2(a^2+c^2)} \qquad B_2 = \frac{2ab-3b^2-c^2}{(a-b)^2(b^2+c^2)^2} \qquad C_2 = \frac{1}{c(b^2+c^2)\sqrt{a^2+c^2}}$$

$$\phi_1 = \tan^{-1}\frac{c}{a} + \tan^{-1}\frac{c}{b} \qquad\qquad \phi_2 = \tan^{-1}\frac{c}{a} + 2\tan^{-1}\frac{c}{b}$$

$f(s)$	$F(t)$
$\dfrac{s^3}{(s-a)(s-b)(s^2+c^2)}$	$a^3 A_1 e^\alpha - b^3 B_1 e^\beta + c^3 C_1 \cos(\gamma+\phi_1)$
$\dfrac{s^2}{(s-a)(s-b)(s^2+c^2)}$	$a^2 A_1 e^\alpha - b^2 B_1 e^\beta + c^2 C_1 \sin(\gamma+\phi_1)$
$\dfrac{s}{(s-a)(s-b)(s^2+c^2)}$	$a A_1 e^\alpha - b B_1 e^\beta + c C_1 \cos(\gamma+\phi_1)$
$\dfrac{1}{(s-a)(s-b)(s^2+c^2)}$	$A_1 e^\alpha - B_1 e^\beta + C_1 \sin(\gamma+\phi_1)$
$\dfrac{1}{s(s-a)(s-b)(s^2+c^2)}$	$\dfrac{A_1}{a}(e^\alpha - 1) - \dfrac{B_1}{b}(e^\beta - 1) - \dfrac{C_1}{c}[\cos(\gamma+\phi_1) + \cos\phi_1]$
$\dfrac{1}{s^2(s-a)(s-b)(s^2+c^2)}$	$\dfrac{A_1}{a^2}(e^\alpha - \alpha - 1) - \dfrac{B_1}{b^2}(e^\beta - \beta - 1) - \dfrac{C_1}{c^2}[\sin(\gamma+\phi_1) + C_3]$
$\dfrac{s^4}{(s-a)(s-b)^2(s^2+c^2)}$	$a^4 A_2 e^\alpha - b^3(4+\beta)e^\beta + b^4 B_2 e^\beta + c^4 C_2 \sin(\gamma+\phi_2)$
$\dfrac{s^3}{(s-a)(s-b)^2(s^2+c^2)}$	$a^3 A_2 e^\alpha - b^2 B_1(3+\beta)e^\beta + b^3 B_2 e^\beta - c^3 C_2 \cos(\gamma+\phi_2)$
$\dfrac{s^2}{(s-a)(s-b)^2(s^2+c^2)}$	$a^2 A_2 e^\alpha - b B_1(2+\beta)e^\beta + b^2 B_2 e^\beta - c^2 C_2 \sin(\gamma+\phi_2)$
$\dfrac{s}{(s-a)(s-b)^2(s^2+c^2)}$	$a A_2 e^\alpha - B_1(1+\beta)e^\beta + b B_2 e^\beta + c C_2 \cos(\gamma+\phi_2)$
$\dfrac{1}{(s-a)(s-b)^2(s^2+c^2)}$	$A_2 e^\alpha - B_1 t e^\beta + B_2 e^\beta + C_2 \sin(\gamma+\phi_2)$
$\dfrac{1}{s(s-a)(s-b)^2(s^2+c^2)}$	$\dfrac{A_2}{a}(e^\alpha - 1) - \dfrac{B_1}{b^2}(e^\beta - \beta - 1) - \dfrac{B_2}{b}(e^\beta - 1) - \dfrac{C_2}{c}[\cos(\gamma+\phi_2) - \cos\phi_2]$

$$\frac{s^m}{s^n(s^2 - 2as + b)^p}$$

a, b = constants $\neq 0$	$\alpha = at$	$\beta = \sqrt{b}\, t$
$R_1 = \sqrt{b - a^2}$	$\omega_1 = R_1 t$	$\phi_1 = \tan^{-1}\dfrac{R_1}{a}$
$R_2 = \sqrt{a^2 - b}$	$\omega_2 = R_2 t$	$\phi_2 = \tan^{-1}\dfrac{R_2}{a}$

$f(s)$	$F(t)$
$\dfrac{s}{s^2 - 2as + b}$	$\dfrac{\sqrt{b}\, e^{\alpha}}{R_1} \sin(\omega_1 + \phi_1)$
$\dfrac{1}{s^2 - 2as + b}$	$\dfrac{e^{\alpha}}{R_1} \sin \omega_1$
$\dfrac{1}{s(s^2 - 2as + b)}$	$\dfrac{e^{\alpha}}{\sqrt{b}\, R_1} \sin(\omega_1 - \phi_1) + \dfrac{1}{b}$
$\dfrac{s^2}{(s^2 - 2as + b)^2}$	$\dfrac{be^{\alpha}}{2R_1{}^3} [\sin \omega_1 - \omega_1 \cos(\omega_1 + 2\phi_1)]$
$\dfrac{s}{(s^2 - 2as + b)^2}$	$\dfrac{\sqrt{b}\, e^{\alpha}}{2R_1{}^3} \left[\dfrac{a}{R_1} \sin \omega_1 - \omega_1 \cos(\omega_1 + \phi_1)\right]$
$\dfrac{1}{(s^2 - 2as + b)^2}$	$\dfrac{e^{\alpha}}{2R_1{}^3} (\sin \omega_1 - \omega_1 \cos \omega_1)$
$\dfrac{1}{s(s^2 - 2as + b)^2}$	$\dfrac{e^{\alpha}}{2\sqrt{b}\, R_1{}^3} \left[\sin(\omega_1 - \phi_1) - \omega_1 \cos(\omega_1 - \phi_1) + \dfrac{R_1}{\sqrt{b}} \cos(\omega_1 - 2\phi_1)\right] + \dfrac{1}{R_1{}^4}$
$\dfrac{s}{s^2 - 2as + b}$	$\dfrac{\sqrt{b}\, e^{\alpha}}{R_2} \sinh(\omega_2 + \phi_2)$
$\dfrac{1}{s^2 - 2as + b}$	$\dfrac{e^{\alpha}}{R_2} \sinh \omega_2$
$\dfrac{1}{s(s^2 - 2as + b)}$	$\dfrac{e^{\alpha}}{\sqrt{b}\, R_1} \sinh(\omega_2 - \phi_2) + \dfrac{1}{b}$
$\dfrac{s^2}{(s^2 - 2as + b)^2}$	$\dfrac{be^{\alpha}}{2R_2{}^3} [\omega_2 \cosh(\omega_2 + 2\phi_2) - \sinh \omega_2]$
$\dfrac{s}{(s^2 - 2as + b)^2}$	$\dfrac{\sqrt{b}\, e^{\alpha}}{2R_2{}^3} \left[\omega_2 \cosh(\omega_2 + \phi_2) - \dfrac{a}{R_2} \sinh \omega_2\right]$
$\dfrac{1}{(s^2 - 2as + b)^2}$	$\dfrac{e^{\alpha}}{2R_2{}^3} (\omega_2 \cosh \omega_2 - \sinh \omega_2)$
$\dfrac{1}{s(s^2 - 2as + b)^2}$	$\dfrac{e^{\alpha}}{2\sqrt{b}\, R_2{}^3} \left[\dfrac{R_2}{\sqrt{b}} \cosh(\omega_2 - 2\phi_2) + \omega_2 \cosh(\omega_2 - \phi_2) - \sinh(\omega_2 - \phi_2)\right] + \dfrac{1}{R_2{}^4}$

The first group of rows corresponds to $b - a^2 > 0$; the second group corresponds to $a^2 - b > 0$.

$$\frac{s^m}{s^n[(s-a)^2 \pm b^2]}$$

$a, b = $ constants $\neq 0$	$\alpha = at$	$\beta = bt$
$R_1 = \sqrt{a^2 + b^2}$		$\phi_1 = \tan^{-1}\dfrac{b}{a}$
$R_2 = \sqrt{a^2 - b^2}$		$\phi_2 = \tanh^{-1}\dfrac{b}{a}$

$f(s)$	$F(t)$
$\dfrac{s}{(s-a)^2 + b^2}$	$\dfrac{R_1 e^\alpha}{b} \sin(\beta + \phi_1)$
$\dfrac{1}{(s-a)^2 + b^2}$	$\dfrac{e^\alpha}{b} \sin \beta$
$\dfrac{1}{s[(s-a)^2 + b^2]}$	$\dfrac{e^\alpha}{bR_1} \sin(\beta - \phi_1) + \dfrac{1}{R_1{}^2}$
$\dfrac{s^2}{[(s-a)^2 + b^2]^2}$	$\dfrac{R_1{}^2 e^\alpha}{2b^3} [\sin\beta - \beta \cos(\beta + 2\phi_1)]$
$\dfrac{s}{[(s-a)^2 + b^2]^2}$	$\dfrac{R_1 e^\alpha}{2b^3} \left[\dfrac{a}{R_1} \sin\beta - \beta \cos(\beta + \phi_1)\right]$
$\dfrac{1}{[(s-a)^2 + b^2]^2}$	$\dfrac{e^\alpha}{2b^3} (\sin\beta - \beta \cos\beta)$
$\dfrac{1}{s[(s-a)^2 + b^2]^2}$	$\dfrac{e^\alpha}{2R_1 b^3} \left[\sin(\beta - \phi_1) - \beta \cos(\beta - \phi_1) + \dfrac{b}{R} \cos(\beta - 2\phi_1)\right] + \dfrac{1}{R_1{}^4}$
$\dfrac{s}{(s-a)^2 - b^2}$	$\dfrac{R_2 e^\alpha}{b} \sinh(\beta + \phi_2)$
$\dfrac{1}{(s-a)^2 - b^2}$	$\dfrac{e^\alpha}{b} \sinh \beta$
$\dfrac{1}{s[(s-a)^2 - b^2]}$	$\dfrac{e^\alpha}{bR_2} \sinh(\beta - \phi_2) + \dfrac{1}{R_2{}^2}$
$\dfrac{s^2}{[(s-a)^2 - b^2]^2}$	$\dfrac{R_2{}^2 e^\alpha}{2b^3} [\beta \cosh(\beta + 2\phi_2) - \sinh\beta]$
$\dfrac{s}{[(s-a)^2 - b^2]^2}$	$\dfrac{R_2 e^\alpha}{2b^3} \left[\beta \cosh(\beta + \phi_2) - \dfrac{a}{R} \sinh\beta\right]$
$\dfrac{1}{[(s-a)^2 - b^2]^2}$	$\dfrac{e^\alpha}{2b^3} (\beta \cosh\beta - \sinh\beta)$
$\dfrac{1}{s[(s-a)^2 - b^2]^2}$	$\dfrac{\alpha^\alpha}{2R_2 b^3} \left[\dfrac{b}{R} \cosh(\beta - 2\phi_2) + \beta \cosh(\beta - \phi_2) - \sinh(\beta - \phi_2)\right] + \dfrac{1}{R_2{}^4}$

$\dfrac{s^m}{s^n[(s-a)^2+b^2](s-c)}$	$\dfrac{s^m}{s^n[(s-a)^2+b^2](s-c)^2}$

$a, b, c =$ constants $\neq 0$ $\qquad\qquad \alpha = at \qquad\qquad \beta = bt \qquad\qquad \gamma = ct$

$$A = \sqrt{a^2+b^2} \qquad B = \sqrt{(a-c)^2+b^2} \qquad \phi_1 = \tan^{-1}\frac{b}{a-c} \qquad \phi_2 = \tan^{-1}\frac{b}{a}$$

$$C_1 = \frac{1}{bB} \qquad E_1 = \frac{e^\gamma - 1}{c} \qquad E_2 = \frac{e^\gamma - \gamma - 1}{c^2} \qquad E_3 = \frac{e^\gamma - \frac{1}{2}\gamma^2 - \gamma - 1}{c^3}$$

$$C_2 = \frac{1}{bB^2} \qquad F_1 = \sin(\phi_1 + \phi_2) \qquad F_2 = \sin(\phi_1 + 2\phi_2) \qquad F_3 = \sin(\phi_1 + 3\phi_2)$$

$$D_1 = \frac{1}{B^2} \qquad G_1 = \sin(2\phi_1 + \phi_2) \qquad G_2 = \sin(2\phi_1 + 2\phi_2)$$

$$D_2 = \frac{2(a-c)}{B^4} \qquad H_1 = \frac{1 - e^\gamma(1-\gamma)}{c^2} \qquad H_2 = \frac{2 + \gamma - e^\gamma(2-\gamma)}{c^3}$$

$f(s)$	$F(t)$
$\dfrac{s^2}{[(s-a)^2+b^2](s-c)}$	$A^2C_1e^\alpha \sin(\beta - \phi_1 + 2\phi_2) + c^2D_1e^\gamma$
$\dfrac{s}{[(s-a)+b^2](s-c)}$	$AC_1e^\alpha \sin(\beta - \phi_1 + \phi_2) + cD_1e^\gamma$
$\dfrac{1}{[(s-a)^2+b^2](s-c)}$	$C_1e^\alpha \sin(\beta - \phi_1) + D_1e^\gamma$
$\dfrac{1}{s[(s-a)^2+b^2](s-c)}$	$\dfrac{C_1}{A}[e^\alpha \sin(\beta - \phi_1 - \phi_2) + F_1] + D_1E_1$
$\dfrac{1}{s^2[(s-a)^2+b^2](s-c)}$	$\dfrac{C_1}{A^2}[e^\alpha \sin(\beta - \phi_1 - 2\phi_2) + tF_1 + F_2] + D_1E_2$
$\dfrac{1}{s^3[(s-a)^2+b^2](s-c)}$	$\dfrac{C_1}{A^3}\left[e^\alpha \sin(\beta - \phi_1 - 3\phi_2) + \dfrac{t^2}{2}F_1 + tF_2 + F_3\right] + D_1E_3$
$\dfrac{s^3}{[(s-a)^2+b^2](s-c)^2}$	$A^3C_2e^\alpha \sin(\beta - 2\phi_1 + 3\phi_2) + c^2(3D_1 + \gamma D_1 + cD_2)e^\gamma$
$\dfrac{s^2}{[(s-a)^2+b^2](s-c)^2}$	$A^2C_2e^\alpha \sin(\beta - 2\phi_1 + 2\phi_2) + c(2D_1 + \gamma D_1 + cD_2)e^\gamma$
$\dfrac{s}{[(s-a)^2+b^2](s-c)^2}$	$AC_2e^\alpha \sin(\beta - 2\phi_1 + \phi_2) + (D_1 + \gamma D_1 + cD_2)e^\gamma$
$\dfrac{1}{[(s-a)^2+b^2](s-c)^2}$	$C_2e^\alpha \sin(\beta - 2\phi_1) + (D_1t + D_2)e^\gamma$
$\dfrac{1}{s[(s-a)^2+b^2](s-c)^2}$	$\dfrac{C_2}{A}[e^\alpha \sin(\beta - 2\phi_1 - \phi_2) + G_1] + D_1iH_1 + D_2E_1$
$\dfrac{1}{s^2[(s-a)^2+b^2](s-c)^2}$	$\dfrac{C_2}{A^2}[e^\alpha \sin(\beta - 2\phi_1 - 2\phi_2) + tG_1 + G_2] + D_1H_2 + D_2E_2$

$\dfrac{s^m}{s''[(s-a)^2-b^2](s-c)}$	$\dfrac{s^m}{s''[(s-a)^2-b^2](s-c)^2}$

a, b, c = constants $\neq 0$ $\alpha = at$ $\beta = bt$ $\gamma = ct$

$A = \sqrt{a^2 - b^2}$ $B = \sqrt{(a-c)^2 - b^2}$ $\phi_1 = \tanh^{-1}\dfrac{b}{a-c}$ $\phi_2 = \tanh^{-1}\dfrac{b}{a}$

$C_1 = \dfrac{1}{bB}$ $E_1 = \dfrac{e^\gamma - 1}{c}$ $E_2 = \dfrac{e^\gamma - \gamma - 1}{c^2}$ $E_3 = \dfrac{e^\gamma - \frac{1}{2}\gamma^2 - \gamma - 1}{c^3}$

$C_2 = \dfrac{1}{bB^2}$ $F_1 = \sinh(\phi_1 + \phi_2)$ $F_2 = \sinh(\phi_1 + 2\phi_2)$ $F_3 = \sinh(\phi_1 + 3\phi_2)$

$D_1 = \dfrac{1}{B^2}$ $G_1 = \sinh(2\phi_1 + \phi_2)$ $G_2 = \sinh(2\phi_1 + 2\phi_2)$

$D_2 = \dfrac{2(a-c)}{B^4}$ $H_1 = \dfrac{1 - e^\gamma(1-\gamma)}{c^2}$ $H_2 = \dfrac{2 + \gamma - e^\gamma(2-\gamma)}{c^3}$

$f(s)$	$F(t)$
$\dfrac{s^2}{[(s-a)^2-b^2](s-c)}$	$A^2C_1e^\alpha \sinh(\beta - \phi_1 + 2\phi_2) + c^2D_1e^\gamma$
$\dfrac{s}{[(s-a)-b^2](s-c)}$	$AC_1e^\alpha \sinh(\beta - \phi_1 + \phi_2) + cD_1e^\gamma$
$\dfrac{1}{[(s-a)^2-b^2](s-c)}$	$C_1e^\alpha \sinh(\beta - \phi_1) + D_1e^\gamma$
$\dfrac{1}{s[(s-a)^2-b^2](s-c)}$	$\dfrac{C_1}{A}[e^\alpha \sinh(\beta - \phi_1 - \phi_2) + F_1] + D_1E_1$
$\dfrac{1}{s^2[(s-a)^2-b^2](s-c)}$	$\dfrac{C_1}{A^2}[e^\alpha \sinh(\beta - \phi_1 - 2\phi_2) + tF_1 + F_2] + D_1E_2$
$\dfrac{1}{s^3[(s-a)^2-b^2](s-c)}$	$\dfrac{C_1}{A^3}\left[e^\alpha \sinh(\beta - \phi_1 - 3\phi_2) + \dfrac{t^2}{2}F_1 + tF_2 + F_1\right] + D_1E_2$
$\dfrac{s^3}{[(s-a)^2-b^2](s-c)^2}$	$A^3C_2e^\alpha \sinh(\beta - 2\phi_1 + 3\phi_2) + c^2(3D_1 + \gamma D_1 + cD_2)e^\gamma$
$\dfrac{s^2}{[(s-a)^2-b^2](s-c)^2}$	$A^2C_2e^\alpha \sinh(\beta - 2\phi_1 + 2\phi_2) + c(2D_1 + \gamma D_1 + cD_2)e^\gamma$
$\dfrac{s}{[(s-a)^2-b^2](s-c)^2}$	$AC_2e^\alpha \sinh(\beta - 2\phi_1 + \phi_2) + (D_1 + \gamma D_1 + cD_2)e^\gamma$
$\dfrac{1}{[(s-a)^2-b^2](s-c)^2}$	$C_2e^\alpha \sinh(\beta - 2\phi_1) + (D_1t + D_2)e^\gamma$
$\dfrac{1}{s[(s-a)^2-b^2](s-c)^2}$	$\dfrac{C_2}{A}[e^\alpha \sinh(\beta - 2\phi_1 - \phi_2) + G_1] + D_1H_1 + D_2E_1$
$\dfrac{1}{s^2[(s-a)^2-b^2](s-c)^2}$	$\dfrac{C_2}{A^2}[e^\alpha \sinh(\beta - 2\phi_1 - 2\phi_2) + tG_1 + G_2] + D_1H_2 + D_2E_2$

$\dfrac{s^m}{s^n(s^2 + a^2)(s^2 + b^2)}$	$\dfrac{s^m}{s^n(s^2 - a^2)(s^2 - b^2)}$

$a, b = $ constants $\neq 0$

$\alpha = at$ $\beta = bt$ $A = \dfrac{1}{a(a^2 - b^2)}$ $B = \dfrac{1}{b(a^2 - b^2)}$

$f(s)$	$F(t)$
$\dfrac{s^3}{(s^2 + a^2)(s^2 + b^2)}$	$a^3 A \cos \alpha - b^3 \cos \beta$
$\dfrac{s^2}{(s^2 + a^2)(s^2 + b^2)}$	$a^2 A \sin \alpha - b^2 B \sin \beta$
$\dfrac{s}{(s^2 + a^2)(s^2 + b^2)}$	$-aA \cos \alpha + bB \cos \beta$
$\dfrac{1}{(s^2 + a^2)(s^2 + b^2)}$	$-A \sin \alpha + B \sin \beta$
$\dfrac{1}{s(s^2 + a^2)(s^2 + b^2)}$	$\dfrac{A}{a}(\cos \alpha - 1) - \dfrac{B}{b}(\cos \beta - 1)$
$\dfrac{1}{s^2(s^2 + a^2)(s^2 + b^2)}$	$\dfrac{A}{a^2}(\sin \alpha - \alpha) - \dfrac{B}{b^2}(\sin \beta - \beta)$
$\dfrac{1}{s^3(s^2 + a^2)(s^2 + b^2)}$	$-\dfrac{A}{a^3}\left(\cos \alpha + \dfrac{\alpha^2}{2} - 1\right) + \dfrac{B}{b^3}\left(\cos \beta + \dfrac{\beta^2}{2} - 1\right)$
$\dfrac{s^3}{(s^2 - a^2)(s^2 - b^2)}$	$a^3 A \cosh \alpha - b^3 B \cosh \beta$
$\dfrac{s^2}{(s^2 - a^2)(s^2 - b^2)}$	$a^2 A \sinh \alpha - b^2 B \sinh \beta$
$\dfrac{s}{(s^2 - a^2)(s^2 - b^2)}$	$aA \cosh \alpha - bB \cosh \beta$
$\dfrac{1}{(s^2 - a^2)(s^2 - b^2)}$	$A \sinh \alpha - B \sinh \beta$
$\dfrac{1}{s(s^2 - a^2)(s^2 - b^2)}$	$\dfrac{A}{a}(\cosh \alpha - 1) - \dfrac{B}{b}(\cosh \beta - 1)$
$\dfrac{1}{s^2(s^2 - a^2)(s^2 - b^2)}$	$\dfrac{A}{a^2}(\sinh \alpha - \alpha) - \dfrac{B}{b^2}(\sinh \beta - \beta)$
$\dfrac{1}{s^3(s^2 - a^2)(s^2 - b^2)}$	$\dfrac{A}{a^3}\left(\cosh \alpha - \dfrac{\alpha^2}{2} - 1\right) - \dfrac{B}{b^3}\left(\cosh \beta - \dfrac{\beta^2}{2} - 1\right)$

$\dfrac{s^m}{s^n(s^2-a^2)(s^2+b^2)}$	$\dfrac{s^m}{s^n(s^2+a^2)(s^2-b^2)}$

a, b = constants $\neq 0$

$\alpha = at$ $\qquad\qquad \beta = bt$ $\qquad\qquad A = \dfrac{1}{a(a^2+b^2)}$ $\qquad\qquad B = \dfrac{1}{b(a^2+b^2)}$

$f(s)$	$F(t)$
$\dfrac{s^3}{(s^2-a^2)(s^2+b^2)}$	$a^3A\cosh\alpha + b^3B\cos\beta$
$\dfrac{s^2}{(s^2-a^2)(s^2+b^2)}$	$a^2A\sinh\alpha + b^2B\sin\beta$
$\dfrac{s}{(s^2-a^2)(s^2+b^2)}$	$aA\cosh\alpha - bB\cos\beta$
$\dfrac{1}{(s^2-a^2)(s^2+b^2)}$	$A\sinh\alpha - B\sin\beta$
$\dfrac{1}{s(s^2-a^2)(s^2+b^2)}$	$\dfrac{A}{a}(\cosh\alpha-1) + \dfrac{B}{b}(\cos\beta-1)$
$\dfrac{1}{s^2(s^2-a^2)(s^2+b^2)}$	$\dfrac{A}{a^2}(\sinh\alpha-\alpha) + \dfrac{B}{b^2}(\sin\beta-\beta)$
$\dfrac{1}{s^3(s^2-a^2)(s^2+b^2)}$	$\dfrac{A}{a^3}\left(\cosh\alpha-\dfrac{\alpha^2}{2}-1\right) - \dfrac{B}{b^3}\left(\cos\beta+\dfrac{\beta^2}{2}-1\right)$
$\dfrac{s^3}{(s^2+a^2)(s^2-b^2)}$	$a^3A\cos\alpha + b^3B\cosh\beta$
$\dfrac{s^2}{(s^2+a^2)(s^2-b^2)}$	$a^2A\sin\alpha + b^2B\sinh\beta$
$\dfrac{s}{(s^2+a^2)(s^2-b^2)}$	$-aA\cos\alpha + bB\cosh\beta$
$\dfrac{1}{(s^2+a^2)(s^2-b^2)}$	$-A\sin\alpha + B\sinh\beta$
$\dfrac{1}{s(s^2+a^2)(s^2-b^2)}$	$\dfrac{A}{a}(\cos\alpha-1) + \dfrac{B}{b}(\cosh\beta-1)$
$\dfrac{1}{s^2(s^2+a^2)(s^2-b^2)}$	$\dfrac{A}{a^2}(\sin\alpha-\alpha) + \dfrac{B}{b^2}(\sinh\beta-\beta)$
$\dfrac{1}{s^3(s^2+a^2)(s^2-b^2)}$	$-\dfrac{A}{a^3}\left(\cos\alpha+\dfrac{\alpha^2}{2}-1\right) + \dfrac{B}{b^3}\left(\cosh\beta-\dfrac{\beta^2}{2}-1\right)$

$$\frac{s^m}{s^n(s^3 + a^3)} \quad \bigg| \quad \frac{s^m}{s^n(s^3 + a^3)(s - d)}$$

$a, b, d = \text{constants} \neq 0 \qquad \alpha = at \qquad \beta = \tfrac{1}{2}at \qquad \gamma = \tfrac{1}{2}\sqrt{3}\,at \qquad \delta = dt$

$$A = \frac{1}{3a^2(a + d)} \qquad\qquad B = \frac{2}{3a^2\sqrt{a^2 - ad + d^2}} \qquad\qquad D = A + B\sin(\phi + \psi)$$

$$\phi = \frac{\pi}{6} \qquad\qquad \psi = \tan^{-1}\frac{a\sqrt{3}}{a - 2d} \qquad\qquad \omega = \frac{1}{3}\cos^{-1}\frac{b^3}{a^3}$$

$$l_1 = 2a\cos\omega \qquad\qquad l_2 = -2a\cos(\omega + 2\phi) \qquad\qquad l_3 = -2a\cos(\omega - 2\phi)$$

$f(s)$	$F(t)$
$\dfrac{s^2}{s^3 + a^3}$	$\dfrac{1}{3}(e^{-\alpha} + 2e^{\beta}\cos\gamma)$
$\dfrac{s}{s^3 + a^3}$	$\dfrac{1}{3a}[-e^{-\alpha} + 2e^{\beta}\sin(\gamma + \phi)]$
$\dfrac{1}{s^3 + a^3}$	$\dfrac{1}{3a^2}[e^{-\alpha} + 2e^{\beta}\sin(\gamma - \phi)]$
$\dfrac{1}{s(s^3 + a^3)}$	$\dfrac{1}{3a^3}(-e^{-\alpha} - 2e^{\beta}\cos\gamma + 3)$
$\dfrac{1}{s^2(s^3 + a^3)}$	$\dfrac{1}{3a^4}[e^{-\alpha} - 2e^{\beta}\cos(\gamma - 2\phi) + 3\alpha]$
$\dfrac{s^3}{(s^3 + a^3)(s - d)}$	$a^3[Ae^{-\alpha} + Be^{\beta}\sin(\gamma - \phi + 2\psi)] + d^3 De^{\delta}$
$\dfrac{s^2}{(s^3 + a^3)(s - d)}$	$a^2[-Ae^{-\alpha} + Be^{\beta}\sin(\gamma - \phi + \psi)] + d^2 De^{\delta}$
$\dfrac{s}{(s^3 + a^3)(s - d)}$	$a[Ae^{-\alpha} + Be^{\beta}\sin(\gamma - \phi)] + dDe^{\delta}$
$\dfrac{1}{(s^3 + a^3)(s - d)}$	$-Ae^{-\alpha} + Be^{\beta}\sin(\gamma - \phi - \psi) + De^{\delta}$
$\dfrac{1}{s(s^3 + a^3)(s - d)}$	$\dfrac{A}{a}(e^{-\alpha} - 1) + \dfrac{D}{d}(e^{\delta} - 1)$ $+ \dfrac{B}{a}[e^{\beta}\sin(\delta - \phi - 2\psi) + \sin(\phi + 2\psi)]$
$\dfrac{1}{s^2(s^3 + a^3)(s - d)}$	$-\dfrac{A}{a^2}(e^{-\alpha} + \alpha - 1) + \dfrac{D}{d^2}(e^{\delta} - \delta - 1)$ $+ \dfrac{B}{a^2}[e^{\beta}\sin(\delta - \phi - 3\psi) + \alpha\sin(\phi + 2\psi) + \sin(\phi + 3\psi)]$
$\dfrac{1}{s^3 - 3a^2s - 2b^3}$	$\dfrac{e^{l_1 t}}{(l_1 + l_2)(l_1 + l_3)} + \dfrac{e^{l_2 t}}{(l_1 + l_2)(l_2 - l_3)} + \dfrac{e^{l_3 t}}{(l_1 + l_3)(l_3 - l_2)} \qquad a > b$

$\dfrac{s^m}{s^n(s^3 - a^3)}$	$\dfrac{s^m}{s^n(s^3 - a^3)(s - d)}$

---

a, b, d = constants $\neq 0$ $\alpha = at$ $\beta = \tfrac{1}{2}at$ $\gamma = \tfrac{1}{2}\sqrt{3}\,at$ $\delta = dt$

$A = \dfrac{1}{3a^2(a - d)}$ $B = \dfrac{2}{3a^2\sqrt{a^2 + ad + d^2}}$ $D = -A + B\sin(\phi - \psi)$

$\phi = \dfrac{\pi}{6}$ $\psi = \tan^{-1}\dfrac{a\sqrt{3}}{a + 2d}$

$f(s)$	$F(t)$
$\dfrac{s^2}{s^3 - a^3}$	$\dfrac{1}{3}(e^\alpha + 2e^{-\beta}\cos\gamma)$
$\dfrac{s}{s^3 - a^3}$	$\dfrac{1}{3a}[e^{-\alpha} + 2e^{-\beta}\sin(\gamma - \phi)]$
$\dfrac{1}{s^3 - a^3}$	$\dfrac{1}{3a^2}[e^\alpha - 2e^{-\beta}\sin(\gamma + \phi)]$
$\dfrac{1}{s(s^3 - a^3)}$	$\dfrac{1}{3a^3}(e^\alpha - 2e^{-\beta}\cos\gamma + 3)$
$\dfrac{1}{s^2(s^3 - a^3)}$	$\dfrac{1}{3a^4}[e^\alpha - 2e^{-\beta}\cos(\gamma - 2\phi) + 3\alpha]$
$\dfrac{s^3}{(s^3 - a^3)(s - d)}$	$a^3[Ae^\alpha + Be^{-\beta}\sin(\gamma + \phi - 4\psi)] + d^3De^\delta$
$\dfrac{s^2}{(s^3 - a^3)(s - d)}$	$a^2[Ae^\alpha + Be^{-\beta}\sin(\gamma + \phi - 3\psi)] + d^2De^\delta$
$\dfrac{s}{(s^3 - a^3)(s - d)}$	$a[Ae^\alpha + Be^{-\beta}\sin(\gamma + \phi - 2\psi)] + dDe^\delta$
$\dfrac{1}{(s^3 - a^3)(s - d)}$	$Ae^\alpha - Be^{-\beta}\sin(\gamma + \phi - \psi) + De^\delta$
$\dfrac{1}{s(s^3 - a^3)(s - d)}$	$\dfrac{A}{a}(e^\alpha - 1) + \dfrac{D}{d}(e^\delta - 1)$ $\quad - \dfrac{B}{a}[e^{-\beta}\sin(\delta + \phi - 2\psi) - \sin(\phi - 2\psi)]$
$\dfrac{1}{s^2(s^3 - a^3)(s - d)}$	$\dfrac{A}{a^2}(e^\alpha - \alpha - 1) + \dfrac{D}{d^2}(e^\delta - \delta - 1)$ $\quad - \dfrac{B}{a^2}[e^{-\beta}\sin(\delta + \phi - 3\psi) - \alpha\sin(\phi - 2\psi) - \sin(\phi - 3\psi)]$
$\dfrac{1}{s^3 - 3a^2s - 2a^3}$	$\dfrac{1}{9a^2}(e^{2\alpha} - e^{-\alpha} - 3\alpha)$

$\dfrac{s^m}{s^n(s^4 + 4a^4)}$	$\dfrac{s^m}{s^n(s^4 - a^4)}$

$a = \text{constant} \neq 0$	$\alpha = at$

$f(s)$	$F(t)$
$\dfrac{s^3}{s^4 + 4a^4}$	$\cosh \alpha \cos \alpha$
$\dfrac{s^2}{s^4 + 4a^4}$	$\dfrac{1}{2a}(\cosh \alpha \sin \alpha + \sinh \alpha \cos \alpha)$
$\dfrac{s}{s^4 + 4a^4}$	$\dfrac{1}{2a^2}\sinh \alpha \sin \alpha$
$\dfrac{1}{s^4 + 4a^4}$	$\dfrac{1}{4a^3}(\cosh \alpha \sin \alpha - \sinh \alpha \cos \alpha)$
$\dfrac{1}{s(s^4 + 4a^4)}$	$\dfrac{1}{4a^4}(1 - \cosh \alpha \cos \alpha)$
$\dfrac{1}{s^2(s^4 + 4a^4)}$	$\dfrac{1}{8a^5}(2\alpha - \cosh \alpha \sin \alpha - \sinh \alpha \cos \alpha)$
$\dfrac{1}{s^3(s^4 + 4a^4)}$	$\dfrac{1}{8a^6}(\alpha^2 - \sinh \alpha \sin \alpha)$
$\dfrac{s^3}{s^4 - a^4}$	$\dfrac{1}{2}(\cosh \alpha + \cos \alpha)$
$\dfrac{s^2}{s^4 - a^4}$	$\dfrac{1}{2a}(\sinh \alpha + \sin \alpha)$
$\dfrac{s}{s^4 - a^4}$	$\dfrac{1}{2a^2}(\cosh \alpha - \cos \alpha)$
$\dfrac{1}{s^4 - a^4}$	$\dfrac{1}{2a^3}(\sinh \alpha - \sin \alpha)$
$\dfrac{1}{s(s^4 - a^4)}$	$\dfrac{1}{2a^4}(\cosh \alpha + \cos \alpha - 2)$
$\dfrac{1}{s^2(s^4 - a^4)}$	$\dfrac{1}{2a^5}(\sinh \alpha + \sin \alpha - 2\alpha)$
$\dfrac{1}{s^3(s^4 - a^4)}$	$\dfrac{1}{2a^6}(\cosh \alpha - \cos \alpha - \alpha^2)$

$p^2 - 4q = 0$	$\dfrac{s^m}{s^n(s^4 + ps^2 + q)}$

p, q = constants $\neq 0$	$\alpha = at$	$a = \sqrt{\left\lvert\dfrac{p}{2}\right\rvert}$

	$f(s)$	$F(t)$
$p > 0, q > 0 \text{ or } q < 0$	$\dfrac{s^3}{s^4 + ps^2 + q}$	$\dfrac{1}{2}(2\cos\alpha - \alpha\sin\alpha)$
	$\dfrac{s^2}{s^4 + ps^2 + q}$	$\dfrac{1}{2a}(\sin\alpha + \alpha\cos\alpha)$
	$\dfrac{s}{s^4 + ps^2 + q}$	$\dfrac{1}{2a^2}\alpha\sin\alpha$
	$\dfrac{1}{s^4 + ps^2 + q}$	$\dfrac{1}{2a^3}(\sin\alpha - \alpha\cos\alpha)$
	$\dfrac{1}{s(s^4 + ps^2 + q)}$	$\dfrac{1}{2a^4}(2 - 2\cos\alpha - \alpha\sin\alpha)$
	$\dfrac{1}{s^2(s^4 + ps^2 + q)}$	$\dfrac{1}{2a^5}(2\alpha + \alpha\cos\alpha - 3\sin\alpha)$
	$\dfrac{1}{s^3(s^4 + ps^2 + q)}$	$\dfrac{1}{2a^6}(\alpha^2 + \alpha\sin\alpha + 4\cos\alpha - 4)$
$p < 0, q > 0 \text{ or } q < 0$	$\dfrac{s^3}{s^4 + ps^2 + q}$	$\dfrac{1}{2}(2\cosh\alpha + \alpha\sinh\alpha)$
	$\dfrac{s^2}{s^4 + ps^2 + q}$	$\dfrac{1}{2a}(\alpha\cosh\alpha + \sinh\alpha)$
	$\dfrac{s}{s^4 + ps^2 + q}$	$\dfrac{1}{2a^2}\alpha\sinh\alpha$
	$\dfrac{1}{s^4 + ps^2 + q}$	$\dfrac{1}{2a^3}(\alpha\cosh\alpha - \sinh\alpha)$
	$\dfrac{1}{s(s^4 + ps^2 + q)}$	$\dfrac{1}{2a^4}(2 - 2\cosh\alpha + \alpha\sinh\alpha)$
	$\dfrac{1}{s^2(s^4 + ps^2 + q)}$	$\dfrac{1}{2a^5}(2\alpha + \alpha\cosh\alpha - 3\sinh\alpha)$
	$\dfrac{1}{s^3(s^4 + ps^2 + q)}$	$\dfrac{1}{2a^6}(\alpha^2 + \alpha\sinh\alpha - 4\cosh\alpha + 4)$

$(p^2 - 4q) < 0$	$\dfrac{s^m}{s^n(s^4 + ps^2 + q)}$

$$p, q = \text{constants} \neq 0 \qquad \alpha = at \qquad \beta = bt \qquad c = \sqrt{|q|}$$

$$\phi = \tan^{-1}\frac{b}{a}$$

$$a = \sqrt{\frac{c}{2} - \left|\frac{p}{4}\right|} \qquad b = \sqrt{\frac{c}{2} + \left|\frac{p}{4}\right|} \qquad r = \sqrt{a^2 + b^2}$$

$$\psi = \tan^{-1}\frac{a}{b}$$

	$f(s)$	$F(t)$
$p > 0, q > 0$ or $q < 0$	$\dfrac{s^3}{s^4 + ps^2 + q}$	$\dfrac{r^4}{4abc}[e^{\alpha}\sin(\beta + 2\phi) - e^{-\alpha}\sin(\beta - 2\phi)]$
	$\dfrac{s^2}{s^4 + ps^2 + q}$	$\dfrac{r^3}{4abc}[e^{\alpha}\sin(\beta + \phi) + e^{-\alpha}\sin(\beta - \phi)]$
	$\dfrac{s}{s^4 + ps^2 + q}$	$\dfrac{r^2}{4abc}(e^{\alpha}\sin\beta - e^{-\alpha}\sin\beta)$
	$\dfrac{1}{s^4 + ps^2 + q}$	$\dfrac{r}{4abc}[e^{\alpha}\sin(\beta - \phi) + e^{-\alpha}\sin(\beta + \phi)]$
	$\dfrac{1}{s(s^4 + ps^2 + q)}$	$\dfrac{1}{4abc}[e^{\alpha}\sin(\beta - 2\phi) - e^{-\alpha}\sin(\beta + 2\phi) + 2\sin 2\phi]$
	$\dfrac{1}{s^2(s^4 + ps^2 + q)}$	$\dfrac{1}{4abcr}[e^{\alpha}\sin(\beta - 3\phi) + e^{-\alpha}\sin(\beta + 3\phi) + 2\sin 3\phi]$
	$\dfrac{1}{s^3(s^4 + ps^2 + q)}$	$\dfrac{1}{4abcr^2}[e^{\alpha}\sin(\beta - 4\phi) - e^{-\alpha}\sin(\beta + 4\phi) + 2\sin 4\phi]$
$p < 0, q > 0$ or $q < 0$	$\dfrac{s^3}{s^4 + ps^2 + q}$	$\dfrac{r^4}{4abc}[e^{\beta}\sin(\alpha + 2\psi) - e^{-\beta}\sin(\alpha - 2\psi)]$
	$\dfrac{s^2}{s^4 + ps^2 + q}$	$\dfrac{r^3}{4abc}[e^{\beta}\sin(\alpha + \psi) + e^{-\beta}\sin(\alpha - \psi)]$
	$\dfrac{s}{s^4 + ps^2 + q}$	$\dfrac{r^2}{4abc}(e^{\beta}\sin\alpha - e^{-\beta}\sin\alpha)$
	$\dfrac{1}{s^4 + ps^2 + q}$	$\dfrac{r}{4abc}[e^{\beta}\sin(\alpha - \psi) + e^{-\beta}\sin(\alpha + \psi)]$
	$\dfrac{1}{s(s^4 + ps^2 + q)}$	$\dfrac{1}{4abc}[e^{\beta}\sin(\alpha - 2\psi) - e^{-\beta}\sin(\alpha + 2\psi) + 2\sin 2\psi]$
	$\dfrac{1}{s^2(s^4 + ps^2 + q)}$	$\dfrac{1}{4abcr}[e^{\beta}\sin(\alpha - 3\psi) + e^{-\beta}\sin(\alpha + 3\psi) + 2\sin 3\psi]$
	$\dfrac{1}{s^3(s^4 + ps^2 + q)}$	$\dfrac{1}{4abcr^2}[e^{\beta}\sin(\alpha - 4\psi) - e^{-\beta}\sin(\alpha + 4\psi) + 2\sin 4\psi]$

$(p^2 - 4q) > 0$	$\dfrac{s^m}{s^n(s^4 + ps^2 + q)}$

$p, q = \text{constants} \neq 0$	$\alpha = at$	$\beta = bt$
$c = \sqrt{|p^2 - 4q|}$	$a = \sqrt{\dfrac{|p| - c}{2}}$	$b = \sqrt{\dfrac{|p| + c}{2}}$

	$f(s)$	$F(t)$
$p > 0, q > 0 \text{ or } q < 0$	$\dfrac{s^3}{s^4 + ps^2 + q}$	$-\dfrac{a^2 \cos \alpha}{c} + \dfrac{b^2 \cos \beta}{c}$
	$\dfrac{s^2}{s^4 + ps^2 + q}$	$-\dfrac{a \sin \alpha}{c} + \dfrac{b \sin \beta}{c}$
	$\dfrac{s}{s^4 + ps^2 + q}$	$\dfrac{\cos \alpha}{c} - \dfrac{\cos \beta}{c}$
	$\dfrac{1}{s^4 + ps^2 + q}$	$\dfrac{\sin \alpha}{ac} - \dfrac{\sin \beta}{bc}$
	$\dfrac{1}{s(s^4 + ps^2 + q)}$	$-\dfrac{\cos \alpha - 1}{a^2 c} + \dfrac{\cos \beta - 1}{b^2 c}$
	$\dfrac{1}{s^2(s^4 + ps^2 + q)}$	$-\dfrac{\sin \alpha - \alpha}{a^3 c} + \dfrac{\sin \beta - \beta}{b^3 c}$
	$\dfrac{1}{s^3(s^4 + ps^2 + q)}$	$-\dfrac{\cos \alpha + \frac{1}{2}\alpha^2 - 1}{a^4 c} - \dfrac{\cos \beta + \frac{1}{2}\beta^2 - 1}{b^4 c}$
$p < 0, q > 0 \text{ or } q < 0$	$\dfrac{s^3}{s^4 + ps^2 + q}$	$-\dfrac{a^2 \cosh \alpha}{c} + \dfrac{b^2 \cosh \beta}{c}$
	$\dfrac{s^2}{s^4 + ps^2 + q}$	$-\dfrac{a \sinh \alpha}{c} + \dfrac{b \sinh \beta}{c}$
	$\dfrac{s}{s^4 + ps^2 + q}$	$-\dfrac{\cosh \alpha}{c} + \dfrac{\cosh \beta}{c}$
	$\dfrac{1}{s^4 + ps^2 + q}$	$-\dfrac{\sinh \alpha}{ac} + \dfrac{\sinh \beta}{bc}$
	$\dfrac{1}{s(s^4 + ps^2 + q)}$	$-\dfrac{\cosh \alpha - 1}{a^2 c} + \dfrac{\cosh \beta - 1}{b^2 c}$
	$\dfrac{1}{s^2(s^4 + ps^2 + q)}$	$-\dfrac{\sinh \alpha - \alpha}{a^3 c} + \dfrac{\sinh \beta - \beta}{b^3 c}$
	$\dfrac{1}{s^3(s^4 + ps^2 + q)}$	$-\dfrac{\cosh \alpha - \frac{1}{2}\alpha^2 - 1}{a^4 c} + \dfrac{\cosh \beta - \frac{1}{2}\beta^2 - 1}{b^4 c}$

a, b, k = constants $\neq 0$	$k, n = 1, 2, 3, \ldots$
erf( ) = error function (Sec. 13.01)	( )!! = double factorial (Sec. 1.03)
$J_k(\ )$ = Bessel function (Sec. 14.24)	$I_k(\ )$ = modified Bessel function (Sec. 14.28)

$f(s)$	$F(t)$	$f(s)$	$F(t)$
$\dfrac{1}{\sqrt{s}}$	$\dfrac{1}{\sqrt{\pi t}}$	$\dfrac{1}{\sqrt{s^{2n+1}}}$	$\dfrac{(2t)^n}{(2n-1)!!\sqrt{\pi t}}$
$\dfrac{1}{s\sqrt{s+a}}$	$\dfrac{\operatorname{erf}\sqrt{at}}{\sqrt{a}}$	$\dfrac{1}{(s-a)\sqrt{s}}$	$e^{at}\dfrac{\operatorname{erf}\sqrt{at}}{\sqrt{a}}$
$\dfrac{1}{\sqrt{(s+a)}}$	$\dfrac{e^{-at}}{\sqrt{\pi t}}$	$\dfrac{1}{\sqrt{(s-a)}}$	$\dfrac{e^{at}}{\sqrt{\pi t}}$
$\dfrac{1}{\sqrt{(s+a)^{2n+1}}}$	$\dfrac{t^n e^{-at}}{[\frac{1}{2}\cdot\frac{3}{2}\cdots(2n-1)/2]\sqrt{\pi t}}$	$\dfrac{1}{\sqrt{(s-a)^{2n+1}}}$	$\dfrac{t^n e^{at}}{[\frac{1}{2}\cdot\frac{3}{2}\cdots(2n-1)/2]\sqrt{\pi t}}$
$\dfrac{s}{\sqrt{(s+a)^3}}$	$\dfrac{(1-2at)e^{-at}}{\sqrt{\pi t}}$	$\dfrac{s}{\sqrt{(s-a)^3}}$	$\dfrac{(1+2at)e^{at}}{\sqrt{\pi t}}$
$\dfrac{1}{\sqrt{s+a}+\sqrt{s+b}}$	$\dfrac{e^{-bt}-e^{-at}}{2(b-a)t\sqrt{\pi t}}$	$\dfrac{1}{(s+a)\sqrt{s+b}}$	$\dfrac{e^{-at}\operatorname{erf}(\sqrt{(b-a)t})}{b-a}$
$\dfrac{1}{(s^2+a^2)^{1/2}}$	$J_0(at)$	$\dfrac{1}{(s^2-a^2)^{1/2}}$	$I_0(at)$
$\dfrac{1}{(s^2+a^2)^{3/2}}$	$\dfrac{tJ_1(at)}{a}$	$\dfrac{1}{(s^2-a^2)^{3/2}}$	$\dfrac{tI_1(at)}{a}$
$\dfrac{1}{(s^2+a^2)^k}$ $k>0$	$\dfrac{\sqrt{\pi}}{\Gamma(k)}\left(\dfrac{t}{2a}\right)^{k-1/2}J_{k-1/2}(at)$	$\dfrac{1}{(s^2-a^2)^k}$ $k>0$	$\dfrac{\sqrt{\pi}}{\Gamma(k)}\left(\dfrac{t}{2a}\right)^{k-1/2}I_{k-1/2}(at)$
$\dfrac{s}{(s^2+a^2)^{3/2}}$	$tJ_0(at)$	$\dfrac{s}{(s^2-a^2)^{3/2}}$	$tI_0(at)$
$\dfrac{s^2}{(s^2+a^2)^{3/2}}$	$J_0(at)-atJ_1(at)$	$\dfrac{s^2}{(s^2-a^2)^{3/2}}$	$I_0(at)+atI_1(at)$
$(\sqrt{s^2+a^2}-s)^k$ $k>0$	$\dfrac{ka^k J_k(at)}{t}$	$(s-\sqrt{s^2-a^2})^k$ $k>0$	$\dfrac{ka^k I_k(at)}{t}$
$\dfrac{(\sqrt{s^2+a^2}-s)^k}{\sqrt{s^2+a^2}}$ $k>-1$	$a^k J_k(at)$	$\dfrac{(s-\sqrt{s^2-a^2})^k}{\sqrt{s^2-a^2}}$ $k>-1$	$a^k I_k(at)$

$a, b, k = $ constants $\neq 0$	$n = 1, 2, 3, \ldots$
erf( ) = error function (Sec. 13.01)	$\Gamma($) = gamma function (Sec. 13.03)
$H_n($) = Hermite polynomial (Sec. 14.23)	$I_k($) = modified Bessel function (Sec. 14.28)

$f(s)$	$F(t)$	$f(s)$	$F(t)$
$\dfrac{1}{\sqrt{s} + \sqrt{a}}$	$\dfrac{1}{\sqrt{\pi t}} - \sqrt{a}\, e^{at}(1 - erf\sqrt{at})$	$\dfrac{1}{\sqrt{s} - \sqrt{a}}$	$\dfrac{1}{\sqrt{\pi t}} + \sqrt{a}\, e^{at}(1 + erf\sqrt{at})$
$\dfrac{\sqrt{s}}{s + a}$	$\dfrac{1}{\sqrt{\pi t}} - \dfrac{2a}{\sqrt{\pi a}}\, e^{-at}\int_0^{\sqrt{at}} e^{\lambda^2}\, d\lambda$	$\dfrac{\sqrt{s}}{s - a}$	$\dfrac{1}{\sqrt{\pi t}} + \sqrt{a}\, e^{at}\, erf\sqrt{at}$
$\dfrac{1}{s(\sqrt{s} + \sqrt{a})}$	$1 - e^{at}(1 - erf\sqrt{at})$	$\dfrac{1}{s(\sqrt{s} - \sqrt{a})}$	$1 - e^{at}(1 + erf\sqrt{at})$
$\dfrac{1}{\sqrt{s}(\sqrt{s} + \sqrt{a})}$	$e^{at}(1 - erf\sqrt{at})$	$\dfrac{1}{\sqrt{s}(\sqrt{s} - \sqrt{a})}$	$e^{at}(1 + erf\sqrt{at})$
$\dfrac{1}{(s + a)\sqrt{s + b}}$	$\dfrac{e^{-at}\,erf\sqrt{(b - a)t}}{\sqrt{b - a}}$	$\dfrac{1}{\sqrt{s + 2a}\sqrt{s + 2b}}$	$e^{-(a+b)t}I_0[(a - b)t]$
$\dfrac{b + a}{(s + a)(\sqrt{s} + \sqrt{b})}$	$e^{-at}(\sqrt{b} - \sqrt{a}\,erf\sqrt{at})$ $-\sqrt{b}\, e^{bt}\,erf\sqrt{bt}$	$\dfrac{b - a}{(s - a)(\sqrt{s} + \sqrt{b})}$	$e^{at}(\sqrt{b} - \sqrt{a}\,erf\sqrt{at})$ $-\sqrt{b}\, e^{bt}\,erf\sqrt{bt}$
$\dfrac{\sqrt{s + 2a} - \sqrt{s}}{\sqrt{s + 2a} + \sqrt{s}}$	$\dfrac{1}{t}\, e^{-at}I_1(at)$	$\dfrac{(\sqrt{s + 2a} + \sqrt{s})^{-2k}}{\sqrt{s}\sqrt{s + 2a}}$	$\dfrac{1}{(2a)^k}\, e^{-at}I_1(at)$
$\dfrac{\sqrt{s + 2a} - \sqrt{s}}{\sqrt{s}}$	$ae^{-at}[I_0(at) + I_1(at)]$	$\dfrac{(2a - 2b)^k}{(\sqrt{s + a} + \sqrt{s + b})^{2k}}$	$\dfrac{k}{t}\, e^{-(a+b)t}I_k[(a - b)t]$ $k > -1$
$A = \sqrt{s + \sqrt{s^2 + a^2}}$ $\dfrac{A}{\sqrt{s^2 + a^2}}$	$\sqrt{\dfrac{2}{\pi t}}\cos at$	$B = \sqrt{s + \sqrt{s^2 - b^2}}$ $\dfrac{B}{\sqrt{s^2 - b^2}}$	$\sqrt{\dfrac{2}{\pi t}}\cosh bt$
$\dfrac{1}{A\sqrt{s^2 + a^2}}$	$\sqrt{\dfrac{2}{\pi t}}\dfrac{\sin at}{a}$	$\dfrac{1}{B\sqrt{s^2 - b^2}}$	$\sqrt{\dfrac{2}{\pi t}}\dfrac{\sinh bt}{b}$
$\dfrac{(1 - s)^n}{s^n\sqrt{s}}$	$\dfrac{n!}{(2n)!\sqrt{\pi t}}H_{2n}(\sqrt{t})$	$\dfrac{(1 - s)^n}{s^{n+1}\sqrt{s}}$	$-\dfrac{n!}{(2n + 1)!\sqrt{\pi}}H_{2n+1}(\sqrt{t})$
$\dfrac{\Gamma(k)}{(s + 2a)^k(s + 2b)^k}$		$\sqrt{\pi}\left(\dfrac{t}{2a - 2b}\right)^{k - 1/2} e^{-(a+b)t}I_{k-1/2}[(a - b)t] \qquad k > 0$	

| a, b = constants $\neq 0$ | Si( ) = integral-sine function (Sec. 13.01) |
| Ber( ) = Ber function (Sec. 14.32) | Bei( ) = Bei function (Sec. 14.32) |

$\dfrac{\sin (a/s)}{s}$	$\text{Bei } (2\sqrt{at})$
$\dfrac{\sin (a/s)}{\sqrt{s}}$	$\dfrac{\sinh \sqrt{2at} \, \sin \sqrt{2at}}{\sqrt{\pi t}}$
$\dfrac{\sin (a/s)}{s\sqrt{s}}$	$\dfrac{\cosh \sqrt{2at} \, \sin \sqrt{2at}}{\sqrt{\pi a}}$
$\dfrac{\cos (a/s)}{s}$	$\text{Ber } (2\sqrt{at})$
$\dfrac{\cos (a/s)}{\sqrt{s}}$	$\dfrac{\cosh \sqrt{2at} \, \cos \sqrt{2at}}{\sqrt{\pi t}}$
$\dfrac{\cos (a/s)}{s\sqrt{s}}$	$\dfrac{\sinh \sqrt{2at} \, \cos \sqrt{2at}}{\sqrt{\pi a}}$
$\dfrac{\sin [a + \tan^{-1} (b/s)]}{\sqrt{a^2 + b^2}}$	$\sin (a + bt)$
$\dfrac{\cos [a + \tan^{-1} (b/s)]}{\sqrt{a^2 + b^2}}$	$\cos (a + bt)$
$\dfrac{e^{-\sqrt{as}} \sin \sqrt{as}}{\sqrt{s}}$	$\dfrac{\sin (a/2t)}{\sqrt{\pi t}}$
$\dfrac{e^{-\sqrt{as}} \cos \sqrt{as}}{\sqrt{s}}$	$\dfrac{\cos (a/2t)}{\sqrt{\pi t}}$
$\tan^{-1} \dfrac{a}{s} = \cot^{-1} \dfrac{s}{a}$	$\dfrac{1}{t} \sin at$
$s\left(\tan^{-1} \dfrac{a}{s} - \dfrac{a}{s}\right)$	$\dfrac{at \cos at - \sin at}{t^2}$
$\dfrac{1}{s} \tan^{-1} \dfrac{a}{s}$	$\text{Si } (at)$

a, k = constants $\neq 0$ $\qquad\qquad J_k(\ \ )$ = Bessel function (Sec. 14.24)

$I_k(\ \ ), K_k(\ \ )$ = modified Bessel functions (Sec. 14.28)

$\dfrac{\sinh\,(a/s)}{s}$	$\dfrac{I_0(2\sqrt{at}) - J_0(2\sqrt{at})}{2}$
$\dfrac{\sinh\,(a/s)}{\sqrt{s}}$	$\dfrac{\cosh\,(2\sqrt{at}) - \cos\,(2\sqrt{at})}{2\sqrt{\pi t}}$
$\dfrac{\sinh\,(a/s)}{s\sqrt{s}}$	$\dfrac{\sinh\,(2\sqrt{at}) - \sin\,(2\sqrt{at})}{2\sqrt{\pi a}}$
$\dfrac{\cosh\,(a/s)}{s}$	$\dfrac{I_0(2\sqrt{at}) + J_0(2\sqrt{at})}{2}$
$\dfrac{\cosh\,(a/s)}{\sqrt{s}}$	$\dfrac{\cosh\,(2\sqrt{at}) + \cos\,(2\sqrt{at})}{2\sqrt{\pi t}}$
$\dfrac{\cosh\,(a/s)}{s\sqrt{s}}$	$\dfrac{\sinh\,(2\sqrt{at}) + \sin\,(2\sqrt{at})}{2\sqrt{\pi a}}$
$\pi e^{-ks} I_0(ks)$	$\begin{cases} [t(2k - t)]^{-1/2} & \text{if } 0 < t < 2k \\ 0 & \text{if } t > 2k \end{cases}$
$e^{-ks} I_1(ks)$	$\begin{cases} \dfrac{k - t}{\pi k \sqrt{t(2k - t)}} & \text{if } 0 < t < 2k \\ 0 & \text{if } t > 2k \end{cases}$
$K_0(ks)$	$\begin{cases} 0 & \text{if } 0 < t < k \\ (t^2 - k^2)^{-1/2} & \text{if } t > k \end{cases}$
$K_0(k\sqrt{s})$	$\dfrac{1}{2t}\,e^{-\alpha} \qquad \alpha = \dfrac{k^2}{4t^2}$
$\dfrac{1}{s}\,e^{ks} K_1(ks)$	$\dfrac{1}{k}\,\sqrt{t(t + 2k)}$
$\dfrac{1}{\sqrt{s}}\,K_1(k\sqrt{s})$	$\dfrac{1}{k}\,e^{-\alpha} \qquad \alpha = \dfrac{k^2}{4t^2}$
$\dfrac{1}{\sqrt{s}}\,e^{k/s} K_0\!\left(\dfrac{k}{s}\right)$	$\dfrac{2}{\sqrt{\pi t}}\,K_0(2\sqrt{2kt})$

EXPONENTIAL TRANSFORMS

$a, b, k, r = $ constants $\neq 0$	$n = 1, 2, 3, \ldots$
erf( ) = error function (Sec. 13.01)	
$J_k($) = Bessel function (Sec. 14.24)	$I_k($) = modified Bessel function (Sec. 14.28)

$f(s)$	$F(t)$	$f(s)$	$F(t)$
$\dfrac{e^{k/s}}{s}$	$I_0(2\sqrt{kt})$	$\dfrac{e^{-k/s}}{s}$	$J_0(2\sqrt{kt})$
$\dfrac{e^{k/s}}{s^{r+1}}$	$\sqrt{\left(\dfrac{t}{k}\right)^r}\, I_r(2\sqrt{kt}) \quad r > 1$	$\dfrac{e^{-k/s}}{s^{r+1}}$	$\sqrt{\left(\dfrac{t}{k}\right)^r}\, J_r(2\sqrt{kt}) \quad r > 1$
$\dfrac{e^{k/s}}{\sqrt{s}}$	$\dfrac{1}{\sqrt{\pi t}} \cosh(2\sqrt{kt})$	$\dfrac{e^{-k/s}}{\sqrt{s}}$	$\dfrac{1}{\sqrt{\pi t}} \cos(2\sqrt{kt})$
$\dfrac{e^{k/s}}{s\sqrt{s}}$	$\dfrac{1}{\sqrt{\pi k}} \sinh(2\sqrt{kt})$	$\dfrac{e^{-k/s}}{s\sqrt{s}}$	$\dfrac{1}{\sqrt{\pi k}} \sin(2\sqrt{kt})$

	$f(s)$	$F(t)$		$f(s)$	$F(t)$
$\alpha = k^2/(4t)$	$\exp(k\sqrt{s})$	$-\dfrac{k}{2t\sqrt{\pi t}} e^{-\alpha} \quad k > 0$	$\alpha = k^2/(4t)$	$\exp(-k\sqrt{s})$	$\dfrac{k}{2t\sqrt{\pi t}} e^{-\alpha} \quad k > 0$
	$\dfrac{\exp(-k\sqrt{s})}{\sqrt{s}}$	$\dfrac{1}{\sqrt{\pi t}} e^{-\alpha} \quad k \geq 0$		$\dfrac{\exp(-k\sqrt{s})}{s\sqrt{s}}$	$2\sqrt{\dfrac{t}{\pi}}\, e^{-\alpha} \quad k \geq 0$
	$\dfrac{\exp(k\sqrt{s})}{s}$	$1 + \operatorname{erf}\sqrt{\alpha} \quad k \geq 0$		$\dfrac{\exp(-k\sqrt{s})}{s}$	$1 - \operatorname{erf}\sqrt{\alpha} \quad k \geq 0$

	$f(s)$	$F(t)$
$\lambda = \sqrt{s(a + \sqrt{a})}$	$\dfrac{a\exp(-k\sqrt{s})}{\lambda}$	$-\exp(k\sqrt{a})\, e^{at}[1 - \operatorname{erf}(\sqrt{at} + \sqrt{\alpha})] + \operatorname{erf}\sqrt{\alpha} \qquad k \geq 0$
	$\dfrac{\exp(-k\sqrt{s})}{\lambda}$	$\exp(k\sqrt{a})\, e^{at}[1 - \operatorname{erf}(\sqrt{at} + \sqrt{\alpha})] \qquad k \geq 0$
	$\dfrac{e^{-k\lambda}}{\lambda}$	$\begin{cases} 0 & \text{if } 0 < t < k \\ e^{-at}I_0[\sqrt{a^2(t^2 - k^2)}] & \text{if } t > k \end{cases}$
$\alpha = \sqrt{s^2 + a^2}$	$\dfrac{e^{-k\alpha}}{\alpha}$	$\begin{cases} 0 & \text{if } 0 < t < k \\ J_0[\sqrt{a^2(t^2 - k^2)}] & \text{if } t > k \end{cases}$
	$e^{-ks} - e^{-k\alpha}$	$\begin{cases} 0 & \text{if } 0 < t < k \\ \dfrac{ak}{\sqrt{t^2 - k^2}} J_1[\sqrt{a^2(t^2 - k^2)}] & \text{if } t > k \end{cases}$
$\beta = \sqrt{s^2 - b^2}$	$\dfrac{e^{-k\beta}}{\beta}$	$\begin{cases} 0 & \text{if } 0 < t < k \\ I_0[\sqrt{a^2(t^2 - k^2)}] & \text{if } t > k \end{cases}$
	$e^{-ks} - e^{-k\beta}$	$\begin{cases} 0 & \text{if } 0 < t < k \\ -\dfrac{ak}{\sqrt{t^2 - k^2}} I_1[\sqrt{a^2(t^2 - k^2)}] & \text{if } t > k \end{cases}$

a, b, k = constants $\neq 0$	$n = 1, 2, 3, \ldots$
erf( ) = error function (Sec. 13.01)	erfc( ) = $1 - $ erf( )
Ci( ), Ei( ), Si( ) = integral functions (Sec. 13.01)	$\psi($) = diagamma function (13.03)

$f(s)$	$F(t)$	$f(s)$	$F(t)$
$\dfrac{\ln s}{s}$	$-C - \ln t$	$\dfrac{\ln s}{s^{n+1}}$	$\dfrac{t^n}{n!}\,[\psi(n) - \ln t]$
$\dfrac{\ln s}{s^2}$	$t(1 - C - \ln t)$	$\dfrac{\ln s}{s^k}$	$\dfrac{t^k}{\Gamma(k)}\,[\psi(k-1) - \ln t]\qquad k > 0$
$\dfrac{\ln (s+b)}{s+a}$	$e^{-at}\{\ln(b-a) + \text{Ei}\,[(b-a)t]\}$	$\dfrac{\ln (s+b)}{s-a}$	$e^{at}\{\ln(b+a) + \text{Ei}\,[(b+a)t]\}$
$\ln \dfrac{s+a}{s+b}$	$\dfrac{1}{t}(e^{-bt} - e^{-at})$	$\ln \dfrac{s-a}{s-b}$	$\dfrac{1}{t}(e^{bt} - e^{at})$
$\dfrac{\ln(1 + k^2 s^2)}{s}$	$-2\,\text{Ci}\left(\dfrac{t}{k}\right)$	$\dfrac{\ln(s^2 + a^2)}{s}$	$2\ln a - 2\,\text{Ci}\,(at)$
$\dfrac{\ln(1 + k^2 s^2)}{s^2}$	$-2\displaystyle\int_0^t \text{Ci}\left(\dfrac{t}{k}\right) dt$	$\dfrac{\ln(s^2 + a^2)}{s^2}$	$\dfrac{2}{a}[at\ln a + \sin \alpha t - at\,\text{Ci}\,(at)]$
$\dfrac{a\ln s}{s^2 + a^2}$	$\cos at\,\text{Si}\,(at)$ $-\sin at\,[\ln a - \text{Ci}\,(at)]$	$\dfrac{\ln s}{s(s^2 + a^2)}$	$\cos at\,[\ln a - \text{Ci}\,(at)]$ $-\sin at\,\text{Si}\,(at)$
$\ln \dfrac{s^2 + a^2}{s^2}$	$\dfrac{2}{t}(1 - \cos at)$	$\ln \dfrac{s^2 - a^2}{s^2}$	$\dfrac{2}{t}(1 - \cosh at)$
$\text{erf}\left(\dfrac{k}{\sqrt{s}}\right)$	$\dfrac{1}{\pi t}\sin(2k\sqrt{t})$	$\text{erfc}\left(\dfrac{k}{\sqrt{s}}\right)$	$\dfrac{1}{s} - \dfrac{1}{\pi t}\sin(2k\sqrt{t})$

The center narrow column spanning the first two rows reads: $C = 0.577\ 215\ 665$

	$f(s)$	$F(t)$		$f(s)$	$F(t)$
$\alpha = k^2 s^2$	$e^{\alpha}\,\text{erfc}\,\sqrt{\alpha}$	$\dfrac{1}{k\sqrt{\pi}}e^{-t^2/4k^2}\qquad k > 0$	$\beta = ks$	$e^{\beta}\,\text{erfc}\,\sqrt{\beta}$	$\dfrac{\sqrt{k}}{\pi(t+k)\sqrt{t}}\qquad k > 0$
	$\dfrac{e^{\alpha}\,\text{erfc}\,\sqrt{\alpha}}{s}$	$\text{erf}\left(\dfrac{t}{2k}\right)\qquad k > 0$		$\dfrac{e^{\beta}\,\text{erfc}\,\sqrt{\beta}}{\sqrt{s}}$	$\dfrac{1}{\sqrt{\pi(t+k)}}\qquad k > 0$
$\lambda = s^2/(4a^2)$	$e^{\lambda}\,\text{erfc}\,\sqrt{\lambda}$	$\dfrac{2a}{\sqrt{\pi}}e^{-a^2 t^2}$	$\kappa = a/s$	$\dfrac{e^{\kappa}\,\text{erf}\,\sqrt{\kappa}}{\sqrt{s}}$	$\dfrac{\sinh(2\sqrt{at})}{\sqrt{\pi t}}$
	$\dfrac{e^{\lambda}\,\text{erfc}\,\sqrt{\lambda}}{s}$	$\text{erf}\,(at)$		$\dfrac{e^{\kappa}\,\text{erfc}\,\sqrt{\kappa}}{\sqrt{s}}$	$\dfrac{\exp(-2\sqrt{at})}{\sqrt{\pi t}}$

a, b, c = constants	$n = 1, 2, 3, \ldots$
$A = e^{-as}$	$C = e^{-cx}$

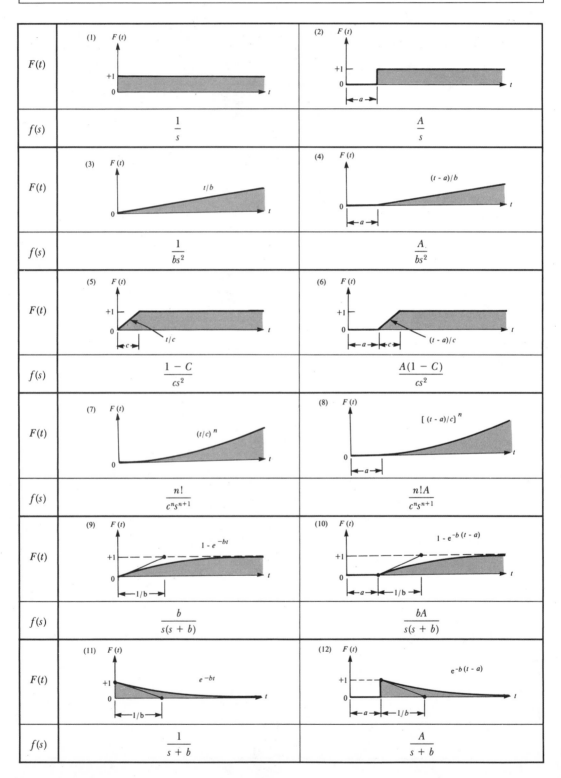

a, b = constants $\qquad\qquad k = 0, 1, 2, \ldots \qquad\qquad\qquad n = 1, 2, 3, \ldots$

$A = e^{-as}$ $\qquad\qquad\qquad\qquad\qquad\qquad\qquad\qquad B = e^{-bs}$

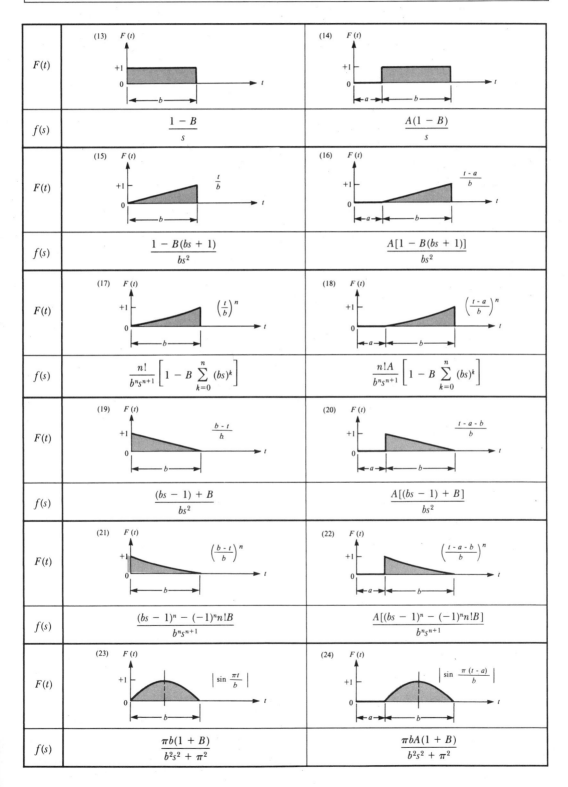

$F(t)$	(13)	(14)
$f(s)$	$\dfrac{1 - B}{s}$	$\dfrac{A(1 - B)}{s}$
$F(t)$	(15) $\dfrac{t}{b}$	(16) $\dfrac{t - a}{b}$
$f(s)$	$\dfrac{1 - B(bs + 1)}{bs^2}$	$\dfrac{A[1 - B(bs + 1)]}{bs^2}$
$F(t)$	(17) $\left(\dfrac{t}{b}\right)^n$	(18) $\left(\dfrac{t - a}{b}\right)^n$
$f(s)$	$\dfrac{n!}{b^n s^{n+1}}\left[1 - B\sum_{k=0}^{n}(bs)^k\right]$	$\dfrac{n!A}{b^n s^{n+1}}\left[1 - B\sum_{k=0}^{n}(bs)^k\right]$
$F(t)$	(19) $\dfrac{b - t}{b}$	(20) $\dfrac{t - a - b}{b}$
$f(s)$	$\dfrac{(bs - 1) + B}{bs^2}$	$\dfrac{A[(bs - 1) + B]}{bs^2}$
$F(t)$	(21) $\left(\dfrac{b - t}{b}\right)^n$	(22) $\left(\dfrac{t - a - b}{b}\right)^n$
$f(s)$	$\dfrac{(bs - 1)^n - (-1)^n n! B}{b^n s^{n+1}}$	$\dfrac{A[(bs - 1)^n - (-1)^n n! B]}{b^n s^{n+1}}$
$F(t)$	(23) $\left\lvert \sin\dfrac{\pi t}{b} \right\rvert$	(24) $\left\lvert \sin\dfrac{\pi(t - a)}{b} \right\rvert$
$f(s)$	$\dfrac{\pi b(1 + B)}{b^2 s^2 + \pi^2}$	$\dfrac{\pi b A(1 + B)}{b^2 s^2 + \pi^2}$

a, b, c, d = constants

$A = e^{-as}$ $\qquad$ $B = e^{-bs}$ $\qquad$ $C = e^{-cs}$ $\qquad$ $D = e^{-ds}$

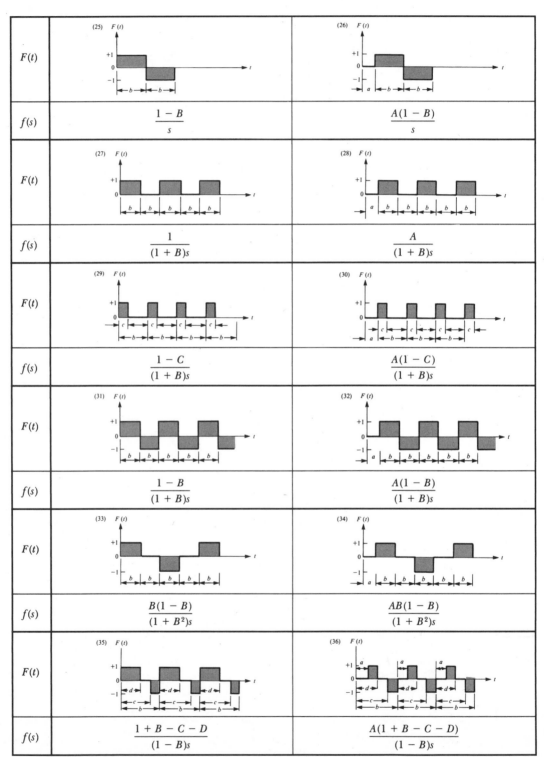

	(25)	(26)
$F(t)$		
$f(s)$	$\dfrac{1 - B}{s}$	$\dfrac{A(1 - B)}{s}$
	(27)	(28)
$F(t)$		
$f(s)$	$\dfrac{1}{(1 + B)s}$	$\dfrac{A}{(1 + B)s}$
	(29)	(30)
$F(t)$		
$f(s)$	$\dfrac{1 - C}{(1 + B)s}$	$\dfrac{A(1 - C)}{(1 + B)s}$
	(31)	(32)
$F(t)$		
$f(s)$	$\dfrac{1 - B}{(1 + B)s}$	$\dfrac{A(1 - B)}{(1 + B)s}$
	(33)	(34)
$F(t)$		
$f(s)$	$\dfrac{B(1 - B)}{(1 + B^2)s}$	$\dfrac{AB(1 - B)}{(1 + B^2)s}$
	(35)	(36)
$F(t)$		
$f(s)$	$\dfrac{1 + B - C - D}{(1 - B)s}$	$\dfrac{A(1 + B - C - D)}{(1 - B)s}$

a, b, c, d = constants		$k, n = 1, 2, 3, \ldots$
$A = e^{-as}$	$B = e^{-bs}$ $D = e^{-ds}$	$N = e^{-2ns}$

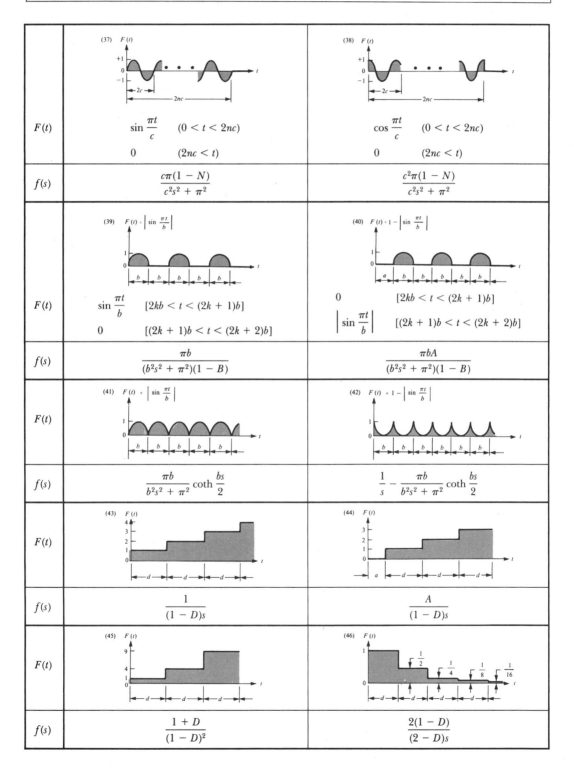

	(37)	(38)
$F(t)$	$\sin \dfrac{\pi t}{c} \quad (0 < t < 2nc)$ $0 \qquad (2nc < t)$	$\cos \dfrac{\pi t}{c} \quad (0 < t < 2nc)$ $0 \qquad (2nc < t)$
$f(s)$	$\dfrac{c\pi(1 - N)}{c^2 s^2 + \pi^2}$	$\dfrac{c^2 \pi(1 - N)}{c^2 s^2 + \pi^2}$
	(39) $F(t) = \left\| \sin \dfrac{\pi t}{b} \right\|$	(40) $F(t) = 1 - \left\| \sin \dfrac{\pi t}{b} \right\|$
$F(t)$	$\sin \dfrac{\pi t}{b} \quad [2kb < t < (2k+1)b]$ $0 \qquad [(2k+1)b < t < (2k+2)b]$	$0 \qquad [2kb < t < (2k+1)b]$ $\left\| \sin \dfrac{\pi t}{b} \right\| \quad [(2k+1)b < t < (2k+2)b]$
$f(s)$	$\dfrac{\pi b}{(b^2 s^2 + \pi^2)(1 - B)}$	$\dfrac{\pi b A}{(b^2 s^2 + \pi^2)(1 - B)}$
	(41) $F(t) = \left\| \sin \dfrac{\pi t}{b} \right\|$	(42) $F(t) = 1 - \left\| \sin \dfrac{\pi t}{b} \right\|$
$F(t)$		
$f(s)$	$\dfrac{\pi b}{b^2 s^2 + \pi^2} \coth \dfrac{bs}{2}$	$\dfrac{1}{s} - \dfrac{\pi b}{b^2 s^2 + \pi^2} \coth \dfrac{bs}{2}$
	(43)	(44)
$F(t)$		
$f(s)$	$\dfrac{1}{(1 - D)s}$	$\dfrac{A}{(1 - D)s}$
	(45)	(46)
$F(t)$		
$f(s)$	$\dfrac{1 + D}{(1 - D)^2}$	$\dfrac{2(1 - D)}{(2 - D)s}$

a, b, c = constants		
$A = e^{-as}$	$a = 2c - b$	$C = e^{-cs}$

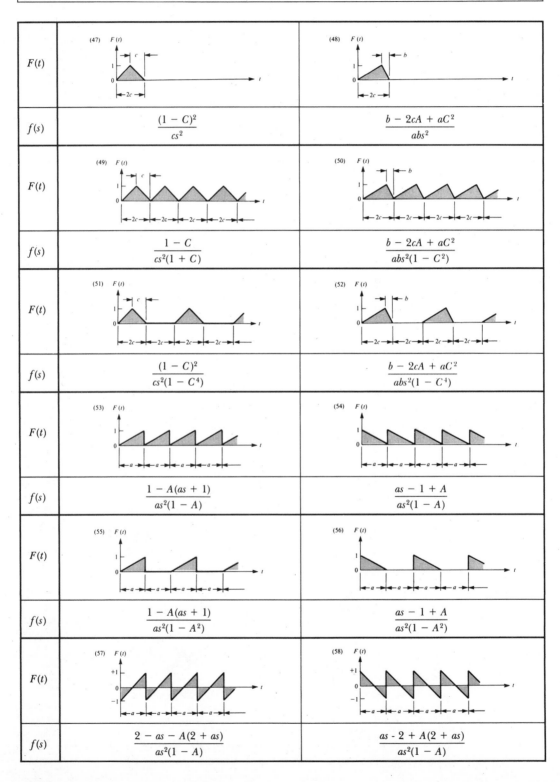

	(47)	(48)
$F(t)$		
$f(s)$	$\dfrac{(1 - C)^2}{cs^2}$	$\dfrac{b - 2cA + aC^2}{abs^2}$
	(49)	(50)
$F(t)$		
$f(s)$	$\dfrac{1 - C}{cs^2(1 + C)}$	$\dfrac{b - 2cA + aC^2}{abs^2(1 - C^2)}$
	(51)	(52)
$F(t)$		
$f(s)$	$\dfrac{(1 - C)^2}{cs^2(1 - C^4)}$	$\dfrac{b - 2cA + aC^2}{abs^2(1 - C^4)}$
	(53)	(54)
$F(t)$		
$f(s)$	$\dfrac{1 - A(as + 1)}{as^2(1 - A)}$	$\dfrac{as - 1 + A}{as^2(1 - A)}$
	(55)	(56)
$F(t)$		
$f(s)$	$\dfrac{1 - A(as + 1)}{as^2(1 - A^2)}$	$\dfrac{as - 1 + A}{as^2(1 - A^2)}$
	(57)	(58)
$F(t)$		
$f(s)$	$\dfrac{2 - as - A(2 + as)}{as^2(1 - A)}$	$\dfrac{as - 2 + A(2 + as)}{as^2(1 - A)}$

$a, b, c, d \neq$ constants	$a + b = 2c$	$k = 0, 1, 2, \ldots$	$n = 1, 2, 3, \ldots$
$A = e^{-as}$	$B = e^{-bs}$	$C = e^{-cs}$	$D = e^{-ds}$

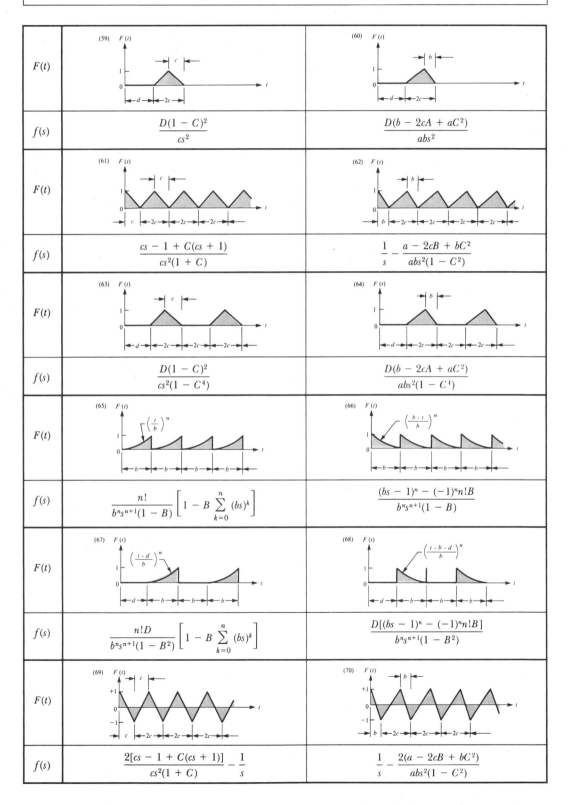

$F(t)$	(59)	(60)
$f(s)$	$\dfrac{D(1 - C)^2}{cs^2}$	$\dfrac{D(b - 2cA + aC^2)}{abs^2}$
$F(t)$	(61)	(62)
$f(s)$	$\dfrac{cs - 1 + C(cs + 1)}{cs^2(1 + C)}$	$\dfrac{1}{s} - \dfrac{a - 2cB + bC^2}{abs^2(1 - C^2)}$
$F(t)$	(63)	(64)
$f(s)$	$\dfrac{D(1 - C)^2}{cs^2(1 - C^4)}$	$\dfrac{D(b - 2cA + aC^2)}{abs^2(1 - C^4)}$
$F(t)$	(65) $\left(\dfrac{t}{b}\right)^n$	(66) $\left(\dfrac{b - t}{b}\right)^n$
$f(s)$	$\dfrac{n!}{b^n s^{n+1}(1 - B)}\left[1 - B\displaystyle\sum_{k=0}^{n}(bs)^k\right]$	$\dfrac{(bs - 1)^n - (-1)^n n! B}{b^n s^{n+1}(1 - B)}$
$F(t)$	(67) $\left(\dfrac{t - d}{b}\right)^n$	(68) $\left(\dfrac{t - b - d}{b}\right)^n$
$f(s)$	$\dfrac{n!D}{b^n s^{n+1}(1 - B^2)}\left[1 - B\displaystyle\sum_{k=0}^{n}(bs)^k\right]$	$\dfrac{D[(bs - 1)^n - (-1)^n n! B]}{b^n s^{n+1}(1 - B^2)}$
$F(t)$	(69)	(70)
$f(s)$	$\dfrac{2[cs - 1 + C(cs + 1)]}{cs^2(1 + C)} - \dfrac{1}{s}$	$\dfrac{1}{s} - \dfrac{2(a - 2cB + bC^2)}{abs^2(1 - C^2)}$

$$\boxed{MY' + NY = g(x)}$$

$$Y(+0) = \text{initial value} \qquad M, N = \text{signed numbers} \qquad p = \frac{N}{M} \qquad \lambda = \left|\frac{N}{M}\right|$$

(1) General Solution

$$\boxed{Y(x) = Y(+0)L_1(x) + G(x)}$$

(2) Shape Functions $(j = 1, 2, \dots, k = 2, 3, \dots)$

$L_j(x)$	$p > 0$	$p = 0$	$p < 0$
$L_1(x)$	$e^{-\lambda x}$	1	$e^{\lambda x}$
$L_2(x)$	$-\dfrac{1}{\lambda}(e^{-\lambda x} - 1)$	x	$\dfrac{1}{\lambda}(e^{\lambda x} - 1)$
$L_3(x)$	$\dfrac{1}{\lambda^2}(e^{-\lambda x} + \lambda x - 1)$	$\dfrac{x^2}{2!}$	$\dfrac{1}{\lambda^2}(e^{\lambda x} - \lambda x - 1)$
$L_k(x)$	$-\dfrac{1}{\lambda}\left[L_{k-1}(x) - \dfrac{(\lambda x)^{k-2}}{(k-2)!}\right]$	$\dfrac{x^{k-1}}{(k-1)!}$	$\dfrac{1}{\lambda}\left[L_{k-1}(x) - \dfrac{(\lambda x)^{k-2}}{(k-2)!}\right]$

(3) Shape Coefficients

$$L_j(x) = L_j \qquad L_j(x - a) = A_j \qquad L_j(x - b) = B_j \qquad L_j(x - d) = D_j$$

For example,

$$B_3 = \begin{cases} \dfrac{1}{\lambda^2}[e^{-\lambda(x-b)} + \lambda(x - b) - 1] & \text{if } p > 0 \\[2mm] \tfrac{1}{2}(x - b)^2 & \text{if } p = 0 \\[2mm] \dfrac{1}{\lambda^2}[e^{\lambda(x-b)} - \lambda(x - b) - 1] & \text{if } p > 0 \end{cases}$$

(4) Convolution Integral

$$G(x) = \frac{1}{M}\int_0^x g(x - \tau)L_1(\tau)\, d\tau$$

Particular forms of convolution integrals are tabulated in Sec. 16.36.

(5) Application

$$\boxed{5Y' + 10Y = g(x)} \qquad \boxed{Y(+0) = 7 \qquad g(x) = \begin{cases} 20x/10 & 0 < x \le 10 \\ 0 & 10 < x \end{cases}}$$

$$p = 10 > 0 \qquad \lambda = \left|\frac{10}{5}\right| = 2 \qquad g(x) \text{ given in Sec. 16.36-2, } w = 20,\ a = 10$$

$$Y(x) = Y(+0)L_1 + \frac{w}{Ma}L_3 = 7e^{-2x} + \tfrac{20}{50}[\tfrac{1}{4}(e^{-2x} + 2x - 1)] \qquad 0 < x \le 10$$

$$Y(x) = Y(+0)L_1 + \frac{w}{Ma}(L_3 - A_3 - aA_2)$$

$$= 7e^{-2x} + \tfrac{20}{50}\{\tfrac{1}{4}(e^{-2x} + 2x - 1) - \tfrac{1}{4}[e^{-2(x-10)} + 2(x - 10) - 1]$$

$$- \tfrac{10}{2}(e^{-2(x-10)} - 1)\} \qquad 10 < x$$

$$MY' + NY = g(x)$$

The analytical expressions for $G(x)$ below apply for all solutions given in Sec. 16.35, provided that the respective shape coefficients are used.

$a, b, c, d, w, \alpha, \beta, P$ = real numbers A_j, B_j, D_j, L_j = shape coefficients (Sec. 16.35-3)

$g(x)$	$G(x) = \dfrac{1}{M} \int_0^x g(x - \tau)L_1(\tau)\, d\tau$	
(1)	$\dfrac{w}{M} L_2$	$0 < x \le a$
	$\dfrac{w}{M}(L_2 - A_2)$	$a < x$
(2)	$\dfrac{w}{Ma} L_3$	$0 < x \le a$
	$\dfrac{w}{Ma}(L_3 - A_3 - aA_2)$	$a < x$
(3)	$\dfrac{2w}{Ma^2} L_4$	$0 < x \le a$
	$\dfrac{2w}{Ma^2}(L_4 - A_4 - aA_3 - \tfrac{1}{2}a^2A_2)$	$a < x$
(4) $n = 1, 2, \ldots$	$\dfrac{n!w}{Ma^n} L_{n+2}$	$0 < x \le a$
	$\dfrac{n!w}{Ma^n}\left(L_{n+2} - \sum_{k=0}^{n} \dfrac{a^k}{k!} A_{n+2-k}\right)$	$a < x$
(5)	$\dfrac{w}{Ma}(aL_2 - L_3)$	$0 < x \le a$
	$\dfrac{w}{Ma}(aL_2 - L_3 + A_3)$	$a < x$
(6)	$\dfrac{2w}{Ma^2}(L_4 - aL_3 + \tfrac{1}{2}a^2L_2)$	$0 < x \le a$
	$\dfrac{2w}{Ma^2}(L_4 - aL_3 + \tfrac{1}{2}a^2L_2 - A_4)$	$a < x$
(7) $n = 1, 2, \ldots$	$\dfrac{n!w}{Ma^n}\left[\sum_{k=0}^{n} \dfrac{(-a)^k}{k!} L_{n+2-k}\right]$	$0 < x \le a$
	$-\dfrac{n!w}{Ma^n}\left[\sum_{k=0}^{n} \dfrac{(-a)^k}{k!} L_{n+2-k} + (-1)^n A_{n+2}\right]$	$a < x$

$$MY' + NY = g(x)$$

$g(x)$	$G(x) = \dfrac{1}{M}\displaystyle\int_0^x g(x-\tau)L_1(\tau)\,d\tau$	
(8)	$\dfrac{w}{M}A_2$	$a < x \le b$
	$\dfrac{w}{M}(A_2 - B_2)$	$b < x$
(9) $w(x-a)c$, $c = b - a$	$\dfrac{w}{Mc}A_3$	$a < x \le b$
	$\dfrac{w}{Mc}(A_3 - B_3 - cB_2)$	$b < x$
(10) $w(x-a)^2/c^2$, $c = b - a$	$\dfrac{2w}{Mc^2}A_4$	$a < x \le b$
	$\dfrac{2w}{Mc^2}(A_4 - B_4 - cB_3 - \tfrac{1}{2}c^2 B_2)$	$b < x$
(11) $w(x-a)^n/c^n$, $c = b - a$, $n = 1, 2, \ldots$	$\dfrac{n!\,w}{Mc^n}A_{n+2}$	$a < x \le b$
	$\dfrac{n!\,w}{Mc^n}\left(A_{n+2} - \displaystyle\sum_{k=0}^{n}\dfrac{c^k}{k!}B_{n+2-k}\right)$	$b < x$
(12) $w(b-x)/c$, $c = b - a$	$\dfrac{w}{Mc}(cA_2 - A_3)$	$a < x \le b$
	$\dfrac{w}{Mc}(cA_2 - A_3 + B_3)$	$b < x$
(13) $w(b-x)^2/c^2$, $c = b - a$	$\dfrac{2w}{Mc^2}(A_4 - cA_3 + \tfrac{1}{2}c^2 A_2)$	$a < x \le b$
	$\dfrac{2w}{Mc^2}(A_4 - cA_3 + \tfrac{1}{2}c^2 A_2 - B_4)$	$b < x$
(14) $w(b-x)^n/c^n$, $c = b - a$, $n = 1, 2, \ldots$	$\dfrac{n!\,w}{Mc^n}\left[\displaystyle\sum_{k=0}^{n}\dfrac{(-c)^k}{k!}A_{n+2-k}\right]$	$a < x \le b$
	$-\dfrac{n!\,w}{Mc^n}\left[\displaystyle\sum_{k=0}^{n}\dfrac{(-c)^k}{k!}A_{n+2-k} + (-1)^n B_{n+2}\right]$	$b < x$

$G(x) = 0 \qquad 0 < x \le a$

16.36 FIRST-ORDER DIFFERENTIAL EQUATION
CONVOLUTION INTEGRALS (Continued) 261

$$MY' + NY = g(x)$$

$g(x)$	$G(x) = \dfrac{1}{M}\int_0^x g(x-\tau)L_1(\tau)\,d\tau$	
(15)	$\dfrac{w}{Mc}A_3$	$a < x \le b$
$w(x-a)c$ $c = b-a$	$\dfrac{w}{Mc}(A_3 - B_3)$	$b < x \le d$
	$\dfrac{w}{Mc}(A_3 - B_3 - cD_2)$	$d < x$
(16)	$\dfrac{w}{M}A_2$	$a < x \le b$
$w(d-x)/c$ $c = d-b$	$\dfrac{w}{Mc}(cA_2 - B_3)$	$b < x \le d$
	$\dfrac{w}{Mc}(cA_2 - B_3 + D_3)$	$d < x$
(17)	$\dfrac{w}{Mc}A_3$	$a < x \le b$
$w(d-x)/h$ $c = b-a$	$\dfrac{w}{Mc}\left(A_3 - B_3 - \dfrac{c}{h}B_3\right)$	$b < x \le d$
	$\dfrac{w}{Mc}\left(A_3 - B_3 - \dfrac{c}{h}B_3 + \dfrac{c}{h}D_3\right)$	$d < x$
(18)	$\dfrac{w}{M}A_2$	$a < x \le b$
w w	$\dfrac{w}{M}(A_2 - 2B_2)$	$b < x \le d$
	$\dfrac{w}{M}(A_2 - 2B_2 + D_2)$	$d < x$
(19)	$\dfrac{w}{Mc}A_3$	$a < x \le b$
$w(x-a)/c$ $c = b-a$	$\dfrac{w}{Mc}(A_3 - 2cB_2)$	$b < x \le d$
	$\dfrac{w}{Mc}(A_3 - 2cB_2 + D_3)$	$d < x$
(20)	$\dfrac{w}{Mc}(cA_2 - A_3)$	$a < x \le d$
$w(b-x)/c$ $c = b-a$	$\dfrac{w}{Mc}(cA_2 - A_3 + cD_2 + D_3)$	$d < x$

$0 < x \le a$ $G(x) = 0$

$$MY' + NY = g(x)$$

$g(x)$	$G(x) = \dfrac{1}{M}\displaystyle\int_0^x g(x-\tau)L_1(\tau)\,d\tau$	
(21)	$\dfrac{w}{Ma}L_3$	$0 < x \le a$
	$\dfrac{w}{Ma}(L_3 - A_3)$	$a < x \le b$
	$\dfrac{w}{Ma}(L_3 - A_3 - aB_2)$	$b < x$
(22)	$\dfrac{w}{M}L_2$	$0 < x \le a$
	$\dfrac{w}{Mc}(cL_2 - A_3)$	$a < x \le b$
	$\dfrac{w}{Mc}(cL_2 - A_3 + B_3)$	$b < x$
(23)	$\dfrac{w}{Ma}L_3$	$0 < x \le a$
	$\dfrac{w}{Ma}\left(L_3 - \dfrac{b}{c}A_3\right)$	$a < x \le b$
	$\dfrac{w}{Ma}\left(L_3 - \dfrac{b}{c}A_3 + \dfrac{a}{c}B_3\right)$	$b < x$
(24)	$\dfrac{w}{M}L_2$	$0 < x \le a$
	$\dfrac{w}{M}(L_2 - 2A_2)$	$a < x \le b$
	$\dfrac{w}{M}(L_2 - 2A_2 + B_2)$	$b < x$
(25)	$\dfrac{w}{Ma}L_3$	$0 < x \le a$
	$\dfrac{w}{Ma}(L_3 - 2aA_2)$	$a < x \le b$
	$\dfrac{w}{Ma}(L_3 - 2aA_2 + B_3)$	$b < x$
(26)	$\dfrac{w}{Ma}(aL_2 - L_3)$	$0 < x \le b$
	$\dfrac{w}{Ma}(aL_2 - L_3 + aB_2 + B_3)$	$b < x$

$$MY' + NY = g(x)$$

$g(x)$	$G(x) = \dfrac{1}{M}\int_0^x g(x-\tau)L_1(\tau)\,d\tau$	
(27)	0	$0 < x \le a$
	$\dfrac{P}{M}A_1$	$a < x$
(28) $w \sin \beta x$, $\beta = \pi/b$	$\dfrac{w[(N/M)\sin\beta x = \beta(\cos\beta x - L_1)]}{M[(N/M)^2 + \beta^2]}$	$0 < x$
(29) $w \cos \beta x$, $\beta = \pi/b$	$\dfrac{w[(N/M)(\cos\beta x - L_1) + \beta\sin\beta x]}{M[(N/M)^2 + \beta^2]}$	$0 < x$
(30) $we^{-\alpha x}\sin\beta x$, $\beta = \pi/b$	$\dfrac{w\{[(N/M)-\alpha]\sin\beta x - \beta(\cos\beta x - L_1)\}}{M\{[(N/M)-\alpha]^2 + \beta^2\}}$	$0 < x$
(31) $we^{-\alpha x}\cos\beta x$, $\beta = \pi/b$	$\dfrac{w\{\beta\sin\beta x + [(N/M)-\alpha](\cos\beta x - L_1)\}}{M\{[(N/M)-\alpha]^2 + \beta^2\}}$	$0 < x$
(32) $we^{-\alpha x}$	$\dfrac{w(e^{-\alpha x} - L_1)}{M[(N/M)-\alpha]}$	$0 < x$
(33) $w(1-e^{-\alpha x})$	$\dfrac{w}{M}L_2 - \dfrac{w(e^{-\alpha x} - L_1)}{M[(N/M)-\alpha]}$	$0 < x$

$$MY'' + NY = g(x)$$

$$Y(+0), Y'(+0) = \text{initial values} \qquad M, N = \text{signed numbers} \qquad p = \frac{N}{M} \qquad \lambda = \sqrt{\left|\frac{N}{M}\right|}$$

(1) General Solution

$$Y(x) = Y(+0)L_1(x) + Y'(+0)L_2(x) + G(x)$$

(2) Shape Functions $(j = 1, 2, \ldots, k = 3, 4, \ldots)$

$L_j(x)$	$p > 0$	$p = 0$	$p < 0$
$L_1(x)$	$\cos \lambda x$	1	$\cosh \lambda x$
$L_2(x)$	$\dfrac{1}{\lambda} \sin \lambda x$	x	$\dfrac{1}{\lambda} \sinh \lambda x$
$L_3(x)$	$-\dfrac{1}{\lambda^2}(\cos \lambda x - 1)$	$\dfrac{x^2}{2!}$	$\dfrac{1}{\lambda^2}(\cosh \lambda x - 1)$
$L_4(x)$	$-\dfrac{1}{\lambda^3}(\sin \lambda x - \lambda x)$	$\dfrac{x^3}{3!}$	$\dfrac{1}{\lambda^3}(\sinh \lambda x - \lambda x)$
$L_5(x)$	$\dfrac{1}{\lambda^4}(\cos \lambda x + \frac{1}{2}\lambda^2 x^2 - 1)$	$\dfrac{x^4}{4!}$	$\dfrac{1}{\lambda^4}(\cosh \lambda x - \frac{1}{2}\lambda^2 x^2 - 1)$
$L_k(x)$	$-\dfrac{1}{\lambda^2}\left[L_{k-2}(x) - \dfrac{(\lambda x)^{k-3}}{(k-3)!}\right]$	$\dfrac{x^{k-1}}{(k-1)!}$	$\dfrac{1}{\lambda^2}\left[L_{k-2}(x) - \dfrac{(\lambda x)^{k-3}}{(k-3)!}\right]$

(3) Shape Coefficients

$$L_j(x) = L_j \qquad L_j(x - a) = A_j \qquad L_j(x - b) = B_j \qquad L_j(x - d) = D_j$$

For example,

$$A_4 = \begin{cases} -\dfrac{1}{\lambda^3}[\sin \lambda(x - a) - \lambda(x - a)] & \text{if } p > 0 \\[2mm] \frac{1}{6}(x - a)^3 & \text{if } p = 0 \\[2mm] \dfrac{1}{\lambda^3}[\sinh \lambda (x - a) - \lambda(x - a)] & \text{if } p < 0 \end{cases}$$

(4) Convolution Integral

$$G(x) = \frac{1}{M} \int_0^x g(x - \tau)L_2(\tau)\, d\tau$$

Particular forms of convolution integrals are tabulated in Sec. 16.38.

(5) Transport Matrix Equations (for all p)

$$\begin{bmatrix} Y(x) \\ Y'(x) \end{bmatrix} = \begin{bmatrix} L_1 & L_2 \\ -pL_2 & L_1 \end{bmatrix}\begin{bmatrix} Y(+0) \\ Y'(+0) \end{bmatrix} + \begin{bmatrix} G(x) \\ G'(x) \end{bmatrix}$$

$$\boxed{MY'' + NY = g(x)}$$

The analytical expressions for $G(x)$ below apply for all solutions given in Sec. 16.37, provided that the respective shape coefficients are used.

$a, b, c, d, w, \alpha, \beta, P$ = real numbers

$g(x)$	$G(x) = \dfrac{1}{M}\displaystyle\int_0^x g(x-\tau)L_2(\tau)\,d\tau$	
(1)	$\dfrac{w}{M}L_3$	$0 < x \le a$
	$\dfrac{w}{M}(L_3 - A_3)$	$a < x$
(2)	$\dfrac{w}{Ma}L_4$	$0 < x \le a$
	$\dfrac{w}{Ma}(L_4 - A_4 - aA_3)$	$a < x$
(3)	$\dfrac{2w}{Ma^2}L_5$	$0 < x \le a$
	$\dfrac{2w}{Ma^2}(L_5 - A_5 - aA_4 - \tfrac{1}{2}a^2A_3)$	$a < x$
(4) $n = 1, 2, \ldots$	$\dfrac{n!\,w}{Ma^n}L_{n+3}$	$0 < x \le a$
	$\dfrac{n!\,w}{Ma^n}\left(L_{n+3} - \displaystyle\sum_{k=0}^{n}\dfrac{a^k}{k!}A_{n+3-k}\right)$	$a < x$
(5)	$\dfrac{w}{Ma}(aL_3 - L_4)$	$0 < x \le a$
	$\dfrac{w}{Ma}(aL_3 - L_4 + A_4)$	$a < x$
(6)	$\dfrac{2w}{Ma^2}(L_5 - aL_4 + \tfrac{1}{2}a^2L_3)$	$0 < x \le a$
	$\dfrac{2w}{Ma^2}(L_5 - aL_4 + \tfrac{1}{2}a^2L_3 - A_5)$	$a < x$
(7) $n = 1, 2, \ldots$	$\dfrac{n!\,w}{Ma^n}\left[\displaystyle\sum_{k=0}^{n}\dfrac{(-a)^k}{k!}L_{n+3-k}\right]$	$0 < x \le a$
	$-\dfrac{n!\,w}{Ma^n}\left[\displaystyle\sum_{k=0}^{n}\dfrac{(-a)^k}{k!}L_{n+3-k} + (-1)^n A_{n+3}\right]$	$a < x$

$$MY'' + NY = g(x)$$

$g(x)$	$G(x) = \dfrac{1}{M}\displaystyle\int_0^x g(x-\tau)L_2(\tau)\,d\tau$	
(8)	$\dfrac{w}{M}A_3$	$a < x \le b$
	$\dfrac{w}{M}(A_3 - B_3)$	$b < x$
(9) $c = b - a$	$\dfrac{w}{Mc}A_4$	$a < x \le b$
	$\dfrac{w}{Mc}(A_4 - B_4 - cB_3)$	$b < x$
(10) $c = b - a$	$\dfrac{2w}{Mc^2}A_5$	$a < x \le b$
	$\dfrac{2w}{Mc^2}(A_5 - B_5 - cB_4 - \tfrac{1}{2}c^2 B_3)$	$b < x$
(11) $c = b - a$ $n = 1, 2, \ldots$	$\dfrac{n!\,w}{Mc^n}A_{n+3}$	$a < x \le b$
	$\dfrac{n!\,w}{Mc^n}\left(A_{n+3} - \displaystyle\sum_{k=0}^{n}\dfrac{c^k}{k!}B_{n+3-k}\right)$	$b < x$
(12) $c = b - a$	$\dfrac{w}{Mc}(cA_3 - A_4)$	$a < x \le b$
	$\dfrac{w}{Mc}(cA_3 - A_4 + B_4)$	$b < x$
(13) $c = b - a$	$\dfrac{2w}{Mc^2}(A_5 - cA_4 + \tfrac{1}{2}c^2 A_3)$	$a < x \le b$
	$\dfrac{2w}{Mc^2}(A_5 - cA_4 + \tfrac{1}{2}c^2 A_3 - B_5)$	$b < x$
(14) $c = b - a$ $n = 1, 2, \ldots$	$\dfrac{n!\,w}{Mc^n}\left[\displaystyle\sum_{k=0}^{n}\dfrac{(-c)^k}{k!}A_{n+3-k}\right]$	$a < x \le b$
	$-\dfrac{n!\,w}{Mc^n}\left[\displaystyle\sum_{k=0}^{n}\dfrac{(-c)^k}{k!}A_{n+3-k} + (-1)^n B_{n+3}\right]$	$b < x$

$G(x) = 0 \qquad 0 < x \le a$

$$MY'' + NY = g(x)$$

$g(x)$	$G(x) = \dfrac{1}{M}\displaystyle\int_0^x g(x-\tau)L_2(\tau)\,d\tau$	
(15)	$\dfrac{w}{Mc}A_4$	$a < x \le b$
	$\dfrac{w}{Mc}(A_4 - B_4)$	$b < x \le d$
	$\dfrac{w}{Mc}(A_4 - B_4 - cD_3)$	$d < x$
(16)	$\dfrac{w}{M}A_3$	$a < x \le b$
	$\dfrac{w}{Mc}(cA_3 - B_4)$	$b < x \le d$
	$\dfrac{w}{Mc}(cA_3 - B_4 + D_4)$	$d < x$
(17)	$\dfrac{w}{Mc}A_4$	$a < x \le b$
	$\dfrac{w}{Mc}\left(A_4 - B_4 - \dfrac{c}{h}B_4\right)$	$b < x \le d$
	$\dfrac{w}{Mc}\left(A_4 - B_4 - \dfrac{c}{h}B_4 + \dfrac{c}{h}D_4\right)$	$d < x$
(18)	$\dfrac{w}{M}A_3$	$a < x \le b$
	$\dfrac{w}{M}(A_3 - 2B_3)$	$b < x \le d$
	$\dfrac{w}{M}(A_3 - 2B_3 + D_3)$	$d < x$
(19)	$\dfrac{w}{Mc}A_4$	$a < x \le b$
	$\dfrac{w}{Mc}(A_4 - 2cB_3)$	$b < x \le d$
	$\dfrac{w}{Mc}(A_4 - 2cB_3 + D_4)$	$d < x$
(20)	$\dfrac{w}{Mc}(cA_3 - A_4)$	$a < x \le d$
	$\dfrac{w}{Mc}(cA_3 - A_4 + cD_3 + D_4)$	$d < x$

$$G(x) = 0 \qquad 0 < x \le a$$

$$MY'' + NY = g(x)$$

$g(x)$	$G(x) = \dfrac{1}{M}\displaystyle\int_0^x g(x-\tau)L_2(\tau)\,d\tau$	
(21)	$\dfrac{w}{Ma}L_1$	$0 < x \le a$
	$\dfrac{w}{Ma}(L_1 - A_1)$	$a < x \le b$
	$\dfrac{w}{Ma}(L_1 - A_1 - aB_3)$	$b < x$
(22)	$\dfrac{w}{M}L_3$	$0 < x \le a$
	$\dfrac{w}{Mc}(cL_3 - A_1)$	$a < x \le b$
	$\dfrac{w}{Mc}(cL_3 - A_1 + B_1)$	$b < x$
(23)	$\dfrac{w}{Ma}L_1$	$0 < x \le a$
	$\dfrac{w}{Ma}\left(L_1 - \dfrac{b}{c}A_1\right)$	$a < x \le b$
	$\dfrac{w}{Ma}\left(L_1 - \dfrac{b}{c}A_1 + \dfrac{a}{c}B_1\right)$	$b < x$
(24)	$\dfrac{w}{M}L_3$	$0 < x \le a$
	$\dfrac{w}{M}(L_3 - 2A_3)$	$a < x \le b$
	$\dfrac{w}{M}(L_3 - 2A_3 + B_3)$	$b < x$
(25)	$\dfrac{w}{Ma}L_1$	$0 < x \le a$
	$\dfrac{w}{Ma}(L_1 - 2aA_3)$	$a < x \le b$
	$\dfrac{w}{Ma}(L_1 - 2aA_3 + B_1)$	$b < x$
(26)	$\dfrac{w}{Ma}(aL_3 - L_1)$	$0 < x \le b$
	$\dfrac{w}{Ma}(aL_3 - L_1 + aB_3 + B_1)$	$b < x$

$$\boxed{MY'' + NY = g(x)}$$

$p = \dfrac{N}{M}$	$u = p + \alpha^2 - \beta^2$	$v = 2\alpha\beta$
$\psi_1 = \tan^{-1}\dfrac{v}{u}$	$F_1 = \sin\psi_1$	$F_2 = \sqrt{\alpha^2 + \beta^2}\,\sin(\psi_1 - \psi_2)$
$\psi_2 = \tan^{-1}\dfrac{\beta}{\alpha}$	$G_1 = \cos\psi_1$	$G_2 = \sqrt{\alpha^2 + \beta^2}\,\cos(\psi_1 - \psi_2)$

$g(x)$	$G(x) = \dfrac{1}{M}\displaystyle\int_0^x g(x - \tau)L_2(\tau)\,d\tau$
(27)	$0 \qquad\qquad\qquad\qquad 0 < x \le a$ $\dfrac{P}{M}A_2 \qquad\qquad\qquad a < x$
(28) $w\sin\beta x$ $\beta = \pi/b$	$\dfrac{w}{M(p - \beta^2)}(\sin\beta x - \beta L_2)$ $0 < x$
(29) $w\cos\beta x$ $\beta = \pi/b$	$\dfrac{w}{M(p - \beta^2)}(\cos\beta x - L_1)$ $0 < x$
(30) $we^{-\alpha x}\sin\beta x$ $\beta = \pi/b$	$\dfrac{w}{M\sqrt{u^2 + v^2}}[e^{-\alpha x}\sin(\beta x + \psi_1) - F_1 L_1 + F_2 L_2]$ $0 < x$
(31) $we^{-\alpha x}\cos\beta x$ $\beta = \pi/b$	$\dfrac{w}{M\sqrt{u^2 + v^2}}[e^{-\alpha x}\cos(\beta x + \psi_1) - G_1 L_1 + G_2 L_2]$ $0 < x$
(32) $we^{-\alpha x}$	$\dfrac{w}{M(p + \alpha^2)}(e^{-\alpha x} - L_1 + \alpha L_2)$ $0 < x$

$$MY'' + NY' + RY = g(x)$$

$Y(+0), Y'(+0) = $ initial values	$M, N, R = $ signed numbers	$p = \dfrac{N}{M}$ $\qquad q = \dfrac{R}{M}$
$\lambda = \frac{1}{2}p$	$\Delta = p^2 - 4q$	$\omega = \frac{1}{2}\sqrt{|\Delta|}$

(1) General Solution

$$Y(x) = Y(+0)L_1(x) + Y'(+0)L_2(x) + G(x)$$

(2) Shape Functions $(j = 1, 2, \ldots)$

$L_j(x)$	$\Delta < 0$	$\Delta = 0$	$\Delta > 0$
$L_1(x)$	$e^{-\lambda x}\left(\cos \omega x + \dfrac{\lambda}{\omega}\sin \omega x\right)$	$(1 + \lambda x)e^{-\lambda x}$	$e^{-\lambda x}\left(\cosh \omega x + \dfrac{\lambda}{\omega}\sinh \omega x\right)$
$L_2(x)$	$e^{-\lambda x}\dfrac{\sin \omega x}{\omega}$	$xe^{-\lambda x}$	$e^{-\lambda x}\dfrac{\sinh \omega x}{\omega}$
$L_3(x)$	$\displaystyle\int_0^x L_2(x)\,dx = \dfrac{1}{q}[1 - L_1(x)]$		
$L_4(x)$	$\displaystyle\int_0^x L_3(x)\,dx = \dfrac{1}{q^2}\{qx - p[1 - L_1(x)] - qL_2(x)\}$		for all Δ
$L_5(x)$	$\displaystyle\int_0^x L_4(x)\,dx = \dfrac{1}{q^3}\{\frac{1}{2}(qx)^2 + (p^2 - q)[1 - L_1(x)] - pq[x - L_2(x)]\}$		

(3) Shape Coefficients

$$L_j(x) = L_j \qquad L_j(x - a) = A_j \qquad L_j(x - b) = B_j \qquad L_j(x - d) = D_j$$

For example,

$$A_3 = \begin{cases} \dfrac{1}{q}\left\{1 - e^{-\lambda(x-a)}\left[\cos \omega(x - a) + \dfrac{\lambda}{\omega}\sin \omega(x - a)\right]\right\} & \text{if } \Delta < 0 \\[3mm] \dfrac{1}{q}\{1 - [1 + \lambda(x - a)e^{-\lambda(x-a)}]\} & \text{if } \Delta = 0 \\[3mm] \dfrac{1}{q}\left\{1 - e^{-\lambda(x-a)}\left[\cosh \omega(x - a) + \dfrac{\lambda}{\omega}\sinh \lambda(x - q)\right]\right\} & \text{if } \Delta < 0 \end{cases}$$

(4) Convolution Integral

$$G(x) = \frac{1}{M}\int_0^x g(x - \tau)L_2(\tau)\,d\tau \qquad \text{(Sec. 16.40)}$$

(5) Transport Matrix Equation   (for all Δ)

$$\begin{bmatrix} Y(x) \\ Y'(x) \end{bmatrix} = \begin{bmatrix} L_1 & L_2 \\ -qL_2 & L_1 - pL_2 \end{bmatrix}\begin{bmatrix} Y(+0) \\ Y'(+0) \end{bmatrix} + \begin{bmatrix} G(x) \\ G'(x) \end{bmatrix}$$

$$MY'' + NY' + RY = g(x)$$

$p = \dfrac{N}{M}$	$u = q - \alpha p + \alpha^2 - \beta^2$	$v = (p - 2\alpha)\beta$	$q = \dfrac{R}{M}$
$\psi_1 = \tan^{-1}\dfrac{v}{u}$	$F_1 = \sin\psi_1$		$F_2 = \sqrt{\alpha^2 + \beta^2}\,\sin(\psi_1 + \psi_2)$
$\psi_2 = \tan^{-1}\dfrac{\beta}{\alpha}$	$G_1 = \cos\psi_1$		$G_2 = \sqrt{\alpha^2 + \beta^2}\,\cos(\psi_1 + \psi_2)$

$g(x)$	$G(x) = \dfrac{1}{M}\displaystyle\int_0^x g(x-\tau)L_2(\tau)\,d\tau$
(1)–(27) Forcing functions shown in Sec. 16.38	Analytical expressions for $G(x)$ corresponding to Cases (1)–(27) are formally identical to those in Sec. 16.38 but must be evaluated in terms of the shape coefficients of Sec. 16.39–3. Since M, N, R are signed numbers, they must be used with their proper sign.
(28)	$\dfrac{w[(q-\beta^2)(\sin\beta x - \beta L_2) - p\beta(\cos\beta x - L_1)]}{M[(q-\beta^2)^2 + (p\beta)^2]}$ $0 < x$
(29)	$\dfrac{w[(q-\beta^2)(\cos\beta x - L_1) - p\beta(\sin\beta x - \beta L_2)]}{M[(q-\beta^2) + (p\beta)^2]}$ $0 < x$
(30)	$\dfrac{w}{1\sqrt{u^2 + v^2}}[e^{-\alpha x}\sin(\beta x - \psi_1) + F_1 L_1 - F_2 L_2]$ $0 < x$
(31)	$\dfrac{w}{M\sqrt{u^2 + v^2}}[e^{-\alpha x}\cos(\beta x - \psi_1) - G_1 L_1 + G_2 L_2]$ $0 < x$
(32)	$\dfrac{w(e^{-\alpha x} - L_1 + \alpha L_2)}{M(q - \alpha p + \alpha^2)}$ $0 < x$

$$\boxed{MY''' + NY = g(x)}$$

$Y(+0), Y'(+0), Y''(+0) =$ initial values $\qquad M, N =$ signed numbers $\qquad p = \dfrac{N}{M}$

$\lambda = \sqrt[3]{\left|\dfrac{N}{M}\right|} \qquad\qquad \eta = \tfrac{1}{2}\lambda \qquad\qquad \omega = \tfrac{1}{2}\sqrt{3}\,\lambda \qquad\qquad \phi = \dfrac{\pi}{3}$

(1) General Solution

$$\boxed{Y(x) = Y(+0)L_1(x) + Y'(+0)L_2(x) + Y''(+0)L_3(x) + G(x)}$$

(2) Shape Functions $\qquad (j = -1, 0, 1, 2, \ldots)$

$L_j(x)$	$p > 0$	$p = 0$	$p < 0$
$L_{-1}(x)$	$\dfrac{\lambda^2}{3}[e^{-\lambda x} + 2e^{\eta x}\cos(\omega x + 2\phi)]$	0	$\dfrac{\lambda^2}{3}[e^{\lambda x} + 2e^{-\eta x}\cos(\omega x - 2\phi)]$
$L_0(x)$	$-\dfrac{\lambda}{3}[e^{-\lambda x} - 2e^{\eta x}\cos(\omega x + \phi)]$	0	$\dfrac{\lambda}{3}[e^{\lambda x} - 2e^{-\eta x}\cos(\omega x - \phi)]$
$L_1(x)$	$\dfrac{1}{3}(e^{-\lambda x} + 2e^{\eta x}\cos\omega x)$	1	$\dfrac{1}{3}(e^{\lambda x} + 2e^{-\eta x}\cos\omega x)$
$L_2(x)$	$-\dfrac{1}{3\lambda}[e^{-\lambda x} - 2e^{\eta x}\cos(\omega x - \phi)]$	x	$\dfrac{1}{3\lambda}[e^{\lambda x} - 2e^{-\eta x}\cos(\omega x + \phi)]$
$L_3(x)$	$\dfrac{1}{3\lambda^2}[e^{-\lambda x} + 2e^{\eta x}\cos(\omega x - 2\phi)]$	$\dfrac{x^2}{2!}$	$\dfrac{1}{3\lambda^2}[e^{\lambda x} + 2e^{-\eta x}\cos(\omega x + 2\phi)]$
$L_4(x)$	$-\dfrac{1}{3\lambda^3}[e^{-\lambda x} - 2e^{\eta x}\cos(\omega x - 3\phi) - 3]$	$\dfrac{x^3}{3!}$	$\dfrac{1}{3\lambda^3}[e^{\lambda x} - 2e^{-\eta x}\cos(\omega x + 3\phi) - 3]$
$L_5(x)$	$\dfrac{1}{3\lambda^4}[e^{-\lambda x} + 2e^{\eta x}\cos(\omega x - 4\phi) + 3\lambda x]$	$\dfrac{x^4}{4!}$	$\dfrac{1}{3\lambda^4}[e^{\lambda x} + 2e^{-\eta x}\cos(\omega x - 4\phi) - 3\lambda x]$
$L_6(x)$	$-\dfrac{1}{3\lambda^5}\left[e^{-\lambda x} - 2e^{\eta x}\cos(\omega x - 5\phi) - \dfrac{3\lambda^2 x^2}{2}\right]$	$\dfrac{x^5}{5!}$	$\dfrac{1}{3\lambda^5}\left[e^{\lambda x} - 2e^{-\eta x}\cos(\omega x + 5\phi) - \dfrac{3\lambda^2 x^2}{2}\right]$

(3) Shape Coefficients

$L_j(x) = L_j \qquad L_j(x - a) = A_j \qquad L_j(x - b) = B_j \qquad L_j(x - d) = D_j$

(4) Convolution Integral

$$G(x) = \dfrac{1}{M}\int_0^x g(x - \tau)L_3(\tau)\,d\tau \qquad \text{(Sec. 16.42)}$$

(5) Transport Matrix Equations $\qquad$ (for all p)

$$\begin{bmatrix} Y(x) \\ Y'(x) \\ Y''(x) \end{bmatrix} = \begin{bmatrix} L_1 & L_2 & L_3 \\ L_0 & L_1 & L_2 \\ L_{-1} & L_0 & L_1 \end{bmatrix} \begin{bmatrix} Y(+0) \\ Y'(+0) \\ Y''(+0) \end{bmatrix} + \begin{bmatrix} G(x) \\ G'(x) \\ G''(x) \end{bmatrix}$$

$$MY''' + NY = g(x)$$

The analytical expressions for $G(x)$ below apply for all solutions given in Sec. 16.41, provided that the respective shape coefficients are used.

$a, b, c, d, w, \alpha, \beta, P$ = real numbers

$g(x)$	$G(x) = \dfrac{1}{M}\displaystyle\int_0^x g(x-\tau)L_3(\tau)\,d\tau$	
(1)	$\dfrac{w}{M}L_4$	$0 < x \le a$
	$\dfrac{w}{M}(L_4 - A_4)$	$a < x$
(2)	$\dfrac{w}{Ma}L_5$	$0 < x \le a$
	$\dfrac{w}{Ma}(L_5 - A_5 - aA_4)$	$a < x$
(3)	$\dfrac{2w}{Ma^2}L_6$	$0 < x \le a$
	$\dfrac{2w}{Ma^2}(L_6 - A_6 - aA_5 - \tfrac{1}{2}a^2A_4)$	$a < x$
(4) $n = 1, 2, \ldots$	$\dfrac{n!\,w}{Ma^n}L_{n+4}$	$0 < x \le a$
	$\dfrac{n!\,w}{Ma^n}\left(L_{n+4} - \displaystyle\sum_{k=0}^{n}\dfrac{a^k}{k!}A_{n+4-k}\right)$	$a < x$
(5)	$\dfrac{w}{Ma}(aL_4 - L_5)$	$0 < x \le a$
	$\dfrac{w}{Ma}(aL_4 - L_5 + A_5)$	$a < x$
(6)	$\dfrac{2w}{Ma^2}(L_6 - aL_5 + \tfrac{1}{2}a^2L_4)$	$0 < x \le a$
	$\dfrac{2w}{Ma^2}(L_6 - aL_5 + \tfrac{1}{2}a^2L_4 - A_6)$	$a < x$
(7) $n = 1, 2, \ldots$	$\dfrac{n!\,w}{Ma^n}\left[\displaystyle\sum_{k=0}^{n}\dfrac{(-a)^k}{k!}L_{n+4-k}\right]$	$0 < x \le a$
	$-\dfrac{n!\,w}{Ma^n}\left[\displaystyle\sum_{k=0}^{n}\dfrac{(-a)^k}{k!}L_{n+4-k} + (-1)^nA_{n+4}\right]$	$a < x$

$$MY''' + NY = g(x)$$

$g(x)$	$G(x) = \dfrac{1}{M}\displaystyle\int_0^x g(x-\tau)L_3(\tau)\,d\tau$	
(8)	$\dfrac{w}{M}A_4$	$a < x \le b$
	$\dfrac{w}{M}(A_4 - B_4)$	$b < x$
(9) $w(x-a)c$ $c = b - a$	$\dfrac{w}{Mc}A_5$	$a < x \le b$
	$\dfrac{w}{Mc}(A_5 - B_5 - cB_4)$	$b < x$
(10) $w(x-a)^2/c^2$ $c = b - a$	$\dfrac{2w}{Mc^2}A_6$	$a < x \le b$
	$\dfrac{2w}{Mc^2}(A_6 - B_6 - cB_5 - \tfrac{1}{2}c^2 B_4)$	$b < x$
(11) $w(x-a)^n/c^n$ $c = b - a$ $n = 1, 2, \ldots$	$\dfrac{n!w}{Mc^n}A_{n+4}$	$a < x \le b$
	$\dfrac{n!w}{Mc^n}\left(A_{n+4} - \displaystyle\sum_{k=0}^{n}\dfrac{c^k}{k!}B_{n+4-k}\right)$	$b < x$
(12) $w(b-x)/c$ $c = b - a$	$\dfrac{w}{Mc}(cA_4 - A_5)$	$a < x \le b$
	$\dfrac{w}{Mc}(cA_4 - A_5 + B_5)$	$b < x$
(13) $w(b-x)^2/c^2$ $c = b - a$	$\dfrac{2w}{Mc^2}(A_6 - cA_5 + \tfrac{1}{2}c^2 A_4)$	$a < x \le b$
	$\dfrac{2w}{Mc^2}(A_6 - cA_5 + \tfrac{1}{2}c^2 A_4 - B_6)$	$b < x$
(14) $w(b-x)^n/c^n$ $c = b - a$ $n = 1, 2, \ldots$	$\dfrac{n!w}{Mc^n}\left[\displaystyle\sum_{k=0}^{n}\dfrac{(-c)^k}{k!}A_{n+4-k}\right]$	$a < x \le b$
	$-\dfrac{n!w}{Mc^n}\left[\displaystyle\sum_{k=0}^{n}\dfrac{(-c)^k}{k!}A_{n+4-k} + (-1)^n B_{n+4}\right]$	$b < x$

$G(x) = 0 \qquad 0 < x \le a$

$$MY''' + NY = g(x)$$

$g(x)$	$G(x) = \dfrac{1}{M}\displaystyle\int_0^x g(x - \tau)L_3(\tau)\,d\tau$	
(15)	$\dfrac{w}{Mc}A_5$	$a < x \le b$
	$\dfrac{w}{Mc}(A_5 - B_5)$	$b < x \le d$
	$\dfrac{w}{Mc}(A_5 - B_5 - cD_4)$	$d < x$
(16)	$\dfrac{w}{M}A_4$	$a < x \le b$
	$\dfrac{w}{Mc}(cA_4 - B_5)$	$b < x \le d$
	$\dfrac{w}{Mc}(cA_4 - B_5 + D_5)$	$d < x$
(17)	$\dfrac{w}{Mc}A_5$	$a < x \le b$
	$\dfrac{w}{Mc}\left(A_5 - B_5 - \dfrac{c}{h}B_5\right)$	$b < x \le d$
	$\dfrac{w}{Mc}\left(A_5 - B_5 - \dfrac{c}{h}B_5 + \dfrac{c}{h}D_5\right)$	$d < x$
(18)	$\dfrac{w}{M}A_4$	$a < x \le b$
	$\dfrac{w}{M}(A_4 - 2B_4)$	$b < x \le d$
	$\dfrac{w}{M}(A_4 - 2B_4 + D_4)$	$d < x$
(19)	$\dfrac{w}{Mc}A_5$	$a < x \le b$
	$\dfrac{w}{Mc}(A_5 - 2cB_4)$	$b < x \le d$
	$\dfrac{w}{Mc}(A_5 - 2cB_4 + D_5)$	$d < x$
(20)	$\dfrac{w}{Mc}(cA_4 - A_5)$	$a < x \le d$
	$\dfrac{w}{Mc}(cA_4 - A_5 + cD_4 + D_5)$	$d < x$

$0 < x \le a$

$G(x) = 0$

$$MY''' + NY = g(x)$$

$g(x)$	$G(x) = \dfrac{1}{M} \displaystyle\int_0^x g(x - \tau)L_3(\tau)\, d\tau$	
(21)	$\dfrac{w}{Ma} L_5$	$0 < x \le a$
	$\dfrac{w}{Ma}(L_5 - A_5)$	$a < x \le b$
	$\dfrac{w}{Ma}(L_5 - A_5 - aB_4)$	$b < x$
(22)	$\dfrac{w}{M} L_4$	$0 < x \le a$
	$\dfrac{w}{Mc}(cL_4 - A_5)$	$a < x \le b$
	$\dfrac{w}{Mc}(cL_4 - A_5 + B_5)$	$b < x$
(23)	$\dfrac{w}{Ma} L_5$	$0 < x \le a$
	$\dfrac{w}{Ma}\left(L_5 - \dfrac{b}{c} A_5\right)$	$a < x \le b$
	$\dfrac{w}{Ma}\left(L_5 - \dfrac{b}{c} A_5 + \dfrac{a}{c} B_5\right)$	$b < x$
(24)	$\dfrac{w}{M} L_4$	$0 < x \le a$
	$\dfrac{w}{M}(L_4 - 2A_4)$	$a < x \le b$
	$\dfrac{w}{M}(L_4 - 2A_4 + B_4)$	$b < x$
(25)	$\dfrac{w}{Ma} L_5$	$0 < x \le a$
	$\dfrac{w}{Ma}(L_5 - 2aA_4)$	$a < x \le b$
	$\dfrac{w}{Ma}(L_5 - 2aA_4 + B_5)$	$b < x$
(26)	$\dfrac{w}{Ma}(aL_4 - L_5)$	$0 < x \le b$
	$\dfrac{w}{Ma}(aL_4 - L_5 + aB_4 + B_5)$	$b < x$

$$\boxed{MY''' + NY = g(x)}$$

$p = \dfrac{N}{M}$	$u = p - \alpha(\alpha^2 - 3\beta^2)$	$v = \beta(3\alpha^2 - \beta^2)$
$\psi_1 = \tan^{-1}\dfrac{v}{u}$ $\quad F_1 = \sin\psi_1$	$F_2 = \sqrt{\alpha^2 + \beta^2}\,\sin(\psi_1 + \psi_2)$	$F_3 = (\alpha^2 + \beta^2)\sin(\psi_1 + 2\psi_2)$
$\psi_2 = \tan^{-1}\dfrac{\beta}{\alpha}$ $\quad G_1 = \cos\psi_1$	$G_2 = \sqrt{\alpha^2 + \beta^2}\,\cos(\psi_1 + \psi_2)$	$G_3 = (\alpha^2 + \beta^2)\cos(\psi_1 + 2\psi_2)$

$g(x)$	$G(x) = \dfrac{1}{M}\displaystyle\int_0^x g(x-\tau)L_3(\tau)\,d\tau$
(27)	$0 \qquad\qquad 0 < x \le a$
	$\dfrac{P}{M}A_3 \qquad\qquad a < x$
(28) $w\sin\beta x$, $\beta = \pi/b$	$\dfrac{w[\beta^3(\cos\beta x - L_1) + p(\sin\beta x - \beta L_2) + \beta^5 L_3}{M(p^2 + \beta^6)}$ $\qquad 0 < x$
(29) $w\cos\beta x$, $\beta = \pi/b$	$\dfrac{w[p(\cos\beta x - L_1) - \beta^3(\sin\beta x - \beta L_2) + p\beta^5 L_3}{M(p^2 + \beta^6)}$ $\qquad 0 < x$
(30) $we^{-\alpha x}\sin\beta x$, $\beta = \pi/b$	$\dfrac{w}{M\sqrt{u^2 + v^2}}[e^{-\alpha x}\sin(\beta x - \psi_1) + F_1 L_1 - F_2 L_2 + F_3 L_3]$ $\qquad 0 < x$
(31) $we^{-\alpha x}\cos\beta x$, $\beta = \pi/b$	$\dfrac{w}{M\sqrt{u^2 + v^2}}[e^{-\alpha x}\cos(\beta x - \psi_1) - G_1 L_1 + G_2 L_2 - G_3 L_3]$ $\qquad 0 < x$
(32) $we^{-\alpha x}$	$\dfrac{w}{M(p - \alpha^3)}(e^{-\alpha x} - L_1 + \alpha L_2 - \alpha^2 L_3)$ $\qquad 0 < x$

$$MY^{iv} + NY = g(x)$$

$Y(+0), Y'(+0), Y''(+0), Y'''(+0)$ = initial values $\qquad$ M, N = signed numbers $\qquad$ $p = \dfrac{N}{M}$

$$\lambda = \sqrt[4]{\frac{N}{4M}} \qquad S_1 = \sin \lambda x \qquad C_1 = \cos \lambda x \qquad \bar{S}_1 = \sinh \lambda x \qquad \bar{C}_1 = \cosh \lambda x$$

$$\eta = \sqrt[4]{\left|\frac{N}{M}\right|} \qquad S_2 = \sin \eta x \qquad C_2 = \cos \eta x \qquad \bar{S}_2 = \sinh \eta x \qquad \bar{C}_2 = \cosh \eta x$$

(1) General Solution

$$Y(x) = Y(+0)L_1(x) + Y'(+0)L_2(x) + Y''(+0)L_3(x) + Y'''(+0)L_4(x) + G(x)$$

(2) Shape Functions $\qquad (j = 1, 2, \ldots, k = 5, 6, \ldots)$

$L_j(x)$	$p > 0$	$p = 0$	$p < 0$
$L_1(x)$	$\bar{C}_1 C_1$	1	$\dfrac{1}{2}(\bar{C}_2 + C_2)$
$L_2(x)$	$\dfrac{1}{2\lambda}(\bar{C}_1 S_1 + \bar{S}_1 C_1)$	x	$\dfrac{1}{2\eta}(\bar{S}_2 + S_2)$
$L_3(x)$	$\dfrac{1}{2\lambda^2}\bar{S}_1 S_1$	$\dfrac{x^2}{2!}$	$\dfrac{1}{2\eta^2}(\bar{C}_2 - C_2)$
$L_4(x)$	$\dfrac{1}{4\lambda^3}(\bar{C}_1 S_1 - \bar{S}_1 C_1)$	$\dfrac{x^3}{3!}$	$\dfrac{1}{2\eta^3}(\bar{S}_2 - S_2)$
$L_5(x)$	$\dfrac{1}{4\lambda^4}[1 - L_1(x)]$	$\dfrac{x^4}{4!}$	$\dfrac{1}{\eta^4}[L_1(x) - 1]$
$L_6(x)$	$\dfrac{1}{4\lambda^4}[x - L_2(x)]$	$\dfrac{x^5}{5!}$	$\dfrac{1}{\eta^4}[L_2(x) - x]$
$L_k(x)$	$\dfrac{1}{4\lambda^4}\left[\dfrac{(x)^{k-5}}{(k-5)!} - L_{k-4}(x)\right]$	$\dfrac{x^{k-1}}{(k-1)!}$	$\dfrac{1}{\eta^4}\left[L_{k-4}(x) - \dfrac{(x)^{k-5}}{(k-5)!}\right]$

(3) Shape Coefficients

$$L_j(x) = L_j \qquad L_j(x - a) = A_j \qquad L_j(x - b) = B_j \qquad L_j(x - d) = D_j$$

(4) Convolution Integral

$$G(x) = \frac{1}{M}\int_0^x g(x - \tau)L_4(\tau)\,d\tau \qquad \text{(Sec. 16.44)}$$

(5) Transport Matrix Equations $\qquad$ (for all p)

$$\begin{bmatrix} Y(x) \\ Y'(x) \\ Y''(x) \\ Y'''(x) \end{bmatrix} = \begin{bmatrix} L_1 & L_2 & L_3 & L_4 \\ -pL_4 & L_1 & L_2 & L_3 \\ -pL_3 & -pL_4 & L_1 & L_2 \\ -pL_2 & -pL_3 & -pL_4 & L_1 \end{bmatrix} \begin{bmatrix} Y(+0) \\ Y'(+0) \\ Y''(+0) \\ Y'''(+0) \end{bmatrix} + \begin{bmatrix} G(x) \\ G'(x) \\ G''(x) \\ G'''(x) \end{bmatrix}$$

$$MY^{iv} + NY = g(x)$$

The analytical expressions for $G(x)$ below apply for all solutions given in Sec. 16.43, provided that the respective shape coefficients are used.

$a, b, c, d, w, \alpha, \beta, P, Q$ = real numbers A_j, B_j, D_j, L_j = shape coefficients (Sec. 16.43–3)

$g(x)$	$G(x) = \dfrac{1}{M}\displaystyle\int_0^x g(x-\tau)L_4(\tau)\,d\tau$	
(1)	$\dfrac{w}{M}L_5$	$0 < x \le a$
	$\dfrac{w}{M}(L_5 - A_5)$	$a < x$
(2)	$\dfrac{w}{Ma}L_6$	$0 < x \le a$
	$\dfrac{w}{Ma}(L_6 - A_6 - aA_5)$	$a < x$
(3)	$\dfrac{2w}{Ma^2}L_7$	$0 < x \le a$
	$\dfrac{2w}{Ma^2}(L_7 - A_7 - aA_6 - \tfrac{1}{2}a^2 A_5)$	$a < x$
(4) $n = 1, 2, \ldots$	$\dfrac{n!\,w}{Ma^n}L_{n+5}$	$0 < x \le a$
	$\dfrac{n!\,w}{Ma^n}\left(L_{n+5} - \displaystyle\sum_{k=0}^{n}\dfrac{a^k}{k!}A_{n+5-k}\right)$	$a < x$
(5)	$\dfrac{w}{Ma}(aL_5 - L_6)$	$0 < x \le a$
	$\dfrac{w}{Ma}(aL_5 - L_6 + A_6)$	$a < x$
(6)	$\dfrac{2w}{Ma^2}(L_7 - aL_6 + \tfrac{1}{2}a^2 L_5)$	$0 < x \le a$
	$\dfrac{2w}{Ma^2}(L_7 - aL_6 + \tfrac{1}{2}a^2 L_5 - A_7)$	$a < x$
(7) $n = 1, 2, \ldots$	$\dfrac{n!\,w}{Ma^n}\left[\displaystyle\sum_{k=0}^{n}\dfrac{(-a)^k}{k!}L_{n+5-k}\right]$	$0 < x \le a$
	$-\dfrac{n!\,w}{Ma^n}\left[\displaystyle\sum_{k=0}^{n}\dfrac{(-a)^k}{k!}L_{n+5-k} + (-1)^n A_{n+5}\right]$	$a < x$

$$MY^{iv} + NY = g(x)$$

$g(x)$	$G(x) = \dfrac{1}{M}\displaystyle\int_0^x g(x-\tau)L_4(\tau)\,d\tau$	
(8) w	$\dfrac{w}{M}A_5$	$a < x \le b$
	$\dfrac{w}{M}(A_5 - B_5)$	$b < x$
(9) $w(x-a)c$ $c = b - a$	$\dfrac{w}{Mc}A_6$	$a < x \le b$
	$\dfrac{w}{Mc}(A_6 - B_6 - cB_5)$	$b < x$
(10) $w(x-a)^2/c^2$ $c = b - a$	$\dfrac{2w}{Mc^2}A_7$	$a < x \le b$
	$\dfrac{2w}{Mc^2}(A_7 - B_7 - cB_6 - \tfrac{1}{2}c^2 B_5)$	$b < x$
(11) $w(x-a)^n/c^n$ $c = b - a$ $n = 1, 2, \ldots$	$\dfrac{n!\,w}{Mc^n}A_{n+5}$	$a < x \le b$
	$\dfrac{n!\,w}{Mc^n}\left(A_{n+5} - \displaystyle\sum_{k=0}^{n}\dfrac{c^k}{k!}B_{n+5-k}\right)$	$b < x$
(12) $w(b-x)/c$ $c = b - a$	$\dfrac{w}{Mc}(cA_5 - A_6)$	$a < x \le b$
	$\dfrac{w}{Mc}(cA_5 - A_6 + B_6)$	$b < x$
(13) $w(b-x)^2/c^2$ $c = b - a$	$\dfrac{2w}{Mc^2}(A_7 - cA_6 + \tfrac{1}{2}c^2 A_5)$	$a < x \le b$
	$\dfrac{2w}{Mc^2}(A_7 - cA_6 + \tfrac{1}{2}c^2 A_5 - B_7)$	$b < x$
(14) $w(b-x)^n/c^n$ $c = b - a$ $n = 1, 2, \ldots$	$\dfrac{n!\,w}{Mc^n}\left[\displaystyle\sum_{k=0}^{n}\dfrac{(-c)^k}{k!}A_{n+5-k}\right]$	$a < x \le b$
	$-\dfrac{n!\,w}{Mc^n}\left[\displaystyle\sum_{k=0}^{n}\dfrac{(-c)^k}{k!}A_{n+5-k} + (-1)^n B_{n+5}\right]$	$b < x$

$G(x) = 0$ $0 > x \le a$

$$MY^{iv} + NY = g(x)$$

$g(x)$	$G(x) = \dfrac{1}{M}\displaystyle\int_0^x g(x-\tau)L_4(\tau)\,d\tau$	
(15)	$\dfrac{w}{Mc}A_6$	$a < x \le b$
	$\dfrac{w}{Mc}(A_6 - B_6)$	$b < x \le d$
	$\dfrac{w}{Mc}(A_6 - B_6 - cD_5)$	$d < x$
(16)	$\dfrac{w}{M}A_5$	$a < x \le b$
	$\dfrac{w}{Mc}(cA_5 - B_6)$	$b < x \le d$
	$\dfrac{w}{Mc}(cA_5 - B_6 + D_6)$	$d < x$
(17)	$\dfrac{w}{Mc}A_6$	$a < x \le b$
	$\dfrac{w}{Mc}\left(A_6 - B_6 - \dfrac{c}{h}B_6\right)$	$b < x \le d$
	$\dfrac{w}{Mc}\left(A_6 - B_6 - \dfrac{c}{h}B_6 + \dfrac{c}{h}D_6\right)$	$d < x$
(18)	$\dfrac{w}{M}A_5$	$a < x \le b$
	$\dfrac{w}{M}(A_5 - 2B_5)$	$b < x \le d$
	$\dfrac{w}{M}(A_5 - 2B_5 + D_5)$	$d < x$
(19)	$\dfrac{w}{Mc}A_6$	$a < x \le b$
	$\dfrac{w}{Mc}(A_6 - 2cB_5)$	$b < x \le d$
	$\dfrac{w}{Mc}(A_6 - 2cB_5 + D_6)$	$d < x$
(20)	$\dfrac{w}{Mc}(cA_5 - A_6)$	$a < x \le d$
	$\dfrac{w}{Mc}(cA_5 - A_6 + cD_5 + D_6)$	$d < x$

Right margin: $G(x) = 0 \qquad 0 < x \le a$

$$MY^{iv} + NY = g(x)$$

$g(x)$	$G(x) = \dfrac{1}{M}\displaystyle\int_0^x g(x-\tau)L_4(\tau)\,d\tau$	
(21)	$\dfrac{w}{Ma}L_6$	$0 < x \le a$
	$\dfrac{w}{Ma}(L_6 - A_6)$	$a < x \le b$
	$\dfrac{w}{Ma}(L_6 - A_6 - aB_5)$	$b < x$
(22)	$\dfrac{w}{M}L_5$	$0 < x \le a$
	$\dfrac{w}{Mc}(cL_5 - A_6)$	$a < x \le b$
	$\dfrac{w}{Mc}(cL_5 - A_6 + B_6)$	$b < x$
(23)	$\dfrac{w}{Ma}L_6$	$0 < x \le a$
	$\dfrac{w}{Ma}\left(L_6 - \dfrac{b}{c}A_6\right)$	$a < x \le b$
	$\dfrac{w}{Ma}\left(L_6 - \dfrac{b}{c}A_6 + \dfrac{a}{c}B_6\right)$	$b < x$
(24)	$\dfrac{w}{M}L_5$	$0 < x \le a$
	$\dfrac{w}{M}(L_5 - 2A_5)$	$a < x \le b$
	$\dfrac{w}{M}(L_5 - 2A_5 + B_5)$	$b < x$
(25)	$\dfrac{w}{Ma}L_6$	$0 < x \le a$
	$\dfrac{w}{Ma}(L_6 - 2aA_5)$	$a < x \le b$
	$\dfrac{w}{Ma}(L_6 - 2aA_5 + B_6)$	$b < x$
(26)	$\dfrac{w}{Ma}(aL_5 - L_6)$	$0 < x \le b$
	$\dfrac{w}{Ma}(aL_5 - L_6 + aB_5 + B_6)$	$b < x$

$$MY^{iv} + NY = g(x)$$

$$p = \frac{N}{M} \qquad u = p + (\alpha^2 - \beta^2)^2 - 4\alpha^2\beta^2 \qquad v = 4\alpha\beta(\beta^2 - \alpha^2) \qquad \psi_1 = \tan^{-1}\frac{v}{u} \qquad \psi_2 = \tan^{-1}\frac{\beta}{\alpha}$$

$$F_1 = \sin\psi_1 \quad F_2 = \sqrt{\alpha^2 + \beta^2}\,\sin(\psi_1 - \psi_2) \quad F_3 = (\alpha^2 + \beta^2)\sin(\psi_1 - 2\psi_2) \quad F_4 = \sqrt{(\alpha^2 + \beta^2)^3}\,\sin(\psi_1 - 3\psi_2)$$

$$G_1 = \cos\psi_1 \quad G_2 = \sqrt{\alpha^2 + \beta^2}\,\cos(\psi_1 - \psi_2) \quad G_3 = (\alpha^2 + \beta^2)\cos(\psi_1 - 2\psi_2) \quad G_4 = \sqrt{(\alpha^2 + \beta^2)^3}\,\cos(\psi_1 - 3\psi_2)$$

$g(x)$	$G(x) = \dfrac{1}{M}\displaystyle\int_0^x g(x-\tau)L_4(\tau)\,d\tau$
(27)	$0 \hfill 0 < x \le a$ $\dfrac{P}{M}A_1 \hfill a < x \le b$ $\dfrac{P}{M}A_4 + \dfrac{Q}{M}B_3 \hfill b < x$
(28)	$\dfrac{w}{M(p + \beta^4)}(\sin\beta x - \beta L_2 + \beta^3 L_1) \hfill 0 < x$
(29)	$\dfrac{w}{M(p + \beta^4)}(\cos\beta x - L_1 + \beta^2 L_3) \hfill 0 < x$
(30)	$\dfrac{w}{M\sqrt{u^2 + v^2}}[e^{-\alpha x}\sin(\beta x + \psi_1)$ $\hfill - F_1 L_1 + F_2 L_2 - F_3 L_3 + F_4 L_4] \quad 0 < x$
(31)	$\dfrac{w}{M\sqrt{u^2 + v^2}}[e^{-\alpha x}\cos(\beta x + \psi_1)$ $\hfill - G_1 L_1 + G_2 L_2 - G_3 L_3 + G_4 L_4] \quad 0 < x$
(32)	$\dfrac{w}{M(p + \alpha^4)}(e^{-\alpha x} - L_1 + \alpha L_2 - \alpha^2 L_3 + \alpha^3 L_4) \hfill 0 < x$

$$\boxed{MY^{iv} + NY'' = g(x)}$$

$Y(+0), Y'(+0), Y''(+0), Y'''(+0) = \text{initial values}$ $p = \dfrac{N}{M}$ $M, N = \text{signed numbers}$

$S = \sin \lambda x$ $C = \cos \lambda x$ $\lambda = \sqrt{\left|\dfrac{N}{M}\right|}$ $\bar{S} = \sinh \lambda x$ $\bar{C} = \cosh \lambda x$

(1) General Solution

$$Y(x) = Y(+0) + Y'(+0)x + Y''(+0)L_3(x) + Y'''(+0)L_4(x) + G(x)$$

(2) Shape Functions $(j = 0, 1, 2, \ldots, k = 5, 6, \ldots)$

$L_j(x)$	$p > 0$	$p = 0$	$p < 0$
$L_0(x)$	$-\lambda S$	0	$\lambda \bar{S}$
$L_1(x)$	C	1	$\bar{C}$
$L_2(x)$	$\dfrac{1}{\lambda} S$	x	$\dfrac{1}{\lambda} \bar{S}$
$L_3(x)$	$\dfrac{1}{\lambda^2}(1 - C)$	$\dfrac{x^2}{2!}$	$\dfrac{1}{\lambda^2}(\bar{C} - 1)$
$L_4(x)$	$\dfrac{1}{\lambda^2}\left(x - \dfrac{S}{\lambda}\right)$	$\dfrac{x^3}{3!}$	$\dfrac{1}{\lambda^2}\left(\dfrac{\bar{S}}{\lambda} - x\right)$
$L_5(x)$	$\dfrac{1}{\lambda^2}\left[\dfrac{x^2}{2} - L_3(x)\right]$	$\dfrac{x^4}{4!}$	$\dfrac{1}{\lambda^2}\left[L_3(x) - \dfrac{x^2}{2}\right]$
$L_6(x)$	$\dfrac{1}{\lambda^2}\left[\dfrac{x^3}{6} - L_4(x)\right]$	$\dfrac{x^5}{5!}$	$\dfrac{1}{\lambda^2}\left[L_4(x) - \dfrac{x^3}{6}\right]$
$L_k(x)$	$\dfrac{1}{\lambda^2}\left[\dfrac{x^{k-3}}{(k-3)!} - L_{k-2}(x)\right]$	$\dfrac{x^{k-1}}{(k-1)!}$	$\dfrac{1}{\lambda^2}\left[L_{k-2}(x) - \dfrac{x^{k-3}}{(k-3)!}\right]$

(3) Shape Coefficients

$L_j(x) = L_j$ $L_j(x - a) = A_j$ $L_j(x - b) = B_j$ $L_j(x - d) = D_j$

(4) Convolution Integral

$$G(x) = \frac{1}{M} \int_0^x g(x - \tau)L_4(\tau)\, d\tau \qquad (\text{Sec. 16.46})$$

(5) Transport Matrix Equations   (for all p)

$$\begin{bmatrix} Y(x) \\ Y'(x) \\ Y''(x) \\ Y'''(x) \end{bmatrix} = \begin{bmatrix} 1 & x & L_3 & L_4 \\ 0 & 1 & L_2 & L_3 \\ 0 & 0 & L_1 & L_2 \\ 0 & 0 & L_0 & L_1 \end{bmatrix} \begin{bmatrix} Y(+0) \\ Y'(+0) \\ Y''(+0) \\ Y'''(+0) \end{bmatrix} + \begin{bmatrix} G(x) \\ G'(x) \\ G''(x) \\ G'''(x) \end{bmatrix}$$

$$\boxed{MY^{iv} + NY'' = g(x)}$$

$p = \dfrac{N}{M}$	$u = (\alpha^2 - \beta^2)^2 - 4\alpha^2\beta^2 + p(\alpha^2 - \beta^2)$	$v = 2\alpha\beta(2\beta^2 - 2\alpha^2 - p)$
$\psi_1 = \tan^{-1}\dfrac{v}{u}$	$F_1 = \sin\psi_1$ $F_{n+1} = (\alpha^2 + \beta^2)^{n/2}\sin(\psi_1 - n\psi_2)$	
$\psi_2 = \tan^{-1}\dfrac{\beta}{\alpha}$	$G_1 = \cos\psi_1$ $G_{n+1} = (\alpha^2 + \beta^2)^{n/2}\cos(\psi_1 - n\psi_2)$	$\Big\}\, n = 1, 2, \ldots$

$g(x)$	$G(x) = \dfrac{1}{M}\displaystyle\int_0^x g(x - \tau)L_4(\tau)\,d\tau$
(1)–(27) Forcing functions shown in Sec. 16.44	Analytical expressions for $G(x)$ corresponding to Cases (1)–(27) are formally identical to those in Sec. 16.44 but must be evaluated in terms of the shape coefficients of Sec. 16.45–3. Since M, N are signed numbers, they must be used with their proper sign.
(28) $w \sin\beta x$, $\beta = \pi/b$	$\dfrac{w}{(\beta^2 - p)\beta^2}(\sin\beta x - \beta x + \beta^3 L_1)$ $0 < x$
(29) $w\cos\beta x$, $\beta = \pi/b$	$\dfrac{w}{(\beta^2 - p)\beta^2}(\cos\beta x - 1 + \beta^2 L_3)$ $0 < x$
(30) $we^{-\alpha x}\sin\beta x$, $\beta = \pi/b$	$\dfrac{w}{M\sqrt{u^2 + v^2}}[e^{-\alpha x}\sin(\beta x + \psi_1)$ $- F_1 + F_2 x - F_3 L_3 + F_4 L_4]$ $0 < x$
(31) $we^{-\alpha x}\cos\beta x$, $\beta = \pi/b$	$\dfrac{w}{M\sqrt{u^2 + v^2}}[e^{-\alpha x}\cos(\beta x + \psi_1)$ $- G_1 + G_2 x - G_3 L_3 + G_4 L_4]$ $0 < x$
(32) $we^{-\alpha x}$	$\dfrac{w(e^{-\alpha x} - 1 + \alpha x - \alpha^2 L_3 + \alpha^3 L_4)}{(p + \alpha^2)\alpha^2}$ $0 < x$

$$MY^{iv} + NY'' + RY = g(x)$$

$Y(+0), Y'(+0), Y''(+0), Y'''(+0)$ = initial values M, N, R = signed numbers $\neq 0$

$$p = \frac{N}{M} \qquad\qquad q = \frac{R}{M} \qquad\qquad \Delta = p^2 - 4q$$

(1) Laplace Transform of Differential Equation

$$y(s) = Y(+0)[f_1(s) + pf_3(s)] + Y'(+0)[f_2(s) + pf_4(s)] + Y''(+0)f_3(s) + Y'''(+0)f_4(s) + f_4(s) \cdot h(s)$$

where

$$f_1(s) = \frac{s^3}{s^4 + ps^2 + q} \qquad f_2(s) = \frac{s^2}{s^4 + ps^2 + q} \qquad f_3(s) = \frac{s}{s^4 + ps^2 + q}$$

$$f_4(s) = \frac{1}{s^4 + ps^2 + s} \qquad h(s) = \frac{1}{M}\mathcal{L}\{g(x)\}$$

(2) Inverse Laplace Transforms

$$\mathcal{L}^{-1}\{f_j(s)\} = L_j(x) \qquad j = 1, 2, 3, 4$$

$$\mathcal{L}^{-1}\{f_4(s) \cdot h(s)\} = \frac{1}{M}\int_0^x g(x - \tau)L_4(\tau)\,d\tau = G(x)$$

where $L_1(x)$, $L_2(x)$, $L_3(x)$, $L_4(x)$ are the *shape functions* given in Sec. 16.48 and $G(x)$ is the *convolution integral* displayed in particular forms in Sec. 16.49.

(3) General Solution

$$Y(x) = Y(+0)[L_1(x) + pL_3(x)] + Y'(+0)[L_2(x) + pL_4(x)] + Y''(+0)L_3(x) + Y'''(+0)L_4(x) + G(x)$$

(4) Properties of Shape Functions $(k = 1, 2, \ldots)$

$$\frac{d^k L_4(x)}{dx^k} = L_{4-k}(x) \qquad\qquad \int_0^x \int_0^x \cdots \int_0^x L_4(x)\,(dx)^k = L_{4+k}(x)$$

(5) Shape Coefficients $(j = -2, -1, 0, 1, 2, \ldots)$

$$L_j(x) = L_j \qquad L_j(x - a) = A_j \qquad L_j(x - b) = B_j \qquad L_j(x - d) = D_j$$

(6) Transport Matrix Equation   (for all Δ)

$$\begin{bmatrix} Y(x) \\ Y'(x) \\ Y''(x) \\ Y'''(x) \end{bmatrix} = \begin{bmatrix} L_1 + pL_3 & L_2 + pL_4 & L_3 & L_4 \\ L_0 + pL_2 & L_1 + pL_3 & L_2 & L_3 \\ L_{-1} + pL_1 & L_0 + pL_2 & L_1 & L_2 \\ L_{-2} + pL_0 & L_{-1} + pL_1 & L_0 & L_1 \end{bmatrix} \begin{bmatrix} Y(+0) \\ Y'(+0) \\ Y''(+0) \\ Y'''(+0) \end{bmatrix} + \begin{bmatrix} G(x) \\ G'(x) \\ G''(x) \\ G'''(x) \end{bmatrix}$$

$$MY^{iv} + NY'' + RY = g(x)$$

(1) Classification of Shape Functions

General solution of the fourth-order differential equation in Sec. 16.47 yields six sets of shape functions depending on the relations of p to q. They are classified as:

Case I:　$p^2 - 4q = 0$　(Sec. 16.48-2)

Case II:　$p^2 - 4q > 0$　(Sec. 16.48-3)

Case III:　$p^2 - 4q < 0$　(Sec. 16.48-4)

(2) Case I

$L_j(x)$	$p > 0 \qquad \lambda = \sqrt{\dfrac{p}{2}}$ — $p^2 - 4q = 0$ — $p < 0 \qquad \eta = \sqrt{\left\|\dfrac{p}{2}\right\|}$	
	$S = \sin \lambda x \qquad C = \cos \lambda x$	$\bar{S} = \sinh \eta x \qquad \bar{C} = \cosh \eta x$
$L_{-2}(x)$	$\dfrac{\lambda^3}{2}(\lambda x C + 5S)$	$\dfrac{\eta^3}{2}(\eta x \bar{C} + 5\bar{S})$
$L_{-1}(x)$	$\dfrac{\lambda^2}{2}(\lambda x S - 4C)$	$\dfrac{\eta^2}{2}(\eta x \bar{S} + 4\bar{C})$
$L_0(x)$	$-\dfrac{\lambda}{2}(\lambda x C + 3S)$	$\dfrac{\eta}{2}(\eta x \bar{C} + 3\bar{S})$
$L_1(x)$	$-\dfrac{1}{2}(\lambda x S - 2C)$	$\dfrac{1}{2}(\eta x \bar{S} + 2\bar{C})$
$L_2(x)$	$\dfrac{1}{2\lambda}(\lambda x C + S)$	$\dfrac{1}{2\eta}(\eta x \bar{C} + \bar{S})$
$L_3(x)$	$\dfrac{1}{2\lambda^2}(\lambda x S)$	$\dfrac{1}{2\eta^2}(\eta x \bar{S})$
$L_4(x)$	$-\dfrac{1}{2\lambda^3}(\lambda x C - S)$	$\dfrac{1}{2\eta^3}(\eta x \bar{C} - \bar{S})$
$L_5(x)$	$-\dfrac{1}{2\lambda^4}(\lambda x S + 2C - 2)$	$\dfrac{1}{2\eta^4}(\eta x \bar{S} - 2\bar{C} + 2)$
$L_6(x)$	$\dfrac{1}{2\lambda^5}(\lambda x C - 3S + 2\lambda x)$	$\dfrac{1}{2\eta^5}(\eta x \bar{C} - 3\bar{S} + 2\eta x)$
$L_7(x)$	$\dfrac{1}{2\lambda^6}(\lambda x S + 4C + (\lambda x)^2 - 4]$	$\dfrac{1}{2\eta^6}(\eta x \bar{S} - 4\bar{C} + (\eta x)^2 + 4]$

For $L_8(x)$, $L_9(x)$, ... use the integrals of Sec. 16.47-4.

$$\boxed{MY^{iv} + NY'' + RY = g(x)}$$

(3) Case II

$$\lambda = \sqrt{\frac{|p| - \omega}{2}} \qquad \omega = \sqrt{p^2 - 4q} \qquad \eta = \sqrt{\frac{|p| + \omega}{2}}$$

$$S_1 = \sin \lambda x \qquad \bar{S}_1 = \sinh \lambda x \qquad S_2 = \sin \eta x \qquad \bar{S}_2 = \sinh \eta x$$

$$C_1 = \cos \lambda x \qquad \bar{C}_1 = \cosh \lambda x \qquad C_2 = \cos \eta x \qquad \bar{C}_2 = \cosh \dot{\eta} x$$

$$K_2 = \frac{1}{\omega \lambda^2} - \frac{1}{\omega \eta^2} \qquad K_4 = \frac{1}{\omega \lambda^4} - \frac{1}{\omega \eta^4} \qquad K_{2n} = \frac{1}{\omega \lambda^{2n}} - \frac{1}{\omega \eta^{2n}}$$

$L_j(x)$	$p > 0$ $p^2 - 4q > 0$	$p < 0$
$L_{-2}(x)$	$-\dfrac{\lambda^5 S_1}{\omega} + \dfrac{\eta^5 S_2}{\omega}$	$-\dfrac{\lambda^5 \bar{S}_1}{\omega} + \dfrac{\eta^5 \bar{S}_2}{\omega}$
$L_{-1}(x)$	$\dfrac{\lambda^4 C_1}{\omega} - \dfrac{\eta^4 C_2}{\omega}$	$-\dfrac{\lambda^4 \bar{C}_1}{\omega} + \dfrac{\eta^4 \bar{C}_2}{\omega}$
$L_0(x)$	$\dfrac{\lambda^3 S_1}{\omega} - \dfrac{\eta^3 S_2}{\omega}$	$-\dfrac{\lambda^3 \bar{S}_1}{\omega} + \dfrac{\eta^3 \bar{S}_2}{\omega}$
$L_1(x)$	$-\dfrac{\lambda^2 C_1}{\omega} + \dfrac{\eta^2 C_2}{\omega}$	$-\dfrac{\lambda^2 \bar{C}_1}{\omega} + \dfrac{\eta^2 \bar{C}_2}{\omega}$
$L_2(x)$	$-\dfrac{\lambda S_1}{\omega} + \dfrac{\eta S_2}{\omega}$	$-\dfrac{\lambda \bar{S}_1}{\omega} + \dfrac{\eta \bar{S}_2}{\omega}$
$L_3(x)$	$\dfrac{C_1}{\omega} - \dfrac{C_2}{\omega}$	$-\dfrac{\bar{C}_1}{\omega} + \dfrac{\bar{C}_2}{\omega}$
$L_4(x)$	$\dfrac{S_1}{\omega \lambda} - \dfrac{S_2}{\omega \lambda}$	$-\dfrac{\bar{S}_1}{\omega \lambda} + \dfrac{\bar{S}_2}{\omega \eta}$
$L_5(x)$	$-\dfrac{C_1}{\omega \lambda^2} + \dfrac{C_2}{\omega \eta^2} + K_2$	$-\dfrac{\bar{C}_1}{\omega \lambda^2} + \dfrac{\bar{C}_2}{\omega \eta^2} + K_2$
$L_6(x)$	$-\dfrac{S_1}{\omega \lambda^3} + \dfrac{S_2}{\omega \eta^3} + xK_2$	$-\dfrac{\bar{S}_1}{\omega \lambda^3} + \dfrac{\bar{S}_2}{\omega \eta^3} + xK_2$
$L_7(x)$	$\dfrac{C_1}{\omega \lambda^4} - \dfrac{C_2}{\omega \eta^4} + \frac{1}{2}x^2 K_2 - K_4$	$-\dfrac{\bar{C}_1}{\omega \lambda^4} + \dfrac{\bar{C}_2}{\omega \eta^4} + \frac{1}{2}x^2 K_2 + K_4$

For $L_8(x)$, $L_9(x)$, ... use the integrals of Sec. 16.47-4.

$$MY^{iv} + NY'' + RY = g(x)$$

(4) Case III $(k = 0, 1, 2, \ldots)$

$$\lambda = \tfrac{1}{2}\sqrt{2\omega^2 - |p|} \qquad\qquad \omega = \sqrt[4]{|q|} \qquad\qquad \eta = \tfrac{1}{2}\sqrt{2\omega^2 + |p|}$$

$$\phi_1 = \tan^{-1}\frac{\eta}{\lambda} \qquad S_{k1} = \sin k\phi_1 \qquad\qquad \phi_2 = \tan^{-1}\frac{\lambda}{\eta} \qquad S_{k2} = \sin k\phi_2$$

$$E_{\lambda 1} = \frac{e^{\lambda x}}{\lambda\eta\omega^2} \qquad S_{\lambda + k} = \sin(\lambda x + k\phi_1) \qquad E_{\eta 1} = \frac{e^{\eta x}}{\lambda\eta\omega^2} \qquad S_{\eta + k} = \sin(\eta x + k\phi_2)$$

$$E_{\lambda 2} = \frac{e^{-\lambda x}}{\lambda\eta\omega^2} \qquad S_{\lambda - k} = \sin(\lambda x - k\phi_1) \qquad E_{\eta 2} = \frac{e^{-\eta x}}{\lambda\eta\omega^2} \qquad S_{\eta - k} = \sin(\eta x - k\phi_2)$$

$L_j(x)$	$p > 0$	$p^2 - 4q < 0$	$p < 0$
$L_{-2}(x)$	$\omega^7(E_{\lambda 1}S_{\eta+5} + E_{\lambda 2}S_{\eta-5})$		$\omega^7(E_{\eta 1}S_{\lambda+5} + E_{\eta 2}S_{\lambda-5})$
$L_{-1}(x)$	$\omega^6(E_{\lambda 1}S_{\eta+4} - E_{\lambda 2}S_{\eta-4})$		$\omega^6(E_{\eta 1}S_{\lambda+4} - E_{\eta 2}S_{\lambda-4})$
$L_0(x)$	$\omega^5(E_{\lambda 1}S_{\eta+3} + E_{\lambda 2}S_{\eta-3})$		$\omega^5(E_{\eta 1}S_{\lambda+3} + E_{\eta 2}S_{\lambda-3})$
$L_1(x)$	$\omega^4(E_{\lambda 1}S_{\eta+2} - E_{\lambda 2}S_{\eta-2})$		$\omega^4(E_{\eta 1}S_{\lambda+2} - E_{\eta 2}S_{\lambda-2})$
$L_2(x)$	$\omega^3(E_{\lambda 1}S_{\eta+1} + E_{\lambda 2}S_{\eta-1})$		$\omega^3(E_{\eta 1}S_{\lambda+1} + E_{\eta 2}S_{\lambda-1})$
$L_3(x)$	$\omega^2(E_{\lambda 1}S_{\eta+0} - E_{\lambda 2}S_{\eta-0})$		$\omega^2(E_{\eta 1}S_{\lambda+0} - E_{\eta 2}S_{\lambda-0})$
$L_4(x)$	$\omega(E_{\lambda 1}S_{\eta-1} + E_{\lambda 2}S_{\eta+1})$		$\omega(E_{\eta 1}S_{\lambda-1} + E_{\eta 2}S_{\lambda+1})$
$L_5(x)$	$(E_{\lambda 1}S_{\eta-2} - E_{\lambda 2}S_{\eta+2} + 2S_{21})$		$(E_{\eta 1}S_{\lambda-2} - E_{\eta 2}S_{\lambda+2} + 2S_{22})$
$L_6(x)$	$\dfrac{1}{\omega}(E_{\lambda 1}S_{\eta-3} + E_{\lambda 2}S_{\eta+3} + 2xS_{21})$		$\dfrac{1}{\omega}(E_{\eta 1}S_{\lambda-3} + E_{\eta 2}S_{\lambda+3} + 2xS_{22})$
$L_7(x)$	$\dfrac{1}{\omega^2}(E_{\lambda 1}S_{\eta-4} - E_{\lambda 2}S_{\eta+4} + x^2S_{21} + S_{41})$		$\dfrac{1}{\omega^2}(E_{\eta 1}S_{\lambda-4} - E_{\eta 2}S_{\lambda+4} + x^2S_{22} + S_{42})$

For $L_8(x)$, $L_9(x)$, ... use the integrals of Sec. 16.47-4.

16.49 FOURTH-ORDER DIFFERENTIAL EQUATION CONVOLUTION INTEGRALS

$$MY^{iv} + NY'' + RY = g(x)$$

$$p = \frac{N}{M} \qquad q = \frac{R}{M} \qquad u = (\alpha^2 - \beta^2)^2 - 4\alpha^2\beta^2 + p(\alpha^2 - \beta^2) + q \qquad v = 2\alpha\beta(2\beta^2 - 2\alpha^2 - p)$$

$$\psi_1 = \tan^{-1}\frac{v}{u} \qquad F_1 = \sin \psi_1 \qquad F_{n+1} = (\alpha^2 + \beta^2)^{n/2} \sin (\psi_1 - n\psi_2)$$

$$\psi_2 = \tan^{-1}\frac{\beta}{\alpha} \qquad G_1 = \cos \psi_1 \qquad G_{n+1} = (\alpha^2 + \beta^2)^{n/2} \cos (\psi_1 - n\psi_2)$$

$$\left.\right\} n = 1, 2, \ldots$$

$g(x)$	$G(x) = \dfrac{1}{M}\displaystyle\int_0^x g(x - \tau)L_4(\tau)\,d\tau$
(1)–(27) Forcing functions shown in Sec. 16.44	Analytical expressions for $G(x)$ corresponding to Cases (1)–(27) are formally identical to those in Sec. 16.44 but must be evaluated in terms of the shape coefficients of Sec. 16.48. Since M, N, R are signed numbers, they must be used with their proper sign.
(28) $w \sin \beta x$ $\beta = \pi/b$	$\dfrac{w[\sin \beta x - \beta L_2 - \beta(p - \beta^2)L_4]}{\beta^4 - p\beta^2 + q}$ $0 < x$
(29) $w \cos \beta x$ $\beta = \pi/b$	$\dfrac{w[\cos \beta x - L_1 - (p - \beta^2)L_3]}{\beta^4 - p\beta^2 + q}$ $0 < x$
(30) $we^{-\alpha x}\sin \beta x$ $\beta = \pi/b$	$\dfrac{w}{M\sqrt{u^2 + v^2}}[e^{\alpha x}\sin (\beta x + \psi_1)$ $- F_1 L_1 + F_2 L_2 - F_3 L_3 + F_4 L_4]$ $0 < x$
(31) $we^{-\alpha x}\cos \beta x$ $\beta = \pi/b$	$\dfrac{w}{M\sqrt{u^2 + v^2}}[e^{\alpha x}\cos (\beta x + \psi_1)$ $- G_1 L_1 + G_2 L_2 - G_3 L_3 + G_4 L_4]$ $0 < x$
(32) $we^{-\alpha x}$	$\dfrac{w(e^{-\alpha x} - L_1 + \alpha L_2 - (p + \alpha^2)(L_3 - \alpha L_4))}{\alpha^4 + p\alpha^2 + q}$ $0 < x$

17
NUMERICAL METHODS

(1) Methods

Whenever the solution of an engineering problem leads to an expression, equation, or system of equations which cannot be evaluated or solved in closed form, *numerical methods* must be employed. The best-known numerical techniques, *approximate evaluation of functions, numerical solution of equations, finite differences*, and *numerical integration*, are outlined in this chapter.

(2) Errors

The *absolute error* ϵ in the result is the difference between the true result a (assumed to be known) and the approximate result $\bar{a}$. The *relative error* $\bar{\epsilon}$ is the absolute error ϵ divided by $\bar{a}$. Aside from possible outright mistakes (blunders) there are three basic types of errors in numerical calculations: *inherent errors* (due to initial data error), *truncation errors* (due to finite approximation of limiting processes), and *round-off errors* (due to use of finite number of digits).

17.02 APPROXIMATIONS BY SERIES EXPANSION

(1) Algebraic Functions

	$\varepsilon_T \leqslant 0.1$ percent	$\varepsilon_T \leqslant 1$ percent
$\dfrac{1}{1+x} \approx 1 - x$	$-0.03 \leqslant x \leqslant +0.03$	$-0.10 \leqslant x \leqslant +0.10$
$\approx 1 - x + x^2$	$-0.10 \leqslant x \leqslant +0.10$	$-0.21 \leqslant x \leqslant +0.21$
$\sqrt{1+x} \approx 1 + \dfrac{x}{2}$	$-0.08 \leqslant x \leqslant +0.10$	$-0.24 \leqslant x \leqslant +0.32$
$\approx 1 + \dfrac{x}{2} - \dfrac{x^2}{8}$	$-0.22 \leqslant x \leqslant +0.27$	$-0.44 \leqslant x \leqslant +0.66$
$\dfrac{1}{\sqrt{1+x}} \approx 1 - \dfrac{x}{2}$	$-0.04 \leqslant x \leqslant +0.06$	$-0.15 \leqslant x \leqslant +0.17$
$\approx 1 - \dfrac{x}{2} + \dfrac{3x^2}{8}$	$-0.14 \leqslant x \leqslant +0.15$	$-0.30 \leqslant x \leqslant +0.32$

(2) Transcendent Functions

	$\varepsilon_T \leqslant 0.1$ percent	$\varepsilon_T \leqslant 1$ percent
$e^x \approx 1 + x$	$-0.04 \leqslant x \leqslant +0.04$	$-0.13 \leqslant x \leqslant +0.14$
$\approx 1 + x + \dfrac{x^2}{2}$	$-0.17 \leqslant x \leqslant +0.19$	$-0.35 \leqslant x \leqslant +0.43$
$\sin x \approx x$	$\left.\begin{array}{r}-0.077 \\ -4.4°\end{array}\right\} \leqslant x \leqslant \left\{\begin{array}{l}+0.077 \\ +4.4°\end{array}\right.$	$\left.\begin{array}{r}-0.244 \\ -14.0°\end{array}\right\} \leqslant x \leqslant \left\{\begin{array}{l}+0.244 \\ +14.0°\end{array}\right.$
$\approx x - \dfrac{x^3}{6}$	$\left.\begin{array}{r}-0.578 \\ -33.1°\end{array}\right\} \leqslant x \leqslant \left\{\begin{array}{l}+0.578 \\ +33.1°\end{array}\right.$	$\left.\begin{array}{r}-1.032 \\ -59.0°\end{array}\right\} \leqslant x \leqslant \left\{\begin{array}{l}+1.032 \\ +59.0°\end{array}\right.$
$\cos x \approx 1$	$\left.\begin{array}{r}-0.045 \\ -2.6°\end{array}\right\} \leqslant x \leqslant \left\{\begin{array}{l}+0.045 \\ +2.6°\end{array}\right.$	$\left.\begin{array}{r}-0.141 \\ -8.1°\end{array}\right\} \leqslant x \leqslant \left\{\begin{array}{l}+0.141 \\ +8.1°\end{array}\right.$
$\approx 1 - \dfrac{x^2}{2}$	$\left.\begin{array}{r}-0.384 \\ -22.0°\end{array}\right\} \leqslant x \leqslant \left\{\begin{array}{l}+0.384 \\ +22.0°\end{array}\right.$	$\left.\begin{array}{r}-0.650 \\ -37.2°\end{array}\right\} \leqslant x \leqslant \left\{\begin{array}{l}+0.650 \\ +37.2°\end{array}\right.$
$\tan x \approx x$	$\left.\begin{array}{r}-0.054 \\ -3.1°\end{array}\right\} \leqslant x \leqslant \left\{\begin{array}{l}+0.054 \\ +3.1°\end{array}\right.$	$\left.\begin{array}{r}-0.183 \\ -10.5°\end{array}\right\} \leqslant x \leqslant \left\{\begin{array}{l}+0.183 \\ +10.5°\end{array}\right.$
$\approx x + \dfrac{x^3}{3}$	$\left.\begin{array}{r}-0.385 \\ -22.0°\end{array}\right\} \leqslant x \leqslant \left\{\begin{array}{l}+0.385 \\ +22.0°\end{array}\right.$	$\left.\begin{array}{r}-0.533 \\ -30.5°\end{array}\right\} \leqslant x \leqslant \left\{\begin{array}{l}+0.533 \\ +30.5°\end{array}\right.$

(Ref. 17.07, p. 30)

k, m, n = positive integers

$P_k(x)$ = Legendre polynomial (Sec. 14.18)

$J_n(x)$ = Bessel function (Sec. 14.24)

$T_k(x)$ = Chebyshev polynomial (Sec. 14.20)

(1) Bessel Series $(0 \le x \le 1)$

(a) Representation

$$f(x) = \sum_{k=1}^{\infty} a_k J_n(\lambda_k x) \quad \text{where } \lambda_k \text{ is the } k\text{th root of } J_n(x) = 0$$

$$a_k = \frac{2}{J_{n+1}^2(\lambda_k)} \int_0^{+1} x f(x) J_n(\lambda_k x)\, dx$$

(b) Integral

$$\int_0^{+1} x J_n(\lambda_k x)\, dx = -\frac{J_{n-1}(\lambda_k)}{\lambda_k} + \frac{n}{\lambda_k} \int_0^{+1} J_{n-1}(\lambda_k x)\, dx$$

(2) Legendre Series $(-1 \le x \le +1)$

(a) Representation

$$f(x) = \sum_{k=0}^{\infty} a_k P_k(x)$$

$$a_k = \frac{2k+1}{2} \int_{-1}^{+1} f(x) P_k(x)\, dx$$

(b) Integrals

$$\int_{-1}^{+1} x P_k(x)\, dx = \frac{\sqrt{2^{k-1}}}{k!!}$$

$$\int_{-1}^{+1} x^m P_k(x)\, dx = \frac{\sqrt{2^{k-m}}}{(m+k-1)!!}$$

(c) Powers of x

$$P_k = P_k(x)$$

$x = P_1$	$x^5 = \frac{1}{63}(27P_1 + 28P_3 + 8P_5)$
$x^2 = \frac{1}{3}(1 + 2P_2)$	$x^6 = \frac{1}{231}(33 + 110P_2 + 72P_4 + 16P_6)$
$x^3 = \frac{1}{5}(P_1 + 2P_3)$	$x^7 = \frac{1}{429}(143P_1 + 182P_3 + 88P_5 + 16P_7)$
$x^4 = \frac{1}{35}(7 + 20P_2 + 8P_4)$	$x^8 = \frac{1}{6,435}(715 + 2,600P_2 + 2,160P_4 + 832P_6 + 128P_8)$

(3) Chebyshev Series $(-1 \le x \le +1)$

(a) Representation

$$f(x) = \frac{a_0}{\pi\sqrt{1-x^2}} + \frac{2}{\pi\sqrt{1-x^2}} \sum_{k=1}^{\infty} a_k T_k(x)$$

$$a_k = \int_{-1}^{+1} f(x) T_k(x)\, dx$$

(b) Integrals

$$\int_{-1}^{+1} x^m T_k(x)\, dx = \begin{cases} 0 & \text{odd } (m+k) \\ 2(m)! \left[\dfrac{T_k(1)}{(m+1)!} - \dfrac{T_k'(1)}{(m+2)!} + \dfrac{T_k''(1)}{(m+3)!} - \cdots \right] & \text{even } (m+k) \end{cases}$$

(c) Powers of x

$$T_k = T_k(x)$$

$x = T_1$	$x^5 = \frac{1}{16}(10T_1 + 5T_3 + T_5)$
$x^2 = \frac{1}{2}(1 - T_2)$	$x^6 = \frac{1}{32}(10 + 15T_2 + 6T_4 + T_6)$
$x^3 = \frac{1}{4}(3T_1 + T_3)$	$x^7 = \frac{1}{64}(35T_1 + 21T_3 + 7T_5 + T_7)$
$x^4 = \frac{1}{8}(3 + 4T_2 + T_4)$	$x^8 = \frac{1}{128}(35 + 56T_2 + 28T_4 + 8T_6 + T_8)$

(1) General Properties

Every algebraic equation of the nth degree (with n real and complex roots),

$$f(x) = a_0 x^n + a_1 x^{n-1} + \cdots + a_{n-1} x + a_n = 0 \qquad a_0 \neq 0$$

can be represented as a product of n linear factors. If $x_1, x_2, \ldots, x_n$ are the roots of this equation, then

$$f(x) = (x - x_1)(x - x_2) \cdots (x - x_n) = 0$$

(2) Relations between the Roots and the Coefficients

$$x_1 + x_2 + \cdots + x_n = \sum_{i=1}^{n} x_i = -\frac{a_1}{a_0}$$

$$x_1 x_2 + x_1 x_3 + \cdots + x_{n-1} x_n = \sum_{i,j=1}^{n} x_i x_j = \frac{a_2}{a_0}$$

$$x_1 x_2 x_3 + x_1 x_2 x_4 + \cdots + x_{n-2} x_{n-1} x_n = \sum_{i,j,k=1}^{n} x_i x_j x_k = -\frac{a_3}{a_0}$$

$$\cdots\cdots\cdots\cdots\cdots\cdots\cdots\cdots\cdots\cdots\cdots\cdots\cdots\cdots$$

$$(-1)^n x_1 x_2 x_3 \cdots x_n = \frac{a_n}{a_0}$$

(3) Methods of Solution

If $n > 4$, there is no formula which gives the roots of this general equation. The following methods are useful:

(a) Roots by trial

Find a number x_1 that satisfies $f(x_1) = 0$. Then divide $f(x)$ by $x - x_1$, thus obtaining an equation of degree one less than that of the original equation. Repeat the same procedure with the reduced equation.

(b) Roots by *regula falsi* approximation

If coefficients $a_0, a_1, \ldots$ are real, introduce $x = a$ and $x = b$ so that $f(a)$ and $f(b)$ have opposite signs. Then

$$c_1 = a - f(a)\frac{b - a}{f(b) - f(a)}$$

is the first approximation of x_1. By repeating the application of this idea, real roots may be obtained to any degree of accuracy.

(c) Roots by Newton's approximation

Assume c_1 as an approximate root, calculate a better approximation by means of

$$c_2 = c_1 - \frac{f(c_1)}{f'(c_1)}$$

and repeat this process to a desired accuracy.

(d) Roots by Steinman's approximation

First locate the roots approximately, and write the equation in the form

$$a_0 x^n = a_1 x^{n-1} + a_2 x^{n-2} + \cdots + a_n$$

Then, with $x = d_1$,

$$d_2 = \frac{a_1 + 2a_2/d_1 + 3a_3/d_1^2 + 4a_4/d_1^3 + \cdots}{a_0 + a_2/d_1^2 + 2a_3/d_1^3 + 3a_4/d_1^4 + \cdots}$$

and repeat this process to a desired accuracy.

(1) General

The solution of a system of linear equations by the method outlined in Sec. 1.13 – 1b becomes impractical if $n > 4$. In such a case the employment of the numerical methods outlined in the following discussion offers a workable solution:

(2) Gauss' Elimination Method

It involves replacing the given system by combination and modification of the initial equations leading to the following *triangular system*.

$$
\begin{aligned}
x_1 + \alpha_{12}x_2 + \cdots + \alpha_{1,n-1}x_{n-1} + \alpha_{1,n}x_n &= \beta_1 \\
x_2 + \cdots + \alpha_{2,n-1}x_{n-1} + \alpha_{2,n}x_n &= \beta_2 \\
\cdots\cdots\cdots\cdots\cdots\cdots\cdots\cdots\cdots\cdots\cdots\cdots \\
x_{n-1} + \alpha_{n-1,n}x_n &= \beta_{n-1} \\
x_n &= \beta_n
\end{aligned}
$$

The last equation (n) yields the value of x_n, which is then substituted in the preceding equation $(n-1)$ from which x_{n-1} is determined, etc.

(3) Gauss-Seidel Iteration Method

This method involves rearrangement and/or modification of the given system to obtain the *largest diagonal coefficients* possible. Then dividing each equation by the respective diagonal coefficient leads to the *carryover form*

$$
x_j = r_{j1}x_1 + r_{j2}x_2 + \cdots + m_j + \cdots + r_{j,n-1}x_{n-1} + r_{jn}x_n
$$

where $r_{j1} = -a_{j1}/a_{jj}$, $r_{j2} = -a_{j2}/a_{jj}$, ... are the *carryover factors* and $m_j = b_j/a_{jj}$ is the *starting value*. Starting with the *trial solution*

$$
x_1^{(1)} = m_1, \; x_2^{(1)} = m_2, \ldots
$$

The successive approximations become

$$
x_1^{(2)} = x_1^{(1)} + \sum_{j=1}^{n} r_{1j}x_j^{(1)}, \; x_2^{(2)} = x_2^{(1)} + \sum_{j=1}^{n} r_{2j}x_j^{(1)}, \ldots
$$

$$
x_1^{(3)} = x_1^{(1)} + x_1^{(2)} + \sum_{j=1}^{n} r_{1j}x_j^{(2)}, \; x_2^{(3)} = x_2^{(1)} + x_2^{(2)} + \sum_{j=1}^{n} r_{2j}x_j^{(2)}, \ldots
$$

$$
\cdots\cdots\cdots\cdots\cdots\cdots\cdots\cdots\cdots\cdots\cdots\cdots\cdots\cdots\cdots\cdots
$$

Under certain conditions this procedure is rapidly convergent, but it may also converge slowly or not at all.

(4) Matrix Iterative Method

The given system is written in a *matrix carryover form* as

$$
X = rX + m
$$

where $X = \{x_1, x_2, \ldots, x_j, \ldots, x_{n-1}, x_n\}$

$$
r = \begin{bmatrix} 0 & r_{12} & \cdots & r_{1,n-1} & r_{1n} \\ r_{21} & 0 & \cdots & r_{2,n-1} & r_{2n} \\ \cdots\cdots\cdots\cdots\cdots\cdots\cdots\cdots \end{bmatrix}
$$

$$
m = \{m_1, m_2, \ldots, m_j, \ldots, m_{n-1}, m_n\}
$$

The successive approximations become

$$
X^{(1)} = m, \; X^{(2)} = m + rm, \; X^{(3)} = m + rm + r^2m, \ldots
$$

and the final solution takes the form of the series,

$$
X = [1 + r + r^2 + \cdots]m
$$

where the sum of the power series in the brackets equals the inverse of A defined in Sec. 1.13 – 1c.

(1) Forward Differences

For a given *discrete function* $y = f(x)$, a set of *equally spaced arguments* $x_n = x_0 + n \, \Delta x (n = 0, \pm 1, \pm 2, \dots; \Delta x = h > 0)$ and a corresponding set of values $y_n = y(x_0 + n \, \Delta x)$ define the *forward differences* as follows:

$\Delta y_n = y_{n+1} - y_n$ First-order difference

$\Delta^2 y_n = \Delta y_{n+1} - \Delta y_n = y_{n+2} - 2y_{n+1} + y_n$ Second-order difference

. . . .

$\Delta^k y_n = \Delta^{k-1} y_{n+1} - \Delta^{k-1} y_n = \sum_{j=0}^{k} (-1)^j \binom{k}{j} y_{n+k-j}$ kth-order difference $k = 2, 3, \dots$

(2) Backward Differences

The same set of values defines the *backward differences* as follows:

$\nabla y_n = y_n - y_{n-1}$ First-order difference

$\nabla^2 y_n = \nabla y_n - \nabla y_{n-1} = y_n - 2y_{n-1} + y_{n-2}$ Second-order difference

. . . .

$\nabla^k y_n = \nabla^{k-1} y_n - \nabla^{k-1} y_{n-1} = \sum_{j=0}^{k} (-1)^j \binom{k}{j} y_{n+j}$ kth-order difference $k = 2, 3, \dots$

(3) Central Differences

For the same discrete function, a similar set of values y_n corresponding to $x_n = x_0 + n \, \Delta x$ defines the *central differences* as follows:

$\delta y_n = y_{n+1/2} - y_{n-1/2}$ First-order difference

$\delta^2 y_n = \delta(y_{n+1/2} - y_{n-1/2}) = y_{n+1} - 2y_n + y_{n-1}$ Second-order difference

. . . .

$\delta^k y_n = \delta^{n-1}(y_{n+1/2} - y_{n-1/2}) = \sum_{j=0}^{k} (-1)^j \binom{k}{j} y_{n-j+k/2}$ kth-order difference

(4) Central Means

For the same discrete functions, the *central means* are defined as follows:

$\mu y_n = \frac{1}{2}(y_{n-1/2} + y_{n+1/2})$ First-order mean

$\mu^2 y_n = \frac{\mu}{2}(y_{n-1/2} + y_{n+1/2}) = \frac{1}{4}(y_{n-1} + 2y_n + y_{n+1})$ Second-order mean

. . . .

$\mu^k y_n = \frac{\mu^{k-1}}{2}(y_{n-1/2} + y_{n+1/2}) = \frac{1}{2^k} \sum_{j=0}^{k} \binom{k}{j} y_{n-j+k/2}$ kth-order mean

(1) Forward Differences

Table					
x_{-2}	y_{-2}				
		Δy_{-2}			
x_{-1}	y_{-1}		$\Delta^2 y_{-2}$		
		Δy_{-1}		$\Delta^3 y_{-2}$	
x_0	y_0		$\Delta^2 y_{-1}$		$\Delta^4 y_{-2}$
		Δy_0		$\Delta^3 y_{-1}$	
x_1	y_1		$\Delta^2 y_0$		
		Δy_1			
x_2	y_2				

Example					
-4	200				
		-180			
-2	20		162		
		-18		654	
0	2		816		$-1,068$
		798		-414	
2	800		402		
		1,200			
4	2,000				

(2) Backward Differences

Table					
x_{-2}	y_{-2}				
		∇y_{-1}			
x_{-1}	y_{-1}		$\nabla^2 y_0$		
		∇y_0		$\nabla^3 y_1$	
x_0	y_0		$\nabla^2 y_1$		$\nabla^4 y_2$
		∇y_1		$\nabla^3 y_2$	
x_1	y_1		$\nabla^2 y_2$		
		∇y_2			
x_2	y_2				

Example					
-4	200				
		-180			
-2	20		162		
		-18		654	
0	2		816		$-1,068$
		798		-414	
2	800		402		
		1,200			
4	2,000				

(3) Central Differences

Table					
x_{-2}	y_{-2}				
		$\delta y_{-3/2}$			
x_{-1}	y_{-1}		$\delta^2 y_{-1}$		
		$\delta y_{-1/2}$		$\delta^3 y_{-1/2}$	
x_0	y_0		$\delta^2 y_0$		$\delta^4 y_0$
		$\delta y_{1/2}$		$\delta^3 y_{1/2}$	
x_1	y_1		$\delta^2 y_1$		
		$\delta y_{3/2}$			
x_2	y_2				

Example					
-4	200				
		-180			
-2	20		162		
		-18		654	
0	2		816		1.068
		798		-414	
2	800		402		
		1,200			
4	2.000				

(4) Relationships $(k, n, p, q = 0, 1, 2, \ldots)$

$$\Delta^k y_n = \nabla^k y_{n+k} = \delta^k y_{n+k/2} \qquad\qquad \nabla^k y_n = \Delta^k y_{n-k} = \delta^k y_{n-k/2}$$

$$\delta^k y_n = \Delta^k y_{n-k/2} = \nabla^k y_{n+k/2} \qquad\qquad \Delta^p \nabla^q y_n = \nabla^q \Delta^p y_n = \Delta^{p+q} y_{n-q}$$

(1) General

The *process of interpolation* is used for finding in-between values of a tabulated function or for the development of a substitute function $\bar{y}(x)$ closely approximating a more complicated function $y(x)$ in a given interval.

(2) Lagrange's Interpolation Formula

From the tabulated $n + 1$ values given as

$$\begin{array}{|l}x_0, x_1, x_2, \ldots, x_n \\ \hline y_0, y_1, y_2, \ldots, y_n\end{array}$$

and not necessarily equally spaced, the *substitute (approximate) function* is

$$\bar{y}(x) = \frac{(x - x_0)(x - x_1) \cdots (x - x_n)}{(x_0 - x_1)(x_0 - x_2) \cdots (x_0 - x_n)} y_0 + \frac{(x - x_0)(x - x_2) \cdots (x - x_n)}{(x_1 - x_0)(x_1 - x_2) \cdots (x_1 - x_n)} y_1$$
$$+ \cdots + \frac{(x - x_0)(x - x_1) \cdots (x - x_{n-1})}{(x_n - x_0)(x_n - x_1) \cdots (x_n - x_{n-1})} y_n$$

If the number of given values is small ($n \le 6$), this formula is a very powerful and convenient approximation model; a special case involving five equally spaced values, called *five-point formula*, is given in Sec. 17.09 − 7.*

(3) Newton's Interpolation Formula

From the same values the *alternate form* of the *substitute function* is

$$\bar{y}(x) = y_0 + (x - x_0) \Delta_1(x_1) + (x - x_0)(x - x_1) \Delta_2(x_2)$$
$$+ \cdots + [(x - x_0)(x - x_1) \cdots (x - x_{n-1})] \Delta_n(x_n)$$

where the *divided differences* are

$$\Delta_1(x_1) = \frac{y_1 - y_0}{x_1 - x_0} \qquad \text{First-order divided difference}$$

$$\Delta_2(x_2) = \frac{\Delta_1(x_2) - \Delta_1(x_1)}{x_2 - x_0} \qquad \text{Second-order divided difference}$$

$$\cdots \cdots \cdots \cdots \cdots \cdots \cdots \cdots$$

$$\Delta_n(x_n) = \frac{\Delta_{n-1}(x_n) - \Delta_{n-1}(x_{n-1})}{x_n - x_0} \qquad n\text{th-order divided difference}$$

Divided differences can be again computed in *tabular form* as shown below.

Table				
x_0	y_0			
x_1	y_1	$\Delta_1(x_1)$		
x_2	y_2	$\Delta_1(x_2)$	$\Delta_2(x_2)$	
x_3	y_3	$\Delta_1(x_3)$	$\Delta_2(x_3)$	$\Delta_3(x_3)$
.	.	.	.	
.	.	.	.	

Example				
6	1,000			
8	2,000	500		
16	6,000	500	0	
26	7,400	320	10	1
.	.	.	.	
.	.	.	.	

*Five-point formula is used frequently for in-table interpolation.

(1) General

If the increment $\Delta x = h$ is a fixed value, the interpolation functions may be taken in a *polynomial form*, the coefficients of which can be expressed in *differences of ascending order* (Sec. 17.06).

(2) Newton's Formula, Forward Interpolation

$$\bar{y}(x) = y_0 + \frac{u}{1!} \Delta y_0 + \frac{u(u-1)}{2!} \Delta^2 y_0 + \frac{u(u-1)(u-2)}{3!} \Delta^3 y_0$$

$$+ \cdots + \frac{u(u-1)(u-2)\cdots(u-n+1)}{n!} \Delta^n y_0 \qquad u = \frac{x-x_0}{h}$$

(3) Newton's Formula, Backward Interpolation

$$\bar{y}(x) = y_n + \frac{u}{1!} \nabla y_n + \frac{u(u+1)}{2!} \nabla^2 y_n + \frac{u(u+1)(u+2)}{3!} \nabla^3 y_n$$

$$+ \cdots + \frac{u(u+1)(u+2)\cdots(u+n-1)}{n!} \nabla^n y_n \qquad u = \frac{x-x_n}{h}$$

(4) Stirling's Interpolation Formula[1]

$$\bar{y}(x) = y_0 + u\frac{\Delta y_0 + \nabla y_0}{2} + \frac{u^2}{2!} \Delta \nabla y_0 + \frac{u(u^2-1)}{3!} \frac{\Delta^2 \nabla y_0 + \Delta \nabla^2 y_0}{2}$$

$$+ \frac{u^2(u^2-1)}{4!} \Delta^2 \nabla^2 y_0 + \cdots \qquad u = \frac{x-x_0}{h}$$

(5) Bessel's Interpolation Formula[1] $(v \neq 0)$

$$\bar{y}(x) = \frac{y_0 + y_1}{2} + v \Delta y_0 + \frac{v^2 - \frac{1}{4}}{2!} \frac{\Delta \nabla y_0 + \Delta \nabla y_1}{2} + \frac{v(v^2 - \frac{1}{4})}{3!} \Delta^2 \nabla y_0$$

$$+ \frac{(v^2 - \frac{1}{4})(v^2 - \frac{9}{4})}{4!} \frac{\Delta^2 \nabla^2 y_0 + \Delta^2 \nabla^2 y_1}{2} + \cdots \qquad v = u - \frac{1}{2} = \frac{x - x_0}{h} - \frac{1}{2}$$

(6) Bessel's Interpolation Formula[1] $(v = 0)$

$$\bar{y}(x) = \frac{1}{2}[(y_0 + y_1) - \frac{1}{8}(\Delta \nabla y_0 + \Delta \nabla y_1) + \frac{3}{128}(\Delta^2 \nabla^2 y_0 + \Delta^2 \nabla^2 y_1) - \frac{5}{1.024}(\Delta^3 \nabla^3 y_0 + \Delta^3 \nabla^3 y_1) + \cdots]$$

(7) Lagrange's Five-point Formula

$$\bar{y}(x) = \frac{f}{24} \left(\frac{y_{-2}}{2+u} - \frac{4y_{-1}}{1+u} + \frac{6y_0}{u} + \frac{4y_1}{1-u} - \frac{y_2}{2-u} \right) \qquad u = \frac{x-x_0}{h}$$

where $\qquad f = (2+u)(1+u)u(1-u)(2-u)$

[1] For relationships of difference operators refer to Sec. 17.07–4.

(1) Concept

Whenever the closed-form integration becomes too involved or is not feasible, the numerical value of a definite integral can be found (to any degree of accuracy) by means of any of several *quadrature formulas* which express the given integral as a linear combination of a selected set of integrands. The basis of these formulas are *difference polynomials* or *orthogonal polynomials*.

(2) Trapezoidal Rule   [n **even or odd**; $h = (b - a)/n$]

$$\int_a^b y(x)\, dx \approx \int_a^b \bar{y}(x)\, dx = \frac{h}{2}(y_0 + 2y_1 + 2y_2 + \cdots + 2y_{n-2} + 2y_{n-1} + y_n) + \varepsilon_T$$

Truncation error: $\varepsilon_T \approx -\dfrac{n(h)^3 f''(\xi)}{12}$ $a \leqslant \xi \leqslant b$

(3) Simpson's Rule   [n **even**; $h = (b - a)/n$]

$$\int_a^b y(x)\, dx \approx \int_a^b \bar{y}(x)\, dx = \frac{h}{3}(y_0 + 4y_1 + 2y_2 + 4y_3 + \cdots + 4y_{n-3} + 2y_{n-2} + 4y_{n-1} + y_n) + \varepsilon_T$$

Truncation error: $\varepsilon_T \approx -\dfrac{n(h)^5 f^{iv}(\xi)}{180}$ $a \leqslant \xi \leqslant b$

(4) Weddle's Rule   [n **must be a multiple of 6**; $h = (b - a)/n$]

$$\int_a^b y(x)\, dx \approx \int_a^b \bar{y}(x)\, dx = \frac{3h}{10}[(y_0 + 5y_1 + y_2 + 6y_3 + y_4 + 5y_5 + y_6)$$

$$+ (y_6 + 5y_7 + y_8 + 6y_9 + y_{10} + 5y_{11} + y_{12})$$

$$+ \cdots + (y_{n-6} + 5y_{n-5} + y_{n-4} + 6y_{n-3} + y_{n-2} + 5y_{n-1} + y_n)] + \epsilon_T$$

Truncation error: $\epsilon_T \approx -\dfrac{n(h)^7 f^{vi}(\xi)}{140}$ $a \leqslant \xi \leqslant b$

(5) Euler's Quadrature Formula[1]   [n **even or odd**; $h = (b - a)/n$]

$$\int_a^b y(x)\, dx \approx \int_a^b \bar{y}(x)\, dx = \frac{h}{2}(y_0 + 2y_1 + 2y_2 + \cdots + 2y_{n-2} + 2y_{n-1} + y_n)$$

$$- \bar{B}_2 \frac{h^2}{2!}[y'(b) - y'(a)] - \bar{B}_4 \frac{h^4}{4!}[y'''(b) - y'''(a)] - \cdots + \epsilon_T$$

Truncation error: $\epsilon_T \approx -n\bar{B}_{2m} \dfrac{h^{2m+1}}{(2m)!} f^{(2m)}(\xi)$ $a \leqslant \xi \leqslant b$

---

[1] $\bar{B}_2, \bar{B}_4, \ldots, \bar{B}_{2m} =$ Bernoulli's numbers, Sec. 8.05.

(1) Gauss-Chebyshev Quadrature Formula

$$\int_a^b y(x)\,dx \approx \frac{b-a}{n}[y(X_1) + y(X_2) + \cdots + y(X_n)] + \varepsilon_T$$

$$X_j = \frac{a+b}{2} + \frac{b-a}{2}x_j \qquad n = 2, 3, 4, 5, 6, 7, 9$$

Truncation error for $n = 3$: $\qquad \varepsilon_T \approx \frac{1}{360}\left(\frac{b-a}{2}\right)^5 y^{ir}(X) \qquad a < X < b$

(2) Gauss-Legendre Quadrature Formula

$$\int_a^b y(x)\,dx \approx (b-a)[C_1 y(X_1) + C_2 y(X_2) + \cdots + C_n y(X_n)] + \varepsilon_T$$

$$X_j = a + (b-a)x_j \qquad n = 1, 2, 3, \ldots$$

Truncation error for $n = n$: $\qquad \varepsilon_T \approx \frac{(n!)^4(b-a)^{2n+1}}{(2n+1)[(2n)!]^3}y^{(2n)}(X) \qquad a < X < b$

(3) Table of x_j for the Gauss-Chebyshev Formula

n	x_1	x_2	x_3	x_4	x_5
2	0.577350	-0.577350			
3	0.707107	0	-0.707107		
4	0.794654	0.187592	-0.187592	-0.794654	
5	0.832497	0.374541	0	0.374541	-0.832497

(4) Table of x_j for the Gauss-Legendre Formula

n	x_1	x_2	x_3	x_4	x_5
1	0.500000				
2	0.211325	0.788675			
3	0.112702	0.500000	0.887298		
4	0.069432	0.330009	0.669991	0.930568	
5	0.046910	0.230765	0.500000	0.769235	0.953090

(5) Table of C_j for the Gauss-Legendre Formula[1]

n	C_1	C_2	C_3	C_4	C_5
1	1.000000				
2	0.500000	0.500000			
3	0.277778	0.444444	0.277778		
4	0.173927	0.326073	0.326073	0.173927	
5	0.118463	0.239314	0.284444	0.239314	0.118463

[1](Ref. 17.09, p. 286)

a, b, c, h = real numbers $\qquad\qquad$ m, n, r = positive integers

$u = f(x)$, $v = g(x)$ $\qquad\qquad\qquad$ $w = g(x + h)$

(1) Derivatives and Differences

(a) Definitions. If $y = f(x)$ satisfies the conditions of Sec. 7.03 and

$$y_1 = y_0 + h, \qquad y_2 = y_0 + 2h, \qquad \ldots, \qquad y_n = y_0 + nh, \qquad \ldots$$

then

$$Dy_n = \lim_{h \to 0} \frac{\Delta y_n}{h}, \qquad D^2 y_n = \lim_{h \to 0} \frac{\Delta^2 y_n}{h^2}, \qquad \ldots, \qquad D^m y_n = \lim_{h \to 0} \frac{\Delta^m y_n}{h^m}$$

where Δ = *forward difference operator* (Sec. 17.06) and D = *derivative operator*, used as

$$D(\ ) = \frac{d}{dx}(\ ), \qquad D^2(\ ) = \frac{d^2}{dx^2}(\ ), \qquad \ldots, \qquad D^m(\ ) = \frac{d^m}{dx^m}(\ )$$

(b) Rules of differential and difference calculus bear close resemblance:

$D[a] = 0 \qquad D[x] = 1$	$\Delta[a] = 0 \qquad \Delta[x] = h$
$D\left[\dfrac{1}{x}\right] = -\dfrac{1}{x^2}$	$\Delta\left[\dfrac{1}{x}\right] = -\dfrac{h}{x(x+h)}$
$D[au] = a\,Du$	$\Delta[au] = a\,\Delta u$
$D[u+v] = Du + Dv$	$\Delta[u+v] = \Delta u + \Delta v$
$D[uv] = v\,Du + u\,Dv$	$\Delta[uv] = v\,\Delta u + u\,\Delta v + \Delta u\,\Delta v$
$D\left[\dfrac{u}{v}\right] = \dfrac{v\,Du - u\,Dv}{v^2}$	$\Delta\left[\dfrac{u}{v}\right] = \dfrac{v\,\Delta u - u\,\Delta v}{vw}$

(2) Relationships of Operators

(a) D in terms of Δ and ∇ $\qquad$ (Sec. 17.06)

$$D = \frac{1}{h}\ln(1 + \Delta) = \frac{1}{h}\left(\Delta - \frac{\Delta^2}{2} + \frac{\Delta^3}{3} - \frac{\Delta^4}{4} + \frac{\Delta^5}{5} - \frac{\Delta^6}{6} + \frac{\Delta^7}{7} - \cdots\right)$$

$$D^2 = \frac{1}{h^2}\ln^2(1 + \Delta) = \frac{1}{h^2}\left(\Delta^2 - \Delta^3 + \tfrac{11}{12}\Delta^4 - \tfrac{5}{6}\Delta^5 + \tfrac{137}{180}\Delta^6 - \cdots\right)$$

$$D^m = \frac{1}{h^m}\ln^m(1 + \Delta) = \frac{1}{h^m}\left[\sum_{r=1}^{\infty}(-1)^r\frac{\Delta^r}{r}\right]^m = \frac{1}{h^m}\ln^m(1 - \nabla) = \frac{1}{h^m}\left[\sum_{r=1}^{\infty}\frac{\nabla^r}{r}\right]^m$$

(b) D in terms of $\lambda = \delta/2$ $\qquad$ (Sec. 17.06)

$$D = \frac{2}{h}\sinh^{-1}\lambda = \frac{2}{h}\left(\lambda - a\lambda^3 + b\lambda^5 - c\lambda^7 + d\lambda^9 - \cdots\right)$$

$$D^2 = \left(\frac{2}{h}\sinh^{-1}\lambda\right)^2 = \frac{4}{h^2}\left[\lambda^2 - 2a\lambda^4 + (a^2 + 2b)\lambda^6 - 2(ab + c)\lambda^8 + \cdots\right]$$

$$D^m = \left(\frac{2}{h}\sinh^{-1}\lambda\right)^m = \left(\frac{2}{h}\right)^m\left[\lambda + \lambda^2\sum_{r=1}^{\infty}(-1)^r\frac{(2r-1)!!}{(2r)!!}\frac{\lambda^{2r-1}}{(2r+1)}\right]^m$$

where $\qquad a = \dfrac{1}{2 \cdot 3} \qquad b = \dfrac{1 \cdot 3}{2 \cdot 4 \cdot 5} \qquad c = \dfrac{1 \cdot 3 \cdot 5}{2 \cdot 4 \cdot 6 \cdot 7} \qquad d = \dfrac{1 \cdot 3 \cdot 5 \cdot 7}{2 \cdot 4 \cdot 6 \cdot 8 \cdot 9}$

18

PROBABILITY AND STATISTICS

(1) Classification of Events

An *observation* (experiment, trial) may have (theoretically) a class (set) S of *possible results* (states, events) $E_1, E_2, \ldots, E_n$ permitting the following classification:

(a) The union $E_1 \cup E_2 \cup \cdots \cup E_n$ of S is the single event of realizing at least one of the events in S.

(b) The intersection $E_i \cap E_j$ of S is the joint event realizing two events E_i and E_j.

(c) The complement $\bar{E}$ of S is the event not included in S.

(2) Simple Probability

If E can occur (happen) in m ways (m times) out of a total of n mutually exclusive and equally likely ways, then

(a) The probability of occurrence (called *success*) of E is

$$P(E) = \frac{m}{n} = p$$

$$0 \le P(E) \le 1$$

(b) The probability of nonoccurrence (called *failure*) of E is

$$P(\bar{E}) = 1 - \frac{m}{n} = q$$

$$0 \le P(\bar{E}) \le 1$$

(3) Special Conditions

(a) Conditional probability of E_1, given that E_2 has occurred, or of E_2, given that E_1 has occurred, are respectively,

$$P\left(\frac{E_1}{E_2}\right) = \frac{P(E_1 \cap E_2)}{P(E_2)}$$

$$P\left(\frac{E_2}{E_1}\right) = \frac{P(E_1 \cap E_2)}{P(E_1)}$$

where $P(E_1) \ne 0$ and $P(E_2) \ne 0$.

(b) Statistical independence. Two events E_1 and E_2 are statistically independent if and only if

$$P\left(\frac{E_1}{E_2}\right) = P(E_1) \ne 0 \qquad\qquad P\left(\frac{E_2}{E_1}\right) = P(E_2) \ne 0$$

(4) Probability Theorems

(a) If E_1 and E_2 are *two mutually not exclusive events*,
$$P(E_1 \cup E_2) = P(E_1) + P(E_2) - P(E_1 \cap E_2)$$

(c) If E and $\bar{E}$ are *two mutually complementary events*,
$$P(E) + P(E) = 1 = p + q$$

(b) If E_1 and E_2 are *two mutually exclusive events*,
$$P(E_1 \cup E_2) = P(E_1) + P(E_2)$$

(d) If E_1 and E_2 are *two mutually independent events*,
$$P(E_1 \cap E_2) = P(E_1)P(E_2)$$

(e) If $E_1, E_2, \ldots, E_k, \ldots, E_n$ form a set S of mutually exclusive events, then for each pair of events E_j, E_k included in this set,

$$P\left(\frac{E_j}{E_k}\right) = \frac{P(E_k \cap E_j)}{P(E_k)} = \frac{P(E_j)P(E_k/E_j)}{\sum\limits_{j=1}^{n} [P(E_j) P(E_k/E_j)]}$$

where $P(E_k) \ne 0$ (Bayes theorem).

(1) Frequency Distribution

A set of numbers $x_1, x_2, \ldots, x_n$ corresponding to the events $E_1, E_2, \ldots, E_n$ which occur with *frequencies* $f_1, f_2, \ldots, f_n$, respectively, so that $f_j \geq 0$ and $x_1 < x_2 < \cdots < x_n$ is called the *frequency distribution*.

(2) Discrete Random Variable

If a variable X can assume a discrete set of values $x_1, x_2, \ldots, x_n$ with respect to probabilities $p_1, p_2, \ldots, p_n$, where $p_1 + p_2 + \cdots + p_n = 1$, then these values define the *discrete probability distribution*. Because x takes specific values with given probabilities, it is designated as the *discrete random variable* (change variable, stochastic variable). The probability that X takes the value x_j is

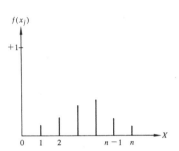

$$\boxed{P(X = x_j) = f(x_j)} \qquad j = 1, 2, \ldots, n$$

where $f(x)$ is the *probability function of the random variable* X, with $f(x_j) \geq 0$, $\sum_1^n f(x_j) = 1$. The *joint probability* that X and Y take the values x_j and y_k, respectively, is

$$\boxed{P(X = x_j, Y = y_j) = f(x_j, y_k)} \qquad \begin{array}{l} j = 1, 2, \ldots, n \\ k = 1, 2, \ldots, r \end{array}$$

where $f(x, y)$ is the *probability function of a two-dimensional random variable.* $\left[f(x_j, y_j) \geq 0, \sum_1^n f(x_j) = 1, \sum_1^r f(y_k) = 1 \right]$.

(3) Continuous Random Variable

The random variable X is denoted as a continuous random variable if $f(x)$ is continuously differentiable.

$$\boxed{\frac{df(x)}{dx} = \lim_{\Delta x \to 0} \frac{P(X < x \leq X + \Delta x)}{\Delta x} = \phi(x)}$$

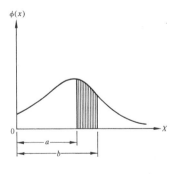

where $\phi(x)$ is the *probability density*, and the *cumulative probability distributions* are

$$\boxed{\begin{aligned} P(x \leq X) &= \int_{-\infty}^{X} \phi(x)\, dx \\[2mm] P(a < x \leq b) &= \int_{a}^{b} \phi(x)\, dx = f(b) - f(a) \\[2mm] P(-\infty < x \leq \infty) &= \int_{-\infty}^{\infty} \phi(x)\, dx = 1 \end{aligned}}$$

A *two-dimensional random variable* is a *continuous random variable* if $f(x, y)$ is continuous for all x_j and y_k, and the *joint probability density*

$$\phi(x, y) = \frac{\partial^2 f(x, y)}{\partial x\, \partial y}$$

exists and is piecewise-continuous.

(1) Measures of Tendency and Frequency

(a) An average is a value which is typical for a set of numbers $X(x_1, x_2, \ldots, x_n)$. Since it tends to lie centrally within the set arranged according to magnitude, it is also a measure of central tendency.

(b) A median of the same set is the average of the two middle values if n is even.

(c) A mode is the value of x_m having a maximum frequency f_m.

(2) Means

(a) The arithmetic mean of a set of n numbers is

$$\overline{X} = \frac{x_1 + x_2 + \cdots + x_n}{n} = \frac{\Sigma x}{n}$$

If the numbers occur $f_1, f_2, \ldots, f_n$ times, respectively, then

$$\overline{X} = \frac{f_1 x_1 + f_2 x_2 + \cdots + f_n x_n}{f_1 + f_2 + \cdots + f_n} = \frac{\Sigma fx}{\Sigma f}$$

If the numbers are associated with a *weighting factor* $w_j \geq 0$, then

$$\overline{X} = \frac{w_1 x_1 + w_2 x_2 + \cdots + w_n x_n}{w_1 + w_2 + \cdots + w_n} = \frac{\Sigma wx}{\Sigma w}$$

(b) The geometric mean of a set of n numbers is

$$\overline{G} = \sqrt[n]{x_1 x_2 \cdots x_n}$$

If the numbers occur $f_1, f_2, \ldots, f_n$ times, respectively, then

$$\overline{G} = \sqrt[n]{x_1^{f_1} x_2^{f_2} \cdots x_n^{f_n}}$$

where $n = f_1 + f_2 + \cdots + f_n$.

(c) The harmonic mean of a set of n numbers is

$$\overline{H} = \frac{1}{(1/n)(1/x_1 + 1/x_2 + \cdots + 1/x_n)} = \frac{n}{\Sigma (1/x)}$$

If the numbers occur $f_1, f_2, \ldots, f_n$ times, respectively, then

$$\overline{H} = \frac{1}{(1/n)(f_1/x_1 + f_2/x_2 + \cdots + f_n/x_n)} = \frac{n}{\Sigma (f/x)}$$

where $n = f_1 + f_2 + \cdots + f_n$

(d) The quadratic mean of a set of n numbers is

$$\overline{Q} = \sqrt{\frac{x_1^2 + x_2^2 + \cdots + x_n^2}{n}} = \sqrt{\frac{\Sigma x^2}{n}}$$

(e) Relations

$$\overline{H} \leq \overline{G} \leq \overline{X}$$

(1) Dispersion

(a) The degree to which a frequency distribution of a set of numbers *tends to spread* about a point of central tendency is called the *dispersion* (spread, variance).

(b) The *range* of a set of numbers is the *difference* between the *largest* and *smallest numbers* in the set.

(2) Deviation

(a) The deviation from the arithmetic mean $\bar{X}$ of each number x_j in a set of numbers $x_1, x_2, \ldots, x_n$ is

$$D_j = x_j - \bar{X} \qquad j = 1, 2, \ldots, n$$

(b) The mean deviation of the same set is

$$\boxed{\bar{D} = \frac{|x_1 + x_2 + \cdots + x_n - n\bar{X}|}{n} = \frac{\Sigma|x - \bar{X}|}{n}}$$

where $|x_j - X|$ is the absolute value of D_j.

    If the numbers occur $f_1, f_2, \ldots, f_n$ times, then $n = f_1 + f_2 + \cdots + f_n$ and

$$\bar{D} = \frac{f_1|x_1 - \bar{X}| + f_2|x_2 - \bar{X}| + \cdots + f_n|x_n - \bar{X}|}{n} = \frac{\Sigma f|x - \bar{X}|}{n}$$

(c) The standard deviation of the same set is

$$\boxed{\sigma = \sqrt{\frac{(x_1 - \bar{X})^2 + (x_2 - \bar{X})^2 + \cdots + (x_n - \bar{X})^2}{n}} = \sqrt{\frac{\Sigma (x - \bar{X})^2}{n}}}$$

If the numbers occur $f_1, f_2, \ldots, f_n$ times, then $n = f_1 + f_2 + \cdots + f_n$ and

$$\sigma = \sqrt{\frac{f_1(x_1 - \bar{X})^2 + f_2(x_2 - \bar{X})^2 + \cdots + f_n(x_n - \bar{X})^2}{n}} = \sqrt{\frac{\Sigma f(x - \bar{X})^2}{n}}$$

(d) The variance of the same set is defined as

$$\sigma^2 = \frac{\Sigma (x - \bar{X})^2}{n} \qquad \text{or} \qquad \sigma^2 = \frac{\Sigma f(x - \bar{X})^2}{n}$$

(e) The covariance of two sets $X(x_1, x_2, \ldots, x_n)$ and $Y(y_1, y_2, \ldots, y_n)$, the arithmetic means of which are $\bar{X}$ and $\bar{Y}$, respectively, is

$$\sigma_{xy} = \frac{\Sigma (x - \bar{X})(y - \bar{Y})}{n}$$

(3) Skewness and Kurtosis

(a) The skewness of the smoothed frequency polygon is the departure of the curve from symmetry.

(b) The kurtosis of the curve is the degree of peakedness of the same curve.

(c) Moments are defined by

$$\boxed{\mu_k = \frac{\Sigma f(x - \bar{X})^k}{n}}$$

where k is the constant indicating the *degree of the moment*. The *first moment* is $\mu_1 = 0$, the *second moment* is $\mu_2 = \sigma^2$, the *coefficient of skewness* is $\gamma_1 = \mu_3/\sigma^3$, the *coefficient of excess* is $\gamma_2 = \mu_4/\sigma^4 - 3$, and the *kurtosis* $\beta_2 = \mu_4/\sigma^4 = \gamma_2 + 3$.

(1) Binomial Distribution

(a) Probability function

$$P(X = x) = f(x) = \frac{n!}{x!(n-x)!} p^x (1-p)^{n-x} \qquad \begin{aligned} & x = 1, 2, \ldots, n \\ & 0 \leq p \leq 1 \end{aligned}$$

(b) Properties

$\mu = np = $ mean

$\sigma^2 = np(1-p) = $ variance

$\sigma = \sqrt{np(1-p)} = $ standard deviation

$\gamma_1 = \dfrac{1-2p}{\sqrt{np(1-p)}} = $ coefficient of skewness

$\gamma_2 = \dfrac{1 - 6p(1-p)}{np(1-p)} = $ coefficient of excess

(2) Poisson Distribution

(a) Probability function

$$P(X = x) = \frac{e^{-\lambda} \lambda^x}{x!} \qquad \begin{aligned} & x = 0, 1, 2, \ldots \\ & \lambda > 0 \\ & e = 2.71828 \cdots \end{aligned}$$

(b) Properties

$\mu = \lambda = $ mean

$\sigma^2 = \lambda = $ variance

$\sigma = \sqrt{\lambda} = $ standard deviation

$\gamma_1 = \dfrac{1}{\sqrt{\lambda}} = $ coefficient of skewness

$\gamma_2 = \dfrac{1}{\lambda} = $ coefficient of excess

(3) Multinomial Distribution

A multinomial distribution is defined by

$$P(X_1 = x_1, X_2 = x_2, \ldots, X_n = x_n) = f(x_1, x_2, \ldots, x_n) = \frac{n!}{x_1! x_2! \cdots x_n!} p_1^{x_1} p_2^{x_2} \cdots p_n^{x_n}$$

where $\qquad \sum_{j=1}^{n} p_j = 1 \qquad \sum_{j=1}^{n} x_j = n \qquad p_j > 0$

(1) Normal Distribution

(a) Density function

$$\phi_N(x) = \frac{1}{\sigma\sqrt{2\pi}}e^{-(x-\mu)^2/2\sigma^2} \qquad \pi = 3.14159\cdots, e = 2.71828\cdots$$

where $\mu =$ mean, $\sigma =$ standard deviation of the random variable X, and $\phi_N(x)$ is the *normal distribution density function* (gaussian function) (Sec. 18.02–3).

(b) Properties

$\mu =$ mean $\gamma_1 = 0 =$ coefficient of skewness

$\sigma^2 =$ variance $\gamma_2 = 0 =$ coefficient of excess

$\sigma =$ standard deviation

Moments about $x = \mu$:   Moments about $x = 0$:

$\mu_1 = 0$ $\nu_1 = \mu$

$\mu_2 = \sigma^2$ $\nu_2 = \mu^2 + \sigma^2$

$\mu_3 = 0$ $\nu_3 = \mu(\mu^2 + 3\sigma^2)$

$\mu_4 = 3\sigma^4$ $\nu_4 = \mu^4 + 6\mu^2\sigma^2 + 3\sigma^4$

(c) Probability function

$$P(X = x) = \int_{-\infty}^{x} \phi_N(x)\,dx = F_N(x)$$

is the *cumulative normal distribution function* (Sec. 18.02–3).

(2) Standard Normal Distribution

(a) Density function

$$\phi_N(t) = \frac{1}{\sqrt{2\pi}}e^{-t^2/2} \qquad t = \frac{x - \mu}{\sigma}$$

is the *standard normal distribution density function,* the ordinates of which are given in Sec. 18.07.

(b) Properties

$\mu = 0 =$ mean

$\sigma^2 = 1 =$ variance

(c) Probability function

$$P(T = t) = \int_{-\infty}^{t} \phi_N(t)\,dt = F_N(t)$$

is the *cumulative standard normal distribution function,* the ordinates of which are given in Sec. 18.08 and are areas under $\phi_N(t)$.

$$\phi_N(t) = \frac{1}{\sqrt{2\pi}} e^{-t^2/2}$$

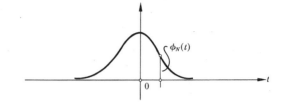

t	0	1	2	3	4	5	6	7	8	9	t
0.0	0.3989	0.3989	0.3989	0.3988	0.3986	0.3984	0.3982	0.3980	0.3977	0.3973	0.0
0.1	0.3970	0.3965	0.3961	0.3956	0.3951	0.3945	0.3939	0.3932	0.3925	0.3918	0.1
0.2	0.3910	0.3902	0.3894	0.3885	0.3876	0.3867	0.3857	0.3847	0.3836	0.3825	0.2
0.3	0.3814	0.3802	0.3790	0.3778	0.3765	0.3752	0.3739	0.3726	0.3712	0.3697	0.3
0.4	0.3683	0.3668	0.3653	0.3637	0.3621	0.3605	0.3589	0.3572	0.3555	0.3538	0.4
0.5	0.3521	0.3503	0.3485	0.3467	0.3448	0.3429	0.3410	0.3391	0.3372	0.3352	0.5
0.6	0.3332	0.3312	0.3292	0.3271	0.3251	0.3230	0.3209	0.3187	0.3166	0.3144	0.6
0.7	0.3123	0.3101	0.3079	0.3056	0.3034	0.3011	0.2989	0.2966	0.2943	0.2920	0.7
0.8	0.2897	0.2874	0.2850	0.2827	0.2803	0.2780	0.2756	0.2732	0.2709	0.2685	0.8
0.9	0.2661	0.2637	0.2613	0.2589	0.2565	0.2541	0.2516	0.2492	0.2468	0.2444	0.9
1.0	0.2420	0.2396	0.2371	0.2347	0.2323	0.2299	0.2275	0.2251	0.2227	0.2203	1.0
1.1	0.2179	0.2155	0.2131	0.2107	0.2083	0.2059	0.2036	0.2012	0.1989	0.1965	1.1
1.2	0.1942	0.1919	0.1895	0.1872	0.1849	0.1826	0.1804	0.1781	0.1758	0.1736	1.2
1.3	0.1714	0.1691	0.1669	0.1647	0.1626	0.1604	0.1582	0.1561	0.1539	0.1518	1.3
1.4	0.1497	0.1476	0.1456	0.1435	0.1415	0.1394	0.1374	0.1354	0.1334	0.1315	1.4
1.5	0.1295	0.1276	0.1257	0.1238	0.1219	0.1200	0.1182	0.1163	0.1145	0.1127	1.5
1.6	0.1109	0.1092	0.1074	0.1057	0.1040	0.1023	0.1006	0.0989	0.0973	0.0957	1.6
1.7	0.0940	0.0925	0.0909	0.0893	0.0878	0.0863	0.0848	0.0833	0.0818	0.0804	1.7
1.8	0.0790	0.0775	0.0761	0.0748	0.0734	0.0721	0.0707	0.0694	0.0681	0.0669	1.8
1.9	0.0656	0.0644	0.0632	0.0620	0.0608	0.0596	0.0584	0.0573	0.0562	0.0551	1.9
2.0	0.0540	0.0529	0.0519	0.0508	0.0498	0.0488	0.0478	0.0468	0.0459	0.0449	2.0
2.1	0.0440	0.0431	0.0422	0.0413	0.0404	0.0396	0.0387	0.0379	0.0371	0.0361	2.1
2.2	0.0355	0.0347	0.0339	0.0332	0.0325	0.0317	0.0310	0.0303	0.0297	0.0290	2.2
2.3	0.0283	0.0277	0.0270	0.0264	0.0258	0.0252	0.0246	0.0241	0.0235	0.0229	2.3
2.4	0.0224	0.0219	0.0213	0.0208	0.0203	0.0198	0.0194	0.0189	0.0184	0.0180	2.4
2.5	0.0175	0.0171	0.0167	0.0163	0.0158	0.0155	0.0151	0.0147	0.0143	0.0139	2.5
2.6	0.0136	0.0132	0.0129	0.0126	0.0122	0.0119	0.0116	0.0113	0.0110	0.0107	2.6
2.7	0.0104	0.0101	0.0099	0.0096	0.0093	0.0091	0.0088	0.0086	0.0084	0.0081	2.7
2.8	0.0079	0.0077	0.0075	0.0073	0.0071	0.0069	0.0067	0.0065	0.0063	0.0061	2.8
2.9	0.0060	0.0058	0.0056	0.0055	0.0053	0.0051	0.0050	0.0048	0.0047	0.0046	2.9
3.0	0.0044	0.0043	0.0042	0.0040	0.0039	0.0038	0.0037	0.0036	0.0035	0.0034	3.0
3.1	0.0033	0.0032	0.0031	0.0030	0.0029	0.0028	0.0027	0.0026	0.0025	0.0025	3.1
3.2	0.0024	0.0023	0.0022	0.0022	0.0021	0.0020	0.0020	0.0019	0.0018	0.0018	3.2
3.3	0.0017	0.0017	0.0016	0.0016	0.0015	0.0015	0.0014	0.0014	0.0013	0.0013	3.3
3.4	0.0012	0.0012	0.0012	0.0011	0.0011	0.0010	0.0010	0.0010	0.0009	0.0009	3.4
t	0	1	2	3	4	5	6	7	8	9	t

18.08 AREAS $F_N(t)$ UNDER THE STANDARD NORMAL CURVE *311*

$$F_N(t) = \frac{1}{\sqrt{2\pi}} \int_{-\infty}^{t} e^{-t^2/2}\, dt$$

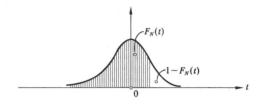

t	0	1	2	3	4	5	6	7	8	9	t
0.0	0.5000	0.5040	0.5080	0.5120	0.5160	0.5199	0.5239	0.5279	0.5319	0.5359	0.0
0.1	0.5398	0.5438	0.5478	0.5517	0.5557	0.5596	0.5636	0.5675	0.5714	0.5754	0.1
0.2	0.5793	0.5832	0.5871	0.5910	0.5948	0.5987	0.6026	0.6064	0.6103	0.6141	0.2
0.3	0.6179	0.6217	0.6255	0.6293	0.6331	0.6368	0.6406	0.6443	0.6480	0.6517	0.3
0.4	0.6554	0.6591	0.6628	0.6664	0.6700	0.6736	0.6772	0.6808	0.6844	0.6879	0.4
0.5	0.6915	0.6950	0.6985	0.7019	0.7054	0.7088	0.7123	0.7157	0.7190	0.7224	0.5
0.6	0.7258	0.7291	0.7324	0.7357	0.7389	0.7422	0.7454	0.7486	0.7517	0.7549	0.6
0.7	0.7580	0.7612	0.7642	0.7673	0.7704	0.7734	0.7764	0.7794	0.7823	0.7852	0.7
0.8	0.7881	0.7910	0.7939	0.7967	0.7996	0.8023	0.8051	0.8078	0.8106	0.8133	0.8
0.9	0.8159	0.8186	0.8212	0.8238	0.8264	0.8289	0.8315	0.8340	0.8365	0.8389	0.9
1.0	0.8413	0.8438	0.8461	0.8485	0.8508	0.8531	0.8554	0.8577	0.8599	0.8621	1.0
1.1	0.8643	0.8665	0.8686	0.8708	0.8729	0.8749	0.8770	0.8790	0.8810	0.8830	1.1
1.2	0.8849	0.8869	0.8888	0.8907	0.8925	0.8944	0.8962	0.8980	0.8997	0.9015	1.2
1.3	0.9032	0.9049	0.9066	0.9082	0.9099	0.9115	0.9131	0.9147	0.9162	0.9177	1.3
1.4	0.9192	0.9207	0.9222	0.9236	0.9251	0.9265	0.9279	0.9292	0.9306	0.9319	1.4
1.5	0.9332	0.9345	0.9357	0.9370	0.9382	0.9394	0.9406	0.9418	0.9429	0.9441	1.5
1.6	0.9452	0.9463	0.9474	0.9484	0.9495	0.9505	0.9515	0.9525	0.9535	0.9545	1.6
1.7	0.9554	0.9564	0.9573	0.9582	0.9591	0.9599	0.9608	0.9616	0.9625	0.9633	1.7
1.8	0.9641	0.9649	0.9656	0.9664	0.9671	0.9678	0.9686	0.9693	0.9699	0.9706	1.8
1.9	0.9713	0.9719	0.9726	0.9732	0.9738	0.9744	0.9750	0.9756	0.9761	0.9767	1.9
2.0	0.9773	0.9778	0.9783	0.9788	0.9793	0.9798	0.9803	0.9808	0.9812	0.9817	2.0
2.1	0.9821	0.9826	0.9830	0.9834	0.9838	0.9842	0.9846	0.9850	0.9854	0.9857	2.1
2.2	0.9861	0.9864	0.9868	0.9871	0.9875	0.9878	0.9881	0.9884	0.9887	0.9890	2.2
2.3	0.9893	0.9896	0.9898	0.9901	0.9904	0.9906	0.9909	0.9911	0.9913	0.9916	2.3
2.4	0.9918	0.9920	0.9922	0.9925	0.9927	0.9929	0.9931	0.9932	0.9934	0.9936	2.4
2.5	0.9938	0.9940	0.9941	0.9943	0.9945	0.9946	0.9948	0.9949	0.9951	0.9952	2.5
2.6	0.9953	0.9955	0.9956	0.9957	0.9959	0.9960	0.9961	0.9962	0.9963	0.9964	2.6
2.7	0.9965	0.9966	0.9967	0.9968	0.9969	0.9970	0.9971	0.9972	0.9973	0.9974	2.7
2.8	0.9974	0.9975	0.9976	0.9977	0.9977	0.9978	0.9979	0.9979	0.9980	0.9981	2.8
2.9	0.9981	0.9982	0.9982	0.9983	0.9984	0.9984	0.9985	0.9985	0.9986	0.9986	2.9
3.0	0.9987	0.9987	0.9987	0.9988	0.9988	0.9989	0.9989	0.9989	0.9990	0.9990	3.0
3.1	0.9990	0.9991	0.9991	0.9991	0.9992	0.9992	0.9992	0.9992	0.9993	0.9993	3.1
3.2	0.9993	0.9993	0.9994	0.9994	0.9994	0.9994	0.9994	0.9995	0.9995	0.9995	3.2
3.3	0.9995	0.9995	0.9996	0.9996	0.9996	0.9996	0.9996	0.9996	0.9996	0.9997	3.3
3.4	0.9997	0.9997	0.9997	0.9997	0.9997	0.9997	0.9997	0.9997	0.9997	0.9998	3.4
t	0	1	2	3	4	5	6	7	8	9	t

$$\binom{n}{0} = 1 \qquad \binom{n}{1} = n \qquad \binom{n}{n} = 1 \qquad \binom{n}{k} = \frac{n!}{k!(n-k)!} = \frac{n(n-1)\cdots(n-k+1)}{k!} = \binom{n}{n-k}$$

n	$\binom{n}{0}$	$\binom{n}{1}$	$\binom{n}{2}$	$\binom{n}{3}$	$\binom{n}{4}$	$\binom{n}{5}$	$\binom{n}{6}$	$\binom{n}{7}$	$\binom{n}{8}$	$\binom{n}{9}$	$\binom{n}{10}$	n
0	1											0
1	1	1										1
2	1	2	1									2
3	1	3	3	1								3
4	1	4	6	4	1							4
5	1	5	10	10	5	1						5
6	1	6	15	20	15	6	1					6
7	1	7	21	35	35	21	7	1				7
8	1	8	28	56	70	56	28	8	1			8
9	1	9	36	84	126	126	84	36	9	1		9
10	1	10	45	120	210	252	210	120	45	10	1	10
11	1	11	55	165	330	462	462	330	165	55	11	11
12	1	12	66	220	495	792	924	792	495	220	66	12
13	1	13	78	286	715	1,287	1,716	1,716	1,287	715	286	13
14	1	14	91	364	1,001	2,002	3,003	3,432	3,003	2,002	1,001	14
15	1	15	105	455	1,365	3,003	5,005	6,435	6,435	5,005	3,003	15
16	1	16	120	560	1,820	4,368	8,008	11,440	12,870	11,440	8,008	16
17	1	17	136	680	2,380	6,188	12,376	19,448	24,310	24,310	19,448	17
18	1	18	153	816	3,060	8,568	18,564	31,824	43,758	48,620	43,758	18
19	1	19	171	969	3,876	11,628	27,132	50,388	75,582	92,378	92,378	19
20	1	20	190	1,140	4,845	15,504	38,760	77,520	125,970	167,960	184,756	20
n	$\binom{n}{0}$	$\binom{n}{1}$	$\binom{n}{2}$	$\binom{n}{3}$	$\binom{n}{4}$	$\binom{n}{5}$	$\binom{n}{6}$	$\binom{n}{7}$	$\binom{n}{8}$	$\binom{n}{9}$	$\binom{n}{10}$	n

Note: For coefficients not given above use $\binom{n}{k} = \binom{n}{n-k}$;

for example, $\binom{19}{15} = \binom{19}{19-15} = \binom{19}{4} = 3,876.$

19
TABLES OF INDEFINITE INTEGRALS

(1) Notation

The more frequently encountered indefinite integrals and their solutions are tabulated in this chapter. Particular symbols used in the following are

a, b, c, d, e, f = constants	α, β, γ = constant equivalents
$m, n\ p, q, r$ = integers	x = independent variable

$A = a + bx$	$M = a^2 + b^2x$
$B = a + bx + cx^2$	$N = a^2 - b^2x$
$C = a^2 + x^2$	$E = a^2 - x^2$
$F = x^2 - a^2$	$G = a^3 \pm x^3$
$H = a^4 + x^4$	$L = a^4 - x^4$
$P = a + bx^q$	$R = ax^q + bx^{q+r}$

In these tables, the constant of integration is omitted but implied, logarithmic expressions are for the absolute value of the respective argument, all angles are in radians, and all inverse functions represent principal values (angles).

(2) Indefinite Integrals Involving

$f(x^m, A^n)$	$A = a + bx$	$a \neq 0$

$$\int x^m\, dx = \frac{x^{m+1}}{m+1} \qquad m \neq -1 \qquad\qquad \int A^n\, dx = \frac{A^{n+1}}{b(n+1)} \qquad n \neq -1$$

$$\int xA\, dx = \frac{x^2}{2}\left(A - \frac{bx}{3}\right) \qquad\qquad \int \frac{A}{x}\, dx = a(\ln x - 1) + A$$

$$\int x^2A\, dx = \frac{x^3}{3}\left(A - \frac{bx}{4}\right) \qquad\qquad \int \frac{A}{x^2}\, dx = b \ln x - \frac{a}{x}$$

$$\int x^mA\, dx = \frac{x^{m+1}}{m+1}\left(A - \frac{bx}{m+2}\right) \qquad m \neq -1, -2$$

$$\int xA^n\, dx = \frac{xA^{n+1}}{(n+1)b}\left(1 + \frac{S}{n+2}\right) \qquad \boxed{S = -\frac{A}{bx}} \qquad n \neq -1, -2$$

$$\int x^2A^n\, dx = \frac{x^2A^{n+1}}{(n+1)b}\left(1 + \frac{2S}{n+2}\left(1 + \frac{S}{n+3}\right)\right) \qquad n \neq -1, -2, -3$$

$$\int x^mA^n\, dx = \frac{x^mA^{n+1}}{(n+1)b}\left(1 + \frac{mS}{n+2}\left(1 + \frac{(m-1)S}{n+3}\left(1 + \cdots + \frac{S}{m+n+1}\right)\right)\right)^* \qquad (n+m+1) > 0$$

*For nested sum refer to Sec. 8.11; S = pocket calculator storage.

(3) Indefinite Integrals Involving*

$$\boxed{f(x) = f(x^m X_p{}^{(r)})} \qquad \boxed{m > 0, \qquad p \neq 0}$$

k, m, r = positive integers	$\mathcal{S}_1{}^{(r)}, \mathcal{S}_2{}^{(r)}, \mathcal{S}_3{}^{(r)}, \ldots$ = Stirling numbers (Sec. A.05)
p = signed number	$X_p{}^{(r)}$ = factorial polynomial (Sec. 1.03)

$$X_1{}^{(r)} = x(x-1)(x-2)(x-3)\cdots(x-r+2)(x-r+1)$$

$$= x\mathcal{S}_1{}^{(r)} + x^2\mathcal{S}_2{}^{(r)} + x^3\mathcal{S}_3{}^{(r)} + \cdots + x^r\mathcal{S}_r{}^{(r)} = \sum_{k=1}^{r} x^k \mathcal{S}_k{}^{(r)}$$

$$= x(\mathcal{S}_1{}^{(r)} + x(\mathcal{S}_2{}^{(r)} + x(\mathcal{S}_3{}^{(r)} + \cdots + x))) = x \bigwedge_{k=1}^{r-1} [\mathcal{S}_k{}^{(r)} + x]^* \qquad \boxed{\mathcal{S}_r{}^{(r)} = 1}$$

$$X_p{}^{(r)} = x(x-p)(x-2p)(x-3p)\cdots(x-pr+2p)(x-pr+p)$$

$$= xp^{r-1}\mathcal{S}_1{}^{(r)} + x^2 p^{r-2}\mathcal{S}_2{}^{(r)} + x^3 p^{r-3}\mathcal{S}_3{}^{(r)} + \cdots + x^r \mathcal{S}_r{}^{(r)} = p^r \sum_{k=1}^{r}\left(\frac{x}{p}\right)^k \mathcal{S}_k{}^{(r)}$$

$$= xp^{r-1}\left(\mathcal{S}_1{}^{(r)} + \frac{x}{p}\left(\mathcal{S}_2{}^{(r)} + \frac{x}{p}\left(\mathcal{S}_3{}^{(r)} + \cdots + \frac{x}{p}\right)\right)\right) = xp^{r-1}\bigwedge_{k=1}^{r-1}\left[\mathcal{S}_k{}^{(r)} + \frac{x}{p}\right]^*$$

$$\int X_1{}^{(2)}\, dx = -\frac{x^2}{2}\left(1 - \frac{2x}{3}\right) \qquad\qquad \int xX_1{}^{(2)}\, dx = -\frac{x^3}{3}\left(1 - \frac{3x}{4}\right)$$

$$\int x^m X_1{}^{(2)}\, dx = -\frac{x^{m+2}}{m+2}\left(1 - \frac{(m+2)x}{m+3}\right)$$

$$\int X_1{}^{(r)}\, dx = \frac{x^2}{2}\left(\mathcal{S}_1{}^{(r)} + \frac{2x}{3}\left(\mathcal{S}_2{}^{(r)} + \frac{3x}{4}\left(\mathcal{S}_3{}^{(r)} + \cdots + \frac{rx}{r+1}\right)\right)\right)$$

$$\int xX_1{}^{(r)}\, dx = \frac{x^3}{3}\left(\mathcal{S}_1{}^{(r)} + \frac{3x}{4}\left(\mathcal{S}_2{}^{(r)} + \frac{4x}{5}\left(\mathcal{S}_3{}^{(r)} + \cdots + \frac{(r+1)x}{r+2}\right)\right)\right)$$

$$\int x^m X_1{}^{(r)}\, dx = \frac{x^{m+2}}{m+2}\left(\mathcal{S}_1{}^{(r)} + \frac{(m+2)x}{m+3}\left(\mathcal{S}_2{}^{(r)} + \frac{(m+3)x}{m+3}\left(\mathcal{S}_3{}^{(r)} + \cdots + \frac{(r+m)x}{r+m+1}\right)\right)\right)^*$$

$$\int X_p{}^{(2)}\, dx = -\frac{px^2}{2}\left(1 - \frac{2x}{3p}\right) \qquad\qquad \int xX_p{}^{(2)}\, dx = -\frac{px^3}{3}\left(1 - \frac{3x}{4p}\right)$$

$$\int x^m X_p{}^{(2)}\, dx = -\frac{px^{m+2}}{m+2}\left(1 - \frac{(m+2)x}{(m+3)p}\right)$$

$$\int X_p{}^{(r)}\, dx = \frac{p^{r-1}x^2}{2}\left(\mathcal{S}_1{}^{(r)} + \frac{2x}{3p}\left(\mathcal{S}_2{}^{(r)} + \frac{3x}{4p}\left(\mathcal{S}_3{}^{(r)} + \cdots + \frac{rx}{(r+1)p}\right)\right)\right)$$

$$\int xX_p{}^{(r)}\, dx = \frac{p^{r-1}x^3}{3}\left(\mathcal{S}_1{}^{(r)} + \frac{3x}{4p}\left(\mathcal{S}_2{}^{(r)} + \frac{4x}{5p}\left(\mathcal{S}_3{}^{(r)} + \cdots + \frac{(r+1)x}{(r+2)p}\right)\right)\right)$$

$$\int x^m X_p{}^{(r)}\, dx = \frac{p^{r-1}x^{m+2}}{m+2}\left(\mathcal{S}_1{}^{(r)} + \frac{(m+2)x}{(m+3)p}\left(\mathcal{S}_2{}^{(r)} + \frac{(m+3)x}{(m+4)p}\left(\mathcal{S}_3{}^{(r)} + \cdots + \frac{(r+m)x}{(r+m+1)p}\right)\right)\right)^*$$

---

*For nested sum refer to Sec. 8.11.

(4) Indefinite Integrals Involving
$$f(x) = \frac{x^m}{A^n} \qquad A = a + bx \qquad a \neq 0$$

$$\int \frac{dx}{A} = \frac{\ln A}{b}$$

$$\int \frac{dx}{A^2} = -\frac{1}{bA}$$

$$\int \frac{dx}{A^3} = -\frac{1}{2bA^2}$$

$$\int \frac{dx}{A^n} = -\frac{1}{(n-1)bA^{n-1}} \qquad n \neq 0, 1$$

$$\int \frac{x\,dx}{A} = \frac{1}{b^2}(A - a\ln A)$$

$$\int \frac{x\,dx}{A^2} = \frac{1}{b^2}\left(\frac{a}{A} + \ln A\right)$$

$$\int \frac{x\,dx}{A^3} = -\frac{2A - a}{2b^2 A^2}$$

$$\int \frac{x\,dx}{A^n} = -\frac{1}{(n-2)b^2 A^{n-1}}\left(A - \frac{n-2}{n-1}a\right) \qquad n \neq 0, 1, 2$$

$$\int \frac{x^2\,dx}{A} = \frac{1}{b^3}\left(\frac{A^2}{2} - 2aA + a^2\ln A\right)$$

$$\int \frac{x^2\,dx}{A^2} = \frac{1}{b^3}\left(A - \frac{a^2}{A} - 2a\ln A\right)$$

$$\int \frac{x^2\,dx}{A^3} = \frac{1}{b^3}\left(\frac{2a}{A} - \frac{a^2}{2A^2} + \ln A\right)$$

$$\int \frac{x^2\,dx}{A^n} = -\frac{1}{(n-3)b^3 A^{n-1}}\left[A^2 - \frac{2(n-3)a}{n-2}A + \frac{n-3}{n-1}a^2\right] \qquad n \neq 0, 1, 2, 3$$

$$\int \frac{x^m}{A}dx = \frac{1}{b}\left[\left(-\frac{a}{b}\right)^m \ln A + x^m \sum_{k=0}^{m-1} \frac{1}{m-k}\left(-\frac{a}{bx}\right)^k\right]$$

$$\int \frac{x^m}{A^n}dx = \frac{1}{b^{m+1}} \sum_{k=0}^{m} \binom{m}{k}\frac{A^{m-n-k+1}(-a)^k}{m-n-k+1}$$

with terms $m - n - k + 1 = 0$ replaced by $\binom{m}{n-1}(-a)^{m-n+1}\ln A$

(5) Indefinite Integrals Involving

$$f(x) = \frac{1}{x^m A^n} \qquad A = a + bx \qquad a \neq 0$$

$$\int \frac{dx}{xA} = -\frac{1}{a} \ln \frac{A}{x}$$

$$\int \frac{dx}{xA^2} = -\frac{1}{a^2} \ln \frac{A}{x} + \frac{1}{aA}$$

$$\int \frac{dx}{xA^n} = -\frac{1}{a^n} \ln \frac{A}{x} + \frac{1}{a^n} \sum_{k=1}^{n-1} \binom{n-1}{k} \frac{(-bx)^k}{kA^k} \qquad n \neq 0$$

$$\int \frac{dx}{x^2 A} = \frac{b}{a^2} \ln \frac{A}{x} - \frac{1}{ax}$$

$$\int \frac{dx}{x^2 A^2} = \frac{2b}{a^3} \ln \frac{A}{x} - \frac{1}{a^2 x} - \frac{b}{a^2 A}$$

$$\int \frac{dx}{x^2 A^3} = \frac{1}{a^4} \left(3b \ln \frac{A}{x} - \frac{A}{x} + \frac{3b^2 x}{A} - \frac{b^3 x^2}{2A^2} \right)$$

$$\int \frac{dx}{x^2 A^n} = \frac{nb}{a^{n+1}} \ln \frac{A}{x} - \frac{A}{a^{n+1} x} + \frac{A}{a^{n+1} x} \sum_{k=2}^{n} \binom{n}{k} \frac{(-bx)^k}{(k-1)A^k} \qquad n \neq 0, 1$$

$$\int \frac{dx}{x^3 A} = -\frac{b^2}{a^3} \ln \frac{A}{x} + \frac{2bA}{a^3 x} - \frac{A^2}{2a^3 x^2}$$

$$\int \frac{dx}{x^3 A^2} = -\frac{3b^2}{a^4} \ln \frac{A}{x} + \frac{3bA}{a^4 x} - \frac{A^2}{2a^4 x^2} - \frac{b^3 x}{a^4 A}$$

$$\int \frac{dx}{x^3 A^3} = -\frac{1}{a^5} \left(6b^2 \ln \frac{A}{x} - \frac{4bA}{x} + \frac{A^2}{2x^2} + \frac{4b^2 x}{A} - \frac{b^4 x^2}{2A^2} \right)$$

$$\int \frac{dx}{x^3 A^n} = -\frac{n(n+1)b^2}{2a^{n+2}} \ln \frac{A}{x} + \frac{(n+1)bA}{a^{n+2} x} - \frac{b^2 A^2}{2a^{n+2} x^2} + \frac{A^2}{a^{n+2} x^2} \sum_{k=3}^{n+1} \binom{n+1}{k} \frac{(-bx)^k}{(k-2)A^k} \qquad n \neq 0, 1, 2$$

$$\int \frac{dx}{x^m A} = \frac{1}{b} \left[\left(-\frac{b}{a} \right)^m \ln A - \frac{1}{x^m} \sum_{k=1}^{m-1} \frac{1}{m+1} \left(-\frac{a}{bx} \right)^k \right]$$

$$\int \frac{dx}{x^m A^n} = \frac{-1}{a^{m+n-1}} \sum_{k=0}^{m+n-2} \binom{m+n-2}{k} \frac{A^{m-k-1}(-b)^k}{(m-k-1)^{m-k-1} x^{m-k-1}}$$

with terms $m-k-1=0$ replaced by $\binom{m+n-2}{n-1}(-b)^{m-1} \ln \frac{A}{x}$

(6) Indefinite Integrals Involving

$f(x) = f(A, D)$	$A = a + x$	$a \neq 0$
$\alpha = a - b \neq 0$	$D = b + x$	$b \neq 0$

$$\int \frac{A}{D}\, dx = \alpha \ln D + x$$

$$\int \frac{dx}{AD} = \frac{1}{\alpha} \ln \frac{D}{A}$$

$$\int \frac{x\, dx}{AD} = \frac{1}{\alpha}(a \ln A - b \ln D)$$

$$\int \frac{x^2\, dx}{AD} = x - \frac{a+b}{2} \ln AD + \frac{a^2 + b^2}{2} \ln \frac{A}{D}$$

$$\int \frac{dx}{AD^2} = \frac{-1}{\alpha D} + \frac{1}{\alpha^2} \ln \frac{A}{D}$$

$$\int \frac{x\, dx}{AD^2} = \frac{b}{\alpha D} - \frac{a}{\alpha^2} \ln \frac{A}{D}$$

$$\int \frac{x^2\, dx}{AD^2} = \frac{b^2}{\alpha D} + \frac{a^2}{\alpha^2} \ln \frac{A}{D} + \ln D$$

$$\int \frac{dx}{A^2 D^2} = -\frac{1}{\alpha^2}\left(\frac{1}{A} + \frac{1}{D}\right) + \frac{2}{\alpha^3} \ln \frac{A}{D}$$

$$\int \frac{x\, dx}{A^2 D^2} = \frac{1}{\alpha^2}\left(\frac{a}{A} + \frac{b}{D}\right) + \frac{a+b}{\alpha^3} \ln \frac{A}{D}$$

$$\int \frac{x^2\, dx}{A^2 D^2} = -\frac{1}{\alpha^2}\left(\frac{a^2}{A} + \frac{b^2}{D}\right) + \frac{2ab}{\alpha^3} \ln \frac{A}{D}$$

$$\int A^m D^n\, dx = \frac{A^{m+1} D^n}{m+n+1} + \frac{n\alpha}{m+n+1} \int A^m D^{n-1}\, dx \qquad m+n \neq -1$$

$$\int \frac{A^m}{D^n}\, dx = \frac{-A^{m+1}}{(n-1)\alpha D^{n-1}} - \frac{n-m-2}{(n-1)\alpha} \int \frac{A^m}{D^{n-1}}\, dx \qquad n \neq 1$$

$$\int \frac{dx}{A^m D^n} = \frac{-1}{(n-1)\alpha A^{m-1} D^{n-1}} - \frac{m+n-2}{(n-1)\alpha} \int \frac{dx}{A^m D^{n-1}} \qquad n \neq 1$$

$$\int \frac{dx}{(c+x)AD} = \frac{\ln A}{\alpha(a-c)} + \frac{\ln D}{\alpha(c-b)} + \frac{\ln (c+x)}{(a-c)(c-b)} \qquad \left.\begin{array}{l} a - c \neq 0 \\[4pt] b - c \neq 0 \end{array}\right.$$

$$\int \frac{x\, dx}{(c+x)AD} = \frac{a \ln A}{\alpha(c-a)} + \frac{b \ln D}{\alpha(b-c)} + \frac{c \ln (c+x)}{(c-a)(b-c)}$$

(7) Indefinite Integrals Involving

$$f(x) = f(x^m, B, D) \qquad \begin{array}{l} B = a + bx + cx^2 \\ D = e + fx \end{array} \qquad a \neq 0$$

$$\omega = b + 2cx = dB/dx \qquad \gamma = 4ac - b^2 \qquad \lambda = ae^2 - bef + cf^2 \qquad m, n > 0$$

$$S = \frac{2Bc}{\gamma} \qquad \gamma \neq 0$$

$$\int B^2 \, dx = \frac{\omega \gamma^2}{60c^3} \left(1 + S(1 + \tfrac{3}{2}S) \right)$$

$$\int B^3 \, dx = \frac{\omega \gamma^3}{600c^4} \left(1 + S(1 + \tfrac{3}{2}S(1 + \tfrac{5}{3}S)) \right)$$

$$\int B^m \, dx = \left(\frac{\gamma}{c} \right)^m \frac{(m!)^2}{(2m+1)!} \frac{\omega}{2c} \overset{m}{\underset{k=1}{\Lambda}} \left[1 + \frac{(2k-1)}{k} S \right]^* \qquad \int x B^m \, dx = \frac{B^{m+1}}{2(m+1)c} - \frac{b}{2c} \int B^m \, dx$$

$$\int \frac{dx}{B} = \begin{cases} + \dfrac{2}{\sqrt{\gamma}} \tan^{-1} \dfrac{\omega}{\sqrt{\gamma}} & \gamma > 0 \\[3mm] - \dfrac{2}{\omega} & \gamma = 0 \\[3mm] - \dfrac{2}{\sqrt{-\gamma}} \tanh^{-1} \dfrac{\omega}{\sqrt{-\gamma}} & \gamma < 0 \end{cases}$$

$$\int \frac{dx}{B^n} = \frac{\omega}{(n-1)\gamma B^{n-1}} + \frac{2(2n-3)c}{(n-1)\gamma} \int \frac{dx}{B^{n-1}} \qquad \int \frac{x \, dx}{B^n} = \frac{-1}{2(n-1)cB^{n-1}} - \frac{b}{2c} \int \frac{dx}{B^n}$$

$$\int \frac{x \, dx}{B} = \frac{1}{2c} \ln B - \frac{b}{2c} \int \frac{dx}{B} \qquad \int \frac{x^2 \, dx}{B} = \frac{x}{c} - \frac{b}{2c^2} \ln B + \frac{b^2 - 2bc}{2c^2} \int \frac{dx}{B}$$

$$\int \frac{x^m \, dx}{B^n} = \frac{1}{(m-2n+1)c} \left[\frac{x^{m-1}}{B^{n-1}} - (m-n)b \int \frac{x^{m-1}}{B^n} \, dx - (m-1)a \int \frac{x^{m-2}}{B^n} \, dx \right]$$

$$\int \frac{dx}{xB} = \frac{1}{2a} \ln \frac{x^2}{B} - \frac{b}{2a} \int \frac{dx}{B} \qquad \int \frac{dx}{x^2 B} = \frac{b}{2a^2} \ln \frac{B}{x^2} - \frac{1}{ax} + \left(\frac{b^2}{2a^2} - \frac{c}{a} \right) \int \frac{dx}{B}$$

$$\int \frac{dx}{x^m B^n} = \frac{-1}{(m-1)a} \left[\frac{1}{x^{m-1}B^{n-1}} + (m+n-2)b \int \frac{dx}{x^{m-1}B^n} + (m+2n-3)c \int \frac{dx}{x^{m-2}B^n} \right]$$

$$\int \frac{D}{B} \, dx = \frac{f}{2c} \ln B + \frac{2ce - bf}{2c} \int \frac{dx}{B} \qquad \int \frac{D}{B^n} \, dx = \frac{f}{2(n-1)aB^{n-1}} + \frac{2ce - bf}{2c} \int \frac{dx}{B^n}$$

$$\int \frac{dx}{DB} = \frac{1}{2\lambda} \left[2e \ln D - e \ln B + (2cf - be) \int \frac{dx}{B} \right]$$

$$\int \frac{dx}{D^m B^n} = \frac{-1}{(m-1)\lambda} \left[\frac{f}{D^{m-1}B^{n-1}} + (m+n-2)(2ce - bf) \int \frac{dx}{D^{m-1}B^n} + (m+2n-3)c \int \frac{dx}{D^{m-2}B^n} \right]$$

*For nested sum refer to Sec. 8.11, S = pocket calculator storage.

(8) Indefinite Integrals Involving

$$f(x) = \frac{x^m}{C^n} \qquad C = a^2 + x^2 \qquad \begin{matrix} a^2 \neq 0 \\ b^2 \neq 0 \end{matrix}$$

$$\int \frac{dx}{C} = \frac{1}{a}\tan^{-1}\frac{x}{a}$$

$$\int \frac{dx}{C^2} = \frac{1}{2a^3}\left(\frac{ax}{C} + \tan^{-1}\frac{x}{a}\right)$$

$$\int \frac{dx}{C^3} = \frac{1}{8a^5}\left(\frac{2a^3x}{C^2} + \frac{3ax}{C} + 3\tan^{-1}\frac{x}{a}\right)$$

$$\int \frac{dx}{C^n} = \frac{x}{2(n-1)a^2C^{n-1}} + \frac{2n-3}{2(n-1)a^2}\int \frac{dx}{C^{n-1}} \qquad n \neq 1$$

$$\int \frac{x\,dx}{C} = \frac{1}{2}\ln C$$

$$\int \frac{x\,dx}{C^2} = \frac{-1}{2C}$$

$$\int \frac{x\,dx}{C^n} = \frac{-1}{2(n-1)C^{n-1}} \qquad n \neq 1$$

$$\int \frac{x^2\,dx}{C} = x - a\tan^{-1}\frac{x}{a}$$

$$\int \frac{x^2\,dx}{C^2} = -\frac{x}{2C} + \frac{1}{2a}\tan^{-1}\frac{x}{a}$$

$$\int \frac{x^2\,dx}{C^n} = \frac{-x}{2(n-1)C^{n-1}} + \frac{1}{2(n-1)}\int \frac{dx}{C^{n-1}} \qquad n \neq 1$$

$$\int \frac{x^3\,dx}{C} = \frac{x^2}{2} - \frac{a^2}{2}\ln C$$

$$\int \frac{x^3\,dx}{C^2} = \frac{a^2}{2C} + \frac{1}{2}\ln C$$

$$\int \frac{x^3\,dx}{C^n} = \frac{1}{2(n-2)C^{n-2}} + \frac{a^2}{2(n-1)C^{n-1}} \qquad n \neq 1, 2$$

$$\int \frac{x^m\,dx}{C^n} = -\frac{x^{m-1}}{2(n-1)C^{n-1}} + \frac{m-1}{2(n-1)}\int \frac{x^{m-2}}{C^{n-1}} \qquad n \neq 1$$

$$\int \frac{e+fx}{C}\,dx = f\ln\sqrt{C} + \frac{e}{a}\tan^{-1}\frac{x}{a}$$

(9) Indefinite Integrals Involving $\boxed{f(x) = \dfrac{1}{x^m C^n} \qquad C = a^2 + x^2 \qquad a^2 \neq 0}$

$$\int \frac{dx}{xC} = \frac{1}{2a^2} \ln \frac{x^2}{C}$$

$$\int \frac{dx}{xC^2} = \frac{1}{2a^2 C} + \frac{1}{2a^4} \ln \frac{x^2}{C}$$

$$\int \frac{dx}{xC^3} = \frac{1}{4a^2 C^2} + \frac{1}{2a^4 C} + \frac{1}{2a^6} \ln \frac{x^2}{C}$$

$$\int \frac{dx}{xC^n} = \frac{1}{2(n-1)a^2 C^{n-1}} + \frac{1}{a^2} \int \frac{dx}{xC^{n-1}} \qquad n \neq 1$$

$$\int \frac{dx}{x^2 C} = \frac{-1}{a^2 x} - \frac{1}{a^3} \tan^{-1} \frac{x}{a}$$

$$\int \frac{dx}{x^2 C^2} = \frac{-1}{a^4 x} - \frac{x}{2a^4 C} - \frac{3}{2a^5} \tan^{-1} \frac{x}{a}$$

$$\int \frac{dx}{x^2 C^n} = \frac{-1}{a^2 x C^{n-1}} - \frac{2n-1}{a^2} \int \frac{dx}{C^n}$$

$$\int \frac{dx}{x^3 C} = \frac{-1}{2a^2 x^2} - \frac{1}{2a^4} \ln \frac{x^2}{C}$$

$$\int \frac{dx}{x^3 C^2} = \frac{-1}{2a^4 x^2} - \frac{1}{2a^4 C} - \frac{1}{a^6} \ln \frac{x^2}{C}$$

$$\int \frac{dx}{x^3 C^n} = \frac{-1}{2a^2 x^2 C^{n-1}} - \frac{n}{a^2} \int \frac{dx}{xC^n}$$

$$\int \frac{dx}{(b^2 + x^2)C} = \frac{1}{b^2 - a^2} \left(\frac{1}{a} \tan^{-1} \frac{x}{a} - \frac{1}{b} \tan^{-1} \frac{x}{b} \right)$$

$$\int \frac{x \, dx}{(b^2 + x^2)C} = \frac{1}{b^2 - a^2} \ln \sqrt{\frac{a^2 + x^2}{b^2 + x^2}}$$

$$\int \frac{x^2 \, dx}{(b^2 + x^2)C} = \frac{1}{b^2 - a^2} \left(b \tan^{-1} \frac{x}{b} - a \tan^{-1} \frac{x}{a} \right)$$

$$\int \frac{dx}{x^m C^n} = \frac{-1}{(m-1)a^2 x^{m-1} C^{n-1}} - \frac{m+2n-3}{(m-1)a^2} \int \frac{dx}{x^{m-2} C^n} \qquad m \neq 1$$

$$\boxed{\int \frac{dx}{(e+fx)C} = \frac{1}{e^2 + a^2 f^2} \left[f \ln (e+fx) - \frac{f}{2} \ln C + \frac{e}{a} \tan^{-1} \frac{x}{a} \right] \qquad e^2 \neq -a^2 f^2}$$

(10) Indefinite Integrals Involving $\qquad\boxed{f(x) = \dfrac{x^m}{E^n} \qquad\qquad E = a^2 - x^2 \qquad a^2 \neq 0}$

$$\int \frac{dx}{E} = \frac{1}{2a}\ln\frac{a+x}{a-x} = \begin{cases} \dfrac{1}{a}\tanh^{-1}\dfrac{x}{a} & x^2 < a^2 \\[2ex] \dfrac{1}{a}\coth^{-1}\dfrac{x}{a} & x^2 > a^2 \end{cases}$$

$$\int \frac{dx}{E^2} = \frac{1}{2a^3}\left(\frac{ax}{E} + \ln\sqrt{\frac{a+x}{a-x}}\right)$$

$$\int \frac{dx}{E^n} = \frac{x}{2(n-1)a^2 E^{n-1}} + \frac{2n-3}{2(n-1)a^2}\int \frac{dx}{E^{n-1}} \qquad n \neq 1$$

$$\int \frac{x\,dx}{E} = -\frac{1}{2}\ln E$$

$$\int \frac{x\,dx}{E^2} = \frac{1}{2E}$$

$$\int \frac{x\,dx}{E^n} = \frac{1}{2(n-1)E^{n-1}} \qquad n \neq 1$$

$$\int \frac{x^2\,dx}{E} = -x + \frac{a}{2}\ln\frac{a+x}{a-x}$$

$$\int \frac{x^2\,dx}{E^2} = \frac{x}{2E} - \frac{1}{4a}\ln\frac{a+x}{a-x}$$

$$\int \frac{x^2\,dx}{E^n} = \frac{x}{2(n-1)E^{n-1}} - \frac{1}{2(n-1)}\int \frac{dx}{E^{n-1}} \qquad n \neq 1$$

$$\int \frac{x^3\,dx}{E} = -\frac{x^2}{2} - \frac{a^2}{2}\ln E$$

$$\int \frac{x^3\,dx}{E^2} = \frac{a^2}{2E} + \frac{1}{2}\ln E$$

$$\int \frac{x^3\,dx}{E^n} = \frac{-1}{2(n-2)E^{n-2}} + \frac{a^2}{2(n-1)E^{n-1}} \qquad n \neq 1, 2$$

$$\int \frac{x^m\,dx}{E^n} = \frac{x^{m-1}}{2(n-1)E^{n-1}} - \frac{m-1}{2(n-1)}\int \frac{x^{m-2}}{E^{n-1}}\,dx \qquad n \neq 1$$

$$\boxed{\int \frac{e + fx}{E}\,dx = -f\ln\sqrt{E} + \frac{e}{a}\ln\sqrt{\frac{a+x}{a-x}}}$$

(11) Indefinite Integrals Involving

$$f(x) = \frac{1}{x^m E^n} \qquad E = a^2 - x^2 \qquad a^2 \neq 0$$

$$\int \frac{dx}{xE} = \frac{1}{2a^2} \ln \frac{x^2}{E}$$

$$\int \frac{dx}{xE^2} = \frac{1}{2a^2 E} + \frac{1}{2a^4} \ln \frac{x^2}{E}$$

$$\int \frac{dx}{xE^3} = \frac{1}{4a^2 E^2} + \frac{1}{2a^4 E} + \frac{1}{2a^6} \ln \frac{x^2}{E}$$

$$\int \frac{dx}{xE^n} = \frac{1}{2(n-1)a^2 E^{n-1}} + \frac{1}{a^2} \int \frac{dx}{xE^{n-1}} \qquad n \neq 1$$

$$\int \frac{dx}{x^2 E} = \frac{-1}{a^2 x} + \frac{1}{a^3} \tanh^{-1} \frac{x}{a}$$

$$\int \frac{dx}{x^2 E^2} = \frac{-1}{a^4 x} + \frac{2}{2a^4 E} + \frac{3}{2a^5} \tanh^{-1} \frac{x}{a}$$

$$\int \frac{dx}{x^2 E^n} = \frac{-1}{a^2 x E^{n-1}} + \frac{2n-1}{a^2} \int \frac{dx}{E^n}$$

$$\int \frac{dx}{x^3 E} = \frac{-1}{2a^2 x^2} + \frac{1}{2a^4} \ln \frac{x^2}{E}$$

$$\int \frac{dx}{x^3 E^2} = \frac{-1}{2a^4 x^2} + \frac{1}{2a^4 E} + \frac{1}{a^6} \ln \frac{x^2}{E}$$

$$\int \frac{dx}{x^3 E^n} = \frac{-1}{2a^2 x^2 E^{n-1}} + \frac{n}{a^2} \int \frac{dx}{xE^n}$$

$$\int \frac{dx}{(b^2 + x^2) E} = \frac{1}{a^2 + b^2} \left(\frac{1}{b} \tan^{-1} \frac{x}{b} + \frac{1}{a} \ln \sqrt{\frac{a+x}{a-x}} \right)$$

$$\int \frac{x \, dx}{(b^2 + x^2) E} = \frac{1}{a^2 + b^2} \ln \sqrt{\frac{b^2 + x^2}{a^2 - x^2}}$$

$$\int \frac{x^2 \, dx}{(b^2 + x^2) E} = \frac{1}{a^2 + b^2} \left(a \ln \sqrt{\frac{a+x}{a-x}} - b \tan^{-1} \frac{x}{b} \right)$$

$$\int \frac{dx}{x^m E^n} = \frac{-1}{(m-1)a^2 x^{m-1}} + \frac{m + 2n - 3}{(m-1)a^2} \int \frac{dx}{x^{m-2} E^n} \qquad m \neq 1$$

$$\int \frac{dx}{(e + fx) E} = \frac{-1}{e^2 - a^2 f^2} \left[f \ln (e + fx) - \frac{f}{2} \ln E - \frac{e}{a} \tanh^{-1} \frac{x}{a} \right] \qquad e^2 \neq a^2 f^2$$

(12) Indefinite Integrals Involving

$$f(x) = \frac{x^m}{G^n} \qquad G = a^3 \pm x^3 \qquad a^3 \neq 0$$

If two signs appear in a formula, the upper sign corresponds to $G = a^3 + x^3$, and the lower sign corresponds to $G = a^3 - x^3$.

$$\int \frac{dx}{G} = \frac{\pm 1}{6a^2} \ln \frac{(a \pm x)^2}{a^2 \mp ax + x^2} + \frac{1}{a^2\sqrt{3}} \tan^{-1} \frac{2x \mp a}{a\sqrt{3}}$$

$$\int \frac{dx}{G^n} = \frac{x}{3a^3(n-1)G^{n-1}} - \frac{4-3n}{3a^3(n-1)} \int \frac{dx}{G^{n-1}} \qquad n \neq 1$$

$$\int \frac{x\,dx}{G} = \frac{1}{6a} \ln \frac{a^2 \mp ax + x^2}{(a \pm x)^2} \pm \frac{1}{a\sqrt{3}} \tan^{-1} \frac{2x \mp a}{a\sqrt{3}}$$

$$\int \frac{x\,dx}{G^2} = \frac{x^2}{3a^3 G} + \frac{1}{3a^3} \int \frac{x\,dx}{G}$$

$$\int \frac{x\,dx}{G^n} = \frac{x^2}{3a^3(n-1)G^{n-1}} - \frac{5-3n}{3a^3(n-1)} \int \frac{x\,dx}{G^{n-1}} \qquad n \neq 1$$

$$\int \frac{x^2\,dx}{G} = \pm \frac{\ln G}{3}$$

$$\int \frac{x^2\,dx}{G^2} = \mp \frac{1}{3G}$$

$$\int \frac{x^2\,dx}{G^n} = \frac{x^3}{3a^3(n-1)G^{n-1}} + \frac{n-2}{a^3(n-1)} \int \frac{x^2\,dx}{G^{n-1}} \qquad n \neq 1$$

$$\int \frac{x^3\,dx}{G} = \pm x \mp a^3 \int \frac{dx}{G}$$

$$\int \frac{x^3\,dx}{G^2} = \mp \frac{x}{3G} \pm \frac{1}{3} \int \frac{dx}{G}$$

$$\int \frac{x^3\,dx}{G^n} = \frac{x^4}{3a^3(n-1)G^{n-1}} + \frac{3n-7}{3a^3(n-1)} \int \frac{x^3\,dx}{G^{n-1}} \qquad n \neq 1$$

$$\int \frac{x^m\,dx}{G^n} = \frac{x^{m+1}}{3a^3(n-1)G^{n-1}} - \frac{m-3n+4}{3a^3(n-1)} \int \frac{x^m\,dx}{G^{n-1}} \qquad n \neq 1$$

$$\int \frac{a+x}{a^3+x^3}\,dx = \int \frac{dx}{a^2-ax+x^2} = \frac{2}{a\sqrt{3}} \tan^{-1} \frac{2x-a}{a\sqrt{3}}$$

$$\int \frac{a-x}{a^3+x^3}\,dx = \frac{1}{3a} \ln \frac{(a+x)^2}{a^2-ax+x^2}$$

(13) Indefinite Integrals Involving

$$f(x) = \frac{1}{x^m G^n} \qquad G = a^3 \pm x^3 \qquad a^3 \neq 0$$

If two signs appear in a formula, the upper sign corresponds to $G = a^3 + x^3$, and the lower corresponds to $G = a^3 - x^3$.

$$\int \frac{dx}{xG} = \frac{1}{3a^3} \ln \frac{x^3}{G}$$

$$\int \frac{dx}{xG^2} = \frac{1}{3a^3 G} + \frac{1}{3a^6} \ln \frac{x^3}{G}$$

$$\int \frac{dx}{xG^3} = \frac{1}{6a^3 G^2} + \frac{1}{a^3} \int \frac{dx}{xG^2}$$

$$\int \frac{dx}{xG^n} = \frac{1}{3a^3(n-1)G^{n-1}} + \frac{1}{a^3} \int \frac{dx}{xG^{n-1}} \qquad n \neq 1$$

$$\int \frac{dx}{x^2 G} = -\frac{1}{a^3 x} \mp \frac{1}{a^3} \int \frac{dx}{G}$$

$$\int \frac{dx}{x^2 G^2} = \frac{-1}{a^6 x} \mp \frac{x^2}{3a^6 G} \mp \frac{4}{3a^6} \int \frac{x\,dx}{G}$$

$$\int \frac{dx}{x^2 G^n} = \frac{1}{3a^3(n-1)xG^{n-1}} + \frac{3n-2}{3a^3(n-1)} \int \frac{dx}{x^2 G^{n-1}} \qquad n \neq 1$$

$$\int \frac{dx}{x^3 G} = -\frac{1}{2a^3 x^2} \mp \frac{1}{a^3} \int \frac{dx}{G}$$

$$\int \frac{dx}{x^3 G^2} = -\frac{1}{2a^6 x^2} \mp \frac{x}{3a^6 G} \mp \frac{5}{3a^6} \int \frac{dx}{G}$$

$$\int \frac{dx}{x^3 G^n} = \frac{1}{3a^3(n-1)x^2 G^{n-1}} + \frac{3n-1}{3a^3(n-1)} \int \frac{dx}{x^3 G^{n-1}} \qquad n \neq 1$$

$$\int \frac{dx}{x^m G^n} = \frac{1}{3a^3(n-1)x^{m-1}G^{n-1}} + \frac{m+3n-4}{3a^3(n-1)} \int \frac{dx}{x^m G^{n-1}} \qquad n \neq 1$$

$$\int \frac{a-x}{a^3-x^3}\,dx = \int \frac{dx}{a^2+ax+x^2} = \frac{2}{a\sqrt{3}} \tan^{-1} \frac{2x+a}{a\sqrt{3}}$$

$$\int \frac{a+x}{a^3-x^3}\,dx = -\frac{1}{3a} \ln \frac{(a-x)^2}{a^2+ax+x^2}$$

(14) Indefinite Integrals Involving

$f(x) = \dfrac{x^n}{H^m}$	$f(x) = \dfrac{1}{x^m H^m}$	$H = a^4 + x^4$	$a^4 \neq 0$

$$\int \frac{dx}{H} = \frac{1}{a^3 \sqrt{8}} \left(\tanh^{-1} \frac{ax\sqrt{2}}{a^2 + x^2} + \tan^{-1} \frac{ax\sqrt{2}}{a^2 - x^2} \right)$$

$$\int \frac{dx}{H^n} = \frac{x}{4(n-1)a^4 H^{n-1}} + \frac{4n-5}{4(n-1)a^4} \int \frac{dx}{H^{n-1}} \qquad n \neq 1$$

$$\int \frac{x\,dx}{H} = \frac{1}{2a^2} \tan^{-1} \left(\frac{x^2}{a^2} \right)$$

$$\int \frac{x^2\,dx}{H} = \frac{1}{a\sqrt{8}} \left(\tanh^{-1} \frac{ax\sqrt{2}}{a^2 - x^2} - \tan^{-1} \frac{ax\sqrt{2}}{a^2 + x^2} \right)$$

$$\int \frac{x^3\,dx}{H} = \tfrac{1}{4} \ln H$$

$$\int \frac{x^4\,dx}{H} = x - \frac{a}{\sqrt{8}} \left(\tanh^{-1} \frac{ax\sqrt{2}}{a^2 + x^2} + \tan^{-1} \frac{ax\sqrt{2}}{a^2 - x^2} \right)$$

$$\int \frac{x^m\,dx}{H} = \frac{x^{m-3}}{m-3} - a^4 \int \frac{x^{m-4}}{H}\,dx \qquad m \neq 3$$

$$\int \frac{x^m\,dx}{H^n} = \frac{x^{m+1}}{4a^4(n-1)H^{n-1}} + \frac{4n-m-5}{4a^4(n-1)} \int \frac{x^m\,dx}{H^{n-1}} \qquad n \neq 1$$

$$\int \frac{dx}{xH} = \frac{1}{2a^4} \ln \frac{x^2}{\sqrt{H}}$$

$$\int \frac{dx}{x^2 H} = -\frac{1}{a^4 x} - \frac{1}{a^5 \sqrt{8}} \left(\tanh^{-1} \frac{ax\sqrt{2}}{a^2 - x^2} - \tan^{-1} \frac{ax\sqrt{2}}{a^2 + x^2} \right)$$

$$\int \frac{dx}{x^3 H} = -\frac{1}{2a^4 x^2} - \frac{1}{2a^6} \tan^{-1} \frac{x^2}{a^2}$$

$$\int \frac{dx}{x^4 H} = -\frac{1}{3a^4 x^3} - \frac{1}{a^7 \sqrt{8}} \left(\tanh^{-1} \frac{ax\sqrt{2}}{a^2 + x^2} + \tan^{-1} \frac{ax\sqrt{2}}{a^2 - x^2} \right)$$

$$\int \frac{dx}{x^m H} = \frac{-1}{(m-1)a^4 x^{m-1}} - \frac{1}{a^4} \int \frac{dx}{x^{m-4} H} \qquad m \neq 1$$

$$\int \frac{dx}{x^m H^n} = \frac{1}{4(n-1)a^4 x^{m-1} H^{n-1}} + \frac{m+4n-5}{4(n-1)a^4} \int \frac{dx}{x^m H^{n-1}} \qquad n \neq 1$$

$$= \frac{-1}{(m-1)a^4 x^{m-1} H^{n-1}} - \frac{m+4n-5}{(m-1)a^4} \int \frac{dx}{x^{m-4} H^n} \qquad m \neq 1$$

(15) Indefinite Integrals Involving $\quad f(x) = \dfrac{x^m}{L^n} \qquad f(x) = \dfrac{1}{x^m L^n} \qquad L = a^4 - x^4 \qquad a^4 \neq 0$

$$\int \frac{dx}{L} = \frac{1}{2a^3} \left(\ln \sqrt{\frac{a+x}{a-x}} + \tan^{-1} \frac{x}{a} \right)$$

$$\int \frac{dx}{L^n} = \frac{x}{4(n-1)a^4 L^{n-1}} + \frac{4n-5}{4(n-1)a^4} \int \frac{dx}{L^{n-1}} \qquad n \neq 1$$

$$\int \frac{x \, dx}{L} = \frac{1}{4a^2} \ln \frac{a^2 + x^2}{a^2 - x^2}$$

$$\int \frac{x^2 \, dx}{L} = \frac{1}{2a} \left(\ln \sqrt{\frac{a+x}{a-x}} - \tan^{-1} \frac{x}{a} \right)$$

$$\int \frac{x^3 \, dx}{L} = -\tfrac{1}{4} \ln L$$

$$\int \frac{x^4 \, dx}{L} = -x + \frac{a}{2} \left(\ln \sqrt{\frac{a+x}{a-x}} + \tan^{-1} \frac{x}{a} \right)$$

$$\int \frac{x^m \, dx}{L} = -\frac{x^{m-3}}{m-3} + a^4 \int \frac{x^{m-4}}{L} \, dx \qquad m \neq 3$$

$$\int \frac{x^m \, dx}{L^n} = \frac{x^{m+1}}{4a^4(n-1)L^{n-1}} + \frac{4n-m-5}{4a^4(n-1)} \int \frac{x^m \, dx}{L^{n-1}} \qquad n \neq 1$$

$$\int \frac{dx}{xL} = \frac{1}{2a^4} \ln \frac{x^2}{\sqrt{L}}$$

$$\int \frac{dx}{x^2 L} = -\frac{1}{a^4 x} + \frac{1}{2a^5} \left(\ln \sqrt{\frac{a+x}{a-x}} - \tan^{-1} \frac{x}{a} \right)$$

$$\int \frac{dx}{x^3 L} = -\frac{1}{2a^4 x^2} + \frac{1}{4a^6} \ln \frac{a^2 + x^2}{a^2 - x^2}$$

$$\int \frac{dx}{x^4 L} = -\frac{1}{3a^4 x^3} + \frac{1}{2a^7} \left(\ln \sqrt{\frac{a+x}{a-x}} + \tan^{-1} \frac{x}{a} \right)$$

$$\int \frac{dx}{x^m L} = \frac{-1}{(m-1)a^4 x^{m-1}} + \frac{1}{a^4} \int \frac{dx}{x^{m-4} L} \qquad m \neq 1$$

$$\int \frac{dx}{x^m L^n} = \frac{1}{4(n-1)a^4 x^{m-1} L^{n-1}} + \frac{m+4n-5}{4(n-1)a^4} \int \frac{dx}{x^m L^{n-1}} \qquad n \neq 1$$

$$= \frac{-1}{(m-1)a^4 x^{m-1} L^{n-1}} + \frac{m+4n-5}{(m-1)a^4} \int \frac{dx}{x^{m-4} L^n} \qquad m \neq 1$$

(16) Indefinite Integrals Involving

$f(x) = \sqrt[p]{x^m}/M^n, N^n, P^n$	$M = a^2 + b^2x$	$a^2 \neq 0$
	$N = a^2 - b^2x$	$a^2 \neq 0$
	$P = a + b\sqrt{x}$	$b \neq 0$

$$\int \sqrt{x^m}\, dx = \frac{2x\sqrt{x^m}}{m+2} \qquad m \neq -2$$

$$\int \sqrt[p]{x^m}\, dx = \frac{px\sqrt[p]{x^m}}{m+p} \qquad m+p \neq 0$$

$$\int \frac{\sqrt{x}\, dx}{M} = \frac{2\sqrt{x}}{b^2} - \frac{2a}{b^3}\tan^{-1}\frac{b\sqrt{x}}{a}$$

$$\int \frac{\sqrt{x}\, dx}{M^2} = \frac{-\sqrt{x}}{b^2M} + \frac{1}{ab^3}\tan^{-1}\frac{b\sqrt{x}}{a}$$

$$\int \frac{\sqrt{x^3}\, dx}{M} = \frac{2\sqrt{x^3}}{3b^2} - \frac{2a^2\sqrt{x}}{b^4} + \frac{2a^3}{b^5}\tan^{-1}\frac{b\sqrt{x}}{a}$$

$$\int \frac{\sqrt{x^3}\, dx}{M^2} = \frac{2\sqrt{x^3}}{b^2M} + \frac{3a^2\sqrt{x}}{b^4M} - \frac{3a}{b^5}\tan^{-1}\frac{b\sqrt{x}}{a}$$

$$\int \frac{\sqrt[p]{x^m}\, dx}{M^n} = \frac{+1}{\sqrt[p]{b^{2+2m}}} \int \frac{(M-a^2)^{m/p}}{M^n}\, dM$$

$$\int \frac{\sqrt{x}\, dx}{N} = -\frac{2\sqrt{x}}{b^2} + \frac{a}{b^3}\ln\frac{a+b\sqrt{x}}{a-b\sqrt{x}}$$

$$\int \frac{\sqrt{x}\, dx}{N^2} = \frac{\sqrt{x}}{b^2M} - \frac{1}{2ab^3}\ln\frac{a+b\sqrt{x}}{a-b\sqrt{x}}$$

$$\int \frac{\sqrt{x^3}\, dx}{N} = -\frac{2\sqrt{x^3}}{3b^2} - \frac{2a^2\sqrt{x}}{b^4} + \frac{a^3}{b^5}\ln\frac{a+b\sqrt{x}}{a-b\sqrt{x}}$$

$$\int \frac{\sqrt{x^3}\, dx}{N^2} = -\frac{2\sqrt{x^3}}{b^2N} + \frac{3a^2\sqrt{x}}{b^4N} - \frac{3a}{2b^5}\ln\frac{a+b\sqrt{x}}{a-b\sqrt{x}}$$

$$\int \frac{\sqrt[p]{x^m}\, dx}{N^n} = \frac{-1}{\sqrt[p]{b^{2+2m}}} \int \frac{(a^2-N)^{m/p}}{N^n}\, dN$$

$$\int \frac{dx}{P} = \frac{2}{b^2}[P - a(1 + \ln P)]$$

$$\int \frac{dx}{P^2} = \frac{2}{Pb^2}(a + P\ln P)$$

$$\int \frac{\sqrt{x}\, dx}{P} = \frac{1}{b}[x - \frac{2a}{b}(\sqrt{x} - \frac{a}{b}\ln P)]$$

(17) Indefinite Integrals Involving

$$f(x) = \frac{M^n, N^n, P^n}{\sqrt[p]{x^m}} \qquad \begin{array}{ll} M = a^2 + b^2x & a^2 \neq 0 \\ N = a^2 - b^2x & a^2 \neq 0 \\ P = a + b\sqrt{x} & b \neq 0 \end{array}$$

$$\int \frac{dx}{\sqrt{x^m}} = \frac{2x}{(2-m)\sqrt{x^m}} \qquad m \neq 2$$

$$\int \frac{dx}{\sqrt[p]{x^m}} = \frac{px}{(p-m)\sqrt[p]{x^m}} \qquad m \neq p$$

$$\int \frac{dx}{M\sqrt{x}} = \frac{2}{ab}\tan^{-1}\frac{b\sqrt{x}}{a}$$

$$\int \frac{dx}{M^2\sqrt{x}} = \frac{\sqrt{x}}{a^2M} + \frac{1}{a^3b}\tan^{-1}\frac{b\sqrt{x}}{a}$$

$$\int \frac{dx}{M\sqrt{x^3}} = \frac{-2}{a^2\sqrt{x}} - \frac{2b}{a^3}\tan^{-1}\frac{b\sqrt{x}}{a}$$

$$\int \frac{dx}{M^2\sqrt{x^3}} = \frac{-2}{a^2M\sqrt{x}} - \frac{3b^2\sqrt{x}}{a^4M} - \frac{3b}{a^5}\tan^{-1}\frac{b\sqrt{x}}{a}$$

$$\int \frac{dx}{M^n\sqrt[p]{x^m}} = \frac{-p}{(m-p)a^2x^{m/p-1}M^{n-1}} - \frac{b^2(m+pn-2p)}{a^2(m-p)}\int \frac{dx}{x^{m/p-1}M^n} \qquad m \neq p$$

$$\int \frac{dx}{N\sqrt{x}} = \frac{1}{ab}\ln\frac{a+b\sqrt{x}}{a-b\sqrt{x}}$$

$$\int \frac{dx}{N^2\sqrt{x}} = \frac{\sqrt{x}}{a^2N} + \frac{1}{2a^3b}\ln\frac{a+b\sqrt{x}}{a-b\sqrt{x}}$$

$$\int \frac{dx}{N\sqrt{x^3}} = \frac{-2}{a^2\sqrt{x}} + \frac{b}{a^3}\ln\frac{a+b\sqrt{x}}{a-b\sqrt{x}}$$

$$\int \frac{dx}{N^2\sqrt{x^3}} = \frac{-2}{a^2N\sqrt{x}} + \frac{3b^2\sqrt{x}}{a^4N} + \frac{3b}{2a^5}\ln\frac{a+b\sqrt{x}}{a-b\sqrt{x}}$$

$$\int \frac{dx}{N^n\sqrt[p]{x^m}} = \frac{-p}{(m-p)a^2x^{m/p-1}N^{n-1}} + \frac{b^2(m+pn-2p)}{a^2(m-p)}\int \frac{dx}{x^{m/p-1}N^n} \qquad m \neq p$$

$$\int \frac{dx}{P\sqrt{x}} = \frac{2}{b}\ln P$$

$$\int \frac{dx}{Px} = \frac{1}{a}(\ln x - 2\ln P)$$

$$\int \frac{dx}{P^2x} = \frac{2}{Pa^2}[a + P(\ln\sqrt{x} - \ln P)]$$

(18) Indefinite Integrals Involving

$$f(x) = x^m \sqrt[p]{A^n} \qquad A = a + bx \qquad a \neq 0$$

$$\int \sqrt{A}\, dx = \frac{2}{3b} \sqrt{A^3}$$

$$S = -\frac{A}{bx}$$

$$\int \sqrt{A^n}\, dx = \frac{2}{(n+2)b} \sqrt{A^{n+2}}$$

$$\int \sqrt[p]{A^n}\, dx = \frac{p}{(n+p)b} \sqrt[p]{A^{n+p}}$$

$$\int x\sqrt{A}\, dx = \frac{2x}{3b} \sqrt{A^3} \left(1 + \frac{2S}{5} \right)$$

$$\int x\sqrt{A^n}\, dx = \frac{2x}{(n+2)b} \sqrt{A^{n+2}} \left(1 + \frac{2S}{n+4} \right)$$

$$\int x\sqrt[p]{A^n}\, dx = \frac{px}{(n+p)b} \sqrt[p]{A^{n+p}} \left(1 + \frac{pS}{n+2p} \right)$$

$$\int x^2\sqrt{A}\, dx = \frac{2x^2}{3b} \sqrt{A^3} \left(1 + \frac{4S}{5} \left(1 + \frac{2S}{7} \right) \right)$$

$$\int x^2\sqrt{A^n} = \frac{2x^2}{(n+2)b} \sqrt{A^{n+2}} \left(1 + \frac{4S}{n+4} \left(1 + \frac{2S}{n+6} \right) \right)$$

$$\int x^2 \sqrt[p]{A^n}\, dx = \frac{px^2}{(n+p)b} \sqrt[p]{A^{n+p}} \left(1 + \frac{2pS}{n+2p} \left(1 + \frac{pS}{n+3p} \right) \right)$$

$$\int x^3\sqrt{A}\, dx = \frac{2x^3}{3b} \sqrt{A^3} \left(1 + \frac{6S}{5} \left(1 + \frac{4S}{7} \left(1 + \frac{2S}{9} \right) \right) \right)$$

$$\int x^3\sqrt{A^n}\, dx = \frac{2x^3}{(n+2)b} \sqrt{A^{n+2}} \left(1 + \frac{6S}{n+4} \left(1 + \frac{4S}{n+6} \left(1 + \frac{2S}{n+8} \right) \right) \right)$$

$$\int x^3\sqrt[p]{A^n}\, dx = \frac{px^3}{(n+p)b} \sqrt[p]{A^{n+p}} \left(1 + \frac{3pS}{n+2p} \left(1 + \frac{2pS}{n+3p} \left(1 + \frac{pS}{n+4p} \right) \right) \right)$$

$$\int x^m\sqrt{A}\, dx = \frac{2x^m}{3b} \sqrt{A^3} \overset{m-1}{\underset{k=0}{\Lambda}} \left[1 + \frac{2(m-k)S}{5+2k} \right]^*$$

$$\int x^m \sqrt{A^n}\, dx = \frac{2x^m}{(n+2)b} \sqrt{A^{n+2}} \overset{m-1}{\underset{k=0}{\Lambda}} \left[1 + \frac{2(m-k)S}{n+4+2k} \right]^*$$

$$\int x^m \sqrt[p]{A^n}\, dx = \frac{px^m}{(n+p)b} \sqrt[p]{A^{n+p}} \overset{m-1}{\underset{k=0}{\Lambda}} \left[1 + \frac{p(m-k)S}{n+2p+kp} \right]^*$$

---

*For nested sum refer to Sec. 8.11; S = pocket calculator storage.

(19) Indefinite Integrals Involving

$$f(x) = \frac{x^m}{\sqrt[p]{A^n}} \qquad A = a + bx \qquad a \neq 0$$

$$\int \frac{dx}{\sqrt{A}} = \frac{2\sqrt{A}}{b} \qquad\qquad\qquad\qquad\qquad\qquad\qquad S = -\frac{A}{bx}$$

$$\int \frac{dx}{\sqrt{A^n}} = \frac{-2}{(n-2)b\sqrt{A^{n-2}}} \qquad\qquad n \neq 2$$

$$\int \frac{dx}{\sqrt[p]{A^n}} = \frac{p}{(p-n)b\sqrt[p]{A^{n-p}}} \qquad\qquad p \neq n$$

$$\int \frac{x\,dx}{\sqrt{A}} = \frac{2x\sqrt{A}}{b}\left(1 + \frac{2S}{3}\right)$$

$$\int \frac{x\,dx}{\sqrt{A^n}} = \frac{-2x}{(n-2)b\sqrt{A^{n-2}}}\left(1 + \frac{2S}{4-n}\right) \qquad n \neq 2, 4$$

$$\int \frac{x\,dx}{\sqrt[p]{A^n}} = \frac{px}{(p-n)b\sqrt[p]{A^{n-p}}}\left(1 + \frac{pS}{2p-n}\right) \qquad n \neq p, 2p$$

$$\int \frac{x^2\,dx}{\sqrt{A}} = \frac{2x^2\sqrt{A}}{b}\left(1 + \frac{4S}{3}\left(1 + \frac{2S}{5}\right)\right)$$

$$\int \frac{x^2\,dx}{\sqrt{A^n}} = \frac{2x^2}{(2-n)b\sqrt{A^{n-2}}}\left(1 + \frac{4S}{4-n}\left(1 + \frac{2S}{6-n}\right)\right) \qquad n \neq 2, 4, 6$$

$$\int \frac{x^2\,dx}{\sqrt[p]{A^n}} = \frac{-px^2}{(n-p)b\sqrt[p]{A^{n-p}}}\left(1 + \frac{2pS}{2p-n}\left(1 + \frac{pS}{3p-n}\right)\right) \qquad n \neq p, 2p, 3p$$

$$\int \frac{x^3\,dx}{\sqrt{A}} = \frac{2x^3\sqrt{A}}{b}\left(1 + 2S\left(1 + \frac{4S}{5}\left(1 + \frac{2S}{7}\right)\right)\right)$$

$$\int \frac{x^3\,dx}{\sqrt{A^n}} = \frac{-2x^3}{(n-2)b\sqrt{A^{n-2}}}\left(1 + \frac{6S}{4-n}\left(1 + \frac{4S}{6-n}\left(1 + \frac{2S}{8-n}\right)\right)\right) \qquad n \neq 2, 4, 6, 8$$

$$\int \frac{x^3\,dx}{\sqrt[p]{A^n}} = \frac{px^3}{(p-n)b\sqrt[p]{A^{n-p}}}\left(1 + \frac{3pS}{2p-n}\left(1 + \frac{2pS}{3p-n}\left(1 + \frac{pS}{4p-n}\right)\right)\right) \qquad n \neq p, 2p, 3p, 4p$$

$$\int \frac{x^m\,dx}{\sqrt{A}} = \frac{2x^m\sqrt{A}}{b}\overset{m-1}{\underset{k=0}{\Lambda}}\left[1 + \frac{2(m-k)S}{3+2k}\right]^*$$

$$\int \frac{x^m\,dx}{\sqrt{A^n}} = \frac{-2x^m}{(n-2)b\sqrt{A^{n-2}}}\overset{m-1}{\underset{k=0}{\Lambda}}\left[1 + \frac{2(m-k)S}{4-n+2k}\right]^* \qquad n \neq 2, 4, 6, \ldots, (m+1)2$$

$$\int \frac{x^m\,dx}{\sqrt[p]{A^n}} = \frac{px^m}{(p-n)b\sqrt[p]{A^{n-p}}}\overset{m-1}{\underset{k=0}{\Lambda}}\left[1 + \frac{p(m-k)S}{2p-n+kp}\right]^* \qquad n \neq p, 2p, \ldots, (m+1)p$$

*For nested sum refer to Sec. 8.11; S = pocket calculator storage.

(20) Indefinite Integrals Involving

$f(x) = \dfrac{1}{x^m \sqrt{A^n}}$	$f(x) = \dfrac{\sqrt{A^n}}{x^m}$	$A = a + bx$	$a \neq 0$

$$\int \frac{dx}{x\sqrt{A}} = \begin{cases} -\dfrac{2}{\sqrt{a}}\tanh^{-1}\sqrt{\dfrac{A}{a}} & a > A > 0 \\[2mm] -\dfrac{2}{\sqrt{a}}\coth^{-1}\sqrt{\dfrac{A}{a}} & A > a > 0 \\[2mm] \dfrac{2}{\sqrt{-a}}\tan^{-1}\sqrt{\dfrac{A}{-a}} & a < 0, A > 0 \\[2mm] \dfrac{1}{\sqrt{a}}\ln\dfrac{\sqrt{A}-\sqrt{a}}{\sqrt{A}+\sqrt{a}} & a > 0, A > 0 \end{cases}$$

$$\int \frac{dx}{x\sqrt{A^n}} = \frac{2}{(n-2)a\sqrt{A^{n-2}}} + \frac{1}{a}\int \frac{dx}{x\sqrt{A^{n-2}}} \qquad n > 2$$

$$\int \frac{dx}{x^2\sqrt{A}} = -\frac{\sqrt{A}}{ax} - \frac{b}{2a}\int \frac{dx}{x\sqrt{A}}$$

$$\int \frac{dx}{x^m\sqrt{A^n}} = -\frac{1}{a(m-1)x^{m-1}\sqrt{A^{n-2}}} - \frac{b(2m+n-4)}{2a(m-1)}\int \frac{dx}{x^{m-1}\sqrt{A^n}} \qquad m \neq 1$$

$$\int \frac{\sqrt{A}}{x}dx = 2\sqrt{A} + a\int \frac{dx}{x\sqrt{A}}$$

$$\int \frac{\sqrt{A^n}}{x}dx = \frac{2\sqrt{A^n}}{n} + a\int \frac{\sqrt{A^{n-2}}}{x}dx$$

$$\int \frac{\sqrt{A}}{x^2}dx = -\frac{\sqrt{A}}{x} + \frac{b}{2}\int \frac{dx}{x\sqrt{A}}$$

$$\int \frac{\sqrt{A^n}}{x^2}dx = -\frac{\sqrt{A^{n+2}}}{ax} + \frac{nb}{2a}\int \frac{\sqrt{A^n}}{x}dx$$

$$\int \frac{\sqrt{A^n}}{x^m}dx = -\frac{\sqrt{A^{n+2}}}{a(m-1)x^{m-1}} - \frac{b(2m-n-4)}{2a(m-1)}\int \frac{\sqrt{A^n}}{x^{m-1}}dx \qquad m \neq 1$$

$$\int \sqrt{\frac{x}{A}}\,dx = \frac{\sqrt{x}}{b}A - \frac{a}{b\sqrt{b}}\ln(A + \sqrt{bx})$$

$$\int \sqrt{\frac{A}{x}}\,dx = \sqrt{x}A + \frac{a}{\sqrt{b}}\ln(A + \sqrt{bx})$$

$$\int \sqrt{xA}\,dx = \frac{A+bx}{4b}\sqrt{xA} - \frac{a^2}{8b\sqrt{b}}\cosh^{-1}\frac{A+bx}{a}$$

(21) Indefinite Integrals Involving

$f(x) = f(\sqrt{A}, \sqrt{D})$	$A = a + bx$	$a \neq 0$
	$D = e + fx$	$e \neq 0$

$\beta = af - be \neq 0$

$$\int \frac{\sqrt{A}}{D} dx = \frac{2\sqrt{A}}{f} + \frac{\beta}{f} \int \frac{dx}{D\sqrt{A}}$$

$$\int \frac{A}{\sqrt{D}} dx = \frac{2\sqrt{D}}{3f} \left(A - \frac{2\beta}{f} \right)$$

$$\int \sqrt{\frac{A}{D}} dx = \frac{\sqrt{AD}}{f} - \frac{\beta}{2f} \int \frac{dx}{\sqrt{AD}}$$

$$\int \frac{\sqrt{A}}{D^n} dx = \frac{-1}{(n-1)f} \left(\frac{\sqrt{A}}{D^{n-1}} - \frac{b}{2} \int \frac{dx}{D^{n-1}\sqrt{A}} \right) \qquad n \neq 1$$

$$\int \frac{A^m}{\sqrt{D}} dx = \frac{2}{(2m+1)f} \left(A^m\sqrt{D} - m\beta \int \frac{A^{m-1}}{\sqrt{D}} dx \right)$$

$$\int \sqrt{AD}\, dx = \frac{\beta + 2bD}{4bf} \sqrt{AD} - \frac{\beta^2}{8bf} \int \frac{dx}{\sqrt{AD}}$$

$$\int A^m\sqrt{D}\, dx = \frac{1}{(2m+3)f} \left(2A^{m+1}\sqrt{D} + \beta \int \frac{A^m\, dx}{\sqrt{D}} \right)$$

$$\int \frac{dx}{\sqrt{AD}} = \begin{cases} \dfrac{2}{\sqrt{-bf}} \tan^{-1} \sqrt{-\dfrac{fA}{bD}} & bf < 0 \\[3mm] \dfrac{2}{\sqrt{bf}} \ln (\sqrt{fA} + \sqrt{bD}) & bf > 0 \end{cases}$$

$$\int \frac{dx}{D\sqrt{A}} = \begin{cases} \dfrac{2}{\sqrt{-\beta f}} \tan^{-1} \sqrt{\dfrac{fA}{-\beta}} & \beta f < 0 \\[3mm] \dfrac{1}{\sqrt{\beta f}} \ln \dfrac{\sqrt{fA} - \sqrt{\beta}}{\sqrt{fA} + \sqrt{\beta}} & \beta f > 0 \end{cases}$$

$$\int \frac{dx}{D^n\sqrt{A}} = -\frac{1}{(n-1)\beta} \left[\frac{\sqrt{A}}{D^{n-1}} + \left(n - \frac{3}{2} \right) b \int \frac{dx}{D^{n-1}\sqrt{A}} \right] \qquad n \neq 1$$

$$\int \sqrt{\frac{A}{a-bx}}\, dx = -\frac{1}{b} \sqrt{(a-bx)A} + \frac{a}{b} \sin^{-1} \frac{bx}{a}$$

$$\int \sqrt{\frac{a-bx}{A}}\, dx = \frac{1}{b} \sqrt{(a-bx)A} + \frac{a}{b} \sin^{-1} \frac{bx}{a}$$

$$\int \frac{dx}{\sqrt{xA}} = \frac{1}{\sqrt{b}} \cosh^{-1} \frac{A+bx}{a}$$

(22) Indefinite Integrals Involving

$$f(x) = x^m \sqrt{B^n} \qquad f(x) = \frac{x^m}{\sqrt{B^n}} \qquad B = a + bx + cx^2 \qquad a \neq 0$$

$$\omega = b + 2cx = dB/dx \qquad \gamma = 4ac - b^2 \qquad \lambda = \gamma/8c$$

$$\int \frac{dx}{\sqrt{B}} = \begin{cases} \dfrac{1}{\sqrt{c}} \ln (\omega + 2\sqrt{cB}) & c > 0 & \gamma \neq 0 \\[2mm] \dfrac{1}{\sqrt{c}} \sinh^{-1} \dfrac{\omega}{\sqrt{\gamma}} & c > 0 & \gamma > 0 \\[2mm] \dfrac{1}{\sqrt{c}} \ln \omega & c > 0 & \gamma = 0 \\[2mm] \dfrac{1}{\sqrt{c}} \cosh^{-1} \dfrac{\omega}{\sqrt{-\gamma}} & c > 0 & \gamma < 0 \\[2mm] \dfrac{-1}{\sqrt{-c}} \sin^{-1} \dfrac{\omega}{\sqrt{-\gamma}} & c < 0 & \gamma < 0 \end{cases}$$

$$\int \frac{dx}{\sqrt{B^3}} = \frac{2\omega}{\gamma \sqrt{B}} \qquad\qquad \int \frac{dx}{\sqrt{B^5}} = \frac{2\omega}{3\gamma \sqrt{B}} \left(\frac{1}{B} + \frac{1}{\lambda} \right)$$

$$\int \frac{dx}{\sqrt{B^n}} = \frac{2\omega}{(n-2)\gamma\sqrt{B^{n-2}}} + \frac{(n-3)}{2(n-2)\lambda} \int \frac{dx}{\sqrt{B^{n-2}}} \qquad n \neq 2$$

$$\int \sqrt{B}\, dx = \frac{\omega}{4c} \sqrt{B} + \lambda \int \frac{dx}{\sqrt{B}} \qquad\qquad \int \sqrt{B^3}\, dx = \frac{\omega}{8c} \sqrt{B^3} + \frac{3\lambda}{2} \int \sqrt{B}\, dx$$

$$\int \sqrt{B^n}\, dx = \frac{\omega\sqrt{B^n}}{2(n+1)c} + \frac{2n\lambda}{n+1} \int \sqrt{B^{n-2}}\, dx \qquad n > 0$$

$$\int x\sqrt{B}\, dx = \frac{\sqrt{B^3}}{3c} - \frac{b}{2c} \int \sqrt{B}\, dx \qquad\qquad \int x\sqrt{B^3}\, dx = \frac{\sqrt{B^5}}{5c} - \frac{b}{2c} \int \sqrt{B^3}\, dx$$

$$\int x^m \sqrt{B^n}\, dx = \frac{1}{(m+n+1)c} \left[x^{m-1}\sqrt{B^{n+2}} - \frac{(2m+n)b}{2} \int x^{m-1}\sqrt{B^n}\, dx \right] \qquad n > 0$$

$$\int \frac{x\, dx}{\sqrt{B}} = \mp \frac{\sqrt{B}}{c} - \frac{b}{2c} \int \frac{dx}{\sqrt{B}} \qquad\qquad \int \frac{x\, dx}{\sqrt{B^3}} = -\frac{2\omega}{\gamma\sqrt{B}}$$

$$\int \frac{x^m\, dx}{\sqrt{B^n}} = \frac{1}{(m-n+1)c} \left[\frac{x^{m-1}}{\sqrt{B^{n-2}}} - \frac{(2m-n)b}{2} \int \frac{x^{m-1}}{\sqrt{B^n}}\, dx - (m-1)a \int \frac{x^{m-2}}{\sqrt{B^n}}\, dx \right] \qquad n > 0$$

$$\int \frac{e + fx}{\sqrt{B}}\, dx = \frac{f}{c}\sqrt{B} + \frac{2ce - bf}{2c} \int \frac{dx}{\sqrt{B}} \qquad\qquad \int \frac{b + 2cx}{\sqrt{B}}\, dx = \int \frac{dB}{\sqrt{B}} = 2\sqrt{B}$$

(23) Indefinite Integrals Involving

$$f(x) = \frac{1}{x^m \sqrt{B^n}} \qquad f(x) = \frac{\sqrt{B^n}}{x^m} \qquad B = a + bx + cx^2 \qquad a \neq 0$$

$$\omega = b + 2cx = dB/dx \qquad \gamma = 4ac - b^2 \qquad \theta = 2a + bx$$

$$\int \frac{dx}{x\sqrt{B}} = \begin{cases} \dfrac{-1}{\sqrt{a}} \ln \dfrac{\theta + 2\sqrt{aB}}{x} & a > 0 \qquad \gamma \neq 0 \\[2ex] \dfrac{-1}{\sqrt{a}} \sinh^{-1} \dfrac{\theta}{x\sqrt{\gamma}} & a > 0 \qquad \gamma > 0 \\[2ex] \dfrac{-1}{\sqrt{a}} \ln \dfrac{\theta}{x} & a > 0 \qquad \gamma = 0 \\[2ex] \dfrac{1}{\sqrt{-a}} \sin^{-1} \dfrac{\theta}{x\sqrt{-\gamma}} & a < 0 \qquad \gamma < 0 \\[2ex] \dfrac{1}{\sqrt{-a}} \tan^{-1} \dfrac{\theta}{2\sqrt{-aB}} & a < 0 \qquad \gamma \leq 0 \end{cases}$$

$$\int \frac{dx}{x^2 \sqrt{B}} = -\frac{\sqrt{B}}{ax} - \frac{b}{2a} \int \frac{dx}{x\sqrt{B}} \qquad\qquad \int \frac{dx}{x^3 \sqrt{B}} = \frac{\sqrt{B}}{a^2 x^2}\left(bx - \frac{\theta}{4}\right) + \frac{2b^2 - \gamma}{8a^2} \int \frac{dx}{x\sqrt{B}}$$

$$\int \frac{dx}{x^m \sqrt{B}} = -\frac{\sqrt{B}}{(m-1)ax^{m-1}} - \frac{b(2m-3)}{2a(m-1)} \int \frac{dx}{x^{m-1}\sqrt{B}} - \frac{c(m-2)}{a(m-1)} \int \frac{dx}{x^{m-2}\sqrt{B}} \qquad m \neq 1$$

$$\int \frac{dx}{x\sqrt{B^n}} = \frac{1}{(n-2)a\sqrt{B^{n-2}}} - \frac{b}{2a} \int \frac{dx}{\sqrt{B^n}} + \frac{1}{a} \int \frac{dx}{x\sqrt{B^{n-2}}} \qquad n \neq 2$$

$$\int \frac{\sqrt{B}}{x} dx = \sqrt{B} + \frac{b}{2} \int \frac{dx}{\sqrt{B}} + a \int \frac{dx}{x\sqrt{B}} \qquad\qquad \int \frac{\sqrt{B}}{x^2} dx = -\frac{\sqrt{B}}{x} + \frac{b}{2} \int \frac{dx}{x\sqrt{B}} + c \int \frac{dx}{\sqrt{B}}$$

$$\int \frac{\sqrt{B}}{x^m} dx = -\frac{\sqrt{B}}{(m-1)x^{m-1}} + \frac{b}{2(m-1)} \int \frac{dx}{x^{m-1}\sqrt{B}} + \frac{c}{m-1} \int \frac{dx}{x^{m-2}\sqrt{B}} \qquad m \neq 1$$

$$\int \frac{\sqrt{B^n}}{x^m} dx = -\frac{\sqrt{B^{n+2}}}{a(m-1)x^{m-1}} + \frac{b(n-2m+4)}{2a(m+1)} \int \frac{\sqrt{B^n}}{x^{m-1}} dx + \frac{c(n-m+3)}{a(m-1)} \int \frac{\sqrt{B^n}}{x^{m-2}} dx \qquad m \neq 1$$

$$\int \frac{dx}{x^m \sqrt{B^n}} = \frac{-1}{a(m-1)x^{m-1}\sqrt{B^{n-2}}} - \frac{b(n+2m-4)}{2a(m-1)} \int \frac{dx}{x^{m-1}\sqrt{B^n}}$$

$$- \frac{c(n+m-3)}{a(m-1)} \int \frac{dx}{x^{m-2}\sqrt{B^n}} \qquad m \neq 1$$

$$\int \frac{dx}{(x-f)\sqrt{B}} = -\int \frac{dz}{\sqrt{c + (b + 2cf)z + (a + bf + cf^2)z^2}} \qquad z = \frac{1}{x-f}$$

(24) Indefinite Integrals Involving

$$f(x) = x^m \sqrt[p]{C^n} \qquad C = a^2 + x^2 \qquad a^2 \neq 0$$

$$\int \sqrt{C}\, dx = \tfrac{1}{2}[x\sqrt{C} + a^2 \ln (x + \sqrt{C})]$$

$$\int \sqrt{C^3}\, dx = \frac{x\sqrt{C^3}}{4} + \frac{3a^2}{4} \int \sqrt{C}\, dx$$

$$\int \sqrt{C^n}\, dx = \frac{x\sqrt{C^n}}{n+1} + \frac{na^2}{n+1} \int \sqrt{C^{n-2}}\, dx \qquad n \neq -1$$

$$\int x\sqrt{C}\, dx = \tfrac{1}{3}\sqrt{C^3}$$

$$\int x\sqrt{C^3}\, dx = \tfrac{1}{5}\sqrt{C^5}$$

$$\int x\sqrt{C^n}\, dx = \frac{1}{n+2}\sqrt{C^{n+2}} \qquad n \neq -2$$

$$\int x^2\sqrt{C}\, dx = \frac{x}{4}\sqrt{C^3} - \frac{a^2 x}{8}\sqrt{C} - \frac{a^4}{8}\ln (x + \sqrt{C})$$

$$\int x^2\sqrt{C^3}\, dx = \frac{x}{6}\sqrt{C^5} - \frac{a^2 x}{24}\sqrt{C^3} - \frac{a^4 x}{16}\sqrt{C} - \frac{a^6}{16}\ln (x + \sqrt{C})$$

$$\int x^2\sqrt{C^n}\, dx = \frac{x\sqrt{C^{n+2}}}{n+3} - \frac{a^2}{n+3} \int \sqrt{C^n} \qquad n \neq -3$$

$$\int x^3\sqrt{C}\, dx = \frac{1}{5}\sqrt{C^5} - \frac{a^2}{3}\sqrt{C^3}$$

$$\int x^3\sqrt{C^3}\, dx = \frac{1}{7}\sqrt{C^7} - \frac{a^2}{5}\sqrt{C^5}$$

$$\int x^3\sqrt{C^n} = \left(x^2 - \frac{2a^2}{n+2}\right)\frac{\sqrt{C^{n+2}}}{n+4} \qquad n \neq -2, -4$$

$$\int x^m\sqrt[p]{C^n}\, dx = \frac{px^{m-1}\sqrt[p]{C^{n+p}}}{2n + mp + p} - \frac{a^2 p(m-1)}{2n + mp + p} \int x^{m-2}\sqrt[p]{C^n}\, dx \qquad 2n \neq -p(1+m)$$

$$\int \sqrt{ax + bx^2}\, dx = \frac{a + 2bx}{4b}\sqrt{ax + bx^2} - \frac{a^2}{8b\sqrt{b}}\cosh^{-1}\frac{a + 2bx}{a}$$

$$\int x^m\sqrt{ax + bx^2}\, dx = \frac{x^{m-1}}{b(m+2)}\sqrt{(ax + bx^2)^3} - \frac{a(2m+1)}{2b(m+2)} \int x^{m-1}\sqrt{ax + bx^2}\, dx$$

(25) Indefinite Integrals Involving

$$f(x) = \frac{x^m}{\sqrt{C^n}} \qquad C = a^2 + x^2 \qquad a^2 \neq 0$$

$$\int \frac{dx}{\sqrt{C}} = \sinh^{-1}\frac{x}{a} = \ln\left(x + \sqrt{C}\right)$$

$$\int \frac{dx}{\sqrt{C^3}} = \frac{x}{a^2\sqrt{C}}$$

$$\int \frac{dx}{\sqrt{C^n}} = \frac{x}{a^2(n-2)\sqrt{C^{n-2}}} - \frac{3-n}{a^2(n-2)}\int \frac{dx}{\sqrt{C^{n-2}}} \qquad n \neq 2$$

$$\int \frac{x\,dx}{\sqrt{C}} = \sqrt{C}$$

$$\int \frac{x\,dx}{\sqrt{C^3}} = \frac{-1}{\sqrt{C}}$$

$$\int \frac{x\,dx}{\sqrt{C^n}} = \frac{x^2}{a^2(n-2)\sqrt{C^{n-2}}} - \frac{4-n}{a^2(n-2)}\int \frac{x\,dx}{\sqrt{C^{n-2}}} \qquad n \neq 2$$

$$\int \frac{x^2\,dx}{\sqrt{C}} = \frac{x}{2}\sqrt{C} - \frac{a^2}{2}\ln\left(x + \sqrt{C}\right)$$

$$\int \frac{x^2\,dx}{\sqrt{C^3}} = \frac{-x}{\sqrt{C}} + \ln\left(x + \sqrt{C}\right)$$

$$\int \frac{x^2\,dx}{\sqrt{C^n}} = \frac{x^3}{a^2(n-2)\sqrt{C^{n-2}}} - \frac{5-n}{a^2(n-2)}\int \frac{x^2\,dx}{\sqrt{C^{n-2}}} \qquad n \neq 2$$

$$\int \frac{x^3\,dx}{\sqrt{C}} = \left(\frac{C}{3} - a^2\right)\sqrt{C}$$

$$\int \frac{x^3\,dx}{\sqrt{C^3}} = \frac{C + a^2}{\sqrt{C}}$$

$$\int \frac{x^3\,dx}{\sqrt{C^n}} = \frac{1}{4-n}\left(x^2 + \frac{2a^2}{n-2}\right)\frac{1}{\sqrt{C^{n-2}}} \qquad n \neq 2, 4$$

$$\int \frac{x^m}{\sqrt{C^n}}\,dx = \frac{px^{m+1}}{2a^2(n-p)\sqrt{C^{n-p}}} + \frac{2n - p(m+3)}{2a^2(n-p)}\int \frac{x^m\,dx}{\sqrt{C^{n-p}}} \qquad n \neq p$$

$$\int \frac{dx}{\sqrt{ax + bx^2}} = \frac{1}{\sqrt{b}}\cosh^{-1}\frac{a + 2bx}{a}$$

$$\int \frac{x^m\,dx}{\sqrt{ax + bx^2}} = \frac{x^{m-1}}{bm}\sqrt{ax + bx^2} - \frac{a(2m-1)}{2bm}\int \frac{x^{m-1}\,dx}{\sqrt{ax + bx^2}}$$

(26) Indefinite Integrals Involving

$$f(x) = \frac{1}{x^m \sqrt[p]{C^n}} \qquad C = a^2 + x^2 \qquad a^2 \neq 0$$

$$\int \frac{dx}{x\sqrt{C}} = -\frac{1}{a} \ln \frac{a + \sqrt{C}}{x}$$

$$\int \frac{dx}{x\sqrt{C^3}} = \frac{1}{a^2\sqrt{C}} - \frac{1}{a^3} \ln \frac{a + \sqrt{C}}{x}$$

$$\int \frac{dx}{x\sqrt{C^n}} = \frac{1}{a^2(n-2)\sqrt{C^{n-2}}} + \frac{1}{a^2} \int \frac{dx}{x\sqrt{C^{n-2}}} \qquad n \neq 2$$

$$\int \frac{dx}{x^2\sqrt{C}} = -\frac{\sqrt{C}}{a^2 x}$$

$$\int \frac{dx}{x^2\sqrt{C^3}} = -\frac{\sqrt{C}}{a^4 x}\left(1 + \frac{x^2}{x}\right)$$

$$\int \frac{dx}{x^2\sqrt{C^n}} = \frac{1}{a^2(n-2)x\sqrt{C^{n-2}}} + \frac{n-1}{a^2(n-2)} \int \frac{dx}{x^2\sqrt{C^{n-2}}} \qquad n \neq 2$$

$$\int \frac{dx}{x^3\sqrt{C}} = -\frac{\sqrt{C}}{2a^2 x^2} + \frac{1}{2a^3} \ln \frac{a + \sqrt{C}}{x}$$

$$\int \frac{dx}{x^3\sqrt{C^3}} = \frac{-1}{2a^2 x^2 \sqrt{C}} - \frac{3}{2a^4\sqrt{C}} + \frac{3}{2a^5} \ln \frac{a + \sqrt{C}}{x}$$

$$\int \frac{dx}{x^3\sqrt{C^n}} = \frac{1}{a^2(n-2)x^2\sqrt{C^{n-2}}} + \frac{n}{a^2(n-2)} \int \frac{dx}{x^3\sqrt{C^{n-2}}} \qquad n \neq 2$$

$$\int \frac{dx}{x^4\sqrt{C}} = \frac{\sqrt{C}}{a^4 x}\left(1 - \frac{C}{3x^2}\right)$$

$$\int \frac{dx}{x^4\sqrt{C^3}} = \frac{x}{a^6\sqrt{C}}\left(1 + \frac{2C}{x^2} - \frac{C^2}{3x^4}\right)$$

$$\int \frac{dx}{x^4\sqrt{C^n}} = \frac{1}{a^2(n-2)x^3\sqrt{C^{n-2}}} + \frac{n+1}{a^2(n-2)} \int \frac{dx}{x^4\sqrt{C^{n-2}}} \qquad n \neq 2$$

$$\int \frac{dx}{x^m\sqrt[p]{C^n}} = \frac{p}{2a^2(n-p)x^{m-1}\sqrt[p]{C^{n-p}}} + \frac{2n + p(m-3)}{2a^2(n-p)} \int \frac{dx}{x^m\sqrt[p]{C^{n-p}}} \qquad n \neq p$$

$$\int \frac{dx}{x\sqrt{ax + bx^2}} = \frac{2}{ax}\sqrt{ax + bx^2}$$

$$\int \frac{dx}{x^m\sqrt{ax + bx^2}} = -\frac{2\sqrt{ax + bx^2}}{a(2m-1)x^m} - \frac{2b(m-1)}{a(2m-1)} \int \frac{dx}{x^{m-1}\sqrt{ax + bx^2}}$$

(27) Indefinite Integrals Involving

$$f(x) = \frac{\sqrt[p]{C^n}}{x^m} \qquad C = a^2 + x^2 \qquad a^2 \neq 0$$

$$\int \frac{\sqrt{C}}{x}\,dx = \sqrt{C} - a \ln \frac{a + \sqrt{C}}{x}$$

$$\int \frac{\sqrt{C^3}}{x}\,dx = \frac{1}{3}\sqrt{C^3} + a^2\sqrt{C} - a^3 \ln \frac{a + \sqrt{C}}{x}$$

$$\int \frac{\sqrt{C^n}}{x}\,dx = \frac{1}{n}\sqrt{C^n} + a^2 \int \frac{\sqrt{C^{n-2}}}{x}\,dx$$

$$\int \frac{\sqrt{C}}{x^2}\,dx = -\frac{1}{x}\sqrt{C} + \ln (x + \sqrt{C})$$

$$\int \frac{\sqrt{C^3}}{x^2}\,dx = -\frac{1}{x}\sqrt{C^3} + \frac{3x}{2}\sqrt{C} + \frac{3a^2}{2} \ln (x + \sqrt{C})$$

$$\int \frac{\sqrt{C^n}}{x^2}\,dx = \frac{\sqrt{C^n}}{(n-1)x} + \frac{a^2 n}{n-1} \int \frac{\sqrt{C^{n-2}}}{x^2}\,dx \qquad n \neq 1$$

$$\int \frac{\sqrt{C}}{x^3}\,dx = -\frac{1}{2x^2}\sqrt{C} - \frac{1}{2a} \ln \frac{a + \sqrt{C}}{x}$$

$$\int \frac{\sqrt{C^3}}{x^3}\,dx = -\frac{1}{2x^2}\sqrt{C^3} + \frac{3}{2}\sqrt{C} - \frac{3a}{2} \ln \frac{a + \sqrt{C}}{x}$$

$$\int \frac{\sqrt{C^n}}{x^3}\,dx = \frac{\sqrt{C^n}}{(n-2)x^2} + \frac{a^2 n}{n-2} \int \frac{\sqrt{C^{n-2}}}{x^3}\,dx \qquad n \neq 2$$

$$\int \frac{\sqrt{C}}{x^4}\,dx = -\frac{\sqrt{C^3}}{3a^2 x^3}$$

$$\int \frac{\sqrt{C^3}}{x^4}\,dx = -\frac{\sqrt{C^3}}{3x^3} - \frac{\sqrt{C}}{x} + \ln (x + \sqrt{C})$$

$$\int \frac{\sqrt{C^n}}{x^4}\,dx = \frac{\sqrt{C^n}}{(n-3)x^3} + \frac{a^2 n}{n-3} \int \frac{\sqrt{C^{n-2}}}{x^4}\,dx \qquad n \neq 3$$

$$\int \frac{\sqrt[p]{C^n}}{x^m}\,dx = \frac{p\sqrt[p]{C^n}}{(2n - mp + p)x^{m-1}} + \frac{2a^2 n}{2n - mp + p} \int \frac{\sqrt[p]{C^{n-p}}}{x^m}\,dx \qquad 2n \neq -p(1 - m)$$

$$\int \frac{\sqrt{ax + bx^2}}{x}\,dx = \sqrt{ax + bx^2} + \frac{a}{2\sqrt{b}} \cosh^{-1} \frac{a + 2bx}{a}$$

$$\int \frac{\sqrt{ax + bx^2}}{x^m}\,dx = -\frac{2\sqrt{(ax + bx^2)^3}}{a(2m - 3)x^{m+1}} - \frac{2b(m - 3)}{a(2m - 3)} \int \frac{\sqrt{ax + bx^2}}{x^{m-1}}\,dx \qquad m \neq 1$$

(28) Indefinite Integrals Involving

$$f(x) = x^m \sqrt[p]{E^n} \qquad E = a^2 - x^2 \qquad a^2 \neq 0$$

$$\int \sqrt{E} \, dx = \frac{1}{2}\left(x\sqrt{E} + a^2 \sin^{-1}\frac{x}{a}\right)$$

$$\int \sqrt{E^3} \, dx = \frac{1}{8}\left(2x\sqrt{E^3} + 3a^2x\sqrt{E} + 3a^4 \sin^{-1}\frac{x}{a}\right)$$

$$\int \sqrt{E^n} \, dx = \frac{x\sqrt{E^n}}{n+1} + \frac{a^2n}{n+1}\int \sqrt{E^{n-2}} \, dx \qquad n \neq -1$$

$$\int x\sqrt{E} \, dx = -\tfrac{1}{3}\sqrt{E^3}$$

$$\int x\sqrt{E^3} \, dx = -\tfrac{1}{5}\sqrt{E^5}$$

$$\int x\sqrt{E^n} \, dx = \frac{x^2\sqrt{E^n}}{n+2} + \frac{a^2n}{n+2}\int x\sqrt{E^{n-2}} \, dx \qquad n \neq -2$$

$$\int x^2\sqrt{E} \, dx = -\frac{x}{4}\sqrt{E^3} + \frac{a^2}{8}\left(x\sqrt{E} + a^2 \sin^{-1}\frac{x}{a}\right)$$

$$\int x^2\sqrt{E^3} \, dx = -\frac{x}{6}\sqrt{E^5} + \frac{a^2x}{24}\sqrt{E^3} + \frac{a^4}{16}\left(x\sqrt{E} + a^2 \sin^{-1}\frac{x}{a}\right)$$

$$\int x^2\sqrt{E^n} \, dx = \frac{x^3\sqrt{E^n}}{n+3} + \frac{a^2n}{n+3}\int x^2\sqrt{E^{n-2}} \, dx \qquad n \neq -3$$

$$\int x^3\sqrt{E} \, dx = \frac{1}{5}\sqrt{E^5} - \frac{a^2}{3}\sqrt{E^3}$$

$$\int x^3\sqrt{E^3} \, dx = \frac{1}{7}\sqrt{E^7} - \frac{a^2}{5}\sqrt{E^5}$$

$$\int x^3\sqrt{E^n} \, dx = \frac{x^4\sqrt{E^n}}{n+4} + \frac{a^2n}{n+4}\int x^3\sqrt{E^{n-2}} \, dx \qquad n \neq -4$$

$$\int x^m\sqrt[p]{E^n} \, dx = \frac{px^{m+1}\sqrt[p]{E^n}}{2n+mp+p} + \frac{2a^2n}{2n+mp+p}\int x^m\sqrt[p]{E^{n-p}} \, dx \qquad 2n \neq -p(1+m)$$

$$\int \sqrt{ax - bx^2} \, dx = \frac{2bx - a}{4b}\sqrt{ax - bx^2} + \frac{a^2}{4b\sqrt{b}}\sin^{-1}\sqrt{\frac{bx}{a}}$$

$$\int x^m\sqrt{ax - bx^2} \, dx = -\frac{x^{m-1}}{b(m+2)}\sqrt{(ax - bx^2)^3} + \frac{a(2m+1)}{2b(m+2)}\int x^{m-1}\sqrt{ax - bx^2} \, dx$$

(29) Indefinite Integrals Involving

$$f(x) = \frac{x^m}{\sqrt[p]{E^n}} \qquad E = a^2 - x^2 \qquad a^2 \neq 0$$

$$\int \frac{dx}{\sqrt{E}} = \sin^{-1}\frac{x}{a}$$

$$\int \frac{dx}{\sqrt{E^3}} = \frac{x}{a^2\sqrt{E}}$$

$$\int \frac{dx}{\sqrt{E^n}} = \frac{x}{a^2(n-2)\sqrt{E^{n-2}}} + \frac{n-3}{a^2(n-2)}\int \frac{dx}{\sqrt{E^{n-2}}} \qquad n \neq 2$$

$$\int \frac{x\,dx}{\sqrt{E}} = -\sqrt{E}$$

$$\int \frac{x\,dx}{\sqrt{E^3}} = \frac{1}{\sqrt{E}}$$

$$\int \frac{x\,dx}{\sqrt{E^n}} = \frac{x^2}{a^2(n-2)\sqrt{E^{n-2}}} + \frac{n-4}{a^2(n-2)}\int \frac{x\,dx}{\sqrt{E^{n-2}}} \qquad n \neq 2$$

$$\int \frac{x^2\,dx}{\sqrt{E}} = -\frac{x}{2}\sqrt{E} + \frac{a^2}{2}\sin^{-1}\frac{x}{a}$$

$$\int \frac{x^2\,dx}{\sqrt{E^3}} = \frac{x}{\sqrt{E}} - \sin^{-1}\frac{x}{a}$$

$$\int \frac{x^2\,dx}{\sqrt{E^n}} = \frac{x^3}{a^2(n-2)\sqrt{E^{n-2}}} + \frac{n-5}{a^2(n-2)}\int \frac{x^2\,dx}{\sqrt{E^{n-2}}} \qquad n \neq 2$$

$$\int \frac{x^3\,dx}{\sqrt{E}} = \left(\frac{E}{3} - a^2\right)\sqrt{E}$$

$$\int \frac{x^3\,dx}{\sqrt{E^3}} = \frac{E + a^2}{\sqrt{E}}$$

$$\int \frac{x^3\,dx}{\sqrt{E^n}} = \frac{x^4}{a^2(n-2)\sqrt{E^{n-2}}} + \frac{n-6}{a^2(n-2)}\int \frac{x^3\,dx}{\sqrt{E^{n-2}}} \qquad n \neq 2$$

$$\int \frac{x^m\,dx}{\sqrt[p]{E^n}} = \frac{px^{m+1}}{2a^2(n-p)\sqrt[p]{E^{n-p}}} + \frac{2n - p(m+3)}{2a^2(n-p)}\int \frac{x^m\,dx}{\sqrt[p]{E^{n-p}}} \qquad n \neq p$$

$$\int \frac{dx}{\sqrt{ax - bx^2}} = \frac{2}{\sqrt{b}}\sin^{-1}\sqrt{\frac{bx}{a}}$$

$$\int \frac{x^m\,dx}{\sqrt{ax - bx^2}} = -\frac{x^{m-1}}{bm}\sqrt{ax - bx^2} + \frac{(2m+1)a}{2bm}\int \frac{x^{m-1}\,dx}{\sqrt{ax - bx^2}}$$

(30) Indefinite Integrals Involving
$$f(x) = \frac{1}{x^m \sqrt[p]{E^n}} \qquad E = a^2 - x^2 \qquad a^2 \neq 0$$

$$\int \frac{dx}{x\sqrt{E}} = -\frac{1}{a} \ln \frac{a + \sqrt{E}}{x}$$

$$\int \frac{dx}{x\sqrt{E^3}} = \frac{1}{a^2 \sqrt{E}} - \frac{1}{a^3} \ln \frac{a + \sqrt{E}}{x}$$

$$\int \frac{dx}{x\sqrt{E^n}} = \frac{1}{a^2 (n-2) \sqrt{E^{n-2}}} + \frac{n-2}{a^2 (n-2)} \int \frac{dx}{x\sqrt{E^{n-2}}} \qquad n \neq 2$$

$$\int \frac{dx}{x^2 \sqrt{E}} = -\frac{\sqrt{E}}{a^2 x}$$

$$\int \frac{dx}{x^2 \sqrt{E^3}} = -\frac{\sqrt{E}}{a^4 x} \left(1 - \frac{x^2}{E} \right)$$

$$\int \frac{dx}{x^2 \sqrt{E^n}} = \frac{1}{a^2 (n-2) x \sqrt{E^{n-2}}} + \frac{n-1}{a^2 (n-2)} \int \frac{dx}{x^2 \sqrt{E^{n-2}}} \qquad n \neq 2$$

$$\int \frac{dx}{x^3 \sqrt{E}} = -\frac{\sqrt{E}}{2a^2 x^2} - \frac{1}{2a^3} \ln \frac{a + \sqrt{E}}{x}$$

$$\int \frac{dx}{x^3 \sqrt{E^3}} = \frac{-1}{2a^2 x^2 \sqrt{E}} + \frac{3}{2a^4 \sqrt{E}} - \frac{3}{2a^5} \ln \frac{a + \sqrt{E}}{x}$$

$$\int \frac{dx}{x^3 \sqrt{E^n}} = \frac{1}{a^2 (n-2) x^2 \sqrt{E^{n-2}}} + \frac{n}{a^2 (n-2)} \int \frac{dx}{x^3 \sqrt{E^{n-2}}} \qquad n \neq 2$$

$$\int \frac{dx}{x^4 \sqrt{E}} = -\frac{\sqrt{E}}{a^4 x} \left(1 + \frac{E}{3x^2} \right)$$

$$\int \frac{dx}{x^4 \sqrt{E^3}} = \frac{1}{a^6 \sqrt{E}} \left(x - \frac{2E}{x} - \frac{E^2}{3x^3} \right)$$

$$\int \frac{dx}{x^4 \sqrt{E^n}} = \frac{1}{a^2 (n-2) x^3 \sqrt{E^{n-2}}} + \frac{n+1}{a^2 (n-2)} \int \frac{dx}{x^4 \sqrt{E^{n-2}}} \qquad n \neq 2$$

$$\int \frac{dx}{x^m \sqrt[p]{E^n}} = \frac{p}{2a^2 (n-p) x^{m-1} \sqrt[p]{E^{n-p}}} + \frac{2n + p(m-3)}{2a^2 (n-p)} \int \frac{dx}{x^m \sqrt[p]{E^{n-p}}} \qquad n \neq p$$

$$\int \frac{dx}{x\sqrt{ax - bx^2}} = -\frac{2}{ax} \sqrt{ax - bx^2}$$

$$\int \frac{dx}{x^m \sqrt{ax - bx^2}} = -\frac{2\sqrt{ax - bx^2}}{a(2m-1)x^m} + \frac{2b(m-1)}{a(2m-1)} \int \frac{dx}{x^{m-1} \sqrt{ax - bx^2}}$$

(31) Indefinite Integrals Involving

$$f(x) = \frac{\sqrt[p]{E^n}}{x^m} \qquad E = a^2 - x^2 \qquad a^2 \neq 0$$

$$\int \frac{\sqrt{E}}{x}\,dx = \sqrt{E} - a\ln\frac{a + \sqrt{E}}{x}$$

$$\int \frac{\sqrt{E^3}}{x}\,dx = \frac{1}{3}\sqrt{E^3} + a^2\sqrt{E} - a^3\ln\frac{a + \sqrt{E}}{x}$$

$$\int \frac{\sqrt{E^n}}{x}\,dx = \frac{1}{n}\sqrt{E^n} + a^2\int\frac{\sqrt{E^{n-2}}}{x}\,dx$$

$$\int \frac{\sqrt{E}}{x^2}\,dx = -\frac{1}{x}\sqrt{E} - \sin^{-1}\frac{x}{a}$$

$$\int \frac{\sqrt{E^3}}{x^2}\,dx = -\frac{1}{x}\sqrt{E^3} - \frac{3x}{2}\sqrt{E} - \frac{3a^2}{2}\sin^{-1}\frac{x}{a}$$

$$\int \frac{\sqrt{E^n}}{x^2}\,dx = \frac{\sqrt{E^n}}{(n-1)x} + \frac{a^2n}{n-1}\int\frac{\sqrt{E^{n-2}}}{x^2}\,dx \qquad n \neq 1$$

$$\int \frac{\sqrt{E}}{x^3}\,dx = -\frac{1}{2x^2}\sqrt{E} + \frac{1}{2a}\ln\frac{a + \sqrt{E}}{x}$$

$$\int \frac{\sqrt{E^3}}{x^3}\,dx = -\frac{1}{2x^2}\sqrt{E^3} - \frac{3}{2}\sqrt{E} + \frac{3a}{2}\ln\frac{a + \sqrt{E}}{x}$$

$$\int \frac{\sqrt{E^n}}{x^3}\,dx = \frac{\sqrt{E^n}}{(n-2)x^2} + \frac{a^2n}{n-2}\int\frac{\sqrt{E^{n-2}}}{x^3}\,dx \qquad n \neq 2$$

$$\int \frac{\sqrt{E}}{x^4}\,dx = -\frac{1}{3a^2x^3}\sqrt{E^3}$$

$$\int \frac{\sqrt{E^3}}{x^4}\,dx = -\frac{1}{3x^3}\sqrt{E^3} + \frac{1}{x}\sqrt{E} + \sin^{-1}\frac{x}{a}$$

$$\int \frac{\sqrt{E^n}}{x^4}\,dx = \frac{\sqrt{E^n}}{(n-3)x^3} + \frac{a^2n}{n-3}\int\frac{\sqrt{E^{n-2}}}{x^4}\,dx \qquad n \neq 3$$

$$\int \frac{\sqrt[p]{E^n}}{x^m}\,dx = \frac{p\sqrt[p]{E^n}}{(2n - mp + p)x^{m-1}} + \frac{2a^2n}{2n - mp + p}\int\frac{\sqrt[p]{E^{n-p}}}{x^m}\,dx \qquad 2n \neq -p(1 - m)$$

$$\int \frac{\sqrt{ax - bx^2}}{x}\,dx = \sqrt{ax - bx^2} + \frac{a}{\sqrt{b}}\sin^{-1}\sqrt{\frac{bx}{a}}$$

$$\int \frac{\sqrt{ax - bx^2}}{x^m}\,dx = \frac{2\sqrt{(ax - bx^2)^3}}{a(3 - 2m)x^m} + \frac{2b(3 - m)}{a(3 - 2m)}\int\frac{\sqrt{ax - bx^2}}{x^{m-1}}\,dx$$

(32) Indefinite Integrals Involving

$$f(x) = x^m \sqrt[p]{F^n} \qquad\qquad F = x^2 - a^2 \qquad a^2 \neq 0$$

$$\int \sqrt{F}\, dx = \frac{x}{2}\sqrt{F} - \frac{a^2}{2}\ln(x + \sqrt{F})$$

$$\int \sqrt{F^3}\, dx = \frac{x\sqrt{F^3}}{4} - \frac{3a^2}{4}\int \sqrt{F}\, dx$$

$$\int \sqrt{F^n}\, dx = \frac{x\sqrt{F^n}}{n+1} - \frac{a^2 n}{n+1}\int \sqrt{F^{n-2}}\, dx \qquad n \neq -1$$

$$\int x\sqrt{F}\, dx = \frac{1}{3}\sqrt{F^3}$$

$$\int x\sqrt{F^3}\, dx = \frac{1}{5}\sqrt{F^5}$$

$$\int x\sqrt{F^n}\, dx = \frac{x^2\sqrt{F^n}}{n+2} - \frac{a^2 n}{n+2}\int x\sqrt{F^{n-2}}\, dx \qquad n \neq -2$$

$$\int x^2\sqrt{F}\, dx = \frac{x}{4}\sqrt{F^3} + \frac{a^2 x}{8}\sqrt{F} - \frac{a^4}{8}\ln(x + \sqrt{F})$$

$$\int x^2\sqrt{F^3}\, dx = \frac{x}{6}\sqrt{F^5} + \frac{a^2 x}{24}\sqrt{F^3} - \frac{a^4 x}{16}\sqrt{F} + \frac{a^6}{16}\ln(x + \sqrt{F})$$

$$\int x^2\sqrt{F^n}\, dx = \frac{x^3\sqrt{F^n}}{n+3} - \frac{a^2 n}{n+3}\int x^2\sqrt{F^{n-2}}\, dx \qquad n \neq -3$$

$$\int x^3\sqrt{F}\, dx = \frac{1}{5}\sqrt{F^5} + \frac{a^2}{3}\sqrt{F^3}$$

$$\int x^3\sqrt{F^3}\, dx = \frac{1}{7}\sqrt{F^7} + \frac{a^2}{5}\sqrt{F^5}$$

$$\int x^3\sqrt{F^n}\, dx = \frac{x^4\sqrt{E^n}}{n+4} - \frac{a^2 n}{n+4}\int x^3\sqrt{F^{n-2}}\, dx \qquad n \neq -4$$

$$\int x^m\sqrt[p]{F^n}\, dx = \frac{px^{m+1}\sqrt[p]{F^n}}{2n+mp+p} - \frac{2a^2 n}{2n+mp+p}\int x^m\sqrt[p]{F^{n-p}}\, dx \qquad 2n \neq -p(m+1)$$

$$\int \sqrt{ax^2 - bx}\, dx = \frac{2ax - b}{4a}\sqrt{ax^2 - bx} - \frac{b^2}{8a\sqrt{a}}\ln(2ax - b + 2\sqrt{a}\sqrt{ax^2 - bx})$$

$$\int x^m\sqrt{ax^2 - bx}\, dx = \frac{x^{m-1}}{a(m+2)}\sqrt{(ax^2 - bx)^3} + \frac{b(2m+1)}{2a(m+1)}\int x^{m-1}\sqrt{ax^2 - bx}\, dx$$

(33) Indefinite Integrals Involving

$$f(x) = \frac{x^m}{\sqrt[p]{F^n}} \qquad F = x^2 - a^2 \qquad a^2 \neq 0$$

$$\int \frac{dx}{\sqrt{F}} = \ln\left(x + \sqrt{F}\right) = \cosh^{-1}\frac{x}{a}$$

$$\int \frac{dx}{\sqrt{F^3}} = -\frac{x}{a^2\sqrt{F}}$$

$$\int \frac{dx}{\sqrt{F^n}} = -\frac{x}{a^2(n-2)\sqrt{F^{n-2}}} - \frac{n-3}{a^2(n-2)} \int \frac{dx}{\sqrt{F^{n-2}}} \qquad n \neq 2$$

$$\int \frac{x\,dx}{\sqrt{F}} = \sqrt{F}$$

$$\int \frac{x\,dx}{\sqrt{F^3}} = \frac{-1}{\sqrt{F}}$$

$$\int \frac{x\,dx}{\sqrt{F^n}} = -\frac{x^2}{a^2(n-2)\sqrt{F^{n-2}}} - \frac{n-4}{a^2(n-2)} \int \frac{x\,dx}{\sqrt{F^{n-2}}} \qquad n \neq 2$$

$$\int \frac{x^2\,dx}{\sqrt{F}} = \frac{x}{2}\sqrt{F} + \frac{a^2}{2}\ln\left(x + \sqrt{F}\right)$$

$$\int \frac{x^2\,dx}{\sqrt{F^3}} = -\frac{x}{\sqrt{F}} + \ln\left(x + \sqrt{F}\right)$$

$$\int \frac{x^2\,dx}{\sqrt{F^n}} = -\frac{x^3}{a^2(n-2)\sqrt{F^{n-2}}} - \frac{n-5}{a^2(n-2)} \int \frac{x^2\,dx}{\sqrt{F^{n-2}}} \qquad n \neq 2$$

$$\int \frac{x^3\,dx}{\sqrt{F}} = \left(\frac{F}{3} + a^2\right)\sqrt{F}$$

$$\int \frac{x^3\,dx}{\sqrt{F^3}} = \frac{F - a^2}{\sqrt{F}}$$

$$\int \frac{x^3\,dx}{\sqrt{F^n}} = -\frac{x^4}{a^2(n-2)\sqrt{F^{n-2}}} - \frac{n-6}{a^2(n-2)} \int \frac{x^3\,dx}{\sqrt{F^{n-2}}} \qquad n \neq 2$$

$$\int \frac{x^m\,dx}{\sqrt[p]{F^n}} = -\frac{px^{m+1}}{2a^2(n-p)\sqrt[p]{F^{n-p}}} - \frac{2n - p(m+3)}{2a^2(n-p)} \int \frac{x^m\,dx}{\sqrt[p]{F^{n-p}}} \qquad n \neq p$$

$$\int \frac{dx}{\sqrt{ax^2 - bx}} = \frac{2}{\sqrt{a}}\ln\left(\sqrt{ax} + \sqrt{ax - b}\right)$$

$$\int \frac{x^m\,dx}{\sqrt{ax^2 - bx}} = \frac{x^{m-1}}{am}\sqrt{ax^2 - bx} + \frac{b(2m-1)}{2am} \int \frac{x^{m-1}\,dx}{\sqrt{ax^2 - bx}}$$

(34) Indefinite Integrals Involving

$$f(x) = \frac{1}{x^m \sqrt[p]{F^n}} \qquad F = x^2 - a^2 \qquad a^2 \neq 0$$

$$\int \frac{dx}{x\sqrt{F}} = \frac{1}{a}\cos^{-1}\frac{a}{x}$$

$$\int \frac{dx}{x\sqrt{F^3}} = \frac{-1}{a^2\sqrt{F}} - \frac{1}{a^3}\cos^{-1}\frac{a}{x}$$

$$\int \frac{dx}{x\sqrt{F^n}} = -\frac{1}{a^2(n-2)\sqrt{F^{n-2}}} - \frac{1}{a^2}\int \frac{dx}{x\sqrt{F^{n-2}}} \qquad n \neq 2$$

$$\int \frac{dx}{x^2\sqrt{F}} = \frac{\sqrt{F}}{a^2 x}$$

$$\int \frac{dx}{x^2\sqrt{F^3}} = -\frac{\sqrt{F}}{a^4 x}\left(1 + \frac{x^2}{F}\right)$$

$$\int \frac{dx}{x^2\sqrt{F^n}} = -\frac{1}{a^2(n-2)x\sqrt{F^{n-2}}} - \frac{n-1}{a^2(n-2)}\int \frac{dx}{x^2\sqrt{F^{n-2}}} \qquad n \neq 2$$

$$\int \frac{dx}{x^3\sqrt{F}} = \frac{\sqrt{F}}{2a^2 x^2} + \frac{1}{2a^3}\cos^{-1}\frac{a}{x}$$

$$\int \frac{dx}{x^3\sqrt{F^3}} = \frac{1}{2a^2 x^2\sqrt{F}} - \frac{3}{2a^4\sqrt{F}} - \frac{3}{2a^5}\cos^{-1}\frac{a}{x}$$

$$\int \frac{dx}{x^3\sqrt{F^n}} = -\frac{1}{a^2(n-2)x^2\sqrt{F^{n-2}}} - \frac{n}{a^2(n-2)}\int \frac{dx}{x^3\sqrt{F^{n-2}}} \qquad n \neq 2$$

$$\int \frac{dx}{x^4\sqrt{F}} = \frac{\sqrt{F}}{a^4 x}\left(1 - \frac{F^2}{3x^2}\right)$$

$$\int \frac{dx}{x^4\sqrt{F^3}} = -\frac{x}{a^6\sqrt{F}}\left(1 + \frac{2F}{x^2} - \frac{F^2}{3x^4}\right)$$

$$\int \frac{dx}{x^4\sqrt{F^n}} = \frac{1}{a^2(n-2)x^3\sqrt{F^{n-2}}} - \frac{n+1}{a^2(n-2)}\int \frac{dx}{x^4\sqrt{F^{n-2}}} \qquad n \neq 2$$

$$\int \frac{dx}{x^m\sqrt[p]{F^n}} = \frac{p}{2a^2(n-p)x^{m-1}\sqrt[p]{F^{n-p}}} - \frac{2n+p(m-3)}{2a^2(n-p)}\int \frac{dx}{x^m\sqrt[p]{F^{n-p}}} \qquad n \neq p$$

$$\int \frac{dx}{x\sqrt{ax^2 - bx}} = \frac{2}{bx}\sqrt{ax^2 - bx}$$

$$\int \frac{dx}{x^m\sqrt{ax^2 - bx}} = \frac{2\sqrt{ax^2 - bx}}{b(2m-1)x^m} + \frac{2a(m-1)}{b(2m-1)}\int \frac{dx}{x^{m-1}\sqrt{ax^2 - bx}}$$

(35) Indefinite Integrals Involving

$$f(x) = \frac{\sqrt[p]{F^n}}{x^m} \qquad F = x^2 - a^2 \qquad a^2 \neq 0$$

$$\int \frac{\sqrt{F}}{x} dx = \sqrt{F} - a \cos^{-1} \frac{a}{x}$$

$$\int \frac{\sqrt{F^3}}{x} dx = \frac{1}{3}\sqrt{F^3} - a^2\sqrt{F} + a^3 \cos^{-1} \frac{a}{x}$$

$$\int \frac{\sqrt{F^n}}{x} dx = \frac{1}{n}\sqrt{F^n} - a^2 \int \frac{\sqrt{F^{n-2}}}{x} dx$$

$$\int \frac{\sqrt{F}}{x^2} dx = -\frac{1}{x}\sqrt{F} + \ln (x + \sqrt{F})$$

$$\int \frac{\sqrt{F^3}}{x^2} dx = -\frac{1}{x}\sqrt{F^3} + \frac{3x}{2}\sqrt{F} - \frac{3a^2}{2} \ln (x + \sqrt{F})$$

$$\int \frac{\sqrt{F^n}}{x^2} dx = \frac{\sqrt{F^n}}{(n-1)x} - \frac{a^2 n}{n-1} \int \frac{\sqrt{F^{n-2}}}{x^2} dx \qquad n \neq 1$$

$$\int \frac{\sqrt{F}}{x^3} dx = \frac{-1}{2x^2}\sqrt{F} + \frac{1}{2a} \cos^{-1} \frac{a}{x}$$

$$\int \frac{\sqrt{F^3}}{x^3} dx = -\frac{1}{2x^2}\sqrt{F^3} + \frac{3}{2}\sqrt{F} - \frac{3a}{2} \cos^{-1} \frac{a}{x}$$

$$\int \frac{\sqrt{F^n}}{x^3} dx = \frac{\sqrt{F^n}}{(n-2)x^2} - \frac{a^2 n}{n-2} \int \frac{\sqrt{F^{n-2}}}{x^3} dx \qquad n \neq 2$$

$$\int \frac{\sqrt{F}}{x^4} dx = \frac{1}{3a^2 x^3}\sqrt{F^3}$$

$$\int \frac{\sqrt{F^3}}{x^4} dx = -\frac{1}{3x^3}\sqrt{F^3} - \frac{1}{x}\sqrt{F} + \ln (x + \sqrt{F})$$

$$\int \frac{\sqrt{F^n}}{x^4} dx = \frac{\sqrt{F^n}}{(n-3)x^3} - \frac{a^2 n}{n-3} \int \frac{\sqrt{F^{n-2}}}{x^4} dx \qquad n \neq 3$$

$$\int \frac{\sqrt[p]{F^n}}{x^m} dx = \frac{p\sqrt[p]{F^n}}{(2n - mp + p)x^{m-1}} - \frac{2a^2 n}{2n - mp + p} \int \frac{\sqrt[p]{F^{n-p}}}{x^m} dx \qquad 2n \neq -p(1 - m)$$

$$\int \frac{\sqrt{ax^2 - bx}}{x} dx = \sqrt{ax^2 - bx} - \frac{b}{\sqrt{a}} \ln (\sqrt{ax} + \sqrt{ax - b})$$

$$\int \frac{\sqrt{ax^2 - bx}}{x^m} dx = \frac{2\sqrt{(ax^2 - bx)^3}}{b(2m - 3)x^{m-1}} + \frac{2a(m - 3)}{b(2m - 3)} \int \frac{\sqrt{ax^2 - bx}}{x^{m-1}} dx$$

(36) Indefinite Integrals Involving

$$f(x) = x^m \sqrt[p]{P^n} \qquad \lambda = \frac{n}{p} \qquad P = a + bx^q \qquad \begin{array}{c} a \neq 0 \\ b \neq 0 \end{array}$$

$$\int P^\lambda \, dx = \frac{1}{1 + q\lambda}\left(xP^\lambda + aq\lambda \int P^{\lambda-1} \, dx\right)$$

$$\int \frac{dx}{P^\lambda} = \frac{-1}{aq(1 - \lambda)}\left[\frac{x}{P^{\lambda-1}} - (1 + q - q\lambda) \int \frac{dx}{P^{\lambda-1}}\right]$$

$$\int x^m P^\lambda \, dx = \frac{1}{1 + m + q\lambda}\left(x^{m+1}P^\lambda + aq\lambda \int x^m P^{\lambda-1} \, dx\right)$$

$$= \frac{-1}{aq(1 + \lambda)}\left[x^{m+1}P^{\lambda+1} - (1 + m + q + q\lambda) \int x^m P^{\lambda+1} \, dx\right]$$

$$= \frac{1}{a(1 + m)}\left[x^{m+1}P^{\lambda+1} - b(1 + m + q + q\lambda) \int x^{m+q}P^\lambda \, dx\right]$$

$$= \frac{1}{b(1 + m + q^\lambda)}\left[x^{m-q+1}P^{\lambda+1} - a(1 + m - q) \int x^{m-q}P^\lambda \, dx\right]$$

(37) Indefinite Integrals Involving

$$f(x) = \frac{x^{p-1}}{x^{2m+1} \pm a^{2m+1}} \qquad a \neq 0$$

$$\int \frac{x^{p-1} \, dx}{x^{2m+1} + a^{2m+1}} = \frac{2(-1)^{p-1}}{(2m + 1)a^{2m-p+1}} \sum_{k=1}^{m} \sin\frac{2kp\pi}{2m + 1} \tan^{-1}\left\{\frac{x + a \cos\left[2k\pi/(2m + 1)\right]}{a \sin\left[2k\pi/(2m + 1)\right]}\right\}$$

$$- \frac{(-1)^{p-1}}{(2m + 1)a^{2m-p+1}} \sum_{k=1}^{m} \cos\frac{2kp\pi}{2m + 1} \ln\left(x^2 + 2ax \cos\frac{2k\pi}{2m + 1} + a^2\right)$$

$$+ \frac{(-1)^{p-1}\ln(x + a)}{(2m + 1)a^{2m-p+1}} \qquad 0 < p \leqslant 2m + 1$$

$$\int \frac{x^{p-1} \, dx}{x^{2m+1} - a^{2m+1}} = \frac{-2}{(2m + 1)a^{2m-p+1}} \sum_{k=1}^{m} \sin\frac{2kp\pi}{2m + 1} \tan^{-1}\left\{\frac{x - a \cos\left[2k\pi/(2m + 1)\right]}{a \sin\left[2k\pi/(2m + 1)\right]}\right\}$$

$$+ \frac{1}{(2m + 1)a^{2m-p+1}} \sum_{k=1}^{m} \cos\frac{2kp\pi}{2m + 1} \ln\left(x^2 - 2ax \cos\frac{2k\pi}{2m + 1} + a^2\right)$$

$$+ \frac{\ln(x - a)}{(2m + 1)a^{2m-p+1}} \qquad 0 < p \leqslant 2m + 1$$

(Ref. 19.08, p. 75).

(38) Indefinite Integrals Involving $\quad \boxed{f(x) = x^m \sqrt[p]{R^n} \qquad \lambda = n/p \qquad R = ax^q + bx^{q+r} \qquad \begin{array}{l} a \neq 0 \\ b \neq 0 \end{array}}$

$$\int R^\lambda \, dx = \frac{\lambda}{1 + \lambda(q + r)}\left(xR^\lambda + ar\lambda \int x^q R^{\lambda-1} \, dx\right)$$

$$\int \frac{dx}{R^\lambda} = \frac{-1}{ar(1 + \lambda)}\left[xR^{\lambda+1} - (1 + r + q\lambda + r\lambda) \int \frac{x^q \, dx}{R^{\lambda-1}}\right]$$

$$\int x^m R^\lambda \, dx = \frac{1}{1 + m + q\lambda + r\lambda}\left(x^{m+1}R^\lambda + ar\lambda \int x^{m+q}R^{\lambda-1} \, dx\right)$$

$$= \frac{-1}{ar(1 + \lambda)}\left[x^{m+1}R^{\lambda+1} - (1 + r + q\lambda + r\lambda) \int x^{m-q}R^{\lambda+1} \, dx\right]$$

(39) Indefinite Integrals Involving $\quad \boxed{f(x) = \frac{x^m}{(x - a_1)(x - a_2) \cdots (x - a_k)} \qquad a_i \neq a_j \neq 0}$

$$\int \frac{x^m \, dx}{(x - a_1)(x - a_2) \cdots (x - a_k)} = \frac{a_1^m \ln(x - a_1)}{(a_1 - a_2)(a_1 - a_3) \cdots (a_1 - a_k)}$$

$$+ \frac{a_2^m \ln(x - a_2)}{(a_2 - a_1)(a_2 - a_3) \cdots (a_2 - a_k)} + \cdots + \frac{a_k^m \ln(x - a_k)}{(a_k - a_1)(a_k - a_2) \cdots (a_k - a_{k-1})}$$

(40) Indefinite Integrals Involving $\quad \boxed{f(x) = \frac{x^{p-1}}{x^{2m} \pm a^{2m}} \qquad a \neq 0}$

$$\int \frac{x^{p-1} \, dx}{x^{2m} + a^{2m}} = \frac{1}{ma^{2m-p}} \sum_{k=1}^{m} \sin\frac{(2k - 1)p\pi}{2m} \tan^{-1}\left\{\frac{x + a\cos[(2k - 1)\pi/2m]}{a\sin[(2k - 1)\pi/2m]}\right\}$$

$$- \frac{1}{2ma^{2m-p}} \sum_{k=1}^{m} \cos\frac{(2k - 1)p\pi}{2m} \ln\left[x^2 + 2ax\cos\frac{(2k - 1)\pi}{2m} + a^2\right] \qquad 0 < p \leq 2m$$

$$\int \frac{x^{p-1} \, dx}{x^{2m} - a^{2m}} = \frac{1}{2ma^{2m-p}} \sum_{k=1}^{m-1} \cos\frac{kp\pi}{m} \ln\left(x^2 - 2ax\cos\frac{k\pi}{m} + a^2\right)$$

$$- \frac{1}{na^{2m-p}} \sum_{k=1}^{m-1} \sin\frac{kp\pi}{m} \tan^{-1}\frac{x - a\cos(k\pi/m)}{a\sin(k\pi/m)} + \frac{1}{2ma^{2m-p}}[\ln(x - a) + (-1)^p \ln(x + a)]$$

$$0 < p \leq 2m$$

(Ref. 19.08, p. 75)

(41) Indefinite Integrals Involving

$f(x) = \sin^n A$	$f(x) = \dfrac{1}{\sin^n A}$	$A = bx$

$$\int \sin A \, dx = -\frac{\cos A}{b}$$

$$\int \sin^2 A \, dx = -\frac{\sin 2A}{4b} + \frac{x}{2}$$

$$\int \sin^3 A \, dx = -\frac{\cos A}{b} + \frac{\cos^3 A}{3b}$$

$$\int \sin^4 A \, dx = -\frac{\sin 2A}{4b} + \frac{\sin 4A}{32b} + \frac{3x}{8}$$

$$\int \sin^5 A \, dx = -\frac{5 \cos A}{8b} + \frac{5 \cos 3A}{48b} - \frac{\cos 5A}{80b}$$

$$\int \sin^6 A \, dx = -\frac{15 \sin 2A}{64b} + \frac{3 \sin 4A}{64b} - \frac{\sin 6A}{192b} + \frac{5x}{16}$$

$$\int \sin^7 A \, dx = -\frac{35 \cos A}{64b} + \frac{7 \cos 3A}{64b} - \frac{7 \cos 5A}{320b} + \frac{\cos 7A}{448b}$$

$$\int \sin^n A \, dx = -\frac{\cos A}{b} \frac{\sin^{n-1} A}{n} + \frac{n-1}{n} \int \sin^{n-2} A \, dx \qquad n > 0$$

$$\int \frac{dx}{\sin A} = \frac{1}{b} \ln \tan \frac{A}{2}$$

$$\int \frac{dx}{\sin^2 A} = -\frac{\cot A}{b}$$

$$\int \frac{dx}{\sin^3 A} = -\frac{\cos A}{2b \sin^2 A} + \frac{1}{2b} \ln \tan \frac{A}{2}$$

$$\int \frac{dx}{\sin^4 A} = -\frac{\cot A}{b} - \frac{\cot^3 A}{3b}$$

$$\int \frac{dx}{\sin^5 A} = -\frac{\cos A}{4b \sin^4 A} - \frac{3 \cos A}{8b \sin^2 A} - \frac{3}{8b} \ln \tan \frac{A}{2}$$

$$\int \frac{dx}{\sin^6 A} = -\frac{\cot A}{b} - \frac{2 \cot^3 A}{3b} - \frac{\cot^5 A}{5b}$$

$$\int \frac{dx}{\sin^7 A} = -\frac{\cos A}{6b \sin^6 A} - \frac{5 \cos A}{24b \sin^4 A} - \frac{5 \cos A}{16b \sin^2 A} + \frac{5}{16b} \ln \tan \frac{A}{2}$$

$$\int \frac{dx}{\sin^n A} = -\frac{\cos A}{b(n-1) \sin^{n-1} A} + \frac{n-2}{n-1} \int \frac{dx}{\sin^{n-2} A} \qquad n > 1$$

(42) Indefinite Integrals Involving

$f(x) = \cos^n A$	$f(x) = \dfrac{1}{\cos^n A}$	$A = bx$

$$\int \cos A \, dx = \frac{\sin A}{b}$$

$$\int \cos^2 A \, dx = \frac{\sin 2A}{4b} + \frac{x}{2}$$

$$\int \cos^3 A \, dx = \frac{\sin A}{b} - \frac{\sin^3 A}{3b}$$

$$\int \cos^4 A \, dx = \frac{\sin 2A}{4b} + \frac{\sin 4A}{32b} + \frac{3x}{8}$$

$$\int \cos^5 A \, dx = \frac{5 \sin A}{8b} + \frac{5 \sin 3A}{48b} + \frac{\sin 5A}{80b}$$

$$\int \cos^6 A \, dx = \frac{15 \sin 2A}{64b} + \frac{3 \sin 4A}{64b} + \frac{\sin 6A}{192b} + \frac{5x}{16}$$

$$\int \cos^7 A \, dx = \frac{35 \sin A}{64b} + \frac{7 \sin 3A}{64b} + \frac{7 \sin 5A}{320b} + \frac{\sin 7A}{448b}$$

$$\int \cos^n A \, dx = \frac{\sin A \, \cos^{n-1} A}{b} + \frac{n-1}{n} \int \cos n^{n-2} A \, dx \qquad n > 0$$

$$\int \frac{dx}{\cos A} = \frac{1}{b} \ln \tan \left(\frac{\pi}{4} + \frac{A}{2} \right)$$

$$\int \frac{dx}{\cos^2 A} = \frac{1}{b} \tan A$$

$$\int \frac{dx}{\cos^3 A} = \frac{\sin A}{2b \cos^2 A} + \frac{1}{2b} \ln \tan \left(\frac{\pi}{4} + \frac{A}{2} \right)$$

$$\int \frac{dx}{\cos^4 A} = \frac{\tan A}{b} + \frac{\tan^3 A}{3b}$$

$$\int \frac{dx}{\cos^5 A} = \frac{\sin A}{4b \cos^4 A} + \frac{3 \sin A}{8 \cos^2 A} + \frac{3}{8b} \ln \tan \left(\frac{\pi}{4} + \frac{A}{2} \right)$$

$$\int \frac{dx}{\cos^6 A} = \frac{\tan A}{b} + \frac{2 \tan^3 A}{3b} + \frac{\tan^5 A}{5b}$$

$$\int \frac{dx}{\cos^7 A} = \frac{\sin A}{6b \cos^6 A} + \frac{5 \sin A}{24b \cos^4 A} + \frac{5 \sin A}{16b \cos^2 A} + \frac{5}{16b} \ln \tan \left(\frac{\pi}{4} + \frac{A}{2} \right)$$

$$\int \frac{dx}{\cos^n A} = \frac{\sin A}{b(n-1) \cos^{n-1} A} + \frac{n-2}{n-1} \int \frac{dx}{\cos^{n-2} A} \qquad n > 1$$

(43) Indefinite Integrals Involving

$$f(x) = \sin^m A \cos^n A \quad \bigg| \quad A = bx$$

$$\int \sin A \cos A \, dx = -\frac{\cos^2 A}{2b}$$

$$\int \sin A \cos^n A \, dx = -\frac{\cos^{n+1} A}{(n+1)b} \qquad n \neq -1 \qquad\qquad \int \sin^m A \cos A \, dx = \frac{\sin^{m+1} A}{(m+1)b} \qquad m \neq -1$$

$$\int \sin^2 A \cos A \, dx = \frac{\sin^3 A}{3b}$$

$$\int \sin^2 A \cos^2 A \, dx = \frac{x}{8} - \frac{\sin 4A}{32b}$$

$$\int \sin^2 A \cos^3 A \, dx = \frac{\sin^3 A \cos^2 A}{5b} + 2\frac{\sin^3 A}{15b}$$

$$\int \sin^2 A \cos^4 A \, dx = \frac{x}{16} + \frac{\sin 2A}{64b} - \frac{\sin 4A}{64b} - \frac{\sin 6A}{192b}$$

$$\int \sin^2 A \cos^n A \, dx = -\frac{\sin A \cos^{n+1} A}{(n+2)b} + \int \frac{\cos^n A \, dx}{n+2} \qquad n \neq -2$$

$$\int \sin^3 A \cos A \, dx = \frac{\sin^4 A}{4b}$$

$$\int \sin^3 A \cos^2 A \, dx = -\frac{\cos^3 A}{3b} + \frac{\cos^5 A}{5b}$$

$$\int \sin^3 A \cos^3 A \, dx = -\frac{3 \cos 2A}{64b} + \frac{\cos 6A}{192b}$$

$$\int \sin^3 A \cos^n A \, dx = -\frac{\cos^{n+1} A}{(n+1)b} + \frac{\cos^{n+3} A}{(n+3)b} \qquad n \neq -1, -3$$

$$\int \sin^4 A \cos A \, dx = \frac{\sin^5 A}{5b}$$

$$\int \sin^4 A \cos^2 A \, dx = \frac{1}{192b}(\sin 6A - 3 \sin 4A - 3 \sin 2A + 12A)$$

$$\int \sin^4 A \cos^3 A \, dx = \frac{\sin^5 A}{5b} - \frac{\sin^7 A}{7b}$$

$$\int \sin^m A \cos^n A \, dx = \frac{\sin^{m+1} A \cos^{n-1} A}{b(m+n)} + \frac{n-1}{m+n} \int \sin^m A \cos^{n-2} A \, dx$$

$$= -\frac{\sin^{m-1} A \cos^{n+1} A}{b(m+n)} + \frac{m-1}{m+n} \int \sin^{m-2} A \cos^n A \, dx \qquad \begin{matrix} m > 0 \\ n > 0 \end{matrix}$$

(44) Indefinite Integrals Involving

$$f(x) = \frac{1}{\sin^m A \cos^n A} \qquad A = bx$$

$$\int \frac{dx}{\sin A \cos A} = \frac{1}{b} \ln (\tan A)$$

$$\int \frac{dx}{\sin A \cos^2 A} = \frac{1}{b} \ln \left(\tan \frac{A}{2} \right) + \frac{1}{b \cos A}$$

$$\int \frac{dx}{\sin A \cos^n A} = \frac{1}{b(n-1) \cos^{n-1} A} + \int \frac{dx}{\sin A \cos^{n-2} A} \qquad n \neq 1$$

$$\int \frac{dx}{\sin^2 A \cos A} = -\frac{1}{b \sin A} + \frac{1}{b} \ln \left[\tan \left(\frac{\pi}{4} + \frac{A}{2} \right) \right]$$

$$\int \frac{dx}{\sin^2 A \cos^2 A} = -\frac{2}{b} \cot 2A$$

$$\int \frac{dx}{\sin^2 A \cos^3 A} = \frac{1}{2b \sin A} \left(\frac{1}{\cos^2 A} - 3 \right) + \frac{3}{2b} \ln \left[\tan \left(\frac{\pi}{4} + \frac{A}{2} \right) \right]$$

$$\int \frac{dx}{\sin^2 A \cos^4 A} = \frac{1}{3b \sin A \cos^3 A} - \frac{8}{3b \tan 2A}$$

$$\int \frac{dx}{\sin^2 A \cos^n A} = \frac{1 - n \cos^2 A}{b(n-1) \sin A \cos^{n-1} A} + \frac{n(n-2)}{n-1} \int \frac{dx}{\cos^{n-2} A} \qquad n \neq 1$$

$$\int \frac{dx}{\sin^3 A \cos A} = \frac{-1}{2b \sin^2 A} + \frac{1}{b} \ln (\tan A)$$

$$\int \frac{dx}{\sin^3 A \cos^2 A} = \frac{1}{b \cos A} - \frac{\cos A}{2b \sin^2 A} + \frac{3}{2b} \ln \left(\tan \frac{A}{2} \right)$$

$$\int \frac{dx}{\sin^3 A \cos^3 A} = -\frac{2 \cos 2A}{b \sin^2 2A} + \frac{2}{b} \ln (\tan A)$$

$$\int \frac{dx}{\sin^3 A \cos^4 A} = \frac{2}{b \cos A} + \frac{1}{3b \cos^3 A} - \frac{\cos A}{2b \sin^2 A} + \frac{5}{2b} \ln \left(\tan \frac{A}{2} \right)$$

$$\int \frac{dx}{\sin^m A \cos A} = \frac{-1}{b(m-1) \sin^{m-1} A} + \int \frac{dx}{\sin^{m-2} A \cos A} \qquad m \neq 1$$

$$\int \frac{dx}{\sin^m A \cos^n A} = \begin{cases} \dfrac{-1}{b(m-1) \sin^{m-1} A \cos^{n-1} A} + \dfrac{m+n-2}{m-1} \displaystyle\int \dfrac{dx}{\sin^{m-2} A \cos^n A} & m \neq 1 \\[4ex] \dfrac{1}{b(n-1) \sin^{m-1} A \cos^{n-1} A} + \dfrac{m+n-2}{n-1} \displaystyle\int \dfrac{dx}{\sin^m A \cos^{n-2} A} & n \neq 1 \end{cases}$$

(45) Indefinite Integrals Involving

$$f(x) = \frac{\sin^m A}{\cos^n A} \qquad A = bx$$

$$\int \frac{\sin A \, dx}{\cos A} = -\frac{\ln(\cos A)}{b}$$

$$\int \frac{\sin A \, dx}{\cos^n A} = \frac{1}{b(n-1)\cos^{n-1} A} \qquad n \neq 1$$

$$\int \frac{\sin^2 A \, dx}{\cos A} = -\frac{\sin A}{b} + \frac{1}{b} \ln\left[\tan\left(\frac{\pi}{4} + \frac{A}{2}\right)\right]$$

$$\int \frac{\sin^m A \, dx}{\cos^{m+2} A} = \frac{\tan^{m+1} A}{b(m+1)} \qquad m \neq -1$$

$$\int \frac{\sin^2 A \, dx}{\cos^2 A} = -x + \frac{1}{b} \tan A$$

$$\int \frac{\sin^2 A \, dx}{\cos^3 A} = \frac{\sin A}{2b \cos^2 A} - \frac{1}{2b} \ln\left[\tan\left(\frac{\pi}{4} + \frac{A}{2}\right)\right]$$

$$\int \frac{\sin^2 A \, dx}{\cos^4 A} = \frac{\tan^3 A}{3b}$$

$$\int \frac{\sin^2 A \, dx}{\cos^5 A \, dx} = \frac{\sin A}{4b \cos^4 A} - \frac{\sin A}{8b \cos^2 A} - \frac{1}{8b} \ln\left[\tan\left(\frac{\pi}{4} + \frac{A}{2}\right)\right]$$

$$\int \frac{\sin^2 A \, dx}{\cos^n A} = \frac{\sin A}{b(n-1)\cos^{n-1} A} - \frac{1}{n-1} \int \frac{dx}{\cos^{n-2} A} \qquad n \neq 1$$

$$\int \frac{\sin^3 A \, dx}{\cos A} = -\frac{\sin^2 A}{2b} - \frac{1}{b} \ln(\cos A)$$

$$\int \frac{\sin^3 A \, dx}{\cos^2 A} = \frac{\cos A}{b} + \frac{1}{b \cos A}$$

$$\int \frac{\sin^3 A \, dx}{\cos^3 A} = \frac{\tan^2 A}{2b} + \frac{1}{b} \ln(\cos A)$$

$$\int \frac{\sin^3 A \, dx}{\cos^4 A} = \frac{1}{3b \cos^3 A} - \frac{1}{b \cos A}$$

$$\int \frac{\sin^3 A \, dx}{\cos^5 A} = \frac{1}{4b \cos^4 A} - \frac{1}{2b \cos^2 A}$$

$$\int \frac{\sin^3 A \, dx}{\cos^n A} = \frac{1}{b(n-1)\cos^{n-1} A} - \frac{1}{b(n-3)\cos^{n-3} A} \qquad n \neq 1, 3$$

$$\int \frac{\sin^m A \, dx}{\cos A} = -\frac{\sin^{m-1} A}{b(m-1)} + \int \frac{\sin^{m-2} A \, dx}{\cos A} \qquad m \neq 1$$

$$\int \frac{\sin^m A \, dx}{\cos^n A} = \begin{cases} -\dfrac{\sin^{m-1} A}{b(m-n)\cos^{n-1} A} + \dfrac{m-1}{m-n} \displaystyle\int \dfrac{\sin^{m-2} A \, dx}{\cos^n A} & m \neq n \\[3ex] \dfrac{\sin^{m-1} A}{b(n-1)\cos^{n-1} A} - \dfrac{m-1}{n-1} \displaystyle\int \dfrac{\sin^{m-2} A \, dx}{\cos^{n-2} A} & n \neq 1 \end{cases}$$

(46) Indefinite Integrals Involving

$$f(x) = \frac{\cos^n A}{\sin^m A} \qquad A = bx$$

$$\int \frac{\cos A \, dx}{\sin A} = \frac{\ln (\sin A)}{b}$$

$$\int \frac{\cos A \, dx}{\sin^m A} = \frac{-1}{b(m-1) \sin^{m-1} A} \qquad m \neq 1$$

$$\int \frac{\cos^2 A \, dx}{\sin A} = \frac{\cos A}{b} + \frac{1}{b} \ln \left(\tan \frac{A}{2} \right)$$

$$\int \frac{\cos^n A \, dx}{\sin^{n+2} A} = -\frac{\cot^{n+1} A}{b(n+1)} \qquad n \neq -1$$

$$\int \frac{\cos^2 A \, dx}{\sin^2 A} = -x - \frac{1}{b} \cot A$$

$$\int \frac{\cos^2 A \, dx}{\sin^3 A} = -\frac{\cos A}{2b \sin^2 A} - \frac{1}{2b} \ln \left(\tan \frac{A}{2} \right)$$

$$\int \frac{\cos^2 A \, dx}{\sin^4 A} = -\frac{\cot^3 A}{3b}$$

$$\int \frac{\cos^2 A \, dx}{\sin^5 A} = -\frac{\cos A}{4b \sin^4 A} + \frac{\cos A}{8b \sin^2 A} - \frac{1}{8b} \ln \left(\tan \frac{A}{2} \right)$$

$$\int \frac{\cos^2 A \, dx}{\sin^m A} = \frac{-\cos A}{b(m-1) \sin^{m-1} A} - \frac{1}{m-1} \int \frac{dx}{\sin^{m-2} A} \qquad m \neq 1$$

$$\int \frac{\cos^3 A \, dx}{\sin A} = \frac{\cos^2 A}{2b} + \frac{1}{b} \ln (\sin A)$$

$$\int \frac{\cos^3 A \, dx}{\sin^2 A} = -\frac{\sin A}{b} - \frac{1}{b \sin A}$$

$$\int \frac{\cos^3 A \, dx}{\sin^3 A} = -\frac{\cot^2 A}{2b} - \frac{1}{b} \ln (\sin A)$$

$$\int \frac{\cos^3 A \, dx}{\sin^4 A} = -\frac{1}{3b \sin^3 A} + \frac{1}{b \sin A}$$

$$\int \frac{\cos^3 A \, dx}{\sin^5 A} = -\frac{1}{4b \sin^4 A} + \frac{1}{2b \sin^2 A}$$

$$\int \frac{\cos^3 A \, dx}{\sin^m A} = \frac{-1}{b(m-1) \sin^{m-1} A} + \frac{1}{b(m-3) \sin^{m-3} A} \qquad m \neq 1, 3$$

$$\int \frac{\cos^n A \, dx}{\sin A} = \frac{\cos^{n-1} A}{b(n-1)} + \int \frac{\cos^{n-2} A \, dx}{\sin A} \qquad n \neq 1$$

$$\int \frac{\cos^n A \, dx}{\sin^m A} = \begin{cases} -\dfrac{\cos^{n-1} A}{b(m-n) \sin^{m-1} A} - \dfrac{n-1}{m-n} \displaystyle\int \dfrac{\cos^{n-2} A \, dx}{\sin^m A} & m \neq n \\[3ex] -\dfrac{\cos^{n-1} A}{b(m-1) \sin^{m-1} A} - \dfrac{n-1}{m-1} \displaystyle\int \dfrac{\cos^{n-2} A \, dx}{\sin^{m-2} A} & m \neq 1 \end{cases}$$

(47) Indefinite Integrals Involving

$$f(x) = x^m \sin^n A \quad | \quad A = bx$$

$$D_1 = \int \sin^n A \, dx \qquad D_2 = \int D_1 \, dx \qquad D_k = \int D_{k-1} \, dx$$

$$\int x \sin A \, dx = -\frac{x \cos A}{b} + \frac{\sin A}{b^2}$$

$$\int x \sin^2 A \, dx = \frac{x^2}{4} - \frac{x \sin 2A}{4b} - \frac{\cos 2A}{8b^2}$$

$$\int x \sin^3 A \, dx = \frac{x \cos 3A}{12b} - \frac{\sin 3A}{36b^2} + \frac{3x \cos A}{4b} + \frac{3 \sin A}{4b^2}$$

$$\int x \sin^n A \, dx = xD_1 - D_2$$

$$\int x^2 \sin A \, dx = -\frac{x^2 \cos A}{b} + \frac{2x \sin A}{b^2} + \frac{2 \cos A}{b^3}$$

$$\int x^2 \sin^2 A \, dx = \frac{1}{2} \int x^2 (1 - \cos 2A) \, dx$$

$$\int x^2 \sin^3 A \, dx = \frac{1}{4} \int x^2 (3 \sin A - \sin 3A) \, dx$$

$$\int x^2 \sin^n A \, dx = x^2 D_1 - 2xD_2 + 2D_3$$

$$\int x^3 \sin A \, dx = -\frac{x^3 \cos A}{b} + \frac{3x^2 \sin A}{b^2} + \frac{6x \cos A}{b^3} - \frac{6 \sin A}{b^4}$$

$$\int x^3 \sin^2 A \, dx = \frac{1}{2} \int x^3 (1 - \cos 2A) \, dx$$

$$\int x^3 \sin^3 A \, dx = \frac{1}{4} \int x^3 (3\sin A - \sin 3A) \, dx$$

$$\int x^3 \sin^n A \, dx = x^3 D_1 - 3x^2 D_2 + 6xD_3 - 6D_4$$

$$\int x^m \sin A \, dx = \frac{m! \sin A}{b} \left[\frac{x^{m-1}}{(m-1)!b} - \frac{x^{m-3}}{(m-3)!b^3} + \frac{x^{m-5}}{(m-5)!b^5} - \cdots \right]$$

$$- \frac{m! \cos A}{b} \left[\frac{x^m}{m!} - \frac{x^{m-2}}{(m-2)!b^2} + \frac{x^{m-4}}{(m-4)!b^4} - \cdots \right]$$

Series terminates with the term involving x^{m-m}

$$\int x^m \sin^n A \, dx = x^m D_1 - mx^{m-1}D_2 + m(m-1)x^{m-2}D_3 - \cdots$$

(48) Indefinite Integrals Involving

$$\boxed{f(x) = x^m \cos^n A} \quad \boxed{A = bx}$$

$$D_1 = \int \cos^n A \, dx \qquad D_2 = \int D_1 \, dx \qquad D_k = \int D_{k-1} \, dx$$

$$\int x \cos A \, dx = \frac{x \sin A}{b} + \frac{\cos A}{b^2}$$

$$\int x \cos^2 A \, dx = \frac{x^2}{4} + \frac{x \sin 2A}{4b} + \frac{\cos 2A}{8b^2}$$

$$\int x \cos^3 A \, dx = \frac{x \sin 3A}{12b} + \frac{\cos 3A}{36b^2} + \frac{3x \sin A}{4b} + \frac{3 \cos A}{4b^2}$$

$$\int x \cos^n A \, dx = xD_1 - D_2$$

$$\int x^2 \cos A \, dx = \frac{x^2 \sin A}{b} + \frac{2x \cos A}{b^2} - \frac{2 \sin A}{b^3}$$

$$\int x^2 \cos^2 A \, dx = \frac{1}{2} \int x^2 (1 + \cos 2A) \, dx$$

$$\int x^2 \cos^3 A \, dx = \frac{1}{4} \int x^2 (3 \cos A + \cos 3A) \, dx$$

$$\int x^2 \cos^n A \, dx = x^2 D_1 - 2xD_2 + D_3$$

$$\int x^3 \cos A \, dx = \frac{x^3 \sin A}{b} + \frac{3x^2 \cos A}{b^2} - \frac{6x \sin A}{b^3} - \frac{6 \cos A}{b^4}$$

$$\int x^3 \cos^2 A \, dx = \frac{1}{2} \int x^3 (1 + \cos 2A) \, dx$$

$$\int x^3 \cos^3 A \, dx = \frac{1}{4} \int x^3 (3 \cos A + \cos 3A) \, dx$$

$$\int x^3 \cos^n A \, dx = x^3 D_1 - 3x^2 D_2 + 6xD_3 - 6D_4$$

$$\int x^m \cos A \, dx = \frac{m! \sin A}{b} \left[\frac{x^m}{m!} - \frac{x^{m-2}}{(m-2)!b^2} + \frac{x^{m-4}}{(m-4)!b^4} - \cdots \right]$$

$$+ \frac{m! \cos A}{b} \left[\frac{x^{m-1}}{(m-1)!b} - \frac{x^{m-3}}{(m-3)!b^3} + \frac{x^{m-5}}{(m-5)!b^5} - \cdots \right] \quad \left. \begin{array}{l} \text{Series terminates} \\ \text{with the term} \\ \text{involving } x^{m-m} \end{array} \right.$$

$$\int x^m \cos^n A \, dx = x^m D_1 - mx^{m-1} D_2 + m(m-1)x^{m-2} D_3 - \cdots$$

(49) Indefinite Integrals Involving

$$f(x) = \frac{x^m}{\sin^n A} \qquad\qquad f(x) = \frac{\sin^n A}{x^m} \qquad A = bx$$

$\text{Si}(A) = $ sine-integral function (Sec. 13.01-1) $B_k = $ auxiliary Bernoulli number (Sec. 8.05-1)

$\text{Ci}(A) = $ cosine-integral function (Sec. 13.01-1) $E_k = $ auxiliary Euler number (Sec. 8.06-1)

$b_0 = 1, \qquad b_k = \dfrac{1}{(2k)!} B_k$ (Sec. 8.15-2) $d_0 = 1, \qquad d_k = \dfrac{4^k - 2}{(2k)!} B_k$ (Sec. 8.15-2)

$$\int \frac{\sin A}{x}\,dx = \text{Si}(A) \qquad\qquad \int \frac{\sin A}{x^2}\,dx = b\,\text{Ci}(A) - \frac{\sin A}{x}$$

$$\int \frac{\sin A}{x^3}\,dx = -\frac{\sin A}{2x^2} - \frac{b\cos A}{2x} - \frac{b^2}{2}\text{Si}(A)$$

$$\int \frac{\sin A}{x^m}\,dx = -\frac{\sin A}{(m-1)x^{m-1}} - \frac{b\cos A}{(m-1)(m-2)x^{m-2}} - \frac{b^2}{(m-1)(m-2)}\int \frac{\sin A}{x^{m-2}}\,dx \qquad m > 2$$

$$\int \frac{x\,dx}{\sin A} = \frac{A}{b^2}\sum_{k=0}^{\infty} d_k \frac{A^{2k}}{2k+1} = \frac{A}{b^2} \Lambda_{k=1}^{\infty}\left[1 + \frac{(2k-1)d_{k+1}}{(2k+1)d_{k-1}}A^2\right]^*$$

$$\int \frac{x\,dx}{\sin^2 A} = -\frac{A}{b^2}\cot A + \frac{1}{b^2}\ln \sin A$$

$$\int \frac{x\,dx}{\sin^3 A} = -\frac{A\cos A}{2b^2 \sin^2 A} - \frac{1}{2b^2 \sin A} + \frac{1}{2}\int \frac{x\,dx}{\sin A}$$

$$\int \frac{x\,dx}{\sin^n A} = -\frac{A\cos A}{(n-1)b^2 \sin^{n-1} A} - \frac{1}{(n-1)(n-2)b^2 \sin^{n-2} A} + \frac{n-2}{n-1}\int \frac{x\,dx}{\sin^{n-2} A} \qquad n > 2$$

$$\int \frac{x^2\,dx}{\sin A} = \frac{A^2}{2b^3}\sum_{k=0}^{\infty} d_k \frac{A^{2k}}{k+1} = \frac{A^2}{2b^3} \Lambda_{k=1}^{\infty}\left[1 + \frac{kd_k}{(k+1)d_{k-1}}A^2\right]^*$$

$$\int \frac{x^2\,dx}{\sin^2 A} = \frac{2A}{b^3}\left[1 - \frac{A}{2}\cot A - \sum_{k=1}^{\infty} b_k \frac{(2A)^{2k}}{2k+1}\right]$$

$$= \frac{2A}{b^3}\left\{-\Lambda_{k=1}^{\infty}\left[1 + \frac{(2k-1)b_k}{(2k+1)b_{k-1}}(2A)^2\right]^* - \frac{A}{2}\cot A + 2\right\}$$

$$\int \frac{x^m\,dx}{\sin A} = \frac{A^m}{b^{m+1}}\sum_{k=0}^{\infty} d_k \frac{A^{2k}}{m+2k} = \frac{A^m}{mb^{m+1}} \Lambda_{k=1}^{\infty}\left[1 + \frac{(m+2k-2)d_k}{(m+2k)d_{k-1}}A^2\right]^*$$

$$\int \frac{\sin^{2n-1} A}{x^m}\,dx = \left(-\frac{1}{4}\right)^{n-1}\int\left\{\sum_{k=0}^{n-1}(-1)^k\binom{2n-1}{k}\sin\left[(2n-2k-1)A\right]\right\}\frac{dx}{x^m}$$

$$\int \frac{\sin^{2n} A}{x^m}\,dx = 2\left(-\frac{1}{4}\right)^n\int\left\{\sum_{k=0}^{n-1}(-1)^k\binom{2n}{k}\cos\left[(2n-2k)A\right]\right\}\frac{dx}{x^m} - \frac{\binom{2n}{n}}{(m-1)4^n x^{m-1}}$$

---

(50) Indefinite Integrals Involving

$$f(x) = \frac{x^m}{\cos^n A} \qquad f(x) = \frac{\cos^n A}{x^m} \qquad A = bx$$

$\text{Si}(A) = $ sine-integral function (Sec. 13.01-1) $B_k = $ auxiliary Bernoulli number (Sec. 8.05-1)

$\text{Ci}(A) = $ cosine-integral function (Sec. 13.01-1) $E_k = $ auxiliary Euler number (Sec. 8.06-1)

$$a_0 = 1, \qquad a_k = \frac{4k-1}{(2k)!} B_k \text{ (Sec. 8.15-2)} \qquad c_0 = 1, \qquad c_k = \frac{1}{(2k)!} E_k \text{ (Sec. 8.15-2)}$$

$$\int \frac{\cos A}{x}\, dx = \text{Ci}(A) \qquad\qquad \int \frac{\cos A}{x^2}\, dx = -b\,\text{Si}(A) - \frac{\cos A}{x}$$

$$\int \frac{\cos A}{x^3}\, dx = -\frac{\cos A}{2x^2} + \frac{b \sin A}{2x} - \frac{b^2}{2}\text{Ci}(A)$$

$$\int \frac{\cos A}{x^m}\, dx = -\frac{\cos A}{(m-1)x^{m-1}} + \frac{b \sin A}{(m-1)(m-2)x^{m-2}} - \frac{b^2}{(m-1)(m-2)} \int \frac{\cos A}{x^{m-2}}\, dx \qquad m > 2$$

$$\int \frac{x\, dx}{\cos A} = \frac{A^2}{2b^2} \sum_{k=0}^{\infty} c_k \frac{A^{2k}}{k+1} = \frac{A^2}{2b^2} \bigwedge_{k=1}^{\infty} \left[1 + \frac{k c_k}{(k+1) c_{k-1}} A^2 \right]^*$$

$$\int \frac{x\, dx}{\cos^2 A} = \frac{A}{b^2} \tan A + \frac{1}{b^2} \ln \cos A$$

$$\int \frac{x\, dx}{\cos^3 A} = \frac{A \sin A}{2b^2 \cos^2 A} - \frac{1}{2b^2 \cos A} + \frac{1}{2} \int \frac{x\, dx}{\cos A}$$

$$\int \frac{x\, dx}{\cos^n A} = \frac{A \sin A}{(n-1) b^2 \cos^{n-1} A} - \frac{1}{(n-1)(n-2) b^2 \cos^{n-2} A} + \frac{n-2}{n-1} \int \frac{x\, dx}{\cos^{n-2} A} \qquad n > 2$$

$$\int \frac{x^2\, dx}{\cos A} = \frac{A^3}{b^3} \sum_{k=0}^{\infty} c_k \frac{A^{2k}}{2k+3} = \frac{A^3}{3b^3} \bigwedge_{k=1}^{\infty} \left[1 + \frac{(2k+1) c_k}{(2k+3) c_{k-1}} A^2 \right]^*$$

$$\int \frac{x^2\, dx}{\cos^2 A} = \frac{2A}{b^3} \left[\frac{A}{2} \tan A - \sum_{k=1}^{\infty} \frac{a_k}{2k+1} (2A)^{2k} \right]$$

$$= \frac{2A}{b^3} \left\{ -\bigwedge_{k=1}^{\infty} \left[1 + \frac{(2k-1) a_k}{(2k+1) a_{k-1}} (2A)^2 \right]^* + \frac{A}{2} \tan A + 1 \right\}$$

$$\int \frac{x^m\, dx}{\cos A} = \frac{A^{m+1}}{b^{m+1}} \sum_{k=0}^{\infty} c_k \frac{A^{2k}}{m+2k+1} = \frac{A^{m+1}}{(m+1) b^{m+1}} \bigwedge_{k=1}^{\infty} \left[1 + \frac{(m+2k-1) c_k}{(m+2k+1) c_{k-1}} A^2 \right]^*$$

$$\int \frac{\cos^{2n-1} A}{x^m}\, dx = \left(\frac{1}{4}\right)^{n-1} \int \left\{ \sum_{k=0}^{n-1} \binom{2n-1}{k} \cos\left[2n - 2k - 1)A \right] \right\} \frac{dx}{x^m}$$

$$\int \frac{\cos^{2n} A}{x^m}\, dx = 2\left(\frac{1}{4}\right)^{n} \int \left\{ \sum_{k=0}^{n-1} \binom{2n}{k} \cos\left[(2n - 2k)A \right] \right\} \frac{dx}{x^m} - \frac{\binom{2n}{n}}{(m-1) 4^n x^{m-1}}$$

---

*For nested sum refer to Sec. 8.11.

(51) Indefinite Integrals Involving

$$f(x) = f(\sin X_k, \ldots, \cos X_k, \ldots)$$

$\alpha_1 = a + b$	$\alpha_2 = a - b$	$\beta_1 = a + 2b$	$\beta_2 = -a + 2b$
$\lambda_1 = a + b + c$	$\lambda_2 = -a + b + c$	$\lambda_3 = a - b + c$	$\lambda_4 = a + b - c$
$A_1 = ax + b$	$A_2 = cx + d$	$B_1 = A_1 + A_2$	$B_2 = A_1 - A_2$
$C_1 = \omega x + a$	$C_2 = \omega x + b$	$D_1 = C_1 + C_2$	$D_2 = C_1 - C_2$

$$\int \sin ax \sin bx \, dx = -\frac{\sin \alpha_1 x}{2\alpha_1} + \frac{\sin \alpha_2 x}{2\alpha_2} \qquad\qquad a \neq b$$

$$\int \sin ax \sin^2 bx \, dx = +\frac{\cos \beta_1 x}{4\beta_1} - \frac{\cos \beta_2 x}{4\beta_2} - \frac{\cos ax}{2a} \qquad\qquad a \neq 2b$$

$$\int \sin ax \sin bx \sin cx \, dx = +\frac{\cos \lambda_1 x}{4\lambda_1} - \frac{\cos \lambda_2 x}{4\lambda_2} - \frac{\cos \lambda_3 x}{4\lambda_3} - \frac{\cos \lambda_4 x}{4\lambda_4} \qquad\qquad \lambda_k \neq 0$$

$$\int \sin ax \cos bx \, dx = -\frac{\cos \alpha_1 x}{2\alpha_1} - \frac{\cos \alpha_2 x}{2\alpha_2} \qquad\qquad a \neq b$$

$$\int \sin ax \cos^2 bx \, dx = -\frac{\cos \beta_1 x}{4\beta_1} + \frac{\cos \beta_2 x}{4\beta_2} - \frac{\cos ax}{2a} \qquad\qquad a \neq 2b$$

$$\int \sin ax \cos bx \cos cx \, dx = -\frac{\cos \lambda_1 x}{4\lambda_1} + \frac{\cos \lambda_2 x}{4\lambda_2} - \frac{\cos \lambda_3 x}{4\lambda_3} - \frac{\cos \lambda_4 x}{4\lambda_4} \qquad\qquad \lambda_k \neq 0$$

$$\int x \sin ax \sin bx \, dx = -x\left(\frac{\sin \alpha_1 x}{2\alpha_1} - \frac{\sin \alpha_2 x}{2\alpha_2}\right) - \left(\frac{\cos \alpha_1 x}{2\alpha_1^2} - \frac{\cos \alpha_2 x}{2\alpha_2^2}\right)$$
$$\qquad\qquad a \neq b$$
$$\int x \sin ax \cos bx \, dx = -x\left(\frac{\cos \alpha_1 x}{2\alpha_1} + \frac{\cos \alpha_2 x}{2\alpha_2}\right) + \left(\frac{\sin \alpha_1 x}{2\alpha_1^2} + \frac{\sin \alpha_2 x}{2\alpha_2^2}\right)$$

$$\int A_1 \sin A_2 \, dx = -\frac{A_1}{c} \cos A_2 + \frac{a}{c^2} \sin A_2 \qquad\qquad a, c \neq 0$$

$$\int \sin A_1 \sin A_2 \, dx = -\frac{\sin B_1}{2(a+c)} + \frac{\sin B_2}{2(a-c)} \qquad\qquad a \neq c$$

$$\int \sin A_1 \cos A_2 \, dx = -\frac{\cos B_1}{2(a+c)} - \frac{\cos B_2}{2(a-c)} \qquad\qquad a \neq c$$

$$\int C_1 \sin C_2 \, dx = -\frac{C_1}{\omega} \cos C_2 + \frac{1}{\omega} \sin C_2$$

$$\int \sin C_1 \sin C_2 \, dx = -\frac{\sin D_1}{4\omega} + \frac{x}{2} \cos D_2 \qquad\qquad \omega \neq 0$$

$$\int \sin C_1 \cos C_2 \, dx = -\frac{\cos D_1}{4\omega} + \frac{x}{2} \sin D_2$$

(52) Indefinite Integrals Involving $\quad\boxed{f(x) = f(\cos X_k, \ldots, \sin X_k, \ldots)}$

$\alpha_1 = a + b$	$\alpha_2 = a - b$	$\beta_1 = a + 2b$	$\beta_2 = 2b - a$
$\lambda_1 = a + b + c$	$\lambda_2 = -a + b + c$	$\lambda_3 = a - b + c$	$\lambda_4 = a + b - c$
$A_1 = ax + b$	$A_2 = cx + d$	$B_1 = A_1 + A_2$	$B_2 = A_1 - A_2$
$C_1 = \omega x + a$	$C_2 = \omega x + b$	$D_1 = C_1 + C_2$	$D_2 = C_1 - C_2$

$$\int \cos ax \cos bx\, dx = +\frac{\sin \alpha_1 x}{2\alpha_1} + \frac{\sin \alpha_2 x}{2\alpha_2} \qquad\qquad a \neq b$$

$$\int \cos ax \cos^2 bx\, dx = +\frac{\sin \beta_1 x}{4\beta_1} + \frac{\sin \beta_2 x}{4\beta_2} + \frac{\sin ax}{2a} \qquad\qquad a \neq 2b$$

$$\int \cos ax \cos bx \cos cx\, dx = +\frac{\sin \lambda_1 x}{4\lambda_1} + \frac{\sin \lambda_2 x}{4\lambda_2} + \frac{\sin \lambda_3 x}{4\lambda_3} + \frac{\sin \lambda_4 x}{4\lambda_4} \qquad\qquad \lambda_k \neq 0$$

$$\int \cos ax \sin bx\, dx = -\frac{\cos \alpha_1 x}{2\alpha_1} + \frac{\cos \alpha_2 x}{2\alpha_2} \qquad\qquad a \neq b$$

$$\int \cos ax \sin^2 bx\, dx = -\frac{\sin \beta_1 x}{4\beta_1} - \frac{\sin \beta_2 x}{4\beta_2} + \frac{\sin ax}{2a} \qquad\qquad a \neq 2b$$

$$\int \cos ax \sin bx \sin cx\, dx = -\frac{\sin \lambda_1 x}{4\lambda_1} - \frac{\sin \lambda_2 x}{4\lambda_2} + \frac{\sin \lambda_3 x}{4\lambda_3} + \frac{\sin \lambda_4 x}{4\lambda_4} \qquad\qquad \lambda_k \neq 0$$

$$\int x \cos ax \cos bx\, dx = +x\left(\frac{\sin \alpha_1 x}{2\alpha_1} + \frac{\sin \alpha_2 x}{2\alpha_2}\right) + \left(\frac{\cos \alpha_1 x}{2\alpha_1{}^2} + \frac{\cos \alpha_2 x}{2\alpha_2{}^2}\right)$$

$$\int x \cos ax \sin bx\, dx = -x\left(\frac{\cos \alpha_1 x}{2\alpha_1} - \frac{\cos \alpha_2 x}{2\alpha_2}\right) + \left(\frac{\sin \alpha_1 x}{2\alpha_1{}^2} - \frac{\sin \alpha_2 x}{2\alpha_2{}^2}\right)$$
$$a \neq b$$

$$\int A_1 \cos A_2\, dx = +\frac{A_1}{c}\sin A_2 + \frac{a}{c^2}\cos A_2 \qquad\qquad a, c \neq 0$$

$$\int \cos A_1 \cos A_2\, dx = +\frac{\sin B_1}{2(a+c)} + \frac{\sin B_2}{2(a-c)} \qquad\qquad a \neq c$$

$$\int \cos A_1 \sin A_2\, dx = -\frac{\cos B_1}{2(a+c)} + \frac{\cos B_2}{2(a-c)} \qquad\qquad a \neq c$$

$$\int C_1 \cos C_2\, dx = +\frac{C_1}{\omega}\sin C_2 + \frac{1}{\omega}\cos C_2$$

$$\int \cos C_1 \cos C_2\, dx = +\frac{\sin D_1}{4\omega} + \frac{x}{2}\cos D_2 \qquad\qquad \omega \neq 0$$

$$\int \cos C_1 \sin C_2\, dx = -\frac{\cos D_1}{4\omega} - \frac{x}{2}\sin D_2$$

(53) Indefinite Integrals Involving

| $f(x) = f(1 \pm \sin A)$ | $A = bx$ |

$$\int \frac{dx}{1 + \sin A} = -\frac{1}{b} \tan \left(\frac{\pi}{4} - \frac{A}{2} \right)$$

$$\int \frac{x \, dx}{1 + \sin A} = -\frac{x}{b} \tan \left(\frac{\pi}{4} - \frac{A}{2} \right) + \frac{2}{b^2} \ln \left[\cos \left(\frac{\pi}{4} - \frac{A}{2} \right) \right]$$

$$\int \frac{\sin A \, dx}{1 + \sin A} = \frac{1}{b} \tan \left(\frac{\pi}{4} - \frac{A}{2} \right) + x$$

$$\int \frac{\cos A \, dx}{1 + \sin A} = \frac{1}{b} \ln \left(1 + \sin A \right)$$

$$\int \frac{dx}{\sin A \, (1 + \sin A)} = \frac{1}{b} \tan \left(\frac{\pi}{4} - \frac{A}{2} \right) + \frac{1}{b} \ln \left(\tan \frac{A}{2} \right)$$

$$\int \frac{dx}{\cos A (1 + \sin A)} = \frac{-1}{2b(1 + \sin A)} + \frac{1}{2b} \ln \left[\tan \left(\frac{\pi}{4} + \frac{A}{2} \right) \right]$$

$$\int \frac{dx}{1 - \sin A} = \frac{1}{b} \tan \left(\frac{\pi}{4} + \frac{A}{2} \right)$$

$$\int \frac{x \, dx}{1 - \sin A} = \frac{x}{b} \cot \left(\frac{\pi}{4} - \frac{A}{2} \right) + \frac{2}{b^2} \ln \left[\sin \left(\frac{\pi}{4} - \frac{A}{2} \right) \right]$$

$$\int \frac{\sin A \, dx}{1 - \sin A} = \frac{1}{b} \tan \left(\frac{\pi}{4} + \frac{A}{2} \right) - x$$

$$\int \frac{\cos A \, dx}{1 - \sin A} = -\frac{1}{b} \ln \left(1 - \sin A \right)$$

$$\int \frac{dx}{\sin A \, (1 - \sin A)} = \frac{1}{b} \tan \left(\frac{\pi}{4} + \frac{A}{2} \right) + \frac{1}{b} \ln \left(\tan \frac{A}{2} \right)$$

$$\int \frac{dx}{\cos A (1 - \sin A)} = \frac{1}{2b(1 - \sin A)} + \frac{1}{2b} \left[\tan \left(\frac{\pi}{4} + \frac{A}{2} \right) \right]$$

$$\int \frac{dx}{(1 + \sin A)^2} = \frac{-1}{2b} \tan \left(\frac{\pi}{4} - \frac{A}{2} \right) - \frac{1}{6b} \tan^3 \left(\frac{\pi}{4} - \frac{A}{2} \right)$$

$$\int \frac{\sin A \, dx}{(1 + \sin A)^2} = \frac{-1}{2b} \tan \left(\frac{\pi}{4} - \frac{A}{2} \right) + \frac{1}{6b} \tan^3 \left(\frac{\pi}{4} - \frac{A}{2} \right)$$

$$\int \frac{dx}{(1 - \sin A)^2} = \frac{1}{2b} \cot \left(\frac{\pi}{4} - \frac{A}{2} \right) + \frac{1}{6b} \cot^3 \left(\frac{\pi}{4} - \frac{A}{2} \right)$$

$$\int \frac{\sin A \, dx}{(1 - \sin A)^2} = -\frac{1}{2b} \cot \left(\frac{\pi}{4} - \frac{A}{2} \right) + \frac{1}{6b} \cot^3 \left(\frac{\pi}{4} - \frac{A}{2} \right)$$

(54) Indefinite Integrals Involving

$f(x) = f(1 \pm \cos A)$	$A = bx$

$$\int \frac{dx}{1 + \cos A} = \frac{1}{b} \tan \frac{A}{2}$$

$$\int \frac{x \, dx}{1 + \cos A} = \frac{x}{b} \tan \frac{A}{2} + \frac{2}{b^2} \ln \left(\cos \frac{A}{2} \right)$$

$$\int \frac{\cos A \, dx}{1 + \cos A} = -\frac{1}{b} \tan \frac{A}{2} + x$$

$$\int \frac{\sin A \, dx}{1 + \cos A} = -\frac{1}{b} \ln \left(1 + \cos A \right)$$

$$\int \frac{dx}{\cos A \, (1 + \cos A)} = -\frac{1}{b} \tan \frac{A}{2} + \frac{1}{b} \ln \left[\tan \left(\frac{\pi}{4} + \frac{A}{2} \right) \right]$$

$$\int \frac{dx}{\sin A \, (1 + \cos A)} = \frac{1}{2b(1 + \cos A)} + \frac{1}{2b} \ln \left(\tan \frac{A}{2} \right)$$

$$\int \frac{dx}{1 - \cos A} = -\frac{1}{b} \cot \frac{A}{2}$$

$$\int \frac{x \, dx}{1 - \cos A} = -\frac{x}{b} \cot \frac{A}{2} + \frac{2}{b^2} \ln \left(\sin \frac{A}{2} \right)$$

$$\int \frac{\cos A \, dx}{1 - \cos A} = -\frac{1}{b} \cot \frac{A}{2} - x$$

$$\int \frac{\sin A \, dx}{1 - \cos A} = \frac{1}{b} \ln \left(1 - \cos A \right)$$

$$\int \frac{dx}{\cos A \, (1 - \cos A)} = -\frac{1}{b} \cot \frac{A}{2} + \frac{1}{b} \ln \left[\tan \left(\frac{\pi}{4} + \frac{A}{2} \right) \right]$$

$$\int \frac{dx}{\sin A \, (1 - \cos A)} = \frac{-1}{2b(1 - \cos A)} + \frac{1}{2b} \ln \left(\tan \frac{A}{2} \right)$$

$$\int \frac{dx}{(1 + \cos A)^2} = \frac{1}{2b} \tan \frac{A}{2} + \frac{1}{6b} \tan^3 \frac{A}{2}$$

$$\int \frac{\cos A \, dx}{(1 + \cos A)^2} = \frac{1}{2b} \tan \frac{A}{2} - \frac{1}{6b} \tan^3 \frac{A}{2}$$

$$\int \frac{dx}{(1 - \cos A)^2} = -\frac{1}{2b} \cot \frac{A}{2} - \frac{1}{6b} \cot^3 \frac{A}{2}$$

$$\int \frac{\cos A \, dx}{(1 - \cos A)^2} = \frac{1}{2b} \cot \frac{A}{2} - \frac{1}{6b} \cot^3 \frac{A}{2}$$

(55) Indefinite Integrals Involving*

| $f(x) = f(1 \pm \alpha \sin A)$ | $A = bx$ |

$$\int \frac{dx}{1 + a \sin A} = \begin{cases} \dfrac{2}{b\sqrt{1-a^2}} \tan^{-1} \dfrac{a + \tan (A/2)}{\sqrt{1-a^2}} & a^2 < 1 \\[3ex] \dfrac{1}{b\sqrt{1-a^2}} \sin^{-1} \dfrac{a + \sin A}{1 + a \sin A} & a^2 < 1 \\[3ex] \dfrac{1}{b\sqrt{a^2-1}} \ln \dfrac{\tan (A/2) + a - \sqrt{a^2-1}}{\tan (A/2) + a + \sqrt{a^2-1}} & a^2 > 1 \end{cases}$$

$$\int \frac{dx}{(1 + a \sin A)^2} = \frac{a \cos A}{b(1 - a^2)(1 + a \sin A)} + \frac{1}{1 - a^2} \int \frac{dx}{1 + a \sin A}$$

$$\int \frac{dx}{(1 + a \sin A)^n} = \frac{a \cos A}{b(n - 1)(1 - a^2)(1 + a \sin A)^{n-1}} + \frac{(2n - 3)}{(n - 1)(1 - a^2)} \int \frac{dx}{(1 + a \sin A)^{n-1}}$$

$$- \frac{(n - 2)}{(n - 1)(1 - a^2)} \int \frac{dx}{(1 + a \sin A)^{n-2}} \qquad \begin{matrix} a^2 \neq 1 \\ n \neq 1 \end{matrix}$$

$$\int \frac{\sin A \, dx}{1 + a \sin A} = \frac{x}{a} - \frac{1}{a} \int \frac{dx}{1 + a \sin A}$$

$$\int \frac{\sin A \, dx}{(1 + a \sin A)^2} = \frac{\cos A}{b(a^2 - 1)(1 + a \sin A)} + \frac{a}{a^2 - 1} \int \frac{dx}{1 + a \sin A} \qquad a^2 \neq 1$$

$$\int \frac{\cos A \, dx}{1 \pm a \sin A} = \pm \frac{\ln (1 \pm a \sin A)}{ab}$$

$$\int \frac{\cos A \, dx}{(1 \pm a \sin A)^2} = \mp \frac{1}{ab(1 \pm a \sin A)}$$

$$\int \frac{\cos A \, dx}{(1 \pm a \sin A)^n} = \mp \frac{1}{ab(n - 1)(1 \pm a \sin A)^{n-1}} \qquad n \neq 1$$

$$\int \frac{dx}{(1 + a \sin A) \sin A} = \frac{1}{b} \ln \left(\tan \frac{A}{2} \right) - a \int \frac{dx}{1 + a \sin A}$$

$$\int \frac{dx}{(1 \pm \sin A) \cos A} = \frac{\mp 1}{2b(1 \pm \sin A)} + \frac{1}{2b} \ln \left[\tan \left(\frac{\pi}{4} + \frac{A}{2} \right) \right]$$

$$\int \frac{\sin A \, dx}{(1 \pm \sin A) \cos A} = \frac{1}{2b(1 \pm \sin A)} \pm \frac{1}{2b} \ln \left[\tan \left(\frac{\pi}{4} + \frac{A}{2} \right) \right]$$

$$\int \frac{\cos A \, dx}{(1 \pm \sin A) \sin A} = -\frac{1}{b} \ln \frac{1 \pm \sin A}{\sin A}$$

$$\int \frac{1 + c \sin A}{1 + a \sin A} \, dx = \frac{cx}{ab} + \frac{a - c}{ab} \int \frac{dx}{1 + a \sin A} \qquad c \neq 0$$

*If $a = 1$ refer to Sec. 19.53.

(56) Indefinite Integrals Involving*

$f(x) = f(1 \pm a \cos A)$	$A = bx$

$$\int \frac{dx}{1 + a \cos A} = \begin{cases} \dfrac{2}{b\sqrt{1 - a^2}} \tan^{-1}\left(\sqrt{\dfrac{1 - a}{1 + a}} \tan \dfrac{A}{2}\right) & a^2 < 1 \\[4mm] \dfrac{1}{b\sqrt{1 - a^2}} \cos^{-1} \dfrac{a + \cos A}{1 + a \cos A} & a^2 < 1 \\[4mm] \dfrac{1}{b\sqrt{a^2 - 1}} \ln \dfrac{(\sqrt{a + 1}) + (\sqrt{a - 1}) \tan (A/2)}{(\sqrt{a + 1}) - (\sqrt{a - 1}) \tan (A/2)} & a^2 > 1 \end{cases}$$

$$\int \frac{dx}{(1 + a \cos A)^2} = \frac{-a \sin A}{b(1 - a^2)(1 + a \cos A)} + \frac{1}{1 - a^2} \int \frac{dx}{1 + a \cos A}$$

$$\int \frac{dx}{(1 + a \cos A)^n} = -\frac{a \sin A}{b(n - 1)(1 - a^2)(1 + a \cos A)^{n-1}} + \frac{2n - 3}{(n - 1)(1 - a^2)} \int \frac{dx}{(1 + a \cos A)^{n-1}}$$
$$- \frac{n - 2}{(n - 1)(1 - a^2)} \int \frac{dx}{(1 + a \cos A)^{n-2}} \qquad \begin{array}{l} a^2 \neq 1 \\ n \neq 1 \end{array}$$

$$\int \frac{\cos A \, dx}{1 + a \cos A} = \frac{x}{a} - \frac{1}{a} \int \frac{dx}{1 + a \cos A}$$

$$\int \frac{\cos A \, dx}{(1 + a \cos A)^2} = \frac{-\sin A}{b(a^2 - 1)(1 + a \cos A)} + \frac{a}{a^2 - 1} \int \frac{dx}{1 + a \cos A} \qquad a^2 \neq 1$$

$$\int \frac{\sin A \, dx}{1 \pm a \cos A} = \mp \frac{\ln (1 \pm a \cos A)}{ab}$$

$$\int \frac{\sin A \, dx}{(1 \pm a \cos A)^2} = \pm \frac{1}{ab(1 \pm a \cos A)}$$

$$\int \frac{\sin A \, dx}{(1 \pm a \cos A)^n} = \pm \frac{1}{ab(n - 1)(1 \pm a \cos A)^{n-1}} \qquad n \neq 1$$

$$\int \frac{dx}{(1 + a \cos A) \cos A} = \frac{1}{b} \ln \left[\tan \left(\frac{A}{2} + \frac{\pi}{4} \right) \right] - a \int \frac{dx}{1 + a \cos A}$$

$$\int \frac{dx}{(1 \pm \cos A) \sin A} = \frac{\pm 1}{2b(1 \pm \cos A)} + \frac{1}{2b} \ln \left(\tan \frac{A}{2} \right)$$

$$\int \frac{\cos A \, dx}{(1 \pm \cos A) \sin A} = \frac{1}{2b(1 \pm \cos A)} \pm \frac{1}{2b} \ln \left(\tan \frac{A}{2} \right)$$

$$\int \frac{\sin A \, dx}{(1 \pm \cos A) \cos A} = \frac{1}{b} \ln \frac{1 \pm \cos A}{\cos A}$$

$$\int \frac{1 + c \cos A}{1 + a \cos A} \, dx = \frac{cx}{ab} + \frac{a - c}{ab} \int \frac{dx}{1 + a \cos A} \qquad c \neq 0$$

*If $a = 1$ refer to Sec. 19.54.

(57) Indefinite Integrals Involving

$$\boxed{f(x) = f(1 \pm a \sin^2 A) \quad \bigg| \quad A = bx}$$

$$\boxed{\alpha = \sqrt{1 + a} \qquad\qquad \beta = \sqrt{1 - a} \qquad\qquad \gamma = \sqrt{a - 1}}$$

$$\int \frac{dx}{1 + \sin^2 A} = \frac{1}{b\sqrt{2}} \tan^{-1}(\sqrt{2}\tan A)$$

$$\int \frac{dx}{1 + a \sin^2 A} = \frac{1}{\alpha b} \tan^{-1}(\alpha \tan A) \qquad\qquad a > 0$$

$$\int \frac{dx}{(1 + a \sin^2 A)^2} = \frac{a \sin 2A}{4\alpha^2 b(1 + a \sin^2 A)} + \frac{2 + a}{2\alpha^3 b} \tan^{-1}(\alpha \tan A) \qquad\qquad a > 0$$

$$\int \frac{dx}{1 - \sin^2 A} = \frac{1}{b} \tan A$$

$$\int \frac{dx}{1 - a \sin^2 A} = \begin{cases} \dfrac{1}{\beta b} \tan^{-1}(\beta \tan A) & 0 < a < 1 \\[2ex] \dfrac{1}{2\gamma b} \ln \dfrac{\gamma \tan A + 1}{\gamma \tan A - 1} & a > 1 \end{cases}$$

$$\int \frac{dx}{(1 - a \sin^2 A)^2} = \frac{-a \sin 2A}{4b\beta^2(1 - a \sin^2 A)} + \frac{2 - a}{2b\beta^2} \begin{cases} \dfrac{1}{\beta} \tan^{-1}(\beta \tan A) & 0 < a < 1 \\[2ex] \dfrac{1}{2\gamma} \ln \dfrac{\gamma \tan A + 1}{\gamma \tan A - 1} & a > 1 \end{cases}$$

$$\int \frac{\sin^2 A\, dx}{1 + a \sin^2 A} = \frac{x}{a} - \frac{1}{\alpha a b} \tan^{-1}(\alpha \tan A) \qquad\qquad a > 0$$

$$\int \frac{\cos^2 A\, dx}{1 + a \sin^2 A} = \frac{\alpha}{ab} \tan^{-1}(\alpha \tan A) - \frac{x}{a} \qquad\qquad a > 0$$

$$\int \frac{\sin A \cos A\, dx}{1 \pm a \sin^2 A} = \pm \frac{1}{ab} \ln \sqrt{1 \pm a \sin^2 A} \qquad\qquad a > 0$$

$$\int \frac{\sin^2 A\, dx}{1 - a \sin^2 A} = \begin{cases} \dfrac{1}{ab\beta} \tan^{-1}(\beta \tan A) - \dfrac{x}{a} & 0 < a < 1 \\[2ex] \dfrac{1}{2ab\gamma} \ln \left(\dfrac{\gamma \tan A + 1}{\gamma \tan A - 1}\right) - \dfrac{x}{a} & a > 0 \end{cases}$$

$$\int \frac{\cos^2 A\, dx}{1 - a \sin^2 A} = \begin{cases} -\dfrac{\beta}{ab} \tan^{-1}(\beta \tan A) + \dfrac{x}{a} & 0 < a < 1 \\[2ex] \dfrac{\gamma}{2ab} \ln \left(\dfrac{\gamma \tan A + 1}{\gamma \tan A - 1}\right) + \dfrac{x}{a} & a > 0 \end{cases}$$

(58) Indefinite Integrals Involving

$f(x) = f(1 \pm a \cos^2 A)$	$A = bx$

$\alpha = \sqrt{1+a}$	$\beta = \sqrt{1-a}$	$\gamma = \sqrt{a-1}$

$$\int \frac{dx}{1+\cos^2 A} = \frac{1}{b\sqrt{2}} \tan^{-1} \frac{\tan A}{\sqrt{2}}$$

$$\int \frac{dx}{1+a\cos^2 A} = \frac{1}{\alpha b} \tan^{-1} \frac{\tan A}{\alpha} \qquad\qquad a > 0$$

$$\int \frac{dx}{(1+a\cos^2 A)^2} = \frac{-a \sin 2A}{4\alpha^2 b(1+a\cos^2 A)} + \frac{2+a}{2\alpha^3 b} \tan^{-1} \frac{\tan A}{\alpha} \qquad\qquad a > 0$$

$$\int \frac{dx}{1-\cos^2 A} = -\frac{1}{b} \cot A$$

$$\int \frac{dx}{1-a\cos^2 A} = \begin{cases} \dfrac{1}{\beta b} \tan^{-1} \dfrac{\tan A}{\beta} & 0 < a < 1 \\[2mm] \dfrac{1}{2\gamma b} \ln \dfrac{\tan A - \gamma}{\tan A + \gamma} & a > 1 \end{cases}$$

$$\int \frac{dx}{(1-a\cos^2 A)^2} = \frac{a \sin 2A}{4b\,\beta^2(1-a\cos^2 A)} + \frac{2-a}{2b\,\beta^2} \begin{cases} \dfrac{1}{\beta} \tan^{-1} \dfrac{\tan A}{\beta} & 0 < a < 1 \\[2mm] \dfrac{1}{2\gamma} \ln \dfrac{\tan A - \gamma}{\tan A + \gamma} & a > 1 \end{cases}$$

$$\int \frac{\cos^2 A\, dx}{1+a\cos^2 A} = \frac{x}{a} - \frac{1}{\alpha a b} \tan^{-1} \frac{\tan A}{\alpha} \qquad\qquad a > 0$$

$$\int \frac{\sin^2 A\, dx}{1+a\cos^2 A} = \frac{\alpha}{ab} \tan^{-1} \left(\frac{\tan A}{\alpha}\right) - \frac{x}{a} \qquad\qquad a > 0$$

$$\int \frac{\sin A \cos A\, dx}{1 \pm a\cos^2 A} = \mp\frac{1}{ab} \ln\sqrt{1 \pm a\cos^2 A} \qquad\qquad a > 0$$

$$\int \frac{\cos^2 A\, dx}{1-a\cos^2 A} = \begin{cases} \dfrac{1}{ab\beta} \tan^{-1} \left(\dfrac{\tan A}{\beta}\right) - \dfrac{x}{a} & 0 < a < 1 \\[2mm] \dfrac{1}{2ab\gamma} \ln \left(\dfrac{\tan A - \gamma}{\tan A + \gamma}\right) - \dfrac{x}{a} & a > 0 \end{cases}$$

$$\int \frac{\sin^2 A\, dx}{1-a\cos^2 A} = \begin{cases} -\dfrac{\beta}{ab} \tan^{-1} \left(\dfrac{\tan A}{\beta}\right) + \dfrac{x}{a} & 0 < a < 1 \\[2mm] \dfrac{\gamma}{2ab} \ln \left(\dfrac{\tan A - \gamma}{\tan A + \gamma}\right) + \dfrac{x}{a} & a > 0 \end{cases}$$

(59) Indefinite Integrals Involving*

$f(x) = f(p \sin A + q \cos A)$	$A = bx$

$$r = \sqrt{p^2 + q^2} \neq 0 \qquad s = \sqrt{a^2 - r^2} \qquad \alpha = \tan^{-1}\frac{q}{p} \qquad \beta = A + \alpha \qquad \lambda = \sin A \pm \cos A$$

$$\int \frac{dx}{\sin A \pm \cos A} = \frac{1}{b\sqrt{2}} \ln \tan\left(\frac{A}{2} \pm \frac{\pi}{8}\right)$$

$$\int \frac{dx}{(\sin A \pm \cos A)^2} = \frac{1}{2b} \tan\left(A \mp \frac{\pi}{4}\right)$$

$$\int \frac{\sin A \, dx}{\sin A \pm \cos A} = \frac{1}{2b}\,(A \mp \ln \lambda)$$

$$\int \frac{\cos A \, dx}{\sin A \pm \cos A} = \frac{1}{2b}\,(\ln \lambda \pm A)$$

$$\int \frac{dx}{p \sin A + q \cos A} = \frac{1}{br} \ln\left(\tan \frac{\beta}{2}\right)$$

$$\int \frac{dx}{(p \sin A + q \cos A)^2} = -\frac{1}{br} \cot \beta = \frac{1}{br}\frac{p \sin A - q \cos A}{p \sin A + q \cos A}$$

$$\int \frac{dx}{(p \sin A + q \cos A)^n} = \frac{1}{r^n}\int \frac{d\beta}{\sin^n \beta} = -\frac{\cos \beta}{br^n(n-1) \sin^{n-1} \beta} + \frac{n-2}{(n-1)r^n}\int \frac{d\beta}{\sin^{n-2} \beta}$$

$$\int \frac{\sin A \, dx}{p \sin A + q \cos A} = \frac{1}{br^2}\,[pA - q \ln (p \sin A + q \cos A)]$$

$$\int \frac{\cos A \, dx}{p \sin A + q \cos A} = \frac{1}{br^2}\,[q \ln (p \sin A + q \cos A) + pA]$$

$$\int \frac{dx}{a + p \sin A + q \cos A} = \begin{cases} \dfrac{2}{bs} \tan^{-1}\left[\dfrac{(a-q) \tan (A/2) + p}{s}\right] & a^2 > r^2 \\[2ex] \dfrac{1}{ab}\left[\dfrac{a - (p+q) \cos A - (p-q) \sin A}{a - (p-q) \cos A + (p+q) \sin A}\right] & a^2 = r^2 \\[2ex] \dfrac{1}{bsi} \ln\left[\dfrac{(a-q) \tan (A/2) + p - si}{(a-q) \tan (A/2) + p + si}\right] & a^2 < r^2 \end{cases}$$

$$\int \frac{(p + q \sin A) \, dx}{\sin A(1 \pm \cos A)} = \frac{p}{2b}\left[\ln\left(\tan \frac{A}{2}\right) \pm \frac{1}{1 \pm \cos A}\right] + q\int \frac{dx}{1 \pm \cos A}$$

$$\int \frac{(p + q \sin A) \, dx}{\cos A (1 \pm \cos A)} = \frac{p}{b}\left[\ln \tan\left(\frac{\pi}{4} + \frac{A}{2}\right)\right] + \frac{q}{b} \ln\left(\frac{1 \pm \cos A}{\cos A}\right) - p\int \frac{dx}{1 \pm \cos A}$$

$$\int \frac{(p + q \cos A) \, dx}{\sin A (1 \pm \sin A)} = \frac{p}{b} \ln\left(\tan \frac{A}{2}\right) - \frac{q}{b} \ln\left(\frac{1 \pm \sin A}{\sin A}\right) - p\int \frac{dx}{1 \pm \sin A}$$

$$\int \frac{(p + q \cos A) \, dx}{\cos A (1 \pm \sin A)} = \frac{p}{2b}\left[\ln \tan\left(\frac{\pi}{4} + \frac{A}{2}\right) \mp \frac{1}{1 \pm \sin A}\right] + q\int \frac{dx}{1 \pm \sin A}$$

*p, q = signed integers or fractions.

(60) Indefinite Integrals Involving*

$f(x) = f(p^2 \sin^2 A \pm q^2 \cos^2 A)$	$A = bx$

$\alpha = p^2 \pm q^2$	$\beta = p^2 \mp q^2$	$\gamma = 4p^2 r^2 - q^4$

$$\int (p^2 \sin^2 A \pm q^2 \cos^2 A)\, dx = \frac{\alpha A}{2b} - \frac{\beta \sin 2A}{4b}$$

$$\int (p^2 \sin^2 A \pm q^2 \cos^2 A) \sin A \cos A\, dx = \frac{1}{4b} (p^2 \sin^4 A \mp q^2 \cos^4 A)$$

$$\int (p^2 \sin^2 A \pm q^2 \cos^2 A)^m \sin A \cos A\, dx = \frac{1}{2(m+1)\beta b} (p^2 \sin^2 A \pm q^2 \cos^2 A)^{m+1} \qquad m > 1 \qquad \beta \neq 0$$

$$\int \frac{dx}{p^2 \sin^2 A + q^2 \cos^2 A} = \frac{1}{bpq} \tan^{-1} \left(\frac{p}{q} \tan A \right)$$

$$\int \frac{dx}{p^2 \sin^2 A - q^2 \cos^2 A} = \frac{1}{bpq} \ln \sqrt{\frac{p \tan A - q}{p \tan A + q}}$$

$$\int \frac{\sin A \cos A\, dx}{p^2 \sin^2 A \pm q^2 \cos^2 A} = \frac{1}{2\beta b} \ln (p^2 \sin^2 A \pm q^2 \cos^2 A) \qquad \beta \neq 0$$

$$\int \frac{\sin A \cos A\, dx}{\sqrt{p^2 \sin^2 A \pm q^2 \cos^2 A}} = \frac{1}{\beta b} \sqrt{p^2 \sin^2 A \pm q^2 \cos^2 A}$$

$$\int \frac{dx}{(p^2 \sin^2 A \pm q^2 \cos^2 A)^2} = \frac{1}{2b(pq)^3} (\alpha z \pm \beta \sin z \cos z)$$

$$\int \frac{dx}{(p^2 \sin^2 A \pm q^2 \cos^2 A)^n} = \frac{1}{b(pq)^{2n-1}} \int (p^2 \sin^2 z \pm q^2 \cos^2 z)^{n-1}\, dz$$

$$\left. \vphantom{\int} \right\} \qquad z = \tan^{-1} \left(\frac{p}{q} \tan A \right)$$

$$\int \frac{dx}{p^2 \sin^2 A + q^2 \sin A \cos A + r^2 \cos^2 A} = \begin{cases} \dfrac{2}{b\sqrt{\gamma}} \tan^{-1} \left(\dfrac{2p^2 \tan A + q^2}{\sqrt{\gamma}} \right) & \gamma > 0 \\[3mm] -\dfrac{2}{b(2p^2 \tan A + q^2)} & \gamma = 0 \\[3mm] \dfrac{1}{b\sqrt{-\gamma}} \ln \left(\dfrac{2p^2 \tan A + q^2 - \sqrt{-\gamma}}{2p^2 \tan A + q^2 + \sqrt{-\gamma}} \right) & \gamma < 0 \end{cases}$$

$$\int \frac{dx}{\sin^2 A (p^2 \pm q^2 \cos^2 A)} = \frac{1}{\alpha b} \left(\int \frac{\pm q^2\, dA}{p^2 \pm q^2 \cos^2 A} - \cot A \right)$$

$$\int \frac{dx}{\cos^2 A (p^2 \pm q^2 \sin^2 A)} = \frac{1}{\alpha b} \left(\int \frac{\pm q^2\, dA}{p^2 \pm q^2 \sin^2 A} + \tan A \right) \qquad \alpha \neq 0$$

---

*p, q = positive integers or fractions.

(61) Indefinite Integrals Involving

$f(x) = f(\sqrt{p \pm q \sin A})$	$A = bx$

$E(k, \phi)$ = incomplete elliptic integral of the second kind (Sec. 13.05)

$F(k, \phi)$ = incomplete elliptic integral of the first kind (Sec. 13.05)

$G(k, \phi) = \tan \phi \sqrt{1 - k^2 \sin \phi}, r = \sqrt{p^2 + q^2}, = p/q, a > r, p > 0, q > 0$

$$\int \sqrt{1 + \sin A}\, dx = -\frac{2\sqrt{2}}{b} \cos\left(\frac{\pi}{4} + \frac{A}{2}\right)$$

$$\int \sqrt{1 - \sin A}\, dx = \frac{2\sqrt{2}}{b} \sin\left(\frac{\pi}{4} + \frac{A}{2}\right)$$

$$\int \sqrt{p + q \sin A}\, dx = -\frac{2}{b} \sqrt{p + q}\, E\left(\sqrt{\frac{2}{1 + \lambda}}, \frac{\cos^{-1}(\sin A)}{2}\right)$$

$$\int \sqrt{p - q \sin A}\, dx = -\frac{2}{b} \sqrt{p + q}\left[E\left(\sqrt{\frac{2}{1 + \lambda}}, \sin^{-1}\sqrt{\frac{\lambda - \sin A}{1 - \sin A}}\right) - G\left(\sqrt{\frac{2}{1 + \lambda}}, \sin^{-1}\sqrt{\frac{\lambda - \sin A}{1 - \sin A}}\right)\right]$$

$$\int \frac{dx}{\sqrt{\sin 2A}} = \frac{\sqrt{2}}{b} F\left(\frac{1}{\sqrt{2}}, \sin^{-1}\sqrt{\frac{2 \sin A}{1 + \sin A + \cos A}}\right)$$

$$\int \frac{dx}{\sqrt{1 + \sin A}} = \frac{\sqrt{2}}{b} \ln\left[\tan\left(\frac{A}{4} + \frac{\pi}{8}\right)\right]$$

$$\int \frac{dx}{\sqrt{1 - \sin A}} = \frac{\sqrt{2}}{b} \ln\left[\tan\left(\frac{A}{4} - \frac{\pi}{8}\right)\right]$$

$$\int \frac{dx}{\sqrt{p + q \sin A}} = \frac{-2}{b\sqrt{p + q}} F\left(\sqrt{\frac{2}{1 + \lambda}}, \sin^{-1}\sqrt{\frac{1 - \sin A}{2}}\right)$$

$$\int \frac{dx}{\sqrt{p - q \sin A}} = \sqrt{\frac{2}{qb^2}} F\left(\sqrt{\lambda}, \sin^{-1}\sqrt{\frac{1 - \sin A}{\lambda - \sin A}}\right)$$

$$\int \frac{\cos A\, dx}{\sqrt{p \pm q \sin A}} = \frac{\pm 1}{2bq} \sqrt{p \pm q \sin A}$$

$$\int \frac{\sin A\, dx}{\sqrt{p + q \sin A}} = \frac{2p}{b\sqrt{p + q}} F\left(\sqrt{\frac{2}{1 + \lambda}}, \sin^{-1}\sqrt{\frac{1 - \sin A}{2}}\right) - \frac{2\sqrt{p + q}}{qb} E\left(\sqrt{\frac{2}{1 + \lambda}}, \sin^{-1}\sqrt{\frac{1 - \sin A}{2}}\right)$$

$$\int \frac{\sin A\, dx}{\sqrt{p - q \sin A}} = -\sqrt{\frac{2}{qb^2}} F\left(\sqrt{\lambda}, \sin^{-1}\sqrt{\frac{1 - \sin A}{\lambda - \sin A}}\right) - \sqrt{\frac{q}{2b^2}} E\left(\sqrt{\lambda}, \sin^{-1}\sqrt{\frac{1 - \sin A}{\lambda - \sin A}}\right)$$

$$\int \sqrt{a + p \sin A + q \cos A}\, dx = -\frac{2}{b} \sqrt{a + r}\, E\left(\sqrt{\frac{2r}{a + r}}, \sin^{-1}\sqrt{\frac{r - p \sin A - q \cos A}{2r}}\right)$$

(62) Indefinite Integrals Involving

$f(x) = f(\sqrt{p \pm q \cos A})$	$A = bx$

$E(k, \phi)$ = incomplete elliptic integral of the second kind (Sec. 13.05)

$F(k, \phi)$ = incomplete elliptic integral of the first kind (Sec. 13.05)

$G(k, \phi) = \tan \phi \sqrt{1 - k^2 \sin \phi}, r = \sqrt{p^2 + q^2}, = p/q, a > r, p > 0, q > 0$

$$\int \sqrt{1 + \cos A}\, dx = \frac{2\sqrt{2}}{b} \sin \frac{A}{2}$$

$$\int \sqrt{1 - \cos A}\, dx = -\frac{2\sqrt{2}}{b} \cos \frac{A}{2}$$

$$\int \sqrt{p + q \cos A}\, dx = \frac{2}{b} \sqrt{p + q}\, E\left(\sqrt{\frac{2}{1 + \lambda}}, \frac{A}{2}\right)$$

$$\int \sqrt{p - q \cos A}\, dx = \frac{2}{b} \sqrt{p + q}\, E\left(\sqrt{\frac{2}{1 + \lambda}}, \sin^{-1}\sqrt{\frac{(p + q)(1 - \cos A)}{2(p - q \cos A)}}\right) - \frac{2q}{b} \frac{\sin A}{\sqrt{p - q \cos A}}$$

$$\int \frac{dx}{\sqrt{\cos 2A}} = \frac{1}{b\sqrt{2}} F\left(\frac{1}{\sqrt{2}}, \sin^{-1}\left(\sqrt{2} \sin A\right)\right)$$

$$\int \frac{dx}{\sqrt{1 + \cos A}} = \frac{\sqrt{2}}{b} \ln\left(\tan \frac{\pi + A}{4}\right)$$

$$\int \frac{dx}{\sqrt{1 - \cos A}} = \frac{\sqrt{2}}{b} \ln\left(\tan \frac{A}{4}\right)$$

$$\int \frac{dx}{\sqrt{p + q \cos A}} = \frac{2}{b\sqrt{p + q}} F\left(\sqrt{\frac{2}{1 + \lambda}}, \frac{A}{2}\right)$$

$$\int \frac{dx}{\sqrt{p - q \cos A}} = \frac{2}{b\sqrt{p + q}} F\left(\sqrt{\frac{2}{1 + \lambda}}, \sin^{-1}\sqrt{\frac{(p + q)(1 - \cos A)}{2(p - q \cos A)}}\right)$$

$$\int \frac{\sin A\, dx}{\sqrt{p \pm q \cos A}} = \frac{\mp 1}{2bq} \sqrt{p \pm q \cos A}$$

$$\int \frac{\cos A\, dx}{\sqrt{p + q \cos A}} = \frac{2\sqrt{p + q}}{bq} E\left(\sqrt{\frac{2}{1 + \lambda}}, \frac{A}{2}\right) - \frac{2\lambda}{b\sqrt{p + q}} F\left(\sqrt{\frac{2}{1 + \lambda}}, \frac{A}{2}\right)$$

$$\int \frac{\cos A\, dx}{\sqrt{p - q \cos A}} = \frac{2}{bq\sqrt{p + q}} E\left(\sqrt{\frac{2}{1 + \lambda}}, \frac{A}{2}\right) - \frac{2\sqrt{p + q}}{b\lambda} F\left(\sqrt{\frac{2}{1 + \lambda}}, \frac{A}{2}\right)$$

$$\int \frac{dx}{\sqrt{a + p \sin A + q \cos A}} = \frac{2}{b\sqrt{a + r}} F\left(\sqrt{\frac{2r}{a + r}}, \sin^{-1}\sqrt{\frac{r - p \sin A - q \cos A}{2r}}\right)$$

(63) Indefinite Integrals Involving

$$f(x) = f(\sqrt{1 \pm a^2 \sin^2 A}) \qquad A = bx$$

$E(k, \phi)$ = incomplete elliptic integral of the second kind (Sec. 13.05)

$F(k, \phi)$ = incomplete elliptic integral of the first kind (Sec. 13.05)

$$\alpha = \sqrt{\frac{1}{1 + a^2}} \qquad \beta = \sqrt{\frac{1}{1 - a^2}} \qquad \lambda = \sqrt{\frac{a^2}{1 + a^2}}$$

$$\int \sqrt{1 + a^2 \sin^2 A} \; dx = -\frac{1}{\alpha b} E\left(\lambda, \frac{\pi}{2} - A\right)$$

$$\int \sin A \sqrt{1 + a^2 \sin^2 A} \; dx = -\frac{\cos A}{2b} \sqrt{1 + a^2 \sin^2 A} - \frac{\sin^{-1}(\lambda \cos A)}{2ab\alpha^2}$$

$$\int \cos A \sqrt{1 + a^2 \sin^2 A} \; dx = \frac{\sin A}{2b} \sqrt{1 + a^2 \sin^2 A} + \frac{\ln(a \sin A + \sqrt{1 + a^2 \sin^2 A})}{2ab}$$

$$\int \sqrt{1 - a^2 \sin^2 A} \; dx = \frac{1}{b} E(a, A)$$

$$\int \sin A \sqrt{1 - a^2 \sin^2 A} \; dx = -\frac{\cos A}{2b} \sqrt{1 - a^2 \sin^2 A} - \frac{\ln(a \cos A + \sqrt{1 - a^2 \sin^2 A})}{2 \, ab\beta^2}$$

$$\int \cos A \sqrt{1 - a^2 \sin^2 A} \; dx = \frac{\sin A}{2b} \sqrt{1 - a^2 \sin^2 A} + \frac{\sin^{-1}(a \sin A)}{2ab}$$

$$\int \frac{dx}{\sqrt{1 + a^2 \sin^2 A}} = -\frac{\alpha}{b} F\left(\lambda, \frac{\pi}{2} - A\right)$$

$$\int \frac{\sin A \; dx}{\sqrt{1 + a^2 \sin^2 A}} = -\frac{1}{ab} \sin^{-1}(\lambda \cos A)$$

$$\int \frac{\cos A \; dx}{\sqrt{1 + a^2 \sin^2 A}} = \frac{1}{ab} \ln(a \sin A + \sqrt{1 + a^2 \sin^2 A})$$

$$\int \frac{dx}{\sqrt{1 - a^2 \sin^2 A}} = \frac{1}{b} F(a, A)$$

$$\int \frac{\sin A \; dx}{\sqrt{1 - a^2 \sin^2 A}} = -\frac{1}{ab} \ln(a \cos A + \sqrt{1 - a^2 \sin^2 A})$$

$$\int \frac{\cos A \; dx}{\sqrt{1 - a^2 \sin^2 A}} = \frac{1}{ab} \sin^{-1}(a \sin A)$$

$$\int \frac{dx}{\sqrt{p^2 \cos^2 A + q^2 \sin^2 A}} = -\frac{1}{bp} F\left(\sqrt{1 - \frac{q^2}{p^2}}, A\right) \qquad q < p$$

(64) Indefinite Integrals Involving

$$\boxed{f(x) = f(\sqrt{1 \pm a^2 \cos^2 A})} \quad \boxed{A = bx}$$

$E(k, \phi)$ = incomplete elliptic integral of the second kind (Sec. 13.05)

$F(k, \phi)$ = incomplete elliptic integral of the first kind (Sec. 13.05)

$$\alpha = \sqrt{\frac{1}{1+a^2}} \qquad \beta = \sqrt{\frac{1}{1-a^2}} \qquad \lambda = \sqrt{\frac{a^2}{1+a^2}} \qquad \kappa = \sqrt{\frac{a^2}{1-a^2}}$$

$$\int \sqrt{1+a^2 \cos^2 A}\, dx = \frac{1}{b\alpha} E(\lambda, A)$$

$$\int \sin A \sqrt{1+a^2 \cos^2 A}\, dx = -\frac{\cos A}{2\alpha b}\sqrt{1-\lambda^2 \sin^2 A} - \frac{\ln(\lambda \cos A + \sqrt{1-\lambda^2 \sin^2 a})}{2ab}$$

$$\int \cos A \sqrt{1+a^2 \cos^2 A}\, dx = \frac{\sin A}{2\alpha b}\sqrt{1-\lambda^2 \sin^2 A} + \frac{\sin^{-1}(\lambda \sin A)}{2ab\alpha^2}$$

$$\int \sqrt{1-a^2 \cos^2 A}\, dx = -\frac{1}{b\beta} E\left(\kappa, \frac{\pi}{2} - A\right)$$

$$\int \sin A \sqrt{1-a^2 \cos^2 A}\, dx = -\frac{\cos A}{2b\beta}\sqrt{1+\kappa^2 \sin^2 A} - \frac{\sin^{-1}(a \cos A)}{2ab}$$

$$\int \cos A \sqrt{1-a^2 \cos^2 A}\, dx = \frac{\sin A}{2b\beta}\sqrt{1+\kappa^2 \sin^2 A} + \frac{\ln(\kappa \sin A + \sqrt{1+\kappa^2 \sin^2 A})}{2ab\beta^2}$$

$$\int \frac{dx}{\sqrt{1+a^2 \cos^2 A}} = \frac{\alpha}{b} F(\lambda, A)$$

$$\int \frac{\sin A\, dx}{\sqrt{1+a^2 \cos^2 A}} = -\frac{1}{ab}\ln(\lambda \cos A + \sqrt{1-\lambda^2 \sin^2 A})$$

$$\int \frac{\cos A\, dx}{\sqrt{1+a^2 \cos^2 A}} = \frac{1}{ab}\sin^{-1}(\lambda \sin A)$$

$$\int \frac{dx}{\sqrt{1-a^2 \cos^2 A}} = -\frac{\beta}{b} F\left(\kappa, \frac{\pi}{2} - A\right)$$

$$\int \frac{\sin A\, dx}{\sqrt{1-a^2 \cos^2 A}} = -\frac{1}{ab}\sin^{-1}(a \cos A)$$

$$\int \frac{\cos A\, dx}{\sqrt{1-a^2 \cos^2 A}} = \frac{1}{ab}\ln(\kappa \sin A + \sqrt{1+\kappa^2 \sin^2 A})$$

$$\int \frac{dx}{\sqrt{p^2 \cos^2 A - q^2 \sin^2 A}} = \frac{1}{b\sqrt{p^2+q^2}} F\left(\sqrt{\frac{p^2}{p^2+q^2}}, \cos^{-1}\frac{\sqrt{p^2 \cos^2 A + q^2 \sin^2 A}}{p}\right)$$

(65) Indefinite Integrals Involving

$f(x) = f(x, \tan A)$	$A = bx$

$$\int \tan A \, dx = -\frac{\ln \, (\cos A)}{b}$$

$$\int \tan^2 A \, dx = \frac{\tan A}{b} - x$$

$$\int \tan^3 A \, dx = \frac{\tan^2 A}{2b} + \frac{\ln \, (\cos A)}{b}$$

$$\int \tan^4 A \, dx = \frac{\tan^3 A}{3b} - \frac{\tan A}{b} + x$$

$$\int \tan^n A \, dx = \frac{\tan^{n-1} A}{(n-1)b} - \int \tan^{n-2} A \, dx \qquad n > 1$$

$$\int \frac{dx}{\tan A} = \frac{\ln \, (\sin A)}{b}$$

$$\int \frac{dx}{\cos^2 A \tan A} = \frac{\ln \, (\tan A)}{b}$$

$$\int \frac{dx}{\tan^2 A} = -\frac{\cot A}{b} - x$$

$$\int \frac{dx}{\tan^3 A} = -\frac{\cot^2 A}{2b} - \frac{\ln \, (\sin A)}{b}$$

$$\int \frac{dx}{\tan^n A} = \frac{-1}{(n-1)b \tan^{n-1} A} - \int \frac{dx}{\tan^{n-2} A} \qquad n > 1$$

$$\int x \tan A \, dx = \frac{A}{b^2} \sum_{k=1}^{\infty} a_k \frac{(2A)^{2k}}{2k+1} = \frac{A^3}{3b^2} \bigwedge_{k=1}^{\infty} \left[1 + \frac{(2k+1)a_{k+1}}{(2k+3)a_k} (2A)^2 \right]^*$$

$$\int \frac{\tan A \, dx}{x} = \frac{1}{A} \sum_{k=1}^{\infty} a_k \frac{(2A)^{2k}}{2k-1} = A \bigwedge_{k=1}^{\infty} \left[1 + \frac{(2k-1)a_{k+1}}{(2k+1)a_k} (2A)^2 \right]^*$$

$$\int \frac{dx}{p + q \tan A} = \frac{1}{b(p^2 + q^2)} \left[pA + q \ln \, (q \sin A + p \cos A) \right]$$

$$\int \frac{\tan A \, dx}{p + q \tan A} = \frac{1}{b(p^2 + q^2)} \left[qA - p \ln \, (q \sin A + p \cos A) \right]$$

$$\int \frac{dx}{\sqrt{p + q \tan^2 A}} = \frac{1}{b\sqrt{p-q}} \sin^{-1} \left(\sqrt{\frac{p-q}{p}} \sin A \right) \qquad p > q$$

$$\int \frac{\tan A \, dx}{\sqrt{p + q \tan^2 A}} = \frac{-1}{b\sqrt{p-q}} \ln \, (\sqrt{p-q} \cos A + \sqrt{p \cos^2 A + q \sin^2 A}) \qquad p > q$$

$$\int \frac{\sin A \, dx}{\sqrt{p + q \tan^2 A}} = \frac{1}{b(q-p)} \sqrt{p \cos^2 A + q \sin^2 A}$$

*a_k = numerical factor (Sec. 8.15−2); for nested sum refer to Sec. 8.11.

(66) Indefinite Integrals Involving

| $f(x) = f(x, \cot A)$ | $A = bx$ |

$$\int \cot A \, dx = \frac{\ln (\sin A)}{b}$$

$$\int \cot^2 A \, dx = -\frac{\cot A}{b} - x$$

$$\int \cot^3 A \, dx = -\frac{\cot^2 A}{2b} - \frac{\ln (\sin A)}{b}$$

$$\int \cot^4 A \, dx = -\frac{\cot^3 A}{3b} + \frac{\cot A}{b} + x$$

$$\int \cot^n A \, dx = -\frac{\cot^{n-1} A}{(n-1)b} - \int \cot^{n-2} A \, dx \qquad n > 1$$

$$\int \frac{dx}{\cot A} = -\frac{\ln (\cos A)}{b}$$

$$\int \frac{dx}{\sin^2 A \cot A} = -\frac{\ln (\cot A)}{b}$$

$$\int \frac{dx}{\cot^2 A} = \frac{\tan A}{b} - x$$

$$\int \frac{dx}{\cot^3 A} = \frac{\tan^2 A}{2b} + \frac{\ln (\cos A)}{b}$$

$$\int \frac{dx}{\cot^n A} = \frac{1}{(n-1)b \cot^{n-1} A} - \int \frac{dx}{\cot^{n-2} A} \qquad n > 1$$

$$\int x \cot A \, dx = \frac{A}{b^2}\left[1 - \sum_{k=1}^{\infty} b_k \frac{(2A)^{2k}}{2k+1}\right] = \frac{A}{b^2}\left\{1 - \frac{A^2}{9} \bigwedge_{k=1}^{\infty}\left[1 + \frac{(2k+1)b_{k+1}}{(2k+3)b_k}(2A)^2\right]\right\}^*$$

$$\int \frac{\cot A}{x} \, dx = -\frac{1}{A}\left[1 + \sum_{k=1}^{\infty} b_k \frac{(2A)^{2k}}{2k-1}\right] = -\frac{1}{A}\left\{1 + \frac{A^2}{3} \bigwedge_{k=1}^{\infty}\left[1 + \frac{(2k-1)b_{k+1}}{(2k+1)b_k}(2A)^2\right]\right\}^*$$

$$\int \frac{dx}{p + q \cot A} = \frac{1}{b(p^2 + q^2)}\left[pA - q \ln (p \sin A + q \cos A)\right]$$

$$\int \frac{\cot A \, dx}{p + q \cot A} = \frac{1}{b(p^2 + q^2)}\left[qA + p \ln (p \sin A + q \cos A)\right]$$

$$\int \frac{dx}{\sqrt{p + q \cot^2 A}} = \frac{1}{b\sqrt{p-q}} \cos^{-1}\left(\sqrt{\frac{p-q}{p}} \cos A\right) \qquad p > q$$

$$\int \frac{\cot A \, dx}{\sqrt{p + q \cot^2 A}} = \frac{1}{b\sqrt{p-q}} \ln (\sqrt{p-q} \sin A + \sqrt{p \sin^2 A + q \cos^2 A}) \qquad p > q$$

$$\int \frac{\cos A \, dx}{\sqrt{p + q \cot^2 A}} = \frac{1}{b(p-q)} \sqrt{p \sin^2 A + q \cos^2 A}$$

$^*b_k$ = numerical factor (Sec. 8.15 − 2); for nested sum refer to Sec. 8.11.

(67) Indefinite Integrals Involving

$$\boxed{f(x) = f(x, \sin^{-1} B) \quad \Big| \quad B = \frac{x}{b}}$$

$$\int \sin^{-1} B \, dx = x \sin^{-1} B + \sqrt{b^2 - x^2}$$

$$\int x \sin^{-1} B \, dx = \left(\frac{x^2}{2} - \frac{b^2}{4}\right) \sin^{-1} B + \frac{x\sqrt{b^2 - x^2}}{4}$$

$$\int x^2 \sin^{-1} B \, dx = \frac{x^3}{3} \sin^{-1} B + \frac{(2b^2 + x^2)\sqrt{b^2 - x^2}}{9}$$

$$\int x^m \sin^{-1} B \, dx = \frac{x^{m+1}}{m+1} \sin^{-1} B - \frac{1}{m+1} \int \frac{x^{m+1}}{\sqrt{b^2 - x^2}} \, dx \qquad m \neq -1$$

$$\int \frac{\sin^{-1} B \, dx}{x} = B + \frac{B^3}{(2)(3)(3)} + \frac{(1)(3)B^5}{(2)(4)(5)(5)} + \frac{(1)(3)(5)B^7}{(2)(4)(6)(7)(7)} + \cdots$$

$$\int \frac{\sin^{-1} B \, dx}{x^2} = -\frac{\sin^{-1} B}{x} - \frac{1}{b} \ln \frac{b + \sqrt{b^2 - x^2}}{x}$$

$$\int \frac{\sin^{-1} B \, dx}{x^m} = -\frac{\sin^{-1} B}{(m-1)x^{m-1}} + \frac{1}{m-1} \int \frac{dx}{x^{m-1}\sqrt{b^2 - x^2}} \qquad m \neq 1$$

$$\int (\sin^{-1} B)^2 \, dx = x(\sin^{-1} B)^2 - 2x + 2\sqrt{b^2 - x^2} \sin^{-1} B$$

(68) Indefinite Integrals Involving

$$\boxed{f(x) = f(x, \tan^{-1} B) \quad \Big| \quad B = \frac{x}{b}}$$

$$\int \tan^{-1} B \, dx = x \tan^{-1} B - b \ln \sqrt{b^2 + x^2}$$

$$\int x \tan^{-1} B \, dx = \frac{b^2 + x^2}{2} \tan^{-1} B - \frac{bx}{2}$$

$$\int x^m \tan^{-1} B \, dx = \frac{x^{m+1}}{m+1} \tan^{-1} B - \frac{b}{m+1} \int \frac{x^{m+1} \, dx}{b^2 + x^2} \qquad m \neq -1$$

$$\int \frac{\tan^{-1} B \, dx}{x} = B - \frac{B^3}{3^2} + \frac{B^5}{5^2} - \frac{B^7}{7^2} + \cdots$$

$$\int \frac{\tan^{-1} B \, dx}{x^2} = -\frac{1}{b}\left(\frac{\tan^{-1} B}{B} + \ln \frac{\sqrt{1 + B^2}}{B}\right)$$

$$\int \frac{\tan^{-1} B \, dx}{x^m} = -\frac{\tan^{-1} B}{(m-1)x^{m-1}} + \frac{b}{m-1} \int \frac{dx}{(b^2 + x^2)x^{m-1}} \qquad m \neq 1$$

(69) Indefinite Integrals Involving

$$f(x) = f(x, \cos^{-1} B) \qquad B = \frac{x}{b}$$

$$\int \cos^{-1} B \, dx = x \cos^{-1} B - \sqrt{b^2 - x^2}$$

$$\int x \cos^{-1} B \, dx = \left(\frac{x^2}{2} - \frac{b^2}{4} \right) \cos^{-1} B - \frac{x\sqrt{b^2 - x^2}}{4}$$

$$\int x^2 \cos^{-1} B \, dx = \frac{x^3}{3} \cos^{-1} B - \frac{(2b^2 + x^2)\sqrt{b^2 - x^2}}{9}$$

$$\int x^m \cos^{-1} B \, dx = \frac{x^{m+1}}{m+1} \cos^{-1} B + \frac{1}{m+1} \int \frac{x^{m+1}}{\sqrt{b^2 - x^2}} \, dx \qquad m \neq -1$$

$$\int \frac{\cos^{-1} B \, dx}{x} = \frac{\pi}{2} \ln x - B - \frac{B^3}{(2)(3)(3)} - \frac{(1)(3)B^5}{(2)(4)(5)(5)} - \frac{(1)(3)(5)B^7}{(2)(4)(6)(7)(7)} - \cdots$$

$$\int \frac{\cos^{-1} B \, dx}{x^2} = -\frac{\cos^{-1} B}{x} + \frac{1}{b} \ln \frac{b + \sqrt{b^2 - x^2}}{x}$$

$$\int \frac{\cos^{-1} B \, dx}{x^m} = -\frac{\cos^{-1} B}{(m-1)x^{m-1}} - \frac{1}{m-1} \int \frac{dx}{x^{m-1}\sqrt{b^2 - x^2}} \qquad m \neq 1$$

$$\int (\cos^{-1} B)^2 \, dx = x(\cos^{-1} B)^2 - 2x - 2\sqrt{b^2 - x^2} \cos^{-1} B$$

(70) Indefinite Integrals Involving

$$f(x) = f(x, \cot^{-1} B) \qquad B = \frac{x}{b}$$

$$\int \cot^{-1} B \, dx = x \cot^{-1} B + b \ln \sqrt{b^2 + x^2}$$

$$\int x \cot^{-1} B \, dx = \frac{b^2 + x^2}{2} \cot^{-1} B + \frac{bx}{2}$$

$$\int x^m \cot^{-1} B \, dx = \frac{x^{m+1}}{m+1} \cot^{-1} B + \frac{b}{m+1} \int \frac{x^{m+1}}{b^2 + x^2} \, dx \qquad m \neq -1$$

$$\int \frac{\cot^{-1} B \, dx}{x} = \frac{\pi}{2} \ln x - B + \frac{B^3}{3^2} - \frac{B^5}{5^2} + \frac{B^7}{7^2} - \cdots$$

$$\int \frac{\cot^{-1} B \, dx}{x^2} = -\frac{1}{b} \left(\frac{\cot^{-1} B}{B} - \ln \frac{\sqrt{1 + B^2}}{B} \right)$$

$$\int \frac{\cot^{-1} B \, dx}{x^m} = -\frac{\cot^{-1} B}{(m-1)x^{m-1}} - \frac{b}{m-1} \int \frac{dx}{(b^2 + x^2)x^{m-1}} \qquad m \neq 1$$

(71) Indefinite Integrals Involving

$f(x) = f(x^m, e^{\pm A})$	$A = bx$

$$\int e^A \, dx = \frac{e^A}{b} \qquad\qquad\qquad \int x e^A \, dx = \frac{(A-1)e^A}{b^2}$$

$$\int x^2 e^A \, dx = \frac{x^3 e^A}{A}\left(1 - \frac{2}{A}\left(1 - \frac{1}{A}\right)\right)$$

$$\int x^3 e^A \, dx = \frac{x^4 e^A}{A}\left(1 - \frac{3}{A}\left(1 - \frac{2}{A}\left(-\frac{1}{A}\right)\right)\right)$$

$$\int x^m e^A \, dx = \frac{x^{m+1} e^A}{A}\left(1 - \frac{m}{A}\left(1 - \frac{m-1}{A}\left(1 - \frac{m-2}{A}\left(1 - \cdots - \frac{1}{A}\right)\right)\right)\right)^*$$

$$\int \frac{e^A}{x} \, dx = \ln x + \frac{A}{1\cdot 1}\left(1 + \frac{A}{2\cdot 2}\left(1 + \frac{2A}{3\cdot 3}\left(1 + \frac{3A}{4\cdot 4}\left(1 + \cdots\right)\right)\right)\right)^*$$

$$\int \frac{e^A}{x^2} \, dx = -\frac{e^A}{x} + b\int \frac{e^A}{x} \, dx$$

$$\int \frac{e^A}{x^3} \, dx = -\frac{e^A}{2x^2}(1 + A) + \frac{b^2}{2}\int \frac{e^A}{x} \, dx$$

$$\int \frac{e^A}{x^m} \, dx = -\frac{e^A}{(m-1)x^{m-1}}\left(1 + \frac{A}{m-2}\left(1 + \frac{A}{m-3}\left(1 + \frac{A}{m-4}\left(1 + \cdots + A\right)\right)\right)\right)^* + \frac{b^{m-1}}{(m-1)!}\int \frac{e^A}{x} \, dx$$

$$\int \frac{dx}{p + q e^A} = \frac{1}{bp}\left[A - \ln(p + q e^A)\right] \qquad\qquad \int \frac{e^A \, dx}{p + q e^A} = \frac{1}{bq}\ln(p + q e^A)$$

$$\int \frac{dx}{p e^A + q e^{-A}} = \begin{cases} -\dfrac{1}{b\sqrt{pq}}\tan^{-1} e^A \sqrt{\dfrac{p}{q}} & pq > 0 \\[3mm] \dfrac{1}{2b\sqrt{-pq}}\ln \dfrac{q + e^A\sqrt{-pq}}{q - e^A\sqrt{-pq}} & pq < 0 \end{cases}$$

$$\int \frac{e^A \, dx}{\sqrt{x}} = 2\sqrt{x}\left(1 + \frac{A}{1\cdot 3}\left(1 + \frac{3A}{2\cdot 5}\left(1 + \frac{5A}{3\cdot 7}\left(1 + \cdots\right)\right)\right)\right)^*$$

$$\int \frac{dx}{\sqrt{p + q e^A}} = \begin{cases} \dfrac{1}{b\sqrt{p}}\ln \dfrac{\sqrt{p + q e^A} - \sqrt{p}}{\sqrt{p + q e^A} + \sqrt{p}} & p > 0 \\[3mm] \dfrac{1}{b\sqrt{-p}}\tan^{-1}\dfrac{\sqrt{p + q e^A}}{\sqrt{-p}} & p < 0 \end{cases}$$

$$\boxed{\int f(x) e^A \, dx = \frac{e^A}{b}\left(f(x) - \frac{1}{b}\left(f'(x) - \frac{1}{b}\left(f''(x) - \frac{1}{b}(f'''(x) - \cdots)\right)\right)\right)^*}$$

*For nested sum refer to Sec. 8.11.

(72) Indefinite Integrals Involving | $f(x) = f(x^m, a^A)$ | $A = bx$ | $B = bx \ln a$

$$\int a^A \, dx = \frac{a^A}{b \ln a}$$

$$\int x a^A \, dx = \frac{(B-1) a^A}{(b \ln a)^2}$$

$$\int x^2 a^A \, dx = \frac{x^3 a^A}{B}\left(1 - \frac{2}{B}\left(1 - \frac{1}{B}\right)\right)$$

$$\int x^3 a^A \, dx = \frac{x^4 a^A}{B}\left(1 - \frac{3}{B}\left(1 - \frac{2}{B}\left(1 - \frac{1}{B}\right)\right)\right)$$

$$\int x^m a^A \, dx = \frac{x^{m+1} a^A}{B}\left(1 - \frac{m}{B}\left(1 - \frac{m-1}{B}\left(1 - \frac{m-2}{B}\left(1 - \cdots - \frac{1}{B}\right)\right)\right)\right)^*$$

$$\int \frac{a^A}{x} \, dx = \ln x + \frac{B}{1 \cdot 1}\left(1 + \frac{B}{2 \cdot 2}\left(1 + \frac{2B}{3 \cdot 3}\left(1 + \frac{3B}{4 \cdot 4}\left(1 + \cdots\right)\right)\right)\right)^*$$

$$\int \frac{a^A}{x^2} \, dx = -\frac{a^A}{x} + (b \ln a)\int \frac{a^A}{x} \, dx$$

$$\int \frac{a^A}{x^3} \, dx = -\frac{a^A}{2x^2}(1 + B) + \frac{(b \ln a)^2}{2}\int \frac{a^A}{x} \, dx$$

$$\int \frac{a^A}{x^m} \, dx = -\frac{a^A}{(m-1)x^{m-1}}\left(1 + \frac{B}{m-2}\left(1 + \frac{B}{m-3}\left(1 + \cdots + B\right)\right)\right)^* + \frac{(b \ln a)^{m-1}}{(m-1)!}\int \frac{a^A}{x} \, dx$$

$$\int \frac{dx}{p + qa^A} = \frac{x}{p} - \frac{x}{pB} \ln (p + qa^A)$$

$$\int \frac{a^A \, dx}{p + qa^A} = \frac{x}{qB} \ln (p + qa^A)$$

$$\int \frac{dx}{pa^A + qa^{-A}} = \begin{cases} \dfrac{x}{B\sqrt{pq}} \tan^{-1}\left(a^A \sqrt{\dfrac{p}{q}}\right) & pq > 0 \\[2ex] \dfrac{x}{2B\sqrt{-pq}} \ln \dfrac{q + a^A\sqrt{-pq}}{q - a^A\sqrt{-pq}} & pq < 0 \end{cases}$$

$$\int \frac{a^A \, dx}{\sqrt{x}} = 2\sqrt{x}\left(1 + \frac{B}{1 \cdot 3}\left(1 + \frac{3B}{2 \cdot 5}\left(1 + \frac{5B}{3 \cdot 7}\left(1 + \cdots\right)\right)\right)\right)^*$$

$$\int \frac{dx}{\sqrt{p + qa^A}} = \begin{cases} \dfrac{x}{B\sqrt{p}} \ln \dfrac{\sqrt{p + qa^A} - \sqrt{p}}{\sqrt{p + qa^A} + \sqrt{p}} & p > 0 \\[2ex] \dfrac{x}{B\sqrt{-p}} \tan^{-1}\dfrac{\sqrt{p + qa^A}}{\sqrt{-p}} & p < 0 \end{cases}$$

$$\boxed{\int f(x) a^A \, dx = \frac{a^A}{b \ln a}\left(f(x) - \frac{1}{b \ln a}\left(f'(x) - \frac{1}{b \ln a}\left(f''(x) - \frac{1}{b \ln a}\left(f'''(x) - \cdots\right)\right)\right)\right)^*}$$

*For nested sum refer to Sec. 8.11.

(73) Indefinite Integrals Involving $\boxed{f(x) = f(x^m, e^A \sin B, \cos b) \quad \Big| \quad A = \alpha x + a, \ B = \beta x + b}$

$|a| \geq 0 \qquad |b| \geq 0 \qquad |\alpha| > 0 \qquad |\beta| > 0 \qquad k, n = 0, 1, 2, \ldots \qquad R_k = \sqrt{[\alpha^2 + (k\beta)^2]^k}$

$$\phi_k = \begin{cases} \tan^{-1}\dfrac{k\beta}{\alpha} & \text{if } \alpha > 0 \\[2mm] \pi + \tan^{-1}\dfrac{k\beta}{\alpha} & \text{if } \alpha < 0 \end{cases} \qquad \begin{array}{l} S_k = \sin(B - k\phi_1) \\[2mm] C_k = \cos(B - k\phi_1) \end{array} \qquad \begin{array}{l} F_k = \dfrac{\sin(kB - \phi_k)}{R_k} \\[2mm] G_k = \dfrac{\cos(kB - \phi_k)}{R_k} \end{array}$$

$$\int e^A \sin B \, dx = e^A F_1 \qquad\qquad \int e^A \cos B \, dx = e^A G_1$$

$$\int e^A \sin^2 B \, dx = \frac{e^A}{2}(\alpha^{-1} - G_2) \qquad\qquad \int e^A \cos^2 B \, dx = \frac{e^A}{2}(\alpha^{-1} + G_2)$$

$$\int e^A \sin^3 B \, dx = \frac{e^A}{4}(3F_1 - F_3) \qquad\qquad \int e^A \cos^3 B \, dx = \frac{e^A}{4}(3G_1 + G_3)$$

$$\int e^A \sin^4 B \, dx = \frac{e^A}{8}(3\alpha^{-1} - 4G_2 + G_4) \qquad\qquad \int e^A \cos^4 B \, dx = \frac{e^A}{8}(3\alpha^{-1} + 4G_2 + G_4)$$

$$\int e^A \sin^{2n} B \, dx = \frac{e^A}{4^n \alpha}\left[\binom{2n}{n} + 2\alpha \sum_{k=1}^{n} (-1)^k \binom{2n}{n-k} G_{2k}\right] \quad \int e^A \cos^{2n} B \, dx = \frac{e^A}{4^n \alpha}\left[\binom{2n}{n} + 2\alpha \sum_{k=1}^{n} \binom{2n}{k-n} G_{2k}\right]$$

$$\int e^A \sin^{2n+1} B \, dx = \frac{e^A}{4^n}\left[\sum_{k=0}^{n}(-1)^k\binom{2n+1}{n-k}F_{2k+1}\right] \qquad \int e^A \cos^{2n+1} B \, dx = \frac{e^A}{4^n}\left[\sum_{k=0}^{n}\binom{2n+1}{n-k}G_{2k+1}\right]$$

$$\int xe^A \sin B \, dx = \frac{xe^A S_1}{R_1}\left(1 - \frac{S_2}{xR_1 S_1}\right) \qquad\qquad \int xe^A \cos B \, dx = \frac{xe^A C_1}{R_1}\left(1 - \frac{C_2}{xR_1 C_1}\right)$$

$$\int x^2 e^A \sin B \, dx = \frac{x^2 e^A S_1}{R_1}\left(1 - \frac{2S_2}{xR_1 S_1}\left(1 - \frac{S_3}{xR_1 S_2}\right)\right) \qquad \int x^2 e^A \cos B \, dx = \frac{x^2 e^A C_1}{R_1}\left(1 - \frac{2C_2}{xR_1 C_1}\left(1 - \frac{C_3}{xR_1 C_2}\right)\right)$$

$$\int x^m e^A \sin B \, dx = \frac{x^m e^A S_1}{R_1}\bigwedge_{k=1}^{m}\left[1 - \frac{(m+1-k)S_{k+1}}{xR_1 S_k}\right]^* \qquad \int x^m e^A \cos B \, dx = \frac{x^m e^A C_1}{R_1}\bigwedge_{k=1}^{m}\left[1 - \frac{(m+1-k)C_{k+1}}{xR_1 C_k}\right]^*$$

$$\int \frac{e^A}{\sin^m B} \, dx = -\frac{\alpha \sin B + (m-2)\beta \cos B}{(m-1)(m-2)\sin^{m-1} B}e^A + \frac{\alpha^2 + (m-2)^2\beta^2}{(m-1)(m-2)\beta^2}\int \frac{e^A}{\sin^{m-2} B}\, dx \qquad m > 2$$

$$\int \frac{e^A}{\cos^m B} \, dx = -\frac{\alpha \cos B - (m-2)\beta \cos B}{(m-1)(m-2)\cos^{m-1} B}e^A + \frac{\alpha^2 + (m-2)^2\beta^2}{(m-1)(m-2)\beta^2}\int \frac{e^A}{\cos^{m-2} B}\, dx \qquad m > 2$$

$$\boxed{\begin{array}{ll} \displaystyle\int \frac{e^A}{\sin B}\, dx = \int \frac{e^a}{\sin B}\left[\sum_{k=0}^{\infty}\frac{(\alpha x)^k}{k!}\right]dx\dagger & \displaystyle\int \frac{e^A}{\cos B}\, dx = \int \frac{e^a}{\cos B}\left[\sum_{k=0}^{\infty}\frac{(\alpha x)^k}{k!}\right]dx\dagger \\[5mm] \displaystyle\int \frac{e^A}{\sin^2 B}\, dx = \int \frac{e^a}{\sin^2 B}\left[\sum_{k=0}^{\infty}\frac{(\alpha x)^k}{k!}\right]dx\dagger & \displaystyle\int \frac{e^A}{\cos^2 B}\, dx = \int \frac{e^a}{\cos^2 B}\left[\sum_{k=0}^{\infty}\frac{(\alpha x)^k}{k!}\right]dx\dagger \end{array}}$$

*For nested sum refer to Sec. 8.11.
†No closed form available; both functions must be expressed in terms of their respective series and the product (Sec. 8.14‒1) integrated term by term.

(74) Indefinite Integrals Involving | $f(x) = f(e^A, \sin B, \cos B, \sin C, \cos C)$ | $A = \alpha x, B = \beta x, C = \gamma x$

$$\omega_1 = \sqrt{\alpha^2 + (\beta + \gamma)^2} \qquad \omega_2 = \sqrt{\alpha^2 + (\beta - \gamma)^2} \qquad \phi_1 = \tan^{-1}\frac{\alpha}{\beta + \gamma} \qquad \phi_2 = \tan^{-1}\frac{\alpha}{\beta - \gamma}$$

$$\int e^A \sin B \sin C \, dx = -\frac{e^A}{2}\left[\frac{\sin (B + C + \phi_1)}{\omega_1} - \frac{\sin (B - C + \phi_2)}{\omega_2}\right]$$

$$\int e^A \sin B \cos C \, dx = -\frac{e^A}{2}\left[\frac{\cos (B + C + \phi_1)}{\omega_1} + \frac{\cos (B - C + \phi_2)}{\omega_2}\right]$$

$$\int e^A \cos B \sin C \, dx = -\frac{e^A}{2}\left[\frac{\cos (B + C + \phi_1)}{\omega_1} - \frac{\cos (B - C + \phi_2)}{\omega_2}\right]$$

$$\int e^A \cos B \cos C \, dx = +\frac{e^A}{2}\left[\frac{\sin (B + C + \phi_1)}{\omega_1} + \frac{\sin (B - C + \phi_2)}{\omega_2}\right]$$

$$\phi_1 = \tan^{-1}\frac{\alpha}{\gamma} \qquad \phi_{2,3} = \tan^{-1}\frac{\alpha}{2\beta \pm \gamma} \qquad \phi_4 = \tan^{-1}\frac{\alpha}{\beta} \qquad \phi_{5,6} = \tan^{-1}\frac{\alpha}{\beta \pm 2\gamma}$$

$$\int e^A \sin^2 B \cos C \, dx = +\frac{e^A}{4}\left[\frac{2 \sin (C + \phi_1)}{\sqrt{\alpha^2 + \gamma^2}} - \frac{\sin (2B + C + \phi_2)}{\sqrt{\alpha^2 + (2\beta + \gamma)^2}} - \frac{\sin (2B - C + \phi_3)}{\sqrt{\alpha^2 + (2\beta - \gamma)^2}}\right]$$

$$\int e^A \sin B \cos^2 C \, dx = -\frac{e^A}{4}\left[\frac{2 \cos (B + \phi_4)}{\sqrt{\alpha^2 + \beta^2}} + \frac{\cos (B + 2C + \phi_5)}{\sqrt{\alpha^2 + (\beta + 2\gamma)^2}} + \frac{\cos (B - 2C + \phi_6)}{\sqrt{\alpha^2 + (\beta - 2\gamma)^2}}\right]$$

$$B + b = \beta x + b \qquad \omega = \sqrt{\alpha^2 + 4\beta^2} \qquad \phi = \tan^{-1}\frac{\alpha}{2\beta}$$

$$\int e^A \sin B \sin (B + b) = +\frac{e^A}{2}\left[\frac{\cos b}{\alpha} - \frac{\sin (2B + b + \phi)}{\omega}\right]$$

$$\int e^A \sin B \cos (B + b) = -\frac{e^A}{2}\left[\frac{\sin b}{\alpha} + \frac{\cos (2B + b + \phi)}{\omega}\right]$$

$$\int e^A \cos B \sin (B + b) = +\frac{e^A}{2}\left[\frac{\sin b}{\alpha} - \frac{\cos (2B + b + \phi)}{\omega}\right]$$

$$\int e^A \cos B \cos (B + b) = +\frac{e^A}{2}\left[\frac{\cos b}{\alpha} + \frac{\sin (2B + b + \phi)}{\omega}\right]$$

If $\alpha > 0$ use ϕ as given by $\tan^{-1}(\ )$; if $\alpha < 0$ replace ϕ by $\pi + \tan^{-1}(\ )$.

(75) Indefinite Integrals Involving | $f(x) = f(e^A, \tan x, \cot x)$ | $A = \alpha x$

$$\int e^A \tan^n x \, dx = +\frac{e^A}{n - 1}\tan^{n-1} x - \frac{\alpha}{n - 1}\int e^A \tan^{n-1} x \, dx - \int e^A \tan^{n-2} x \, dx$$

$$\int e^A \cot^n x \, dx = -\frac{e^A}{n - 1}\cot^{n-1} x + \frac{\alpha}{n - 1}\int e^A \cot^{n-1} x \, dx - \int e^A \cot^{n-2} x \, dx$$

(76) Indefinite Integrals Involving

$f(x) = f(x^m, \sinh A)$	$A = bx$

$$D_1 = \int \sinh^n A \, dx \qquad D_2 = \int D_1 \, dx \qquad D_k = \int D_{k-1} \, dx$$

$$\int \sinh A \, dx = \frac{\cosh A}{b}$$

$$\int \sinh^2 A \, dx = \frac{\sinh A \cosh A}{2b} - \frac{x}{2}$$

$$\int \sinh^n A \, dx = \frac{\sinh^{n-1} A \cosh A}{bn} - \frac{n-1}{n} \int \sinh^{n-2} A \, dx$$

$$\int x \sinh A \, dx = \frac{x \cosh A}{b} - \frac{\sinh A}{b^2}$$

$$\int x^2 \sinh A \, dx = \left(\frac{x^2}{b} + \frac{2}{b^3}\right) \cosh A - \frac{2x}{b^2} \sinh A$$

$$\int x^m \sinh A \, dx = \frac{x^m \cosh A}{b} - \frac{m}{b} \int x^{m-1} \cosh A \, dx$$

$$\int x^m \sinh^n A \, dx = x^m D_1 - m x^{m-1} D_2 + m(m-1) x^{m-2} D_3 - \cdots (-1)^m m! D_{m+1}$$

$$\int \frac{\sinh A \, dx}{x} = A + \frac{A^3}{(3)(3!)} + \frac{A^5}{(5)(5!)} + \cdots$$

$$\int \frac{\sinh A \, dx}{x^2} = -\frac{\sinh A}{x} + b\left(\ln x + \frac{A^2}{(2)(2!)} + \frac{A^4}{(4)(4!)} + \cdots\right)$$

$$\int \frac{\sinh A \, dx}{x^m} = \frac{-\sinh A}{(m-1)x^{m-1}} + \frac{b}{m-1} \int \frac{\cosh A}{x^{m-1}} \, dx \qquad m \neq 1$$

$$\int \frac{dx}{\sinh A} = \frac{\ln\left[\tanh\left(A/2\right)\right]}{b}$$

$$\int \frac{dx}{\sinh^2 A} = -\frac{\coth A}{b}$$

$$\int \frac{dx}{\sinh^n A} = \frac{-\cosh A}{b(n-1)\sinh^{n-2} A} - \frac{n-2}{n-1} \int \frac{dx}{\sinh^{n-2} A} \qquad n \neq 1$$

$$\int \frac{x \, dx}{\sinh^n A} = \frac{-x \cosh A}{b(n-1)\sinh^{n-1} A} - \frac{1}{b^2(n-1)(n-2)\sinh^{n-2} A} - \frac{n-2}{n-1} \int \frac{x \, dx}{\sinh^{n-2} A} \qquad n \neq 1, 2$$

(77) Indefinite Integrals Involving | $f(x) = f(x^m, \cosh A)$ | $A = bx$

$$D_1 = \int \cosh^n A \, dx \qquad D_2 = \int D_1 \, dx \qquad D_k = \int D_{k-1} \, dx$$

$$\int \cosh A \, dx = \frac{\sinh A}{b}$$

$$\int \cosh^2 A \, dx = \frac{\sinh A \cosh A}{2b} + \frac{x}{2}$$

$$\int \cosh^n A \, dx = \frac{\cosh^{n-1} A \sinh A}{bn} + \frac{n-1}{n} \int \cosh^{n-2} A \, dx$$

$$\int x \cosh A \, dx = \frac{x \sinh A}{b} - \frac{\cosh A}{b^2}$$

$$\int x^2 \cosh A \, dx = \left(\frac{x^2}{b} + \frac{2}{b^3}\right) \sinh A - \frac{2x}{b^2} \cosh A$$

$$\int x^m \cosh A \, dx = \frac{x^m \sinh A}{b} - \frac{m}{b} \int x^{m-1} \sinh A \, dx$$

$$\int x^m \cosh^n A \, dx = x^m D_1 - m x^{m-1} D_2 + m(m-1) x^{m-1} D_3 - \cdots (-1)^m m! D_{m+1}$$

$$\int \frac{\cosh A \, dx}{x} = \ln x + \frac{A^2}{(2)(2!)} + \frac{A^4}{(4)(4!)} + \cdots$$

$$\int \frac{\cosh A \, dx}{x^2} = -\frac{\cosh A}{x} + b\left[A + \frac{A^3}{(3)(3!)} + \frac{A^5}{(5)(5!)} + \cdots\right]$$

$$\int \frac{\cosh A \, dx}{x^m} = \frac{-\cosh A}{(m-1)x^{m-1}} + \frac{b}{m-1} \int \frac{\sinh A}{x^{m-1}} \, dx \qquad m \neq 1$$

$$\int \frac{dx}{\cosh A} = \frac{\tan^{-1}(\sinh A)}{b}$$

$$\int \frac{dx}{\cosh^2 A} = \frac{\tanh A}{b}$$

$$\int \frac{dx}{\cosh^n A} = \frac{\sinh A}{b(n-1)\cosh^{n-1} A} + \frac{n-2}{n-1} \int \frac{dx}{\cosh^{n-2} A} \qquad n \neq 1$$

$$\int \frac{x \, dx}{\cosh^n A} = \frac{x \sinh A}{b(n-1)\cosh^{n-1} A} + \frac{1}{b^2(n-1)(n-2)\cosh^{n-2} A} + \frac{n-2}{n-1} \int \frac{x \, dx}{\cosh^{n-2} A} \qquad n \neq 1, 2$$

(78) Indefinite Integrals Involving

$f(x) = f(\sinh A, \cosh A)$	$A = bx$

$$\int \sinh A \cosh A \, dx = \frac{\sinh^2 A}{2b}$$

$$\int \sinh^2 A \cosh^2 A \, dx = \frac{\sinh 4A}{32b} - \frac{x}{8}$$

$$\int \sinh^n A \cosh A \, dx = \frac{\sinh^{n+1} A}{(n+1)b} \qquad n \neq -1$$

$$\int \sinh A \cosh^n A \, dx = \frac{\cosh^{n+1} A}{(n+1)b} \qquad n \neq -1$$

$$\int \frac{dx}{\sinh A \cosh A} = \frac{\ln(\tanh A)}{b}$$

$$\int \frac{dx}{\sinh^2 A \cosh A} = -\frac{\tan^{-1}(\sinh A) + \operatorname{csch} A}{b}$$

$$\int \frac{dx}{\sinh A \cosh^2 A} = \frac{\ln[\tanh(A/2)] + \operatorname{sech} A}{b}$$

$$\int \frac{dx}{\sinh^2 A \cosh^2 A} = -\frac{2 \coth 2A}{b}$$

$$\int \frac{\sinh^2 A}{\cosh A} \, dx = \frac{\sinh A - \tan^{-1}(\sinh A)}{b}$$

$$\int \frac{\cosh^2 A}{\sinh A} \, dx = \frac{\cosh A + \ln[\tanh(A/2)]}{b}$$

$$\int \frac{\sinh A}{\cosh^n A} \, dx = -\frac{1}{(n-1)b \cosh^{n-1} A} \qquad n \neq 1$$

$$\int \frac{\cosh A}{\sinh^n A} \, dx = -\frac{1}{(n-1)b \sinh^{n-1} A} \qquad n \neq 1$$

$$\int \frac{\sinh A}{\cosh A \pm 1} \, dx = \frac{1}{b} \ln(\cosh A \pm 1) \qquad \int \frac{\cosh A}{1 \pm \sinh A} \, dx = \pm \frac{1}{b} \ln(1 \pm \sinh A)$$

$$\int e^{ax} \sinh A \, dx = \frac{(a \sinh A - b \cosh A)e^{ax}}{a^2 - b^2}$$

$$a^2 \neq b^2$$

$$\int e^{ax} \cosh A \, dx = \frac{(a \cosh A - b \sinh A)e^{ax}}{a^2 - b^2}$$

(79) Indefinite Integrals Involving $\boxed{f(x) = f(\sinh \alpha x, \cosh \alpha x, \sinh \beta x, \cosh \beta x)}$

$$\left. \begin{aligned} \int \sinh \alpha x \sinh \beta x &= \frac{\sinh (\alpha + \beta)x}{2(\alpha + \beta)} - \frac{\sinh (\alpha - \beta)x}{2(\alpha - \beta)} \\[2mm] \int \sinh \alpha x \cosh \beta x &= \frac{\cosh (\alpha + \beta)x}{2(\alpha + \beta)} + \frac{\cosh (\alpha - \beta)x}{2(\alpha - \beta)} \\[2mm] \int \cosh \alpha x \cosh \beta x &= \frac{\sinh (\alpha + \beta)x}{2(\alpha + \beta)} + \frac{\sinh (\alpha - \beta)x}{2(\alpha - \beta)} \end{aligned} \right\} \alpha^2 \neq \beta^2$$

(80) Indefinite Integrals Involving $\boxed{f(x) = f(\sinh \alpha x, \cosh \alpha x, \sin \beta x, \cos \beta x)}$

$$\int \sinh \alpha x \sin \beta x = \frac{\alpha \cosh \alpha x \sin \beta x - \beta \sinh \alpha x \cos \beta x}{\alpha^2 + \beta^2}$$

$$\int \sinh \alpha x \cos \beta x = \frac{\alpha \cosh \alpha x \cos \beta x + \beta \sinh \alpha x \sin \beta x}{\alpha^2 + \beta^2}$$

$$\int \cosh \alpha x \sin \beta x = \frac{\alpha \sinh \alpha x \sin \beta x - \beta \cosh \alpha x \cos \beta x}{\alpha^2 + \beta^2}$$

$$\int \cosh \alpha x \cos \beta x = \frac{\alpha \sinh \alpha x \cos \beta x + \beta \cosh \alpha x \sin \beta x}{\alpha^2 + \beta^2}$$

(81) Indefinite Integrals Involving $\boxed{f(x) = f(\tanh A, \coth A)}$ $\boxed{A = bx}$

$$\int \tanh A \, dx = \frac{\ln (\cosh A)}{b} \qquad\qquad \int \tanh^2 A \, dx = x - \frac{\tanh A}{b}$$

$$\int \tanh^n A \, dx = -\frac{\tanh^{n-1} A}{b(n-1)} + \int \tanh^{n-2} A \, dx \qquad n \neq 1$$

$$\int \coth A \, dx = \frac{\ln (\sinh A)}{b} \qquad\qquad \int \coth^2 A \, dx = x - \frac{\coth A}{b}$$

$$\int \coth^n A \, dx = -\frac{\coth^{n-1} A}{b(n-1)} + \int \coth^{n-2} A \, dx \qquad n \neq 1$$

$$\int x^m \tanh x \, dx = -\sum_{k=1}^{\infty} \frac{(-4)^k a_k x^{m+2k}}{m + 2k} \qquad\qquad \int x^m \coth x \, dx = \frac{x^m}{m} - \sum_{k=1}^{\infty} \frac{(-4)^k b_k x^{m+2k}}{m + 2k}$$

a_k, b_k = numerical factors (Sec. 8.15-2)

(82) Indefinite Integrals Involving

$$f(x) = f(x, \sinh^{-1} B) \qquad B = \frac{x}{b}$$

$$\int \sinh^{-1} B \, dx = x \sinh^{-1} B - \sqrt{x^2 + b^2}$$

$$\int x \sinh^{-1} B \, dx = \left(\frac{x^2}{2} + \frac{b^2}{4}\right) \sinh^{-1} B - \frac{x\sqrt{x^2 + b^2}}{4}$$

$$\int x^2 \sinh^{-1} B \, dx = \frac{x^3}{3} \sinh^{-1} B + \frac{(2b^2 - x^2)\sqrt{x^2 + b^2}}{9}$$

$$\int x^m \sinh^{-1} B \, dx = \frac{x^{m+1}}{m+1} \sinh^{-1} B - \frac{1}{m+1} \int \frac{x^{m+1}}{\sqrt{x^2 + b^2}} \, dx \qquad m \neq -1$$

$$\int \frac{\sinh^{-1} B \, dx}{x} = B - \frac{B^3}{(2)(3)(3)} + \frac{(1)(3)B^5}{(2)(4)(5)(5)} - \frac{(1)(3)(5)B^7}{(2)(4)(6)(7)(7)} + \cdots \qquad x^2 < b^2$$

$$\int \frac{\sinh^{-1} B \, dx}{x^2} = -\frac{\sinh^{-1} B}{x} - \frac{1}{b} \ln \frac{b + \sqrt{x^2 + b^2}}{x}$$

(83) Indefinite Integrals Involving

$$f(x) = f(x, \tanh^{-1} B) \qquad B = \frac{x}{b}$$

$$\int \tanh^{-1} B \, dx = x \tanh^{-1} B + b \ln \sqrt{b^2 - x^2}$$

$$\int x \tanh^{-1} B \, dx = \frac{x^2 - b^2}{2} \tanh^{-1} B + \frac{bx}{2}$$

$$\int x^m \tanh^{-1} B \, dx = \frac{x^{m+1}}{m+1} \tanh^{-1} B - \frac{b}{m+1} \int \frac{x^{m+1}}{b^2 - x^2} \, dx \qquad m \neq 1$$

$$\int \frac{\tanh^{-1} B \, dx}{x} = B + \frac{B^3}{3^2} + \frac{B^5}{5^2} + \frac{B^7}{7^2} + \cdots$$

$$\int \frac{\tanh^{-1} B \, dx}{x^2} = -\frac{1}{b}\left(\frac{\tanh^{-1} B}{B} + \ln \frac{\sqrt{1 - B^2}}{B}\right)$$

$$\int \frac{\tanh^{-1} B \, dx}{x^3} = -\frac{1}{2x^2}\left[B - (B^2 - 1) \tanh^{-1} B\right]$$

$$\int \frac{\tanh^{-1} B \, dx}{x^m} = -\frac{\tanh^{-1} B}{(m-1)x^{m-1}} + \frac{b}{m-1} \int \frac{dx}{(b^2 - x^2)x^{m-1}}$$

(84) Indefinite Integrals Involving $\qquad$ $f(x) = f(x, \cosh^{-1} B)$ $\quad$ $B = \dfrac{x}{b}$

$$\int \cosh^{-1} B \, dx = x \cosh^{-1} B \mp \sqrt{x^2 - b^2}$$

$$\int x \cosh^{-1} B \, dx = \left(\frac{x^2}{2} - \frac{b^2}{4}\right) \cosh^{-1} B \mp \frac{x\sqrt{x^2 - b^2}}{4}$$

$$\int x^2 \cosh^{-1} B \, dx = \frac{x^3}{3} \cosh^{-1} B \mp \frac{(2b^2 + x^2)\sqrt{x^2 - b^2}}{9}$$

$$\int x^m \cosh^{-1} B \, dx = \frac{x^{m+1}}{m+1} \cosh^{-1} B \mp \frac{1}{m+1} \int \frac{x^{m+1}\, dx}{\sqrt{x^2 - b^2}} \qquad m \neq -1$$

$\left. \begin{array}{l} \\ \\ \\ \\ \end{array} \right\}$ $-$ if $\cosh^{-1} B > 0$ $+$ if $\cosh^{-1} B < 0$

$$\int \frac{\cosh^{-1} B \, dx}{x} = \mp \left[\frac{1}{2}(\ln 2B)^2 + \frac{B^2}{(2)(2)(2)} + \frac{(1)(3)B^4}{(2)(4)(4)(4)} \right.$$
$$\left. + \frac{(1)(3)(5)B^6}{(2)(4)(6)(6)(6)} + \cdots \right]$$

$$\int \frac{\cosh^{-1} B \, dx}{x^2} = -\frac{\cosh^{-1} B}{x} \mp \frac{1}{b} \ln \frac{b + \sqrt{x^2 + b^2}}{x}$$

$\left. \begin{array}{l} \\ \\ \\ \end{array} \right\}$ $-$ if $\cosh^{-1} B < 0$ $+$ if $\cosh^{-1} B > 0$

(85) Indefinite Integrals Involving $\qquad$ $f(x) = f(x, \coth^{-1} B)$ $\quad$ $B = \dfrac{x}{b}$

$$\int \coth^{-1} B \, dx = x \coth^{-1} B + b \ln \sqrt{x^2 - b^2}$$

$$\int x \coth^{-1} B \, dx = \frac{x^2 - b^2}{2} \coth^{-1} B + \frac{bx}{2}$$

$$\int x^m \coth^{-1} B \, dx = \frac{x^{m+1}}{m+1} \coth^{-1} B + \frac{b}{m+1} \int \frac{x^{m+1}}{x^2 - b^2} \, dx \qquad m \neq -1$$

$$\int \frac{\coth^{-1} B \, dx}{x} = -B - \frac{B^3}{3^2} - \frac{B^5}{5^2} - \frac{B^7}{7^2} - \cdots$$

$$\int \frac{\coth^{-1} B \, dx}{x^2} = -\frac{1}{b}\left(\frac{\coth^{-1} B}{B} + \ln \frac{\sqrt{B^2 - 1}}{B}\right)$$

$$\int \frac{\coth^{-1} B \, dx}{x^3} = -\frac{1}{2x^2}[B - (B^2 - 1)\coth^{-1} B]$$

$$\int \frac{\coth^{-1} B \, dx}{x^m} = -\frac{\coth^{-1} B}{(m-1)x^{m-1}} + \frac{b}{m-1} \int \frac{dx}{(b^2 - x^2)x^{m-1}}$$

(86) Indefinite Integrals Involving

$f(x) = f(x, \ln A)$	$A = bx$

$$\int \ln A \, dx = x(\ln A - 1)$$

$$\int x \ln A \, dx = \left(\frac{x}{2}\right)^2 [\ln (A^2) - 1]$$

$$\int x^m \ln A \, dx = \frac{x^{m+1}}{(m+1)^2}(\ln A^{m+1} - 1) \qquad m \neq -1$$

$$\int (\ln A)^n \, dx = x(\ln A)^n - n \int (\ln A)^{n-1} \, dx \qquad n \neq -1$$

$$\int x^m (\ln A)^n \, dx = \frac{x^{m+1}(\ln A)^n - n \int x^m (\ln A)^{n-1} \, dx}{m+1} \qquad m, n \neq -1$$

$$\int \frac{\ln A \, dx}{x} = \frac{\ln^2 A}{2}$$

$$\int \frac{(\ln A)^n \, dx}{x} = \frac{(\ln A)^{n+1}}{n+1} \qquad n \neq -1$$

$$\int \frac{\ln A \, dx}{x^m} = -\frac{1 + (m-1)\ln A}{(m-1)^2 x^{m-1}} \qquad m \neq 1$$

$$\int \frac{(\ln A)^n \, dx}{x^m} = -\frac{(\ln A)^n}{(m-1)x^{m-1}} + \frac{n}{m-1} \int \frac{(\ln A)^{n-1}}{x^m} \, dx \qquad m \neq 1$$

$$\int \frac{dx}{\ln A} = \frac{1}{b}\left[\ln (\ln A) + \ln A + \frac{(\ln A)^2}{(2)(2!)} + \frac{(\ln A)^3}{(3)(3!)} + \cdots\right]$$

$$\int \frac{x^m \, dx}{\ln A} = \frac{1}{b^{m+1}}\left[\ln (\ln A) + (m+1)\ln A + \frac{(m+1)^2(\ln A)^2}{(2)(2!)} + \frac{(m+1)^3(\ln A)^3}{(3)(3!)} + \cdots\right] \qquad m > 0$$

$$\int \sin (\ln A) \, dx = \frac{x}{2}[\sin (\ln A) - \cos (\ln A)]$$

$$\int \cos (\ln A) \, dx = \frac{x}{2}[\sin (\ln A) + \cos (\ln A)]$$

$$\int \ln (\sin A) \, dx = x\left[\ln A - 1 - \sum_{k=1}^{\infty} \frac{b_k(2A)^{2k}}{2k(2k+1)}\right] \qquad |A| < \pi$$

$$\int \ln (\cos A) \, dx = -x \sum_{k=1}^{\infty} \frac{a_k(2A)^{2k}}{2k(2k+1)} \qquad |A| < \frac{\pi}{2}$$

a_k, b_k = numerical factors (Sec. 8.15-2)

20
TABLES OF DEFINITE INTEGRALS

(1) Definite Integrals

The more frequently encountered definite integrals of elementary functions are tabulated in this chapter. Particular symbols used in the following are:

a, b, c = constants	$\alpha, \beta, \gamma, \lambda$ = constant equivalents
k, m, n, p, q, r = integers	x = independent variable

In these tables, logarithmic expressions are for the absolute value of the respective argument, all angles are in radians, and all inverse functions represent principal values (angles). Special theorems and rules useful in the evaluation of the definite integrals are given in Secs. (2) through (8).

(2) Improper Integrals

If either or both of the *limits* of a definite integral are *infinitely large* or if the *integrand becomes infinite* in the interval of integration, the integral is called an *improper integral*.

$$\int_a^{+\infty} f(x)\, dx = \lim_{b \to +\infty} \int_a^b f(x)\, dx \qquad\qquad \int_{-\infty}^b f(x)\, dx = \lim_{a \to -\infty} \int_a^b f(x)\, dx$$

$$\int_{-\infty}^{+\infty} f(x)\, dx = \lim_{\substack{a \to -\infty \\ b \to +\infty}} \int_a^b f(x)\, dx$$

If $\lim_{x \to c} f(x) = \infty$, then

$$\int_a^b f(x)\, dx = \int_a^c f(x)\, dx + \int_c^b f(x)\, dx = \lim_{\epsilon_1 \to 0} \int_a^{c-\epsilon_1} f(x)\, dx + \lim_{\epsilon_2 \to 0} \int_{c+\epsilon_2}^b f(x)\, dx \qquad \begin{array}{l} \epsilon_1 > 0 \\ \epsilon_2 > 0 \end{array}$$

(3) First Mean-Value Theorem

If $f(x)$ and $g(x)$ are continuous in $[a, b]$, and $g(x)$ is integrable over this interval and *does not change sign* anywhere in it, then there exists at least one point c in (a, b) such that

$$\int_a^b f(x)g(x)\, dx = f(c) \int_a^b g(x)\, dx \qquad \text{and for } g(x) = 1, \qquad \int_a^b f(x)\, dx = (b-a)f(c)$$

(4) Second Mean-Value Theorem

If $f(x)$ and $g(x)$ are continuous in $[a, b]$, $f(x)$ is a *positive, monotonic, decreasing function* in this interval, and $g(x)$ is integrable in it, then there exists at least one point c in (a, b) such that

$$\int_a^b f(x)g(x)\, dx = f(a) \int_a^c g(x)\, dx$$

If the same conditions are valid but $f(x)$ is a *positive, monotonic, increasing function,* then there exists at least one point c in (a, b) such that

$$\int_a^b f(x)g(x)\, dx = f(b) \int_c^b g(x)\, dx$$

In general,

$$\int_a^b f(x)g(x)\, dx = f(a) \int_a^c g(x)\, dx + f(b) \int_c^b g(x)\, dx$$

where $f(x)$ is *monotonic increasing or decreasing* and *is not necessarily always positive* in (a, b).

(5) Even and Odd Functions

If $f(x)$ and $g(x)$ are integrable in $[-a, a]$, $f(x)$ is an *even function* such that $f(-x) = f(x)$, and $g(x)$ is an odd function such that $g(-x) = -g(x)$, then

$$\int_{-a}^{a} f(x)\, dx = 2 \int_{0}^{a} f(x)\, dx$$

$$\int_{-a}^{a} g(x)\, dx = 0$$

(6) Periodic and Antiperiodic Functions

If $n =$ integer, $T =$ positive number (period), $f(x)$ and $g(x)$ are integrable in $[0, l]$, $f(x)$ is a periodic function such that $f(nT + x) = f(x)$, and $g(x)$ is an antiperiodic function such that $g[(2n-1)T + x] = -g(x)$, $g(2nT + x) = g(x)$, then with

$$l = a = nT + t: \qquad \int_{0}^{a} f(x)\, dx = nF(T) + F(t) - (n+1)F(0)$$

$$l = b = (2n-1)T + t: \qquad \int_{0}^{b} g(x)\, dx = G(T) - G(t)$$

$$l = c = 2nT + t: \qquad \int_{0}^{c} g(x)\, dx = G(t) - G(0)$$

where $dF(x)/dx = f(x)$, $dG(x)/dx = g(x)$, and $t < T$.

(7) Trigonometric Identities

If $f(\sin x)$ and $f(\cos x)$ are, respectively and exclusively, functions of $\sin x$ and $\cos x$, then

$$\int_{0}^{\pi/2} f(\sin x)\, dx = \int_{0}^{\pi/2} f(\cos x)\, dx = \frac{1}{2} \int_{0}^{\pi} f(\sin x)\, dx$$

$$\int_{0}^{\lambda\pi/2} f\left(\sin \frac{x}{\lambda}\right) dx = \int_{0}^{\lambda\pi/2} f\left(\cos \frac{x}{\lambda}\right) dx = \frac{\lambda}{2} \int_{0}^{\pi} f(\sin x)\, dx$$

where $\lambda =$ integer or fraction.

(8) Change in Limits and Variables

If $f(x)$ is integrable in $[a, b]$ and $x = g(t)$ and its derivative $dg(t)/dt$ are continuous in $\alpha \le t \le \beta$, then with $a = g(\alpha)$, $b = g(\beta)$,

$$\int_{x=a}^{x=b} f(x)\, dx = \int_{t=\alpha}^{t=\beta} f(g(t)) \frac{dg(t)}{dt}\, dt$$

Particular cases of this transformation are:

$$\int_{0}^{b} f(x)\, dx = \int_{0}^{b} f(b - x)\, dx \qquad\qquad \int_{a}^{b} f(x)\, dx = \int_{a}^{b} f(a + b - x)\, dx$$

$$\int_{a}^{b} f(x)\, dx = \frac{1}{\lambda} \int_{\lambda a}^{\lambda b} f\left(\frac{x}{\lambda}\right) dx \qquad\qquad \int_{a}^{b} f(x)\, dx = \int_{a \pm \lambda}^{b \pm \lambda} f(x \mp \lambda)\, dx$$

$$\int_{a}^{b} f(x)\, dx = \int_{0}^{b-a} f(x + a)\, dx = (b - a) \int_{0}^{1} f((b - a)x + a)\, dx$$

(9) Definite Integrals Involving

$$\boxed{f(x) = f(a^p \pm x^p)} \quad \boxed{[a, b]}$$

m, n, p = positive integers $\neq 0$
a, b, c, α, β = real numbers $\neq 0$

$\Gamma(\ )$ = gamma function (Sec. 13.03)
$B(\ )$ = beta function (Sec. 13.03)

$$\int_0^a x^m (a-x)^n \, dx = \frac{m!\,n!\,a^{m+n+1}}{(m+n+1)!}$$

$$\int_a^b (x-a)^m (b-x)^n \, dx = \frac{m!\,n!\,(b-a)^{m+n+1}}{(m+n+1)!}$$

$$\int_0^a x^\alpha (a-x)^\beta \, dx = \frac{\Gamma(\alpha+1)\Gamma(\beta+1)}{\Gamma(\alpha+\beta+2)} a^{\alpha+\beta+1} = B(\alpha+1, \beta+1) a^{\alpha+\beta+1}$$

$$\int_a^b (x-a)^\alpha (b-x)^\beta \, dx = \frac{\Gamma(\alpha+1)\Gamma(\beta+1)}{\Gamma(\alpha+\beta+2)} (b-a)^{\alpha+\beta+1} = B(\alpha+1, \beta+1)(b-a)^{\alpha+\beta+1}$$

$$\int_0^a x^m (a^n - x^n)^p \, dx = \frac{p!\,n^p a^{np+m+1}}{(m+1)(m+1+p)(m+1+2p) \cdots (m+1+np)}$$

$$\int_0^a x^\alpha (a^n - x^n)^\beta \, dx = \frac{\Gamma\left(\dfrac{\alpha+1}{n}\right)\Gamma(\beta+1)}{n\Gamma\left(\dfrac{\alpha+1}{n}+\beta+1\right)} a^{\alpha+n\beta+1} = \frac{1}{n} B\left(\frac{\alpha+1}{n}, \beta+1\right) a^{\alpha+n\beta+1}$$

$$\int_a^b \frac{1}{x-c} \left(\frac{x-a}{b-x}\right)^\alpha dx = \frac{\pi}{\sin \alpha\pi} \left[1 - \left(\frac{c-a}{b-c}\right)^\alpha \cos \alpha\pi\right] \qquad a < c < b, |\alpha| < 1$$

$$\int_0^a \frac{x^m}{a+x} \, dx = (-a)^m \left[\ln 2 + \sum_{k=1}^m (-1)^k \frac{1}{k}\right]$$

$$\int_0^a \frac{x^m}{a^n+x^n} \, dx = a^{m-n+1} \left[\sum_{k=0}^\infty (-1)^k \frac{1}{m+1+kn}\right]$$

$$\int_0^a \frac{x^b}{(a-x)^b} \, dx = \frac{\pi ab}{\sin b\pi} \qquad |b| < 1$$

$$\int_0^a \frac{x^b}{(a-x)^{b+1}} \, dx = \frac{\pi}{\sin b\pi} \qquad 0 < b < 1$$

$$\int_0^1 \frac{x^\alpha - x^{-\alpha}}{x \pm 1} \, dx = \begin{cases} \dfrac{1}{\alpha} - \dfrac{\pi}{\sin \alpha\pi} & \text{for } + \\[2mm] \dfrac{1}{\alpha} - \dfrac{\pi}{\tan \alpha\pi} & \text{for } - \end{cases} \qquad |\alpha| < 1$$

$$\int_0^1 \frac{x^\alpha - x^{-\alpha}}{x^2 \pm 1} \, dx = \begin{cases} \dfrac{1}{\alpha} - \dfrac{\pi}{2\sin(\alpha\pi/2)} & \text{for } + \\[2mm] \dfrac{1}{\alpha} - \dfrac{\pi}{2\tan(\alpha\pi/2)} & \text{for } - \end{cases} \qquad |\alpha| < 1$$

$$\int_0^a \frac{(x/a)^{m-1} + (x/a)^{n-1}}{(a+x)^{m+n}} \, dx = \frac{a(m-1)!(n-1)!}{a^{m+n}(m+n-1)!}$$

$$\int_0^a \frac{(x/a)^{\alpha-1} + (x/a)^{\beta-1}}{(a+x)^{\alpha+\beta}} \, dx = \frac{B(\alpha, \beta)}{a^{\alpha+\beta-1}}$$

$$\int_0^a \frac{dx}{a^2 + ax + x^2} = \frac{\pi}{3a\sqrt{3}}$$

$$\int_0^a \frac{dx}{a^2 - ax + x^2} = \frac{2\pi}{3a\sqrt{3}}$$

(10) Definite Integrals Involving

$$\boxed{f(x) = f(\sqrt{a^p \pm x^p})} \quad \boxed{[0, a]}$$

m, n, p = positive integers $\neq 0$ $\Gamma(\ )$ = gamma function (Sec. 13.03)
a, α, β = real numbers $\neq 0$ $B(\ )$ = beta function (Sec. 13.03)

$$\int_0^a \sqrt{a^2 + x^2}\, dx = \frac{a^2}{2}\left[\sqrt{2} + \ln\left(\sqrt{2}+1\right)\right]$$
$$\int_0^a \sqrt{a^2 - x^2}\, dx = \frac{\pi a^2}{4}$$

$$\int_0^a x\sqrt{a^2 + x^2}\, dx = \frac{a^3}{3}\left(\sqrt{8}-1\right)$$
$$\int_0^a x\sqrt{a^2 - x^2}\, dx = \frac{a^3}{3}$$

$$\int_0^a x^{2m+1}\sqrt{a^2 - x^2}\, dx = \frac{(2m)(2m-2)\cdots 6\cdot 4\cdot 2}{(2m+1)(2m-1)\cdots 5\cdot 3\cdot 1}\frac{a^{2m+3}}{2m+3} = \frac{(2m)!!}{(2m+3)!!}a^{2m+3}$$

$$\int_0^a x^{2m}\sqrt{a^2 - x^2}\, dx = \frac{(2m-1)(2m-3)\cdots 5\cdot 3\cdot 1}{(2m)(2m-2)\cdots 6\cdot 4\cdot 2}\frac{\pi a^{2m+2}}{(2m+2)2} = \frac{(2m-1)!!}{(2m+2)!!}\frac{\pi a^{2m+2}}{2}$$

$$\int_0^a x^\alpha \sqrt{a^2 - x^2}\, dx = \frac{\sqrt{\pi}}{2}\frac{\Gamma\left(\dfrac{\alpha+1}{2}\right)}{\Gamma\left(\dfrac{\alpha+2}{2}\right)}\frac{a^{\alpha+2}}{\alpha+2}$$
$$\int_0^a x^\alpha \sqrt[p]{a^n - x^n}\, dx = a^{\alpha+1}\frac{\sqrt[p]{a^n}}{n}B\left(\frac{\alpha+1}{n},\frac{p+1}{p}\right]$$

$$\int_0^a \frac{x^m\, dx}{\sqrt{a-x}} = \frac{2m(2m-2)\cdots 6\cdot 4\cdot 2}{(2m+1)(2m-1)\cdots 5\cdot 3\cdot 1}\frac{2a^{m+1}}{\sqrt{a}} = \frac{(2m)!!}{(2m+1)!!}\frac{2a^{m+1}}{\sqrt{a}}$$

$$\int_0^a \frac{x\, dx}{\sqrt{a-x}} = \frac{4a^2}{3\sqrt{a}}$$
$$\int_0^a \frac{x^2\, dx}{\sqrt{a-x}} = \frac{16a^3}{15\sqrt{a}}$$

$$\int_0^a \frac{dx}{\sqrt{a^2 + x^2}} = \ln\left(\sqrt{2}+1\right)$$
$$\int_0^a \frac{dx}{\sqrt{a^2 - x^2}} = \frac{\pi}{2}$$

$$\int_0^a \frac{x\, dx}{\sqrt{a^2 + x^2}} = \left(\sqrt{2}-1\right)a$$
$$\int_0^a \frac{x\, dx}{\sqrt{a^2 - x^2}} = a$$

$$\int_0^a \frac{x^{2m+1}\, dx}{\sqrt{a^2 - x^2}} = \frac{2m(2m-2)\cdots 6\cdot 4\cdot 2}{(2m+1)(2m-1)\cdots 5\cdot 3\cdot 1}a^{2m+1} = \frac{(2m)!!\,a^{2m+1}}{(2m+1)!!}$$

$$\int_0^a \frac{x^{2m}\, dx}{\sqrt{a^2 - x^2}} = \frac{(2m-1)(2m-3)\cdots 5\cdot 3\cdot 1}{(2m)(2m-2)\cdots 6\cdot 4\cdot 2}\frac{\pi a^{2m}}{2} = \frac{(2m-1)!!}{(2m)!!}\frac{\pi a^{2m}}{2}$$

$$\int_0^a \frac{dx}{\sqrt{a^3 - x^3}} = \frac{1.403\,160\cdots}{\sqrt{a}}$$
$$\int_0^a \frac{dx}{\sqrt{a^4 - x^4}} = \frac{5.244\,115\cdots}{a}$$

$$\int_0^a \frac{dx}{\sqrt{a^n - x^n}} = \frac{a}{n}\sqrt{\frac{\pi}{a^n}}\frac{\Gamma(1/n)}{\Gamma(1/n+\frac{1}{2})}$$
$$\int_0^a \frac{dx}{\sqrt[p]{a^n - x^n}} = \frac{a}{n\sqrt[p]{a^n}}B\left(\frac{p-1}{p},\frac{1}{n}\right)$$

$$\int_0^a \frac{x^m\, dx}{\sqrt{a^n - x^n}} = \frac{a^{m+1}}{n}\sqrt{\frac{\pi}{a^n}}\frac{\Gamma\left(\dfrac{m+1}{n}\right)}{\Gamma\left(\dfrac{m+1}{n}+\dfrac{1}{2}\right)}$$
$$\int_0^a \frac{x^m\, dx}{\sqrt[p]{a^n - x^n}} = \frac{a^{m+1}}{n\sqrt[p]{a^n}}B\left(\frac{p-1}{p},\frac{m+1}{n}\right)$$

---

(11) Definite Integrals Involving

$$\boxed{f(x) = f(a^p + x^p)} \quad \boxed{[0, \infty]}$$

m, n, p = positive integers $\neq 0$
a, b, λ = positive real numbers $\neq 0$
m, n as arguments in beta function may also be fractions.

$(\ )!!$ = double factorial (Sec. 1.03)
$B(\ )$ = beta function (Sec. 13.03)

$$\int_0^\infty \frac{x^\lambda \, dx}{a+x} = \frac{\pi a^\lambda}{\sin{(\lambda+1)\pi}} \qquad 0 < \lambda < 1$$

$$\int_0^\infty \frac{x^{-\lambda} \, dx}{a+x} = \frac{\pi a^{-\lambda}}{\sin{\lambda\pi}} \qquad 0 < \lambda < 1$$

$$\int_0^\infty \frac{dx}{a^2+x^2} = \frac{\pi}{2a}$$

$$\int_0^\infty \frac{x \, dx}{a^2+x^2} = \infty$$

$$\int_0^\infty \frac{dx}{a^3+x^3} = \frac{2\pi}{3a^2\sqrt{3}}$$

$$\int_0^\infty \frac{x \, dx}{a^3+x^3} = \frac{2\pi}{3a\sqrt{3}}$$

$$\int_0^\infty \frac{dx}{a^4+x^4} = \frac{\pi}{2a^3\sqrt{2}}$$

$$\int_0^\infty \frac{x \, dx}{a^4+x^4} = \frac{\pi}{4a^2}$$

$$\int_0^\infty \frac{dx}{a^n+x^n} = \frac{a\pi}{na^n \sin{(\pi/n)}}$$

$$\int_0^\infty \frac{x \, dx}{a^n+x^n} = \frac{\pi}{na^{n-2} \sin{(2\pi/n)}}$$

$$\int_0^\infty \frac{x^m \, dx}{a^n+x^n} = \frac{\pi a^{m+1}}{na^n \sin{[(m+1)\pi/n]}}$$

$$\int_0^\infty \frac{x^2 \, dx}{a^n+x^n} = \frac{\pi}{na^{n-3} \sin{(3\pi/n)}}$$

$$\int_0^\infty \frac{dx}{(a+bx)^2} = \frac{1}{ab}$$

$$\int_0^\infty \frac{x \, dx}{(a+bx)^3} = \frac{1}{2ab^2}$$

$$\int_0^\infty \frac{dx}{(a+bx)^n} = \frac{B(1, n-1)}{a^{n-1}b}$$

$$\int_0^\infty \frac{x^m \, dx}{(a+bx)^n} = \frac{B(m+1, n-m-1)}{a^{n-m-1}b^{m+1}}$$

$$\int_0^\infty \frac{x^{m-1} \, dx}{(1+ax)^n} = \frac{B(m, n-m)}{a^m} \qquad |a| < \pi$$

$$n > m > 0$$

$$\int_0^\infty \frac{dx}{(a^2+x^2)(b^2+x^2)} = \frac{\pi}{2ab(a+b)}$$

$$\int_0^\infty \frac{dx}{(a^2+x^2)(a^n+x^n)} = \frac{\pi}{4a^{n+1}}$$

$$\int_0^\infty \frac{dx}{(a^2+x^2)^n} = \frac{(2n-3)(2n-5)\cdots 5\cdot 3\cdot 1}{(2n-2)(2n-4)\cdots 6\cdot 4\cdot 2}\frac{\pi}{2a^{2n-1}} = \frac{(2n-3)!!}{(2n-2)!!}\frac{\pi}{2a^{2n-1}}$$

$$\int_0^\infty \frac{x^{2m} \, dx}{(a+bx^2)^n} = \frac{(2m-1)!!(2n-2m-3)!!\pi}{(2n-2)!!2a^{n-m-1}b^m\sqrt{ab}} \qquad n > (m+1)$$

$$\int_0^\infty \frac{x^{2m+1} \, dx}{(a+bx^2)^n} = \frac{m!(n-m-2)!}{(n-1)!2a^{n-m-1}b^{m+1}} \qquad n > (m+1) \geq 1$$

$$\int_0^\infty \frac{dx}{a+2bx+cx^2} = \frac{1}{\sqrt{ac-b^2}}\cot^{-1}\frac{b}{\sqrt{ac-b^2}} \qquad (ac-b^2) > 0$$

(12) Definite Integrals Involving

$$f(x) = f(x^m, e^{-ax}, e^{-ax^2}) \qquad [0, \infty]$$

Ei( ) = exponential integral function (Sec. 13.01)   Γ( ) = gamma function (Sec. 13.03)
erf( ) = error function (Sec. 13.01)   Z( ) = zeta function (Sec. A.06)
For definitions of a, b, m, n, and ( )!! see opposite page.

$$\int_0^\infty e^{-ax}\,dx = \frac{1}{a}$$

$$\int_0^\infty xe^{-x}\,dx = 1$$

$$\int_0^\infty x^m e^{-x}\,dx = \Gamma(m+1) = m!$$

$$\int_0^\infty x^m e^{-ax}\,dx = \frac{\Gamma(m+1)}{a^{m+1}} = \frac{m!}{a^{m+1}}$$

$$\int_0^\infty \frac{e^{-ax}\,dx}{x} = \infty$$

$$\int_0^\infty \frac{e^{-ax}\,dx}{b+x} = -e^{ab}\,\mathrm{Ei}(ab)$$

$$\int_0^\infty \frac{e^{-ax}\,dx}{\sqrt{x}} = \sqrt{\frac{\pi}{a}}$$

$$\int_0^\infty \frac{e^{-ax}\,dx}{\sqrt{b+x}} = \sqrt{\frac{\pi}{a}}\,e^{ab}[1 - \mathrm{erf}(\sqrt{ab})]$$

$$\int_0^\infty \frac{dx}{e^{ax}+1} = \frac{\ln 2}{a}$$

$$\int_0^\infty \frac{dx}{e^{ax}-1} = \infty$$

$$\int_0^\infty \frac{x\,dx}{e^{ax}+1} = \frac{\pi^2}{12a^2}$$

$$\int_0^\infty \frac{x\,dx}{e^{ax}-1} = \frac{\pi^2}{6a^2}$$

$$\int_0^\infty \frac{x^2\,dx}{e^{ax}+1} = \frac{3}{2a^3}\,Z(3)$$

$$\int_0^\infty \frac{x^2\,dx}{e^{ax}-1} = \frac{2}{a^3}\,Z(3)$$

$$\int_0^\infty \frac{x^b\,dx}{e^{ax}+1} = \frac{\Gamma(b+1)}{a^{b+1}}\sum_{k=0}^\infty \frac{1}{(2k+1)^{b+1}} = \frac{\Gamma(b+1)}{a^{b+1}}\left(1 - \frac{1}{2^b}\right)Z(b+1)$$

$$\int_0^\infty \frac{x^b\,dx}{e^{ax}-1} = \frac{\Gamma(b+1)}{a^{b+1}}\sum_{k=1}^\infty \frac{1}{k^{b+1}} = \frac{\Gamma(b+1)}{a^{b+1}}\,Z(b+1)$$

$$\int_0^\infty e^{-x^2}\,dx = \frac{\sqrt{\pi}}{2}$$

$$\int_0^\infty xe^{-x^2}\,dx = \frac{1}{2}$$

$$\int_0^\infty e^{-ax^2}\,dx = \frac{1}{2}\sqrt{\frac{\pi}{a}}$$

$$\int_0^\infty xe^{-ax^2}\,dx = \frac{1}{2a}$$

$$\int_0^\infty x^{2m+1}e^{-ax^2}\,dx = \frac{m!}{2a^{m+1}}$$

$$\int_0^\infty x^b e^{-ax^2}\,dx = \frac{\Gamma\left(\dfrac{b+1}{2}\right)}{2\sqrt{a^{b+1}}}$$

$$\int_0^\infty x^{2m}e^{-ax^2}\,dx = \frac{(2m-1)(2m-3)\cdots 5\cdot 3\cdot 1}{2(2a)^m}\sqrt{\frac{\pi}{a}} = \frac{(2m-1)!!}{2(2a)^m}\sqrt{\frac{\pi}{a}}$$

$$\int_0^\infty \frac{e^{-ax}-e^{-bx}}{x}\,dx = \ln\frac{b}{a}$$

$$\int_0^\infty \frac{e^{-ax^2}-e^{-bx^2}}{x}\,dx = \ln\sqrt{\frac{b}{a}}$$

(13) Definite Integrals Involving

$$f(x) = f(\sin^m x, \cos^m x, \sin^n x, \cos^n x) \quad \left[0, \frac{\pi}{2} \right]$$

m, n = positive integers $\neq 0$ \
( )!! = double factorial (Sec. 1.03)

$\Gamma( )$ = gamma function (Sec. 13.03) \
B( ) = beta function (Sec. 13.03)

$$\int_0^{\pi/2} \sin x \, dx = \int_0^{\pi/2} \cos x \, dx = 1$$

$$\int_0^{\pi/2} \sin^2 x \, dx = \int_0^{\pi/2} \cos^2 x \, dx = \frac{\pi}{4}$$

$$\int_0^{\pi/2} \sin^3 x \, dx = \int_0^{\pi/2} \cos^3 x \, dx = \frac{2}{3}$$

$$\int_0^{\pi/2} \sin^4 x \, dx = \int_0^{\pi/2} \cos^4 x \, dx = \frac{3\pi}{16}$$

$$\int_0^{\pi/2} \sin^{2m+1} x \, dx = \int_0^{\pi/2} \cos^{2m+1} x \, dx = \frac{(2m)(2m-2)\cdots 6 \cdot 4 \cdot 2}{(2m+1)(2m-1)\cdots 5 \cdot 3 \cdot 1} = \frac{(2m)!!}{(2m+1)!!}$$

$$\int_0^{\pi/2} \sin^{2m} x \, dx = \int_0^{\pi/2} \cos^{2m} x \, dx = \frac{(2m-1)(2m-3)\cdots 5 \cdot 3 \cdot 1}{(2m)(2m-2)\cdots 6 \cdot 4 \cdot 2} \frac{\pi}{2} = \frac{(2m-1)!!}{(2m)!!} \frac{\pi}{2}$$

$$\int_0^{\pi/2} \sin x \cos x \, dx = \frac{1}{2}$$

$$\int_0^{\pi/2} \sin^2 x \cos^2 x \, dx = \frac{\pi}{16}$$

$$\int_0^{\pi/2} \sin^3 x \cos^3 x \, dx = \frac{1}{12}$$

$$\int_0^{\pi/2} \sin^4 x \cos^4 x \, dx = \frac{3\pi}{256}$$

$$\int_0^{\pi/2} \sin^5 x \cos^5 x \, dx = \frac{1}{60}$$

$$\int_0^{\pi/2} \cos^6 x \sin^6 x \, dx = \frac{15\pi}{6144}$$

$$\int_0^{\pi/2} \sin^{2m+1} x \cos^{2n+1} x \, dx = \frac{m!\,n!}{(m+n+1)!} \frac{1}{2} = \frac{\Gamma(m+1)\Gamma(n+1)}{2\Gamma(m+n+2)} = \frac{1}{2}\,\mathrm{B}(m+1, n+1)$$

$$\int_0^{\pi/2} \sin^{2m} x \cos^{2n} x \, dx = \frac{(2m-1)!!(2n-1)!!}{(2m+2n)!!} \frac{\pi}{2} = \frac{\Gamma(m+\frac{1}{2})\Gamma(n+\frac{1}{2})}{2\Gamma(m+n+1)} = \tfrac{1}{2}\,\mathrm{B}(m+\tfrac{1}{2}, n+\tfrac{1}{2})$$

$$\int_0^{\pi/2} \sin x \cos^2 x \, dx = \frac{1}{3}$$

$$\int_0^{\pi/2} \cos x \sin^2 x \, dx = \frac{1}{3}$$

$$\int_0^{\pi/2} \sin x \cos^3 x \, dx = \frac{1}{4}$$

$$\int_0^{\pi/2} \cos x \sin^3 x \, dx = \frac{1}{4}$$

$$\int_0^{\pi/2} \sin x \cos^n x \, dx = \frac{1}{n+1}$$

$$\int_0^{\pi/2} \cos x \sin^n x \, dx = \frac{1}{n+1}$$

$$\int_0^{\pi/2} \sin^{2m+1} x \cos^{2n} x \, dx = \frac{(2m)!!(2n-1)!!}{(2m+2n+1)!!} \frac{\Gamma(m+1)\Gamma(n+\frac{1}{2})}{2\Gamma(m+n+\frac{3}{2})} = \tfrac{1}{2}\mathrm{B}(m+1, n+\tfrac{1}{2})$$

$$\int_0^{\pi/2} \sin^{2m} x \cos^{2n+1} x \, dx = \frac{(2n)!!(2m-1)!!}{(2m+2n+1)!!} = \frac{\Gamma(n+1)\Gamma(m+\frac{1}{2})}{2\Gamma(m+n+\frac{3}{2})} = \tfrac{1}{2}\mathrm{B}(m+\tfrac{1}{2}, n+1)$$

(14) Definite Integrals Involving

$$f(x) = f(a \sin x \pm b \cos x) \qquad \left[0, \frac{\pi}{2}\right]$$

a, b = positive real numbers $\neq 0$

k = modulus ($|k| < 1$) (Sec. 13.05–1)

$G = \bar{Z}(2) = 0.915\ 965\ 594$ (Sec. A.06)

K, E = complete elliptic integrals (Sec. 13.05–2)

$$\int_0^{\pi/2} \frac{dx}{1 + \sin x} = \int_0^{\pi/2} \frac{dx}{1 + \cos x} = 1$$

$$\int_0^{\pi/2} \frac{\sin x\ dx}{1 + \sin x} = \int_0^{\pi/2} \frac{\cos x\ dx}{1 + \cos x} = \frac{\pi}{2} - 1$$

$$\int_0^{\pi/2} \frac{x\ dx}{1 + \sin x} = \ln 2$$

$$\int_0^{\pi/2} \frac{x\ dx}{1 + \cos x} = \frac{\pi}{2} - \ln 2$$

$$\int_0^{\pi/2} \frac{x \cos x\ dx}{1 + \sin x} = \pi \ln 2 - 4G$$

$$\int_0^{\pi/2} \frac{x \sin x\ dx}{1 + \cos x} = -\frac{\pi}{2} \ln 2 + 2G$$

$$\int_0^{\pi/2} \frac{dx}{\sin x \pm \cos x} = \mp \frac{1}{\sqrt{2}} \ln\left(\tan \frac{\pi}{8}\right)$$

$$\int_0^{\pi/2} \frac{dx}{(\sin x \pm \cos x)^2} = \pm 1$$

$$\int_0^{\pi/2} \frac{dx}{1 \pm a \sin x} = \int_0^{\pi/2} \frac{dx}{1 \pm a \cos x} = \frac{\pi \mp 2 \sin^{-1} a}{2\sqrt{1 - a^2}} \qquad 0 < a < 1$$

$$\int_0^{\pi/2} \frac{dx}{(1 \pm a \sin x)^2} = \int_0^{\pi/2} \frac{dx}{(1 \pm a \cos x)^2} = \frac{\pi \mp 2 \sin^{-1} a}{2\sqrt{(1 - a^2)^3}} \mp \frac{a}{1 - a^2} \qquad 0 < \sin^{-1} a < \frac{\pi}{2}$$

$$\int_0^{\pi/2} \frac{dx}{(a \sin x + b \cos x)^2} = \frac{1}{ab}$$

$$\int_0^{\pi/2} \frac{x\ dx}{(a \sin x + b \cos x)^2} = \frac{ab}{a^2 + b^2} \frac{\pi}{2} - \frac{\ln ab}{a^2 + b^2}$$

$$\int_0^{\pi/2} \frac{dx}{1 \pm a^2 \sin^2 x} = \int_0^{\pi/2} \frac{dx}{1 \pm a^2 \cos^2 x} = \frac{\pi}{2\sqrt{1 \pm a^2}}$$

$$\int_0^{\pi/2} \frac{dx}{(1 \pm a^2 \sin^2 x)^2} = \int_0^{\pi/2} \frac{d}{(1 \pm a^2 \cos^2 x)^2} = \frac{\pi (2 \pm a^2)}{4\sqrt{(1 \pm a^2)^3}} \qquad 0 < a^2 < 1$$

$$\int_0^{\pi/2} \frac{dx}{a^2 \sin^2 x + b^2 \cos^2 x} = \frac{\pi}{2ab}$$

$$\int_0^{\pi/2} \frac{dx}{(a^2 \sin^2 x + b^2 \cos^2 x)^2} = \frac{2\pi (a^2 + b^2)}{(2ab)^3}$$

$$\int_0^{\pi/2} \frac{\cos^2 x\ dx}{a^2 \sin^2 x + b^2 \cos^2 x} = \frac{\pi}{2b(a + b)}$$

$$\int_0^{\pi/2} \frac{\cos^2 x\ dx}{(a^2 \sin^2 x + b^2 \cos^2 x)^2} = \frac{\pi}{4ab^3}$$

$$\int_0^{\pi/2} \frac{\sin^2 x\ dx}{a^2 \sin^2 x + b^2 \cos^2 x} = \frac{\pi}{2a(a + b)}$$

$$\int_0^{\pi/2} \frac{\sin^2 x\ dx}{(a^2 \sin^2 x + b^2 \cos^2 x)^2} = \frac{\pi}{4a^3 b}$$

$$\int_0^{\pi/2} \frac{\sin x\ dx}{\sqrt{1 - k^2 \sin^2 x}} = \frac{1}{2k} \ln \frac{1 + k}{1 - k}$$

$$\int_0^{\pi/2} \frac{\cos x\ dx}{\sqrt{1 - k^2 \sin^2 x}} = \frac{1}{k} \sin^{-1} k$$

$$\int_0^{\pi/2} \frac{\sin^2 x\ dx}{\sqrt{1 - k^2 \sin^2 x}} = \frac{1}{k^2} (K - E)$$

$$\int_0^{\pi/2} \frac{\cos^2 x\ dx}{\sqrt{1 - k^2 \sin^2 x}} = \frac{1}{k^2} [E - (1 - k^2)K]$$

(15) Definite Integrals Involving

| $f(x) = f(\sin^p x, \cos^p x, \sin qx, \cos qx)$ | $[0, \pi], [-a, a]$ |

m, n, p, q = positive integers $\neq 0$
a = positive real number $\neq 0$

( )!! = double factorial (Sec. 1.03)
B( ) = beta function (Sec. 13.03)

$$\int_0^\pi \sin x \, dx = 2 \qquad \int_0^\pi \sin^2 x \, dx = \int_0^\pi \cos^2 x \, dx = \frac{\pi}{2} \qquad \int_0^\pi \cos x \, dx = 0$$

$$\int_0^\pi \sin^{2m+1} x \, dx = 2 \frac{(2m)!!}{(2m+1)!!} \qquad\qquad \int_0^\pi \cos^{2m+1} x \, dx = 0$$

$$\int_0^\pi \sin^{2m} x \, dx = \pi \frac{(2m-1)!!}{(2m)!!} \qquad\qquad \int_0^\pi \cos^{2m} x \, dx = \pi \frac{(2m-1)!!}{(2m)!!}$$

$$\int_0^\pi \sin x \cos x \, dx = 0 \qquad\qquad \int_0^\pi \sin^2 x \cos^2 x \, dx = \frac{\pi}{8}$$

$$\int_0^\pi \sin^3 x \cos^3 x \, dx = 0 \qquad\qquad \int_0^\pi \sin^4 x \cos^4 x \, dx = \frac{3\pi}{128}$$

$$\int_0^\pi \sin^{2m+1} x \cos^{2m+1} x \, dx = 0 \qquad\qquad \int_0^\pi \sin^{2m} x \cos^{2m} x \, dx = B\left(m + \tfrac{1}{2}, m + \tfrac{1}{2}\right)$$

$$\int_0^\pi \sin mx \sin nx \, dx = \begin{cases} 0 & \text{if } m \neq n \\ \dfrac{\pi}{2} & \text{if } m = n \end{cases}$$

$$\int_0^\pi \sin mx \cos nx \, dx = \begin{cases} 0 & \text{if } m \neq n \text{ and } m + n \text{ is even} \\ \dfrac{2m}{m^2 - n^2} & \text{if } m \neq n \text{ and } m + n \text{ is odd} \\ 0 & \text{if } m = n \end{cases}$$

$$\int_0^\pi \cos mx \cos nx \, dx = \begin{cases} 0 & \text{if } m \neq n \\ \dfrac{\pi}{2} & \text{if } m = n \end{cases}$$

$$\int_{-a}^a \sin \frac{m\pi x}{a} \sin \frac{n\pi x}{a} \, dx = \int_{-a}^a \cos \frac{m\pi x}{a} \cos \frac{n\pi x}{a} \, dx = \begin{cases} 0 & \text{if } m \neq n \\ a & \text{if } m = n \end{cases}$$

$$\int_{-a}^a \sin \frac{m\pi x}{a} \cos \frac{n\pi x}{a} \, dx = \int_{-a}^a \cos \frac{m\pi x}{a} \sin \frac{n\pi x}{a} \, dx = 0 \qquad\qquad \text{all } m, n$$

(16) Definite Integrals Involving

$f(x) = f(x, \sin^p x, \cos^p x)$	$[0, \pi]$

m, n, p = positive integers $\neq 0$	$\alpha = \sqrt{1 + a^2}$	$1 < \alpha^2 < 2$
a = positive real number $\neq 0$	$\beta = \sqrt{1 - a^2}$	$0 < \beta^2 < 1$

$$\int_0^\pi x \sin x \, dx = \pi$$

$$\int_0^\pi x \cos x \, dx = -2$$

$$\int_0^\pi x \sin^2 x \, dx = \frac{\pi^2}{4}$$

$$\int_0^\pi x \cos^2 x \, dx = \frac{\pi^2}{4}$$

$$\int_0^\pi x \sin^{2n+1} x \, dx = \frac{(2n)!!}{(2n+1)!!} \pi$$

$$\int_0^\pi x \cos^{2n+1} x \, dx = -\frac{2}{4^n} \sum_{k=0}^{n} \binom{2n+1}{k} \frac{1}{(2n-2k-1)^2}$$

$$\int_0^\pi x \sin^{2n} x \, dx = \frac{(2n-1)!!}{(2n)!!} \frac{\pi^2}{2}$$

$$\int_0^\pi x \cos^{2n} x \, dx = \frac{(2n-1)!!}{(2n)!!} \frac{\pi^2}{2}$$

$$\int_0^\pi \frac{dx}{1 \pm a \sin x} = \frac{\pi \mp 2 \sin^{-1} a}{\beta}$$

$$\int_0^\pi \frac{dx}{1 \pm a \cos x} = \frac{\pi}{\beta}$$

$$\int_0^\pi \frac{dx}{(1 \pm a \sin x)^2} = \frac{\pi \mp 2 \sin^{-1} a}{\beta^3} \pm \frac{2a}{\beta^2}$$

$$\int_0^\pi \frac{dx}{(1 \pm a \cos x)^2} = \frac{\pi}{\beta^3}$$

$$\int_0^\pi (\alpha^2 - 2a \cos x)^n \, dx = \pi \left[1 + \binom{n}{1}^2 a^2 + \binom{n}{2}^2 a^4 + \cdots + \binom{n}{n}^2 a^{2n} \right]$$

$$\int_0^\pi \frac{dx}{\alpha^2 \pm 2a \cos x} = \frac{\pi}{|\beta^2|}$$

$$\int_0^\pi \frac{dx}{\alpha^2 \pm 2a \sin x} = \frac{\pi \mp 2 \sin^{-1} (2a/\alpha^2)}{|\beta^2|}$$

$$\int_0^\pi \frac{dx}{(\alpha^2 - 2a \cos x)^n} = \frac{\pi}{\beta^{2n}} \sum_{k=0}^{n-1} \binom{2k}{k} \binom{n+k-1}{n-k-1} \left(\frac{a}{\beta}\right)^{2k}$$

$$\int_0^\pi \frac{\sin x \, dx}{\alpha^2 - 2a \cos x} = \frac{2}{a} \tanh^{-1} a$$

$$\int_0^\pi \frac{\cos x \, dx}{\alpha^2 - 2a \cos x} = \frac{\pi a}{\beta^2}$$

$$\int_0^\pi \frac{\sin^2 x \, dx}{\alpha^2 - 2a \cos x} = \frac{\pi}{2}$$

$$\int_0^\pi \frac{\cos^2 x \, dx}{\alpha^2 - 2a \cos x} = \frac{\pi}{2} \left(\frac{\alpha}{\beta}\right)^2$$

$$\int_0^\pi \frac{\sin x \sin mx \, dx}{\alpha^2 - 2a \cos x} = \frac{\pi a^m}{2a}$$

$$\int_0^\pi \frac{\cos x \cos mx \, dx}{\alpha^2 - 2a \cos x} = \frac{\pi a^m}{2a} \left(\frac{\alpha}{\beta}\right)^2$$

$$\int_0^\pi \frac{\cos mx \, dx}{(\alpha^2 - 2a \cos x)^n} = \left(\frac{a}{\beta^2}\right)^{2n-1} a^{m-1} \pi \sum_{k=0}^{n-1} \binom{m+n-1}{k} \binom{2n-k-2}{n-1} \left(\frac{\beta}{a}\right)^{2k}$$

$$\int_0^\pi \frac{dx}{a^2 \sin^2 x + b^2 \cos^2 x} = \frac{\pi^2}{2ab}$$

$$\int_0^\pi \frac{x \sin x \cos x \, dx}{a^2 \sin^2 x - b^2 \cos^2 x} = \frac{\pi}{b^2 - a^2} \ln \frac{a+b}{2b}$$

(17) Definite Integrals Involving

$$\boxed{f(x) = f(\sin nx) \qquad [0,\, \pi]}$$

---

n = positive integer $\neq 0$ $\qquad$ a, b = real numbers $\neq 0$ $\qquad$ $\lambda = (-1)^n$

$\alpha_1 = a + b$ $\qquad$ $\beta_1 = (b+n)\pi$ $\qquad$ $\omega_1 = 2\sqrt{a^2 + (b+n)^2}$ $\qquad$ $\phi_1 = \tan^{-1}\dfrac{a}{b+n}$

$\alpha_2 = a - b$ $\qquad$ $\beta_2 = (b-n)\pi$ $\qquad$ $\omega_2 = 2\sqrt{a^2 + (b-n)^2}$ $\qquad$ $\phi_2 = \tan^{-1}\dfrac{a}{b-n}$

$$\int_0^\pi \sin nx \, dx = \frac{1-\lambda}{n}$$

$$\int_0^\pi x \sin nx \, dx = -\frac{\lambda\pi}{n}$$

$$\int_0^\pi x^2 \sin nx \, dx = \frac{2(\lambda - 1)}{n^3} - \frac{\lambda\pi^2}{n}$$

$$\int_0^\pi x^3 \sin nx \, dx = \frac{6\lambda\pi}{n^3} - \frac{\lambda\pi^3}{n}$$

$$\int_0^\pi e^{ax} \sin nx \, dx = \frac{n(1 - \lambda e^{a\pi})}{a^2 + n^2}$$

$$\int_0^\pi a^{bx} \sin nx \, dx = \frac{n(1 - \lambda a^{b\pi})}{(b \ln a)^2 + n^2}$$

$$\int_0^\pi x e^{ax} \sin nx \, dx = -\frac{n\lambda\pi e^{a\pi}}{a^2 + n^2} + \frac{2an}{(a^2 + n^2)^2}(\lambda e^{a\pi} - 1)$$

$$\int_0^\pi \sin ax \sin nx \, dx = \frac{n\lambda \sin a\pi}{a^2 - n^2}$$

$$\int_0^\pi \sinh ax \sin nx \, dx = -\frac{n\lambda \sinh a\pi}{a^2 + n^2}$$

$$\int_0^\pi \cos ax \sin nx \, dx = \frac{n(\lambda \cos a\pi - 1)}{a^2 - n^2}$$

$$\int_0^\pi \cosh ax \sin nx \, dx = -\frac{n(\lambda \cosh a\pi - 1)}{a^2 + n^2}$$

$$\int_0^\pi x \sin bx \sin nx \, dx = -\frac{\pi^2}{2}\left[(\lambda \sin b\pi)\left(\frac{1}{\beta_1} - \frac{1}{\beta_2}\right) + (\lambda \cos b\pi - 1)\left(\frac{1}{\beta_1^2} - \frac{1}{\beta_2^2}\right) \right]$$

$$\int_0^\pi x \cos bx \sin nx \, dx = -\frac{\pi^2}{2}\left[(\lambda \cos b\pi)\left(\frac{1}{\beta_1} - \frac{1}{\beta_2}\right) - (\lambda \sin b\pi)\left(\frac{1}{\beta_1^2} - \frac{1}{\beta_2^2}\right) \right]$$

$$\int_0^\pi e^{ax} \sin bx \sin nx \, dx = -e^{a\pi}\left[\frac{\sin(\phi_1 + \beta_1)}{\omega_1} - \frac{\sin(\phi_2 + \beta_2)}{\omega_2} \right] + \frac{\sin \phi_1}{\omega_1} - \frac{\sin \phi_2}{\omega_2}$$

$$\int_0^\pi e^{ax} \cos bx \sin nx \, dx = -e^{a\pi}\left[\frac{\cos(\phi_1 + \beta_1)}{\omega_1} - \frac{\cos(\phi_2 + \beta_2)}{\omega_2} \right] + \frac{\cos \phi_1}{\omega_1} - \frac{\cos \phi_2}{\omega_2}$$

$$\int_0^\pi \sin ax \sin bx \sin nx \, dx = +\frac{n(1 - \lambda \cos \alpha_1 \pi)}{2(\alpha_1^2 - n^2)} - \frac{n(1 - \lambda \cos \alpha_2 \pi)}{2(\alpha_2^2 - n^2)}$$

$$\int_0^\pi \cos ax \sin bx \sin nx \, dx = +\frac{n\lambda \sin \alpha_1 \pi}{2(\alpha_1^2 - n^2)} - \frac{n\lambda \sin \alpha_2 \pi}{2(\alpha_2^2 - n^2)}$$

$$\int_0^\pi \cos ax \cos bx \sin nx \, dx = +\frac{n(\lambda \cos \alpha_1 \pi - 1)}{2(\alpha_1^2 - n^2)} + \frac{n(\lambda \cos \alpha_2 \pi - 1)}{2(\alpha_2^2 - n^2)}$$

(18) Definite Integrals Involving

$$\boxed{f(x) = f(\cos nx) \qquad [0, \pi]}$$

n = positive integer $\neq 0$		a, b = real numbers $\neq 0$	$\lambda = (-1)^n$
$\alpha_1 = a + b$	$\beta_1 = (b + n)\pi$	$\omega_1 = 2\sqrt{a^2 + (b+n)^2}$	$\phi_1 = \tan^{-1}\dfrac{a}{b+n}$
$\alpha_2 = a - b$	$\beta_2 = (b - n)\pi$	$\omega_2 = 2\sqrt{a^2 + (b-n)^2}$	$\phi_2 = \tan^{-1}\dfrac{a}{b-n}$

$$\int_0^\pi \cos nx \, dx = 0 \qquad\qquad \int_0^\pi x \cos nx \, dx = \frac{\lambda - 1}{n^2}$$

$$\int_0^\pi x^2 \cos nx \, dx = \frac{2\lambda\pi}{n^2} \qquad\qquad \int_0^\pi x^3 \cos nx \, dx = \frac{3\lambda\pi^2}{n^2} - \frac{6(\lambda - 1)}{nx}$$

$$\int_0^\pi e^{ax} \cos nx \, dx = \frac{a(\lambda e^{a\pi} - 1)}{a^2 + n^2} \qquad\qquad \int_0^\pi a^{bx} \cos nx \, dx = \frac{(b \ln a)(\lambda a^{b\pi} - 1)}{(b \ln a)^2 + n^2}$$

$$\int_0^\pi x e^{ax} \cos nx \, dx = + \frac{n\lambda\pi e^{a\pi}}{a^2 + n^2} + \frac{a^2 - n^2}{(a^2 + n^2)^2}(1 - \lambda e^{a\pi})$$

$$\int_0^\pi \sin ax \cos nx \, dx = \frac{a(1 - \lambda \cos a\pi)}{a^2 - n^2} \qquad\qquad \int_0^\pi \sinh ax \cos nx \, dx = -\frac{a(1 - \lambda \cosh a\pi)}{a^2 + n^2}$$

$$\int_0^\pi \cos ax \cos nx \, dx = \frac{a\lambda \sin a\pi}{a^2 - n^2} \qquad\qquad \int_0^\pi \cosh ax \cos nx \, dx = +\frac{a\lambda \sinh a\pi}{a^2 + n^2}$$

$$\int_0^\pi x \sin bx \cos nx \, dx = -\frac{\pi^2}{2}\left[(\lambda \cos b\pi)\left(\frac{1}{\beta_1} + \frac{1}{\beta_2}\right) - (\lambda \sin b\pi)\left(\frac{1}{\beta_1^2} + \frac{1}{\beta_2^2}\right)\right]$$

$$\int_0^\pi x \cos bx \cos nx \, dx = +\frac{\pi^2}{2}\left[(\lambda \sin b\pi)\left(\frac{1}{\beta_1} + \frac{1}{\beta_2}\right) - (\lambda \cos b\pi - 1)\left(\frac{1}{\beta_1^2} + \frac{1}{\beta_2^2}\right)\right]$$

$$\int_0^\pi e^{ax} \sin bx \cos nx \, dx = -e^{a\pi}\left[\frac{\cos(\phi_1 + \beta_1)}{\omega_1} + \frac{\cos(\phi_2 + \beta_2)}{\omega_2}\right] + \frac{\cos \phi_1}{\omega_1} + \frac{\cos \phi_2}{\omega_2}$$

$$\int_0^\pi e^{ax} \cos bx \cos nx \, dx = +e^{a\pi}\left[\frac{\sin(\phi_1 + \beta_1)}{\omega_1} + \frac{\sin(\phi_2 + \beta_2)}{\omega_2}\right] - \frac{\sin \phi_1}{\omega_1} - \frac{\sin \phi_2}{\omega_2}$$

$$\int_0^\pi \sin ax \sin bx \cos nx \, dx = -\frac{\lambda\alpha_1 \sin \alpha_1\pi}{2(\alpha_1^2 - n^2)} + \frac{\lambda\alpha_2 \sin \alpha_2\pi}{2(\alpha_2^2 - n^2)}$$

$$\int_0^\pi \sin ax \cos bx \cos nx \, dx = +\frac{\alpha_1(1 - \lambda \cos \alpha_1\pi)}{2(\alpha_1^2 - n^2)} + \frac{\alpha_2(1 - \lambda \cos \alpha_2\pi)}{2(\alpha_2^2 - n^2)}$$

$$\int_0^\pi \cos ax \cos bx \cos nx \, dx = +\frac{\lambda\alpha_1 \sin \alpha_1\pi}{2(\alpha_1^2 - n^2)} + \frac{\lambda\alpha_2 \sin \alpha_2\pi}{2(\alpha_2^2 - n^2)}$$

(19) Definite Integrals Involving

| $f(x) = f(\sin nx)$ | $[-\pi, +\pi]$ |

$n = $ positive integer $\neq 0$		$a, b = $ real numbers $\neq 0$	$\lambda = (-1)^n$
$\alpha_1 = a + b$	$\beta_1 = (b + n)\pi$	$\omega_1 = 2\sqrt{a^2 + (b + n)^2}$	$\phi_1 = \tan^{-1}\dfrac{a}{b + n}$
$\alpha_2 = a - b$	$\beta_2 = (b - n)\pi$	$\omega_2 = 2\sqrt{a^2 + (b - n)^2}$	$\phi_2 = \tan^{-1}\dfrac{a}{b - n}$

$$\int_{-\pi}^{+\pi} \sin nx \, dx = 0$$

$$\int_{-\pi}^{+\pi} x \sin nx \, dx = -\frac{2\lambda\pi}{n}$$

$$\int_{-\pi}^{+\pi} x^2 \sin nx \, dx = 0$$

$$\int_{-\pi}^{+\pi} x^3 \sin nx \, dx = \frac{12\lambda\pi}{n^3} - \frac{2\lambda\pi^3}{n}$$

$$\int_{-\pi}^{+\pi} e^{ax} \sin nx \, dx = -\frac{2n\lambda \sinh a\pi}{a^2 + n^2}$$

$$\int_{-\pi}^{+\pi} a^{bx} \sin nx \, dx = \frac{2n\lambda \sinh (\pi b \ln a)}{(b \ln a)^2 + n^2}$$

$$\int_{-\pi}^{+\pi} xe^{ax} \sin nx \, dx = -\frac{2n\lambda \sinh a\pi}{(a^2 + n^2)^2} \left[\pi(a^2 + n^2) - 2a \right]$$

$$\int_{-\pi}^{+\pi} \sin ax \sin nx \, dx = \frac{2n\lambda \sin a\pi}{a^2 - n^2}$$

$$\int_{-\pi}^{+\pi} \sinh ax \sin nx \, dx = -\frac{2n\lambda \sinh a\pi}{a^2 + n^2}$$

$$\int_{-\pi}^{+\pi} \cos ax \sin nx \, dx = 0$$

$$\int_{-\pi}^{+\pi} \cosh ax \sin nx \, dx = 0$$

$$\int_{-\pi}^{+\pi} x \sin bx \sin nx \, dx = 0$$

$$\int_{-\pi}^{+\pi} x \cos bx \sin nx \, dx = -\pi^2 \left[(\lambda \cos b\pi)\left(\frac{1}{\beta_1} - \frac{1}{\beta_2}\right) - (\lambda \sin b\pi)\left(\frac{1}{\beta_1^2} - \frac{1}{\beta_2^2}\right) \right]$$

$$\int_{-\pi}^{+\pi} e^{ax} \sin bx \sin nx \, dx = -e^{a\pi} \left[\frac{\sin (\phi_1 + \beta_1)}{\omega_1} - \frac{\sin (\phi_2 + \beta_2)}{\omega_2} \right] + e^{-a\pi} \left[\frac{\sin (\phi_1 - \beta_1)}{\omega_1} - \frac{\sin (\phi_2 - \beta_2)}{\omega_2} \right]$$

$$\int_{-\pi}^{+\pi} e^{ax} \cos bx \sin nx \, dx = -e^{a\pi} \left[\frac{\cos (\phi_1 + \beta_1)}{\omega_1} - \frac{\cos (\phi_2 + \beta_2)}{\omega_2} \right] + e^{-a\pi} \left[\frac{\cos (\phi_1 - \beta_1)}{\omega_1} - \frac{\cos (\phi_2 - \beta_2)}{\omega_2} \right]$$

$$\int_{-\pi}^{+\pi} \sin ax \sin bx \sin nx \, dx = 0$$

$$\int_{-\pi}^{+\pi} \cos ax \sin bx \sin nx \, dx = +\frac{n\lambda \sin \alpha_1 \pi}{\alpha_1^2 - n^2} - \frac{n\lambda \sin \alpha_2 \pi}{\alpha_2^2 - n^2}$$

$$\int_{-\pi}^{+\pi} \cos ax \cos bx \sin nx \, dx = 0$$

(20) Definite Integrals Involving

| $f(x) = f(\cos nx)$ | $[-\pi, +\pi]$ |

---

n = positive integer $\neq 0$ $\qquad$ a, b = real numbers $\neq 0$ $\qquad$ $\lambda = (-1)^n$

$\alpha_1 = a + b$ $\qquad$ $\beta_1 = (b + n)\pi$ $\qquad$ $\omega_1 = 2\sqrt{a^2 + (b+n)^2}$ $\qquad$ $\phi_1 = \tan^{-1}\dfrac{a}{b+n}$

$\alpha_2 = a - b$ $\qquad$ $\beta_2 = (b - n)\pi$ $\qquad$ $\omega_2 = 2\sqrt{a^2 + (b-n)^2}$ $\qquad$ $\phi_2 = \tan^{-1}\dfrac{a}{b-n}$

$$\int_{-\pi}^{+\pi} \cos nx\, dx = 0 \qquad\qquad \int_{-\pi}^{+\pi} x \cos nx\, dx = 0$$

$$\int_{-\pi}^{+\pi} x^2 \cos nx\, dx = \frac{4\lambda\pi}{n^2} \qquad\qquad \int_{-\pi}^{+\pi} x^3 \cos nx\, dx = 0$$

$$\int_{-\pi}^{+\pi} e^{ax} \cos nx\, dx = \frac{2a\lambda \sinh a\pi}{a^2 + n^2} \qquad\qquad \int_{-\pi}^{+\pi} a^{bx} \cos nx\, dx = \frac{2(b \ln a)\lambda \sinh (\pi b \ln a)}{(b \ln a)^2 + n^2}$$

$$\int_{-\pi}^{+\pi} x e^{ax} \cos nx\, dx = \frac{2\lambda \sinh a\pi}{(a^2 + n^2)^2} \left[n\pi(a^2 + n^2) - a^2 + n^2 \right]$$

$$\int_{-\pi}^{+\pi} \sin ax \cos nx\, dx = 0 \qquad\qquad \int_{-\pi}^{+\pi} \sinh ax \cos nx\, dx = 0$$

$$\int_{-\pi}^{+\pi} \cos ax \cos nx\, dx = \frac{2a\lambda \sin a\pi}{a^2 - n^2} \qquad\qquad \int_{-\pi}^{+\pi} \cosh ax \cos nx\, dx = \frac{2a\lambda \sinh a\pi}{a^2 + b^2}$$

$$\int_{-\pi}^{+\pi} x \sin ax \cos nx\, dx = -\pi^2 \left[(\lambda \cos a\pi - 1)\left(\frac{1}{\beta_1} + \frac{1}{\beta_2}\right) - (\lambda \sin a\pi)\left(\frac{1}{\beta_1{}^2} + \frac{1}{\beta_2{}^2}\right) \right]$$

$$\int_{-\pi}^{+\pi} x \cos ax \cos nx\, dx = 0$$

$$\int_{-\pi}^{+\pi} e^{ax} \sin bx \cos nx\, dx = -e^{a\pi}\left[\frac{\cos(\phi_1 + \beta_1)}{\omega_1} + \frac{\cos(\phi_2 + \beta_2)}{\omega_2} \right] + e^{-a\pi}\left[\frac{\cos(\phi_1 - \beta_1)}{\omega_1} + \frac{\cos(\phi_2 - \beta_2)}{\omega_2} \right]$$

$$\int_{-\pi}^{+\pi} e^{ax} \cos bx \cos nx\, dx = +e^{a\pi}\left[\frac{\sin(\phi_1 + \beta_1)}{\omega_1} + \frac{\sin(\phi_2 + \beta_2)}{\omega_2} \right] - e^{-a\pi}\left[\frac{\sin(\phi_1 - \beta_1)}{\omega_1} + \frac{\sin(\phi_2 - \beta_2)}{\omega_2} \right]$$

$$\int_{-\pi}^{+\pi} \sin ax \sin bx \cos nx\, dx = -\frac{\lambda\alpha_1 \sin \alpha_1\pi}{\alpha_1{}^2 - n^2} + \frac{\lambda\alpha_2 \sin \alpha_2\pi}{\alpha_2{}^2 - n^2}$$

$$\int_{-\pi}^{+\pi} \sin ax \cos bx \cos nx\, dx = 0$$

$$\int_{-\pi}^{+\pi} \cos ax \cos bx \cos nx\, dx = +\frac{\lambda\alpha_1 \sin \alpha_1\pi}{\alpha_1{}^2 - n^2} + \frac{\lambda\alpha_2 \sin \alpha_2\pi}{\alpha_2{}^2 - n^2}$$

(21) Definite Integrals Involving

$$f(x) = f\left(\frac{1}{x^m}, \sin^p ax, \cos^p bx\right)$$ $[0 \ \ \infty]$

---

m, n, p = positive integers $\neq 0$ ( )!! = double factorial (Sec. 1.03)

a, b, c = positive real numbers $\neq 0$ $\Gamma( )$ = gamma function (Sec. 13.03)

$$\int_0^\infty \frac{\sin(\pm ax)}{x}\, dx = \pm\frac{\pi}{2}$$

$$\int_0^\infty \frac{\tan(\pm ax)}{x}\, dx = \pm\frac{\pi}{2}$$

$$\int_0^\infty \frac{\sin^{2m} ax}{x^2}\, dx = \frac{(2m-3)!!}{(2m-2)!!}\frac{a\pi}{2} \qquad m > 1$$

$$\int_0^\infty \frac{\sin^{2m+1} ax}{x}\, dx = \frac{(2m-1)!!}{(2m)!!}\frac{\pi}{2} \qquad m > 0$$

$$\int_0^\infty \frac{\sin^2 ax}{x^2}\, dx = \frac{\pi a}{2}$$

$$\int_0^\infty \frac{\sin^{2m} ax}{x}\, dx = \int_0^\infty \frac{\cos^{2m} ax}{x}\, dx = \infty$$

$$\int_0^\infty \frac{\sin^4 ax}{x^2}\, dx = \frac{\pi a}{4}$$

$$\int_0^\infty \frac{\sin^3 ax}{x}\, dx = \frac{\pi}{4}$$

$$\int_0^\infty \frac{\sin^4 ax}{x^3}\, dx = a^2 \ln 2$$

$$\int_0^\infty \frac{\sin^3 ax}{x^2}\, dx = \frac{3}{4}a \ln 3$$

$$\int_0^\infty \frac{\sin ax}{\sqrt{x}}\, dx = \sqrt{\frac{\pi}{2a}}$$

$$\int_0^\infty \frac{\cos ax}{\sqrt{x}}\, dx = \sqrt{\frac{\pi}{2a}}$$

$$\int_0^\infty \frac{\sin ax}{\sqrt[b]{x}}\, dx = \frac{\pi\sqrt[b]{a}}{2a\Gamma(1/b)\sin(\pi/2b)}$$

$$\int_0^\infty \frac{\cos ax}{\sqrt[b]{x}}\, dx = \frac{\pi\sqrt[b]{a}}{2a\Gamma(1/b)\cos(\pi/2b)}$$

$$\int_0^\infty \frac{\sin ax \sin bx}{x}\, dx = \ln\sqrt{\frac{a+b}{a-b}}$$

$$\int_0^\infty \frac{\cos ax \cos bx}{x}\, dx = \infty$$

$$\int_0^\infty \frac{\sin ax \cos bx}{x}\, dx = \begin{cases} 0 & \text{if } b > a \geq 0 \\ \pi/2 & \text{if } a > b \geq 0 \\ \pi/4 & \text{if } a = b > 0 \end{cases}$$

$$\int_0^\infty \frac{\sin^2 ax}{b^2 + x^2}\, dx = \frac{\pi}{4b}(1 - e^{-2ab})$$

$$\int_0^\infty \frac{\cos^2 ax}{b^2 + x^2}\, dx = \frac{\pi}{4b}(1 + e^{-2ab})$$

$$\int_0^\infty \frac{\sin ax \sin bx}{c^2 + x^2}\, dx = \frac{\pi}{2c}e^{-ac}\sinh bc \qquad a \geq b$$

$$\int_0^\infty \frac{\cos ax \cos bx}{c^2 + x^2}\, dx = \frac{\pi}{2c}e^{-ac}\cosh bc \qquad a \geq b$$

$$\int_0^\infty \sin(x^{m+1})\, dx = \Gamma\left(1 + \frac{1}{m+1}\right)\sin\frac{\pi}{2(m+1)}$$

$$\int_0^\infty \cos(x^{m+1})\, dx = \Gamma\left(1 + \frac{1}{m+1}\right)\cos\frac{\pi}{2(m+1)}$$

(22) Definite Integrals Involving

$$f(x) = f(e^{-ax}, \sin^p ax, \cos^p bx) \quad [0, \infty]$$

For definition of a, b, c, m, n, p, and $\Gamma(\ )$ see opposite page.

$$\alpha = a^2 + (b + c)^2 \qquad \beta = a^2 + (b - c)^2 \qquad \gamma = a^2 + b^2 - c^2$$

$$\int_0^\infty e^{-ax} \sin bx \, dx = \frac{b}{a^2 + b^2}$$

$$\int_0^\infty e^{-ax} \cos bx \, dx = \frac{a}{a^2 + b^2}$$

$$\int_0^\infty xe^{-ax} \sin bx \, dx = \frac{2ab}{(a^2 + b^2)^2}$$

$$\int_0^\infty xe^{-ax} \cos bx \, dx = \frac{a^2 - b^2}{(a^2 + b^2)^2}$$

$$\int_0^\infty x^a e^{-bx} \sin cx \, dx = \frac{\Gamma(a+1) \sin \omega}{\sqrt{(b^2 + c^2)^{a+1}}}$$

$$\omega = (a+1) \tan^{-1} \frac{c}{b}$$

$$\int_0^\infty x^a e^{-bx} \cos cx \, dx = \frac{\Gamma(a+1) \cos \omega}{\sqrt{(b^2 + c^2)^{a+1}}}$$

$$\int_0^\infty e^{-ax} \sin (bx + c) \, dx = \frac{1}{a^2 + b^2} (a \sin c + b \cos c)$$

$$\int_0^\infty e^{-ax} \cos (bx + c) \, dx = \frac{1}{a^2 + b^2} (a \cos c - b \sin c)$$

$$\int_0^\infty e^{-ax} \sin^2 bx \, dx = \frac{2b^2}{a(a^2 + 4b^2)}$$

$$\int_0^\infty e^{-ax} \cos^2 bx \, dx = \frac{a^2 + 2b^2}{a(a^2 + 4b^2)}$$

$$\int_0^\infty \frac{e^{-ax}}{x} \sin bx \, dx = \tan^{-1} \frac{b}{a}$$

$$\int_0^\infty \frac{e^{-ax}}{x} \cos bx \, dx = \infty$$

$$\int_0^\infty \frac{e^{-ax}}{x} \sin^2 bx \, dx = \ln \sqrt[4]{\frac{a^2 + 4b^2}{a^2}}$$

$$\int_0^\infty \frac{e^{-ax}}{x^m} \cos^n bx \, dx = \infty$$

$$\int_0^\infty \frac{e^{-ax}}{x^2} \sin^2 bx \, dx = b \tan^{-1} \frac{2b}{a} - a \ln \sqrt[4]{\frac{a^2 + 4b^2}{a^2}}$$

$$\int_0^\infty \frac{e^{-ax} - e^{-bx}}{x} \sin cx \, dx = \tan^{-1} \frac{c(b-a)}{ab + c^2}$$

$$\int_0^\infty \frac{e^{-ax} - e^{-bx}}{x} \cos cx \, dx = \ln \sqrt{\frac{b^2 + c^2}{a^2 + c^2}}$$

$$\int_0^\infty \frac{e^{-ax} - e^{-bx}}{x^2} \sin cx \, dx = -a \tan^{-1} \frac{c}{a} + b \tan^{-1} \frac{c}{b} + c \ln \sqrt{\frac{b^2 + c^2}{a^2 + c^2}}$$

$$\int_0^\infty e^{-ax} \sin bx \sin cx \, dx = \frac{2abc}{\alpha\beta}$$

$$\int_0^\infty e^{-ax} \cos bx \cos cx \, dx = \frac{a(\alpha + \beta)}{2\alpha\beta}$$

$$\int_0^\infty e^{-ax} \sin bx \cos cx \, dx = \frac{b\gamma}{\alpha\beta}$$

$$\int_0^\infty \frac{e^{-ax}}{x} \sin bx \sin cx \, dx = \frac{1}{4} \ln \frac{\alpha}{\beta}$$

(23) Definite Integrals Involving

$$f(x) = f\left(\frac{1}{x}, \sqrt{1 \pm a^2 \sin^2 x}, \sqrt{1 \pm a^2 \cos^2 x}\right) \qquad [0, \infty]$$

Complete elliptic integral of the first kind (Sec. 13.05–2):

$$K(a) = f\left(a, \frac{\pi}{2}\right) = \int_0^{\pi/2} \frac{d\phi}{\sqrt{1 - a^2 \sin^2 \phi}} = \int_0^{\pi/2} \frac{d\phi}{\sqrt{b^2 + a^2 \cos^2 \phi}}$$

Complete elliptic integral of the second kind (Sec. 13.05–2):

$$E(a) = E\left(a, \frac{\pi}{2}\right) = \int_0^{\pi/2} \sqrt{1 - a^2 \sin^2 \phi}\, d\phi = \int_0^{\pi/2} \sqrt{b^2 + a^2 \cos^2 \phi}\, d\phi$$

Modulus: $k = a,\ 0 < a < 1$ $\qquad\qquad$ Complementary modulus: $k' = b = \sqrt{1 - a^2}$

$$\int_0^\infty \frac{\sin x}{x} \sqrt{1 - a^2 \sin^2 x}\, dx = E(a) \qquad\qquad \int_0^\infty \frac{\sin x}{x} \sqrt{1 - a^2 \cos^2 x}\, dx = E(a)$$

$$\int_0^\infty \frac{\sin x\, dx}{x\sqrt{1 - a^2 \sin^2 x}} = K(a) \qquad\qquad \int_0^\infty \frac{\sin x\, dx}{x\sqrt{1 - a^2 \cos^2 x}} = K(a)$$

$$\int_0^\infty \frac{\sin x \cos x\, dx}{x\sqrt{1 - a^2 \sin^2 x}} = \frac{E(a) - b^2 K(a)}{a^2} \qquad\qquad \int_0^\infty \frac{\sin x \cos x\, dx}{x\sqrt{1 - a^2 \cos^2 x}} = \frac{K(a) - E(a)}{a^2}$$

$$\int_0^\infty \frac{\sin x \cos^2 x\, dx}{x\sqrt{1 - a^2 \sin^2 x}} = \frac{E(a) - b^2 K(a)}{a^2} \qquad\qquad \int_0^\infty \frac{\sin x \cos^2 x\, dx}{x\sqrt{1 - a^2 \cos^2 x}} = \frac{K(a) - E(a)}{a^2}$$

$$\int_0^\infty \frac{\sin x\, dx}{x\sqrt{1 + \sin^2 x}} = \sqrt{\tfrac{1}{2}}\, K(\sqrt{\tfrac{1}{2}}) \qquad\qquad \int_0^\infty \frac{\sin x\, dx}{x\sqrt{1 + \cos^2 x}} = \sqrt{\tfrac{1}{2}}\, K(\sqrt{\tfrac{1}{2}})$$

$$\int_0^\infty \frac{\sin x \cos x\, dx}{x\sqrt{1 + \sin^2 x}} = \sqrt{2}\,[K(\sqrt{\tfrac{1}{2}}) - E(\sqrt{\tfrac{1}{2}})] \qquad\qquad \int_0^\infty \frac{\sin x \cos x\, dx}{x\sqrt{1 + \cos^2 x}} = \sqrt{2}\,[E(\sqrt{\tfrac{1}{2}}) - \tfrac{1}{2}K(\sqrt{\tfrac{1}{2}})]$$

$$\int_0^\infty \frac{\sin x \cos^2 x\, dx}{x\sqrt{1 + \sin^2 x}} = \sqrt{2}\,[K(\sqrt{\tfrac{1}{2}}) - E(\sqrt{\tfrac{1}{2}})] \qquad\qquad \int_0^\infty \frac{\sin x \cos^2 x\, dx}{x\sqrt{1 + \cos^2 x}} = \sqrt{2}\,[E(\sqrt{\tfrac{1}{2}}) - \tfrac{1}{2}K(\sqrt{\tfrac{1}{2}})$$

$$\int_0^\infty \frac{\sin^3 x\, dx}{x\sqrt{1 + \sin^2 x}} = \sqrt{\tfrac{1}{2}}\,[2E(\sqrt{\tfrac{1}{2}}) - K(\sqrt{\tfrac{1}{2}})] \qquad\qquad \int_0^\infty \frac{\sin^3 x\, dx}{x\sqrt{1 + \cos^2 x}} = \sqrt{\tfrac{1}{2}}\,[K(\sqrt{\tfrac{1}{2}}) - E(\sqrt{\tfrac{1}{2}})]$$

$$\int_0^\infty \frac{\tan x}{x} \sqrt{1 - a^2 \sin^2 x}\, dx = E(a) \qquad\qquad \int_0^\infty \frac{\tan x}{x} \sqrt{1 - a^2 \cos^2 x}\, dx = E(a)$$

$$\int_0^\infty \frac{\tan x\, dx}{x\sqrt{1 - a^2 \sin^2 x}} = K(a) \qquad\qquad \int_0^\infty \frac{\tan x\, dx}{x\sqrt{1 - a^2 \cos^2 x}} = K(a)$$

$$\int_0^\infty \frac{\tan x \sin^2 x\, dx}{x\sqrt{1 - a^2 \sin^2 x}} = \frac{K(a) - E(a)}{a^2} \qquad\qquad \int_0^\infty \frac{\tan x \sin^2 x\, dx}{x\sqrt{1 - a^2 \cos^2 x}} = \frac{E(a) - b^2 K(a)}{a^2}$$

$$\int_0^\infty \frac{\tan x\, dx}{x\sqrt{1 + \sin^2 x}} = \sqrt{\tfrac{1}{2}}\, K(\sqrt{\tfrac{1}{2}}) \qquad\qquad \int_0^\infty \frac{\tan x\, dx}{x\sqrt{1 + \cos^2 x}} = \sqrt{\tfrac{1}{2}}\, K(\sqrt{\tfrac{1}{2}})$$

(24) Definite Integrals Involving

$f(x) = f(\ln x)$	$[0, 1]$

m, n = positive integers $\neq 0$ $C = 0.577\ 215\ 665$ = Euler's constant (Sec. 13.01)
a, b = positive real numbers $\neq 0$ $\Gamma(\ )$ = gamma function (Sec. 13.03)
Exponential integral functions (Sec. 13.01):

$$\text{Ei}(x) = C + \ln x + \sum_{k=1}^{\infty} \frac{(-x)^k}{k(k)!} \qquad\qquad \bar{\text{Ei}}(x) = C + \ln x + \sum_{k=1}^{\infty} \frac{x^k}{k(k)!}$$

$$\int_0^1 \ln \frac{1}{x}\, dx = 1 \qquad\qquad \int_0^1 \left(\ln \frac{1}{x}\right)^a dx = \Gamma(a+1)$$

$$\int_0^1 x \ln \frac{1}{x}\, dx = \frac{1}{4} \qquad\qquad \int_0^1 x^a \left(\ln \frac{1}{x}\right)^b dx = \frac{\Gamma(b+1)}{(a+1)^{b+1}}$$

$$\int_0^1 \sqrt{\ln \frac{1}{x}}\, dx = \frac{\sqrt{\pi}}{2} \qquad\qquad \int_0^1 \frac{dx}{\sqrt{\ln(1/x)}} = \sqrt{\pi}$$

$$\int_0^1 (\ln x)^m\, dx = (-1)^m m! \qquad\qquad \int_0^1 x^m (\ln x)^n\, dx = \frac{(-1)^n n!}{(m+1)^{n+1}}$$

$$\int_0^1 \ln(1+x)\, dx = 2\ln 2 - 1 \qquad\qquad \int_0^1 \ln(1-x)\, dx = -1$$

$$\int_0^1 x \ln(1+x)\, dx = \tfrac{1}{4} \qquad\qquad \int_0^1 x \ln(1-x)\, dx = -\tfrac{3}{4}$$

$$\int_0^1 \frac{\ln x}{1+x}\, dx = -\frac{\pi^2}{12} \qquad\qquad \int_0^1 \frac{\ln x}{1-x}\, dx = -\frac{\pi^2}{6}$$

$$\int_0^1 \frac{\ln(1+x^a)}{x}\, dx = +\frac{\pi^2}{12a} \qquad\qquad \int_0^1 \frac{\ln(1-x^a)}{x}\, dx = -\frac{\pi^2}{6a}$$

$$\int_0^1 \frac{dx}{a+\ln x} = +e^{-a}\, \bar{\text{Ei}}(a) \qquad\qquad \int_0^1 \frac{dx}{a-\ln x} = -e^a\, \text{Ei}(a)$$

$$\int_0^1 \frac{dx}{(a+\ln x)^2} = e^{-a}\, \bar{\text{Ei}}(a) - \frac{1}{a} \qquad\qquad \int_0^1 \frac{dx}{(a-\ln x)^2} = e^a\, \text{Ei}(a) + \frac{1}{a}$$

$$\int_0^1 \frac{dx}{(a+\ln x)^m} = \frac{1}{(m-1)!}\left[\frac{1}{e^a}\bar{\text{Ei}}(a) - \frac{1}{a^m}\sum_{k=1}^{m-1} (m-k-1)!\, a^k\right]$$

$$\int_0^1 \frac{dx}{(a-\ln x)^m} = \frac{(-1)^m}{(m-1)!}\left[e^a\, \text{Ei}(a) - \frac{1}{a^m}\sum_{k=1}^{m-1} (m-k-1)!\, (-a)^k\right]$$

$$\int_0^1 \ln x[\ln(1+x)]\, dx = 2 - \ln 4 - \frac{\pi^2}{12} \qquad\qquad \int_0^1 \ln x[\ln(1-x)]\, dx = 2 - \frac{\pi^2}{6}$$

(25) Definite Integrals Involving $\boxed{f(x) = f[\ln (\sin x), \ln (\cos x), \ldots]}$ $\boxed{\left[0, \dfrac{\pi}{k}\right]}$

---

m, k = positive integers $\neq 0$ $L = \dfrac{\pi \ln 2}{8}$ $M = \dfrac{\bar{Z}(2)}{2} = 0.457\ 982\ 797$

$\bar{Z}(m) = \displaystyle\sum_{k=1}^{\infty} \dfrac{(-1)^{k+1}}{(2k-1)^m}$ = complementary zeta function (Sec. A.06)

$\displaystyle\int_0^{\pi/4} \ln (\sin x)\ dx = -2L - M$ $\displaystyle\int_0^{\pi/4} \ln (\cos x)\ dx = -2L + M$

$\displaystyle\int_0^{\pi/4} \ln (\sin x + \cos x)\ dx = -(L - M)$ $\displaystyle\int_0^{\pi/4} \ln (\cos x - \sin x)\ dx = -(L + M)$

$\displaystyle\int_0^{\pi/4} \ln (\tan x)\ dx = -2M$ $\displaystyle\int_0^{\pi/4} \ln (\cot x)\ dx = +2M$

$\displaystyle\int_0^{\pi/4} \ln (1 + \tan x)\ dx = L$ $\displaystyle\int_0^{\pi/4} \ln (\cot x + 1)\ dx = L + 2M$

$\displaystyle\int_0^{\pi/4} \ln (1 - \tan x)\ dx = L - 2M$ $\displaystyle\int_0^{\pi/4} \ln (\cot x - 1)\ dx = L$

$\displaystyle\int_0^{\pi/4} \ln (\tan x + \cot x)\ dx = 4L$ $\displaystyle\int_0^{\pi/4} \ln (\cot x - \tan x)\ dx = 2L$

$\displaystyle\int_0^{\pi/4} [\ln (\tan x)]^2\ dx = \dfrac{\pi^3}{16}$ $\displaystyle\int_0^{\pi/4} [\ln (\tan x)]^m\ dx = m!\,(-1)^m\,\bar{Z}(m+1)$

$\displaystyle\int_0^{\pi/2} \ln (\sin x)\ dx = \int_0^{\pi/2} \ln (\cos x)\ dx = -4L$

$\displaystyle\int_0^{\pi/2} [\ln (\sin x)]^2\ dx = \int_0^{\pi/2} [\ln (\cos x)]^2\ dx = \dfrac{\pi}{2}\left[(\ln 2)^2 + \dfrac{\pi^2}{12}\right]$

$\displaystyle\int_0^{\pi/2} [\ln (\sin x)]\sin x\ dx = \int_0^{\pi/2} [\ln (\cos x)]\cos x\ dx = \ln 2 - 1$

$\displaystyle\int_0^{\pi/2} [\ln (\sin x)]\cos x\ dx = \int_0^{\pi/2} [\ln (\cos x)]\sin x\ dx = -1$

$\displaystyle\int_0^{\pi/2} \ln (1 \pm \sin x)\ dx = \int_0^{\pi/2} \ln (1 \pm \cos x)\ dx = -4(L \mp M)$

$\displaystyle\int_0^{\pi} \ln (\sin x)\ dx = -8L$ $\displaystyle\int_0^{\pi} x \ln (\sin x)\ dx = -4\pi L$

$\displaystyle\int_0^{\pi} \ln (1 \pm \sin x)\ dx = -8(L \mp M)$ $\displaystyle\int_0^{\pi} \ln (1 \pm \cos x)\ dx = -8L$

(26) Definite Integrals Involving

$$f(x) = f[\ln (x, e^{-ax}, \sin bx, \cos cx) \qquad [0, \infty]$$

a, b, c, λ = positive real numbers $\neq 0$ For definition of L and M see opposite page

$C = 0.577\ 215\ 665$ = Euler's constant (Sec. 13.01)

$$\int_0^\infty \frac{\ln x}{1+x^2}\, dx = 0$$

$$\int_0^\infty \frac{\ln x}{1-x^2}\, dx = -\frac{\pi^2}{4}$$

$$\int_0^\infty \frac{\ln x}{a^2+x^2}\, dx = \frac{\pi}{2a} \ln a$$

$$\int_0^\infty \frac{\ln x\, dx}{(a+x)(b+x)} = \frac{(\ln a)^2 - (\ln b)^2}{2(a-b)} \qquad a \neq b$$

$$\int_0^\infty \frac{\ln x}{a^2+b^2x^2}\, dx = \frac{\pi}{2ab} \ln \frac{a}{b}$$

$$\int_0^\infty \frac{\ln x}{a^2-b^2x^2}\, dx = -\frac{\pi^2}{4ab}$$

$$\int_0^\infty \frac{\ln (x+1)}{1+x^2}\, dx = 2(L+M)$$

$$\int_0^\infty \frac{\ln (x-1)}{1+x^2}\, dx = L$$

$$\int_0^\infty \ln \frac{a^2+x^2}{b^2+x^2}\, dx = (a-b)\pi$$

$$\int_0^\infty \ln \frac{a^2-x^2}{b^2-x^2}\, dx = (a+b)\pi$$

$$\int_0^\infty \frac{\ln (a^2+b^2x^2)}{c^2+x^2}\, dx = \frac{\pi}{c} \ln (ac+b)$$

$$\int_0^\infty \frac{\ln (a^2+b^2x^2)}{c^2-x^2}\, dx = -\frac{\pi}{c} \tan^{-1}\frac{bc}{a}$$

$$\int_0^\infty \frac{\ln (\sin ax)}{b^2+x^2}\, dx = \frac{\pi}{2b} \ln \frac{\sinh ab}{e^{ab}}$$

$$\int_0^\infty \frac{\ln (\sin ax)}{b^2-x^2}\, dx = -\frac{\pi^2}{4b} + \frac{a\pi}{2}$$

$$\int_0^\infty \frac{\ln (\cos ax)}{b^2+x^2}\, dx = \frac{\pi}{2b} \ln \frac{\cosh ab}{e^{ab}}$$

$$\int_0^\infty \frac{\ln (\cos x)}{b^2-x^2}\, dx = \frac{a\pi}{2}$$

$$\int_0^\infty \ln (1+e^{-x}) = \frac{\pi^2}{12}$$

$$\int_0^\infty \ln (1-e^{-x})\, dx = -\frac{\pi^2}{6}$$

$$\int_0^\infty e^{-ax} \ln x\, dx = \frac{-1}{a}(C+\ln a)$$

$$\int_0^\infty \frac{e^{-ax}}{\ln x}\, dx = 0$$

$$\int_0^\infty e^{-ax} \ln \frac{1}{x}\, dx = \frac{1}{a}(C+\ln a)$$

$$\int_0^\infty xe^{-ax} \ln \frac{1}{x}\, dx = \frac{2}{a^2}(C+\ln a - 1)$$

$$\int_0^\infty e^{-ax}(\ln x) \sin bx\, dx = \frac{b}{a^2+b^2}\left[\ln \sqrt{a^2+b^2} + \frac{a}{b} \tan^{-1}\frac{b}{a} - C\right]$$

$$\int_0^\infty e^{-ax}(\ln x) \cos bx\, dx = \frac{-a}{a^2+b^2}\left[\ln\sqrt{a^2+b^2} + \frac{b}{a} \tan^{-1}\frac{b}{a} + C\right]$$

$$\int_0^\infty \frac{\ln (a^2 \sin^2 \lambda x + b^2 \cos^2 \lambda x)\, dx}{c^2+x^2} = \frac{\pi}{c} [\ln (a \sinh \lambda c + b \cosh \lambda c) - \lambda c]$$

(27) Definite Integrals Involving

| $f(x) = f(\sinh x, \cosh x)$ | $[0, \infty]$ |

---

m = positive integer $\neq 0$ a, b, c = positive real numbers $\neq 0$

$\alpha = \dfrac{a\pi}{2c}$ $A = \cosh 2\alpha + \cosh 2\beta$ $Z(\ ) = $ zeta function (Sec. A.06)

$\beta = \dfrac{b\pi}{2c}$ $B = \cos 2\alpha + \cosh 2\beta$ $\bar{Z}(\ ) = $ complementary zeta function (Sec. A.06)

$$\int_0^\infty \frac{dx}{\sinh ax}\, dx = \infty$$

$$\int_0^\infty \frac{x\, dx}{\sinh ax}\, dx = \left(\frac{\pi}{2a}\right)^2$$

$$\int_0^\infty \frac{x^m}{\sinh bx}\, dx = \frac{2^{m+1}-1}{a(2a)^m}\, m!\, Z(m+1)$$

$$\int_0^\infty \frac{dx}{\cosh ax} = \frac{\pi}{2a}$$

$$\int_0^\infty \frac{x\, dx}{\cosh ax} = \frac{1.831\,329\,803\,\cdots}{a^2}$$

$$\int_0^\infty \frac{x^m}{\cosh ax}\, dx = \frac{2^{m+1}}{a(2a)^m}\, m!\, \bar{Z}(m+1)$$

$$\int_0^\infty \frac{dx}{a+\sinh bx} = \frac{1}{b\sqrt{1+a^2}} \ln\frac{1+a+\sqrt{1+a^2}}{1+a-\sqrt{1+a^2}} \qquad (1+a) > \sqrt{1+a^2}$$

$$\int_0^\infty \frac{dx}{a+\cosh bx} = \frac{1}{b\sqrt{a^2-1}} \ln\frac{1+a+\sqrt{a^2-1}}{1+a-\sqrt{a^2-1}} \qquad a^2 > 1$$

$$\int_0^\infty \frac{dx}{a\sinh cx+b\cosh cx} = \frac{1}{c\sqrt{a^2-b^2}} \ln\frac{a+b+\sqrt{a^2-b^2}}{a+b-\sqrt{a^2-b^2}} \qquad a^2 > b^2$$

$$\int_0^\infty \frac{\sin ax}{\sinh cx}\, dx = \frac{\alpha}{a}\tanh \alpha$$

$$\int_0^\infty \frac{\sinh ax}{\sinh cx}\, dx = \frac{\alpha}{a}\tan \alpha$$

$$\int_0^\infty \frac{x\sin ax}{\cosh cx}\, dx = \left(\frac{\alpha}{a}\right)^2 \frac{\tanh \alpha}{\cosh \alpha}$$

$$\int_0^\infty \frac{\cos ax}{\cosh cx}\, dx = \frac{\alpha}{a}\operatorname{sech} \alpha$$

$$\int_0^\infty \frac{\cosh ax}{\cosh cx}\, dx = \frac{\alpha}{a}\sec \alpha$$

$$\int_0^\infty \frac{x\cos ax}{\sinh cx}\, dx = \left(\frac{\alpha}{a}\right)^2 \operatorname{sech}^2 \alpha$$

$$\int_0^\infty \frac{\sin ax \sin bx}{\cosh cx}\, dx = \frac{\pi}{cA}\sinh \alpha \sinh \beta$$

$$\int_0^\infty \frac{\sin ax \cos bx}{\sinh cx}\, dx = \frac{\pi}{2cA}\sinh 2\alpha$$

$$\int_0^\infty \frac{\cos ax \cos bx}{\cosh cx}\, dx = \frac{\pi}{cA}\cosh \alpha \cosh \beta$$

$$\int_0^\infty \frac{\sinh ax \sin bx}{\cosh cx}\, dx = \frac{\pi}{cB}\sin \alpha \sinh \beta$$

$$\int_0^\infty \frac{\sinh ax \cos bx}{\sinh cx}\, dx = \frac{\pi}{2cB}\sin 2\alpha$$

$$\int_0^\infty \frac{\cosh ax \cos bx}{\cosh cx}\, dx = \frac{\pi}{cB}\cos \alpha \cosh \beta$$

$$\int_0^\infty e^{-ax} \sinh bx\, dx = \frac{b}{a^2-b^2}$$

$$\int_0^\infty e^{-ax} \cosh bx\, dx = \frac{a}{a^2-b^2}$$

21
PLANE CURVES
AND AREAS

(1) Equations of a Plane Curve

A plane curve is defined analytically in one of the following forms:

Form	Cartesian coordinates		Polar coordinates	
Explicit	$y = f(x)$ or	$x = g(y)$	$r = f(\theta)$ or	$\theta = g(r)$
Implicit	$F(x, y) = 0$		$F(r, \theta) = 0$	
Parametric	$x = x(t)$	$y = y(t)$	$r = r(t)$	$\theta = \theta(t)$

where x, y are the *cartesian coordinates*, r, θ are the *polar coordinates*, and t is the *parameter of variation*. The basic terms used in the analysis of plane curves are defined below.

(2) Derivatives

$$y' = d[f(x)]/dx \qquad x' = d[g(y)]/dy \qquad r' = d[f(\theta)]/d\theta \qquad \theta' = d[g(r)]/dr$$

$$F_x = d[F(x, y)]/dx \qquad F_y = d[F(x, y)]/dy \qquad F_r = d[F(r, \theta)]/d\theta \qquad F_\theta = d[F(r, \theta)]/dr$$

$$\dot{x} = d[x(t)]/dt \qquad \dot{y} = d[y(t)]/dt \qquad \dot{r} = d[r(t)]/dt \qquad \dot{\theta} = d[\theta(t)]/dt$$

$$y'' = d^2[f(x)]/dx^2 \qquad x'' = d^2[g(y)]/dy^2 \qquad r'' = d^2[f(\theta)]/d\theta^2 \qquad \theta'' = d^2[g(r)]/dr^2$$

$$F_{xx} = d^2[F(x, y)]/dx^2 \qquad F_{yy} = d^2[F(x, y)]/dy^2 \qquad F_{\theta\theta} = d^2[F(r, \theta)]/d\theta^2 \qquad F_{rr} = d^2[F(r, \theta)]/dr^2$$

$$F_{xy} = d^2[F(x, y)]/dx\,dy \qquad\qquad F_{\theta r} = d^2[F(r, \theta)]/d\theta\,dr$$

$$\ddot{x} = d^2[x(t)]/dt^2 \qquad \ddot{y} = d^2[y(t)]/dt^2 \qquad \ddot{r} = d^2[r(t)]/dt^2 \qquad \ddot{\theta} = d^2[\theta(t)]/dt^2$$

(3) Differential Element

$$ds = \sqrt{1 + (y')^2}\, dx = \sqrt{1 + (x')^2}\, dy$$

$$= \sqrt{r^2 + (r')^2}\, d\theta = \sqrt{1 + (r\theta')^2}\, dr$$

$$= \sqrt{1 + \left(\frac{F_x}{F_y}\right)^2}\, dx = \sqrt{1 + \left(\frac{F_r}{F_\theta}\right)^2}\, d\theta$$

$$= \sqrt{(\dot{x})^2 + (\dot{y})^2}\, dt = \sqrt{(\dot{r})^2 + (r\dot{\theta})^2}\, dt$$

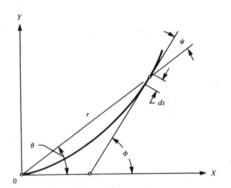

(4) Direction Functions

$\tan \phi$	$\cos \phi$	$\sin \phi$	$\tan \psi$	$\cos \psi$	$\sin \psi$
$\dfrac{dy}{dx}$	$\dfrac{dx}{ds}$	$\dfrac{dy}{ds}$	$\dfrac{r}{r'}$	$\dfrac{dr}{ds}$	$\dfrac{r\,d\theta}{ds}$
$\dfrac{-F_x}{F_y}$	$\dfrac{F_y}{\sqrt{F_x^2 + F_y^2}}$	$\dfrac{-F_x}{\sqrt{F_x^2 + F_y^2}}$	$\dfrac{-rF_r}{F_\theta}$	$\dfrac{r'F_\theta}{\sqrt{F_r^2 + F_\theta^2}}$	$\dfrac{rF_\theta}{\sqrt{F_r^2 + F_\theta^2}}$
$\dfrac{\dot{y}}{\dot{x}}$	$\dfrac{\dot{x}}{\sqrt{(\dot{x})^2 + (\dot{y})^2}}$	$\dfrac{\dot{y}}{\sqrt{(\dot{x})^2 + (\dot{y})^2}}$	$\dfrac{r\dot{\theta}}{\dot{r}}$	$\dfrac{\dot{r}}{\sqrt{(\dot{r})^2 + (r\dot{\theta})^2}}$	$\dfrac{r\dot{\theta}}{\sqrt{(\dot{r})^2 + (r\dot{\theta})^2}}$

ϕ = direction angle of tangent
ψ = angle between tangent and radius vector.

(1) Characteristics of a Function

(a) The rates of change of $f(x)$ at $x = x$ are

$f'(x) > 0$; $f(x)$ is increasing $f'(x) < 0$; $f(x)$ is decreasing

$f'(x) = 0$; $f(x)$ has a tangent parallel to the X axis

(b) The curvature of $f(x)$ at $x = x$ is as follows:

$f''(x) > 0$; $f(x)$ is convex $f''(x) < 0$; $f(x)$ is concave

$f''(x) = 0$ and changes in sign; $f(x)$ has an inflection point

$f''(x) = 0$ and does not change in sign; $f(x)$ has a flat point

(c) The necessary condition for a maximum or a minimum of $f(x)$ at $x = a$ is

$$f'(a) = 0$$

(d) Sufficient conditions for extrema of $f(x)$ at $x = a$ are

$f'(a) = 0$	$f''(a) < 0$	(maximum)
$f'(a) = 0$	$f''(a) > 0$	(minimum)

(e) Necessary and sufficient conditions for extrema of $f(x)$ at $x = a$ are if $f'(a) = f''(a) = \cdots = f^{(n-1)}(a) = 0$ and $f^{(n)}(a) \neq 0$, then

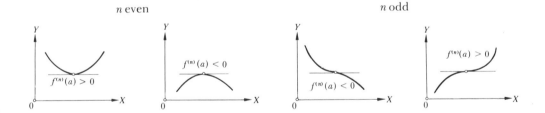

n even n odd

(2) Shape of a Curve

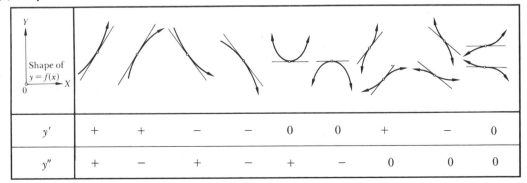

Shape of $y = f(x)$									
y'	+	+	−	−	0	0	+	−	0
y''	+	−	+	−	+	−	0	0	0

(1) Notation

S = point of contact   K = center of curvature

x, y = coordinates of contact point   X, Y = coordinates of running point

$\overline{AS}$ = tangent $\overline{DS}$ = normal

$\overline{AB}$ = subtangent $\overline{DB}$ = subnormal

ρ = radius of curvature

For definition of y', $\dot{x}$, $\dot{y}$, y'', $\ddot{x}$, $\ddot{y}$, F_x, F_y, F_{xx}, F_{xy}, F_{yy}, ds refer to Sec. 21.01.

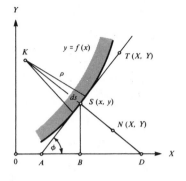

(2) Equivalents

$$P = F_x \begin{vmatrix} F_{xy} & F_x \\ F_{yy} & F_y \end{vmatrix} - F_y \begin{vmatrix} F_{xx} & F_x \\ F_{xy} & F_y \end{vmatrix} \qquad Q = \begin{vmatrix} \dot{x} & \dot{y} \\ \ddot{x} & \ddot{y} \end{vmatrix}$$

(3) Equations of Geometry

Curve		$y = f(x)$	$F(x, y) = 0$	$x = x(t) \quad y = y(t)$
Slope, $\tan \phi$		$y' = \dfrac{dy}{dx}$	$y' = -\dfrac{F_x}{F_y}$	$\dfrac{\dot{y}}{\dot{x}}$
Length ds		$\sqrt{1 + (y')^2}\, dx$	$\sqrt{1 + \left(\dfrac{F_x}{F_y}\right)^2}\, dx$	$\sqrt{(\dot{x})^2 + (\dot{y})^2}\, dt$
Tangent at S		$Y - y = y'(X - x)$	$F_x(X - x) = -F_y(Y - y)$	$(Y - y)\dot{x} = (X - x)\dot{y}$
Normal at S		$X - x = -y'(Y - y)$	$F_x(Y - y) = F_y(X - x)$	$(X - x)\dot{x} = -(Y - y)\dot{y}$
Length $\overline{AS}$		$\left\| \dfrac{y}{y'} \sqrt{1 + (y')^2} \right\|$	$\left\| \dfrac{y}{F_x} \sqrt{F_x^2 + F_y^2} \right\|$	$\left\| \dfrac{y\sqrt{(\dot{x})^2 + (\dot{y})^2}}{\dot{y}} \right\|$
Length $\overline{AB}$		$\left\| \dfrac{y}{y'} \right\|$	$\left\| \dfrac{yF_y}{F_x} \right\|$	$\left\| \dfrac{y\dot{x}}{\dot{y}} \right\|$
Length $\overline{DS}$		$\left\| y\sqrt{1 + (y')^2} \right\|$	$\left\| \dfrac{y}{F_y} \sqrt{F_x^2 + F_y^2} \right\|$	$\left\| \dfrac{y\sqrt{(\dot{x})^2 + (\dot{y})^2}}{\dot{x}} \right\|$
Length $\overline{DB}$		$\|yy'\|$	$\left\| \dfrac{yF_x}{F_y} \right\|$	$\left\| \dfrac{y\dot{y}}{\dot{x}} \right\|$
Radius of curvature	ρ	$\dfrac{[\sqrt{1 + (y')^2}]^3}{y''}$	$\dfrac{[\sqrt{F_x^2 + F_y^2}]^3}{P}$	$\dfrac{[\sqrt{(\dot{x})^2 + (\dot{y})^2}]^3}{Q}$
Coordinates of center of curvature K	x_K	$x - \dfrac{y'[1 + (y')^2]}{y''}$	$x + F_x\dfrac{F_x^2 + F_y^2}{P}$	$x - \dot{y}\dfrac{(\dot{x})^2 + (\dot{y})^2}{Q}$
	y_K	$y + \dfrac{1 + (y')^2}{y''}$	$y + F_y\dfrac{F_x^2 + F_y^2}{P}$	$y + \dot{x}\dfrac{(\dot{x})^2 + (\dot{y})^2}{Q}$

(1) Notation

S = point of contact $\qquad$ K = center of curvature

r, θ = coordinates of contact point $\qquad$ R, τ = coordinates of running point

$\overline{AS}$ = tangent $\qquad$ $\overline{DS}$ = normal

$\overline{A0}$ = subtangent $\qquad$ $\overline{D0}$ = subnormal

ρ = radius of curvature

For definition of r', θ', r'', θ'', . . . , ds refer to Sec. 21.01.

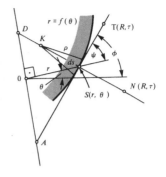

(2) Equivalents

$$u = R \cos \tau - r \cos \theta \qquad\qquad v = R \sin \tau + r \sin \theta$$

$$K_1 = \frac{r' \sin \theta + r \cos \theta}{r' \cos \theta - r \sin \theta} \qquad\qquad K_2 = \frac{\dot{r} \sin \theta + r\dot{\theta} \cos \theta}{\dot{r} \cos \theta - r\dot{\theta} \sin \theta}$$

$$L_1 = \sqrt{r^2 + (r')^2} \qquad\qquad L_2 = \sqrt{(\dot{r})^2 + (r\dot{\theta})^2}$$

(3) Equations of Geometry

Curve		$r = f(\theta)$	$r = r(t) \qquad \theta = \theta(t)$
Slope		$\tan \phi = K_1 \qquad \tan \psi = \dfrac{r}{r'}$	$\tan \phi = K_2 \qquad \tan \psi = \dfrac{r\dot{\theta}}{\dot{r}}$
Length ds		$L_1\, d\theta$	$L_2\, dt$
Tangent at S		$v = K_1 u$	$v = K_2 u$
Normal at S		$u = -K_1 v$	$u = -K_2 v$
Length $\overline{AS}$		$\left\| \dfrac{r}{r'} L_1 \right\|$	$\left\| \dfrac{r}{\dot{r}} L_2 \right\|$
Length $\overline{A0}$		$\left\| \dfrac{r^2}{r'} \right\|$	$\left\| \dfrac{r^2 \dot{\theta}}{\dot{r}} \right\|$
Length $\overline{DS}$		$\|L_1\|$	$\left\| \dfrac{1}{\dot{\theta}} L_2 \right\|$
Length $\overline{D0}$		$\|r'\|$	$\left\| \dfrac{\dot{r}}{\dot{\theta}} \right\|$
Radius of curvature	ρ	$\dfrac{L_1{}^3}{r^2 + 2rr' - rr''}$	$\dfrac{L_2{}^3 (\dot{\theta})^2}{r^2(\dot{\theta})^2 + 2r\dot{r}(\dot{\theta}) - r\ddot{\theta}}$
Coordinates of center of curvature K	x_K	$r \cos \theta - \rho\, \dfrac{K_1}{\sqrt{1 + K_1{}^2}}$	$r \cos \theta - \rho\, \dfrac{K_2}{\sqrt{1 + K_2{}^2}}$
	y_K	$r \sin \theta + \rho\, \dfrac{1}{\sqrt{1 + K_1{}^2}}$	$r \sin \theta + \rho\, \dfrac{1}{\sqrt{1 + K_2{}^2}}$

(1) Segment AB

The geometry of a plane curve is defined in the forms introduced in Sec. 21.01-1. The coordinates of the end points of the segment AB of this curve are x_A, y_A, x_B, y_B, r_A, θ_A, r_B, θ_B and t_A, t_B, respectively.

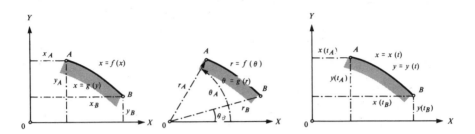

The static functions of the segment AB shown in the three coordinate systems above are defined analytically below.

(2) Length of Segment AB

$$L = \int_{x_A}^{x_B} \sqrt{1 + [f'(x)]^2}\, dx = \int_{y_A}^{y_B} \sqrt{1 + [g'(y)]^2}\, dy = \int_{t_A}^{t_B} \sqrt{[\dot{x}(t)]^2 + [\dot{y}(t)]^2}\, dt$$

$$= \int_{\theta_B}^{\theta_A} \sqrt{r^2 + [f'(\theta)]^2}\, d\theta = \int_{r_B}^{r_A} \sqrt{1 + r^2[g'(r)]^2}\, dr$$

(3) Static Moments of Segment AB about X, Y

$$M_x = \int_{x_A}^{x_B} f(x)\sqrt{1 + [f'(x)]^2}\, dx = \int_{y_A}^{y_B} y\sqrt{1 + [g'(y)]^2}\, dy = \int_{t_A}^{t_B} y(t)\sqrt{[\dot{x}(t)]^2 + [\dot{y}(t)]^2}\, dt$$

$$= \int_{\theta_B}^{\theta_A} [f(\theta)] \sin\theta \sqrt{r^2 + [f'(\theta)]^2}\, d\theta = \int_{r_B}^{r_A} r \sin[g(r)]\sqrt{1 + r^2[g'(r)]^2}\, dr$$

$$M_y = \int_{x_A}^{x_B} x\sqrt{1 + [f'(x)]^2}\, dx = \int_{y_A}^{y_B} g(y)\sqrt{1 + [g'(y)]^2}\, dy = \int_{t_A}^{t_B} x(t)\sqrt{[\dot{x}(t)]^2 + [\dot{y}(t)]^2}\, dt$$

$$= \int_{\theta_B}^{\theta_A} [f(\theta)] (\cos\theta)\sqrt{r^2 + [f'(\theta)]^2}\, d\theta = \int_{r_B}^{r_A} r \cos[g(r)]\sqrt{1 + r^2[g'(r)]^2}\, dr$$

(4) Coordinates of Centroid C of Segment AB

$$x_C = \frac{M_y}{L} \qquad y_C = \frac{M_x}{L}$$

where L and M_x, M_y are given analytically in (2) and (3) above.

(5) Static Moments of Segment AB about U, V

$$M_u = M_x \cos\omega + M_y \sin\omega$$

$$M_v = -M_x \sin\omega + M_y \cos\omega$$

where M_x, M_y are given analytically in (3) above and ω is the angle of rotation.

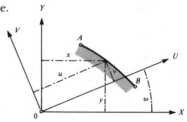

(1) Segment AB

All functions tabulated below are related to the plane segment shown in the three coordinate systems in Sec. 21.05-1.

(2) Moments of Inertia of Segment AB about X, Y

$$I_{xx} = \int_{x_A}^{x_B} [f(x)]^2 \sqrt{1 + [f'(x)]^2}\, dx = \int_{y_A}^{y_B} y^2 \sqrt{1 + [g'(y)]^2}\, dy = \int_{t_A}^{t_B} [y(t)]^2 \sqrt{[\dot{x}(t)]^2 + [\dot{y}(t)]^2}\, dt$$

$$= \int_{\theta_B}^{\theta_A} [f(\theta)]^2 \sin^2 \theta \sqrt{r^2 + [f'(\theta)]^2}\, d\theta = \int_{r_B}^{r_A} r^2 \sin^2 [g(r)] \sqrt{1 + r^2[g'(r)]^2}\, dr$$

$$I_{yy} = \int_{x_A}^{x_B} x^2 \sqrt{1 + [f'(x)]^2}\, dx = \int_{y_A}^{y_B} [g(y)]^2 \sqrt{1 + [g'(y)]^2}\, dy = \int_{t_A}^{t_B} [x(t)]^2 \sqrt{[\dot{x}(t)]^2 + [\dot{y}(t)]^2}\, dt$$

$$= \int_{\theta_B}^{\theta_A} [f(\theta)]^2 \cos^2 \theta \sqrt{r^2 + [f'(\theta)]^2}\, d\theta = \int_{r_B}^{r_A} r^2 \cos^2 [g(r)] \sqrt{1 + r^2[g'(r)]^2}\, dr$$

(3) Products of Inertia of Segment AB in X, Y

$$I_{xy} = \int_{x_A}^{x_B} xf(x) \sqrt{1 + [f'(x)]^2}\, dx = \int_{y_A}^{y_B} yg(y) \sqrt{1 + [g'(y)]^2}\, dy = \int_{t_B}^{t_A} x(t)y(t) \sqrt{[\dot{x}(t)]^2 + [\dot{y}(t)]^2}\, dt$$

$$= \int_{\theta_B}^{\theta_A} [f(\theta)]^2 \sin \theta \cos \theta \sqrt{r^2 + [f'(\theta)]^2}\, d\theta$$

$$= \int_{r_B}^{r_A} r^2 \sin [g(r)] \cos [g(r)] \sqrt{1 + r^2[g'(r)]^2}\, dr$$

(4) Moments and Product of Inertia of Segment AB in U, V

In rotated axes U, V (Sec. 21.05-5),

$$I_{uu} = \tfrac{1}{2}(I_{xx} + I_{yy}) + \tfrac{1}{2}(I_{xx} - I_{yy}) \cos 2\omega - I_{xy} \sin 2\omega$$

$$I_{vv} = \tfrac{1}{2}(I_{xx} + I_{yy}) - \tfrac{1}{2}(I_{xx} - I_{yy}) \cos 2\omega + I_{xy} \sin 2\omega$$

$$I_{uv} = \tfrac{1}{2}(I_{xx} - I_{yy}) \sin 2\omega + I_{xy} \cos 2\omega$$

where I_{xx}, I_{yy}, I_{xy} are the inertia functions given in (2) and (3) above and ω is the angle of rotation.

(5) Principal Moments of Inertia of Segment AB

If

$$\omega = -\tfrac{1}{2} \tan^{-1} \frac{I_{xy}}{I_{xx} - I_{yy}}$$

the moments of inertia in (4) above become the principal moments of inertia and the product of inertia becomes zero.

(6) Inertia Functions of Segment AB in Parallel Axes

$$I_{xx} = I_{Cxx} + y_C^2 L \qquad\qquad I_{yy} = I_{Cyy} + x_C^2 L$$

$$I_{xy} = I_{Cxy} + x_C y_C L$$

where I_{xx}, I_{yy}, I_{xy} are the inertia functions in X, Y, x_C, y_C are the coordinates of the centroid, $I_{Cxx}, I_{Cyy}, I_{Cxy}$ are the inertia functions in the centroidal axes, and L is the length of the segment.

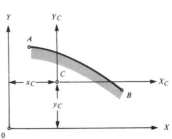

(1) General Equation

The general equation of the second degree,

$$a_{11}x^2 + 2a_{12}xy + a_{22}y^2 + 2a_{13}x + 2a_{23}y + a_{33} = 0$$

defines the following conic sections: a circle, an ellipse, a hyperbola, a parabola, a pair of straight lines, a straight line, or a point (the last three cases are degenerate curves of the second degree).

(2) Invariants

The quantities (note that $a_{ik} = a_{ki}$)

$$I_3 = \begin{vmatrix} a_{11} & a_{12} & a_{13} \\ a_{21} & a_{22} & a_{23} \\ a_{31} & a_{32} & a_{33} \end{vmatrix} \qquad I_2 = \begin{vmatrix} a_{11} & a_{12} \\ a_{21} & a_{22} \end{vmatrix} \qquad I_1 = a_{11} + a_{22}$$

and the sign of the quantity

$$A = \begin{vmatrix} a_{22} & a_{23} \\ a_{32} & a_{33} \end{vmatrix} + \begin{vmatrix} a_{11} & a_{13} \\ a_{31} & a_{33} \end{vmatrix}$$

are invariants of transformation (they are not affected by the translation or the rotation of coordinate axes) and define properties of conics.

(3) Principal Axis

The direction angle of the principal axis of a real ellipse, hyperbola, and parabola is

$$\omega = \frac{1}{2}\tan^{-1}\frac{2a_{12}}{a_{11} - a_{22}} \qquad \leq \frac{\pi}{2}$$

(4) Classification (CS = conic section)

Type	$I_2 \neq 0$		$I_2 = 0$		
	Central CS		Noncentral CS		
	$I_2 > 0$	$I_2 < 0$			
Proper $I_3 \neq 0$	Real ellipse $I_1I_3 < 0$	Hyperbola	Parabola		
	Imaginary ellipse $I_1I_3 > 0$				
Improper $I_3 = 0$	Two nonparallel straight lines		Two parallel straight lines		One straight line
	Imaginary	Real	Real $A < 0$	Imaginary $A > 0$	Real $A = 0$

a_{ij} = constants in X, Y axes (Sec. 21.07) b_{ij} = constants in U, V axes (Sec. 21.08)

$i, j = 1, 2, 3$ ω = position angle of $\bar{X}$ axis

I_1, I_2, I_3 = invariants (Sec. 21.07) a, b = semiaxes in U, V axes

(1) Central Conic Section $(I_2 \neq 0)$

(a) Coordinates of center *M*

$$x_M = -\frac{a_{13}a_{22} - a_{12}a_{23}}{I_2} \qquad y_M = -\frac{a_{11}a_{23} - a_{12}a_{13}}{I_2}$$

(b) Equation of CS in *U, V* axes

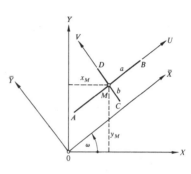

$$\boxed{b_{11}u^2 + b_{22}v^2 + b_{33} = 0}$$

$$b_{11}, b_{22} = \frac{I_1}{2} \pm \sqrt{\left(\frac{I_1}{2}\right)^2 - I_2} \qquad b_{33} = \frac{I_3}{I_2}$$

(c) Principal semiaxes

$$a^2 = \left|-\frac{b_{33}}{b_{11}}\right| \qquad b^2 = \left|-\frac{b_{33}}{b_{22}}\right|$$

For ellipse $\dfrac{b_{33}}{b_{11}} < 0$ $\dfrac{b_{33}}{b_{22}} < 0$. For hyperbola $\dfrac{b_{33}}{b_{11}} < 0$ $\dfrac{b_{33}}{b_{22}} > 0$

(2) Noncentral Conic Section $(I_2 = 0)$

(a) Equation of principal axis *U* or *V*

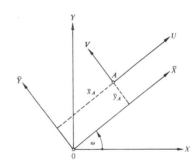

$$a_{11}x + a_{12}y + \frac{a_{11}a_{13} + a_{12}a_{23}}{I_1} = 0$$

which is tangent to the conic section at the vertex A.

(b) Equation of CS in *U, V* axes

$$\boxed{b_{11}u^2 + b_{23}v = 0 \qquad b_{22}v^2 + 2b_{13}u = 0}$$

$$b_{11} = I_1 \qquad b_{13} = a_{13}\sqrt{\frac{a_{11}}{I_1}} + a_{23}\sqrt{\frac{a_{22}}{I_1}}$$

$$b_{22} = I_1 \qquad b_{23} = a_{23}\sqrt{\frac{a_{11}}{I_1}} - a_{13}\sqrt{\frac{a_{22}}{I_1}}$$

(c) Coordinates of vertex *A* in $\bar{X}, \bar{Y}$ axes

	Axis of symmetry	
	$\bar{X}$ axis	$\bar{Y}$ axis
$\bar{x}_A$	$\dfrac{b_{23}}{b_{22}}$	$\dfrac{b_{13}}{b_{11}}$
$\bar{y}_A$	$\dfrac{a_{33}}{2b_{23}} - \dfrac{b_{23}{}^2}{2b_{22}b_{13}}$	$\dfrac{a_{33}}{2b_{23}} - \dfrac{b_{13}{}^2}{2b_{11}b_{23}}$

$2\phi_0$ = central angle, radians $\qquad$ α = angle at A, radians $\qquad$ a = radius

$L = 2\phi_0 a$ = segment length $\qquad$ h = segment height $\qquad$ $\alpha = \phi_0 = 2\tan^{-1}\dfrac{h}{c}$

$S_k = \sin\phi_k$ $\qquad\qquad\qquad$ $C_K = \cos\phi_k$ $\qquad\qquad\qquad$ $k = 0, 1, 2$

For definitions of $x_C, y_C, I_{xx}, I_{yy}, I_{xy}, M_x, M_y$ see Secs. 21.05–21.06.
For other properties of circular segment see Secs. 4.05–4.06.

(1) Segment AB in X_A, Y_A

$$M_{ax} = 2(S_0 - \phi_0 C_0)a^2 \qquad M_{Ay} = 2\phi_0 S_0^2 a^2$$

$$x_C = aS_0 \qquad\qquad y_C = \left(\frac{S_0}{\phi_0} - C_0\right)a$$

$$I_{Axx} = (\phi_0 - 3S_0 C_0 - 2\phi_0 C_0)a^3$$

$$I_{Ayy} = 2S_0(S_0 - \phi_0 C_0)a^3$$

$$I_{Axy} = (\phi_0 - S_0 C_0 + 2\phi_0 S_0^2)a^3$$

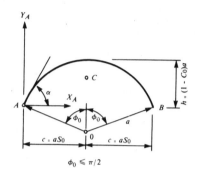

$$\phi_0 \leqslant \pi/2$$

(2) Segment AB in X_0, Y_0

$$M_{0x} = 2S_0 a^2 \qquad M_{0y} = 0$$

$$x_C = 0 \qquad\qquad y_C = \frac{aS_0}{\phi_0}$$

$$I_{0xx} = (\phi_0 + S_0 C_0)a^3$$

$$I_{0yy} = (\phi_0 - S_0 C_0)a^3$$

$$I_{0xy} = 0$$

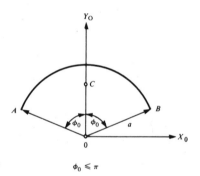

$$\phi_0 \leqslant \pi$$

(3) Segment AB in X_C, Y_C

$$M_{Cx} = 0 \qquad M_{Cy} = 0$$

$$x_C = 0 \qquad y_C = 0$$

$$I_{Cxx} = \left(\phi_0 + S_0 C_0 - \frac{2S_0^2}{\phi_0}\right)a^3$$

$$I_{Cyy} = (\phi_0 - S_0 C_0)a^3$$

$$I_{Cxy} = 0$$

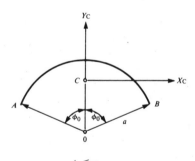

$$\phi_0 \leqslant \pi$$

(4) Segment AB in X_V, Y_V

$$M_{Vx} = -2(\phi_0 - S_0)a^2 \qquad M_{Vy} = 0$$

$$x_C = 0 \qquad\qquad y_C = \left(1 - \frac{S_0}{\phi_0}\right)a$$

$$I_{Vxx} = (2\phi_0 - 4S_0 + S_0 C_0)a^3$$

$$I_{Vyy} = (\phi_0 - S_0 C_0)a^3$$

$$I_{Vxy} = 0$$

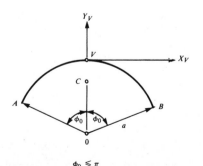

$$\phi_0 \leqslant \pi$$

(1) General Cases   **(Notation Sec. 21.09)**

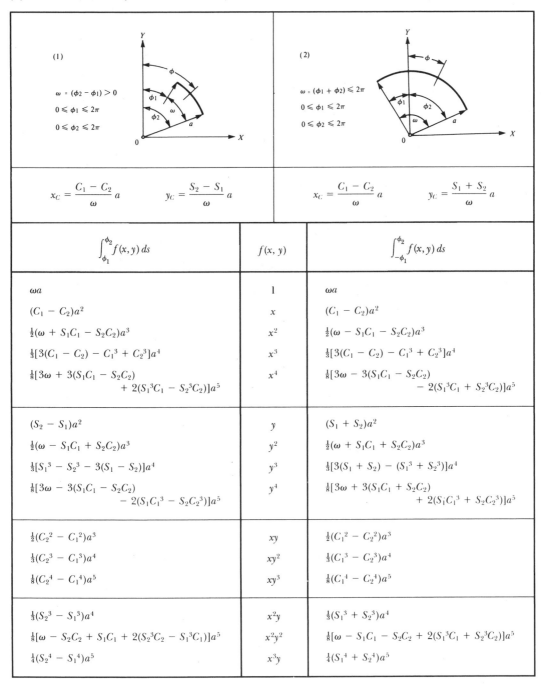

$$x_C = \frac{C_1 - C_2}{\omega} a \qquad y_C = \frac{S_2 - S_1}{\omega} a \qquad\qquad x_C = \frac{C_1 - C_2}{\omega} a \qquad y_C = \frac{S_1 + S_2}{\omega} a$$

$\displaystyle\int_{\phi_1}^{\phi_2} f(x,y)\,ds$	$f(x,y)$	$\displaystyle\int_{-\phi_1}^{\phi_2} f(x,y)\,ds$
ωa	1	ωa
$(C_1 - C_2)a^2$	x	$(C_1 - C_2)a^2$
$\frac{1}{2}(\omega + S_1C_1 - S_2C_2)a^3$	x^2	$\frac{1}{2}(\omega - S_1C_1 - S_2C_2)a^3$
$\frac{1}{3}[3(C_1 - C_2) - C_1^3 + C_2^3]a^4$	x^3	$\frac{1}{3}[3(C_1 - C_2) - C_1^3 + C_2^3]a^4$
$\frac{1}{8}[3\omega + 3(S_1C_1 - S_2C_2) + 2(S_1^3C_1 - S_2^3C_2)]a^5$	x^4	$\frac{1}{8}[3\omega - 3(S_1C_1 - S_2C_2) - 2(S_1^3C_1 + S_2^3C_2)]a^5$
$(S_2 - S_1)a^2$	y	$(S_1 + S_2)a^2$
$\frac{1}{2}(\omega - S_1C_1 + S_2C_2)a^3$	y^2	$\frac{1}{2}(\omega + S_1C_1 + S_2C_2)a^3$
$\frac{1}{3}[S_1^3 - S_2^3 - 3(S_1 - S_2)]a^4$	y^3	$\frac{1}{3}[3(S_1 + S_2) - (S_1^3 + S_2^3)]a^4$
$\frac{1}{8}[3\omega - 3(S_1C_1 - S_2C_2) - 2(S_1C_1^3 - S_2C_2^3)]a^5$	y^4	$\frac{1}{8}[3\omega + 3(S_1C_1 + S_2C_2) + 2(S_1C_1^3 + S_2C_2^3)]a^5$
$\frac{1}{2}(C_2^2 - C_1^2)a^3$	xy	$\frac{1}{2}(C_1^2 - C_2^2)a^3$
$\frac{1}{3}(C_2^3 - C_1^3)a^4$	xy^2	$\frac{1}{3}(C_1^3 - C_2^3)a^4$
$\frac{1}{8}(C_2^4 - C_1^4)a^5$	xy^3	$\frac{1}{8}(C_1^4 - C_2^4)a^5$
$\frac{1}{3}(S_2^3 - S_1^3)a^4$	x^2y	$\frac{1}{3}(S_1^3 + S_2^3)a^4$
$\frac{1}{8}[\omega - S_2C_2 + S_1C_1 + 2(S_2^3C_2 - S_1^3C_1)]a^5$	x^2y^2	$\frac{1}{8}[\omega - S_1C_1 - S_2C_2 + 2(S_1^3C_1 + S_2^3C_2)]a^5$
$\frac{1}{4}(S_2^4 - S_1^4)a^5$	x^3y	$\frac{1}{4}(S_1^4 + S_2^4)a^5$

**(2) Static and Inertia Functions
of a Complete Circular Curve**

$$L = 2\pi a \qquad M_{Cx} = M_{Cy} = 0$$

$$I_{Cxx} = 2\pi a^3 \qquad I_{Cyy} = 2\pi a^3 \qquad I_{Cxy} = 0$$

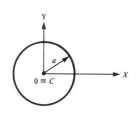

α = position angle, radians	a, b = semiaxes	$e = \sqrt{a^2 - b^2}$
ϕ = tangent angle, radians	$D = \sqrt{1 - k^2 \sin^2 \alpha}$	$k = \dfrac{e}{a} \qquad r = \dfrac{b}{a}$
$S = \sin \alpha \qquad C = \cos \alpha$	$x = a \sin \alpha$	$y = b \cos \alpha$

$E(k, \alpha), F(k, \alpha)$ = incomplete elliptic integrals (Sec. 13.05)
For definition of other symbols refer to Secs. 4.07, 21.03–21.04.

(1) Differential Geometry

$$\frac{x^2}{a^2} + \frac{y^2}{b^2} = 1$$

Slope at 1: $\quad y' = \tan \phi = -\dfrac{r^2 x_1}{y_1}$

Tangent at 1: $\quad y - y_1 = (x - x_1) \tan \phi$

Normal at 1: $\quad y - y_1 = (x_1 - x) \cot \phi$

Radius of curvature

at 1: $\quad \rho_1 = \dfrac{(b^4 x_1{}^2 + a^4 y_1{}^2)^{3/2}}{a^4 b^4}$

at H, V: $\quad \rho_H = \dfrac{b^2}{a} \qquad \rho_V = \dfrac{a^2}{b}$

Center of curvature

at 1: $\quad x_{K1} = \dfrac{e^2}{a^4} x_1{}^3 \qquad y_{K1} = \dfrac{e^2}{b^4} y_1{}^3$

at H: $\quad x_{KH} = \pm \dfrac{e^2}{a} \qquad y_{KH} = 0$

at V: $\quad x_{KV} = 0 \qquad y_{KV} = \pm \dfrac{e^2}{b}$

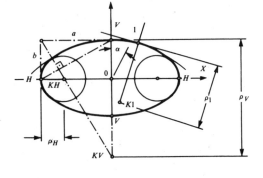

(2) Basic Integrals

$$ds = aD\, d\alpha$$

$I_0(\alpha) = \displaystyle\int D \, d\alpha \quad = E(k, \alpha)$

$I_1(\alpha) = \displaystyle\int DS \, d\alpha \quad = \dfrac{1}{2k} [r^2 \ln (kC + D) + kCD]$

$I_2(\alpha) = \displaystyle\int DS^2 \, d\alpha \quad = \dfrac{1}{3k^2} [(3k^2 - 1)E(k, \alpha) + r^2 F(k, \alpha) - k^2 CDS]$

$I_3(\alpha) = \displaystyle\int DS^3 \, d\alpha \quad = \dfrac{1}{8k^3} [(3k^4 - 2k^2 - 1) \ln (kC + D) - (2k^3 S^2 - 3k^3 - k)CD]$

$I_4(\alpha) = \displaystyle\int CD \, d\alpha \quad = \dfrac{1}{2k} [\sin^{-1}(kS) + kDS]$

$I_5(\alpha) = \displaystyle\int C^2 D \, d\alpha \quad = \dfrac{1}{3k^2} [(k^2 + 1)E(k, \alpha) - r^2 F(k, \alpha) + k^2 CDS]$

$I_6(\alpha) = \displaystyle\int C^3 D \, d\alpha \quad = \dfrac{1}{8k^3} [(4k^2 - 1) \sin^{-1} (kS) + (2k^3 C^2 + 2k^3 + 1)DS]$

$I_7(\alpha) = \displaystyle\int CDS \, d\alpha \quad = -\dfrac{1}{3k^2} D^3$

$I_8(\alpha) = \displaystyle\int CDS^2 \, d\alpha = \dfrac{1}{8k^3} [\sin^{-1} (kS) + (2k^3 S^2 - k)DS]$

$I_9(\alpha) = \displaystyle\int C^2 DS \, d\alpha = \dfrac{1}{8k^3} [r^4 \ln (kC + D) - (2k^3 C^2 + kr^2)CD]$

(1) General Cases   (Notation Sec. 21.11)

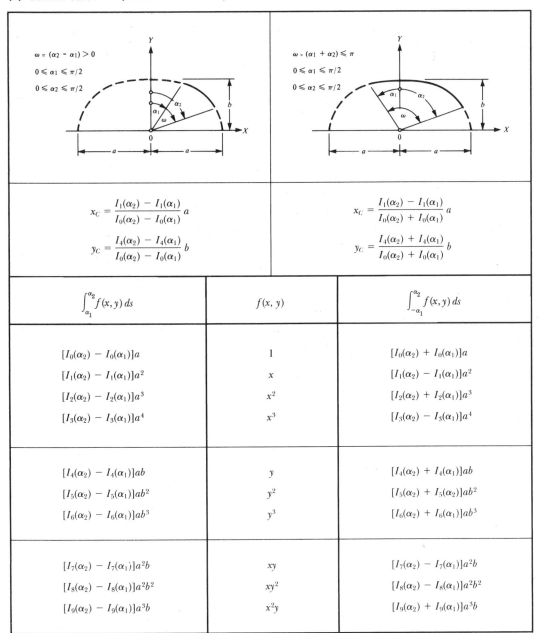

$$x_C = \frac{I_1(\alpha_2) - I_1(\alpha_1)}{I_0(\alpha_2) - I_0(\alpha_1)} a$$

$$y_C = \frac{I_4(\alpha_2) - I_4(\alpha_1)}{I_0(\alpha_2) - I_0(\alpha_1)} b$$

$$x_C = \frac{I_1(\alpha_2) - I_1(\alpha_1)}{I_0(\alpha_2) + I_0(\alpha_1)} a$$

$$y_C = \frac{I_4(\alpha_2) + I_4(\alpha_1)}{I_0(\alpha_2) + I_0(\alpha_1)} b$$

$\displaystyle\int_{\alpha_1}^{\alpha_2} f(x, y)\, ds$	$f(x, y)$	$\displaystyle\int_{-\alpha_1}^{\alpha_2} f(x, y)\, ds$
$[I_0(\alpha_2) - I_0(\alpha_1)]a$	1	$[I_0(\alpha_2) + I_0(\alpha_1)]a$
$[I_1(\alpha_2) - I_1(\alpha_1)]a^2$	x	$[I_1(\alpha_2) - I_1(\alpha_1)]a^2$
$[I_2(\alpha_2) - I_2(\alpha_1)]a^3$	x^2	$[I_2(\alpha_2) + I_2(\alpha_1)]a^3$
$[I_3(\alpha_2) - I_3(\alpha_1)]a^4$	x^3	$[I_3(\alpha_2) - I_3(\alpha_1)]a^4$
$[I_4(\alpha_2) - I_4(\alpha_1)]ab$	y	$[I_4(\alpha_2) + I_4(\alpha_1)]ab$
$[I_5(\alpha_2) - I_5(\alpha_1)]ab^2$	y^2	$[I_5(\alpha_2) + I_5(\alpha_2)]ab^2$
$[I_6(\alpha_2) - I_6(\alpha_1)]ab^3$	y^3	$[I_6(\alpha_2) + I_6(\alpha_1)]ab^3$
$[I_7(\alpha_2) - I_7(\alpha_1)]a^2b$	xy	$[I_7(\alpha_2) - I_7(\alpha_1)]a^2b$
$[I_8(\alpha_2) - I_8(\alpha_1)]a^2b^2$	xy^2	$[I_8(\alpha_2) - I_8(\alpha_1)]a^2b^2$
$[I_9(\alpha_2) - I_9(\alpha_1)]a^3b$	x^2y	$[I_9(\alpha_2) + I_9(\alpha_1)]a^3b$

(2) Static and Inertia Functions of a Complete Elliptic Curve

$$L = 4aI_0\left(\frac{1}{2}\pi\right) = 2\pi a\left[1 - \left(\frac{1}{2}\right)^2\left(\frac{k^2}{1}\right) - \left(\frac{1\cdot 3}{2\cdot 4}\right)^2\left(\frac{k^4}{3}\right)\right.$$

$$\left. -\left(\frac{1\cdot 3\cdot 5}{2\cdot 4\cdot 6}\right)^2\left(\frac{k^6}{5}\right) - \cdots\right]$$

$$\cong \pi[1.5(a + b) - \sqrt{ab}] \qquad M_{Cx} = M_{Cy} = 0$$

$$I_{Cxx} = 4ab^2 I_5(\tfrac{1}{2}\pi)$$

$$I_{Cyy} = 4a^3 I_3(\tfrac{1}{2}\pi) \qquad I_{Cxy} = 0$$

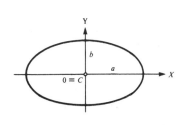

α = position angle, radians　　　　a, b = semiaxes　　　　$e = \sqrt{a^2 + b^2}$

ϕ = tangent angle, radians　　$D = \sqrt{1 - k^2 \sin^2 \alpha}$　　$k = \dfrac{a}{e}$　　$r = \dfrac{b}{a}$

$S = \sin \alpha$　　　　　　　　　$C = \cos \alpha$　　　　$x = \dfrac{a}{\sin \alpha}$　　$y = \pm \dfrac{b \cos \alpha}{\sin \alpha}$

$E(k, \alpha), F(k, \alpha)$ = incomplete elliptic integrals (Sec. 13.05)
For other symbols refer to Secs. 4.08, 21.03–21.06.

(1) Differential Geometry, Normal Position

$$\boxed{\dfrac{x^2}{a^2} - \dfrac{y^2}{b^2} = 1}$$

Slope at 1:　$y_1' = \tan \phi = \dfrac{r^2 x_1}{y_1}$

Tangent at 1:　$y - y_1 = (x - x_1) \tan \phi$

Normal at 1:　$y - y_1 = (x_1 - x) \cot \phi$

Radius of curvature

at 1:　$\rho_1 = \dfrac{(b^4 x_1^2 + a^4 y_1^2)^{3/2}}{a^4 b^4}$　　　　at H:　$\rho_H = rb$

Center of curvature

at 1:　$x_{K1} = \dfrac{e^2}{a^4} x_1^3$　　　　at H:　$x_{KH} = \pm \dfrac{e^2}{a}$

$y_{K1} = -\dfrac{e^2}{b^4} y_1^3$　　　　　　$y_{KH} = 0$

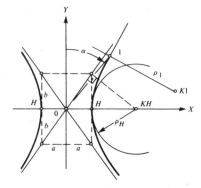

(2) Differential Geometry, Asymptotic Position

$$\boxed{xy = a^2}$$

Slope at $1 = y_1' = \tan \phi = -\dfrac{y_1}{x_1}$

Radius of curvature

at 1:　$\rho_1 = -\dfrac{2 x_1 y_1}{\sqrt{x_1^2 + y_1^2}}$　　　at H:　$\rho_H = a\sqrt{2}$

Center of curvature

at 1:　$x_{K1} = \dfrac{a^2}{2 x_1} \left(\dfrac{3 x_1^2}{a^2} + \dfrac{a^2}{x_1^2} \right)$　　　at H:　$x_{KH} = \pm 2a$

$y_{K1} = \dfrac{a^2}{2 x_1} \left(\dfrac{x_1^4}{a^4} + 3 \right)$　　　　　　$y_{KH} = \pm 2a$

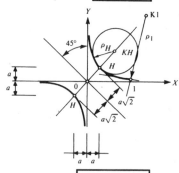

(3) Basic Integrals

$$\boxed{ds = \dfrac{eD}{S^2} d\alpha}$$

$I_0(\alpha) = \displaystyle\int \dfrac{D}{S^2} d\alpha = r^2 F(k, \alpha) - E(k, \alpha) - \dfrac{CD}{S}$

$I_1(\alpha) = \displaystyle\int \dfrac{D}{S^3} d\alpha = \dfrac{1}{4} \left(r^2 \ln \dfrac{D + C}{D - C} - \dfrac{2CD}{S^2} \right)$

$I_2(\alpha) = \displaystyle\int \dfrac{D}{S^4} d\alpha = \dfrac{1}{3} \left[2r^2 F(k, \alpha) + (k^2 - 2) E(k, \alpha) + (k^2 - 3) \dfrac{CD}{S} - \dfrac{C^3 D}{S^3} \right]$

$I_3(\alpha) = \displaystyle\int \dfrac{D}{S^5} d\alpha = \dfrac{1}{16} \left[r^2 (k^2 + 3) \ln \dfrac{D + C}{D - C} + \dfrac{3(k^3 - 3) S^2 + 2}{S^4} CD \right]$

$I_4(\alpha) = \displaystyle\int \dfrac{CD}{S^3} d\alpha = \dfrac{1}{4} \left(k^2 \ln \dfrac{1 + D}{1 - D} - \dfrac{2D}{S^2} \right)$

$I_5(\alpha) = \displaystyle\int \dfrac{CD}{S^4} d\alpha = -\dfrac{1}{3} \left(\dfrac{D}{S} \right)^3$

$I_6(\alpha) = \displaystyle\int \dfrac{CD}{S^5} d\alpha = \dfrac{1}{16} \left[k^4 \ln \dfrac{D + 1}{D - 1} + 2(k^2 S^2 - 2) \dfrac{D}{S^4} \right]$

(1) General Cases   (Notation Sec. 21.13)

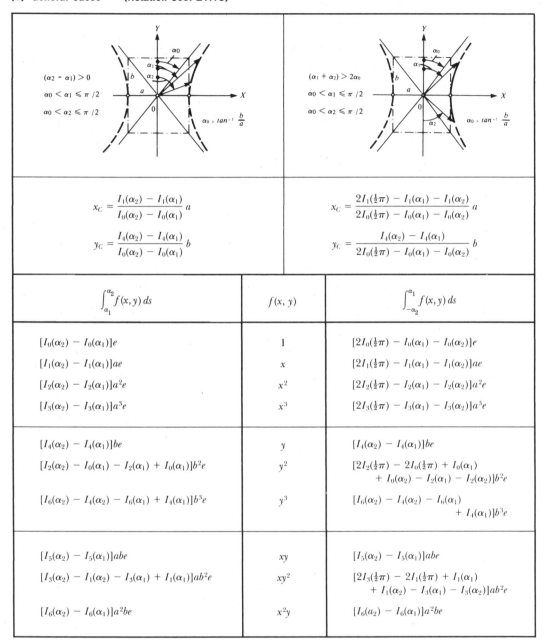

$$x_C = \frac{I_1(\alpha_2) - I_1(\alpha_1)}{I_0(\alpha_2) - I_0(\alpha_1)}\,a$$

$$y_C = \frac{I_4(\alpha_2) - I_4(\alpha_1)}{I_0(\alpha_2) - I_0(\alpha_1)}\,b$$

$$x_C = \frac{2I_1(\tfrac{1}{2}\pi) - I_1(\alpha_1) - I_1(\alpha_2)}{2I_0(\tfrac{1}{2}\pi) - I_0(\alpha_1) - I_0(\alpha_2)}\,a$$

$$y_C = \frac{I_4(\alpha_2) - I_4(\alpha_1)}{2I_0(\tfrac{1}{2}\pi) - I_0(\alpha_1) - I_0(\alpha_2)}\,b$$

$\displaystyle\int_{\alpha_1}^{\alpha_2} f(x,y)\,ds$	$f(x,y)$	$\displaystyle\int_{-\alpha_2}^{\alpha_1} f(x,y)\,ds$
$[I_0(\alpha_2) - I_0(\alpha_1)]e$	1	$[2I_0(\tfrac{1}{2}\pi) - I_0(\alpha_1) - I_0(\alpha_2)]e$
$[I_1(\alpha_2) - I_1(\alpha_1)]ae$	x	$[2I_1(\tfrac{1}{2}\pi) - I_1(\alpha_1) - I_1(\alpha_2)]ae$
$[I_2(\alpha_2) - I_2(\alpha_1)]a^2e$	x^2	$[2I_2(\tfrac{1}{2}\pi) - I_2(\alpha_1) - I_2(\alpha_2)]a^2e$
$[I_3(\alpha_2) - I_3(\alpha_1)]a^3e$	x^3	$[2I_3(\tfrac{1}{2}\pi) - I_3(\alpha_1) - I_3(\alpha_2)]a^3e$
$[I_4(\alpha_2) - I_4(\alpha_1)]be$	y	$[I_4(\alpha_2) - I_4(\alpha_1)]be$
$[I_2(\alpha_2) - I_0(\alpha_1) - I_2(\alpha_1) + I_0(\alpha_1)]b^2e$	y^2	$[2I_2(\tfrac{1}{2}\pi) - 2I_0(\tfrac{1}{2}\pi) + I_0(\alpha_1)$ $+ I_0(\alpha_2) - I_2(\alpha_1) - I_2(\alpha_2)]b^2e$
$[I_6(\alpha_2) - I_4(\alpha_2) - I_6(\alpha_1) + I_4(\alpha_1)]b^3e$	y^3	$[I_6(\alpha_2) - I_4(\alpha_2) - I_6(\alpha_1)$ $+ I_4(\alpha_1)]b^3e$
$[I_5(\alpha_2) - I_5(\alpha_1)]abe$	xy	$[I_5(\alpha_2) - I_5(\alpha_1)]abe$
$[I_3(\alpha_2) - I_1(\alpha_2) - I_3(\alpha_1) + I_1(\alpha_1)]ab^2e$	xy^2	$[2I_3(\tfrac{1}{2}\pi) - 2I_1(\tfrac{1}{2}\pi) + I_1(\alpha_1)$ $+ I_1(\alpha_2) - I_3(\alpha_1) - I_3(\alpha_2)]ab^2e$
$[I_6(\alpha_2) - I_6(\alpha_1)]a^2be$	x^2y	$[I_6(a_2) - I_6(\alpha_1)]a^2be$

(2) Static and Inertia Functions of a Complete Hyperbolic Curve

$$L = 4e[I_0(\tfrac{1}{2}\pi) - I_0(\alpha)]$$

$$M_{Cx} = 0 \qquad\qquad M_{Cy} = 0$$

$$I_{Cxx} = 4b^2e[I_2(\tfrac{1}{2}\pi) - I_0(\tfrac{1}{2}\pi) + I_2(\alpha) + I_0(\alpha)]$$

$$I_{Cyy} = 4a^2e[I_2(\tfrac{1}{2}\pi) - I_2(\alpha)] \qquad\qquad I_{Cxy} = 0$$

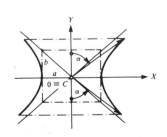

$2p$ = parameter $\qquad$ a, b, c = segments $\qquad$ $\bar{x} = x/p$ $\qquad$ $\bar{y} = y/p$

ϕ = tangent angle, radians $\qquad$ $D = \sqrt{1 + \bar{x}^2}$ $\qquad$ $H = \sinh^{-1} \bar{x}$

$\tan \phi = \bar{x}$ $\qquad$ $S = \sin \phi = \bar{x}/D$ $\qquad$ $C = \cos \phi = 1/D$

For other symbols, refer to Secs. 4.09, 21.03–21.06.

(1) Differential Geometry

$$y = \frac{1}{2p}x^2 \qquad \bar{y} = \frac{1}{2}\bar{x}^2$$

Slope at 1: $\quad y' = \tan \phi = \dfrac{x_1}{p}$

Tangent at 1: $\quad y - y_1 = \dfrac{x_1}{p}(x - x_1)$

Normal at 1: $\quad y - y_1 = \dfrac{p}{x_1}(x_1 - x)$

Radius of curvature

at 1: $\quad \rho_1 = \dfrac{(p^2 + x_1^2)^{3/2}}{p^2}$ $\qquad$ at 0: $\quad \rho_0 = p$

Center of curvature

at 1: $\quad x_{K1} = -\dfrac{x_1^3}{p^2}$ $\qquad y_{K1} = p + \dfrac{3x_1^2}{p}$ $\qquad$ at 0: $\quad x_{K0} = 0$ $\qquad y_{K0} = p$

(2) Basic Integrals $\quad (m, n = 0, 1, 2, \ldots)$

$$ds = \left(\sqrt{1 + \frac{x^2}{p}}\right) dx = D\, dx$$

$\displaystyle\int x^n\, ds \quad = p^{n+1} \int \bar{x}^n D\, d\bar{x} = p^{n+1} I_n(x)$

$\displaystyle\int x^n C\, ds \quad = p^{n+1} \int \bar{x}^n\, d\bar{x} \quad = \frac{x^{n+1}}{n+1}$

$\displaystyle\int x^n S\, ds \quad = p^{n+1} \int \bar{x}^{n+1}\, d\bar{x} = \frac{x^{n+2}}{(n+2)p}$

$\displaystyle\int x^n C^2\, ds = p^{n+1} \int \frac{\bar{x}^n}{D}\, d\bar{x} \quad = p^{n+1} J_n(x)$

$\displaystyle\int x^n CS\, ds = p^{n+1} \int \frac{\bar{x}^{n+1}}{D}\, d\bar{x} = p^{n+1} J_{n+1}(x)$

$\displaystyle\int x^n S^2\, ds = p^{n+1} \int \frac{\bar{x}^{n+2}}{D}\, d\bar{x} = p^{n+1} J_{n+2}(x)$

For integrals of y^n and $x^m y^n$, use

$$y^n = \frac{x^{2n}}{(2p)^n}$$

$$x^m y^n = \frac{x^{m+2n}}{(2p)^n}$$

(3) Table of $I_n(x)$, $J_n(x)$

$I_0(x) = \dfrac{1}{2}\left(\dfrac{x}{p}D + H\right)$	$J_0(x) = H$
$I_1(x) = \dfrac{1}{3}(D^3 - 1)$	$J_1(x) = D - 1$
$I_2(x) = \dfrac{1}{4}\left[\dfrac{xD^3}{p} - I_0(x)\right]$	$J_2(x) = I_0(x) - J_0(x)$
$I_3(x) = \dfrac{1}{5}\left[\left(\dfrac{x}{p}\right)^2 D^3 - 2I_1(x)\right]$	$J_3(x) = I_1(x) - J_1(x)$
$I_n(x) = \dfrac{1}{n+2}\left[\left(\dfrac{x}{p}\right)^{n-1} D^3 - (n-1)I_{n-2}(x)\right]$	$J_n(x) = I_{n-2}(x) - J_{n-2}(x)$

(1) General Cases   (Notation Sec. 21.15)

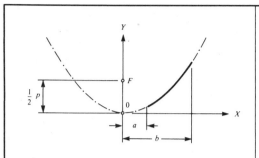

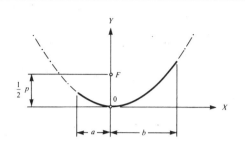

$$x_C = \frac{[I_1(b) - I_1(a)]p}{I_0(b) - I_0(a)}$$

$$y_C = \frac{\frac{1}{2}[I_2(b) - I_2(a)]p}{I_0(b) - I_0(a)}$$

$$x_C = \frac{[I_1(b) - I_1(a)]p}{I_0(b) + I_0(a)}$$

$$y_C = \frac{\frac{1}{2}[I_2(b) + I_2(a)]p}{I_0(b) + I_0(a)}$$

$\int_a^b f(x)\,ds$	$f(x)$	$\int_{-a}^b f(x)\,ds$
$[I_0(b) - I_0(a)]p$	1	$[I_0(b) + I_0(a)]p$
$[I_1(b) - I_1(a)]p^2$	x	$[I_1(b) - I_1(a)]p^2$
$[I_2(b) - I_2(a)]p^3$	x^2	$[I_2(b) + I_2(a)]p^3$
$[I_n(b) - I_n(a)]p^{n+1}$	x^n	$[I_n(b) - I_n(-a)]p^{n+1}$
$[J_0(b) - J_0(a)]p$	C^2	$[J_0(b) + J_0(a)]p$
$[J_n(b) - J_n(a)]p^{n+1}$	$x^n C^2$	$[J_n(b) - J_n(-a)]p^{n+1}$
$[J_1(b) - J_1(a)]p$	CS	$[J_1(b) - J_1(a)]p$
$[J_{n+1}(b) - J_{n+1}(a)]p^{n+1}$	$x^n CS$	$[J_{n+1}(b) - J_{n+1}(-a)]p^{n+1}$
$[J_2(b) - J_2(a)]p$	S^2	$[J_2(b) + J_2(a)]p$
$[J_{n+2}(b) - J_{n+2}(a)]p^{n+1}$	$x^n S^2$	$[J_{n+2}(b) - J_{n+2}(-a)]p^{n+1}$

(2) Static and Inertia Functions of a Parabolic Curve Segment

$$L = [I_0(c)]p \qquad M_{0x} = \tfrac{1}{2}[I_2(c)]p^2 \qquad M_{0y} = [I_1(c)]p^2$$

$$x_C = \frac{[I_1(c)]p}{[I_0(c)]} \qquad y_C = \frac{[I_2(c)]p}{2[I_0(c)]}$$

$$I_{Cxx} = \tfrac{1}{4}[I_4(c)]p^3 \qquad I_{cxy} = \tfrac{1}{2}[I_3(c)]p^3$$

$$I_{Cyy} = [I_2(c)]p^3$$

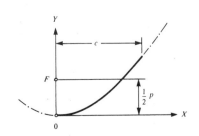

(1) Notation

a, b, c = constants	e = eccentricity	f = equivalent
k = exponent	m, n = integers	t = parameter
x, y = cartesian coordinates		r, θ = polar coordinates
x_C, y_C = coordinates of centroid of a plane curve segment		$x_K\, y_K$ = coordinates of center of curvature
A_C = area between a curve and its asymptote		A_X = area between a curve and X axis
A_Y = area between a curve and Y axis		A_0 = area of loop
$C = \cos\theta$	$S = \sin\theta$ $T = \tan\theta$	ϕ = tangent angle (or rolling angle)
F = focus	R = radius V = vertex	ρ = radius of curvature

(2) Properties

A selected set of special plane curves is displayed in Secs. 21.18–21.23. All curves are defined by their equations, and their shapes are depicted by the adjacent graphs. The remaining properties can be readily determined by the relations introduced in Secs. 21.01–21.06.

21.18 POWER FUNCTIONS

(1) Exponent $k = 2n$

$$y = (ax)^{2n} \qquad r^{2n-1} = \frac{S}{(aC)^{2n}}$$

$$x = \frac{t}{a} \qquad v = t^{2n}$$

(2) Exponent $k = 2n + 1$

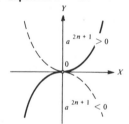

$$y = (ax)^{2n+1} \qquad r^{2n} = \frac{S}{(aC)^{2n+1}}$$

$$x = \frac{t}{a} \qquad y = t^{2n+1}$$

(3) Exponent $k = -2n$

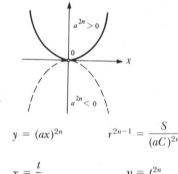

$$y = \left(\frac{a}{x}\right)^{2n} \qquad r^{2n+1} = \frac{C}{(aC)^{2n}}$$

$$x = \frac{a}{t} \qquad y = t^{2n}$$

(4) Exponent $k = -(2n - 1)$

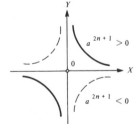

$$y = \left(\frac{a}{x}\right)^{2n+1} \qquad r^{2n+1} = \frac{a^{2n+1}}{SC^{2n+1}}$$

$$x = \frac{a}{t} \qquad y = t^{2n+1}$$

(5) Exponent $k = \dfrac{2m + 1}{2n + 1}$

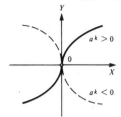

$$y = (ax)^{(2m+1)/(2n+1)}$$

$$rS = (arC)^{(2m+1)/(2n+1)}$$

$$x = \frac{t}{a} \qquad y = t^{(2m+1)/(2n+1)}$$

(6) Exponent $k = -\dfrac{2m + 1}{2n + 1}$

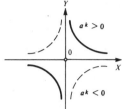

$$y = \left(\frac{a}{x}\right)^{(2m+1)/(2n+1)}$$

$$rS = \left(\frac{a}{rC}\right)^{(2m+1)/(2n+1)}$$

$$x = \frac{a}{t} \qquad y = t^{(2m+1)/(2n+1)}$$

(7) Exponent $k = \dfrac{2m + 1}{2n}$

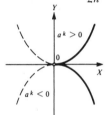

$$y = (ax)^{(2m+1)/2n}$$

$$rS = (arC)^{(2m+1)/2n}$$

$$x = \frac{t}{a} \qquad y = t^{(2m+1)/2n}$$

(8) Exponent $k = -\dfrac{2m + 1}{2n}$

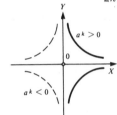

$$y = \left(\frac{a}{x}\right)^{(2m+1)/2n}$$

$$rS = \left(\frac{a}{rC}\right)^{(2m+1)/2n}$$

$$x = \frac{a}{t} \qquad y = t^{(2m+1)/2n}$$

(9) Exponent $k = \dfrac{2m}{2n + 1}$

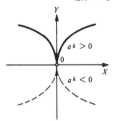

$$y = (ax)^{2m/(2n+1)}$$

$$rS = (arC)^{2m/(2n+1)}$$

$$x = \frac{t}{a} \qquad y = t^{2m/(2n+1)}$$

(10) Exponent $k = -\dfrac{2m}{2n + 1}$

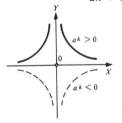

$$y = \left(\frac{a}{x}\right)^{2m/(2n+1)}$$

$$rS = \left(\frac{a}{rC}\right)^{2m/(2n+1)}$$

$$x = \frac{a}{t} \qquad y = t^{2m/(2n+1)}$$

Notation (Sec. 21.17)$\qquad\qquad\qquad\qquad R = \tfrac{1}{2}a = $ radius

(1) Witch of Agnesi (Verniera)

$$y = \frac{a^3}{x^2 + a^2} \qquad r = \frac{a}{S}\sqrt{C^2 + S^6}$$

$$x = \frac{a}{T} \qquad y = \frac{aT^2}{1 + T^2}$$

$$A_C = 4\pi R^2$$

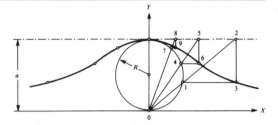

Construction sequence:

$\quad$ 1, 2, 3; 4, 5, 6; . . .

Asymptote: X axis

(2) Cissoid of Diocles

$$x^2 = \frac{y^3}{a - y} \qquad \rho = \frac{aC}{T}$$

$$x = \frac{a}{(1 + T^2)T} \qquad y = \frac{a}{1 + T^2}$$

$$A_C = 3\pi R^2$$

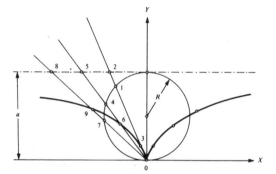

Construction sequence:

$\quad \overline{12} = \overline{03};\ \overline{45} = \overline{06};$ . . .

Asymptote: $y = a$

(3) Strofoid

$$y^2 = x^2 \frac{a - x}{a + x} \qquad r = \frac{a(C^2 - S^2)}{C}$$

$$x = \frac{a(1 - T^2)}{1 + T^2} \qquad y = \frac{a(1 - T^2)T}{1 + T^2}$$

$$A_0 = a^2(1 - \tfrac{1}{4}\pi) \qquad A_C = a^2(1 + \tfrac{1}{4}\pi)$$

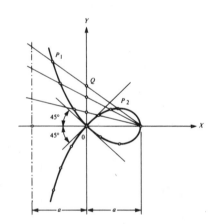

Construction sequence:

$\quad \overline{0Q} = \overline{QP_1} = \overline{QP_2};$ . . .

Asymptote: $x = -a$

(4) Folium of Descartes

$$x^3 + y^3 - 3axy = 0 \qquad r = \frac{3aCS}{S^3 + C^3}$$

$$x = \frac{3aT}{1 + T^3} \qquad y = \frac{3aT^2}{1 + T^3}$$

$$A_0 = A_C = \tfrac{3}{2}a^2$$

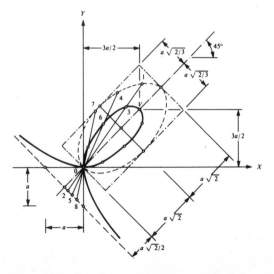

Construction sequence:

$\quad \overline{01} = \overline{23};\ \overline{04} = \overline{56};$ . . .

Asymptote: $x + y = -a$

> Notation (Sec. 21.17) $f^4 = (a^4 - e^4)/e^4$

(1) Conchoids of Nicomedes

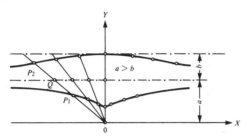

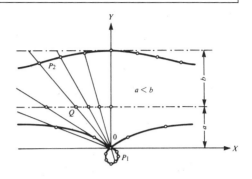

$$(x^2 + y^2)(y - a)^2 - b^2 y^2 = 0 \qquad x = \frac{a}{T} \pm \frac{b}{\sqrt{1 + T^2}} \qquad y = a \pm \frac{bT}{\sqrt{1 + T^2}}$$

$r = \dfrac{a}{S} \pm b$ Construction sequence: $\overline{0P_1} = \overline{0Q} - b,\ \overline{0P_2} = \overline{0Q} + b;\ \ldots$

(2) Pascal's Snails

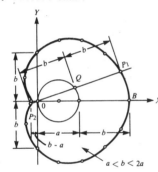

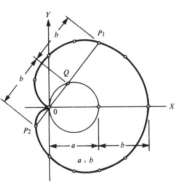

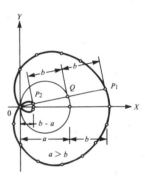

$$(x^2 + y^2 - ax^2) - b^2(x^2 + y^2) = 0 \qquad x = \frac{a \pm b\sqrt{1 + T^2}}{1 + T^2} \qquad y = \frac{(a \pm b\sqrt{1 + T^2})T}{1 + T^2}$$

$r = aC \pm b$ Construction sequence: $\overline{0P_1} = \overline{0Q} + b,\ \overline{0P_2} = \overline{0Q} - b;\ \ldots$

For curves without the loop, the area is $\pi(b^2 + \tfrac{1}{2}a^2)$.

(3) Ovals of Cassini

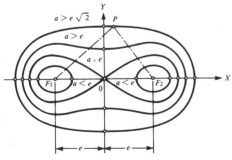

$$(x^2 + y^2)^2 - 2e^2(x^2 - y^2) = a^4 - e^4$$

$$r = e\sqrt{C^2 - S^2 \pm \sqrt{[(C^2 - S^2)^2 + f^4]}}$$

Construction sequence: $\overline{F_1 P} \cdot \overline{F_2 P} = a^2$

(4) Lemniscate of Bernoulli

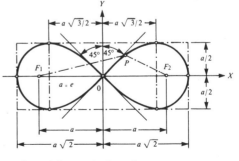

$$(x^2 + y^2)^2 - 2a^2(x^2 - y^2) = 0$$

$$r = a\sqrt{2(C^2 - S^2)} \qquad A_0 = a^2$$

Radius of curvature $= \dfrac{2a}{3\sqrt{2(C^2 - S^2)}}$

Notation (Sec. 21.17)	a = radius of rolling circle	b = radius of fixed circle

(1) Ordinary Cycloid

$x = a(\phi - \sin \phi)$ $y = a(1 - \cos \phi)$

Length of one arc = $8a$

Area under one arc = $3\pi a^2$

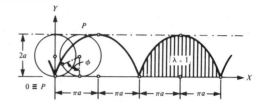

The curve is generated by a fixed point P on a circle of radius a which rolls without slipping along the X axis.

(2) Trochoids

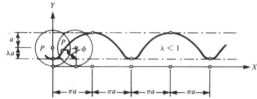

$x = a(\phi - \lambda \sin \phi)$ If $\lambda < 1$, the curve is called a *curtate cycloid*.

$y = a(1 - \lambda \cos \phi)$ If $\lambda > 1$, the curve is called a *prolate cycloid*.

The curve is generated by a fixed point P at the distance λa from the center of a circle of radius a which rolls without slipping along the X axis.

(3) Ordinary Epicycloid

$x = (a + b) \cos \phi - a \cos \left(\dfrac{a + b}{a} \phi \right)$

$y = (a + b) \sin \phi - a \sin \left(\dfrac{a + b}{a} \phi \right)$

Length of one arc = $8(a + b)/N$

Area of one sector = $(a\pi/b)(a + b)(2a + b)$

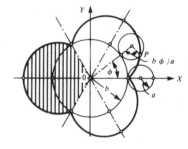

The curve is generated by a fixed point P on a circle of radius a which rolls without slipping on the outside of a fixed circle of radius b. If $b/a = N$ is an integer, the curve has N equal branches. If N is a fraction, the branches cross one other.

(4) Epitrochoids

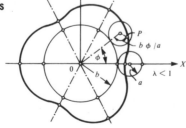
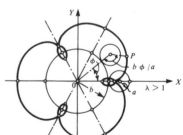

$x = (a + b) \cos \phi - \lambda a \cos \left(\dfrac{a + b}{a} \phi \right)$ If $\lambda < 1$, the curve is called a *curtate epicycloid*.

$y = (a + b) \sin \phi - \lambda a \sin \left(\dfrac{a + b}{a} \phi \right)$ If $\lambda > 1$, the curve is called a *prolate epicycloid*.

The curve is generated by a fixed point P at a distance λa from the center of a circle of radius a which rolls without slipping on the outside of a fixed circle of radius b. If $b/a = N$ is an integer, the curve has N equal branches. If N is a fraction, the branches cross one other.

(5) Ordinary Hypocycloid

$$x = (b - a) \cos \phi + a \cos \left(\frac{b - a}{a} \phi \right)$$

$$y = (b - a) \sin \phi - a \sin \left(\frac{b - a}{a} \phi \right)$$

Length of one arc $= 8(b - a)/N$

Area of one sector $= (a\pi/b)(b - a)(b - 2a)$

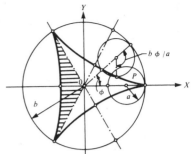

The curve is generated by a fixed point P on a circle of radius a which rolls without slipping on the inside of a fixed circle of radius b. If $b/a = N$ is an integer, the curve has N equal branches. If N is a fraction, the branches cross one other. If $N = 2$, the curve reduces to a straight vertical segment.

(6) Hypotrochoids

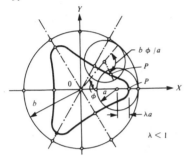

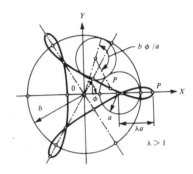

$$x = (b - a) \cos \phi + \lambda a \cos \left(\frac{b - a}{a} \phi \right) \qquad \text{If } \lambda < 1, \text{ the curve is called a } \textit{curtate hypocycloid.}$$

$$y = (b - a) \sin \phi - \lambda a \sin \left(\frac{b - a}{a} \phi \right) \qquad \text{If } \lambda > 1, \text{ the curve is called a } \textit{prolate hypocycloid.}$$

The curve is generated by a fixed point P at a distance λa from the center of a circle of radius a which rolls without slipping on the inside of a fixed circle of radius b. If $b/a = N$ is an integer, the curve has N equal branches. If N is a fraction, the branches cross one other.

(7) Astroid

$$x^{2/3} + y^{2/3} = b^{2/3}$$

$$x = b \cos^3 \phi \qquad\qquad y = b \sin^3 \phi$$

Length of one arc $= 3b/2$

Area of one sector $= 3\pi b^2/32$

The curve is a hypocycloid of $N = 4$.

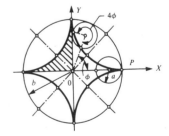

(8) Cardioid

$$(x^2 + y^2 - a^2)^2 = 4a^2[(x - a)^2 + y^2]$$

$$x = a(2 \cos \phi - \cos 2\phi)$$

$$y = a(2 \sin \phi - \sin 2\phi)$$

Length of curve $= 16a$

Area enclosed by curve $= 7\pi a^2$

The curve is a epicycloid of $N = 1$, $(a = b)$.

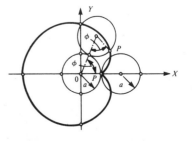

Notation (Sec. 21.17) A = constant $B = \ln A$

(1) Ordinary Exponential Curves

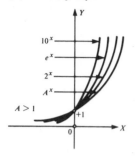
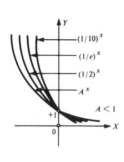

$$y = A^x = e^{Bx} \qquad r = \sqrt{x^2 + A^{2x}}$$

$$y' = \tan \phi = BA^x \qquad \tan \theta = \frac{A^x}{x}$$

(2) Reciprocal Exponential Curves

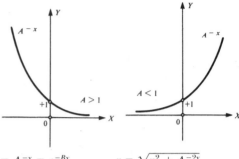

$$y = A^{-x} = e^{-Bx} \qquad r = \sqrt{x^2 + A^{-2x}}$$

$$y' = \tan \phi = BA^{-x} \qquad \tan \theta = \frac{A^{-x}}{x}$$

(3) Distribution Curve

$$y = e^{-(Ax)^2} \qquad r = \sqrt{x^2 + e^{-2(Ax)^2}}$$

$$y' = \tan \phi = -2A^2 x e^{-(2x)^2}$$

$$\tan \theta = x^{-1} e^{-(Ax)^2}$$

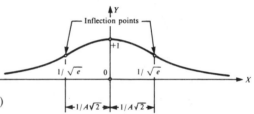

Particular case: Gauss' curve (Secs. 18.06, 20.12)

(4) General Logarithmic Curves

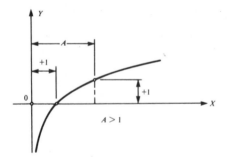
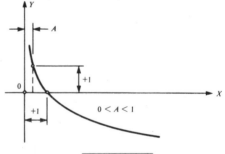

$$y = \log_A x = \frac{\ln x}{\ln A} \qquad y' = \tan \phi = \frac{1}{x \ln A}$$

$$r = \sqrt{x^2 + (\log_A x)^2}$$

$$\tan \theta = x^{-1} \log_A x$$

(5) Catenary

$$y = A \cosh \frac{x}{A} \qquad r = \sqrt{x^2 + A^2 \cosh^2 \frac{x}{A}}$$

$$\rho = A \cosh^2 \frac{x}{A} \qquad \tan \theta = \frac{A \cosh (x/A)}{x}$$

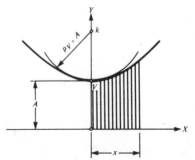

A heavy flexible cable hanging between two fixed points has the shape of the *catenary*.

Length of arc = $A \sinh \dfrac{x}{A}$ (one branch)   Area below one branch = $A^2 \sinh \dfrac{x}{A}$

Notation (Sec. 21.17)	a, b, c = constants	α, β = phase angles

(1) Exponential Curves Magnified by x^b

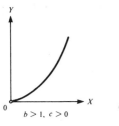 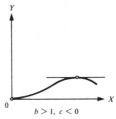 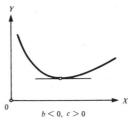 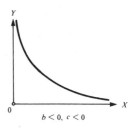

$b > 1, c > 0$ $b > 1, c < 0$ $b < 0, c > 0$ $b < 0, c < 0$

$$y = ax^b e^{cx} \qquad y' = \tan\phi = a\left(\frac{b}{x} + c\right)x^b e^{cx} \qquad r = \sqrt{x^2 + (ax^b e^{cx})^2} \qquad \tan\theta = ax^{b-1}e^{cx}$$

(2) Spiral of Archimedes $(a > 0)$

$$r = a\theta \qquad x = a\theta\cos\theta \qquad y = a\theta\sin\theta$$

$$y' = \tan\phi = \frac{1 + \tan\theta}{1 - \tan\theta}$$

Length of arc = $\frac{1}{2}a(\theta\sqrt{\theta^2 + 1} + \sinh^{-1}\theta)$

Area of sector bounded by θ_1 and θ_2 is

$\frac{1}{6}a^2(\theta_2^3 - \theta_1^3)$

The curve is generated by the point moving with a constant linear velocity along a straight line rotating with constant angular velocity about the origin 0.

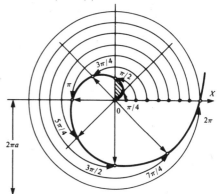

(3) Hyperbolic Spiral $(a > 0)$

$$r = \frac{a}{\theta} \qquad x = \frac{a}{\theta}\cos\theta \qquad y = \frac{a}{\theta}\sin\theta$$

$$y' = \tan\theta = \frac{\theta - \tan\theta}{1 + \theta\tan\theta}$$

Area of sector bounded by θ_1 and θ_2: $\frac{1}{2}a^2\left(\frac{1}{\theta_1} - \frac{1}{\theta_2}\right)$

Asymptote: $y = a$

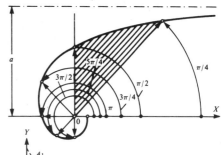

(4) Free Damped Vibration Curve $(k = 1, 2, 3, \ldots)$

$$y = ae^{-bx}\sin(cx + \alpha) \qquad R = \sqrt{b^2 + c^2}$$

$$y' = aRe^{-bx}\sin(cx + \alpha + \beta) \qquad \beta = \tan^{-1}\frac{b}{c}$$

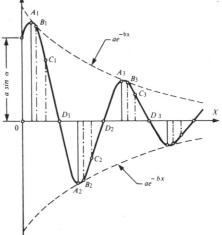

Amplitudes A_k: $x_k = (k\pi - \alpha + \beta)/c$
Tangent points B_k: $x_k = [(k + \frac{1}{2})\pi - \alpha]/c$
Inflection points C_k: $x_k = (k\pi - \alpha + 2\beta)/c$
Zero points D_k: $x_k = (k\pi - \alpha)/c$
Asymptote: x axis
The number $\lambda = \ln(A_k/A_{k+1}) = a\pi/c$, where A_k, A_{k+1} are two consecutive amplitudes, is called the *logarithmic decrement* of damping.

$$
S = \sin\theta \qquad\qquad r = f(\theta) \qquad\qquad d\theta_1 = \left[\frac{df(\theta)}{d\theta}\cos\theta - f(\theta)\sin\theta\right] d\theta
$$

$$
C = \cos\theta \qquad\qquad r' = \frac{df(\theta)}{d\theta} \qquad\qquad d\theta_2 = \left[\frac{df(\theta)}{d\theta}\sin\theta + f(\theta)\cos\theta\right] d\theta
$$

(1) Geometry

The geometry of a curvilinear, trapezoidal area in terms of notation introduced in Sec. 21.01-1 is shown below. The coordinates of end points are $x_A = a$, $y_A = d$, $x_B = b$, $y_B = c$ and r_A, α, r_B, β.

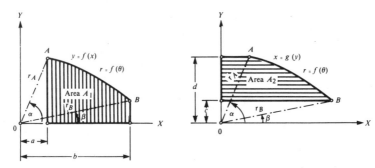

(2) Areas

$$
A_1 = \int_a^b f(x)\,dx = \int_\beta^\alpha Sf(\theta)\,d\theta_1 \qquad\qquad A_2 = \int_c^d g(y)\,dy = \int_\beta^\alpha Cf(\theta)\,d\theta_2
$$

(3) Moments of Areas about X, Y

$$
M_{x1} = \frac{1}{2}\int_a^b [f(x)]^2\,dx = \frac{1}{2}\int_\beta^\alpha S^2[f(\theta)]^2\,d\theta_1 \qquad M_{x2} = \int_c^d yg(y)\,dy = \int_\beta^\alpha CS[f(\theta)]^2\,d\theta_2
$$

$$
M_{y1} = \int_a^b xf(x)\,dx = \int_\beta^\alpha CS[f(\theta)]^2\,d\theta_1 \qquad M_{y2} = \frac{1}{2}\int_c^d [g(y)]^2\,dy = \frac{1}{2}\int_\beta^\alpha C^2[f(\theta)]^2\,d\theta_2
$$

(4) Coordinates of Centroid of Areas in X, Y

$$
x_{C1} = M_{y1}/A_1 \qquad\qquad y_{C1} = M_{x1}/A_1 \qquad\qquad x_{C2} = M_{y2}/A_2 \qquad\qquad y_{C2} = M_{x2}/A_2
$$

(5) Inertia Functions of Areas in X, Y

$$
I_{xx1} = \frac{1}{3}\int_a^b [f(x)]^3\,dx = \frac{1}{3}\int_\beta^\alpha S^3[f(\theta)]^3\,d\theta_1 \qquad I_{xx2} = \int_a^b y^2 g(y)\,dy = \int_\beta^\alpha S^2 C[f(\theta)]^3\,d\theta_2
$$

$$
I_{yy1} = \int_a^b x^2 f(x)\,dx = \int_\beta^\alpha SC^2[f(\theta)]^3\,d\theta_1 \qquad I_{yy2} = \frac{1}{3}\int_a^b [g(y)]^3\,dy = \frac{1}{3}\int_\beta^\alpha C^3[f(\theta)]^3\,d\theta_2
$$

$$
I_{xy1} = \frac{1}{2}\int_a^b x[f(x)]^2\,dx = \frac{1}{2}\int_\beta^\alpha CS^2[f(\theta)]^3\,d\theta_1 \qquad I_{yx2} = \frac{1}{2}\int_a^b y[g(y)]^2\,dy = \frac{1}{2}\int_\beta^\alpha C^2 S[f(\theta)]^3\,d\theta_2
$$

(6) Transformations

The analytical relations for the calculation of static and inertia functions in rotated axes and the calculation of extreme values (principal moments of inertia) are formally identical to the relations given in Secs. 21.05-5 and 21.06-4, 5, 6, respectively.

Notation (Sec. 21.24) $A_1 = A_2 = A$ (for triangle only)

(1) Geometry

The geometry of a curvilinear, triangular area in terms of notation introduced in Sec. 21.01-1 is shown below. The coordinates of end points are $x_A = a$, $y_A = d$, $x_B = b$, $y_B = c$ and r_A, α, r_B, β.

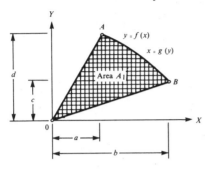
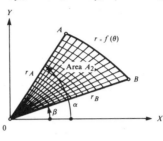

(2) Area

$$A = \tfrac{1}{2}(ad - bc) + \int_a^b f(x)\, dx = \tfrac{1}{2}(bc - ad) + \int_c^d g(y)\, dy = \tfrac{1}{2}\int_\beta^\alpha [f(\theta)]^2\, d\theta$$

(3) Moments of Area about X, Y

$$M_x = \tfrac{1}{6}(ad^2 - bc^2) + \tfrac{1}{2}\int_a^b [f(x)]^2\, dx$$

$$M_y = \tfrac{1}{6}(a^2d - b^2c) + \int_a^b xf(x)\, dx$$

$$= \tfrac{1}{6}(bc^2 - ad^2) + \int_c^d yg(y)\, dy$$

$$= \tfrac{1}{6}(b^2x - a^2d) + \tfrac{1}{2}\int_c^d [g(y)]^2\, dy$$

$$= \tfrac{1}{3}\int_\beta^\alpha S[f(\theta)]^3\, d\theta$$

$$= \tfrac{1}{3}\int_\beta^\alpha C[f(\theta)]^3\, d\theta$$

(4) Coordinates of Centroid of Area in X, Y

$$x_C = \frac{M_y}{A} \qquad y_C = \frac{M_x}{A}$$

(5) Inertia Functions of Area in X, Y

$$I_{xx} = \tfrac{1}{12}(ad^3 - bc^3) + \tfrac{1}{3}\int_a^b [f(x)]^3\, dx = \tfrac{1}{12}(bc^3 - ad^3) + \int_c^d y^2 g(y)\, dy = \tfrac{1}{4}\int_\beta^\alpha S^2[f(\theta)]^3\, d\theta$$

$$I_{yy} = \tfrac{1}{12}(a^3d - b^3c) + \tfrac{1}{3}\int_a^b x^2 f(x)\, dx = \tfrac{1}{12}(b^3c - a^3d) + \tfrac{1}{3}\int_c^d [g(y)]^3\, dy = \tfrac{1}{4}\int_\beta^\alpha C^2[f(\theta)]^3\, d\theta$$

$$I_{xy} = \tfrac{1}{8}(a^2d^2 - b^2c^2) + \tfrac{1}{2}\int_a^b x[f(x)]^2\, dx = \tfrac{1}{8}(b^2c^2 - a^2d^2) + \tfrac{1}{2}\int_c^d y[g(y)]^2\, dy$$

$$= \tfrac{1}{4}\int_\beta^\alpha CS[f(\theta)]^3\, d\theta$$

(6) Transformations

The analytical relations for the calculation of static and inertia functions in rotated and in parallel axes and the calculation of principal moments of inertia are formally identical to those given in Secs. 21.05-5 and 21.06-4, 5, 6, respectively.

| A = area | $I_{AA}, I_{BB}, I_{TT}, I_{xx}, I_{yy}$ = moments of inertia |
| x_C, y_C = coordinates of centroid | I_{AB}, I_{xy} = products of inertia |

Square

$A = a^2$

$x_C = \dfrac{a}{2}$

$y_C = \dfrac{a}{2}$

$I_{AA} = \dfrac{a^4}{12}$

$I_{BB} = \dfrac{a^4}{12}$

$I_{AB} = 0$

$I_{xx} = \dfrac{a^4}{3}$

$I_{yy} = \dfrac{a^4}{3}$

$I_{xy} = \dfrac{a^4}{4}$

Rectangle

$A = ab$

$x_C = \dfrac{a}{2}$

$y_C = \dfrac{b}{2}$

$I_{AA} = \dfrac{ab^3}{12}$

$I_{BB} = \dfrac{a^3 b}{12}$

$I_{AB} = 0$

$I_{xx} = \dfrac{ab^3}{3}$

$I_{yy} = \dfrac{a^3 b}{3}$

$I_{xy} = \dfrac{a^2 b^2}{4}$

Triangle

$A = \dfrac{bh}{2}$

$x_C = \dfrac{v - u}{3}$

$y_C = \dfrac{h}{3}$

$I_{TT} = \dfrac{bh^3}{4}$

$I_{AA} = \dfrac{bh^3}{36}$

$I_{AB} = \dfrac{(u^2 - v^2) h^2}{72}$

$I_{xx} = \dfrac{bh^3}{12}$

$I_{yy} = \dfrac{(u^3 + v^3) h}{12}$

$I_{xy} = \dfrac{(v^2 - u^2) h^2}{24}$

Triangle

$A = \dfrac{1}{2} \begin{vmatrix} x_1 & y_1 & 1 \\ x_2 & y_2 & 1 \\ x_3 & y_3 & 1 \end{vmatrix}$

$x_C = \dfrac{x_1 + x_2 + x_3}{3}$

$y_C = \dfrac{y_1 + y_2 + y_3}{3}$

x, y = coordinates in X, Y axes
$\bar{x}, \bar{y}$ = coordinates in A, B axes

$I_{AA} = \dfrac{A}{12} (\bar{y}_1^2 + \bar{y}_2^2 + \bar{y}_3^2)$

$I_{BB} = \dfrac{A}{12} (\bar{x}_1^2 + \bar{x}_2^2 + \bar{x}_3^2)$

$I_{AB} = \dfrac{A}{12} (\bar{x}_1 \bar{y}_1 + \bar{x}_2 \bar{y}_2 + \bar{x}_3 \bar{y}_3)$

Trapezoid

$A = (a + b) h$

$e = \dfrac{h}{3} \dfrac{a + 2b}{a + b}$

$f = \dfrac{h}{3} \dfrac{2a + b}{a + b}$

$I_{AA} = \dfrac{h^3}{18} \dfrac{a^2 + 4ab + b^2}{a + b}$

$I_{BB} = \dfrac{h}{6} \dfrac{a^4 - b^4}{a - b}$

$I_{AB} = 0$

$I_{xx} = \dfrac{h^3}{6} (a + 3b)$

$I_{TT} = \dfrac{h^3}{6} (3a + b)$

Regular polygon

$A = na^2 \cot \alpha$

$r = a \cot \alpha$

$R = \dfrac{a}{\sin \alpha}$

$I_{AA} = \dfrac{A}{12} [3r^2 + a^2]$

$\quad = \dfrac{A}{12} [3R^2 - 2a^2]$

$\quad = I_{BB} = I_{CC}$

A = area	$C = \cos\theta$	$I_{AA}, I_{BB}, I_{xx}, I_{yy}$ = moments of inertia
x_C, y_C = coordinates of centroid	$S = \sin\theta$	I_{AB}, I_{xy} = products of inertia

Ellipse

For circular areas use $a = b$

$A = \pi ab$

$x_C = 0$

$y_C = 0$

$I_{AA} = \dfrac{\pi ab^3}{4}$

$I_{BB} = \dfrac{\pi a^3 b}{4}$

$I_{AB} = 0$

$I_{xx} = \dfrac{5\pi ab^3}{4}$

$I_{yy} = \dfrac{5\pi a^3 b}{4}$

$I_{xy} = \pi a^2 b^2$

Half ellipse

$A = \dfrac{\pi ab}{2}$

$x_C = a$

$y_C = \dfrac{4b}{3\pi}$

$I_{AA} = \dfrac{(9\pi^2 - 64)ab^3}{72\pi}$

$I_{BB} = \dfrac{\pi a^3 b}{8}$

$I_{AB} = 0$

$I_{xx} = \dfrac{\pi ab^3}{8}$

$I_{yy} = \dfrac{5\pi a^3 b}{8}$

$I_{xy} = \dfrac{2a^2 b^2}{3}$

Quarter ellipse

$A = \dfrac{\pi ab}{4}$

$x_C = \dfrac{4a}{3\pi}$

$y_C = \dfrac{4b}{3\pi}$

$I_{AA} = \dfrac{(9\pi^2 - 64)ab^3}{144\pi}$

$I_{BB} = \dfrac{(9\pi^2 - 64)a^3 b}{144\pi}$

$I_{AB} = \dfrac{(9\pi^2 - 64)a^2 b^2}{72\pi}$

$I_{xx} = \dfrac{\pi ab^3}{16}$

$I_{yy} = \dfrac{\pi a^3 b}{16}$

$I_{xy} = \dfrac{a^2 b^2}{8}$

Elliptic complement

$\alpha = 1 - \dfrac{\pi}{4}$ $\beta = \dfrac{10 - 3\pi}{12 - 3\pi}$

$A = \alpha ab$

$x_C = \beta a$

$y_C = \beta b$

$I_{AA} = \left(1 - \dfrac{5\pi}{16} - \alpha\beta^2\right)ab^3$

$I_{BB} = \left(1 - \dfrac{5\pi}{16} - \alpha^2\beta\right)a^3 b$

$I_{AB} = \left(\dfrac{9\alpha - 2}{72\alpha} - \alpha\beta^2\right)a^2 b^2$

$I_{xx} = \left(1 - \dfrac{5\pi}{16}\right)ab^3$

$I_{yy} = \left(1 - \dfrac{5\pi}{16}\right)a^3 b$

$I_{xy} = \dfrac{(9\alpha - 2)a^2 b^2}{72\alpha}$

Elliptic sector

$\beta = \dfrac{2S}{3\theta}$

$A = \theta ab$

$x_C = \beta a$

$y_C = 0$

$I_{AA} = \dfrac{(\theta - CS)ab^3}{4}$

$I_{BB} = \dfrac{(\theta + CS - 4\theta\beta^2)a^3 b}{4}$

$I_{AB} = 0$

$I_{xx} = \dfrac{(\theta - CS)ab^3}{4}$

$I_{yy} = \dfrac{(\theta + CS)a^3 b}{4}$

$I_{xy} = 0$

Elliptic segment

$\alpha = \theta - CS$ $\beta = \dfrac{2S^3}{3(\theta - CS)}$

$A = \alpha ab$

$x_C = \beta a$

$y_C = 0$

$I_{AA} = \dfrac{\alpha(1 - \beta C)ab^3}{4}$

$I_{BB} = \dfrac{\alpha(1 + 3\beta C - 4\beta^2)a^3 b}{4}$

$I_{AB} = 0$

$I_{xx} = \dfrac{\alpha(1 - \beta C)ab^3}{4}$

$I_{yy} = \dfrac{\alpha(1 + 3\beta C)a^3 b}{4}$

$I_{xy} = 0$

A = area	$\lambda = c/b$	$I_{AA}, I_{BB}, I_{TT}, I_{xx}, I_{yy}$ = moments of inertia
x_C, y_C = coordinates of centroid	V = vertex	I_{AB}, I_{xy} = products of inertia

Parabola

$A = \dfrac{4ab}{3}$

$x_C = a$

$y_C = \dfrac{2b}{5}$

$I_{TT} = \dfrac{4ab^3}{7}$

$I_{AA} = \dfrac{16ab^3}{175}$

$I_{BB} = \dfrac{4a^3b}{15}$

$I_{AB} = 0$

$I_{xx} = \dfrac{32ab^3}{105}$

$I_{yy} = \dfrac{8a^3b}{5}$

$I_{xy} = \dfrac{8a^2b^2}{15}$

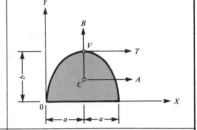

Half parabola

$A = \dfrac{2ab}{3}$

$x_C = \dfrac{3a}{8}$

$y_C = \dfrac{2b}{5}$

$I_{TT} = \dfrac{2ab^3}{7}$

$I_{AA} = \dfrac{8ab^3}{175}$

$I_{BB} = \dfrac{19a^3b}{480}$

$I_{AB} = \dfrac{a^2b^2}{60}$

$I_{xx} = \dfrac{8ab^3}{105}$

$I_{yy} = \dfrac{2a^3b}{15}$

$I_{xy} = \dfrac{a^2b^2}{8}$

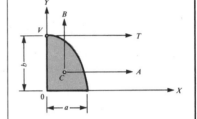

Parabolic complement

$A = \dfrac{ab}{3}$

$x_C = \dfrac{3a}{4}$

$y_C = \dfrac{3b}{10}$

$I_{AA} = \dfrac{37ab^3}{2100}$

$I_{BB} = \dfrac{a^3b}{80}$

$I_{AB} = \dfrac{a^2b^2}{120}$

$I_{xx} = \dfrac{ab^3}{21}$

$I_{yy} = \dfrac{a^3b}{5}$

$I_{xy} = \dfrac{a^2b^2}{12}$

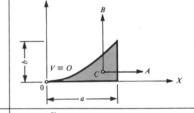

Parabolic fillet

$A = \dfrac{a^2}{6}$

$x_C = \dfrac{2a}{5}$

$y_C = \dfrac{2a}{5}$

$I_{AA} = \dfrac{11a^4}{2100}$

$I_{BB} = \dfrac{11a^4}{2100}$

$I_{AB} = \dfrac{13a^4}{4200}$

$I_{xx} = \dfrac{13a^4}{700}$

$I_{yy} = \dfrac{13a^4}{700}$

$I_{xy} = \dfrac{43a^4}{4200}$

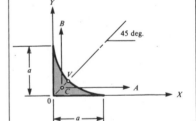

Parabolic sector

$A = \dfrac{(4 + 3\lambda)ab}{3}$

$x_C = \dfrac{(8 + 20\lambda + 10\lambda^2)b}{4 + 3\lambda}$

$y_C = 0$

$I_{xx} = \dfrac{(5\lambda + 8)a^3b}{15}$

$I_{yy} = \dfrac{(64 + 224\lambda + 280\lambda^2 + 105\lambda^3)ab^3}{210}$

$I_{xy} = 0$

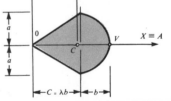

Parabolic segment

$A = \dfrac{4ab}{3}$

$x_C = \dfrac{(2 + 5\lambda)b}{5}$

$y_C = 0$

$I_{xx} = \dfrac{4a^3b}{15}$

$I_{yy} = \dfrac{(64 + 224\lambda + 280\lambda^2)ab^3}{210}$

$I_{xy} = 0$

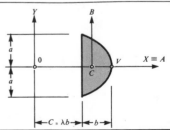

22

SPACE CURVES
AND SURFACES

(1) Equations of a Space Curve

A space curve is defined analytically in one of the following forms:

(A) Intersection form	(B) Projection form
$F(x, y, z) = 0 \qquad G(x, y, z) = 0$	$y = f(x) \qquad z = g(x)$
(C) Parametric form in t	(D) Parametric form in s
$x = x(t) \qquad y = y(t) \qquad z = z(t)$	$x = x(s) \qquad y = y(s) \qquad z = z(s)$
(E) Cartesian vector form in t	(F) Cartesian vector form in s
$\mathbf{r}(t) = x(t)\mathbf{i} + y(t)\mathbf{j} + z(t)\mathbf{k}$	$\mathbf{r}(s) = x(s)\mathbf{i} + y(s)\mathbf{j} + z(s)\mathbf{k}$

where in (A), $F(x, y)$, $G(x, y)$ are the implicit functions representing two surfaces whose intersection is the curve; in (B), $f(x)$, $g(x)$ are the projections of the curve in XY and YZ planes, respectively; in (C), t is a selected parameter; in (D), s is the curvilinear coordinate measured along the curve; and in (E, F), the parametric equations of (C, D) are used as scalar functions multiplied by the respective cartesian unit vectors $\mathbf{i}$, $\mathbf{j}$, $\mathbf{k}$.

(2) Derivatives, Form (A)

$$\frac{dy}{dx} = y_x = \frac{F_z G_x - F_x G_z}{F_y G_z - F_z G_y} \qquad \frac{dx}{dy} = x_y = \frac{F_y G_z - F_z G_y}{F_z G_x - F_x G_z} \qquad \frac{d^2y}{dx^2} = y_{xx} \qquad \frac{d^2x}{dy^2} = x_{yy}$$

$$\frac{dz}{dx} = z_x = \frac{F_x G_y - F_y G_x}{F_y G_z - F_z G_y} \qquad \frac{dx}{dz} = x_z = \frac{F_y G_z - F_z G_y}{F_x G_y - F_y G_x} \qquad \frac{d^2z}{dx^2} = z_{xx} \qquad \frac{d^2x}{dz^2} = x_{zz}$$

where F_x, F_y, . . . , G_x, G_y are the partial derivatives of $F(x, y, z)$, $G(x, y, z)$ defined in Sec. 7.03.

(3) Derivatives, Form (B)

$$dy/dx = f_x \qquad dx/dy = 1/f_x \qquad d^2y/dx^2 = f_{xx} \qquad d^2x/dy^2 = -f_{xx}/f_x^3$$

$$dz/dx = g_x \qquad dx/dz = 1/g_x \qquad d^2z/dx^2 = g_{xx} \qquad d^2x/dz^2 = -g_{xx}/g_x^3$$

(4) Derivatives, Forms (C, D)

$\dot{x} = dx(t)/dt$	$\ddot{x} = d^2x(t)/dt^2$	$x' = dx(s)/ds$	$x'' = d^2x(s)/ds^2$
$\dot{y} = dy(t)/dt$	$\ddot{y} = d^2y(t)/dt^2$	$y' = dy(s)/ds$	$y'' = d^2y(s)/ds^2$
$\dot{z} = dz(t)/dt$	$\ddot{z} = d^2z(t)/dt^2$	$z' = dz(s)/ds$	$z'' = d^2z(s)/ds^2$

(5) Derivatives, Forms (E, F)

$$\dot{\mathbf{r}}(t) = \frac{d\mathbf{r}(t)}{dt} = \dot{x}\mathbf{i} + \dot{y}\mathbf{j} + \dot{z}\mathbf{k} \qquad\qquad \mathbf{r}'(s) = \frac{d\mathbf{r}(s)}{ds} = x'\mathbf{i} + y'\mathbf{j} + z'\mathbf{k}$$

$$\ddot{\mathbf{r}}(t) = \frac{d^2\mathbf{r}(t)}{dt^2} = \ddot{x}\mathbf{i} + \ddot{y}\mathbf{j} + \ddot{z}\mathbf{k} \qquad\qquad \mathbf{r}''(s) = \frac{d^2\mathbf{r}(s)}{ds^2} = x''\mathbf{i} + y''\mathbf{j} + z''\mathbf{k}$$

(6) Differential Elements, Forms (A, B, C, D)

$$ds_A = \sqrt{1 + (y')^2 + (z')^2}\, dx \qquad\qquad ds_C = \sqrt{(\dot{x})^2 + (\dot{y})^2 + (\dot{z})^2}\, dt$$

$$ds_B = \sqrt{1 + (f_x)^2 + (g_x)^2}\, dx \qquad\qquad ds_D = \sqrt{(x')^2 + (y')^2 + (z')^2}\, ds$$

where the subscripts A, B, C, D identify the respective forms.

(1) Moving Trihedral

At every point of the curve, three othogonal planes intersect in three straight lines and form the moving trihedral X^s, Y^s, Z^s shown in the adjacent figure. The planes are the *osculating plane* X^sY^s, the *normal plane* Y^sZ^s, and the *rectifying plane* Z^sX^s. Their intersections are the *tangent* X^s of unit vector $\mathbf{e}_t$, the *principal normal* Y^s of unit vector $\mathbf{e}_n$, and the *binormal* Z^s of unit vector $\mathbf{e}_b$. The relationship between these unit vectors and their cartesian counterparts are

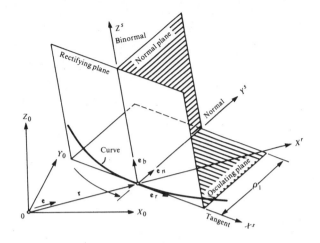

$$
\begin{bmatrix} \mathbf{e}_t \\ \mathbf{e}_n \\ \mathbf{e}_b \end{bmatrix} =
\begin{bmatrix} x_t & \beta_t & \gamma_t \\ x_n & \beta_n & \gamma_n \\ x_b & \beta_b & \gamma_b \end{bmatrix}
\begin{bmatrix} \mathbf{i} \\ \mathbf{j} \\ \mathbf{k} \end{bmatrix}
\qquad\qquad
\begin{bmatrix} \mathbf{i} \\ \mathbf{j} \\ \mathbf{k} \end{bmatrix} =
\begin{bmatrix} \alpha_t & \alpha_n & \alpha_b \\ \beta_t & \beta_n & \beta_b \\ \gamma_t & \gamma_n & \gamma_b \end{bmatrix}
\begin{bmatrix} \mathbf{e}_t \\ \mathbf{e}_n \\ \mathbf{e}_b \end{bmatrix}
$$

where $(\alpha_t, \beta_t, \gamma_t)$, $(\alpha_n, \beta_n, \gamma_n)$, $(\alpha_b, \beta_b, \gamma_b)$ are the direction cosines of the X^s, Y^s, Z^s, respectively. Their analytical forms are given below.

(2) Direction Cosines

$$
\begin{array}{lll}
\Delta_{A1} = \sqrt{1 + y_x^2 + z_x^2} & \Delta_{A2} = \sqrt{y_x^4 x_{yy}^2 + y_{xx}^2 + z_{xx}^2} & \Delta_{A3} = \Delta_{A1}\Delta_{A2} \\[6pt]
\Delta_{B1} = \sqrt{1 + f_x^2 + g_x^2} & \Delta_{B2} = \sqrt{f_{xx}^2/f_x^2 + f_{xx}^2 + g_{xx}^2} & \Delta_{B3} = \Delta_{B1}\Delta_{B2} \\[6pt]
\Delta_{C1} = \sqrt{(\dot{x})^2 + (\dot{y})^2 + (\dot{z})^2} & \Delta_{C2} = \sqrt{(\ddot{x})^2 + (\ddot{y})^2 + (\ddot{z})^2} & \Delta_{C3} = \Delta_{C1}\Delta_{C2} \\[6pt]
\Delta_{D1} = 1 & \Delta_{D2} = \sqrt{(x'')^2 + (y'')^2 + (z'')^2} & \Delta_{D3} = \Delta_{D2}
\end{array}
$$

	Form (A)	Form (B)	Form (C)	Form (D)
α_t	$1/\Delta_{A1}$	$1/\Delta_{B1}$	$\dot{x}/\Delta_{C1}$	x'
β_t	y_x/Δ_{A1}	f_x/Δ_{B1}	$\dot{y}/\Delta_{C1}$	y'
γ_t	z_x/Δ_{A1}	g_x/Δ_{B1}	$\dot{z}/\Delta_{C1}$	z'
α_n	$-x_{yy}y_x^2/\Delta_{A2}$	$-f_{xx}f_x^{-2}/\Delta_{B2}$	$\ddot{x}/\Delta_{C2}$	x''/Δ_{D2}
β_n	y_{xx}/Δ_{A2}	f_{xx}/Δ_{B2}	$\ddot{y}/\Delta_{C2}$	y''/Δ_{D2}
γ_n	z_{xx}/Δ_{A2}	g_{xx}/Δ_{B2}	$\ddot{z}/\Delta_{C2}$	z''/Δ_{D2}
α_b	$\dfrac{1}{\Delta_{A3}}\begin{vmatrix} y_x & z_x \\ y_{xx} & z_{xx} \end{vmatrix}$	$\dfrac{1}{\Delta_{B3}}\begin{vmatrix} f_x & g_x \\ f_{xx} & g_{xx} \end{vmatrix}$	$\dfrac{1}{\Delta_{C3}}\begin{vmatrix} \dot{y} & \dot{z} \\ \ddot{y} & \ddot{z} \end{vmatrix}$	$\dfrac{1}{\Delta_{D3}}\begin{vmatrix} y' & z' \\ y'' & z'' \end{vmatrix}$
β_b	$\dfrac{1}{\Delta_{A3}}\begin{vmatrix} z_x & 1 \\ z_{xx} & x_{yy}y_x^2 \end{vmatrix}$	$\dfrac{1}{\Delta_{B3}}\begin{vmatrix} g_x & 1 \\ g_{xx} & \dfrac{f_{xx}}{f_x} \end{vmatrix}$	$\dfrac{1}{\Delta_{C3}}\begin{vmatrix} \dot{z} & \dot{x} \\ \ddot{z} & \ddot{x} \end{vmatrix}$	$\dfrac{1}{\Delta_{D3}}\begin{vmatrix} z' & x' \\ z'' & x'' \end{vmatrix}$
γ_b	$\dfrac{1}{\Delta_{A3}}\begin{vmatrix} 1 & y_x \\ -x_{yy}y_x^2 & y_{xx} \end{vmatrix}$	$\dfrac{1}{\Delta_{B3}}\begin{vmatrix} 1 & f_x \\ -\dfrac{f_{xx}}{f_x} & f_{xx} \end{vmatrix}$	$\dfrac{1}{\Delta_{C3}}\begin{vmatrix} \dot{x} & \dot{y} \\ \ddot{x} & \ddot{y} \end{vmatrix}$	$\dfrac{1}{\Delta_{D3}}\begin{vmatrix} x' & y' \\ x'' & y'' \end{vmatrix}$

x, y, z = coordinates of running point　　　　x_1, y_1, z_1 = coordinates of contact point

$\mathbf{r}$ = position vector of running point　　　　$\mathbf{r}_1$ = position vector of contact point

For definitions of forms (A, B, C, D, E, F) and unit vectors $\mathbf{e}_t$, $\mathbf{e}_u$, $\mathbf{e}_b$, refer to Secs. 22.01 and 22.02, respectively.

(1) Equations of Axes in Algebraic Forms (A, B, C, D)

$$\text{Tangent } X^s: \quad \frac{x - x_1}{\alpha_t} = \frac{y - y_1}{\beta_t} = \frac{z - z_1}{\gamma_t}$$

$$\text{Normal } Y^s: \quad \frac{x - x_1}{\alpha_n} = \frac{y - y_1}{\beta_n} = \frac{z - z_1}{\gamma_n}$$

$$\text{Binormal } Z^s: \quad \frac{x - x_1}{\alpha_b} = \frac{y - y_1}{\beta_b} = \frac{z - z_1}{\gamma_b}$$

where $\alpha_t, \beta_t, \gamma_t, \alpha_n, \beta_n, \gamma_n, \alpha_b, \beta_b, \gamma_b$ are the direction cosines of the tangent, normal, and binormal, respectively (Sec. 22.02-2).

(2) Equations of Axes in Vector Forms (E, F)

Tangent X^s:　$(\mathbf{r} - \mathbf{r}_1) \times (\alpha_t \mathbf{i} + \beta_t \mathbf{j} + \gamma_t \mathbf{k}) = 0$

Normal Y^s:　$(\mathbf{r} - \mathbf{r}_1) \times (\alpha_n \mathbf{i} + \beta_n \mathbf{j} + \gamma_n \mathbf{k}) = 0$

Binormal Z^s:　$(\mathbf{r} - \mathbf{r}_1) \times (\alpha_b \mathbf{i} + \beta_b \mathbf{j} + \gamma_b \mathbf{k}) = 0$

where $\mathbf{r} - \mathbf{r}_1 = (x - x_1)\mathbf{i} + (y - y_1)\mathbf{j} + (z - z_1)\mathbf{k}$.

(3) Equations of Planes in Algebraic Forms (A, B, C, D)

Plane $X^s Y^s$:　$(x - x_1)\alpha_b + (y - y_1)\beta_b + (z - z_1)\gamma_b = 0$

Plane $Y^s Z^s$:　$(x - x_1)\alpha_t + (y - y_1)\beta_t + (z - z_1)\gamma_t = 0$

Plane $Z^s X^s$:　$(x - x_1)\alpha_n + (y - y_1)\beta_n + (z - z_1)\gamma_n = 0$

where $\alpha_t, \beta_t, \ldots, \gamma_b$ are the same as in (1) above.

(4) Equations of Planes in Vector Forms (E, F)

Plane $X^s Y^s$:　$(\mathbf{r} - \mathbf{r}_1) \cdot (\alpha_b \mathbf{i} + \beta_b \mathbf{j} + \gamma_b \mathbf{k}) = 0$

Plane $Y^s Z^s$:　$(\mathbf{r} - \mathbf{r}_1) \cdot (\alpha_t \mathbf{i} + \beta_t \mathbf{j} + \gamma_t \mathbf{k}) = 0$

Plane $Z^s X^s$:　$(\mathbf{r} - \mathbf{r}_1) \cdot (\alpha_n \mathbf{i} + \beta_n \mathbf{j} + \gamma_n \mathbf{k}) = 0$

where $\mathbf{r} - \mathbf{r}_1$ is the same as in (2) above.

(5) Unit Vectors

Tangent unit vector:　$\mathbf{e}_t = \alpha_t \mathbf{i} + \beta_t \mathbf{j} + \gamma_t \mathbf{k} = \mathbf{e}_n \times \mathbf{e}_b$

Normal unit vector:　$\mathbf{e}_n = \alpha_n \mathbf{i} + \beta_n \mathbf{j} + \gamma_n \mathbf{k} = \mathbf{e}_b \times \mathbf{e}_t$

Binormal unit vector:　$\mathbf{e}_b = \alpha_b \mathbf{i} + \beta_b \mathbf{j} + \gamma_b \mathbf{k} = \mathbf{e}_t \times \mathbf{e}_n$

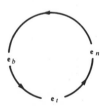

K = curvature $\qquad\qquad$ ρ_K = radius of curvature $\qquad\qquad$ $D_1 = \sqrt{(\dot{x})^2 + (\dot{y})^2 + (\dot{z})^2}$

T = torsion $\qquad\qquad$ ρ_T = radius of torsion $\qquad\qquad$ $D_2 = \sqrt{(x'')^2 + (y'')^2 + (z'')^2}$

$D_3 = \sqrt{(\dot{y}\ddot{z} - \ddot{y}\dot{z})^2 + (\dot{z}\ddot{x} - \ddot{z}\dot{x})^2 + (\dot{x}\ddot{y} - \ddot{x}\dot{y})^2}$

For definition of forms (C, D, E, F) refer to Sec. 22.01.

(1) Curvature in Algebraic Forms (C, D)

$$K_C = \frac{D_3}{D_1^3} = \frac{1}{\rho_{KC}} \qquad\qquad\qquad K_D = D_2 = \frac{1}{\rho_{KD}}$$

The curvature and radius of curvature are always positive.

(2) Curvature in Vector Forms (E, F)

$$K_E = \sqrt{\frac{(\dot{\mathbf{r}} \times \ddot{\mathbf{r}})^2}{(\dot{\mathbf{r}}^2)^3}} = \frac{1}{\rho_{KE}} \qquad\qquad K_F = \sqrt{\mathbf{r}'' \cdot \mathbf{r}''} = \frac{1}{\rho_{KF}}$$

(3) Torsion in Algebraic Forms (C, D)

$$T_C = \frac{1}{D_3^2}\begin{vmatrix} \dot{x} & \ddot{x} & \dddot{x} \\ \dot{y} & \ddot{y} & \dddot{y} \\ \dot{z} & \ddot{z} & \dddot{z} \end{vmatrix} = \frac{1}{\rho_{TC}} \qquad\qquad T_D = \frac{1}{D_2^2}\begin{vmatrix} x' & x'' & x''' \\ y' & y'' & y''' \\ z' & z'' & z''' \end{vmatrix} = \frac{1}{\rho_{TD}}$$

where $\dddot{x} = \dfrac{d^3 x(t)}{dt^3}, \ldots, x''' = \dfrac{d^3 x(s)}{ds^3}, \ldots$, and the torsion is positive if the binormal Z^s rotates in the right-hand direction about the tangent X^s and is negative in the opposite direction.

(4) Torsion in Vector Forms (E, F)

$$T_E = \rho_{KE}^2 \frac{\dot{\mathbf{r}} \cdot \ddot{\mathbf{r}} \times \dddot{\mathbf{r}}}{(\dot{\mathbf{r}} \cdot \dot{\mathbf{r}})^3} = \frac{1}{\rho_{TE}} \qquad\qquad T_F = \rho_{KF}^2(\mathbf{r}' \cdot \mathbf{r}'' \times \mathbf{r}''') = \frac{1}{\rho_{TF}}$$

where $\dddot{\mathbf{r}} = \dfrac{d^3 \mathbf{r}(t)}{dt^3}$ $\quad$ and $\quad$ $\mathbf{r}''' = \dfrac{d^3 \mathbf{r}(s)}{ds^3}$.

(5) Coordinates of Centers

The *position vector of the center of curvature* in all forms is

$$\mathbf{r}_K = (x_1 + \rho_K\alpha_n)\mathbf{i} + (y_1 + \rho_K\beta_n)\mathbf{j} + (z_1 + \rho_K\gamma_n)\mathbf{k}$$

The *position vector of center of torsion* in all forms is

$$\mathbf{r}_T = (x_1 + \rho_T\alpha_b)\mathbf{i} + (y_1 + \rho_T\beta_b)\mathbf{j} + (z_1 + \rho_T\gamma_b)\mathbf{k}$$

(6) Serret-Frenet Matrix Equation

The relation of the unit vectors and their derivatives are

$$\begin{bmatrix} \dot{\mathbf{e}}_t \\ \dot{\mathbf{e}}_n \\ \dot{\mathbf{e}}_b \end{bmatrix} = \dot{s}\begin{bmatrix} 0 & K & 0 \\ -K & 0 & T \\ 0 & -T & 0 \end{bmatrix}\begin{bmatrix} \mathbf{e}_t \\ \mathbf{e}_n \\ \mathbf{e}_b \end{bmatrix} \qquad\qquad \begin{bmatrix} \mathbf{e}_t' \\ \mathbf{e}_n' \\ \mathbf{e}_b' \end{bmatrix} = \begin{bmatrix} 0 & K & 0 \\ -K & 0 & T \\ 0 & -T & 0 \end{bmatrix}\begin{bmatrix} \mathbf{e}_t \\ \mathbf{e}_n \\ \mathbf{e}_b \end{bmatrix}$$

where $\dot{\mathbf{e}}_t = \dfrac{d\mathbf{e}_t}{dt}, \ldots, \mathbf{e}_t' = \dfrac{d\mathbf{e}_t}{ds}, \ldots$, and

$$\dot{s} = \sqrt{(\dot{x})^2 + (\dot{y})^2 + (\dot{z})^2}.$$

$R, h = $ constants	$S_1 = \sin t$	$S_2 = \sin (\ln t)$
$\lambda = h/R = \tan \phi$	$C_1 = \cos t$	$C_2 = \cos (\ln t)$
$a = \sqrt{1 + \lambda^2}$	$b = \sqrt{2 + \lambda^2}$	$t = $ parameter

For definitions of other symbols refer to Secs 22.01–22.04.

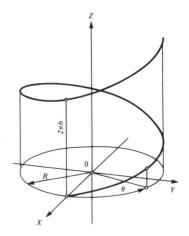

(1) Cylindrical Helix in Form (C)

(a) Geometry of the helix is defined by the circular cylinder of radius R on the surface of which the curve rises with $\phi = \tan^{-1} \lambda$. The pitch of the helix is $p = 2\pi h$.

(b) Governing equations and derivatives

$$x = RC_1 \qquad y = RS_1 \qquad z = \lambda Rt$$

$$\dot{x} = -RS_1 \qquad \dot{y} = RC_1 \qquad \dot{z} = \lambda R$$

$$\ddot{x} = -RC_1 \qquad \ddot{y} = -RS_1 \qquad \ddot{z} = 0$$

(c) Direction cosines, curvature and torsion

$\alpha_t = -\dfrac{S_1}{a}$	$\beta_t = \dfrac{C_1}{a}$	$\gamma_t = \dfrac{\lambda}{a}$	$\dfrac{K}{T} = \lambda$
$\alpha_n = -C_1$	$\beta_n = -S_1$	$\gamma_n = 0$	$\rho_K = Ra^2$
$\alpha_b = \dfrac{\lambda S_1}{a}$	$\beta_b = -\dfrac{\lambda C_1}{a}$	$\gamma_b = \dfrac{1}{a}$	$\rho_T = \dfrac{Ra^2}{\lambda}$

(2) Conical Helix in Form (C)

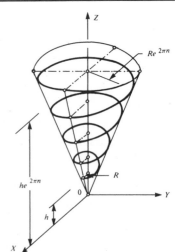

(a) Geometry of the helix is defined by the circular cone with vertex at 0 on the surface of which the curve rises with $\phi = \tan^{-1} [\lambda/\sqrt{1 + \lambda^2}]$. The pitch is

$$p_n = h[e^{2\pi n} - e^{2\pi(n-1)}]$$

with $n = 1, 2, 3, \ldots$.

(b) Governing equations and derivatives

$$x = RtC_2 \qquad y = RtS_2 \qquad z = \lambda Rt$$

$$\dot{x} = R(C_2 - S_2) \qquad \dot{y} = R(C_2 + S_2) \qquad \dot{z} = \lambda R$$

$$\ddot{x} = -\frac{R}{t}(C_2 + S_2) \qquad \ddot{y} = \frac{R}{t}(C_2 - S_2) \qquad \ddot{z} = 0$$

(c) Direction cosines, curvature, torsion

$\alpha_t = \dfrac{C_2 - S_2}{b}$	$\beta_t = \dfrac{C_2 + S_2}{b}$	$\gamma_t = \dfrac{\lambda}{b}$	$\dfrac{K}{T} = \dfrac{\lambda}{\sqrt{2}}$
$\alpha_n = -\dfrac{C_2 + S_2}{2}$	$\beta_n = \dfrac{C_2 - S_2}{2}$	$\gamma_n = 0$	$\rho_K = \dfrac{b^2 Rt}{\sqrt{2}}$
$\alpha_b = -\dfrac{(C_2 - S_2)}{b\sqrt{2}}$	$\beta_b = -\dfrac{(C_2 + S_2)}{b\sqrt{2}}$	$\gamma_b = \dfrac{\sqrt{2}}{c}$	$\rho_T = \dfrac{b^2 Rt}{\lambda}$

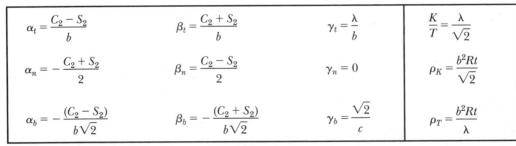

$$\Delta_1 = \sqrt{a^2b^2 - x^4}$$

$$\Delta_2 = \sqrt{a^6b^6 + b^4x^2y^6}$$

$$\Delta_3 = \sqrt{(a^2 - b^2)^2x^6 + b^4y^6 + a^4z^6} \qquad t = \text{parameter}$$

$$\Delta_4 = R\sqrt{1 + \cos^2 t}$$

$$\Delta_5 = R\sqrt{4 + \sin^2 t}$$

$$C = \cos t \qquad S = \sin t$$

For definitions of other symbols refer to Secs. 22.01–22.04.

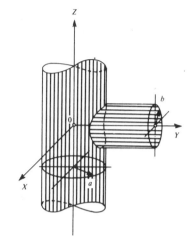

(1) Cylinder-Cylinder Intersection Curve in Form (*B*)

(a) Curve of intersection of two cylinders of radii a, b,

$$x^2 + y^2 = a^2 \qquad x^2 + z^2 = b^2$$

(b) Governing equations and derivatives

$$y = \sqrt{a^2 - x^2} = f \qquad\qquad z = \sqrt{b^2 - x^2} = g$$

$$\frac{dy}{dx} = -\frac{x}{y} = f_x \qquad\qquad \frac{dz}{dx} = -\frac{x}{z} = g_x$$

$$\frac{d^2y}{dx^2} = -\frac{a^2}{y^3} = f_{xx} \qquad\qquad \frac{d^2z}{dx^2} = -\frac{b^2}{z^3} = g_{xx}$$

$$\frac{d^3y}{dx^3} = -\frac{3a^2x}{y^5} = f_{xxx} \qquad\qquad \frac{d^3z}{dx^3} = -\frac{3b^2x}{z^5} = g_{xxx}$$

(c) Direction cosines, curvature, torsion

$\alpha_t = \dfrac{yz}{\Delta_1}$	$\beta_t = \dfrac{zx}{\Delta_1}$	$\gamma_t = \dfrac{xy}{\Delta_1}$	$\rho_K = \dfrac{xy^3z^3}{\Delta_2}$
$\alpha_n = \dfrac{a^2yz^3}{\Delta_2}$	$\beta_n = -\dfrac{a^2xz^3}{\Delta_2}$	$\gamma_n = -\dfrac{b^2xy^3}{\Delta_2}$	$\rho_T = \dfrac{\Delta_1{}^2\Delta_2}{3a^2b^2x^3(y^3 - z^2)}$
$\alpha_b = -\dfrac{(a^2 - b^2)x^3}{\Delta_3}$	$\beta_b = \dfrac{b^2y^3}{\Delta_3}$	$\gamma_b = -\dfrac{a^2z^3}{\Delta_3}$	

(2) Sphere-Cylinder Intersection Curve in Form (*C*)

(a) Curve of intersection of sphere of radius R and cylinder of radius $R/2$,

$$x^2 + y^2 + z^2 = R^2 \qquad x^2 + y^2 - Rx = 0$$

(b) Governing equations and derivatives

$$x = RC^2 \qquad\qquad y = RCS \qquad\qquad z = RS$$

$$\dot{x} = -2RCS \qquad\qquad \dot{y} = R(C^2 - S^2) \qquad\qquad \dot{z} = RC$$

$$\ddot{x} = -2R(C^2 - S^2) \qquad\qquad \ddot{y} = -4RCS \qquad\qquad \ddot{z} = -RS$$

$$\dddot{x} = 8RCS \qquad\qquad \dddot{y} = -4R(C^2 - S^2) \qquad\qquad \dddot{z} = -RC$$

(c) Direction cosines, curvature, torsion

$\alpha_t = -\dfrac{2RCS}{\Delta_4}$	$\beta_t = \dfrac{R(C^2 - S^2)}{\Delta_4}$	$\gamma_t = \dfrac{RC}{\Delta_4}$	$\rho_K = R\sqrt{\dfrac{(1 + C^2)^3}{5 + 3C^2}}$
$\alpha_n = -\dfrac{2R(C^2 - S^2)}{\Delta_5}$	$\beta_n = -\dfrac{4RCS}{\Delta_5}$	$\gamma_n = -\dfrac{RS}{\Delta_5}$	$\rho_T = R\dfrac{5 + 3C^2}{6C}$
$\alpha_b = \dfrac{R^2S(2C^2 + 1)}{\Delta_4\Delta_5}$	$\beta_b = -\dfrac{2R^2C^3}{\Delta_4\Delta_5}$	$\gamma_b = \dfrac{2R^2}{\Delta_4\Delta_5}$	

(1) Equations of a Surface

A surface is defined analytically in one of the following forms:

(A) Implicit form	(B) Explicit form
$F(x, y, z) = 0$	$z = f(x, y)$
(C) Parametric form	(D) Vector form
$X = X(u, v)$ $\qquad$ $Y = Y(u, v)$ $\qquad$ $Z = Z(u, v)$	$R = X(u, v)\mathbf{i} + Y(u, v)\mathbf{j} + Z(u, v)\mathbf{k}$

where X, Y, Z are the cartesian coordinates and u, v are parameters called *surface coordinates*. Because the vector form allows a concise representation of the analytical properties of the surface, all subsequent relations are shown in this form.

(2) Position Vector and Parametric Derivative

$$\mathbf{R} = X\mathbf{i} + Y\mathbf{j} + Z\mathbf{k}$$

$$\frac{\partial R}{\partial u} = \mathbf{R}_u = X_u\mathbf{i} + Y_u\mathbf{j} + Z_u\mathbf{k} \qquad\qquad \frac{\partial R}{\partial u} = \mathbf{R}_v = X_v\mathbf{i} + Y_v\mathbf{j} + Z_v\mathbf{k}$$

$$\frac{\partial^2 R}{\partial u^2} = \mathbf{R}_{uv} = X_{uu}\mathbf{i} + Y_{uu}\mathbf{j} + Z_{uu}\mathbf{k} \qquad\qquad \frac{\partial^2 R}{\partial v^2} = \mathbf{R}_{vv} = X_{vv}\mathbf{i} + Y_{vv}\mathbf{j} + Z_{vv}\mathbf{k}$$

$$\frac{\partial^2 \mathbf{R}}{\partial u\,\partial v} = \mathbf{R}_{uv} = X_{uv}\mathbf{i} + Y_{uv}\mathbf{j} + Z_{uv}\mathbf{k}$$

where $X_u = \dfrac{\partial X}{\partial u}$, $X_{uu} = \dfrac{\partial^2 X}{\partial u^2}$, $X_{uv} = \dfrac{\partial^2 X}{\partial u\,\partial v}$, . . . , are the partial derivatives of X, Y, Z with respect to u and v. $\mathbf{R}$ designates the position vector of a running point, and $\mathbf{R}_1$ defines the position of a particular point.

(3) Unit Vectors

The first derivatives of $\mathbf{R}$ are

$\mathbf{R}_u$ = unit tangent vector in u direction

$\mathbf{R}_v$ = unit tangent vector in v direction

$$\mathbf{n} = \frac{\mathbf{R}_u \times \mathbf{R}_v}{|\mathbf{R}_u \times \mathbf{R}_v|} = \text{unit normal vector}$$

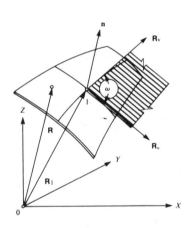

(4) Angle of Surface Coordinates

The angle ω between two curves u and v is given by

$$\cos \omega = \mathbf{R}_u \cdot \mathbf{R}_v$$

and for $\omega = \pi/2$, the *surface coordinates are orthogonal.*

(5) Tangent Plane of Surface

$$\boxed{(\mathbf{R} - \mathbf{R}_1) \cdot (\mathbf{R}_u \times \mathbf{R}_v) = 0}$$

$$\begin{vmatrix} X - X_1 & Y - Y_1 & Z - Z_1 \\ X_u & Y_u & Z_u \\ X_v & Y_v & Z_v \end{vmatrix} = 0$$

(6) Normal of Surface

$$\boxed{\frac{\mathbf{R} - \mathbf{R}_1}{|\mathbf{R} - \mathbf{R}_1|} \times \mathbf{n} = 0}$$

$$\frac{X - X_1}{\begin{vmatrix} Y_u & Z_u \\ Y_v & Z_v \end{vmatrix}} = \frac{Y - Y_1}{\begin{vmatrix} Z_u & X_u \\ Z_v & X_v \end{vmatrix}} = \frac{Z - Z_1}{\begin{vmatrix} X_u & Y_u \\ X_v & Y_v \end{vmatrix}}$$

(1) First Fundamental Quantity in Form (C)

$$\Phi_1 = ds^2 = d\mathbf{R} \cdot d\mathbf{R} = E\,du^2 + 2F\,du\,dv + G\,dv^2$$

where

$$E = \mathbf{R}_u \cdot \mathbf{R}_u \qquad F = \mathbf{R}_u \cdot \mathbf{R}_v \qquad G = \mathbf{R}_v \cdot \mathbf{R}_v$$

and ds^2 is the elemental area of the surface. For $\omega = \pi/2$, $F = 0$.

(2) Second Fundamental Quantity in Form (C)

$$\Phi_2 = d\mathbf{R} \cdot d\mathbf{n} = L\,du^2 + 2M\,du\,dv + N\,dv^2$$

where

$$L = \frac{\mathbf{R}_{uu} \times \mathbf{R}_u \cdot \mathbf{R}_v}{\sqrt{EG - F^2}} \qquad M = -\frac{\mathbf{R}_u \times \mathbf{R}_{uv} \cdot \mathbf{R}_v}{\sqrt{EG - F^2}} \qquad N = \frac{\mathbf{R}_u \cdot \mathbf{R}_v \times \mathbf{R}_{vv}}{\sqrt{EG - F^2}}$$

and E, F, G are the same as in (1) above.

(3) Fundamental Quantities in Form (B)

For $z = f(x, y)$ written as $x = u$, $y = v$, $z = f(u, v)$ and with the notation

$$f_x = \frac{\partial z}{\partial x} \qquad f_y = \frac{\partial z}{\partial y} \qquad f_{xx} = \frac{\partial^2 z}{\partial x^2} \qquad f_{xy} = \frac{\partial^2 z}{\partial x\,\partial y} \qquad f_{yy} = \frac{\partial^2 x}{\partial y^2}$$

the parameters of Φ_1 and Φ_2 reduce to

$$E = 1 + f_x^2 \qquad\qquad F = 1 + f_x f_y \qquad\qquad G = 1 + f_y^2$$

$$L = \frac{f_{xx}}{\sqrt{1 + f_x^2 + f_y^2}} \qquad M = \frac{f_{xy}}{\sqrt{1 + f_x^2 + f_y^2}} \qquad N = \frac{f_{yy}}{\sqrt{1 + f_x^2 + f_y^2}}$$

(4) Curvature in Form (C)

The curvature at a point of the surface is

$$K_n = \frac{1}{\rho_n} = \frac{\Phi_2}{\Phi_1} = L\left(\frac{du}{ds}\right)^2 + 2M\left(\frac{du\,dv}{ds\,ds}\right) + N\left(\frac{dv}{ds}\right)^2$$

where ρ_n is the radius of curvature of the curve cut by the normal plane at that point. When the normal plane rotates about $\mathbf{n}$, ρ varies between between two extreme values ρ_1 and ρ_2, called the *principal radii of curvature*, given by the solution of

$$\frac{1}{\rho_1} + \frac{1}{\rho_2} = \frac{GL - 2FM - EN}{EG - F^2} \qquad\qquad \frac{1}{\rho_1}\frac{1}{\rho_2} = \frac{LN - M^2}{EG - F^2}$$

The directions of the principal planes of curvature are angles,

$$\alpha_{1,2} = \tan^{-1}\frac{-(EN - GL) \pm \sqrt{(EN - GL)^2 - 4(EM - FL)(FN - GM)}}{2(EM - FL)}$$

measured from $\mathbf{R}_u$ in the right-hand direction.

(5) Classification of Surfaces

According to the sign of the product of curvatures, the surface at a point is classified as *synclastic* $\left(\dfrac{1}{\rho_1\rho_2} > 0\right)$, *developable* $\left(\dfrac{1}{\rho_1\rho_2} = 0\right)$, or *anticlastic* $\left(\dfrac{1}{\rho_1\rho_2} < 0\right)$, and the points are called elliptic, parabolic, hyperbolic, respectively.

(1) Definition of a Surface

A surface of revolution is generated by the rotation of a plane curve (meridian) about an axis in the plane of this curve. The surface is defined analytically in:

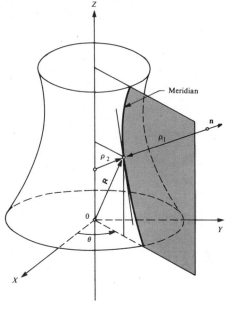

— Meridian

Form (E)	Form (F)
$x = r \cos \theta$	$x = g(z) \cos \theta$
$y = r \sin \theta$	$y = g(z) \sin \theta$
$z = f(r)$	$z = z$

where r, θ, z are cylindrical coordinates and $f(r)$, $g(z)$ are abbreviated below as f, g, respectively.

(2) Position Vector and Derivatives in Form (E)

$$\mathbf{R} = r \cos \theta \mathbf{i} + r \sin \theta \mathbf{j} + f \mathbf{k}$$

$$\mathbf{R}_\theta = -r \sin \theta \mathbf{i} + r \cos \theta \mathbf{j}$$

$$\mathbf{R}_{\theta\theta} = -r \cos \theta \mathbf{i} + r \sin \theta \mathbf{j}$$

$$\mathbf{R}_r = \cos \theta \mathbf{i} + \sin \theta \mathbf{j} + f_r \mathbf{k}$$

$$\mathbf{R}_{rr} = f_{rr} \mathbf{k} \qquad \mathbf{R}_{\theta r} = -\sin \theta \mathbf{i} + \cos \theta \mathbf{j}$$

where f_r, f_{rr} are the derivatives of $f = f(r)$.

(3) Position Vector and Derivatives in Form (F)

$$\mathbf{R} = g \cos \theta \mathbf{i} + g \sin \theta \mathbf{j} + z \mathbf{k} \qquad \mathbf{R}_{\theta z} = 0$$

$$\mathbf{R}_\theta = -g \sin \theta \mathbf{i} + g \cos \theta \mathbf{j} \qquad \mathbf{R}_z = g_z \cos \theta \mathbf{i} + g_z \sin \theta \mathbf{j} + \mathbf{k}$$

$$\mathbf{R}_{\theta\theta} = -g \cos \theta \mathbf{i} - g \sin \theta \mathbf{j} \qquad \mathbf{R}_{zz} = g_{zz} \cos \theta \mathbf{i} + g_{zz} \sin \theta \mathbf{j}$$

where g_z, g_{zz} are the derivatives of $g = g(z)$.

(4) Parameters of Fundamental Quantities in Forms (E, F)

(E)	$E = r^2$	$L = -\dfrac{r f_r}{\sqrt{G}}$	(F)	$E = g^2$	$L = -\dfrac{g}{\sqrt{G}}$
	$F = 0$	$M = 0$		$F = 0$	$M = 0$
	$G = 1 + f_r^2$	$N = -\dfrac{f_{rr}}{\sqrt{G}}$		$G = 1 + g_z^2$	$N = -\dfrac{g_{zz}}{\sqrt{G}}$

(5) Normal Unit Vectors and Radii of Curvature in Forms (E, F)

(E)	$\mathbf{n} = \dfrac{f_r \cos \theta \mathbf{i} + f_r \sin \theta \mathbf{j} - \mathbf{k}}{\sqrt{G}}$	(F)	$\mathbf{n} = \dfrac{\cos \theta \mathbf{i} + \sin \theta \mathbf{j} - g_z \mathbf{k}}{\sqrt{G}}$	
	$\rho_1 = \dfrac{G \sqrt{G}}{f_{rr}} \qquad \rho_2 = \dfrac{r \sqrt{G}}{f_r}$		$\rho_1 = \dfrac{G \sqrt{G}}{g_{zz}} \qquad \rho_2 = g \sqrt{G}$	

(1) General Equation

The general equation of the second degree.

$$a_{11}x^2 + a_{22}y^2 + a_{33}z^2 + 2a_{12}xy + 2a_{13}xz + 2a_{23}yz + 2a_{14}x + 2a_{24}y + 2a_{34}z + a_{44} = 0$$

defines the following quadratic surfaces: a sphere, an ellipsoid, a hyperboloid, a paraboloid, a cone, a cylinder, two planes, a line, and a point (the last three are the degenerated surfaces of the second degree).

(2) Invariants

The quantities ($a_{ik} = a_{ki}$)

$$J_4 = \begin{vmatrix} a_{11} & a_{12} & a_{13} & a_{14} \\ a_{21} & a_{22} & a_{23} & a_{24} \\ a_{31} & a_{32} & a_{33} & a_{34} \\ a_{41} & a_{42} & a_{43} & a_{44} \end{vmatrix} \qquad J_3 = \begin{vmatrix} a_{11} & a_{12} & a_{13} \\ a_{21} & a_{22} & a_{23} \\ a_{31} & a_{32} & a_{33} \end{vmatrix}$$

$$J_2 = \begin{vmatrix} a_{11} & a_{12} \\ a_{21} & a_{22} \end{vmatrix} + \begin{vmatrix} a_{11} & a_{13} \\ a_{31} & a_{33} \end{vmatrix} + \begin{vmatrix} a_{22} & a_{23} \\ a_{32} & a_{33} \end{vmatrix} \qquad J_1 = a_{11} + a_{22} + a_{33}$$

are invariants of the general equation; they remain unchanged under transformation of coordinates.

(3) Center

The center of a central quadratic surface ($J_3 \neq 0$) is the point of intersection of the three planes

$$\begin{aligned} a_{11}x + a_{12}y + a_{13}z + a_{14} &= 0 \\ a_{21}x + a_{22}y + a_{23}z + a_{24} &= 0 \\ a_{31}x + a_{32}y + a_{33}z + a_{34} &= 0 \end{aligned}$$

(4) Classification

Type		$J_3 \neq 0$		$J_3 = 0$
		Central surface		Noncentral surface
		$J_3 J_1 > 0$ $J_2 > 0$	$J_3 J_1$ or $J_2 > 0$	
$J_4 \neq 0$	$J_4 < 0$	Real ellipsoid	Hyperboloid of two sheets	Elliptic paraboloid
	$J_4 > 0$	Imaginary ellipsoid	Hyperboloid of one sheet	Hyperbolic paraboloid
$J_4 = 0$		Imaginary cone	Real cone	Cylinder or two planes

R = radius $\qquad$ a, b, h = semiaxes	x_C, y_C, z_C = coordinates of centroid
r, θ, z = cylindrical coordinates (Sec. 5.01)	r, θ, ϕ = spherical coordinates (Sec. 5.01)
$S_\theta = \sin\theta, \quad C_\theta = \cos\theta \quad (0 \leq \theta \leq 2\pi)$	$S_\phi = \sin\phi, \quad C_\phi = \cos\phi \quad (-\tfrac{1}{2}\pi \leq \phi \leq \tfrac{1}{2}\pi)$

For definition of forms (A, B, C) and other relations refer to Secs. 22.07–22.09.

Sphere

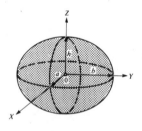

Center at 0

(A): $x^2 + y^2 + z^2 - R^2 = 0$

(B): $z = \pm\sqrt{R^2 - x^2 - y^2}$

(C): $x = RC_\theta S_\phi \qquad y = RS_\theta S_\phi \qquad z = RC_\phi$

Center at C

(A): $(x - x_C)^2 + (y - y_C)^2 + (z - z_C)^2 - R^2 = 0$

(B): $z = z_C \pm\sqrt{R^2 - (x - x_C)^2 - (y - y_C)^2}$

(C): $x = x_C + RC_\theta S_\phi \qquad y = y_C + RS_\phi S_\phi \qquad z = z_C + RC$

Ellipsoid

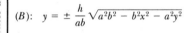

Center at 0

(A): $\dfrac{x^2}{a^2} + \dfrac{y^2}{b^2} + \dfrac{z^2}{h^2} - 1 = 0$

(B): $y = \pm\dfrac{h}{ab}\sqrt{a^2 b^2 - b^2 x^2 - a^2 y^2}$

(C): $x = aC_\theta S_\phi \qquad y = bS_\theta S_\phi \qquad z = hC_\phi$

Center at C

(A): $\left(\dfrac{x - x_C}{a}\right)^2 + \left(\dfrac{y - y_C}{b}\right)^2 + \left(\dfrac{z - z_C}{h}\right)^2 - 1 = 0$

(B): $z = z_C \pm h\sqrt{1 - \left(\dfrac{x - x_C}{a}\right)^2 - \left(\dfrac{y - y_C}{b}\right)^2}$

(C): $x = x_C + aC_\theta S_\phi \qquad y = y_C + bS_\theta S_\phi \qquad z = z_C + hC_\phi$

Rotational ellipsoid

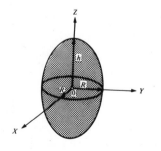

Center at 0

(A): $\dfrac{x^2}{R^2} + \dfrac{y^2}{R^2} + \dfrac{z^2}{h^2} - 1 = 0$

(B): $z = \pm\dfrac{h}{R}\sqrt{R^2 - x^2 - y^2}$

(C): $x = RC_\theta S_\phi \qquad y = RS_\theta S_\phi \qquad z = hC_\phi$

Center at C

(A): $\left(\dfrac{x - x_C}{R}\right)^2 + \left(\dfrac{y - y_C}{R}\right)^2 + \left(\dfrac{z - z_C}{h}\right)^2 - 1 = 0$

(B): $z = z_C \pm h\sqrt{1 - \left(\dfrac{x - x_C}{R}\right)^2 - \left(\dfrac{y - y_C}{R}\right)^2}$

(C): $x = x_C + RC_\theta S_\phi \qquad y = y_C + RS_\theta S_\phi \qquad z = z_C + hC_\phi$

a, b, h = semiaxes	t, τ = parameters	x_C, y_C, z_C = coordinates of centroid

r, θ, z = cylindrical coordinates (Sec. 5.01) $\qquad S_\theta = \sin \theta, \quad C_\theta = \cos \theta \qquad (0 \leq \theta \leq 2\pi)$

For definition of forms (A, B, C) and other relations refer to Secs. 22.07–22.09.

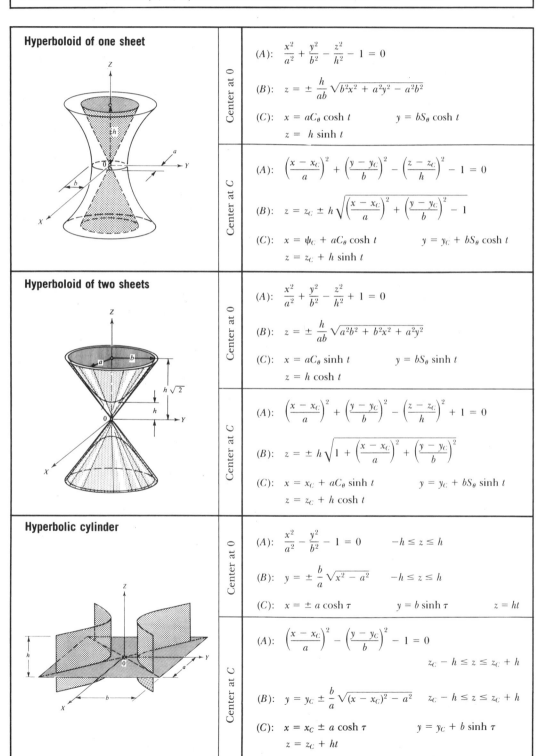

Hyperboloid of one sheet

Center at 0

$(A): \quad \dfrac{x^2}{a^2} + \dfrac{y^2}{b^2} - \dfrac{z^2}{h^2} - 1 = 0$

$(B): \quad z = \pm \dfrac{h}{ab} \sqrt{b^2 x^2 + a^2 y^2 - a^2 b^2}$

$(C): \quad x = a C_\theta \cosh t \qquad y = b S_\theta \cosh t$
$\qquad\quad z = h \sinh t$

Center at C

$(A): \quad \left(\dfrac{x - x_C}{a}\right)^2 + \left(\dfrac{y - y_C}{b}\right)^2 - \left(\dfrac{z - z_C}{h}\right)^2 - 1 = 0$

$(B): \quad z = z_C \pm h \sqrt{\left(\dfrac{x - x_C}{a}\right)^2 + \left(\dfrac{y - y_C}{b}\right)^2 - 1}$

$(C): \quad x = \psi_C + a C_\theta \cosh t \qquad y = y_C + b S_\theta \cosh t$
$\qquad\quad z = z_C + h \sinh t$

Hyperboloid of two sheets

Center at 0

$(A): \quad \dfrac{x^2}{a^2} + \dfrac{y^2}{b^2} - \dfrac{z^2}{h^2} + 1 = 0$

$(B): \quad z = \pm \dfrac{h}{ab} \sqrt{a^2 b^2 + b^2 x^2 + a^2 y^2}$

$(C): \quad x = a C_\theta \sinh t \qquad y = b S_\theta \sinh t$
$\qquad\quad z = h \cosh t$

Center at C

$(A): \quad \left(\dfrac{x - x_C}{a}\right)^2 + \left(\dfrac{y - y_C}{b}\right)^2 - \left(\dfrac{z - z_C}{h}\right)^2 + 1 = 0$

$(B): \quad z = \pm h \sqrt{1 + \left(\dfrac{x - x_C}{a}\right)^2 + \left(\dfrac{y - y_C}{b}\right)^2}$

$(C): \quad x = x_C + a C_\theta \sinh t \qquad y = y_C + b S_\theta \sinh t$
$\qquad\quad z = z_C + h \cosh t$

Hyperbolic cylinder

Center at 0

$(A): \quad \dfrac{x^2}{a^2} - \dfrac{y^2}{b^2} - 1 = 0 \qquad -h \leq z \leq h$

$(B): \quad y = \pm \dfrac{b}{a} \sqrt{x^2 - a^2} \qquad -h \leq z \leq h$

$(C): \quad x = \pm a \cosh \tau \qquad y = b \sinh \tau \qquad z = ht$

Center at C

$(A): \quad \left(\dfrac{x - x_C}{a}\right)^2 - \left(\dfrac{y - y_C}{b}\right)^2 - 1 = 0$
$\qquad\qquad\qquad\qquad\qquad\qquad z_C - h \leq z \leq z_C + h$

$(B): \quad y = y_C \pm \dfrac{b}{a} \sqrt{(x - x_C)^2 - a^2} \qquad z_C - h \leq z \leq z_C + h$

$(C): \quad x = x_C \pm a \cosh \tau \qquad y = y_C + b \sinh \tau$
$\qquad\quad z = z_C + ht$

R = radius p, t, τ = parameters x_C, y_C, z_C = coordinates of centroid

r, θ, z = cylindrical coordinates (Sec. 5.01) x_V, y_V, z_V = coordinates of vertex

$S_\theta = \sin\theta$ $C_\theta = \cos\theta$ $(\phi \le \theta \le 2\pi)$ a, b, h = semiaxes

For definition of forms (A, B, C) and other relations refer to Secs. 22.07–22.09.

Circular cylinder

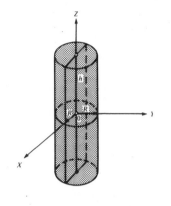

Center at 0

(A): $x^2 + y^2 - R^2 = 0$ $-h \le z \le h$

(B): $y = \pm\sqrt{R^2 - x^2}$ $-h \le z \le h$

(C): $x = RC_\theta$ $y = RS_\theta$ $z = ht$

Center at C

(A): $(x - x_C)^2 + (y - y_C)^2 - R^2 = 0$ $z_C - h \le z \le z_C + h$

(B): $y = y_C \pm\sqrt{R^2 - (x - x_C)^2}$ $z_C - h \le z \le z_C + h$

(C): $x = x_C + RC_\theta$ $y = y_C + RS_\theta$ $z = z_C + ht$

Elliptic cylinder

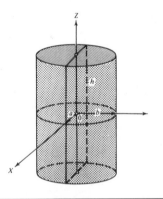

Center at 0

(A): $\dfrac{x^2}{a^2} + \dfrac{y^2}{b^2} - 1 = 0$ $-h \le z \le h$

(B): $y = \pm\dfrac{b}{a}\sqrt{a^2 - x^2}$ $-h \le z \le h$

(C): $x = aC_\theta$ $y = bS_\theta$ $z = ht$

Center at C

(A): $\left(\dfrac{x - x_C}{a}\right)^2 + \left(\dfrac{y - y_C}{b}\right)^2 - 1 = 0$ $z_C - h \le z \le z_C + h$

(B): $y = y_C \pm\dfrac{b}{a}\sqrt{a^2 - (x - x_C)^2}$ $z_C - h \le z \le z_C + h$

(C): $x = x_C + aC_\theta$ $y = y_C + RS_\theta$ $z = z_C + ht$

Parabolic cylinder

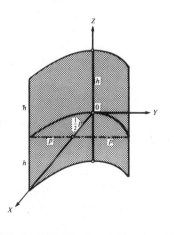

Vertex at 0

(A): $y^2 - 2px = 0$ $-h \le z \le h$

(B): $y = \pm\sqrt{2px}$ $-h \le z \le h$

(C): $x = \dfrac{\tau}{2p}$ $y = \tau$ $z = ht$

Vertex at V

(A): $(y - y_V)^2 - 2p(x - x_V) = 0$ $z_V - h \le z \le z_V + h$

(B): $y = y_V \pm\sqrt{2p(x - x_V)}$ $z_V - h \le z \le z_V + h$

(C): $x = x_V + \dfrac{\tau}{2p}$ $y = y_V + \tau$ $z = z_V + ht$

R = radius	a, b, h = semiaxes in cones or segments in conoids
r, θ, z = cylindrical coordinates (Sec. 5.01)	x_V, y_V, z_V = coordinates of vertex
t = parameter	$S_\theta = \sin\theta, \quad C_\theta = \cos\theta \qquad (0 \le \theta \le 2\pi)$

For definition of forms (A, B, C) and other relations refer to Secs. 22.07–22.09.

<table>
<tr>
<td colspan="2">

Circular cone

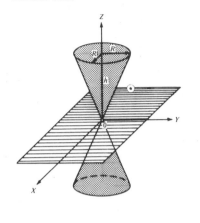
</td>
<td>Circular</td>
<td>

$(A): \quad \dfrac{x^2}{R^2} + \dfrac{y^2}{R^2} - \dfrac{z^2}{h^2} = 0$

$(B): \quad z = \pm\dfrac{h}{R}\sqrt{x^2 + y^2}$

$(C): \quad x = RtC_\theta \quad , \qquad y = RtS_\theta \qquad z = ht$
</td>
</tr>
<tr>
<td colspan="2"></td>
<td>Parabolic</td>
<td>

$(A): \quad \left(\dfrac{x - x_V}{R}\right)^2 + \left(\dfrac{y - y_V}{R}\right)^2 - \left(\dfrac{z - z_V}{h}\right)^2 = 0$

$(B): \quad z = z_V \pm \dfrac{h}{R}\sqrt{(x - x_V)^2 + (y - y_V)^2}$

$(C): \quad x = x_V + RtC_\theta \qquad y = y_V + RtS_\theta$
$\qquad\quad z = z_V + ht$
</td>
</tr>
<tr>
<td colspan="2">

Elliptic cone

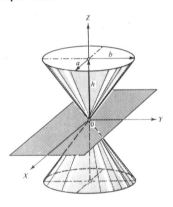
</td>
<td>Vertex at 0</td>
<td>

$(A): \quad \dfrac{x^2}{a^2} + \dfrac{y^2}{b^2} - \dfrac{z^2}{h^2} = 0$

$(B): \quad z = \pm\dfrac{h}{ab}\sqrt{b^2x^2 + a^2y^2}$

$(C): \quad x = atC_\theta \qquad y = btS_\theta \qquad z = ht$
</td>
</tr>
<tr>
<td colspan="2"></td>
<td>Vertex at V</td>
<td>

$(A): \quad \left(\dfrac{x - x_V}{a}\right)^2 + \left(\dfrac{y - y_V}{b}\right)^2 - \left(\dfrac{z - z_V}{h}\right)^2 = 0$

$(B): \quad z = z_V \pm h\sqrt{\left(\dfrac{x - x_V}{a}\right)^2 + \left(\dfrac{y - y_V}{b}\right)^2}$

$(C): \quad x = x_V + atC_\theta \qquad y = y_V + btS_\theta$
$\qquad\quad z = z_V \pm ht$
</td>
</tr>
<tr>
<td colspan="2">

Conoid

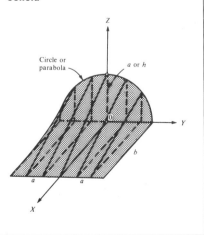

Circle or parabola a or h
</td>
<td>Vertex at 0</td>
<td>

A ruled surface determined by the directrix curve $y^2 + z^2 = a^2$ (circle), directrix straight line $x = b$, $z = 0$ (line parallel to Y axis), and directrix plane $y = 0$ (XZ plane).

$(A): \quad (a^2 - y^2)(b - x)^2 - b^2z^2 = 0$

$(B): \quad z = \dfrac{b - x}{b}\sqrt{a^2 - y^2}$
</td>
</tr>
<tr>
<td colspan="2"></td>
<td>Vertex at V</td>
<td>

A ruled surface determined by the directrix curve $z = h(a^2 - y^2)/a^2$ (parabola), directrix straight line $x = b$, $z = 0$ (line parallel to Y axis), and directrix plane $y = 0$ (XZ plane).

$(A): \quad hxy^2 - a^2bz = 0$

$(B): \quad z = \dfrac{hxy^2}{a^2b}$
</td>
</tr>
</table>

R = radius	h = height
θ = polar angle of R	
$S_\theta = \sin\theta$, $C_\theta = \cos\theta$ $\quad (0 \le \theta \le 2\pi)$	
x_V, y_V, z_V = coordinates of vertex	

a = minor radius of torus
ω = polar angle of a
$S_\omega = \sin\omega$, $C_\omega = \cos\omega$ $\quad (0 \le \omega \le 2\pi)$
x_C, y_C, z_C = coordinates of center

For definition of forms (A, B, C) and other relations refer to Secs. 22.07–22.09.

Circular paraboloid

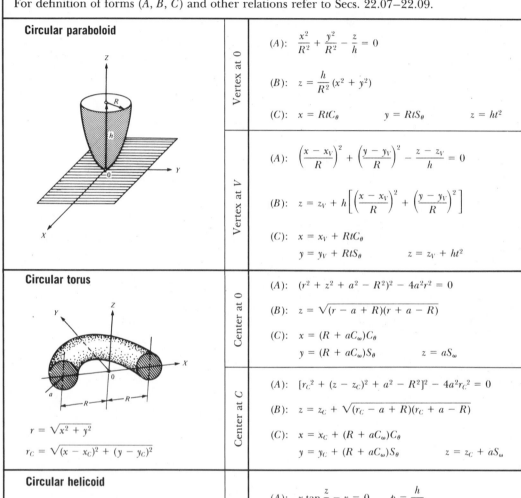

Vertex at 0

(A): $\dfrac{x^2}{R^2} + \dfrac{y^2}{R^2} - \dfrac{z}{h} = 0$

(B): $z = \dfrac{h}{R^2}(x^2 + y^2)$

(C): $x = RtC_\theta \qquad y = RtS_\theta \qquad z = ht^2$

Vertex at V

(A): $\left(\dfrac{x - x_V}{R}\right)^2 + \left(\dfrac{y - y_V}{R}\right)^2 - \dfrac{z - z_V}{h} = 0$

(B): $z = z_V + h\left[\left(\dfrac{x - x_V}{R}\right)^2 + \left(\dfrac{y - y_V}{R}\right)^2\right]$

(C): $x = x_V + RtC_\theta$
$\quad\;\; y = y_V + RtS_\theta \qquad z = z_V + ht^2$

Circular torus

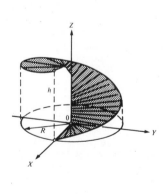

$r = \sqrt{x^2 + y^2}$

$r_C = \sqrt{(x - x_C)^2 + (y - y_C)^2}$

Center at 0

(A): $(r^2 + z^2 + a^2 - R^2)^2 - 4a^2r^2 = 0$

(B): $z = \sqrt{(r - a + R)(r + a - R)}$

(C): $x = (R + aC_\omega)C_\theta$
$\quad\;\; y = (R + aC_\omega)S_\theta \qquad z = aS_\omega$

Center at C

(A): $[r_C^2 + (z - z_C)^2 + a^2 - R^2]^2 - 4a^2r_C^2 = 0$

(B): $z = z_C + \sqrt{(r_C - a + R)(r_C + a - R)}$

(C): $x = x_C + (R + aC_\omega)C_\theta$
$\quad\;\; y = y_C + (R + aC_\omega)S_\theta \qquad z = z_C + aS_\omega$

Circular helicoid

$0 \le \sqrt{x^2 + y^2} \le R$

Axis at 0

(A): $x \tan\dfrac{z}{p} - y = 0 \qquad p = \dfrac{h}{2\pi}$

(B): $z = p \tan^{-1}\dfrac{y}{x}$

(C): $x = RtC_\theta \qquad y = RtS_\theta \qquad z = p\theta$

Axis at C

(A): $(x - x_C)\tan\dfrac{z - z_C}{p} - y + y_C = 0$

(B): $z = z_C + p \tan^{-1}\dfrac{y - y_C}{x - x_C}$

(C): $x = x_C + RtC_\theta$
$\quad\;\; y = y_C + RtS_\theta \qquad z = z_C + p\theta$

a, b, h = semiaxes	x_V, y_V, z_V = coordinates of vertex
t = parameter	$S_\theta = \sin\theta,\quad C_\theta = \cos\theta\qquad(0 \le \theta \le 2\pi)$

For definition of forms (A, B, C) and other relations refer to Secs. 22.07–22.09.

Elliptic paraboloid

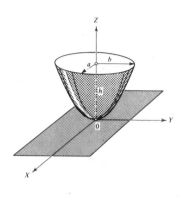

Vertex at 0

$(A):\quad \dfrac{x^2}{a^2} + \dfrac{y^2}{b^2} - \dfrac{z}{h} = 0$

$(B):\quad z = h\left(\dfrac{x^2}{a^2} + \dfrac{y^2}{b^2}\right)$

$(C):\quad x = atC_\theta \qquad y = btS_\theta \qquad z = ht^2$

Vertex at V

$(A):\quad \left(\dfrac{x - x_V}{a}\right)^2 + \left(\dfrac{y - y_V}{b}\right)^2 - \dfrac{z - z_V}{h} = 0$

$(B):\quad z = z_V + h\left[\left(\dfrac{x - x_V}{a}\right)^2 + \left(\dfrac{y - y_V}{b}\right)^2\right]$

$(C):\quad x = x_V + atC_\theta$

$y = y_V + btS_\theta \qquad z = z_V + ht^2$

Hyperbolic paraboloid of the first kind

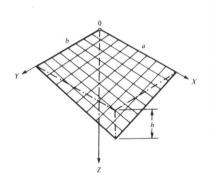

Vertex at 0

$(A):\quad hxy - abz = 0$

$(B):\quad z = \dfrac{h}{ab}xy$

$(C):\quad x = t\sqrt{ab} \qquad y = t\sqrt{ab} \qquad z = ht^2$

Vertex at V

$(A):\quad h(x - x_V)(y - y_V) - ab(z - z_V) = 0$

$(B):\quad z = z_V + \dfrac{h}{ab}(x - x_V)(y - y_V)$

$(C):\quad x = x_V + t\sqrt{ab}$

$y = y_V + t\sqrt{ab} \qquad z = z_V + ht^2$

Hyperbolic paraboloid of the second kind

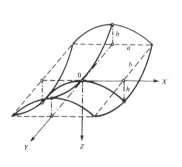

Vertex at 0

$(A):\quad \dfrac{x^2}{a^2} - \dfrac{y^2}{b^2} - \dfrac{z}{h} = 0$

$(B):\quad z = h\left(\dfrac{x^2}{a^2} - \dfrac{y^2}{b^2}\right)$

$(C):\quad x = \dfrac{at}{b} \qquad y = \dfrac{bt}{a} \qquad z = \left(\dfrac{h}{b^2} - \dfrac{h}{a^2}\right)t^2$

Vertex at V

$(A):\quad \left(\dfrac{x - x_V}{a}\right)^2 - \left(\dfrac{y - y_V}{b}\right)^2 - \dfrac{z - z_V}{h} = 0$

$(B):\quad z = z_V + h\left[\left(\dfrac{x - x_V}{a}\right)^2 - \left(\dfrac{y - y_V}{b}\right)^2\right]$

$(C):\quad x = x_V + \dfrac{at}{b}$

$y = y_V + \dfrac{bt}{a} \qquad z = z_V + \left(\dfrac{h}{b^2} - \dfrac{h}{a^2}\right)t^2$

(1) Geometry

The analytical expressions for the static and inertia functions of surfaces (Secs. 22.07–22.09) are defined in forms (B, C, E, F) below in terms of the following equivalents

$$\Delta_B = \sqrt{1 + f_x^2 + f_y^2} \qquad \Delta_C = \sqrt{EG - F} \qquad \Delta_E = \sqrt{1 + f_r^2} \qquad \Delta_F = \sqrt{1 + g_z^2}$$

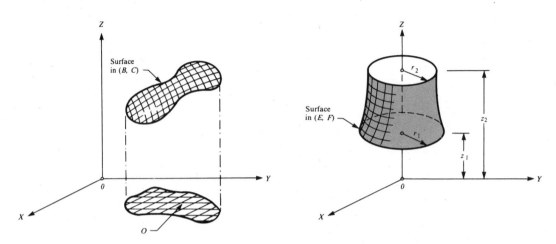

(2) Area of Surface in X, Y, Z and Moments of Surface about XY, YZ, ZX Planes

$$A = \iint_O \Delta_B \, dx \, dy = \iint_O \Delta_C \, du \, dv = 2\pi \int_{r_1}^{r_2} r\Delta_E \, dr = 2\pi \int_{z_1}^{z_2} g(z)\Delta_F \, dz$$

$$M_{xy} = \iint_O z\Delta_B \, dx \, dy = \iint_O z\Delta_C \, du \, dv = 2\pi \int_{r_1}^{r_2} rf\Delta_E \, dr = 2\pi \int_{z_1}^{z_2} zg\Delta_F \, dz$$

$$\left.\begin{array}{l} M_{yz} = \displaystyle\iint_O x\Delta_B \, dx \, dy = \iint_O x\Delta_C \, du \, dv \\[2mm] M_{zx} = \displaystyle\iint_O y\Delta_B \, dx \, dy = \iint_O y\Delta_C \, du \, dv \end{array}\right\} \quad \text{In forms } (E, F), M_{yz} = M_{zx} = 0$$

(3) Coordinates of Centroid of Surface in X, Y, Z

$$x_C = M_{yz}/A \qquad y_C = M_{zx}/A \qquad z_C = M_{xy}/A$$

(4) Moments and Products of Inertia of Surface in X, Y, Z

$$I_{xx} = \iint_O (y^2 + z^2)\Delta_B \, dx \, dy = \iint_O (y^2 + z^2)\Delta_C \, du \, dv \qquad\qquad I_{xy} = \iint_O xy\Delta_B \, dx \, dy$$

$$= 2\pi \int_{r_1}^{r_2} (r^2 + \tfrac{1}{2}f^2)\Delta_E \, dr = 2\pi \int_{z_1}^{z_2} (z^2 + \tfrac{1}{2}g^2)\Delta_F \, dz \qquad\qquad = \iint_O xy\Delta_C \, du \, dv$$

$$I_{yy} = \iint_O (z^2 + x^2)\Delta_B \, dx \, dy = \iint_O (z^2 + x^2)\Delta_C \, du \, dv \qquad\qquad I_{yz} = \iint_O yz\Delta_B \, dx \, dy$$

$$= 2\pi \int_{r_1}^{r_2} (r^2 + \tfrac{1}{2}f^2)\Delta_E \, dr = 2\pi \int_{z_1}^{z_2} (z^2 + \tfrac{1}{2}g^2)\Delta_F \, dz \qquad\qquad = \iint_O yz\Delta_C \, du \, dv$$

$$I_{zz} = \iint_O (x^2 + y^2)\Delta_B \, dx \, dy = \iint_O (x^2 + y^2)\Delta_C \, du \, dv \qquad\qquad I_{zx} = \iint_O zx\Delta_B \, dx \, dy$$

$$= 2\pi \int_{r_1}^{r_2} r^3\Delta_E \, dr = 2\pi \int_{z_1}^{z_2} g^3\Delta_F \, dx \qquad\qquad = \iint_O zx\Delta_C \, du \, dv$$

In forms (E, F), $I_{xy} = I_{yz} = I_{zx} = 0$. For parallel axes use the relations of Sec. 22.18–2.

(1) System X, Y ($0 =$ origin, $C =$ centroid)

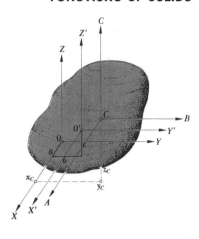

Volume
$\delta = $ density

$$dV = dx\, dy\, dz\, \delta \qquad V = \iiint_V dV$$

Static
moments

$$M_{xy} = \iiint_V z\, dV \qquad M_{yz} = \iiint_V x\, dV$$

$$M_{zx} = \iiint_V y\, dV$$

Coordinates
of centroid

$$x_C = \frac{M_{yz}}{V} \qquad z_C = \frac{M_{xy}}{V}$$

$$y_C = \frac{M_{zx}}{V}$$

Moments of inertia

$$I_{xx} = \iiint_V (y^2 + z^2)\, dV$$

$$I_{yy} = \iiint_V (x^2 + z^2)\, dV$$

$$I_{zz} = \iiint_V (x^2 + y^2)\, dV$$

Products of inertia

$$I_{yz} = I_{zy} = \iiint_V yz\, dV$$

$$I_{xz} = I_{zx} = \iiint_V xz\, dV$$

$$I_{xy} = I_{yx} = \iiint_V xy\, dV$$

Polar moment of inertia

$$J_0 = \iiint_V (x^2 + y^2 + z^2)\, dV$$

$$= \frac{I_{xx} + I_{yy} + I_{zz}}{2}$$

Radii of gyration

$$k_x = \sqrt{\frac{I_{xx}}{V}} \qquad k_y = \sqrt{\frac{I_{yy}}{V}} \qquad k_z = \sqrt{\frac{I_{zz}}{V}} \qquad k_0 = \sqrt{\frac{J_0}{V}}$$

(2) System $X'Y'$ ($0' =$ origin, $C =$ centroid)

Static moments $M'_{yz} = M_{yz} - aV \qquad M'_{zx} = M_{zx} - bV \qquad M'_{xy} = M_{xy} - cV$

Coordinates
of centroid $x'_C = x_C - a \qquad\qquad y'_C = y_C - b \qquad\qquad z'_C = z_C - c$

Moments of inertia

$$I'_{xx} = I_{xx} - 2bM_{zx} - 2cM_{xy} + (b^2 + c^2)V$$

$$I'_{yy} = I_{yy} - 2aM_{yz} - 2cM_{xy} + (a^2 + c^2)V$$

$$I'_{zz} = I_{zz} - 2aM_{yz} - 2bM_{zx} + (a^2 + b^2)V$$

Products of inertia

$$I'_{yz} = I_{yz} - bM_{zx} - cM_{xy} + bcV$$

$$I'_{xz} = I_{xz} - aM_{yz} - cM_{xy} + acV$$

$$I'_{xy} = I_{xy} - aM_{yz} - bM_{zx} + abV$$

Polar moment of inertia   ($r^2 = a^2 + b^2 + c^2$)

$$J'_0 = \frac{I_{xx} + I_{yy} + I_{zz}}{2} - 2aM_{yz} - 2bM_{zx} - 2cM_{xy} + r^2 V$$

Radii of gyration

$$k'_x = \sqrt{k_x^2 - 2by_C - 2cz_C + (b^2 + c^2)} \qquad k'_z = \sqrt{k_z^2 - 2ax_C - 2by_C + (a^2 + b^2)}$$

$$k'_y = \sqrt{k_y^2 - 2ax_C - 2cz_C + (a^2 + c^2)} \qquad k'_0 = \sqrt{k_0^2 - 2(ax_C + by_C + cz_C) + r^2}$$

V = volume

δ = density per unit length

$m = V\delta$ = mass

$I_{xx}, I_{yy}, I_{zz}, I_{AA}, I_{BB}, I_{CC}$ = moments of inertia

$I_{xy}, I_{yz}, I_{zx}, I_{AB}, I_{BC}, I_{CA}$ = products of inertia

x_C, y_C, z_C = coordinates of centroid

Straight bar

$m = l\delta$

$x_C = 0$

$y_C = l/2$

$z_C = 0$

$I_{AA} = \dfrac{ml^2}{12}$

$I_{BB} = 0$

$I_{CC} = I_{AA}$

$I_{AB} = 0, \ldots$

$I_{xx} = \dfrac{ml^2}{3}$

$I_{yy} = 0$

$I_{zz} = I_{xx}$

$I_{xy} = 0, \ldots$

Straight bar

$m = l\delta$

$x_C = \dfrac{a}{2}$

$y_C = \dfrac{b}{2}$

$z_C = \dfrac{c}{2}$

$I_{AA} = m\dfrac{b^2 + c^2}{12}$

$I_{BB} = m\dfrac{c^2 + a^2}{12}$

$I_{CC} = m\dfrac{a^2 + b^2}{12}$

$I_{AB} = m\dfrac{ab}{12}, \ldots$

$I_{xx} = m\dfrac{b^2 + c^2}{3}$

$I_{yy} = m\dfrac{c^2 + a^2}{3}$

$I_{zz} = m\dfrac{a^2 + b^2}{3}$

$I_{xy} = m\dfrac{ab}{3}, \ldots$

Circular ring

$m = 2\pi a\delta$

$x_C = 0$

$y_C = a$

$z_C = a$

$I_{AA} = ma^2$

$I_{BB} = \dfrac{ma^2}{2}$

$I_{CC} = I_{BB}$

$I_{AB} = 0, \ldots$

$I_{xx} = 3ma^2$

$I_{yy} = \tfrac{3}{2}ma^2$

$I_{zz} = I_{yy}$

$I_{xy} = I_{zx} = 0, I_{yz} = ma^2$

Circular bar

$m = 2\alpha a\delta$

$x_C = 0$

$y_C = 0$

$z_C = \dfrac{a}{\alpha}\sin\alpha$

$I_{AA} = ma^2\left(1 - \dfrac{\sin^2\alpha}{\alpha^2}\right)$

$I_{BB} = I_{yy} - mz_C^2$

$I_{CC} = I_{zz}$

$I_{AB} = 0, \ldots$

$I_{xx} = ma^2$

$I_{yy} = \dfrac{ma^2}{2}\left(1 + \dfrac{\sin 2\alpha}{2\alpha}\right)$

$I_{zz} = \dfrac{ma^2}{2}\left(1 - \dfrac{\sin 2\alpha}{2\alpha}\right)$

$I_{xy} = 0, \ldots$

Parabolic bar

$\delta(\phi) = \dfrac{\delta}{\cos\phi}$

$m = 2a\delta$

$x_C = y_C = 0$

$z_C = \dfrac{2h}{3}$

$I_{AA} = m\dfrac{15a^2 + 4h^2}{45}$

$I_{BB} = \dfrac{4mh^2}{45}$

$I_{CC} = \dfrac{ma^2}{3}$

$I_{AB} = 0, \ldots$

$I_{xx} = m\dfrac{5a^2 + 8h^2}{15}$

$I_{yy} = \dfrac{8mh^2}{15}$

$I_{zz} = \dfrac{ma^2}{3}$

$I_{xy} = 0, \ldots$

V = volume $\qquad$ $I_{xx}, I_{yy}, I_{zz}, I_{AA}, I_{BB}, I_{CC}$ = moments of inertia
δ = density $\qquad$ $I_{xy}, I_{yz}, I_{zx}, I_{AB}, I_{BC}, I_{CA}$ = products of inertia
m = mass $\qquad\qquad\qquad$ x_C, y_C, z_C = coordinates of centroid
$m = V\delta$

Cube			
$m = a^3\delta$	$I_{AA} = \dfrac{ma^2}{6}$	$I_{xx} = \dfrac{2ma^2}{3}$	
$x_C = \dfrac{a}{2}$	$I_{BB} = I_{CC} = I_{AA}$	$I_{yy} = I_{xx}$	
$y_C = \dfrac{a}{2}$	$I_{TT} = \dfrac{5ma^2}{12}$	$I_{zz} = I_{xx}$	
$z_C = \dfrac{a}{2}$	$I_{AB} = 0$	$I_{xy} = I_{yz} = I_{zx} = \dfrac{ma^2}{4}$	

Prism			
$m = abc\delta$	$I_{AA} = m\dfrac{b^2 + c^2}{12}$	$I_{xx} = m\dfrac{b^2 + c^2}{3}$	
$x_C = \dfrac{a}{2}$	$I_{BB} = m\dfrac{a^2 + c^2}{12}$	$I_{yy} = m\dfrac{a^2 + c^2}{3}$	
$y_C = \dfrac{b}{2}$	$I_{TT} = m\dfrac{a^2 + 4c^2}{12}$	$I_{zz} = m\dfrac{a^2 + b^2}{3}$	
$z_C = \dfrac{c}{2}$	$I_{AB} = 0$	$I_{xy} = \dfrac{mab}{4} \quad \cdots$	

Rectangular pyramid			
$m = \dfrac{abh\delta}{3}$	$I_{AA} = \dfrac{m}{80}(4b^2 + 3h^2)$	$I_{xx} = \dfrac{m}{20}(b^2 + 2h^2)$	
$x_C = 0$	$I_{BB} = \dfrac{m}{80}(4a^2 + 3h^2)$	$I_{yy} = \dfrac{m}{20}(a^2 + 2h^2)$	
$y_C = 0$	$I_{TT} = \dfrac{m}{20}(b^2 + 12h^2)$	$I_{zz} = \dfrac{m}{20}(a^2 + b^2)$	
$z_C = \dfrac{h}{4}$	$I_{AB} = 0$	$I_{xy} = I_{yz} = I_{zx} = 0$	

Tetrahedron			
$m = \dfrac{abh\delta}{6}$	$I_{AA} = 3m\dfrac{b^2 + h^2}{80}$	$I_{xx} = m\dfrac{b^2 + h^2}{10}$	
$x_C = \dfrac{a}{4}$	$I_{BB} = 3m\dfrac{a^2 + h^2}{80}$	$I_{yy} = m\dfrac{a^2 + h^2}{10}$	
$y_C = \dfrac{b}{4}$	$I_{CC} = 3m\dfrac{a^2 + b^2}{80}$	$I_{zz} = m\dfrac{a^2 + b^2}{10}$	
$z_C = \dfrac{h}{4}$	$I_{AB} = -\dfrac{mab}{80}, \ \cdots$	$I_{xy} = \dfrac{mab}{20}, \ \cdots$	

Wedge			
$m = \dfrac{abh\delta}{2}$	$I_{AA} = m\dfrac{3b^2 + 2h^2}{36}$	$I_{xx} = m\dfrac{3b^2 + 11h^2}{36}$	
$x_C = 0$	$I_{BB} = m\dfrac{3a^2 + 4h^2}{72}$	$I_{yy} = m\dfrac{3a^2 + 22h^2}{72}$	
$y_C = 0$	$I_{CC} = m\dfrac{a^2 + 4b^2}{24}$	$I_{zz} = m\dfrac{a^2 + 4b^2}{24}$	
$z_C = \dfrac{h}{3}$	$I_{AB} = I_{BC} = I_{CA} = 0$	$I_{xy} = I_{yz} = I_{zx} = 0$	

V = volume $\qquad I_{xx}, I_{yy}, I_{zz}, I_{AA}, I_{BB}, I_{CC}$ = moments of inertia

δ = density $\qquad I_{xy}, I_{yz}, I_{zx}, I_{AB}, I_{BC}, I_{CA}$ = products of inertia

m = mass $\qquad\qquad\qquad x_C, y_C, z_C$ = coordinates of centroid

$m = V\delta$

$$\alpha = 1 - \frac{64}{9\pi^2}$$

$$\beta = 12 - \frac{80}{3\pi^2}$$

Elliptic cylinder

For circular base $a = b$

$m = \pi abh\delta$

$x_C = 0$

$y_C = 0$

$z_C = \dfrac{h}{2}$

$I_{AA} = m\dfrac{3b^2 + h^2}{12}$

$I_{BB} = m\dfrac{3a^2 + h^2}{12}$

$I_{CC} = m\dfrac{a^2 + b^2}{4}$

$I_{AB} = I_{BC} = I_{CA} = 0$

$I_{xx} = m\dfrac{3b^2 + 4h^2}{12}$

$I_{yy} = m\dfrac{3a^2 + 4h^2}{12}$

$I_{TT} = m\dfrac{5a^2 + b^2}{5}$

$I_{xy} = I_{yz} = I_{zx} = 0$

Half elliptic cylinder

$m = \dfrac{\pi abh\delta}{2}$

$x_C = -\dfrac{4a}{3\pi}$

$y_C = 0$

$z_C = \dfrac{h}{2}$

$I_{AA} = m\dfrac{3b^2 + h^2}{12}$

$I_{BB} = m\dfrac{h^2 + \alpha a^2}{12}$

$I_{CC} = m\dfrac{a^2 + \alpha b^2}{4}$

$I_{AB} = I_{BC} = I_{CA} = I_{xy} = I_{yz} = 0$

$I_{xx} = m\dfrac{3b^2 + 4h^2}{12}$

$I_{yy} = m\dfrac{3a^2 + h^2}{12}$

$I_{zz} = m\dfrac{a^2 + h^2}{4}$

$I_{zx} = -\dfrac{2mah}{3\pi}$

Elliptic cone

$m = \dfrac{\pi abh\delta}{3}$

$x_C = 0$

$y_C = 0$

$z_C = \dfrac{3h}{4}$

$I_{AA} = 3m\dfrac{4b^2 + h^2}{80}$

$I_{BB} = 3m\dfrac{4a^2 + h^2}{80}$

$I_{CC} = 3m\dfrac{a^2 + b^2}{20}$

$I_{AB} = I_{BC} = I_{CA} = 0$

$I_{xx} = 3m\dfrac{b^2 + 4h^2}{20}$

$I_{yy} = 3m\dfrac{a^2 + 4h^2}{20}$

$I_{TT} = m\dfrac{3b^2 + 2h^2}{20}$

$I_{xy} = I_{yz} = I_{zx} = 0$

Half elliptic cone

$m = \dfrac{\pi abh\delta}{6}$

$x_C = -\dfrac{a}{\pi}$

$y_C = 0$

$z_C = \dfrac{3h}{4}$

$I_{AA} = 3m\dfrac{4b^2 + h^2}{80}$

$I_{BB} = 3m\dfrac{h^2 + \beta a^2}{80}$

$I_{CC} = 3m\dfrac{b^2 + \lambda a^2}{20}$

$I_{AB} = I_{BC} = I_{CA} = I_{xy} = I_{yz} = 0$

$I_{xx} = 3m\dfrac{b^2 + 4h^2}{20}$

$I_{yy} = 3m\dfrac{a^2 + 4h^2}{20}$

$I_{zz} = 3m\dfrac{a^2 + b^2}{20}$

$I_{zx} = -\dfrac{3mah}{4\pi}$

$$\lambda = 3 - \frac{20}{\pi^2}$$

Half torus

$m = \pi Rr^2\delta$

$x_C = -\dfrac{4R^2 + r^2}{2\pi R}$

$y_C = 0$

$z_C = 0$

$I_{AA} = m\dfrac{4R^2 + 5r^2}{8}$

$I_{BB} = m\left(\dfrac{4R^2 + 5r^2}{8} - x_C{}^2\right)$

$I_{CC} = m\left(\dfrac{4R^2 + 3r^2}{4} - x_C{}^2\right)$

$I_{AB} = I_{BC} = I_{CA} = I_{xy} = I_{yz} = I_{zx} = 0$

$I_{xx} = m\dfrac{4R^2 + 5r^2}{8}$

$I_{yy} = m\dfrac{4R^2 + 5r^2}{8}$

$I_{zz} = m\dfrac{4R^2 + 3r^2}{4}$

For a complete torus use $m = 2\pi Rr^2\delta$ and $x_C = 0$.

V = volume $I_{xx}, I_{yy}, I_{zz}, I_{AA}, I_{BB}, I_{CC}$ = moments of inertia

δ = density $I_{xy}, I_{yz}, I_{zx}, I_{AB}, I_{BC}, I_{CA}$ = products of inertia

m = mass x_C, y_C, z_C = coordinates of centroid

$m = V\delta$

Sphere

$m = \dfrac{4\pi a^3 \delta}{3}$

$x_C = 0$

$y_C = 0$

$z_C = 0$

$I_{TT} = \dfrac{7ma^2}{5}$

$I_{NN} = I_{TT}$

$I_{TN} = 0$

$I_{xx} = \dfrac{2ma^2}{5}$

$I_{xx} = I_{yy} = I_{zz}$

$I_{xy} = I_{yz} = I_{zx} = 0$

Hemisphere

$m = \dfrac{2\pi a^3 \delta}{3}$

$x_C = 0$

$y_C = 0$

$z_C = \dfrac{3a}{8}$

$I_{AA} = \dfrac{83ma^2}{320}$

$I_{TT} = \dfrac{208ma^2}{320}$

$I_{TN} = 0$

$I_{xx} = \dfrac{2ma^2}{5}$

$I_{xx} = I_{yy} = I_{zz}$

$I_{xy} = I_{yz} = I_{zx} = 0$

Ellipsoid

$m = \dfrac{4\pi abc\delta}{3}$

$x_C = 0$

$y_C = 0$

$z_C = 0$

$I_{TT} = m\dfrac{b^2 + 6c^2}{5}$

$I_{NN} = m\dfrac{a^2 + 6c^2}{5}$

$I_{TN} = 0$

$I_{xx} = m\dfrac{b^2 + c^2}{5}$

$I_{yy} = m\dfrac{a^2 + c^2}{5}$

$I_{zz} = m\dfrac{a^2 + b^2}{5}$

$I_{xy} = I_{yz} = I_{zx} = 0$

Semiellipsoid

$m = \dfrac{2\pi abc}{3}$

$x_C = 0$

$y_C = 0$

$z_C = \dfrac{3c}{8}$

$I_{xx} = m\dfrac{b^2 + c^2}{5}$

$I_{yy} = m\dfrac{a^2 + c^2}{5}$

$I_{zz} = m\dfrac{a^2 + b^2}{5}$

$I_{AB} = I_{BC} = I_{CA} = 0$

$I_{AA} = m\dfrac{64b^2 + 19c^2}{320}$

$I_{BB} = m\dfrac{64a^2 + 19c^2}{320}$

$I_{CC} = 3m\dfrac{a^2 + b^2}{5}$

$I_{xy} = I_{yz} = I_{zx} = 0$

Elliptic paraboloid

$m = \dfrac{\pi abh\delta}{2}$

$x_C = 0$

$y_C = 0$

$z_C = \dfrac{2h}{3}$

$I_{AA} = \dfrac{m}{18}(3b^2 + h^2)$

$I_{BB} = \dfrac{m}{18}(3a^2 + h^2)$

$I_{AB} = 0$

$I_{xx} = \dfrac{m}{6}(b^2 + 3h^2)$

$I_{yy} = \dfrac{m}{6}(a^2 + 3h^2)$

$I_{zz} = \dfrac{m}{6}(a^2 + b^2)$

$I_{xy} = I_{yz} = I_{zx} = 0$

V = volume	$I_{xx}, I_{yy}, I_{zz}, I_{AA}, I_{BB}, I_{CC}$ = moments of inertia
δ = density	$I_{xy}, I_{yz}, I_{zx}, I_{AB}, I_{BC}, I_{CA}$ = products of inertia
$m = V\delta$ = mass	x_C, y_C, z_C = coordinates of centroid
$t = 1$ = thickness	

Spherical shell

$m = 4\pi a^2 \delta$	$I_{AA} = \dfrac{2ma^2}{3}$	$I_{xx} = \dfrac{5ma^2}{3}$
$x_C = 0$	$I_{BB} = I_{AA}$	$I_{yy} = I_{xx}$
$y_C = 0$	$I_{CC} = I_{AA}$	$I_{zz} = \dfrac{2ma^2}{3}$
$z_C = a$	$I_{AB} = 0, \dots$	$I_{xy} = 0, \dots$

Hemispherical shell

$m = 2\pi a^2 \delta$	$I_{AA} = \dfrac{5ma^2}{12}$	$I_{xx} = \dfrac{2ma^2}{3}$
$x_C = 0$	$I_{BB} = I_{AA}$	$I_{yy} = I_{xx}$
$y_C = 0$	$I_{CC} = \dfrac{2ma^2}{3}$	$I_{zz} = I_{yy}$
$z_C = -\dfrac{a}{2}$	$I_{AB} = 0, \dots$	$I_{xy} = 0, \dots$

Circular cylindrical shell

$m = 2\pi a h \delta$	$I_{AA} = \dfrac{m}{12}(6a^2 + h^2)$	$I_{xx} = \dfrac{m}{6}(3a^2 + 2h^2)$
$x_C = 0$	$I_{BB} = I_{AA}$	$I_{yy} = I_{xx}$
$y_C = 0$	$I_{CC} = ma^2$	$I_{zz} = ma^2$
$z_C = \dfrac{h}{2}$	$I_{AB} = 0, \dots$	$I_{xy} = 0, \dots$

Circular conical shell

$m = \pi a \sqrt{a^2 + h^2}\, \delta$	$I_{AA} = \dfrac{m}{18}(9a^2 + 10h^2)$	$I_{xx} = \dfrac{m}{2}(a^2 + 2h^2)$
$x_C = 0$	$I_{BB} = I_{AA}$	$I_{yy} = I_{xx}$
$y_C = 0$	$I_{CC} = \dfrac{ma^2}{2}$	$I_{zz} = \dfrac{ma^2}{2}$
$z_C = \dfrac{2h}{3}$	$I_{AB} = 0, \dots$	$I_{xy} = 0, \dots$

Semicircular cylindrical shell

$m = \pi a h \delta$	$I_{AA} = ma^2\left(1 - \dfrac{4}{\pi^2}\right)$	$I_{xx} = ma^2$
$x_C = \dfrac{h}{2}$	$I_{BB} = \dfrac{m}{12}\left(6a^2 - \dfrac{48a^2}{\pi^2} + h^2\right)$	$I_{yy} = \dfrac{m}{6}(3a^2 + 2h^2)$
$y_C = 0$	$I_{CC} = I_{NN} = \dfrac{m}{12}(6a^2 + h^2)$	$I_{zz} = I_{yy}$
$z_C = -\dfrac{2a}{\pi}$	$I_{AB} = 0, \dots$	$I_{xy} = 0, \dots$

Appendix A
NUMERICAL
TABLES

| | $n! = n$ factorial (Sec. 1.03–1) | $n!! = n$ double factorial (Sec. 1.03–7) | | $n = 1$–100 | |

n	$n!$	$n!!$	n	$n!$	$n!!$
1	1.000 000 000 (00)	1.000 000 000 (00)	51	1.551 118 753 (066)	2.980 227 914 (33)
2	2.000 000 000 (00)	2.000 000 000 (00)	52	8.065 817 517 (067)	2.706 443 182 (34)
3	6.000 000 000 (00)	3.000 000 000 (00)	53	4.274 883 284 (069)	1.579 520 794 (35)
4	2.400 000 000 (01)	8.000 000 000 (00)	54	2.308 436 973 (071)	1.461 479 318 (36)
5	1.200 000 000 (02)	1.500 000 000 (01)	55	1.269 640 335 (073)	8.687 364 368 (36)
6	7.200 000 000 (02)	4.800 000 000 (01)	56	7.109 985 878 (074)	8.184 284 181 (37)
7	5.040 000 000 (03)	1.050 000 000 (02)	57	4.052 691 950 (076)	4.951 797 690 (38)
8	4.032 000 000 (04)	3.840 000 000 (02)	58	2.350 561 331 (078)	4.746 884 825 (39)
9	3.628 800 000 (05)	9.450 000 000 (02)	59	1.386 831 185 (080)	2.921 560 637 (40)
10	3.628 800 000 (06)	3.840 000 000 (03)	60	8.320 987 112 (081)	2.848 130 895 (41)
11	3.991 680 000 (07)	1.039 500 000 (04)	61	5.075 802 139 (083)	1.782 151 989 (42)
12	4.790 016 000 (08)	4.608 000 000 (04)	62	3.146 997 326 (085)	1.765 841 155 (43)
13	6.227 020 800 (09)	1.351 350 000 (05)	63	1.982 608 315 (087)	1.122 755 753 (44)
14	8.717 829 120 (10)	6.451 200 000 (05)	64	1.268 869 322 (089)	1.130 138 339 (45)
15	1.307 674 368 (12)	2.027 025 000 (06)	65	8.247 650 562 (090)	7.297 912 393 (46)
16	2.092 278 989 (13)	1.032 192 000 (07)	66	5.443 449 391 (092)	7.458 913 039 (46)
17	3.556 874 281 (14)	3.445 942 500 (07)	67	3.647 111 092 (094)	4.889 601 304 (47)
18	6.402 373 706 (15)	1.857 945 600 (08)	68	2.480 035 542 (096)	5.072 060 866 (48)
19	1.216 451 004 (17)	6.547 290 750 (08)	69	1.711 224 524 (098)	3.373 824 899 (49)
20	2.432 902 008 (18)	3.715 891 200 (09)	70	1.197 857 167 (100)	3.550 442 606 (50)
21	5.109 094 217 (19)	1.374 931 058 (10)	71	8.504 785 855 (101)	2.395 415 679 (51)
22	1.124 000 728 (21)	8.174 960 640 (10)	72	6.123 445 837 (103)	2.556 318 677 (52)
23	2.585 201 674 (22)	3.162 341 432 (11)	73	4.470 115 461 (105)	1.748 653 445 (53)
24	6.204 484 017 (23)	1.961 990 554 (12)	74	3.307 885 441 (107)	1.891 675 821 (54)
25	1.551 121 004 (25)	7.905 853 581 (12)	75	2.480 914 081 (109)	1.311 490 084 (55)
26	4.032 914 611 (26)	5.101 175 439 (13)	76	1.885 494 702 (111)	1.437 673 624 (56)
27	1.088 886 945 (28)	2.134 580 467 (14)	77	1.451 830 920 (113)	1.009 847 365 (57)
28	3.048 883 446 (29)	1.428 329 123 (15)	78	1.132 428 118 (115)	1.121 385 427 (58)
29	8.841 761 994 (30)	6.190 283 354 (15)	79	8.946 182 130 (116)	7.977 794 181 (58)
30	2.652 528 598 (32)	4.284 987 369 (16)	80	7.156 945 704 (118)	8.971 083 412 (59)
31	8.222 838 654 (33)	1.918 987 840 (17)	81	5.797 126 020 (120)	6.462 013 287 (60)
32	2.631 308 369 (35)	1.371 195 858 (18)	82	4.753 643 337 (122)	7.356 228 389 (61)
33	8.683 317 619 (36)	6.332 659 871 (18)	83	3.945 523 970 (124)	5.363 471 028 (62)
34	2.952 327 990 (38)	4.662 066 258 (19)	84	3.314 240 134 (126)	6.179 282 254 (63)
35	1.033 314 797 (40)	2.216 430 955 (20)	85	2.817 104 114 (128)	4.558 950 374 (64)
36	3.719 933 268 (41)	1.678 343 853 (21)	86	2.422 709 638 (130)	5.314 182 739 (65)
37	1.376 375 309 (43)	8.200 794 533 (21)	87	2.107 757 298 (132)	3.966 286 825 (66)
38	5.230 226 175 (44)	6.377 706 640 (22)	88	1.854 826 422 (134)	4.676 480 810 (67)
39	2.039 788 208 (46)	3.198 309 868 (23)	89	1.650 795 516 (136)	3.529 995 274 (68)
40	8.159 152 832 (47)	2.551 082 656 (24)	90	1.485 715 964 (138)	4.208 832 729 (69)
41	3.345 252 661 (49)	1.311 307 046 (25)	91	1.352 001 528 (140)	3.212 295 700 (70)
42	1.405 006 118 (51)	1.071 454 716 (26)	92	1.243 841 405 (142)	3.872 126 111 (71)
43	5.041 526 306 (52)	5.638 620 297 (26)	93	1.156 772 507 (144)	3.987 435 001 (72)
44	2.658 271 575 (54)	4.714 400 749 (27)	94	1.087 366 157 (146)	3.639 798 544 (73)
45	1.196 222 209 (56)	2.537 379 134 (28)	95	1.032 997 849 (148)	2.838 063 251 (74)
46	5.502 622 160 (57)	2.168 624 344 (29)	96	9.916 779 348 (149)	3.494 206 602 (75)
47	2.586 232 415 (59)	1.192 568 193 (30)	97	9.619 275 968 (151)	2.752 921 353 (76)
48	1.241 391 553 (61)	1.040 939 685 (31)	98	9.426 890 448 (153)	3.424 322 470 (77)
49	6.082 818 640 (62)	5.843 585 145 (31)	99	9.332 621 544 (155)	2.725 393 140 (78)
50	3.041 409 320 (64)	5.204 698 426 (32)	100	9.332 621 544 (157)	3.424 322 470 (79)

[1]$\Gamma(n+1) = n! =$ gamma function (Sec. 13.03–1) $\Pi(n) = n! =$ pi function (Sec. 13.03–2)

| $\Gamma(u + 1) = u!$ (Sec. 13.03−1) | | $\Pi(u) = u!$ (Sec. 13.03−2) | | | $u = 0.005{-}1.000$ | |

u	$u!$	u	$u!$	u	$u!$	u	$u!$
0.005	0.997 138 535	0.255	0.905 385 766	0.505	0.886 398 974	0.755	0.920 209 222
0.010	0.994 325 851	0.260	0.904 397 118	0.510	0.886 591 685	0.760	0.921 374 885
0.015	0.991 561 289	0.265	0.903 426 295	0.515	0.886 804 980	0.765	0.922 559 518
0.020	0.988 844 203	0.270	0.902 503 065	0.520	0.887 038 783	0.770	0.923 763 128
0.025	0.986 173 963	0.275	0.901 597 199	0.525	0.887 293 023	0.775	0.924 985 721
0.030	0.983 549 951	0.280	0.900 718 477	0.530	0.887 567 838	0.780	0.926 227 306
0.035	0.980 971 561	0.285	0.899 866 677	0.535	0.887 862 529	0.785	0.927 487 893
0.040	0.978 438 201	0.290	0.899 041 586	0.540	0.888 177 659	0.790	0.928 767 490
0.045	0.975 949 292	0.295	0.898 242 995	0.545	0.888 512 953	0.795	0.930 066 112
0.050	0.973 504 266	0.300	0.897 470 696	0.550	0.888 868 348	0.800	0.931 383 771
0.055	0.971 102 566	0.305	0.896 724 490	0.555	0.889 243 783	0.805	0.932 720 481
0.060	0.968 743 650	0.310	0.896 004 177	0.560	0.889 638 199	0.810	0.934 076 259
0.065	0.966 426 982	0.315	0.895 309 564	0.565	0.890 054 539	0.815	0.935 451 120
0.070	0.964 152 043	0.320	0.894 640 463	0.570	0.890 489 746	0.820	0.936 845 083
0.075	0.961 918 319	0.325	0.893 996 687	0.575	0.890 944 769	0.825	0.938 258 168
0.080	0.959 725 311	0.330	0.893 378 054	0.580	0.891 419 554	0.830	0.939 690 395
0.085	0.957 572 527	0.335	0.892 784 385	0.585	0.891 914 052	0.835	0.941 141 786
0.090	0.955 459 488	0.340	0.892 215 507	0.590	0.892 428 214	0.840	0.942 612 363
0.095	0.953 385 723	0.345	0.891 671 249	0.595	0.892 961 995	0.845	0.944 102 152
0.100	0.951 350 770	0.350	0.891 151 442	0.600	0.893 515 349	0.850	0.945 611 176
0.105	0.949 541 178	0.355	0.890 655 924	0.605	0.894 098 234	0.855	0.947 129 464
0.110	0.947 955 504	0.360	0.890 184 532	0.610	0.894 680 609	0.860	0.948 687 042
0.115	0.945 474 315	0.365	0.889 737 112	0.615	0.895 292 433	0.865	0.950 253 939
0.120	0.943 590 186	0.370	0.889 313 807	0.620	0.895 923 669	0.870	0.951 840 186
0.125	0.941 742 700	0.375	0.888 913 569	0.625	0.896 574 280	0.875	0.953 445 813
0.130	0.939 931 450	0.380	0.888 537 149	0.630	0.897 244 233	0.880	0.955 070 853
0.135	0.939 138 036	0.385	0.888 184 104	0.635	0.897 933 493	0.885	0.956 715 340
0.140	0.936 416 066	0.390	0.887 854 292	0.640	0.898 642 030	0.890	0.958 379 308
0.145	0.934 711 134	0.395	0.887 547 575	0.645	0.899 369 814	0.895	0.960 062 793
0.150	0.933 040 931	0.400	0.887 263 818	0.650	0.900 116 816	0.900	0.961 765 832
0.155	0.931 405 022	0.405	0.887 002 888	0.655	0.900 983 010	0.905	0.963 488 463
0.160	0.929 803 967	0.410	0.886 764 658	0.660	0.901 668 371	0.910	0.965 230 726
0.165	0.928 234 712	0.415	0.886 548 999	0.665	0.902 472 875	0.915	0.966 992 661
0.170	0.926 699 611	0.420	0.886 355 790	0.670	0.903 296 500	0.920	0.968 774 309
0.175	0.925 197 423	0.425	0.886 184 908	0.675	0.904 139 224	0.925	0.970 575 713
0.180	0.923 727 814	0.430	0.886 036 236	0.680	0.905 001 030	0.930	0.972 396 918
0.185	0.922 290 459	0.435	0.885 909 659	0.685	0.905 881 900	0.935	0.974 237 967
0.190	0.920 885 037	0.440	0.885 805 064	0.690	0.906 781 816	0.940	0.976 098 908
0.195	0.919 511 234	0.445	0.885 722 340	0.695	0.907 700 765	0.945	0.977 979 786
0.200	0.918 168 742	0.450	0.885 661 380	0.700	0.908 638 733	0.950	0.979 880 651
0.205	0.916 857 261	0.455	0.885 622 080	0.705	0.909 595 708	0.955	0.981 801 552
0.210	0.915 576 493	0.460	0.885 604 336	0.710	0.910 571 680	0.960	0.983 742 540
0.215	0.914 326 150	0.465	0.885 608 050	0.715	0.911 566 639	0.965	0.985 703 666
0.220	0.913 105 948	0.470	0.885 633 122	0.720	0.912 580 578	0.970	0.987 684 984
0.225	0.911 915 607	0.475	0.885 679 458	0.725	0.913 613 490	0.975	0.989 686 546
0.230	0.910 754 856	0.480	0.885 746 965	0.730	0.914 665 373	0.980	0.991 708 408
0.235	0.909 623 427	0.485	0.885 835 552	0.735	0.915 736 217	0.985	0.993 750 627
0.240	0.908 521 058	0.490	0.885 945 132	0.740	0.916 826 025	0.990	0.995 813 260
0.245	0.907 447 492	0.495	0.886 075 617	0.745	0.917 934 795	0.995	0.997 896 364
0.250	0.906 402 477	0.500	0.886 226 926	0.750	0.919 062 527	1.000	1.000 000 000

(1) Polynomials $\bar{B}_m(x) = \sum\limits_{k=0}^{m} b_k x^k$ (Sec. 8.05−2)

$\bar{B}_m(x)$ \ b_k	b_0	b_1	b_2	b_3	b_4	b_5	b_6	b_7	b_8	b_9	b_{10}
$\bar{B}_0(x)$	1										
$\bar{B}_1(x)$	$-\frac{1}{2}$	1									
$\bar{B}_2(x)$	$\frac{1}{6}$	-1	1								
$\bar{B}_3(x)$	0	$\frac{1}{2}$	$-\frac{3}{2}$	1							
$\bar{B}_4(x)$	$-\frac{1}{30}$	0	1	-2	1						
$\bar{B}_5(x)$	0	$-\frac{1}{6}$	0	$\frac{5}{3}$	$-\frac{5}{2}$	1					
$\bar{B}_6(x)$	$\frac{1}{42}$	0	$-\frac{1}{2}$	0	$\frac{5}{2}$	-3	1				
$\bar{B}_7(x)$	0	$\frac{1}{6}$	0	$-\frac{7}{6}$	0	$\frac{7}{2}$	$-\frac{7}{2}$	1			
$\bar{B}_8(x)$	$-\frac{1}{30}$	0	$\frac{2}{3}$	0	$-\frac{7}{3}$	0	$\frac{14}{3}$	-4	1		
$\bar{B}_9(x)$	0	$-\frac{3}{10}$	0	2	0	$-\frac{21}{5}$	0	6	$-\frac{9}{2}$	1	
$\bar{B}_{10}(x)$	$\frac{5}{66}$	0	$-\frac{3}{2}$	0	5	0	-7	0	$\frac{15}{2}$	-5	1

(2) Numbers B_m **and** $\bar{B}_m$ (Sec. 8.05−1)

B_m	$\bar{B}_m$	Fractions	Decimals
	$\bar{B}_0$	$1:1$	$1.000\ 000\ 000\ 000$ $(+00)$
	$-\bar{B}_1$	$1:2$	$5.000\ 000\ 000\ 000$ (-01)
B_1	$\bar{B}_2$	$1:6$	$1.666\ 666\ 666\ 667$ (-01)
B_2	$-\bar{B}_4$	$1:30$	$3.333\ 333\ 333\ 333$ (-02)
B_3	$\bar{B}_6$	$1:42$	$2.380\ 952\ 380\ 952$ (-02)
B_4	$-\bar{B}_8$	$1:30$	$3.333\ 333\ 333\ 333$ (-02)
B_5	$\bar{B}_{10}$	$5:66$	$7.575\ 757\ 575\ 758$ (-02)
B_6	$-\bar{B}_{12}$	$691:2\ 730$	$2.531\ 135\ 531\ 136$ (-01)
B_7	$\bar{B}_{14}$	$7:6$	$1.166\ 666\ 666\ 667$ $(+00)$
B_8	$-\bar{B}_{16}$	$3\ 617:510$	$7.092\ 156\ 862\ 745$ $(+00)$
B_9	$\bar{B}_{18}$	$43\ 867:798$	$5.497\ 117\ 794\ 486$ $(+01)$
B_{10}	$-\bar{B}_{20}$	$174\ 611:330$	$5.291\ 242\ 424\ 242$ $(+02)$
$\bar{B}_3 = \bar{B}_5 = \bar{B}_7 = \cdots = 0$			

(1) Polynomials $\bar{E}_m(x) = \sum_{k=0}^{m} e_k x^k$ (Sec. 8.06–2)

$\bar{E}_m(x)$ \ e_k	e_0	e_1	e_2	e_3	e_4	e_5	e_6	e_7	e_8	e_9	e_{10}
$\bar{E}_0(x)$	1										
$\bar{E}_1(x)$	$-\frac{1}{2}$	1									
$\bar{E}_2(x)$	0	-1	1								
$\bar{E}_3(x)$	$\frac{1}{4}$	0	$-\frac{3}{2}$	1							
$\bar{E}_4(x)$	0	1	0	-2	1						
$\bar{E}_5(x)$	$-\frac{1}{2}$	0	$\frac{5}{2}$	0	$-\frac{5}{2}$	1					
$\bar{E}_6(x)$	0	-3	0	5	0	-3	1				
$\bar{E}_7(x)$	$\frac{17}{8}$	0	$-\frac{21}{2}$	0	$\frac{35}{4}$	0	$-\frac{7}{2}$	1			
$\bar{E}_8(x)$	0	17	0	-28	0	14	0	-4	1		
$\bar{E}_9(x)$	$-\frac{31}{2}$	0	$\frac{153}{2}$	0	-65	0	21	0	$-\frac{9}{2}$	1	
$\bar{E}_{10}(x)$	0	-155	0	255	0	-126	0	30	0	-5	1

(2) Numbers E_m **and** $\bar{E}_m$ (Sec. 8.06–1)

E_m	$\bar{E}_m$	Integers
	$\bar{E}_0$	1
	$\bar{E}_1$	0
E_1	$-\bar{E}_2$	1
E_2	$\bar{E}_4$	5
E_3	$-\bar{E}_6$	61
E_4	$\bar{E}_8$	$1\,385$
E_5	$-\bar{E}_{10}$	$50\,521$
E_6	$\bar{E}_{12}$	$2\,702\,765$
E_7	$-\bar{E}_{14}$	$199\,360\,981$
E_8	$\bar{E}_{16}$	$19\,391\,512\,145$
E_9	$-\bar{E}_{18}$	$2\,404\,879\,675\,441$
E_{10}	$\bar{E}_{20}$	$370\,371\,188\,237\,525$
$\bar{E}_3 = \bar{E}_5 = \bar{E}_7 = \cdots = 0$		

$\mathscr{S}_k^{(p)}$ = Stirling number	$\Gamma(x)$ = gamma function (Sec. 13.03–1)
$X_h^{(p)}$ = factorial polynomial	$\dbinom{x}{p}$ = binomial coefficient (Sec. 1.04–2)
k, p = positive integers	h, x = real numbers

(1) Relations

$$X_1^{(p)} = x(x-1)(x-2) \cdots (x-p+1) = \binom{x}{p} p! = \sum_{k=1}^{p} x^k \mathscr{S}_k^{(p)}$$

$$X_h^{(p)} = x(x-h)(x-2h) \cdots (x-ph+h) = \frac{h^p \Gamma\left(\dfrac{x}{h}+1\right)}{\Gamma\left(\dfrac{x}{h}-p+1\right)} = h^p \sum_{k=1}^{p} \left(\frac{x}{h}\right)^k \mathscr{S}_k^{(p)}$$

$$X_1^{(-p)} = \frac{1}{(x+1)(x+2)(x+3) \cdots (x+p)} = \left[\binom{x+p}{p} p!\right]^{-1} = \left[\sum_{k=1}^{p} (x+p)^k \mathscr{S}_k^{(p)}\right]^{-1}$$

$$X_h^{(-p)} = \frac{1}{(x+h)(x+2h)(x+3h) \cdots (x+ph)} = \frac{\Gamma\left(\dfrac{x}{h}+1\right)}{h^p \Gamma\left(\dfrac{x}{h}+p+1\right)} = \left[h^p \sum_{k=1}^{p} \left(\frac{x}{h}+p\right)^k \mathscr{S}_k^{(p)}\right]^{-1}$$

(2) Numerical Values

k	$\mathscr{S}_k^{(1)}$	$\mathscr{S}_k^{(2)}$	$\mathscr{S}_k^{(3)}$	$\mathscr{S}_k^{(4)}$	$\mathscr{S}_k^{(5)}$	$\mathscr{S}_k^{(6)}$	$\mathscr{S}_k^{(7)}$	$\mathscr{S}_k^{(8)}$	$\mathscr{S}_k^{(9)}$
1	1	-1	2	-6	24	-120	720	$-5\,040$	40 320
2		1	-3	11	-50	274	$-1\,764$	13 068	$-109\,584$
3			1	-6	35	-225	1 624	$-13\,132$	118 121
4				1	-10	85	-735	6 769	$-67\,284$
5					1	-15	175	$-1\,960$	22 449
6						1	-21	322	$-4\,536$
7							1	-28	546
8								1	-36
9									1

k	$\mathscr{S}_k^{(10)}$	$\mathscr{S}_k^{(11)}$	$\mathscr{S}_k^{(12)}$	$\mathscr{S}_k^{(13)}$
1	$-362\,880$	3 628 800	$-39\,916\,800$	479 001 600
2	1 026 576	$-10\,628\,640$	120 543 840	$-1\,486\,442\,880$
3	$-1\,172\,700$	12 753 576	$-150\,917\,976$	1 931 559 552
4	723 680	$-8\,409\,500$	105 258 076	$-1\,414\,014\,888$
5	$-269\,325$	3 416 930	$-45\,995\,730$	657 206 836
6	63 273	$-902\,055$	13 339 535	$-206\,070\,150$
7	$-9\,450$	157 773	$-2\,637\,558$	44 990 231
8	870	$-18\,150$	357 423	$-6\,926\,634$
9	-45	1 320	$-32\,670$	749 463
10	1	-55	1 925	$-55\,770$
11		1	-66	2 717
12			1	-78
13				1

[1]For applications see Secs. 1.03 and 19.03.

$Z(m)$ = zeta function	$\bar{Z}(m)$ = complementary zeta function
$\bar{B}_m$ = Bernoulli number (Sec. A.03)	$\bar{E}_m$ = Euler number (Sec. A.04)
m, k = positive integers	$\beta = (-1)^{k+1}$

(1) Relations

$$\sum_{k=1}^{\infty} \frac{1}{k^m} = Z(m)$$

$$\sum_{k=1}^{\infty} \frac{1}{(2k)^m} = 2^{-m} Z(m)$$

$$\sum_{k=1}^{\infty} \frac{1}{(2k-1)^m} = (2^m - 1) 2^{-m} Z(m)$$

$$\sum_{k=1}^{\infty} \frac{1}{k^{2m}} = \frac{(2\pi)^{2m}}{(2m)!2} |\bar{B}_{2m}|$$

$$\sum_{k=1}^{\infty} \frac{1}{(2k)^{2m}} = \frac{(\pi)^{2m}}{(2m)!2} |\bar{B}_{2m}|$$

$$\sum_{k=1}^{\infty} \frac{1}{(2k-1)^{2m}} = \frac{(\pi)^{2m}(2^{2m}-1)}{(2m)!2} |\bar{B}_{2m}|$$

$$\sum_{k=1}^{\infty} \frac{\beta}{k^m} = (2^m - 2) 2^{-m} Z(m)$$

$$\sum_{k=1}^{\infty} \frac{\beta}{(2k)^m} = (2^m - 2) 4^{-m} Z(m)$$

$$\sum_{k=1}^{\infty} \frac{\beta}{(2k-1)^m} = \bar{Z}(m)$$

$$\sum_{k=1}^{\infty} \frac{\beta}{k^{2m}} = \frac{(\pi)^{2m}(2^{2m}-2)}{(2m)!2} |\bar{B}_{2m}|$$

$$\sum_{k=1}^{\infty} \frac{\beta}{(2k)^{2m}} = \frac{(\pi/2)^{2m}(2^{2m}-2)}{(2m)!2} |\bar{B}_{2m}|$$

$$\sum_{k=1}^{\infty} \frac{\beta}{(2k-1)^{2m-1}} = \frac{(\pi/2)^{2m-1}}{(2m)!2} |\bar{E}_{2m}|$$

(2) Numerical Values

m	$Z(m)$	$(2^m - 1) 2^{-m} Z(m)$	$(2^m - 2) 2^{-m} Z(m)$	$\bar{Z}(m)$
1	∞	∞	0.693 147 181	0.785 398 163
2	1.644 934 067	1.233 700 550	0.822 467 033	0.915 965 594
3	1.202 056 903	1.051 799 790	0.901 542 677	0.968 946 146
4	1.082 323 234	1.014 678 032	0.947 032 829	0.988 944 552
5	1.036 927 755	1.004 523 763	0.972 119 770	0.996 157 828
6	1.017 343 062	1.001 447 077	0.985 551 091	0.998 685 222
7	1.008 349 277	1.000 471 549	0.992 593 820	0.999 554 508
8	1.004 077 356	1.000 155 179	0.996 233 002	0.999 849 990
9	1.002 008 393	1.000 051 345	0.998 094 298	0.999 949 684
10	1.000 994 575	1.000 017 041	0.999 039 508	0.999 983 164
11	1.000 494 189	1.000 005 666	0.999 517 143	0.999 994 375
12	1.000 246 087	1.000 001 886	0.999 757 685	0.999 998 122
13	1.000 122 713	1.000 000 628	0.999 878 543	0.999 999 374
14	1.000 061 248	1.000 000 209	0.999 939 170	0.999 999 791
15	1.000 030 589	1.000 000 070	0.999 969 551	0.999 999 930
16	1.000 015 282	1.000 000 023	0.999 984 764	0.999 999 977
17	1.000 007 637	1.000 000 008	0.999 923 782	0.999 999 992
18	1.000 003 817	1.000 000 003	0.999 996 188	0.999 999 997
19	1.000 001 908	1.000 000 001	0.999 998 094	0.999 999 999
20	1.000 000 956	1.000 000 000	0.999 999 047	1.000 000 000
21	1.000 000 477	1.000 000 000	0.999 999 523	1.000 000 000
22	1.000 000 238	1.000 000 000	0.999 999 762	1.000 000 000
23	1.000 000 119	1.000 000 000	0.999 999 881	1.000 000 000
24	1.000 000 060	1.000 000 000	0.999 999 440	1.000 000 000
25	1.000 000 030	1.000 000 000	0.999 999 970	1.000 000 000

[1]For applications see Secs. 8.08, 13.04, 20.12, 20.14, 20.25, and 20.27.

$$x = 0.01 - 0.50$$

x, rad	$\sin x$	$\cos x$	$\tan x$	e^x	e^{-x}	$\sinh x$	$\cosh x$	$\tanh x$	x, deg
0.01	0.01000	0.99995	0.01000	1.01005	0.99005	0.01000	1.00005	0.01000	0.57
0.02	0.02000	0.99980	0.02000	1.02020	0.98020	0.02000	1.00020	0.02000	1.15
0.03	0.03000	0.99955	0.03001	1.03045	0.97045	0.03000	1.00045	0.02999	1.72
0.04	0.03999	0.99920	0.04002	1.04081	0.96079	0.04001	1.00080	0.03998	2.29
0.05	0.04998	0.99875	0.05004	1.05127	0.95123	0.05002	1.00125	0.04996	2.86
0.06	0.05996	0.99820	0.06007	1.06184	0.94176	0.06004	1.00180	0.05993	3.44
0.07	0.06994	0.99755	0.07011	1.07251	0.93239	0.07006	1.00245	0.06989	4.01
0.08	0.07991	0.99680	0.08017	1.08329	0.92312	0.08009	1.00320	0.07983	4.58
0.09	0.08988	0.99595	0.09024	1.09417	0.91393	0.09012	1.00405	0.08976	5.16
0.10	0.09983	0.99500	0.10033	1.10517	0.90484	0.10017	1.00500	0.09967	5.73
0.11	0.10978	0.99396	0.11045	1.11628	0.89583	0.11022	1.00606	0.10956	6.30
0.12	0.11971	0.99281	0.12058	1.12750	0.88692	0.12029	1.00721	0.11943	6.88
0.13	0.12963	0.99156	0.13074	1.13883	0.87810	0.13037	1.00846	0.12927	7.45
0.14	0.13954	0.99022	0.14092	1.15027	0.86936	0.14046	1.00982	0.13909	8.02
0.15	0.14944	0.98877	0.15114	1.16183	0.86071	0.15056	1.01127	0.14889	8.59
0.16	0.15932	0.98723	0.16138	1.17351	0.85214	0.16068	1.01283	0.15865	9.17
0.17	0.16918	0.98558	0.17166	1.18530	0.84366	0.17082	1.01448	0.16838	9.74
0.18	0.17903	0.98384	0.18197	1.19722	0.83527	0.18097	1.01624	0.17808	10.31
0.19	0.18886	0.98200	0.19232	1.20925	0.82696	0.19115	1.01810	0.18775	10.89
0.20	0.19867	0.98007	0.20271	1.22140	0.81873	0.20134	1.02007	0.19738	11.46
0.21	0.20846	0.97803	0.21314	1.23368	0.81058	0.21155	1.02213	0.20697	12.03
0.22	0.21823	0.97590	0.22362	1.24608	0.80252	0.22178	1.02430	0.21652	12.61
0.23	0.22798	0.97367	0.23414	1.25860	0.79453	0.23203	1.02657	0.22603	13.18
0.24	0.23770	0.97134	0.24472	1.27125	0.78663	0.24231	1.02894	0.23550	13.75
0.25	0.24740	0.96891	0.25534	1.28403	0.77880	0.25261	1.03141	0.24492	14.32
0.26	0.25708	0.96639	0.26602	1.29693	0.77105	0.26294	1.03399	0.25430	14.90
0.27	0.26673	0.96377	0.27676	1.30996	0.76338	0.27329	1.03667	0.26362	15.47
0.28	0.27636	0.96106	0.28755	1.32313	0.75578	0.28367	1.03946	0.27291	16.04
0.29	0.28595	0.95824	0.29841	1.33643	0.74826	0.29408	1.04235	0.28213	16.62
0.30	0.29552	0.95534	0.30934	1.34986	0.74082	0.30452	1.04534	0.29131	17.19
0.31	0.30506	0.95233	0.32033	1.36343	0.73345	0.31499	1.04844	0.30044	17.76
0.32	0.31457	0.94924	0.33139	1.37713	0.72615	0.32549	1.05164	0.30951	18.33
0.33	0.32404	0.94604	0.34252	1.39097	0,71892	0.33602	1.05495	0.31852	18.91
0.34	0.33349	0.94275	0.35374	1.40495	0.71177	0.34659	1.05836	0.32748	19.48
0.35	0.34290	0.93937	0.36503	1.41907	0.70469	0.35719	1.06188	0.33638	20.05
0.36	0.35227	0.93590	0.37640	1.43333	0.69768	0.36783	1.06550	0.34521	20.63
0.37	0.36162	0.93233	0.38786	1.44773	0.69073	0.37850	1.06923	0.35399	21.20
0.38	0.37092	0.92866	0.39941	1.46228	0.68386	0.38921	1.07307	0.36271	21.77
0.39	0.38019	0.92491	0.41105	1.47698	0.67706	0.39996	1.07702	0.37136	22.35
0.40	0.38942	0.92106	0.42279	1.49182	0.67032	0.41075	1.08107	0.37995	22.92
0.41	0.39861	0.91712	0.43463	1.50682	0.66365	0.42158	1.08523	0.38847	23.49
0.42	0.40776	0.91309	0.44657	1.52196	0.65705	0.43246	1.08950	0.39693	24.06
0.43	0.41687	0.90897	0.45862	1.53726	0.65051	0.44337	1.09388	0.40532	24.64
0.44	0.42594	0.90475	0.47078	1.55271	0.64404	0.45434	1.09837	0.41364	25.21
0.45	0.43497	0.90045	0.48306	1.56831	0.63763	0.46534	1.10297	0.42190	25.78
0.46	0.44395	0.89605	0.49545	1.58407	0.63128	0.47640	1.10768	0.43008	26.36
0.47	0.45289	0.89157	0.50797	1.59999	0.62500	0.48750	1.11250	0.43820	26.93
0.48	0.46178	0.88699	0.52061	1.61607	0.61878	0.49865	1.11743	0.44624	27.50
0.49	0.47063	0.88233	0.53339	1.63232	0.61263	0.50984	1.12247	0.45422	28.07
0.50	0.47943	0.87758	0.54630	1.64872	0.60653	0.52110	1.12763	0.46212	28.65

$$x = 0.51 - 1.00$$

x, rad	$\sin x$	$\cos x$	$\tan x$	e^x	e^{-x}	$\sinh x$	$\cosh x$	$\tanh x$	x, deg
0.51	0.48818	0.87274	0.55936	1.66529	0.60050	0.53240	1.13289	0.46995	29.22
0.52	0.49688	0.86782	0.57256	1.68203	0.59452	0.54375	1.13827	0.47770	29.79
0.53	0.50553	0.86281	0.58592	1.69893	0.58860	0.55516	1.14377	0.48538	30.37
0.54	0.51414	0.85771	0.59943	1.71601	0.58275	0.56663	1.14938	0.49299	30.94
0.55	0.52269	0.85252	0.61311	1.73325	0.57695	0.57815	1.15510	0.50052	31.51
0.56	0.53119	0.84726	0.62695	1.75067	0.57121	0.58973	1.16094	0.50798	32.09
0.57	0.53963	0.84190	0.64097	1.76827	0.56553	0.60137	1.16690	0.51536	32.66
0.58	0.54802	0.83646	0.65517	1.78604	0.55990	0.61307	1.17297	0.52267	33.23
0.59	0.55636	0.83094	0.66956	1.80399	0.55433	0.62483	1.17916	0.52990	33.80
0.60	0.56464	0.82534	0.68414	1.82212	0.54881	0.63665	1.18547	0.53705	34.38
0.61	0.57287	0.81965	0.69892	1.84043	0.54335	0.64854	1.19189	0.54413	34.95
0.62	0.58104	0.81388	0.71391	1.85893	0.53794	0.66049	1.19844	0.55113	35.52
0.63	0.58914	0.80803	0.72911	1.87761	0.53259	0.67251	1.20510	0.55805	36.10
0.64	0.59720	0.80210	0.74454	1.89648	0.52729	0.68459	1.21189	0.56490	36.67
0.65	0.60519	0.79608	0.76020	1.91554	0.52205	0.69675	1.21879	0.57167	37.24
0.66	0.61312	0.78999	0.77610	1.93479	0.51685	0.70897	1.22582	0.57836	37.82
0.67	0.62099	0.78382	0.79225	1.95424	0.51171	0.72126	1.23297	0.58498	38.39
0.68	0.62879	0.77757	0.80866	1.97388	0.50662	0.73363	1.24025	0.59152	38.96
0.69	0.63654	0.77125	0.82534	1.99372	0.50158	0.74607	1.24765	0.59798	39.53
0.70	0.64422	0.76484	0.84229	2.01375	0.49659	0.75858	1.25517	0.60437	40.11
0.71	0.65183	0.75836	0.85953	2.03399	0.49164	0.77117	1.26282	0.61068	40.68
0.72	0.65938	0.75181	0.87707	2.05443	0.48675	0.78384	1.27059	0.61691	41.25
0.73	0.66687	0.74517	0.89492	2.07508	0.48191	0.79659	1.27849	0.62307	41.83
0.74	0.67429	0.73847	0.91309	2.09594	0.47711	0.80941	1.28652	0.62915	42.40
0.75	0.68164	0.73169	0.93160	2.11700	0.47237	0.82232	1.29468	0.63515	42.97
0.76	0.68892	0.72484	0.95045	2.13828	0.46767	0.83530	1.30297	0.64108	43.54
0.77	0.69614	0.71791	0.96967	2.15977	0.46301	0.84838	1.31139	0.64693	44.12
0.78	0.70328	0.71091	0.98926	2.18147	0.45841	0.86153	1.31994	0.65271	44.69
0.79	0.71035	0.70385	1.00925	2.20340	0.45384	0.87478	1.32862	0.65841	45.26
0.80	0.71736	0.69671	1.02964	2.22554	0.44933	0.88811	1.33743	0.66404	45.84
0.81	0.72429	0.68950	1.05046	2.24791	0.44486	0.90152	1.34638	0.66959	46.41
0.82	0.73115	0.68222	1.07171	2.27050	0.44043	0.91503	1.35547	0.67507	46.98
0.83	0.73793	0.67488	1.09343	2.29332	0.43605	0.92863	1.36468	0.68048	47.56
0.84	0.74464	0.66746	1.11563	2.31637	0.43171	0.94233	1.37404	0.68581	48.13
0.85	0.75128	0.65998	1.13833	2.33965	0.42741	0.95612	1.38353	0.69107	48.70
0.86	0.75784	0.65244	1.16156	2.36316	0.42316	0.97000	1.39316	0.69626	49.27
0.87	0.76433	0.64483	1.18532	2.38691	0.41895	0.98398	1.40293	0.70137	49.85
0.88	0.77074	0.63715	1.20966	2.41090	0.41478	0.99806	1.41284	0.70642	50.42
0.89	0.77707	0.62941	1.23460	2.43513	0.41066	1.01224	1.42289	0.71139	50.99
0.90	0.78333	0.62161	1.26016	2.45960	0.40657	1.02652	1.43309	0.71630	51.57
0.91	0.78950	0.61375	1.28637	2.48432	0.40252	1.04090	1.44342	0.72113	52.14
0.92	0.79560	0.60582	1.31326	2.50929	0.39852	1.05539	1.45390	0.72590	52.71
0.93	0.80162	0.59783	1.34087	2.53451	0.39455	1.06998	1.46453	0.73059	53.29
0.94	0.80756	0.58979	1.36923	2.55998	0.39063	1.08468	1.47530	0.73522	53.86
0.95	0.81342	0.58168	1.39838	2.58571	0.38674	1.09948	1.48623	0.73978	54.43
0.96	0.81919	0.57352	1.42836	2.61170	0.38289	1.11440	1.49729	0.74428	55.00
0.97	0.82489	0.56530	1.45920	2.63794	0.37908	1.12943	1.50851	0.74870	55.58
0.98	0.83050	0.55702	1.49096	2.66446	0.37531	1.14457	1.51988	0.75307	56.15
0.99	0.83603	0.54869	1.52368	2.69123	0.37158	1.15983	1.53141	0.75736	56.72
1.00	0.84147	0.54030	1.55741	2.71828	0.36788	1.17520	1.54308	0.76159	57.30

$$x = 1.01 - 1.50$$

x, rad	$\sin x$	$\cos x$	$\tan x$	e^x	e^{-x}	$\sinh x$	$\cosh x$	$\tanh x$	x, deg
1.01	0.84683	0.53186	1.59221	2.74560	0.36422	1.19069	1.55491	0.76576	57.87
1.02	0.85211	0.52337	1.62813	2.77319	0.36059	1.20630	1.56689	0.76987	58.44
1.03	0.85730	0.51482	1.66524	2.80107	0.35701	1.22203	1.57904	0.77391	59.01
1.04	0.86240	0.50622	1.70361	2.82922	0.35345	1.23788	1.59134	0.77789	59.59
1.05	0.86742	0.49757	1.74332	2.85765	0.34994	1.25386	1.60379	0.78181	60.16
1.06	0.87236	0.48887	1.78442	2.88637	0.34646	1.26996	1.61641	0.78566	60.73
1.07	0.87720	0.48012	1.82703	2.91538	0.34301	1.28619	1.62919	0.78946	61.31
1.08	0.88196	0.47133	1.87122	2.94468	0.33960	1.30254	1.64214	0.79320	61.88
1.09	0.88663	0.46249	1.91709	2.97427	0.33622	1.31903	1.65525	0.79688	62.45
1.10	0.89121	0.45360	1.96476	3.00417	0.33287	1.33565	1.66852	0.80050	63.03
1.11	0.89570	0.44466	2.01434	3.03436	0.32956	1.35240	1.68196	0.80406	63.60
1.12	0.90010	0.43568	2.06596	3.06485	0.32628	1.36929	1.69557	0.80757	64.17
1.13	0.90441	0.42666	2.11975	3.09566	0.32303	1.38631	1.70934	0.81102	64.74
1.14	0.90863	0.41759	2.17588	3.12677	0.31982	1.40347	1.72329	0.81441	65.32
1.15	0.91276	0.40849	2.23450	3.15819	0.31664	1.42078	1.73741	0.81775	65.89
1.16	0.91680	0.39934	2.29580	3.18993	0.31349	1.43822	1.75171	0.82104	66.46
1.17	0.92075	0.39015	2.35994	3.22199	0.31037	1.45581	1.76618	0.82427	67.04
1.18	0.92461	0.38092	2.42727	3.25437	0.30728	1.47355	1.78083	0.82745	67.61
1.19	0.92837	0.37166	2.49790	3.28708	0.30422	1.49143	1.79565	0.83058	68.18
1.20	0.93204	0.36236	2.57215	3.32012	0.30119	1.50946	1.81066	0.83365	68.75
1.21	0.93562	0.35302	2.65032	3.35348	0.29820	1.52764	1.82584	0.83668	69.33
1.22	0.93910	0.34365	2.73275	3.38719	0.29523	1.54598	1.84121	0.83965	69.90
1.23	0.94249	0.33424	2.81982	3.42123	0.29229	1.56447	1.85676	0.84258	70.47
1.24	0.94578	0.32480	2.91193	3.45561	0.28938	1.58311	1.87250	0.84546	71.05
1.25	0.94898	0.31532	3.00957	3.49034	0.28650	1.60192	1.88842	0.84828	71.62
1.26	0.95209	0.30582	3.11327	3.52542	0.28365	1.62088	1.90454	0.85106	72.19
1.27	0.95510	0.29628	3.22363	3.56085	0.28083	1.64001	1.92084	0.85380	72.77
1.28	0.95802	0.28672	3.34135	3.59664	0.27804	1.65930	1.93734	0.85648	73.34
1.29	0.96084	0.27712	3.46721	3.63279	0.27527	1.67876	1.95403	0.85913	73.91
1.30	0.96356	0.26750	3.60210	3.66930	0.27253	1.69838	1.97091	0.86172	74.48
1.31	0.96618	0.25785	3.74708	3.70617	0.26982	1.71818	1.98800	0.86428	75.06
1.32	0.96872	0.24818	3.90335	3.74342	0.26714	1.73814	2.00528	0.86678	75.63
1.33	0.97115	0.23848	4.07231	3.78104	0.26448	1.75828	2.02276	0.86925	76.20
1.34	0.97348	0.22875	4.25562	3.81904	0.26185	1.77860	2.04044	0.87167	76.78
1.35	0.97572	0.21901	4.45522	3.85743	0.25924	1.79909	2.05833	0.87405	77.35
1.36	0.97786	0.20924	4.67344	3.89619	0.25666	1.81977	2.07643	0.87639	77.92
1.37	0.97991	0.19945	4.91306	3.93535	0.25411	1.84062	2.09473	0.87869	78.50
1.38	0.98185	0.18964	5.17744	3.97490	0.25158	1.86166	2.11324	0.88095	79.07
1.39	0.98370	0.17981	5.47069	4.01485	0.24908	1.88289	2.13196	0.88317	79.64
1.40	0.98545	0.16997	5.79788	4.05520	0.24660	1.90430	2.15090	0.88535	80.21
1.41	0.98710	0.16010	6.16536	4.09596	0.24414	1.92591	2.17005	0.88749	80.79
1.42	0.98865	0.15023	6.58112	4.13712	0.24171	1.94770	2.18942	0.88960	81.36
1.43	0.99010	0.14033	7.05546	4.17870	0.23931	1.96970	2.20900	0.89167	81.93
1.44	0.99146	0.13042	7.60183	4.22070	0.23693	1.99188	2.22881	0.89370	82.51
1.45	0.99271	0.12050	8.23809	4.26311	0.23457	2.01427	2.24884	0.89569	83.08
1.46	0.99387	0.11057	8.98861	4.30596	0.23224	2.03686	2.26910	0.89765	83.65
1.47	0.99492	0.10063	9.88737	4.34924	0.22993	2.05965	2.28958	0.89958	84.22
1.48	0.99588	0.09067	10.98338	4.39295	0.22764	2.08265	2.31029	0.90147	84.80
1.49	0.99674	0.08071	12.34986	4.43710	0.22537	2.10586	2.33123	0.90332	85.37
1.50	0.99749	0.07074	14.10142	4.48169	0.22313	2.12928	2.35241	0.90515	85.94

$$x = 1.51 - 2.00$$

x, rad	$\sin x$	$\cos x$	$\tan x$	e^x	e^{-x}	$\sinh x$	$\cosh x$	$\tanh x$	x, deg
1.51	0.99815	0.06076	16.42809	4.52673	0.22091	2.15291	2.37382	0.90694	86.52
1.52	0.99871	0.05077	19.66953	4.57223	0.21871	2.17676	2.39547	0.90870	87.09
1.53	0.99917	0.04079	24.49841	4.61818	0.21654	2.20082	2.41736	0.91042	87.66
1.54	0.99953	0.03079	32.46114	4.66459	0.21438	2.22510	2.43949	0.91212	88.24
1.55	0.99978	0.02079	48.07848	4.71147	0.21225	2.24961	2.46186	0.91379	88.81
1.56	0.99994	0.01080	92.62050	4.75882	0.21014	2.27434	2.48448	0.91542	89.38
1.57	1.00000	0.00080	1255.76559	4.80665	0.20805	2.29930	2.50735	0.91703	89.95
1.58	0.99996	−0.00920	−108.64920	4.85496	0.20598	2.32449	2.53047	0.91860	90.53
1.59	0.99982	−0.01920	−52.06697	4.90375	0.20393	2.34991	2.55384	0.92015	91.10
1.60	0.99957	−0.02920	−34.23253	4.95303	0.20190	2.37557	2.57746	0.92167	91.67
1.61	0.99923	−0.03919	−25.49474	5.00281	0.19989	2.40146	2.60135	0.92316	92.25
1.62	0.99879	−0.04918	−20.30728	5.05309	0.19790	2.42760	2.62549	0.92462	92.82
1.63	0.99825	−0.05917	−16.87110	5.10387	0.19593	2.45397	2.64990	0.92606	93.39
1.64	0.99761	−0.06915	−14.42702	5.15517	0.19398	2.48059	2.67457	0.92747	93.97
1.65	0.99687	−0.07912	−12.59926	5.20698	0.19205	2.50746	2.69951	0.92886	94.54
1.66	0.99602	−0.08909	−11.18055	5.25931	0.19014	2.53459	2.72472	0.93022	95.11
1.67	0.99508	−0.09904	−10.04718	5.31217	0.18825	2.56196	2.75021	0.93155	95.68
1.68	0.99404	−0.10899	−9.12077	5.36556	0.18637	2.58959	2.77596	0.93286	96.26
1.69	0.99290	−0.11892	−8.34923	5.41948	0.18452	2.61748	2.80200	0.93415	96.83
1.70	0.99166	−0.12884	−7.69660	5.47395	0.18268	2.64563	2.82832	0.93541	97.40
1.71	0.99033	−0.13875	−7.13726	5.52896	0.18087	2.67405	2.85491	0.93665	97.98
1.72	0.98889	−0.14865	−6.65244	5.58453	0.17907	2.70273	2.88180	0.93786	98.55
1.73	0.98735	−0.15853	−6.22810	5.64065	0.17728	2.73168	2.90897	0.93906	99.12
1.74	0.98572	−0.16840	−5.85353	5.69734	0.17552	2.76091	2.93643	0.94023	99.69
1.75	0.98399	−0.17825	−5.52038	5.75460	0.17377	2.79041	2.96419	0.94138	100.27
1.76	0.98215	−0.18808	−5.22209	5.81244	0.17204	2.82020	2.99224	0.94250	100.84
1.77	0.98022	−0.19789	−4.95341	5.87085	0.17033	2.85026	3.02059	0.94361	101.41
1.78	0.97820	−0.20768	−4.71009	5.92986	0.16864	2.88061	3.04925	0.94470	101.99
1.79	0.97607	−0.21745	−4.48866	5.98945	0.16696	2.91125	3.07821	0.94576	102.56
1.80	0.97385	−0.22720	−4.28626	6.04965	0.16530	2.94217	3.10747	0.94681	103.13
1.81	0.97153	−0.23693	−4.10050	6.11045	0.16365	2.97340	3.13705	0.94783	103.71
1.82	0.96911	−0.24663	−3.92937	6.17186	0.16203	3.00492	3.16694	0.94884	104.28
1.83	0.96659	−0.25631	−3.77118	6.23389	0.16041	3.03674	3.19715	0.94983	104.85
1.84	0.96398	−0.26596	−3.62449	6.29654	0.15882	3.06886	3.22768	0.95080	105.42
1.85	0.96128	−0.27559	−3.48806	6.35982	0.15724	3.10129	3.25853	0.95175	106.00
1.86	0.95847	−0.28519	−3.36083	6.42374	0.15567	3.13403	3.28970	0.95268	106.57
1.87	0.95557	−0.29476	−3.24187	6.48830	0.15412	3.16709	3.32121	0.95359	107.14
1.88	0.95258	−0.30430	−3.13038	6.55350	0.15259	3.20046	3.35305	0.95449	107.72
1.89	0.94949	−0.31381	−3.02566	6.61937	0.15107	3.23415	3.38522	0.95537	108.29
1.90	0.94630	−0.32329	−2.92710	6.68589	0.14957	3.26816	3.41773	0.95624	108.86
1.91	0.94302	−0.33274	−2.83414	6.75309	0.14808	3.30250	3.45058	0.95709	109.43
1.92	0.93965	−0.34215	−2.74630	6.82096	0.14661	3.33718	3.48378	0.95792	110.01
1.93	0.93618	−0.35153	−2.66316	6.88951	0.14515	3.37218	3.51733	0.95873	110.58
1.94	0.93262	−0.36087	−2.58433	6.95875	0.14370	3.40752	3.55123	0.95953	111.15
1.95	0.92896	−0.37018	−2.50948	7.02869	0.14227	3.44321	3.58548	0.96032	111.73
1.96	0.92521	−0.37945	−2.43828	7.09933	0.14086	3.47923	3.62009	0.96109	112.30
1.97	0.92137	−0.38868	−2.37048	7.17068	0.13946	3.51561	3.65507	0.96185	112.87
1.98	0.91744	−0.39788	−2.30582	7.24274	0.13807	3.55234	3.69041	0.96259	113.45
1.99	0.91341	−0.40703	−2.24408	7.31553	0.13670	3.58942	3.72611	0.96331	114.02
2.00	0.90930	−0.41615	−2.18504	7.38906	0.13534	3.62686	3.76220	0.96403	114.59

$$x = 2.01 - 2.50$$

x, rad	$\sin x$	$\cos x$	$\tan x$	e^x	e^{-x}	$\sinh x$	$\cosh x$	$\tanh x$	x, deg
2.01	0.90509	−0.42522	−2.12853	7.46332	0.13399	3.66466	3.79865	0.96473	115.16
2.02	0.90079	−0.43425	−2.07437	7.53832	0.13266	3.70283	3.83549	0.96541	115.74
2.03	0.89641	−0.44323	−2.02242	7.61409	0.13134	3.74138	3.87271	0.96609	116.31
2.04	0.89193	−0.45218	−1.97252	7.69061	0.13003	3.78029	3.91032	0.96675	116.88
2.05	0.88736	−0.46107	−1.92456	7.76790	0.12873	3.81958	3.94832	0.96740	117.46
2.06	0.88271	−0.46992	−1.87841	7.84597	0.12745	3.85926	3.98671	0.96803	118.03
2.07	0.87796	−0.47873	−1.83396	7.92482	0.12619	3.89932	4.02550	0.96865	118.60
2.08	0.87313	−0.48748	−1.79111	8.00447	0.12493	3.93977	4.06470	0.96926	119.18
2.09	0.86821	−0.49619	−1.74977	8.08492	0.12369	3.98061	4.10430	0.96986	119.75
2.10	0.86321	−0.50485	−1.70985	8.16617	0.12246	4.02186	4.14431	0.97045	120.32
2.11	0.85812	−0.51345	−1.67127	8.24824	0.12124	4.06350	4.18474	0.97103	120.89
2.12	0.85294	−0.52201	−1.63396	8.33114	0.12003	4.10555	4.22558	0.97159	121.47
2.13	0.84768	−0.53051	−1.59785	8.41487	0.11884	4.14801	4.26685	0.97215	122.04
2.14	0.84233	−0.53896	−1.56288	8.49944	0.11765	4.19089	4.30855	0.97269	122.61
2.15	0.83690	−0.54736	−1.52898	8.58486	0.11648	4.23419	4.35067	0.97323	123.19
2.16	0.83138	−0.55570	−1.49610	8.67114	0.11533	4.27791	4.39323	0.97375	123.76
2.17	0.82578	−0.56399	−1.46420	8.75828	0.11418	4.32205	4.43623	0.97426	124.33
2.18	0.82010	−0.57221	−1.43321	8.84631	0.11304	4.36663	4.47967	0.97477	124.90
2.19	0.81434	−0.58039	−1.40310	8.93521	0.11192	4.41165	4.52356	0.97526	125.48
2.20	0.80850	−0.58850	−1.37382	9.02501	0.11080	4.45711	4.56791	0.97574	126.05
2.21	0.80257	−0.59656	−1.34534	9.11572	0.10970	4.50301	4.61271	0.97622	126.62
2.22	0.79657	−0.60455	−1.31761	9.20733	0.10861	4.54936	4.65797	0.97668	127.20
2.23	0.79048	−0.61249	−1.29061	9.29987	0.10753	4.59617	4.70370	0.97714	127.77
2.24	0.78432	−0.62036	−1.26429	9.39333	0.10646	4.64344	4.74989	0.97759	128.34
2.25	0.77807	−0.62817	−1.23863	9.48774	0.10540	4.69117	4.79657	0.97803	128.92
2.26	0.77175	−0.63592	−1.21359	9.58309	0.10435	4.73937	4.84372	0.97846	129.49
2.27	0.76535	−0.64361	−1.18916	9.67940	0.10331	4.78804	4.89136	0.97888	130.06
2.28	0.75888	−0.65123	−1.16530	9.77668	0.10228	4.83720	4.93948	0.97929	130.63
2.29	0.75233	−0.65879	−1.14200	9.87494	0.10127	4.88684	4.98810	0.97970	131.21
2.30	0.74571	−0.66628	−1.11921	9.97418	0.10026	4.93696	5.03722	0.98010	131.78
2.31	0.73901	−0.67370	−1.09694	10.07442	0.09926	4.98758	5.08684	0.98049	132.35
2.32	0.73223	−0.68106	−1.07514	10.17567	0.09827	5.03870	5.13697	0.98087	132.93
2.33	0.72538	−0.68834	−1.05381	10.27794	0.09730	5.09032	5.18762	0.98124	133.50
2.34	0.78146	−0.69556	−1.03293	10.38124	0.09633	5.14245	5.23878	0.98161	134.07
2.35	0.71147	−0.70271	−1.01247	10.48557	0.09537	5.19510	5.29047	0.98197	134.65
2.36	0.70441	−0.70979	−0.99242	10.59095	0.09442	5.24827	5.34269	0.98233	135.22
2.37	0.69728	−0.71680	−0.97276	10.69739	0.09348	5.30196	5.39544	0.98267	135.79
2.38	0.69007	−0.72374	−0.95349	10.80490	0.09255	5.35618	5.44873	0.98301	136.36
2.39	0.68280	−0.73060	−0.93458	10.91349	0.09163	5.41093	5.50256	0.98335	136.94
2.40	0.67546	−0.73739	−0.91601	11.02318	0.09072	5.46623	5.55695	0.98367	137.51
2.41	0.66806	−0.74411	−0.89779	11.13396	0.08982	5.52207	5.61189	0.98400	138.08
2.42	0.66058	−0.75075	−0.87989	11.24586	0.08892	5.57847	5.66739	0.98431	138.66
2.43	0.65304	−0.75732	−0.86230	11.35888	0.08804	5.63542	5.72346	0.98462	139.23
2.44	0.64543	−0.76382	−0.84501	11.47304	0.08716	5.69294	5.78010	0.98492	139.80
2.45	0.63776	−0.77023	−0.82802	11.58835	0.08629	5.75103	5.83732	0.98522	140.37
2.46	0.63003	−0.77657	−0.81130	11.70481	0.08543	5.80969	5.89512	0.98551	140.95
2.47	0.62223	−0.78283	−0.79485	11.82245	0.08458	5.86893	5.95352	0.98579	141.52
2.48	0.61437	−0.78901	−0.77866	11.94126	0.08374	5.92876	6.01250	0.98607	142.09
2.49	0.60645	−0.79512	−0.76272	12.06128	0.08291	5.98918	6.07209	0.98635	142.67
2.50	0.59847	−0.80114	−0.74702	12.18249	0.08208	6.05020	6.13229	0.98661	143.24

$$x = 2.51 - 3.00$$

x, rad	$\sin x$	$\cos x$	$\tan x$	e^x	e^{-x}	$\sinh x$	$\cosh x$	$\tanh x$	x, deg
2.51	0.59043	−0.80709	−0.73156	12.30493	0.08127	6.11183	6.19310	0.98688	143.81
2.52	0.58233	−0.81295	−0.71632	12.42860	0.08046	6.17407	6.25453	0.98714	144.39
2.53	0.57417	−0.81873	−0.70129	12.55351	0.07966	6.23692	6.31658	0.98739	144.96
2.54	0.56596	−0.82444	−0.68648	12.67967	0.07887	6.30040	6.37927	0.98764	145.53
2.55	0.55768	−0.83005	−0.67186	12.80710	0.07808	6.36451	6.44259	0.98788	146.10
2.56	0.54936	−0.83559	−0.65745	12.93582	0.07730	6.42926	6.50656	0.98812	146.68
2.57	0.54097	−0.84104	−0.64322	13.06582	0.07654	6.49464	6.57118	0.98835	147.25
2.58	0.53253	−0.84641	−0.62917	13.19714	0.07577	6.56068	6.63646	0.98858	147.82
2.59	0.52404	−0.85169	−0.61530	13.32977	0.07502	6.62738	6.70240	0.98881	148.40
2.60	0.51550	−0.85689	−0.60160	13.46374	0.07427	6.69473	6.76901	0.98903	148.97
2.61	0.50691	−0.86200	−0.58806	13.59905	0.07353	6.76276	6.83629	0.98924	149.54
2.62	0.49826	−0.86703	−0.57468	13.73572	0.07280	6.83146	6.90426	0.98946	150.11
2.63	0.48957	−0.87197	−0.56145	13.87377	0.07208	6.90085	6.97292	0.98966	150.69
2.64	0.48082	−0.87682	−0.54837	14.01320	0.07136	6.97092	7.04228	0.98987	151.26
2.65	0.47203	−0.88158	−0.53544	14.15404	0.07065	7.04169	7.11234	0.99007	151.83
2.66	0.46319	−0.88626	−0.52264	14.29629	0.06995	7.11317	7.18312	0.99026	152.41
2.67	0.45431	−0.89085	−0.50997	14.43997	0.06925	7.18536	7.25461	0.99045	152.98
2.68	0.44537	−0.89534	−0.49743	14.58509	0.06856	7.25827	7.32683	0.99064	153.55
2.69	0.43640	−0.89975	−0.48502	14.73168	0.06788	7.33190	7.39978	0.99083	154.13
2.70	0.42738	−0.90407	−0.47273	14.87973	0.06721	7.40626	7.47347	0.99101	154.70
2.71	0.41832	−0.90830	−0.46055	15.02928	0.06654	7.48137	7.54791	0.99118	155.27
2.72	0.40921	−0.91244	−0.44848	15.18032	0.06587	7.55722	7.62310	0.99136	155.84
2.73	0.40007	−0.91648	−0.43653	15.33289	0.06522	7.63383	7.69905	0.99153	156.42
2.74	0.39088	−0.92044	−0.42467	15.48699	0.06457	7.71121	7.77578	0.99170	156.99
2.75	0.38166	−0.92430	−0.41292	15.64263	0.06393	7.78935	7.85328	0.99186	157.56
2.76	0.37240	−0.92807	−0.40126	15.79984	0.06329	7.86828	7.93157	0.99202	158.14
2.77	0.36310	−0.93175	−0.38970	15.95863	0.06266	7.94799	8.01065	0.98218	158.71
2.78	0.35376	−0.93533	−0.37822	16.11902	0.06204	8.02849	8.09053	0.99233	159.28
2.79	0.34439	−0.93883	−0.36683	16.28102	0.06142	8.10980	8.17122	0.99248	159.86
2.80	0.33499	−0.94222	−0.35553	16.44465	0.06081	8.19192	8.25273	0.99263	160.43
2.81	0.32555	−0.94553	−0.34431	16.60992	0.06020	8.27486	8.33506	0.99278	161.00
2.82	0.31608	−0.94873	−0.33316	16.77685	0.05961	8.35862	8.41823	0.99292	161.57
2.83	0.30657	−0.95185	−0.32208	16.94546	0.05901	8.44322	8.50224	0.99306	162.15
2.84	0.29704	−0.95486	−0.31108	17.11577	0.05843	8.52867	8.58710	0.99320	162.72
2.85	0.28748	−0.95779	−0.30015	17.28778	0.05784	8.61497	8.67281	0.99333	163.29
2.86	0.27789	−0.96061	−0.28928	17.46153	0.05727	8.70213	8.75940	0.99346	163.87
2.87	0.26827	−0.96334	−0.27847	17.63702	0.05670	8.79016	8.84686	0.99359	164.44
2.88	0.25862	−0.96598	−0.26773	17.81427	0.05613	8.87907	8.93520	0.99372	165.01
2.89	0.24895	−0.96852	−0.25704	17.99331	0.05558	8.96887	9.02444	0.99384	165.58
2.90	0.23925	−0.97096	−0.24641	18.17415	0.05502	9.05956	9.11458	0.99396	166.16
2.91	0.22953	−0.97330	−0.23582	18.35680	0.05448	9.15116	9.20564	0.99408	166.73
2.92	0.21978	−0.97555	−0.22529	18.54129	0.05393	9.24368	9.29761	0.99420	167.30
2.93	0.21002	−0.97770	−0.21481	18.72763	0.05340	9.33712	9.39051	0.99431	167.88
2.94	0.20023	−0.97975	−0.20437	18.91585	0.05287	9.43149	9.48436	0.99443	168.45
2.95	0.19042	−0.98170	−0.19397	19.10595	0.05234	9.52681	9.57915	0.99494	169.02
2.96	0.18060	−0.98356	−0.18362	19.29797	0.05182	9.62308	9.67490	0.99464	169.60
2.97	0.17075	−0.98531	−0.17330	19.49192	0.05130	9.72031	9.77161	0.99475	170.17
2.98	0.16089	−0.98697	−0.16301	19.68782	0.05079	9.81851	9.86930	0.99485	170.74
2.99	0.15101	−0.98853	−0.15276	19.88568	0.05029	9.91770	9.96798	0.99496	171.31
3.00	0.14112	−0.98999	−0.14255	20.08554	0.04979	10.01787	10.06766	0.99505	171.89

$$x = 3.05 - 5.00$$

x, rad	sin x	cos x	tan x	e^x	e^{-x}	sinh x	cosh x	tanh x	x, deg
3.05	0.09146	− 0.99581	− 0.09185	21.11534	0.04736	10.53399	10.58135	0.99552	174.75
3.10	0.04158	− 0.99914	− 0.04162	22.19795	0.04505	11.07645	11.12150	0.99595	177.62
3.15	− 0.00841	− 0.99996	0.00841	23.33606	0.04285	11.64661	11.68946	0.99633	180.48
3.20	− 0.05837	− 0.99829	0.05847	24.53253	0.04076	12.24588	12.28665	0.99668	183.35
3.25	− 0.10820	− 0.99413	0.10883	25.79034	0.03877	12.87578	12.91456	0.99700	186.21
3.30	− 0.15775	− 0.98748	0.15975	27.11264	0.03688	13.53788	13.57476	0.99728	189.08
3.35	− 0.20690	− 0.97836	0.21148	28.50273	0.03508	14.23382	14.26891	0.99754	191.94
3.40	− 0.25554	− 0.96680	0.26432	29.96410	0.03337	14.96536	14.99874	0.99777	194.81
3.45	− 0.30354	− 0.95282	0.31857	31.50039	0.03175	15.73432	15.76607	0.99799	197.67
3.50	− 0.35078	− 0.93646	0.37459	33.11545	0.03020	16.54263	16.57282	0.99818	200.54
3.55	− 0.39715	− 0.91775	0.43274	34.81332	0.02872	17.39230	17.42102	0.99835	203.40
3.60	− 0.44252	− 0.89676	0.49347	36.59823	0.02732	18.28546	18.31278	0.99851	206.26
3.65	− 0.48679	− 0.87352	0.55727	38.47467	0.02599	19.22434	19.25033	0.99865	209.13
3.70	− 0.52984	− 0.84810	0.62473	40.44730	0.02472	20.21129	20.23601	0.99878	211.99
3.75	− 0.57156	− 0.82056	0.69655	42.52108	0.02352	21.24878	21.27230	0.99889	214.86
3.80	− 0.61186	− 0.79097	0.77356	44.70118	0.02237	22.33941	22.36178	0.99900	217.72
3.85	− 0.65063	− 0.75940	0.85676	46.99306	0.02128	23.48589	23.50717	0.99909	220.59
3.90	− 0.68777	− 0.72593	0.94742	49.40245	0.02024	24.69110	24.71135	0.99918	223.45
3.95	− 0.72319	− 0.69065	1.04711	51.93537	0.01925	25.95806	25.97731	0.99926	226.32
4.00	− 0.75680	− 0.65364	1.15782	54.59815	0.01832	27.28992	27.30823	0.99933	229.18
4.05	− 0.78853	− 0.61500	1.28215	57.39746	0.01742	28.69002	28.70744	0.99939	232.05
4.10	− 0.81828	− 0.57482	1.42353	60.34029	0.01657	30.16186	30.17843	0.99945	234.91
4.15	− 0.84598	− 0.53321	1.58659	63.43400	0.01576	31.70912	31.72488	0.99950	237.78
4.20	− 0.87158	− 0.49026	1.77778	66.68633	0.01500	33.33567	33.35066	0.99955	240.64
4.25	− 0.89499	− 0.44609	2.00631	70.10541	0.01426	35.04557	35.05984	0.99959	243.51
4.30	− 0.91617	− 0.40080	2.28585	73.69979	0.01357	36.84311	36.85668	0.99963	246.37
4.35	− 0.93505	− 0.35451	2.63760	77.47846	0.01291	38.73278	38.74568	0.99967	249.24
4.40	− 0.95160	− 0.30733	3.09632	81.45087	0.01228	40.71930	40.73157	0.99970	252.10
4.45	− 0.96577	− 0.25939	3.72327	85.62694	0.01168	42.80763	42.81931	0.99973	254.97
4.50	− 0.97753	− 0.21080	4.63733	90.01713	0.01111	45.00301	45.01412	0.99975	257.83
4.55	− 0.98684	− 0.16168	6.10383	94.63241	0.01057	47.31092	47.32149	0.99978	260.70
4.60	− 0.99369	− 0.11215	8.86017	99.48432	0.01005	49.73713	49.74718	0.99980	263.56
4.65	− 0.99805	− 0.06235	16.00767	104.58499	0.00956	52.28771	52.29727	0.99982	266.43
4.70	− 0.99992	− 0.01239	80.71276	109.94717	0.00910	54.96904	54.97813	0.99983	269.29
4.75	− 0.99929	0.03760	− 26.57541	115.58428	0.00865	57.78782	57.79647	0.99985	272.15
4.80	− 0.99616	0.08750	− 11.38487	121.51042	0.00823	60.75109	60.75932	0.99986	275.02
4.85	− 0.99055	0.13718	− 7.22093	127.74039	0.00783	63.86628	63.87411	0.99988	277.88
4.90	− 0.98245	0.18651	− 5.26749	134.28978	0.00745	67.14117	67.14861	0.99989	280.75
4.95	− 0.97190	0.23538	− 4.12906	141.17496	0.00708	70.58394	70.59102	0.99990	283.61
*5.00	− 0.95892	0.28366	− 3.38052	148.41316	0.00674	74.20321	74.20995	0.99991	286.48

*For $x > 5.00$ and other numerical data, refer to:

Abramowitz, M., and I. A. Stegun: "Handbook of Mathematical Functions," National Bureau of Standards, Washington, D.C., 1964.

Dwight, H. B.: "Mathematical Tables," 3d ed., Dover, New York, 1961.

Flecher, A., J. C. P. Miller, L. Rosenhead, and L. J. Comrie: "An Index of Mathematical Tables," 2d ed., Addison-Wesley, Reading, Mass., 1962.

Lebedev, A. V., and R. M. Fedorova: "A Guide to Mathematical Tables," Pergamon, New York, 1960.

(1) Legendre Polynomials $P_n(x)$ (Sec. 14.18) $\boxed{P_0(x) = 1}$

$$P_1(x) = x$$

$$P_4(x) = \frac{1}{8}(35x^4 - 30x^2 + 3)$$

$$P_2(x) = \frac{1}{2}(3x^2 - 1)$$

$$P_5(x) = \frac{1}{8}(63x^5 - 70x^3 + 15x)$$

$$P_3(x) = \frac{1}{2}(5x^3 - 3x)$$

$$\boxed{P_n(x) = \frac{2n-1}{n}(x)P_{n-1}(x) - \frac{n-1}{n}P_{n-2}(x)}$$

(2) Legendre Polynomials $Q_n(x)$ (Sec. 14.18) $\boxed{Q_0(x) = \ln\sqrt{\frac{1+x}{1-x}}}$

$$Q_1(x) = xQ_0(x) - 1$$

$$Q_4(x) = P_4(x)\,Q_0(x) - \frac{35}{8}x^3 + \frac{55}{24}x$$

$$Q_2(x) = P_2(x)Q_0(x) - \frac{3}{2}x$$

$$Q_5(x) = P_5(x)\,Q_0(x) - \frac{63}{8}x^4 + \frac{49}{8}x^2 - \frac{8}{15}$$

$$Q_3(x) = P_3(x)Q_0(x) - \frac{5}{2}x^2 + \frac{2}{3}$$

$$\boxed{Q_n(x) = \frac{2n-1}{n}(x)Q_{n-1}(x) - \frac{n-1}{n}Q_{n-2}(x)}$$

(3) Chebyshev Polynomials $T_n(x)$ (Sec. 14.20) $\boxed{T_0(x) = 1}$

$$T_1(x) = x$$

$$T_4(x) = 8x^4 - 8x^2 + 1$$

$$T_2(x) = 2x^2 - 1$$

$$T_5(x) = 16x^5 - 20x^3 + 5x$$

$$T_3(x) = 4x^3 - 3x$$

$$\boxed{T_n(x) = 2xT_{n-1}(x) - T_{n-2}(x)}$$

(4) Chebyshev Polynomials $U_n(x)$ (Sec. 14.20) $\boxed{U_0(x) = \sin^{-1}x}$

$$U_1(x) = \sqrt{1-x^2}$$

$$U_4(x) = (8x^3 - 4x)\sqrt{1-x^2}$$

$$U_2(x) = 2x\sqrt{1-x^2}$$

$$U_5(x) = (16x^4 - 12x^2 + 1)\sqrt{1-x^2}$$

$$U_3(x) = (4x^2 - 1)\sqrt{1-x^2}$$

$$\boxed{U_n(x) = 2xU_{n-1}(x) - U_{n-2}(x)}$$

(5) Laguerre Polynomials $L_n(x)$ (Sec. 14.22) $\boxed{L_0(x) = 1}$

$$L_1(x) = 1 - x$$

$$L_4(x) = 24 - 96x + 72x^2 - 16x^3 + x^4$$

$$L_2(x) = 2 - 4x + x^2$$

$$L_5(x) = 120 - 600x + 600x^2 - 200x^3 + 25x^4 - x^5$$

$$L_3(x) = 6 - 18x + 9x^2 - x^3$$

$$\boxed{L_n(x) = (2n-1-x)L_{n-1}(x) - (n-1)^2 L_{n-2}(x)}$$

(6) Hermite Polynomials $H_n(x)$ (Sec. 14.23) $\boxed{H_0(x) = 1}$

$$H_1(x) = 2x$$

$$H_4(x) = 16x^4 - 48x^2 + 12$$

$$H_2(x) = 4x^2 - 2$$

$$H_5(x) = 32x^5 - 160x^3 + 120x$$

$$H_3(x) = 8x^3 - 12x$$

$$\boxed{H_n(x) = 2xH_{n-1}(x) - 2(n-1)\,H_{n-2}(x)}$$

(1) Archimedes Number π

(a) Definition. The symbol π denotes the ratio of the circumference of a circle to its diameter,

$$\pi = 3.141\ 592\ 653\ 589\ 793\ 238\ 462\ 643\ldots$$

In 1882 C.L.P. Lindemann proved that π is both an irrational and a transcendental number and thus has shown that the problem of rectification and squaring of circle with ruler and compass alone is a mathematical impossibility.

(b) Approximation by fraction

$$\pi = \frac{22}{7} - \epsilon = 3.142\ 857\ 143 - \epsilon \qquad \epsilon < 1.3 \times 10^{-3}$$

$$\pi = \frac{355}{113} - \epsilon = 3.141\ 592\ 920 - \epsilon \qquad \epsilon < 2.7 \times 10^{-7}$$

(c) Evaluation by series

$$\pi = 4 \sum_{k=1}^{\infty} \frac{(-1)^{k-1}}{2k-1} \left(\frac{4}{5^{2k-1}} - \frac{1}{239^{2k-1}} \right) \qquad n=7, \epsilon < 5 \times 10^{-10}$$

(2) Base of Natural Logarithms e

(a) Definition. The symbol e denotes the limit

$$e = \lim_{m \to \infty} \left(1 + \frac{1}{m} \right)^{m} = \lim_{n \to 0} (1 + n)^{1/n} = 2.718\ 281\ 828\ 459\ 045\ 235\ 360\ldots$$

and is the base of the natural system of logarithms. In 1873, C. Hermite proved that e is both an irrational and a transcendental number.

(b) Approximation by fraction

$$e = \frac{19}{7} + \epsilon = 2.714\ 285\ 714 + \epsilon \qquad \epsilon < 4 \times 10^{-3}$$

$$e = \frac{1264}{465} + \epsilon = 2.718\ 279\ 570 + \epsilon \qquad \epsilon < 2.3 \times 10^{-6}$$

(c) Evaluation by series

$$e = 2 \left(\frac{1}{1!} + \frac{2}{3!} + \frac{3}{5!} + \cdots \right) = 2 \sum_{k=1}^{\infty} \frac{k}{(2k-1)!} \qquad n=7, \epsilon < 4.6 \times 10^{-10}$$

(3) Euler Constant C

(a) Definition. The symbol C denotes the limit

$$C = \lim_{n \to \infty} \left(1 + \frac{1}{2} + \frac{1}{3} + \cdots + \frac{1}{n} - \ln n \right) = 0.577\ 215\ 664\ 901\ 532\ 860\ 606 \ldots$$

No proof is yet available whether C is an irrational number.

(b) Approximation by fraction

$$C = \sqrt{\frac{1}{3}} - \epsilon = 0.577\ 350\ 269 - \epsilon \qquad \epsilon < 1.3 \times 10^{-4}$$

$$C = \frac{228}{395} - \epsilon = 0.577\ 215\ 190 - \epsilon \qquad \epsilon < 4.8 \times 10^{-7}$$

(c) Evaluation by series

$$C = \sum_{k=1}^{n} \frac{1}{k} - \ln n - \epsilon$$

$n = 10^3, \epsilon < 5 \times 10^{-4}$

$n = 10^6, \epsilon < 5 \times 10^{-7}$

$n = 10^9, \epsilon < 5 \times 10^{-10}$

(4) Selected Eigenvalues

Characteristic equation	Eigenvalues					
	x_1	x_2	x_3	x_4	x_5	$x_n{}^*$
$\cosh x \cos x + 1 = 0$	1.8751	4.6941	7.8548	10.9955	14.1372	$\frac{1}{2}(2n - 1)\pi$
$\cosh x \cos x - 1 = 0$	4.7300	7.8532	10.9956	14.1372	17.2788	$\frac{1}{2}(2n + 1)\pi$
$\tanh x + \tan x = 0$	2.3650	5.4978	8.6394	11.7810	14.9226	$\frac{1}{4}(4n - 1)\pi$
$\tanh x - \tan x = 0$	3.9266	7.0686	10.2102	13.3518	16.4934	$\frac{1}{4}(4n + 1)\pi$
$\tan x - x = 0$	4.4934	7.7253	10.9041	14.0662	17.2208	$\frac{1}{2}(2n + 1)\pi$

*For $n > 5$.

(1) Definition

The Dirac Delta function $\mathscr{D}(x - t)$ is not a function in the true sense but a definition of a distribution.

$$\mathscr{D}(x - t) = \lim_{\Delta \to 0} \begin{cases} 0 & \text{for } x < \left(t - \dfrac{\Delta}{2}\right) \\[2mm] \dfrac{1}{\Delta} & \text{for } \left(t - \dfrac{\Delta}{2}\right) < x < \left(t + \dfrac{\Delta}{2}\right) \\[2mm] 0 & \text{for } x > \left(t + \dfrac{\Delta}{2}\right) \end{cases}$$

(2) Basic Relations

$$\mathscr{D}(x) = \mathscr{D}(-x)$$

$$a\mathscr{D}(x) = \mathscr{D}\left(\frac{x}{a}\right)$$

$$f(a \pm x)\mathscr{D}(x) = f(a)\mathscr{D}(x)$$

$$\mathscr{D}(x \neq 0) = 0$$

$$x\mathscr{D}(x) = 0$$

$$f(x)\mathscr{D}(x) = f(0)\mathscr{D}(x)$$

(3) Properties

$$\int_{-\infty}^{+\infty} \mathscr{D}(x - t)\, dt = 1$$

$$\int_{-\infty}^{+\infty} f(t)\mathscr{D}(x - t)\, dt = f(x)$$

$$\int_{0}^{+\infty} \mathscr{D}(t)\, dt = 1$$

$$\int_{0}^{+\infty} f(t)\mathscr{D}(t)\, dt = f(0)$$

$$\int_{-\infty}^{+\infty} \mathscr{D}(x - t)\mathscr{D}(y - t)\, dt = \mathscr{D}(x - y)$$

(4) Derivatives

$$\frac{d\mathscr{D}(x)}{dx} = -\frac{\mathscr{D}(x)}{x}$$

$$\frac{d^n\mathscr{D}(x)}{dx^n} = (-1)^n \frac{n!\,\mathscr{D}(x)}{x^n}$$

(5) Integrals

$$\int_{-\infty}^{+\infty} f(t)\frac{d\mathscr{D}(x - t)}{dx}\, dt = \frac{df(x)}{dx}$$

$$\int_{-\infty}^{+\infty} f(t)\frac{d^n\mathscr{D}(x - t)}{dx^n}\, dt = (-1)^n \frac{d^n f(x)}{dx^n}$$

(6) Laplace Transforms

$$\mathscr{L}\{[\mathscr{D}(t)]\} = 1 \qquad\qquad \mathscr{L}\{[\mathscr{D}(t - a)]\} = e^{-as}$$

Appendix B
GLOSSARY OF SYMBOLS AND MATHEMATICS TERMS

$=$ or $::$	Equals		$\neq$ or $\neq$	Does not equal
$>$	Greater than		$<$	Less than
$\geq$	Greater than or equal		$\leq$	Less than or equal
$\equiv$	Identical		$\approx$	Approximately equal

B.2 ALGEBRA

$+$	Plus or positive		$-$	Minus or negative
$\pm$	Plus or minus / Positive or negative		$\mp$	Minus or plus / Negative or positive
$\times$	Multiplied by		$\div$ or $:$	Divided by
a^n	nth power of a		$\sqrt[n]{a}$	nth root of a
$\log$ / $\log_{10}$	Common logarithm or Briggs' logarithm		$\ln$ / $\log_e$	Natural logarithm or Napier's logarithm
$(\ )$	Parentheses	$[\ ]$ Brackets	$\{\ \}$	Braces

$$\begin{vmatrix} a_1 & a_2 & \cdots \\ b_1 & b_2 & \cdots \\ \cdots & \cdots & \cdots \end{vmatrix} \quad \text{Determinant} \qquad \begin{bmatrix} a_1 & a_2 & \cdots \\ b_1 & b_2 & \cdots \\ \cdots & \cdots & \cdots \end{bmatrix} \quad \text{Matrix}$$

I	Unit matrix	Adj	Adjoint matrix
A^{-1}	Inverse of the A matrix	A^T	Transpose of the A matrix
$n!$	n factorial	$\binom{n}{k}$	Binomial coefficient
$n!!$	n double factorial	$X_h^{(p)}$	Factorial polynomial

$k^P n$ The number of all possible permutations of n elements, among which there are k elements of equal value.

$k^V n$ The number of all possible permutations of n elements taken k at a time.

$k^C n$ The number of all possible permutations (without repetition) of n elements taken k at a time.

B.3 COMPLEX NUMBERS

$i = \sqrt{-1}$	Unit imaginary number	$z = x + iy$	Complex variable
$|z|$	Absolute value of z	$\bar{z}$	Conjugate of z
$\text{Re}\,(z), \mathscr{R}(z)$	Real part of z	$\text{Im}\,(z), \mathscr{I}(z)$	Imaginary part of z

$\parallel$	Parallel to	$\alpha°$	α in degrees
$\perp$	Perpendicular to	α'	α in minutes
$\angle$	Angle	α''	α in seconds
$\cong$	Congruent to	$\sim$	Similar to

$\triangle$	Triangle	$\bigcirc$	Circle
$\square$	Parallelogram	$\square$	Square

$\overline{AB}$　　The line segment between A and B　　　　$\overparen{AB}$　　The arc segment between A and B

$P(x, y, z)$　　　Point P given by the cartesian coordinates x, y, z

$P(r, \theta, z)$　　　Point P given by the cylindrical coordinates r, θ, z

$P(\rho, \theta, \phi)$　　　Point P given by the spherical coordinates ρ, θ, ϕ

B.5　CIRCULAR AND HYPERBOLIC FUNCTIONS

sin	Sine	sinh	Hyperbolic sine
cos	Cosine	cosh	Hyperbolic cosine
tan	Tangent	tanh	Hyperbolic tangent
cot	Cotangent	coth	Hyperbolic cotangent
sec	Secant	sech	Hyperbolic secant
csc	Cosecant	csch	Hyperbolic cosecant

vers	Versine	covers	Coversine

$\sin^{-1}$	Inverse sine	$\sinh^{-1}$	Inverse hyperbolic sine
$\cos^{-1}$	Inverse cosine	$\cosh^{-1}$	Inverse hyperbolic cosine
$\tan^{-1}$	Inverse tangent	$\tanh^{-1}$	Inverse hyperbolic tangent
$\cot^{-1}$	Inverse cotangent	$\coth^{-1}$	Inverse hyperbolic cotangent
$\sec^{-1}$	Inverse secant	sech^{-1}	Inverse hyperbolic secant
$\csc^{-1}$	Inverse cosecant	csch^{-1}	Inverse hyperbolic cosecant

B.6　VECTOR ANALYSIS

$\mathbf{i}, \mathbf{j}, \mathbf{k}$　　Unit vectors, cartesian system of coordinates　　　　$\mathbf{e}_s$　　Unit vector in s direction

$\mathbf{r} = \mathbf{i}x + \mathbf{j}y + \mathbf{k}z$　　　Position vector, cartesian coordinates

$\mathbf{r} = r_a\mathbf{e}_a + r_\theta\mathbf{e}_\theta + r_z\mathbf{e}_z$　　　Position vector, cylindrical coordinates

$\mathbf{r} = r_b\mathbf{e}_b + r_\theta\mathbf{e}_\theta + r_\phi\mathbf{e}_\phi$　　　Position vector, spherical coordinates

$\mathbf{r}_1 \bullet \mathbf{r}_2$　　Scalar product　　　　$\mathbf{r}_1 \times \mathbf{r}_2$　　Vector product

∇　　Vector differential operator　　　　∇^2　　Laplacian operator

(a, b)	The bounded open interval	$[a, b]$	The bounded closed interval
$f(x), F(x)$	The function of x	$f^{-1}(x), F^{-1}(x)$	The inverse function of x
$\sum\limits_{i=1}^{n} u_i$	The sum of n terms	$\prod\limits_{i=1}^{n} u_i$	The product of n terms

$\bigwedge\limits_{k=1}^{n} [1 \pm a_k s]$ The nested sum of $n + 1$ terms

Δu	The increment of u	du	The differential of u

$\left.\begin{array}{l} \dfrac{dy}{dx}, y', D_x y \\[2ex] \dfrac{df(x)}{dx}, f'(x), D_x f(x) \end{array}\right\}$ The first-order derivative of $y = f(x)$ with respect to x

$\left.\begin{array}{l} \dfrac{d^n y}{dx^n}, y^{(n)}, D_x^{(n)} y \\[2ex] \dfrac{d^n f(x)}{dx}, f^{(n)}(x), D_x^{(n)} f(x) \end{array}\right\}$ The nth-order derivative of $y = f(x)$ with respect to x

$\left.\begin{array}{l} \dfrac{\partial w}{\partial x}, w_x, D_x w \\[2ex] \dfrac{\partial f}{\partial x}, f_x, F_x \end{array}\right\}$ The first-order partial derivative of $w = f(x, y, \ldots)$ with respect to x

$\left.\begin{array}{l} \dfrac{\partial^2 w}{\partial x \partial y}, w_{xy}, D_{xy} w \\[2ex] \dfrac{\partial^2 f}{\partial x \partial y}, f_{xy}, F_{xy} \end{array}\right\}$ The second-order partial derivative of $w = f(x, y, \ldots)$ with respect to x and then with respect to y

$f'(a+)$	The derivative on the right of $x = a$	$f'(a-)$	The derivative on the left of $x = a$

$\dfrac{\partial (f_1, f_2, \ldots, f_n)}{\partial (x_1, x_2, \ldots, x_n)}, J \dfrac{(f_1, f_2, \ldots, f_n)}{(x_1, x_2, \ldots, x_n)}$ Jacobian determinant (Sec. 7.08)

$\int f(x)\, dx$	The indefinite integral of $y = f(x)$	$\int_a^b f(x)\, dx$	The definite integral of $y = f(x)$ between limits a and b
$\int\!\int$	The double integral	$\int\!\int\!\int$	The triple integral
$\int_C$	The line integral	$\int_S$ The surface integral	$\int_V$ The volume integral

A	Area	M	Static moment
V	Volume	I	Moment of inertia
k	Radius of gyration	J	Polar moment of inertia
ρ	Radius of curvature	κ	Curvature
L	Length of curve	x_C, y_C, z_C	Coordinates of centroid

π	Archimedes number	$\bar{B}_m$	Bernoulli number
e	Euler's number	B_m	Auxiliary Bernoulli number
C	Euler's constant	$\bar{E}_m$	Euler number
$\mathscr{S}_h^{(p)}$	Stirling number	E_m	Auxiliary Euler number

B.9 SPECIAL FUNCTIONS

$\Gamma(x)$	Gamma function	$\Pi(x)$	Pi function
$\mathrm{B}(x)$	Beta function	$\mathrm{erf}(x)$	Error function
$\Psi(x)$	Digamma function	$\Psi^{(m)}(x)$	Polygamma function
$Z(x)$	Zeta function	$\bar{Z}(x)$	Complementary zeta function

$\mathrm{Si}(x), \mathrm{Ci}(x), \mathrm{Ei}(x), \bar{\mathrm{Ei}}(x)$ Integral functions

$S(x), C(x)$ Fresnel integrals

$\left.\begin{array}{l} E(k, x), F(k, x), \Pi(n, k, x) \\ E(k, \phi), F(k, \phi), \Pi(n, k, \phi) \end{array}\right\}$ Elliptic integrals

$\mathrm{sn}\, u, \mathrm{cn}\, u, \mathrm{dn}\, u$ Jacobi's elliptic functions

$\mathscr{L} f(t) = f(s)$ Laplace transform

$\mathscr{D}(x - t)$ **Dirac delta functions**

B.10 BESSEL FUNCTIONS

J_n	Bessel function of the first kind of order n
Y_n	Bessel function of the second kind of order n
I_n	Modified Bessel function of the first kind of order n
K_n	Modified Bessel function of the second kind of order n

$H_n^{(1)}$	Hankel function of the first kind of order n
$H_n^{(2)}$	Hankel function of the second kind of order n

Ber_n	Ber function of order n	Ker_n	Ker function of order n
Bei_n	Bei function of order n	Kei_n	Kei function of order n

B.11 ORTHOGONAL POLYNOMIALS

$p_n(x)$	Orthogonal polynomial in n	$w(x)$	Weight function
$F(\alpha, \beta, \gamma, x)$	Hypergeometric series		
$P_n(x)$	Legendre polynomial in n		
$T_n(x)$	Chebyshev polynomial in n		
$L_n(x)$	Laguerre polynomial in n		
$H_n(x)$	Hermite polynomial in n		

ϵ	Absolute error		$\bar{\epsilon}$	Relative error
ϵ_T	Truncation error		$\bar{y}$	Substitute function
r_{ij}	Carryover value		m_j	Starting value
Δy_n	Forward difference		∇y_n	Backward difference
δy_n	Central difference		Δx_n	Divided difference

B.13 PROBABILITY AND STATISTICS

$\cup$	Union		$\cap$	Intersection
$P(E)$	Probability of occurrence		$P(\bar{E})$	Probability of nonoccurrence
$\phi(x)$	Probability density		$\phi(x, y)$	Joint probability density
$\bar{X}$	Arithmetic means		$\bar{G}$	Geometric means
$\bar{H}$	Harmonic means		$\bar{Q}$	Quadratic means
D	Deviation		$\bar{D}$	Mean deviation
σ	Standard deviation		σ^2	Variance
μ_k	Moment of degree k		β_2	Kurtosis
γ_1	Coefficient of skewness		γ_2	Coefficient of excess
$\phi_N(t)$	Ordinate of the standard normal curve		$F_N(t)$	Area under the standard normal curve

B.14 GREEK ALPHABET

A	α	Alpha	H	η	Eta	N	ν	Nu	T	τ	Tau
B	β	Beta	Θ	θ	Theta	Ξ	ξ	Xi	Υ	υ	Upsilon
Γ	γ	Gamma	I	ι	Iota	O	o	Omicron	Φ	ϕ	Phi
Δ	δ	Delta	K	κ	Kappa	Π	π	Pi	X	χ	Chi
E	ϵ	Epsilon	Λ	λ	Lambda	P	ρ	Rho	Ψ	ψ	Psi
Z	ζ	Zeta	M	μ	Mu	Σ	σ	Sigma	Ω	ω	Omega

B.15 GERMAN ALPHABET

Aa	Bb	Cc	Dd	Ee	Ff	Gg	Hh	Ii	Jj	Kk	Ll	Mm
𝔄𝔞	𝔅𝔟	ℭ𝔠	𝔇𝔡	𝔈𝔢	𝔉𝔣	𝔊𝔤	ℌ𝔥	ℑ𝔦	𝔍𝔧	𝔎𝔩	𝔏𝔩	𝔐𝔪

Nn	Oo	Pp	Qq	Rr	Ss	Tt	Uu	Vv	Ww	Xx	Yy	Zz
𝔑𝔫	𝔒𝔬	𝔓𝔭	𝔔𝔮	𝔑𝔯	𝔖𝔰	𝔗𝔱	𝔘𝔲	𝔙𝔳	𝔚𝔴	𝔛𝔵	𝔜𝔶	𝔷𝔷

B.16 RUSSIAN ALPHABET

Upright	Cursive	Transliteration (pronunciation)	Upright	Cursive	Transliteration (pronunciation)
А а	*А а*	a	Р р	*Р р*	r
Б б	*Б б*	b	С с	*С с*	s
В в	*В в*	v	Т т	*Т т*	t
Г г	*Г г*	g	У у	*У у*	u (as in moon)
Д д	*Д д*	d	Ф ф	*Ф ф*	f
Е е	*Е е*	ye (as in yell)	Х х	*Х х*	kh
Ё ё	*Ё ё*	yё (yo as in yore)	Ц ц	*Ц ц*	ts
Ж ж	*Ж ж*	zh	Ч ч	*Ч ч*	ch (church)
З з	*З з*	z	Ш ш	*Ш ш*	sh
И и	*И и*	i	Щ щ	*Щ щ*	shch
Й й	*Й й*	y	Ъ ъ	*Ъ ъ*	[double apostrophe, no sound]
К к	*К к*	k	Ы ы	*Ы ы*	y (rhythm)
Л л	*Л л*	l	Ь ь	*Ь ь*	[apostrophe; palatalizes preceding consonant]
М м	*М м*	m			
Н н	*Н н*	n	Э э	*Э э*	e (elder)
О о	*О о*	o	Ю ю	*Ю ю*	yu (union)
П п	*П п*	p	Я я	*Я я*	ya (yard)

Symbol	Name	Quantity	Relations
cd	candela	luminous intensity	basic unit
kg	kilogram	mass	basic unit
lm	lumen	luminous flux	$cd \cdot sr$
lx	lux	illumination	$lm \cdot m^{-2}$
m	meter	length	basic unit
mol	mole	amount of substance	basic unit
rad	radian	plane angle	supplementary unit
s	second	time	basic unit
sr	steradian	solid angle	supplementary unit
A	ampere	electric current	basic unit
C	coulomb	electric charge	$A \cdot s$
F	farad	electric capacitance	$A \cdot s \cdot V^{-1}$
H	henry	electric inductance	$V \cdot s \cdot A^{-1}$
Hz	hertz	frequency	s^{-1}
J	joule	work, energy	$N \cdot m$
K	kelvin	temperature degree	basic unit
N	newton	force	$kg \cdot m \cdot s^{-2}$
Pa	pascal	pressure, stress	$N \cdot m^{-2}$
S	siemens	electric conductance	Ω^{-1}
T	tesla	magnetic flux density	$Wb \cdot m^{-2}$
V	volt	voltage, electromotive force	$W \cdot A^{-1}$
W	watt	power	$J \cdot s^{-1}$
Wb	weber	magnetic flux	$V \cdot s$
Ω	ohm	electric resistance	$V \cdot A^{-1}$

B.18   DECIMAL MULTIPLES AND FRACTIONS OF UNITS

Factor	Prefix	Symbol	Factor	Prefix	Symbol
10^1	deka	D*	10^{-1}	deci	d
10^2	hecto	h	10^{-2}	centi	c
10^3	kilo	k	10^{-3}	milli	m
10^6	mega	M	10^{-6}	micro	μ
10^9	giga	G	10^{-9}	nano	n
10^{12}	tera	T	10^{-12}	pico	p
10^{15}	peta	P	10^{-15}	femto	f
10^{18}	exa	E	10^{-18}	atto	a

*In some literature, the symbol "da" is used for deka

Symbol*	Name	Quantity	Relations
deg	degree	plane angle	supplementary unit
ft	foot	length	basic unit
g	standard gravity	acceleration	$32.174 \text{ ft} \cdot \text{sec}^{-2}$
hp	horsepower	power	$550 \text{ lbf} \cdot \text{ft} \cdot \text{sec}^{-1}$
lb	pound-mass	mass	$\text{lbf} \cdot\cdot \text{g}^{-1}$
lbf	pound-force	force	basic unit
pd	poundal	force	$\text{lb} \cdot \text{ft} \cdot \text{s}^{-2}$
sec	second	time	basic unit
sl	slug	mass	$\text{lbf} \cdot \text{ft}^{-1} \cdot \text{sec}^2$
Btu	British thermal unit	work, energy	$778.128 \text{ ft} \cdot \text{lbf}$
°F	degree Fahrenheit	temperature	basic unit

*For A, C, F, ... refer to Sec. B.17.

B.20 METRIC SYSTEM OF UNITS (MKS SYSTEM)

Symbol*	Name	Quantity	Relations
cal	calorie	work, energy	$0.42665 \text{ m} \cdot \text{kgf}$
deg	degree	plane angle	supplementary unit
g	standard gravity	acceleration	$9.80665 \text{ m} \cdot \text{sec}^{-2}$
hp	horsepower	power	$75 \text{ kgf} \cdot \text{m} \cdot \text{sec}^{-1}$
kg	kilogram-mass	mass	$\text{kgf} \cdot \text{g}^{-1}$
kgf	kilogram-force	force	basic unit
m	meter	length	basic unit
sec	second	time	basic unit
°C	degree Celsius	temperature	basic unit

*For A, C, F, . . . refer to Sec. B.17.

abscissa   The x ordinate in the Cartesian system. The abscissa of the point (a, b) is a.

algorithm   A sequence of instructions that describe how to exactly accomplish a specific task.

analog   Describing a system in which numbers are represented by an indicating device such as a slide rule, nondigital voltmeter, ammeter, etc.

analytic geometry   The branch of mathematics that uses algebra to help in the study of geometry. It allows you to draw pictures of algebraic equations, and it helps you understand geometry by allowing you to describe geometric figures by means of algebraic equations.

antiderivative   An antiderivative of a function $f(x)$ is a function $F(x)$ whose derivative is $f(x)$; $dF(x)/dx = f(x)$. Also $F(x)$ is called the *indefinite integral* of $f(x)$.

asymptote   A straight line that is a close approximation to a particular curve as the curve goes to infinity in one direction. The curve comes very close to the asymptote line, but never touches it.

axiom   A statement that is assumed to be true without proof. *Axiom* is a synonym for *postulate*.

Bessel function   The differential equation

$$x^2 y'' + xy' + (x^2 - n^2)y = 0 \qquad n = \text{constant}$$

is called Bessel's equation of order n. Certain of its solutions are called *Bessel functions*.

binomial   The sum of two terms, such as $ax + b$.

binomial theorem   This theorem explains how to expand the expression $(a + b)^n$.

Boolean algebra   The study of operations carried out on variables that can have only two values: 1 (true) or 0 (false).

calculus   This branch of mathematics is divided into two basic or general areas: differential calculus and integral calculus. In differential calculus, the basic problem is to find the rate of change of a function. The reverse process of differentiation is integration (or antidifferentiation).

Cartesian coordinates   A system in which points on a plane are identified by an ordered pair of numbers, representing the distances to two perpendicular axes. The horizontal axis is the X axis, and the vertical axis is the Y axis.

characteristic   The integer part of a common logarithm. The decimal part is called the *mantissa*.

closed interval   An interval that contains its endpoints. The interval $0 \le x \le 1$ is a closed interval because the two endpoints (0 and 1) are included.

coefficient   A technical term for something that multiplies something else (usually a constant multiplying a variable).

complex number   A complex number is formed by adding a purely imaginary number to a real number. The general form of a complex number is $a + bi$, where a and b are real numbers and i is the imaginary unit: $i^2 = -1$.

conic sections   Circles, ellipses, parabolas, and hyperbolas are called conic sections because they can be formed by the intersection of a plane with a right circular cone.

constant   A quantity that does not change in value or form.

coplanar   A set of points is coplanar if all points lie in the same plane. Any three points are always coplanar.

critical point   critical point for a function is a point where the first derivative(s) is (are) zero.

definite integral   If $f(x)$ represents a function of x that is always nonnegative, then the definite integral of $f(x)$ between a and b represents the area under the curve $y = f(x)$, above the x axis to the right of the line $x = a$ and to the left of the line $x = b$.

delta   The Greek capital letter delta (Δ) is used to represent "change in." So Δx represents the change in x.

dependent variable   The dependent variable stands for the output number of a function. In the equation $y = f(x)$, y is the dependent variable and x the independent variable.

derivative   The derivative of a function is the rate of change of that function. It is the slope of the tangent line at the point $(x, f(x))$, or the velocity of a moving point.

differential   Refers to an infinitesimal change in a variable. It is symbolized by d, as in dx.

differential equation   An equation containing the derivatives of a function with respect to one or more independent variables. The order of the equation is the highest derivative that appears in the equation. Second-order differential equations commonly appear in physics.

differentiation   The process of finding a derivative.

direction cosines   Direction cosines of a line are the cosines of the angles that the line makes with the three coordinate axes.

directrix   A line that helps define a geometric figure; commonly used in the definition of conic sections.

double integral   The double integral of a two-variable function $f(x, y)$ represents the volume under the surface $z = f(x, y)$ and above the xy plane in a specified region.

e   Used to represent a fundamental irrational number with the decimal approximation of $e = 2.7182818\cdots$.

eigenvalue   If a square matrix $\mathbf{A}$ multiplies a vector $\mathbf{x}$, and the resulting vector is proportional to $\mathbf{x}$:

$$\mathbf{A}\mathbf{x} = \lambda\mathbf{x}$$

In this case, λ is called an eigenvalue of matrix $\mathbf{A}$, and $\mathbf{x}$ is the corresponding eigenvector. In order to find the eigenvalues, rewrite the equation as

$$(\lambda\mathbf{I} - \mathbf{A})\mathbf{x} = 0$$

where $\mathbf{I}$ is the appropriate identity matrix.

equation   A mathematical statement that says that two mathematical expressions have the same value.

exponent   A number that indicates the operation of repeated multiplication.

exponential function   A function of the form $f(x) = a^x$, where a is a constant known as the *base*.

finite   Something is finite if it doesn't take forever to count or measure it.

Fourier series   Any periodic function that can be expressed as a series involving sines and cosines known as a Fourier series.

function   A rule that turns each member of one set into a member of another set. The most common functions are those that turn one number into another number. An important property of functions is that for each value of the independent variable, there is only one value of the dependent variable. An inverse function does exactly the opposite of the original function.

fundamental theorem of calculus

$$\int_a^b f(x)\,dx = F(b) - F(a)$$

The theorem tells how to find the area under a curve by taking an integral.

imaginary number   An imaginary number is of the form ni, where n is a real number that is being multiplied by the imaginary unit i, and i is defined by $i^2 = -1$.

indefinite integral   The indefinite integral of a function f is symbolized as follows:

$$\int f(x)\, dx = F(x) + C$$

where F is an antiderivative function for x, that is, $dF/dx = f(x)$, and C is called the *arbitrary constant* of integration.

independent variable   The input number to a function; i.e., in $y = f(x)$, x is the independent variable and y is the dependent variable.

infinitesimal   Description of a variable quantity that approaches very close to zero.

infinity ∞ is a symbol representing a limitless quantity.

inverse function   A function that does exactly the opposite of the original function.

irrational number   A real number that is not a rational number; it cannot be expressed as the ratio of two integers.

Laplace transform   A class of integral transforms that have many applications in modern technology. The behavior of ac electric circuits is readily expressed in terms of integrodifferential equations. Also, many types of nonelectric systems lend themselves conveniently to analysis by analogy to electric systems, thus providing a convenient means of generalizing and unifying the analysis of complex systems involving subsystems of various kinds. The adaptability of the Laplace transform method to the solution of differential and integral equations, and the analysis of discontinuous functions, has brought about its wide application to control problems.

limit   The limit of a function is the value that the dependent variable approaches as the independent variable approaches some fixed value.

logarithm   An inverse of an exponential. Any positive number, except 1, can be used as the base for a logarithm function. The two most useful bases are 10 and e.

matrix   A table of numbers arranged in rows and columns. The plural of *matrix* is *matrices*.

modulus   Absolute value of a complex number.

natural logarithm   The natural logarithm of a positive number x (written as $\ln x$) is the logarithm of x to the base e (2.7182818 $\cdots$). The natural logarithm function can also be defined by the definite integral:

$$\ln x = \int_1^x t^{-1}\, dt$$

Newton's method   Newton's method provides a way to estimate the points where complicated functions cross the x axis. The use of the Newton method equation may involve iteration, so that greater accuracy may be obtained in the solution of the intractable equation. The Newton method equation is represented by

$$x_2 = x_1 - f(x_1)/(f'(x_1))$$

where $f'(x_1)$ is the derivative of the function f at point x_1.

numerical integration   This method is used when it is not possible to find a formula that can be evaluated to give the value of a definite integral.

ordinate   Another name for the y coordinate in the Cartesian coordinate system.

partial derivative   The partial derivative of $y = f(x_1, x_2, \ldots, x_n)$ with respect to x_i is found by taking the derivative of y with respect to x_i, while all the other independent variables are held constant.

polar coordinates   Any point in a plane can be identified by its distance from the origin r and its angle of inclination θ. It is an alternative to the Cartesian coordinate system.

Polish notation   In this system, operators are written before their operands.

reverse Polish notation (RPN)   This notation system is the same as Polish notation except it is written in reverse order—operators come after operands.

scalar   A quantity having size but not direction. Real numbers are scalars. A vector has both size and direction.

stochastic   A stochastic variable is the same as a random variable.

transcendental numbers   Numbers that cannot occur as the roots of polynomial equations with rational coefficients. The transcendental numbers are a subset of the irrational numbers. Most values for trigonometric functions, and the number e, are transcendental.

triple integral   A triple integral integrates a function over an entire volume.

vector   A quantity that has both magnitude and direction.

Y intercept   The Y intercept of a curve is the value of y at the point where the curve crosses the Y axis.

Appendix C
UNITS OF MEASUREMENT AND CONVERSIONS BETWEEN INTERNATIONAL AND U.S. CUSTOMARY SYSTEMS

C.01 CONVERSIONS FOR LENGTH, PRESSURE, VELOCITY, VOLUME, AND WEIGHT

To Convert from:	To:	Multiply by:
Length		
Centimeters	Inches	0.393 7
Centimeters	Yards	0.010 94
Feet	Inches	12.0
Feet	Meters	0.304 81
Feet	Yards	0.333
Inches	Centimeters	2.540
Inches	Feet	0.083 33
Inches	Meters	0.025 40
Inches	Micrometers	.25 400.
Inches	Millimeters	25.400
Inches	Yards	0.027 78
Kilometers	Feet	3 281.
Kilometers	Miles (nautical)	0.533 6
Kilometers	Miles (statute)	0.621 4
Kilometers	Yards	1 094.
Meters	Feet	3.280 9
Meters	Yards	1.093 6
Micrometers	Inches	0.000 039 4
Micrometers	Meters	0.000 001
Miles (statute)	Feet	5 280.
Miles (statute)	Kilometers	1.609 3
Miles (statute)	Meters	1 609.34
Miles (statute)	Yards	1 760.
Miles (nautical)	Feet	6 080.2
Miles (nautical)	Kilometers	1.852 0
Miles (nautical)	Meters	1 852.0
Millimeters	Inches	0.039 37
Rods	Meters	5.029 2
Yards	Centimeters	91.44
Yards	Feet	3.0
Yards	Inches	36.0
Yards	Meters	0.914 4
Pressure		
Dynes per square centimeter	Pascals	0.100 0
Grams per cubic centimeter	Ounces per cubic inch	0.578 0
Kilograms per square centimeter	Pounds per square inch	14.223
Kilograms per square centimeter	Pascals	98 066.5
Kilograms per square meter	Pascals	9.806 6
Kilograms per square meter	Pounds per square foot	0.204 8
Kilograms per square meter	Pounds per square yard	1.843 3
Kilograms per cubic meter	Pounds per cubic foot	0.062 43
Ounces per cubic inch	Grams per cubic centimeter	1.730 0
Pounds per cubic feet	Kilograms per cubic meter	16.019
Pounds per square feet	Kilograms per square meter	4.882 4
Pounds per square feet	Pascals	47.880

To Convert from:	To:	Multiply by:
Pounds per square inch	Kilograms per square centimeter	0.070 3
Pounds per square inch	Pascals	6 894.76
Pounds per square yard	Kilograms per square meter	0.542 5

Velocity

Feet per minute	Meters per second	0.005 08
Feet per second	Meters per second	0.304 8
Inches per second	Meters per second	0.025 4
Kilometers	Meters per second	0.277 8
Knots	Meters per second	0.514 4
Miles per hour	Meters per second	0.447 0
Miles per minute	Meters per second	26.822 4

Volume

Cubic centimeters	Cubic inches	0.061 02
Cubic feet	Cubic inches	1 728.0
Cubic feet	Cubic meters	0.028 3
Cubic feet	Cubic yards	0.037 0
Cubic feet	Gallons	7.481
Cubic feet	Liters	28.32
Cubic feet	Quarts	29.922 2
Cubic inches	Cubic centimeters	16.39
Cubic inches	Cubic feet	0.000 578 7
Cubic inches	Cubic meters	0.000 016 39
Cubic inches	Liters	0.016 4
Cubic inches	Gallons	0.004 329
Cubic inches	Quarts	0.017 32
Cubic meters	Cubic feet	35.31
Cubic meters	Cubic inches	61 023.
Cubic meters	Cubic yards	1.308 7
Cubic yards	Cubic feet	27.0
Cubic yards	Cubic meters	0.764 1
Gallons	Cubic feet	0.133 7
Gallons	Cubic inches	231.0
Gallons	Cubic meters	0.003 785
Gallons	Liters	3.785
Gallons	Quarts	4.0
Liters	Cubic feet	0.035 31
Liters	Cubic inches	61.017
Liters	Gallons	0.264 2
Liters	Pints	2.113 3
Liters	Quarts	1.057
Liters	Cubic meters	0.001 0
Pints	Cubic meters	0.004 732
Pints	Liters	0.473 2
Pints	Quarts	0.50
Quarts	Cubic feet	0.033 42
Quarts	Cubic inches	57.75
Quarts	Cubic meters	0.000 946 4
Quarts	Gallons	0.25
Quarts	Liters	0.946 4
Quarts	Pints	2.0

To Convert from:	To:	Multiply by:
	Weight	
Grams	Kilograms	0.001
Grams	Ounces	0.035 27
Grams	Pounds	0.002 205
Kilograms	Ounces	35.274
Kilograms	Pounds	2.204 6
Ounces	Grams	28.35
Ounces	Kilograms	0.028 35
Ounces	Pounds	0.062 5
Pounds	Grams	453.6
Pounds	Kilograms	0.453 6
Pounds	Ounces	16.0

C.02 STANDARD CONVERSION TABLE: MEASURES ARE FOUND FROM THE TABLE

Multiply:	By:	To obtain:
Acres	43 560	Square feet
Acres	4 047	Square meters
Acres	$1\ 562 \times 10^{-3}$	Square miles
Acres	4 840	Square yards
Acre-feet	43 560	Cubic feet
Acre-feet	325 851	Gallons
Acre-feet	1 233.49	Cubic meters
Atmospheres	76.0	Centimeters of mercury
Atmospheres	29.92	Inches of mercury
Atmospheres	33.90	Feet of water
Atmospheres	10 333	Kilograms per square meter
Atmospheres	14.70	Pounds per square inch
Atmospheres	1.058	Tons per square foot
Barrels—oil	42	Gallons—oil
Barrels—cement	376	Pounds—cement
Bags or sacks—cement	94	Pounds—cement
Board feet	$144\ \text{in}^2 \times 1\ \text{in}$	Cubic inches
British thermal units	0.252 0	Kilogram-calories
British thermal units	777.5	Foot-pounds
British thermal units	3.927×10^{-4}	Horsepower-hours
British thermal units	107.5	Kilogram-meters
British thermal units	2.928×10^{-4}	Kilowatthours
British thermal units per minute	12.96	Foot-pounds per second
British thermal units per minute	0.023 56	Horsepower
British thermal units per minute	0.017 57	Kilowatts
British thermal units per minute	17.57	Watts
Centares (centiares)	1	Square meters
Centigrams	0.01	Grams
Centiliters	0.01	Liters
Centimeters	0.393 7	Inches
Centimeters	0.01	Meters
Centimeters	10	Millimeters

Multiply:	By:	To obtain:
Centimeters of mercury	0.013 16	Atmospheres
Centimeters of mercury	0.446 1	Feet of water
Centimeters of mercury	136.0	Kilograms per square meter
Centimeters of mercury	27.85	Pounds per square foot
Centimeters of mercury	0.193 4	Pounds per square inch
Centimeters per second	1.969	Feet per minute
Centimeters per second	0.032 81	Feet per second
Centimeters per second	0.036	Kilometers per hour
Centimeters per second	0.6	Meters per minute
Centimeters per second	0.022 37	Miles per hour
Centimeters per second	3.728×10^{-4}	Miles per minute
Centimeters per second per second	0.032 81	Feet per second per second
Cubic centimeters	3.531×10^{-5}	Cubic feet
Cubic centimeters	6.102×10^{-2}	Cubic inches
Cubic centimeters	10^{-6}	Cubic meters
Cubic centimeters	1.308×10^{-6}	Cubic yards
Cubic centimeters	2.642×10^{-4}	Gallons
Cubic centimeters	10^{-3}	Liters
Cubic centimeters	2.113×10^{-3}	Pints (liq.)
Cubic centimeters	1.057×10^{-3}	Quarts (liq.)
Cubic feet	2.832×10^{4}	Cubic centimeters
Cubic feet	1 728	Cubic inches
Cubic feet	0.028 32	Cubic meters
Cubic feet	0.037 04	Cubic yards
Cubic feet	7.480 52	Gallons
Cubic feet	28.32	Liters
Cubic feet	59.84	Pints (liq.)
Cubic feet	29.92	Quarts (liq.)
Cubic feet per minute	472.0	Cubic centimeters per second
Cubic feet per minute	0.124 7	Gallons per second
Cubic feet per minute	0.472 0	Liters per second
Cubic feet per minute	62.43	Pounds of water per minute
Cubic feet per second	0.646 317	Million gallons per day
Cubic feet per second	448.831	Gallons per minute
Cubic inches	16.39	Cubic centimeters
Cubic inches	5.787×10^{-4}	Cubic feet
Cubic inches	1.639×10^{-5}	Cubic inches
Cubic inches	2.143×10^{-5}	Cubic yards
Cubic inches	4.329×10^{-3}	Gallons
Cubic inches	1.639×10^{-2}	Liters
Cubic inches	0.034 63	Pints (liq.)
Cubic inches	0.017 32	Quarts (liq.)
Cubic meters	10^{6}	Cubic centimeters
Cubic meters	35.31	Cubic feet
Cubic meters	61.023×10^{3}	Cubic inches
Cubic meters	1.308	Cubic yards
Cubic meters	264.2	Gallons
Cubic meters	10^{3}	Liters
Cubic meters	2 113	Pints (liq.)
Cubic meters	1 057	Quarts (liq.)
Cubic yards	7.646×10^{5}	Cubic centimeters
Cubic yards	27	Cubic feet
Cubic yards	46 656	Cubic inches
Cubic yards	0.764 6	Cubic meters
Cubic yards	202.0	Gallons
Cubic yards	764.6	Liters
Cubic yards	1 616	Pints (liq.)

Multiply:	By:	To obtain:
Cubic yards	807.9	Quarts (liq.)
Cubic yards per minute	0.45	Cubic feet per second
Cubic yards per minute	3.367	Gallons per second
Cubic yards per minute	12.74	Liters per second
Decigrams	0.1	Grams
Deciliters	0.1	Liters
Decimeters	0.1	Meters
Degrees (angle)	60	Minutes
Degrees (angle)	0 017 45	Radians
Degrees (angle)	3 600	Seconds
Degrees per second	0.017 45	Radians per second
Degrees per second	0.166 7	Revolutions per minute
Degrees per second	0.002 778	Revolutions per second
Dekagrams	10	Grams
Dekaliters	10	Liters
Dekameters	10	Meters
Drams	27.343 75	Grains
Drams	0.062 5	Ounces
Drams	1.771 845	Grams
Fathoms	6	Feet
Feet	30.48	Centimeters
Feet	12	Inches
Feet	0.304 8	Meters
Feet	$\frac{1}{3}$	Yards
Feet of water	0.029 50	Atmospheres
Feet of water	0.882 6	Inches of mercury
Feet of water	304.8	Kilograms per square meter
Feet of water	62.43	Pounds per square foot
Feet of water	0.433 5	Pounds per square inch
Feet per minute	0.508 0	Centimeters per second
Feet per minute	0.016 67	Feet per second
Feet per minute	0.018 29	Kilometers per hour
Feet per minute	0.304 8	Meters per minute
Feet per minute	0.011 36	Miles per hour
Feet per second	30.48	Centimeters per second
Feet per second	1.097	Kilometers per hour
Feet per second	0.592 1	Knots
Feet per second	18.29	Meters per minute
Feet per second	0.681 8	Miles per hour
Feet per second	0.011 36	Miles per minute
Feet per second per second	30.48	Centimeters per second per second
Feet per second per second	0.304 8	Meters per second per second
Foot-pounds	1.286×10^{-3}	British thermal units
Foot-pounds	5.050×10^{-7}	Horsepower-hours
Foot-pounds	3.241×10^{-4}	Kilogram-calories
Foot-pounds	0.138 3	Kilogram-meters
Foot-pounds	3.766×10^{-7}	Kilowatt hours
Foot-pounds per minute	1.286×10^{-3}	British thermal units per minute
Foot-pounds per minute	0.016 67	Foot pounds per second
Foot-pounds per minute	3.030×10^{-5}	Horsepower
Foot-pounds per minute	3.241×10^{-4}	Kilogram-calories per minute
Foot-pounds per minute	2.260×10^{-5}	Kilowatts
Foot-pounds per second	7.717×10^{-2}	British thermal units per minute
Foot-pounds per second	1.818×10^{-3}	Horsepower
Foot-pounds per second	1.945×10^{-2}	Kilogram-calories per minute
Foot-pounds per second	1.356×10^{-3}	Kilowatts
Gallons	3 785	Cubic centimeters

Multiply:	By:	To obtain:
Gallons	0.133 7	Cubic feet
Gallons	231	Cubic inches
Gallons	3.785×10^{-3}	Cubic meters
Gallons	4.95×10^{-3}	Cubic yards
Gallons	3.785	Liters
Gallons	8	Pints (liq.)
Gallons	4	Quarts (liq.)
Gallons—Imperial	1.200 95	U.S. gallons
Gallons—U.S.	0.832 67	Imperial gallons
Gallons of water	8.345 3	Pounds of water
Gallons per minute	2.228×10^{-3}	Cubic feet per second
Gallons per minute	0.063 08	Liters per second
Gallons per minute	8.020 8	Cubic feet per hour
Gallons per minute	8.020 8 area (square feet)	Overflow rate (feet per hour)
Gallons of water per minute	6.008 6	Tons of water per 24 hours
Grains (troy)	1	Grains (avoir.)
Grains (troy)	0.064 80	Grams
Grains (troy)	0.041 67	Pennyweights (troy)
Grains (troy)	$2.083\ 3 \times 10^{-3}$	Ounces (troy)
Grains per U.S. gallon	17.118	Parts per million
Grains per U.S. gallon	142.86	Pounds per million gallons
Grains per Imperial gallon	14.254	Parts per million
Grams	980.7	Dynes
Grams	15.43	Grains
Grams	10^{-3}	Kilograms
Grams	10^{3}	Milligrams
Grams	0.035 27	Ounces
Grams	0.032 15	Ounces (troy)
Grams	2.205×10^{-3}	Pounds
Grams per centimeter	5.600×10^{-3}	Pounds per inch
Grams per cubic centimeter	62.43	Pounds per cubic foot
Grams per cubic centimeter	0.036 13	Pounds per cubic inch
Grams per liter	58.417	Grains per gallon
Grams per liter	8.345	Pounds per 1000 gallons
Grams per liter	0 062 427	Pounds per cubic foot
Grams per liter	1 000	Parts per million
Hectares	2.471	Acres
Hectares	1.076×10^{5}	Square feet
Hectograms	100	Grams
Hectoliters	100	Liters
Hectometers	100	Meters
Hectowatts	100	Watts
Horsepower	42.44	British thermal units per minute
Horsepower	33 000	Foot-pounds per minute
Horsepower	550	Foot-pounds per second
Horsepower	1.014	Horsepower (metric)
Horsepower	10.70	Kilogram-calories per minute
Horsepower	0.745 7	Kilowatts
Horsepower	745.7	Watts
Horsepower (boiler)	33 479	British thermal units per hour
Horsepower (boiler)	9 803	Kilowatts
Horsepower-hours	2 547	British thermal units
Horsepower-hours	1.98×10^{6}	Foot-pounds
Horsepower-hours	641.7	Kilogram-calories
Horsepower-hours	2.737×10^{5}	Kilogram-meters
Horsepower-hours	0.745 7	Kilowatthours
Inches	2.540	Centimeters

Multiply:	By:	To obtain:
Inches of mercury	0.033 42	Atmospheres
Inches of mercury	1.133	Feet of water
Inches of mercury	345.3	Kilograms per square meter
Inches of mercury	70.73	Pounds per square foot
Inches of mercury	0.491 2	Pounds per square inch
Inches of water	0.002 458	Atmospheres
Inches of water	0.073 55	Inches of mercury
Inches of water	25.40	Kilograms per square meter
Inches of water	0.578 1	Ounces per square inch
Inches of water	5.202	Pounds per square foot
Inches of water	0.036 13	Pounds per square inch
Kilograms	980 665	Dynes
Kilograms	2.205	Pounds
Kilograms	1.102×10^{-3}	Tons (short)
Kilograms	10^3	Grams
Kilogram-calories	3.968	British thermal units
Kilogram-calories	3 087	Foot-pounds
Kilogram-calories	1.558×10^{-3}	Horsepower-hours
Kilogram-calories	1.162×10^{-3}	Kilowatthours
Kilogram-calories per minute	51.43	Foot-pounds per second
Kilogram-calories per minute	0.093 51	Horsepower
Kilogram-calories per minute	0.069 72	Kilowatts
Kilograms per meter	0.672 0	Pounds per foot
Kilograms per square meter	9.678×10^{-5}	Atmospheres
Kilograms per square meter	3.281×10^{-3}	Feet of water
Kilograms per square meter	2.896×10^{-3}	Inches of mercury
Kilograms per square meter	0.204 8	Pounds per square foot
Kilograms per square meter	1.422×10^{-3}	Pounds per square inch
Kilograms per square millimeter	10^6	Kilograms per square meter
Kiloliters	10^3	Liters
Kilometers	10^5	Centimeters
Kilometers	3 281	Feet
Kilometers	10^3	Meters
Kilometers	0.621 4	Miles
Kilometers	1 094	Yards
Kilometers per hour	27.78	Centimeters per second
Kilometers per hour	54.68	Feet per minute
Kilometers per hour	0.911 3	Feet per second
Kilometers per hour	0.539 6	Knots
Kilometers per hour	16.67	Meters per minute
Kilometers per hour	0.621 4	Miles per hour
Kilograms per hour per second	27.78	Centimeters per second per second
Kilograms per hour per second	0.911 3	Feet per second per second
Kilograms per hour per second	0.277 8	Meters per second per second
Kilowatts	56.92	British thermal units per minute
Kilowatts	4.425×10^4	Foot-pounds per minute
Kilowatts	737.6	Foot-pounds per second
Kilowatts	1.341	Horsepower
Kilowatts	14.34	Kilogram-calories per minute
Kilowatts	10^3	Watts
Kilowatthours	3 415	British thermal units
Kilowatthours	2.655×10^6	Foot-pounds
Kilowatthours	1.341	Horsepower-hours
Kilowatthours	860.5	Kilogram-calories
Kilowatthours	3.671×10^5	Kilogram-meters
Liters	10^3	Cubic centimeters
Liters	0.035 31	Cubic feet

Multiply:	By:	To obtain:
Liters	61.02	Cubic inches
Liters	10^3	Cubic meters
Liters	1.308×10^{-3}	Cubic yards
Liters	0.264 2	Gallons
Liters	2.113	Pints (liq.)
Liters	1.057	Quarts (liq.)
Liters per minute	5.886×10^{-4}	Cubic feet per second
Liters per minute	4.403×10^{-3}	Gallons per second
Lumber width (inches) $\times \dfrac{\text{thickness (inches)}}{12}$	Length (feet)	Board feet
Meters	100	Centimeters
Meters	3.281	Feet
Meters	39.37	Inches
Meters	10^{-3}	Kilometers
Meters	10^3	Millimeters
Meters	1.094	Yards
Meters per minute	1.667	Centimeters per second
Meters per minute	3.281	Feet per minute
Meters per minute	0.054 68	Feet per second
Meters per minute	0.06	Kilometers per hour
Meters per minute	0.037 28	Miles per hour
Meters per second	196.8	Feet per minute
Meters per second	3.281	Feet per second
Meters per second	3.6	Kilometers per hour
Meters per second	0.06	Kilometers per minute
Meters per second	2.237	Miles per hour
Meters per second	0.037 28	Miles per minute
Micrometers	10^{-6}	Meters
Miles	$1\,609 \times 10^5$	Centimeters
Miles	5 280	Feet
Miles	1.609	Kilometers
Miles	1 760	Yards
Miles per hour	44.70	Centimeters per second
Miles per hour	88	Feet per minute
Miles per hour	1.467	Feet per second
Miles per hour	1.609	Kilometers per hour
Miles per hour	0.868 4	Knots
Miles per hour	26.82	Meters per minute
Miles per minute	2 682	Centimeters per second
Miles per minute	88	Feet per second
Miles per minute	1.609	Kilometers per minute
Miles per minute	60	Miles per hour
Milligrams	10^{-6}	Kilograms
Milligrams	10^{-3}	Grams
Milliliters	10^{-3}	Liters
Millimeters	0.1	Centimeters
Millimeters	0.039 37	Inches
Milligrams per liter	1	Parts per million
Million gallons per day	1.547 23	Cubic feet per second
Miner's inches	1.5	Cubic feet per minute
Minutes (angle)	2.909×10^{-4}	Radians
Ounces	16	Drams
Ounces	437.5	Grains
Ounces	0.062 5	Pounds
Ounces	28.349 527	Grams
Ounces	0.911 5	Ounces (troy)
Ounces	2.790×10^{-5}	Tons (long)

Multiply:	By:	To obtain:
Ounces	2.835×10^{-5}	Tons (metric)
Ounces (troy)	480	Grains
Ounces (troy)	20	Pennyweights (troy)
Ounces (troy)	0.083 33	Pounds (troy)
Ounces (troy)	31.103 481	Grams
Ounces (troy)	1.097 14	Ounces (avoir.)
Ounces (fluid)	1.805	Cubic inches
Ounces (fluid)	0.029 57	Liters
Ounces per square inch	0.062 5	Pounds per square inch
Overflow rate (feet per hour)	$0.124\ 68 \times$ area (feet^2)	Gallons per minute
$\dfrac{1}{\text{Overflow rate (feet per hour)}}$	8.020 8	Square feet per gallon per minute
Parts per million	0.058 4	Grains per U.S. gallon
Parts per million	0.070 16	Grains per Imperial gallon
Parts per million	8.345	Pounds per million gallons
Pennyweights (troy)	24	Grains
Pennyweights (troy)	1.555 17	Grams
Pennyweights (troy)	0.05	Ounces (troy)
Pennyweights (troy)	$4.166\ 7 \times 10^{-3}$	Pounds (troy)
Pounds	16	Ounces
Pounds	256	Drams
Pounds	7 000	Grains
Pounds	0.000 5	Tons (short)
Pounds	453.592 4	Grams
Pounds	1.215 28	Pounds (troy)
Pounds	14.583 3	Ounces (troy)
Pounds (troy)	5 760	Grains
Pounds (troy)	240	Pennyweights (troy)
Pounds (troy)	12	Ounces (troy)
Pounds (troy)	373.241 77	Grams
Pounds (troy)	0.822 857	Pounds (avoir.)
Pounds (troy)	13.165 7	Ounces (avoir.)
Pounds (troy)	$3.673\ 5 \times 10^{-4}$	Tons (long)
Pounds (troy)	$4.114\ 3 \times 10^{-4}$	Tons (short)
Pounds (troy)	$3.732\ 4 \times 10^{-4}$	Tons (metric)
Pounds of water	0.016 02	Cubic feet
Pounds of water	27.68	Cubic inches
Pounds of water	0.119 8	Gallons
Pounds of water per minute	2.670×10^{-4}	Cubic feet per second
Pounds per cubic foot	0.016 02	Grams per cubic centimeter
Pounds per cubic foot	16.02	Kilograms per cubic meter
Pounds per cubic foot	5.787×10^{-4}	Pounds per cubic inch
Pounds per cubic inch	27.68	Grams per cubic centimeter
Pounds per cubic inch	2.768×10^{4}	Kilograms per cubic meter
Pounds per cubic inch	1 728	Pounds per cubic foot
Pounds per foot	1.488	Kilograms per meter
Pounds per inch	178.6	Grams per centimeter
Pounds per square foot	0.016 02	Feet of water
Pounds per square foot	4.883	Kilograms per square meter
Pounds per square foot	6.945×10^{-3}	Pounds per square inch
Pounds per square inch	0.068 04	Atmospheres
Pounds per square inch	2.307	Feet of water
Pounds per square inch	2.036	Inches of mercury
Pounds per square inch	703.1	Kilograms per square meter
Quadrants (angle)	90	Degrees
Quadrants (angle)	5 400	Minutes
Quadrants (angle)	1.571	Radians

Multiply:	By:	To obtain:
Quarts (dry)	67.20	Cubic inches
Quarts (liq.)	57.75	Cubic inches
Quintals, Argentine	101.28	Pounds
Quintals, Brazil	129.54	Pounds
Quintals, Castile, Peru	101.43	Pounds
Quintals, Chile	101.41	Pounds
Quintals, Mexico	101.47	Pounds
Quintals, metric	220.46	Pounds
Quires	25	Sheets
Radians	57.30	Degrees
Radians	3 438	Minutes
Radians	0.637	Quadrants
Radians per second	57.30	Degrees per second
Radians	0.159 2	Revolutions per second
Radians per second	9.549	Revolutions per minute
Radians per second per second	573.0	Revolutions per minute per minute
Radians per second per second	0.159 2	Revolutions per second per second
Revolutions	360	Degrees
Revolutions	4	Quadrants
Revolutions	6.283	Radians
Revolutions per minute	6	Degrees per second
Revolutions per minute	0.104 7	Radians per second
Revolutions per minute	0.016 67	Revolutions per second
Revolutions per minute per minute	1.745×10^{-3}	Radians per second per second
Revolutions per minute per minute	2.778×10^{-4}	Revolutions per second per second
Revolutions per second	360	Degrees per second
Revolutions per second	6.283	Radians per second
Revolutions per second	60	Revolutions per minute
Revolutions per second per second	6.283	Radians per second per second
Revolutions per second per second	3 600	Revolutions per minute per minute
Seconds (angle)	4.848×10^{-6}	Radians
Square centimeters	1.076×10^{-3}	Square feet
Square centimeters	0.155 0	Square inches
Square centimeters	10^{-4}	Square meters
Square centimeters	100	Square millimeters
Square feet	2.296×10^{-5}	Acres
Square feet	929.0	Square centimeters
Square feet	144	Square inches
Square feet	0.092 90	Square meters
Square feet	3.587×10^{-8}	Square miles
Square feet	$\frac{1}{9}$	Square yards
$\dfrac{1}{\text{Square feet per gallon per minute}}$	8.020 8	Overflow rate (feet per hour)
Square inches	6.452	Square centimeters
Square inches	6.944×10^{-3}	Square feet
Square inches	645.2	Square millimeters
Square kilometers	247.1	Acres
Square kilometers	10.76×10^{6}	Square feet
Square kilometers	10^{6}	Square meters
Square kilometers	0.386 1	Square miles
Square kilometers	1.196×10^{6}	Square yards
Square meters	2.471×10^{-4}	Acres
Square meters	10.76	Square feet
Square meters	3.861×10^{-7}	Square miles
Square meters	1.196	Square yards
Square miles	640	Acres

Multiply:	By:	To obtain:
Square miles	27.88×10^6	Square feet
Square miles	2.590	Square kilometers
Square miles	3.098×10^6	Square yards
Square millimeters	0.01	Square centimeters
Square millimeters	1.550×10^{-3}	Square inches
Square yards	2.066×10^{-4}	Acres
Square yards	9	Square feet
Square yards	0.836 1	Square meters
Square yards	3.228×10^{-7}	Square miles
Temperature, °C + 273	1	Absolute temperature (°C)
Temperature, °C + 17.78	1.8	Temperature (°F)
Temperature, °F + 460	1	Absolute temperature (°F)
Temperature, °F − 32	$\frac{5}{9}$	Temperature (°C)
Tons (long)	1 016	Kilograms
Tons (long)	2 240	Pounds
Tons (long)	1.120 00	Tons (short)
Tons (metric)	10^3	Kilograms
Tons (metric)	2 205	Pounds
Tons (short)	2 000	Pounds
Tons (short)	32 000	Ounces
Tons (short)	907.184 86	Kilograms
Tons (short)	2 430.56	Pounds (troy)
Tons (short)	0.892 87	Tons (long)
Tons (short)	29 166.66	Ounces (troy)
Tons (short)	0.90718	Tons (metric)
Tons dry solids per 24 hours $\dfrac{1}{\text{Area (square feet)}}$		Square feet per ton per 24 hours
Tons of water per 24 hours	83.333	Pounds of water per hour
Tons of water per 24 hours	0.166 43	Gallons per minute
Tons of water per 24 hours	1.334 9	Cubic feet per hour
Watts	0.056 92	British thermal units per minute
Watts	44.26	Foot-pounds per minute
Watts	0.737 6	Foot-pounds per second
Watts	1.341×10^{-3}	Horsepower
Watts	0.014 34	Kilogram-calories per minute
Watts	10^{-3}	Kilowatts
Watthours	3.415	British thermal units
Watthours	2 655	Foot-pounds
Watthours	1.341×10^{-3}	Horsepower-hours
Watthours	0.860 5	Kilogram-calories
Watthours	367.1	Kilogram-meters
Watthours	10^{-3}	Kilowatthours
Yards	91.44	Centimeters
Yards	3	Feet
Yards	36	Inches
Yards	0.914 4	Meters

Note: In the preceding conversions for SI, kilograms, when used as a weight or force, must be multiplied by 9.81 to arrive at newtons. Many modern (SI) torque and force listings are given in newton-meters (N·m) or kilograms, instead of kilogram-meters (kg·m) or gram-centimeters (g·cm), etc. This confusion was created when the kilogram was used as a force or weight. In modern SI applications, the kilogram is considered only as a unit of mass, while the newton is considered as a force. A 1-kg mass used as a force must be specified as $1 \times 9.81 = 9.81$ N. The kilogram is multiplied by the metric gravitational constant of 9.81 m/s² to arrive at the newton measure of force or weight. One newton is equal to 0.101 97 kg force.

To convert from:	To:	Multiply by:	To convert from:	To:	Multiply by:
angstrom	m	$1.000\,0 \times 10^{-10}$ [a]	hp[e]	W	$7.457\,0 \times 10^{2}$
atm	Pa	$1.013\,3 \times 10^{5}$	hp[f]	W	$7.460\,0 \times 10^{2}$
Btu[b]	J	1.054×10^{3}	in	m	$2.540\,0 \times 10^{-2}$
Btu[b]/ft²·h	W/m²	$3.152\,6$	in²	m²	$6.451\,6 \times 10^{-4}$
Btu[b]/ft²·h·°F	W/m²·K	$5.674\,5$	in³	m³	$1.638\,7 \times 10^{-5}$
Btu[b]/ft/h·ft²·°F	W/m·K	$1.729\,6$	in of Hg[h]	Pa	$3.386\,4 \times 10^{3}$
Btu[b]/ft²·s	W/m²	1.135×10^{4}	in of water[c]	Pa	$2.490\,8 \times 10^{2}$
Btu[b]/in/ft²·h·°F	W/m·K	$1.441\,3 \times 10^{-1}$	K	°C	$t_C = t_K - 273.15$
Btu[b]/in/s·ft²·°F	W/m·K	$5.188\,7 \times 10^{2}$	kgf	N	$9.806\,65$ [a]
Btu[b]/lbm·°F	J/kg·K	$4.184\,0 \times 10^{3}$	kgf/mm²	Pa	$9.806\,65 \times 10^{6}$ [a]
cal[b]	J	$4.184\,0$ [a]	ksi	MPa	$6.894\,8$
cal[b]/cm·s·°C	W/m·K	$4.184\,0 \times 10^{2}$ [a]	ksi	Pa	$6.894\,8 \times 10^{6}$
cal[b]/g	J/kg	$4.184\,0 \times 10^{3}$ [a]	ksi$\sqrt{\text{in}}$	MPa$\sqrt{\text{m}}$	$1.098\,8$
cal[b]/g·°C	J/kg·K	$4.184\,0 \times 10^{3}$ [a]	lb[h]	kg	$4.535\,9 \times 10^{-1}$
circ mil	m²	$5.067\,1 \times 10^{-10}$	lb/in³	kg/m³	$2.768\,0 \times 10^{4}$
°C	K	$t_K = t_C + 273.15$	lbf	N	4.4482
degree	rad	$1.745\,3 \times 10^{-2}$	lbf·in	N·m	1.1298×10^{-1}
dyne/cm²	Pa	$1.000\,0 \times 10^{-1}$ [a]	lbf·ft	N·m	$1.355\,8$
°F	°C	$t_C = (t_F - 32)/1.8$	MPa$\sqrt{\text{m}}$	$\text{MNm}^{-3/2}$	$1.000\,0$ [a]
°F	K	$t_K = (t_F + 459.67)/1.8$	μin	m	$2.540\,0 \times 10^{-8}$ [a]
ft	m	$3.048\,0 \times 10^{-1}$	mil	m	$2.540\,0 \times 10^{-5}$ [a]
ft²	m²	$9.290\,3 \times 10^{-2}$	N/m²	Pa	$1.000\,0$ [a]
ft³	m³	$2.831\,7 \times 10^{-2}$	oersted	A/m	79.578
ft of water[c]	Pa	$2.989\,0 \times 10^{3}$	oz/ft²	kg/m²	$3.051\,5 \times 10^{-1}$
ft²/h (thermal diffusivity)	m²/s	$2.580\,64 \times 10^{-5}$ [a]	psi	Pa	$6.894\,8 \times 10^{3}$
ft·lbf	J	$1.355\,8$	°R	K	$t_K = t_R/1.8$
ft·lbf/s	W	$1.355\,8$	ton[i]	kg	$9.071\,8 \times 10^{2}$
ft/s	m/s	$3.048\,0 \times 10^{-1}$	ton[j]	kg	$1.016\,0 \times 10^{3}$
gauss	T	1.0000×10^{-4} [a]	ton/in²	Pa	$1.378\,6 \times 10^{4}$
gallon[d]	m³	$3.785\,4 \times 10^{-3}$	tonne	kg	$1.000\,0 \times 10^{3}$ [a]
g/cm³	kg/m³	$1.000\,0 \times 10^{3}$ [a]	torr	Pa	$1.333\,2 \times 10^{2}$
g/cm³	Mg/m³	$1.000\,0$ [a]	Ω/circ mil·ft	Ω·m	$1.662\,4 \times 10^{-9}$

[a] Exactly. [b] Thermochemical. [c] At 4 °C (39.2 °F). [d] U.S. liquid. [e] Mechanical (1 hp = 550 ft·lbf/s). [f] Electrical. [g] At 0 °C (32 °F). [h] Avoirdupois. [i] Short; equal to 2000 lbm. [j] Long; 2240 lbm.

C.04 UNITS, NOT PART OF COHERENT SYSTEM, BUT GENERALLY ACCEPTED FOR USE WITH SI UNITS

Quantity	Name of unit	Unit symbol	Magnitude in SI units
Time	Minute	min	60 s
	Hour	h	3 600 s
	Day	d	86 400 s
Plane angle	Degree	°	$\pi/180$ rad
	Minute	′	$\pi/10\,800$ rad
	Second	″	$\pi/648\,000$ rad
Volume	Liter	l	1 l = 1 dm³
Mass	Tonne	t	1 t = 10^{3} kg
Energy	Electronvolt	eV	approx. $1.602\,19 \times 10^{-19}$ J
Mass of an atom	Atomic mass unit	u	approx. $1.660\,53 \times 10^{-27}$ kg
Length	Astronomical unit	AU	$149\,600 \times 10^{6}$ m
	Parsec	pc	approx. $30\,857 \times 10^{12}$ m

Quantity	Derived Units	Abbreviation	
Acceleration	Meter per second squared	m/s^2	
Activity (of radioactive source)	1 per second	s^{-1}	
Angular acceleration	Radian per second squared	rad/s^{-2}	
Angular velocity	Radian per second	rad/s	
Area	Square meter	m^2	
Density	Kilogram per cubic meter	kg/m^3	
Dynamic viscosity	Newton-second per square meter	$N \cdot s/m^2$	
Electric capacitance	Farad	F	$(A \cdot s/V)$
Electric charge	Coulomb	C	$(A \cdot s)$
Electric field strength	Volt per meter	V/m	
Electric resistance	Ohm	Ω	(V/A)
Entropy	Joule per kelvin	J/K	
Force	Newton	N	$(kg \cdot m/s^2)$
Frequency	Hertz	Hz	(s^{-1})
Illumination	Lux	lx	(lm/m^2)
Inductance	Henry	H	$(V \cdot s/A)$
Kinematic viscosity	Square meter per second	m^2/s	
Luminance	Candela per square meter	cd/m^2	
Luminous flux	Lumen	lm	$(cd \cdot sr)$
Magnetomotive force	Ampere	A	
Magnetic field strength	Ampere per meter	A/m	
Magnetic flux	Weber	Wb	$(V \cdot s)$
Magnetic flux density	Tesla	T	(Wb/m^2)
Power	Watt	W	(J/s)
Pressure	Newton per square meter	N/m^2	
Radiant intensity	Watt per steradian	W/sr	
Specific heat	Joule per kilogram-kelvin	$J/kg \cdot K$	
Thermal conductivity	Watt per meter-kelvin	$W/m \cdot K$	
Velocity	Meter per second	m/s	
Volume	Cubic meter	m^3	
Voltage, potential difference, electromotive force	Volt	V	(W/A)
Wave number	1 per meter	m^{-1}	
Work, energy, quantity of heat	Joule	J	$(N \cdot m)$

The following lists of physical constants are recommended by the National Academy of Sciences and have been adopted by the National Institute of Standards and Technology (NIST). The lists are taken from the NIST *Technical News Bulletin*.

Adjusted Values of Constants

Constant	Symbol	Value	Est.‡ error limit	Système International (MKSA)		Centimeter-gram-second (CGS)	
Speed of light in vacuum	c	2.997 925	3	$\times 10^8$	m s^{-1}	$\times 10^{10}$	cm s^{-1}
Elementary charge	e	1.602 10	7	10^{-19}	C	10^{-20}	cm$^{1/2}$ g$^{1/2}$ *
		4.802 98	20		. . .	10^{-10}	cm$^{3/2}$ g$^{1/2}$ s^{-1} †
Avogadro constant	N_A	6.022 52	28	10^{23}	mol^{-1}	10^{23}	mol^{-1}
Electron rest mass	m_e	9.109 1	4	10^{-31}	kg	10^{-28}	g
		5.485 97	9	10^{-4}	u	10^{-4}	u
Proton rest mass	m_p	1.672 52	8	10^{-27}	kg	10^{-24}	g
		1.007 276 63	24	10^0	u	10^0	u
Neutron rest mass	m_n	1.674 82	8	10^{-27}	kg	10^{-24}	g
		1.008 665 4	13	10^0	u	10^0	u
Faraday constant	$F\cdot$	9.648 70	16	10^4	C mol^{-1}	10^3	cm$^{1/2}$ g$^{1/2}$ mol^{-1} *
		2.892 61	5		. . .	10^{14}	cm$^{3/2}$ g$^{1/2}$ s^{-1} mol^{-1} †
Planck constant	h	6.625 6	5	10^{-34}	J s	10^{-27}	erg s
	$\hbar$	1.054 50	7	10^{-34}	J s	10^{-27}	erg s
Fine structure constant	α	7.297 20	10	10^{-3}	. . .	10^{-3}	
	$1/\alpha$	1.370 388	19	10^2	. . .	10^2	
	$\alpha/2\pi$	1.161 385	16	10^{-3}	. . .	10^{-3}	
	α^2	5.324 92	14	10^{-5}	. . .	10^{-5}	
Charge to mass ratio for electron	e/m_e	1.758 796	19	10^{11}	C kg^{-1}	10^7	cm$^{1/2}$ g$^{-1/2}$ *
		5.272 74	6		. . .	10^{17}	cm$^{3/2}$ g$^{-1/2}$ s^{-1} †
Quantum-charge ratio	h/e	4.135 56	12	10^{-15}	J s C^{-1}	10^{-7}	cm$^{3/2}$ g$^{1/2}$ s^{-1} *
		1.379 47	4		. . .	10^{-17}	cm$^{1/2}$ g$^{1/2}$ †
Compton wavelength of electron	λ_C	2.426 21	6	10^{-12}	m	10^{-10}	cm
	$\lambda_C/2\pi$	3.861 44	9	10^{-13}	m	10^{-11}	cm
Compton wavelength of proton	$\lambda_{C,p}$	1.321 40	4	10^{-15}	m	10^{-13}	cm
	$\lambda_{C,p}/2\pi$	2.103 07	6	10^{-16}	m	10^{-14}	cm
Rydberg constant	R_∞	1.097 373 1	3	10^7	m^{-1}	10^5	cm^{-1}
Bohr radius	a_0	5.291 67	7	10^{-11}	m	10^{-9}	cm
Electron radius	r_e	2.817 77	11	10^{-15}	m	10^{-13}	cm
	r_e^2	7.939 8	6	10^{-30}	m^2	10^{-26}	cm^2
Thomson cross section	$8\pi r_e^2/3$	6.651 6	5	10^{-29}	m^2	10^{-25}	cm^2
Gyromagnetic ratio of proton	γ	2.675 19	2	10^8	rad s^{-1} T^{-1}	10^4	rad s^{-1} G^{-1} *
	$\gamma/2\pi$	4.257 70	3	10^7	Hz T^{-1}	10^3	s^{-1} G^{-1} *
(uncorrected for diamagnetism, H$_2$O)	γ'	2.675 12	2	10^8	rad s^{-1} T^{-1}	10^4	rad s^{-1} G^{-1} *
	$\gamma'/2\pi$	4.257 59	3	10^7	Hz T^{-1}	10^3	s^{-1} G^{-1} *
Bohr magneton	μ_B	9.273 2	6	10^{-24}	J T^{-1}	10^{-21}	erg G^{-1} *
Nuclear magneton	μ_N	5.050 5	4	10^{-27}	J T^{-1}	10^{-24}	erg G^{-1} *
Proton moment	μ_p	1.410 49	13	10^{-26}	J T^{-1}	10^{-23}	erg G^{-1} *
	μ_p/μ_N	2.792 76	7	10^0	. . .	10^0	
(uncorrected for diamagnetism, H$_2$O)	μ'_p/μ_N	2.792 68	7	10^0	. . .	10^0	
Anomalous electron moment correction	$(\mu_e/\mu_0) - 1$	1.159 615	15	10^{-3}	. . .	10^{-3}	
Zeeman splitting constant	μ_B/hc	4.668 58	4	10^1	m^{-1} T^{-1}	10^{-5}	cm^{-1} G^{-1} *
Gas constant	R	8.314 3	12	10^0	J K^{-1} mol^{-1}	10^7	erg K^{-1} mol^{-1}
Normal volume perfect gas	V_0	2.241 36	30	10^{-2}	m^3 mol^{-1}	10^4	cm^3 mol^{-1}
Boltzmann constant	k	1.380 54	18	10^{-23}	J K^{-1}	10^{-16}	erg K^{-1}
First radiation constant ($2\pi hc^2$)	c_1	3.741 5	3	10^{-16}	W m^2	10^{-5}	erg cm^2 s^{-1}
Second radiation constant	c_2	1.438 79	19	10^{-2}	m K	10^0	cm K
Wien displacement constant	b	2.897 8	4	10^{-3}	m K	10^{-1}	cm K
Stefan-Boltzmann constant	σ	5.669 7	29	10^{-8}	W m^{-2} K^{-4}	10^{-5}	erg cm^{-2} s^{-1} K^{-4}
Gravitational constant	G	6.670	15	10^{-11}	N m^2 kg^{-2}	10^{-8}	dyn cm^2 g^{-2}
Electron-volt	eV	1.602 10	7	$\times 10^{-19}$	J $(eV)^{-1}$	$\times 10^{-12}$	erg $(eV)^{-1}$
Energy associated with:							
Unified atomic mass unit	c^2/Ne	9.314 78	15	10^8	eV u^{-1}	10^8	eV u^{-1}
Proton mass	$m_p c^2/e$	9.382 56	15	10^8	eV m_p^{-1}	10^8	eV m_p^{-1}
Neutron mass	$m_n c^2/e$	9.395 50	15	10^8	eV m_n^{-1}	10^8	eV m_n^{-1}
Electron mass	$m_e c^2/e$	5.110 06	5	10^5	eV m_e^{-1}	10^5	eV m_e^{-1}
Cycle	e/h	2.418 04	7	10^{14}	Hz $(eV)^{-1}$	10^{14}	s^{-1} $(eV)^{-1}$
Wavelength	ch/e	1.239 81	4	10^{-6}	eV m	10^{-4}	eV cm
Wavenumber	e/ch	8.065 73	23	10^5	m^{-1} $(eV)^{-1}$	10^3	cm^{-1} $(eV)^{-1}$
K	e/k	1.160 49	16	10^4	K $(eV)^{-1}$	10^4	K $(eV)^{-1}$

‡ Based on 3 standard deviations; applied to last digits in preceding column.

* Electromagnetic system.

† Electrostatic system.

Abbreviations: C—coulomb; J—joule; Hz—hertz; W—watt; N—newton; T—tesla, G—gauss.

U.S. Customary System	Système International
Length—foot	Length—meter
Time—second	Time—second
Force—pound-force	Force—newton, $kg \cdot m/s^2$
Weight—pound	Weight—newton
Mass—slug, $lb \cdot s^2/ft$	Mass—kilogram
Work or energy—foot-pounds	Work or energy—newton-meter, $kg \cdot m^2/s^2$, joule
Power—foot-pounds per minute; 1 horsepower, 550 ft · lb/s	Power—watt, joules per second; newton-meters per second, $kg \cdot m^2/s^3$
Torque (moment)—pounds-feet	Torque (moment)—newton-meters
Velocity—feet per minute, feet per second	Velocity—meters per minute, meters per second
Acceleration—unit length per s^2	Acceleration—unit length per s^2
Stress—pounds per in^2; 1 pound per in^2 = 6 894.757 Pa 1 lb/in^2 = 6.894 757 kPa	Stress—pascals; N/m^2; 1 Pa = 0.000 145 lb/in^2 1 kPa = 0.145 037 7 lb/in^2

(1) Equivalent Units for Energy, Work, Power, Force, and Mass

Units of energy or work and heat energy are as follows:

U.S. Customary System	SI (metric)
Foot-pound	Newton-meter
British thermal unit (Btu) [the amount of heat energy required to raise the temperature of 1 lb of water (H_2O) at 72°F by 1°F]	Joule (1 J = 1 N · m)
	Watthour, kilowatthour (Wh, kWh)
Horsepower-hour	Gram-calorie (the amount of heat energy required to raise the temperature of 1 g of water at 22°C by 1°C)

Units of power are

U.S. Customary System	SI (metric)
Horsepower	Watt, kilowatt (1 W = 1 J/s = 1 N · m/s
Btu per second, minute, or hour	Joule per second, minute, or hour
Foot-pounds per second, minute, or hour	Newton-meters per second, minute, or hour

Units of force are

U.S. Customary System	SI (metric)
Pound-force, lbf (force)	Newton (force)
Pound-foot, pound-inch (torque or turning moment)	Newton-meter (torque or turning moment)
	Kilogram (when used as weight or force must be specified as mg · N; i.e., 5-kg weight expressed as a force is equal to 5 × 9.81 = 49.05 N)

Units of mass are

U.S. Customary System	SI (metric)
Pounds, slugs ($lb \cdot s^2/ft$) (see previous text)	Kilograms

By definition, *energy* is a force acting through a distance or heat equivalent (work is the expenditure of energy), and *power* is the time rate of expending energy or heat equivalent. Power and energy are thus related by

$$\text{Power, W or Btu/h} = \frac{\text{energy, Wh or Btu}}{\text{time, h}}$$

Occasionally in the literature you may notice errors in specification of the correct units in energy, power, force, mass, or torque (e.g., foot-pounds used for torque instead of pounds-feet, watts used as an energy unit instead of watthours, Btu used as a power unit instead of Btu/h).

(2) Torque Conversion Units

It is often necessary to convert torque from one system of units to another. The following charts will be useful and time-saving and should also help prevent conversion errors. (See *Note* following these tables for explanation of asterisks.)

To convert pounds-feet to:	Multiply by:
Gram-inches	5 443.108 8
Ounce-inches	192.0
Pound-inches	12.0
Kilogram-centimeters*	13.825 7
Kilogram-meters*	0.138 257

To convert pounds-inches to:	Multiply by:
Gram-inches	453.592 4
Ounce-inches	16.0
Pound-feet	0.083 34
Kilogram-centimeters*	1.152
Kilogram-meters*	1.152×10^{-2}

To convert ounce-inches to:	Multiply by:
Gram-inches	28.349 5
Pound-inches	0.062 5
Pound-feet	$5.208\ 7 \times 10^{-3}$
Kilogram-centimeters*	72.808×10^{-3}
Kilogram-meters*	728.08×10^{-6}

To convert gram-inches to:	Multiply by:
Ounce-inches	0.035 27
Pound-inches	2.205×10^{-3}
Pound-feet	$1.837\ 6 \times 10^{-4}$
Kilogram-centimeters*	2.54×10^{-3}
Kilogram-meters*	2.54×10^{-5}

To convert kilogram-centimeters to:	Multiply by:
Gram-inches	393.7
Ounce-inches	13.885 8
Pound-inches	$85.810\ 8 \times 10^{-2}$
Pound-feet	72.346×10^{-3}
Kilogram-meters*	0.01

To convert kilogram-meters to:	Multiply by:
Gram-inches	39 370.0
Ounce-inches	1 388.58
Pound-inches	85.810 8
Pound-feet	7.234 6
Kilogram-centimeters*	100.0

Note: In the preceding conversions for SI, kilograms, when used as a weight or force, must be multiplied by 9.81 to arrive at the newton conversion. Many modern SI torque listings are given as newton-meters (N · m) instead of kilogram-meters (kg · m), or gram-centimeters (g · cm). This confusion was created when the kilogram was previously used as a force or weight. In modern SI systems, the kilogram is considered only as a unit of mass, while the newton is considered as a force. A 1-kg mass used as a force may be specified as $1 \times 9.81 = 9.81$ N. The kilogram is multiplied by the metric gravitational constant of 9.81 m/s^2 to arrive at the newton measure of force. One newton is equal to 0.101 97 kg force.

C.08 WEIGHTS AND MEASURES—U.S. SYSTEM

(1) Linear Measure

Inches	Feet	Yards	Rods	Furlongs	Miles
1.0 =	0.083 33 =	0.027 78 =	0.005 050 5 =	0.000 126 26 =	0.000 015 78
12.0 =	1.0 =	0.333 33 =	0.060 606 1 =	0.001 515 15 =	0.000 189 39
36.0 =	3.0 =	1.0 =	0.181 818 2 =	0.004 545 45 =	0.000 568 18
198.0 =	16.5 =	5.5 =	1.0	= 0.025	= 0.003 125
7 920.0 =	660.0 =	220.0 =	40.0	= 1.0	= 0.125
63 360.0 =	5 280.0 =	1 760.0 =	320.0	= 8.0	= 1.0

(2) Square and Land Measure

Square inches	Square feet	Square yards	Square rods	Acres	Square miles
1.0 =	0.006 944 =	0.000 772			
144.0 =	1.0 =	0.111 111			
1.296.0 =	9.0 =	1.0 =	0.033 06 =	0.000 207	
39 204.0 =	272.25 =	30.25 =	1.0 =	0.006 25 =	0.000 009 8
	43 560.0 =	4 840.0 =	160.0 =	1.0 =	0.001 562 5
		3 097 600.0 =	102 400.0 =	640.0 =	1.0

(3) Avoirdupois Weight

Grains	Drams	Ounces	Pounds	Tons
1.0 =	0.036 57 =	0.002 286 =	0.000 143 =	0.000 000 071 4
27.343 75 =	1.0 =	0.062 5 =	0.003 906 =	0.000 001 95
437.5 =	16.0 =	1.0 =	0.062 5 =	0.000 031 25
7 000.0 =	256.0 =	16.0 =	1.0 =	0.000 5
14 000 000.0 =	512 000.0 =	32 000.0 =	2 000.0 =	1.0

(4) Dry Measure

Pints		Quarts		Pecks		Cubic feet		Bushels
1.0	=	0.5	=	0.062 5	=	0.019 45	=	0.015 63
2.0	=	1.0	=	0.125	=	0.038 91	=	0.031 25
16.0	=	8.0	=	1.0	=	0.311 12	=	0.25
51.426 27	=	25.713 14	=	3.214 14	=	1.0	=	0.803 54
64.0	=	32.0	=	4.0	=	1.244 5	=	1.0

(5) Liquid Measure

Gills		Pints		Quarts		U.S. gallons		Cubic feet
1.0	=	0.25	=	0.125	=	0.031 25	=	0.004 18
4.0	=	1.0	=	0.5	=	0.125	=	0.016 71
8.0	=	2.0	=	1.0	=	0.250	=	0.033 42
32.0	=	8.0	=	4.0	=	1.0	=	0.133 78
						7.480 52	=	1.0

C.09 WEIGHT EQUIVALENT REFERENCE (U.S. CUSTOMARY SYSTEM AND SI)

1 gram	= 15.43 grains
1 gram	= 15 430 milligrains
1 gram	= 15 430 000 micrograins
1 pound	= 7 000 grains
1 ounce	= 437.5 grains
1 ounce	= 28.35 grams
1 grain	= 0.002 286 ounce
1 grain	= 0.064 8 gram
1 grain	= 64.8 milligrams
0.1 grain	= 0.006 48 gram
0.1 grain	= 6.48 milligrams
1 micrograin	= 0.000 000 064 8 gram
1 micrograin	= 0.000 064 8 milligram
10 grains	= 0.022 86 ounce or 0.648 gram
100 grains	= 0.228 6 ounce or 6.48 grams

Material	Relative resistivity* at 20°C	Density, g/cm³	Thermal conductivity at 20°C, W/(cm·°C)	Thermal expansion × 10⁻⁶°C, in	Melting point, °C
Aluminum	1.54	2.70	2.22	23.6	660
Beryllium	2.3	1.85	1.46	11.6	1 277
Bismuth	67.0	9.80	0.084	13.3	271
Brass, yellow	3.7	8.47	1.17	20.3	930
Cadmium	4.3	8.65	0.92	29.8	321
Carbon, graphite	790	2.25	0.24	0.6–4.3	Sublimes
Chromium	7.4	7.19	0.67	6.2	1 875
Cobalt	3.6	8.85	0.69	13.8	1 495
Columbium (see niobium)					
Constantan	29.0	8.9	0.21	14.9	1 290
Copper, hard-drawn	1.03	8.94	3.91	16.8	1 083
Gallium	4.7	5.91	0.29	18.0	30
Germanium	2.7×10^6	5.33	0.59	5.75	937
Gold	1.36	19.32	2.96	14.2	1 063
Inconel, 17–16–8	56.9	8.51	0.15	11.5	1 425
Indium	4.9	7.31	0.24	33.0	156
Invar, 64–36	46.0	8.00	0.11	0–2	1 425
Iron	5.6	7.87	0.75	11.8	1 536
Lead	12.0	11.34	0.35	29.3	327
Magnesium	2.58	1.74	1.53	27.1	650
Mercury	55.6	13.55	0.082	—	−38.9
Molybdenum	3.3	10.22	1.42	4.9	2 610
Monel, 67–30	27.9	8.84	0.26	14.0	1 325
Nichrome, 80–20	62.5	8.4	0.134	13.0	1 400
Nickel	5.5	8.89	0.61	13.3	1 440
Niobium	7.2	8.57	0.52	7.31	2 468
Palladium	6.3	12.02	0.70	11.8	1 552
Phosphor-bronze 95–5	6.4	8.86	0.71	17.8	1 000
Platinum, 99,9%	6.16	21.45	0.69	8.9	1 769
Silicon	10^{11}	2.33	1.25	2.5	1 420
Silver	0.922	10.49	4.18	19.7	961
Steel, .4–.5 C	7–12	7.8	0.5	11.0	1 480
Steel, stainless 304	42	7.9	0.16	17.0	1 430
Steel, stainless 410	33	7.7	0.24	11	1 500
Tantalum	7.4	16.6	0.54	6.6	3 000
Thorium	8.1	11.6	0.37	12.5	1 750
Tin	7.0	7.30	0.63	23	232
Titanium	24.2	4.51	0.41	8.4	1 670
Tungsten	3.2	19.3	1.67	4.6	3 410
Uranium	17.5	18.7	0.3	6.8–14.1	1 132
Zinc	3.5	7.14	1.10	27	420
Zirconium	23	6.5	0.21	5.8	1 852

Note: Standard resistivity of 100% IACS copper at 20°C = 1.7241×10^{-6} $\Omega \cdot$cm.

Element name	Symbol	Atomic number	Atomic weight	Melting point, °C	Electrical resistivity, $\Omega \cdot$ cm
Actinium	Ac	89	227.028 ·	1 050	—
Aluminum	Al	13	26.981 5	660.37	2.655×10^{-6}
Americium	Am	95	(243)●	994 ± 4	—
Antimony	Sb	51	121.75	630.74	39.0×10^{-6}
Argon	A	18	39.948	−189.2	—
Arsenic	As	33	74.921 6	817■	35.0×10^{-6}
Astatine	At	85	(210)●	302	—
Barium	Ba	56	137.33	725	—
Berkelium	Bk	97	(247)●	—	—
Beryllium	Be	4	9.012 2	1 278 ± 5	4.0×10^{-6}
Bismuth	Bi	83	208.980	271.3	106.8×10^{-6}
Boron	B	5	10.81	2 079	1.8×10^{-6}
Bromine	Br	35	79.904	−7.2	—
Cadmium	Cd	48	112.41	320.9	6.83×10^{-6}
Calcium	Ca	20	40.08	839 ± 2	3.91×10^{-6}
Californium	Cf	98	(251)●	—	—
Carbon	C	6	12.011	3 652▲	$1 375 \times 10^{-6}$
Cerium	Ce	58	140.12	798 ± 2	75.0×10^{-6}
Cesium	Cs	55	132.905 4	28.40	20.0×10^{-6}
Chlorine	Cl	17	35.453	−100.98	—
Chromium	Cr	24	51.996	1 857 ± 20	129×10^{-6}
Cobalt	Co	27	51.933 2	1 495	6.24×10^{-6}
Copper	Cu	29	63.546	1 083.4	1.673×10^{-6}
Curium	Cm	96	(247)●	1 340 ± 40	—
Dysprosium	Dy	66	162.50	1 409	90×10^{-6}
Einsteinium	Es	99	(252)●	—	—
Erbium	Er	68	167.26	1 522	107.0×10^{-6}
Europium	Eu	63	151.96	822 ± 5	90.0×10^{-6}
Fermium	Fm	100	(257)●	—	—
Fluorine	F	9	18.998 4	−219.62	—
Francium	Fr	87	(223)●	27 *	—
Gadolinium	Gd	64	157.25	1 311 ± 1	134×10^{-6}
Gallium	Ga	31	69.72	29.78	56.8×10^{-6}
Germanium	Ge	32	72.59	937.4	89.0×10^{-6}
Gold	Au	79	196.967	1 064.4	2.44×10^{-6}
Hafnium	Hf	72	178.49	2 227 ± 20	35.1×10^{-6}
Helium	He	2	4.002 6	−272.2◆	—
Holmium	Ho	67	164.93	1 470	87.0×10^{-6}
Hydrogen	H	1	1.007 9	−259.14	—
Indium	In	49	114.82	156.61	8.37×10^{-6}
Iodine	I	53	126.905	113.5	—
Iridium	Ir	77	192.22	2 410	4.71×10^{-6}
Iron	Fe	26	55.847	1 535	9.71×10^{-6}
Krypton	Kr	36	83.80	−156.6	—
Lanthanum	La	57	138.906	920 ± 5	57.0×10^{-6}
Lawrencium	Lw	103	(260)●	—	—
Lead	Pb	82	207.2	327.5	20.65×10^{-6}
Lithium	Li	3	6.941	180.54	8.55×10^{-6}
Lutetium	Lu	71	174.967	1 656 ± 5	79.0×10^{-6}
Magnesium	Mg	12	24.305	648.8	4.45×10^{-6}
Manganese	Mn	25	54.938	1 244 ± 3	185.0×10^{-6}
Mendelevium	Md	101	(258)●	—	—
Mercury	Hg	80	200.59	−38.87	95.78×10^{-6}

Element name	Symbol	Atomic number	Atomic weight	Melting point, °C	Electrical resistivity, $\Omega \cdot$ cm
Molybdenum	Mo	42	95.94	2 617	5.2×10^{-6}
Neodymium	Nd	60	144.24	1 010	64.0×10^{-6}
Neon	Ne	10	20.118	−248.67	—
Neptunium	Np	93	237.048	640 ± 1	—
Nickel	Ni	28	58.69	1 453	6.84×10^{-6}
Niobium	Nb	41	92.906	2 468 ± 10	14.6×10^{-6}
Nitrogen	N	7	14.007	−209.86	—
Nobelium	No	102	(259)●	—	—
Osmium	Os	76	190.2	3 045 ± 30	9.5×10^{-6}
Oxygen	O	8	15.999	−218.4	—
Palladium	Pd	46	106.42	1 554	9.93×10^{-6}
Phosphorus	P	15	30.974	44.1	1×10^{11}
Platinum	Pt	78	195.08	1 772	9.85×10^{-6}
Plutonium	Pu	94	(244)●	641	146.45×10^{-6}
Polonium	Po	84	(209)●	254	42×10^{-6}
Potassium	K	19	39.098	63.25	6.15×10^{-6}
Praseodymium	Pr	59	140.908	931 ± 4	68.0×10^{-6}
Promethium	Pm	61	(145)●	1 080 *	—
Protactinium	Pa	91	231.036	1 600	—
Radium	Ra	88	226.025	700	—
Radon	Rn	86	(222)●	−71.0	—
Rhenium	Re	75	186.207	3 180	19.3×10^{-6}
Rhodium	Rh	45	102.906	1 965 ± 3	4.33×10^{-6}
Rubidium	Rb	37	85.468	38.89	12.5×10^{-6}
Ruthenium	Ru	44	101.07	2 310	7.6×10^{-6}
Samarium	Sm	62	150.36	1 072 ± 5	90.0×10^{-6}
Scandium	Sc	21	44.956	1 539	61.0×10^{-6}
Selenium	Se	34	78.96	217	10.0×10^{6}
Silicon	Si	14	28.086	1 410	10
Silver	Ag	47	107.868	961.9	1.59×10^{-6}
Sodium	Na	11	22.99	97.8	4.2×10^{-6}
Strontium	Sr	38	87.62	769	23.0×10^{-6}
Sulfur	S	16	32.06	112.8	2.0×10^{18}
Tantalum	Ta	73	180.948	2 996	12.45×10^{-6}
Technetium	Tc	43	(98)●	2 172	—
Tellurium	Te	52	127.60	449.5	0.43
Terbium	Tb	65	158.925	1 360 ± 4	116×10^{-6}
Thallium	Tl	81	204.383	303.5	18.0×10^{-6}
Thorium	Th	90	232.038	1 750	13.0×10^{-6}
Thulium	Tm	69	168.934	1 545 ± 15	79.0×10^{-6}
Tin	Sn	50	118.71	231.97	11.5×10^{-6}
Titanium	Ti	22	47.88	1 660 ± 10	42.0×10^{-6}
Tungsten	W	74	183.85	3 410 ± 20	5.5×10^{-6}
Uranium	U	92	238.029	1 132	30.0×10^{-6}
Vanadium	V	23	50.942	1 890 ± 10	24.8×10^{-6}
Xenon	Xe	54	131.29	−111.9	—
Ytterbium	Yb	70	173.04	824 ± 5	28.0×10^{-6}
Yttrium	Y	39	88.906	1 523 ± 8	57.0×10^{-6}
Zinc	Zn	30	65.39	419.6	5.92×10^{-6}
Zirconium	Zr	40	91.224	1 852 ± 2	

Note: ● Values in parentheses are for the mass number of the most stable isotope known; ■ at 28 atm; ▲ sublimates; * approximate; ◆ at 26 atm.

n	0	1	2	3	4	5	6	7	8	9
0					2^2		$2\cdot3$		2^3	3^2
1	$2\cdot5$		$2^2\cdot3$		$2\cdot7$	$3\cdot5$	2^4		$2\cdot3^2$	
2	$2^2\cdot5$	$3\cdot7$	$2\cdot11$		$2^3\cdot3$	5^2	$2\cdot13$	3^3	$2^2\cdot7$	
3	$2\cdot3\cdot5$		2^5	$3\cdot11$	$2\cdot17$	$5\cdot7$	$2^2\cdot3^2$		$2\cdot19$	$3\cdot13$
4	$2^3\cdot5$		$2\cdot3\cdot7$		$2^2\cdot11$	$3^2\cdot5$	$2\cdot23$		$2^4\cdot3$	7^2
5	$2\cdot5^2$	$3\cdot17$	$2^2\cdot13$		$2\cdot3^3$	$5\cdot11$	$2^3\cdot7$	$3\cdot19$	$2\cdot29$	
6	$2^2\cdot3\cdot5$		$2\cdot31$	$3^2\cdot7$	2^6	$5\cdot13$	$2\cdot3\cdot11$		$2^2\cdot17$	$3\cdot23$
7	$2\cdot5\cdot7$		$2^3\cdot3^2$		$2\cdot37$	$3\cdot5^2$	$2^2\cdot19$	$7\cdot11$	$2\cdot3\cdot13$	
8	$2^4\cdot5$	3^4	$2\cdot41$		$2^2\cdot3\cdot7$	$5\cdot17$	$2\cdot43$	$3\cdot29$	$2^3\cdot11$	
9	$2\cdot3^2\cdot5$	$7\cdot13$	$2^2\cdot23$	$3\cdot31$	$2\cdot47$	$5\cdot19$	$2^5\cdot3$		$2\cdot7^2$	$3^2\cdot11$
10	$2^2\cdot5^2$		$2\cdot3\cdot17$		$2^3\cdot13$	$3\cdot5\cdot7$	$2\cdot53$		$2^2\cdot3^3$	
11	$2\cdot5\cdot11$	$3\cdot37$	$2^4\cdot7$		$2\cdot3\cdot19$	$5\cdot23$	$2^2\cdot29$	$3^2\cdot13$	$2\cdot59$	$7\cdot17$
12	$2^3\cdot3\cdot5$	11^2	$2\cdot61$	$3\cdot41$	$2^2\cdot31$	5^3	$2\cdot3^2\cdot7$		2^7	$3\cdot43$
13	$2\cdot5\cdot13$		$2^2\cdot3\cdot11$	$7\cdot19$	$2\cdot67$	$3^3\cdot5$	$2^3\cdot17$		$2\cdot3\cdot23$	
14	$2^2\cdot5\cdot7$	$3\cdot47$	$2\cdot71$	$11\cdot13$	$2^4\cdot3^2$	$5\cdot29$	$2\cdot73$	$3\cdot7^2$	$2^2\cdot37$	
15	$2\cdot3\cdot5^2$		$2^3\cdot19$	$3^2\cdot17$	$2\cdot7\cdot11$	$5\cdot31$	$2^2\cdot3\cdot13$		$2\cdot79$	$3\cdot53$
16	$2^5\cdot5$	$7\cdot23$	$2\cdot3^4$		$2^2\cdot41$	$3\cdot5\cdot11$	$2\cdot83$		$2^3\cdot3\cdot7$	13^2
17	$2\cdot5\cdot17$	$3^2\cdot19$	$2^2\cdot43$		$2\cdot3\cdot29$	$5^2\cdot7$	$2^4\cdot11$	$3\cdot59$	$2\cdot89$	
18	$2^2\cdot3^2\cdot5$		$2\cdot7\cdot13$	$3\cdot61$	$2^3\cdot23$	$5\cdot37$	$2\cdot3\cdot31$	$11\cdot17$	$2^2\cdot47$	$3^3\cdot7$
19	$2\cdot5\cdot19$		$2^6\cdot3$		$2\cdot97$	$3\cdot5\cdot13$	$2^2\cdot7^2$		$2\cdot3^2\cdot11$	
20	$2^3\cdot5^2$	$3\cdot67$	$2\cdot101$	$7\cdot29$	$2^2\cdot3\cdot17$	$5\cdot41$	$2\cdot103$	$3^2\cdot23$	$2^4\cdot13$	$11\cdot19$
21	$2\cdot3\cdot5\cdot7$		$2^2\cdot53$	$3\cdot71$	$2\cdot107$	$5\cdot43$	$2^3\cdot3^3$	$7\cdot31$	$2\cdot109$	$3\cdot73$
22	$2^2\cdot5\cdot11$	$13\cdot17$	$2\cdot3\cdot37$		$2^5\cdot7$	$3^2\cdot5^2$	$2\cdot113$		$2^3\cdot3\cdot19$	
23	$2\cdot5\cdot23$	$3\cdot7\cdot11$	$2^3\cdot29$		$2\cdot3^2\cdot13$	$5\cdot47$	$2^2\cdot59$	$3\cdot79$	$2\cdot7\cdot17$	
24	$2^4\cdot3\cdot5$		$2\cdot11^2$	3^5	$2^2\cdot61$	$5\cdot7^2$	$2\cdot3\cdot41$	$13\cdot19$	$2^3\cdot31$	$3\cdot83$
25	$2\cdot5^3$		$2^2\cdot3^2\cdot7$	$11\cdot23$	$2\cdot127$	$3\cdot5\cdot17$	2^8		$2\cdot3\cdot43$	$7\cdot37$
26	$2^2\cdot5\cdot13$	$3^2\cdot29$	$2\cdot131$		$2^3\cdot3\cdot11$	$5\cdot53$	$2\cdot7\cdot19$	$3\cdot89$	$2^2\cdot67$	
27	$2\cdot3^3\cdot5$		$2^4\cdot17$	$3\cdot7\cdot13$	$2\cdot137$	$5^2\cdot11$	$2^2\cdot3\cdot23$		$2\cdot139$	$3^2\cdot31$
28	$2^3\cdot5\cdot7$		$2\cdot3\cdot47$		$2^2\cdot71$	$3\cdot5\cdot19$	$2\cdot11\cdot13$	$7\cdot41$	$2^5\cdot3^2$	17^2
29	$2\cdot5\cdot29$	$3\cdot97$	$2^2\cdot73$		$2\cdot3\cdot7^2$	$5\cdot59$	$2^3\cdot37$	$3^3\cdot11$	$2\cdot149$	$13\cdot23$
30	$2^2\cdot3\cdot5^2$	$7\cdot43$	$2\cdot151$	$3\cdot101$	$2^4\cdot19$	$5\cdot61$	$2\cdot3^2\cdot17$		$2^2\cdot7\cdot11$	$3\cdot103$
31	$2\cdot5\cdot31$		$2^3\cdot3\cdot13$		$2\cdot157$	$3^2\cdot5\cdot7$	$2^2\cdot79$		$2\cdot3\cdot53$	$11\cdot29$
32	$2^6\cdot5$	$3\cdot107$	$2\cdot7\cdot23$	$17\cdot19$	$2^2\cdot3^4$	$5^2\cdot13$	$2\cdot163$	$3\cdot109$	$2^3\cdot41$	$7\cdot47$
33	$2\cdot3\cdot5\cdot11$		$2^2\cdot83$	$3^2\cdot37$	$2\cdot167$	$5\cdot67$	$2^4\cdot3\cdot7$		$2\cdot13^2$	$3\cdot113$
34	$2^2\cdot5\cdot17$	$11\cdot31$	$2\cdot3^2\cdot19$	7^3	$2^3\cdot43$	$3\cdot5\cdot23$	$2\cdot173$		$2^2\cdot3\cdot29$	
35	$2\cdot5^2\cdot7$	$3^3\cdot13$	$2^5\cdot11$		$2\cdot3\cdot59$	$5\cdot71$	$2^2\cdot89$	$3\cdot7\cdot17$	$2\cdot179$	
36	$2^3\cdot3^2\cdot5$	19^2	$2\cdot181$	$3\cdot11^2$	$2^2\cdot7\cdot13$	$5\cdot73$	$2\cdot3\cdot61$		$2^4\cdot23$	$3^2\cdot41$
37	$2\cdot5\cdot37$	$7\cdot53$	$2^2\cdot3\cdot31$		$2\cdot11\cdot17$	$3\cdot5^3$	$2^3\cdot47$	$13\cdot29$	$2\cdot3^3\cdot7$	
38	$2^2\cdot5\cdot19$	$3\cdot127$	$2\cdot191$		$2^7\cdot3$	$5\cdot7\cdot11$	$2\cdot193$	$3^2\cdot43$	$2^2\cdot97$	
39	$2\cdot3\cdot5\cdot13$	$17\cdot23$	$2^3\cdot7^2$	$3\cdot131$	$2\cdot197$	$5\cdot79$	$2^2\cdot3^2\cdot11$		$2\cdot199$	$3\cdot7\cdot19$
40	$2^4\cdot5^2$		$2\cdot3\cdot67$	$13\cdot31$	$2^2\cdot101$	$3^4\cdot5$	$2\cdot7\cdot29$	$11\cdot37$	$2^3\cdot3\cdot17$	
41	$2\cdot5\cdot41$	$3\cdot137$	$2^2\cdot103$	$7\cdot59$	$2\cdot3^2\cdot23$	$5\cdot83$	$2^5\cdot13$	$3\cdot139$	$2\cdot11\cdot19$	
42	$2^2\cdot3\cdot5\cdot7$		$2\cdot211$	$3^2\cdot47$	$2^3\cdot53$	$5^2\cdot17$	$2\cdot3\cdot71$	$7\cdot61$	$2^2\cdot107$	$3\cdot11\cdot13$
43	$2\cdot5\cdot43$		$2^4\cdot3^3$		$2\cdot7\cdot31$	$3\cdot5\cdot29$	$2^2\cdot109$	$19\cdot23$	$2\cdot3\cdot73$	
44	$2^3\cdot5\cdot11$	$3^2\cdot7^2$	$2\cdot13\cdot17$		$2^2\cdot3\cdot37$	$5\cdot89$	$2\cdot223$	$3\cdot149$	$2^6\cdot7$	
45	$2\cdot3^2\cdot5^2$	$11\cdot41$	$2^2\cdot113$	$3\cdot151$	$2\cdot227$	$5\cdot7\cdot13$	$2^3\cdot3\cdot19$		$2\cdot229$	$3^3\cdot17$
46	$2^2\cdot5\cdot23$		$2\cdot3\cdot7\cdot11$		$2^4\cdot29$	$3\cdot5\cdot31$	$2\cdot233$		$2^2\cdot3^2\cdot13$	$7\cdot67$
47	$2\cdot5\cdot47$	$3\cdot157$	$2^3\cdot59$	$11\cdot43$	$2\cdot3\cdot79$	$5^2\cdot19$	$2^2\cdot7\cdot17$	$3^2\cdot53$	$2\cdot239$	
48	$2^5\cdot3\cdot5$	$13\cdot37$	$2\cdot241$	$3\cdot7\cdot23$	$2^2\cdot11^2$	$5\cdot97$	$2\cdot3^5$		$2^3\cdot61$	$3\cdot163$
49	$2\cdot5\cdot7^2$		$2^2\cdot3\cdot41$	$17\cdot29$	$2\cdot13\cdot19$	$3^2\cdot5\cdot11$	$2^4\cdot31$	$7\cdot71$	$2\cdot3\cdot83$	
50	$2^2\cdot5^3$	$3\cdot167$	$2\cdot251$		$2^3\cdot3^2\cdot7$	$5\cdot101$	$2\cdot11\cdot23$	$3\cdot13^2$	$2^2\cdot127$	
51	$2\cdot3\cdot5\cdot17$	$7\cdot73$	2^9	$3^3\cdot19$	$2\cdot257$	$5\cdot103$	$2^2\cdot3\cdot43$	$11\cdot47$	$2\cdot7\cdot37$	$3\cdot173$
52	$2^3\cdot5\cdot13$		$2\cdot3^2\cdot29$		$2^2\cdot131$	$3\cdot5^2\cdot7$	$2\cdot263$	$17\cdot31$	$2^4\cdot3\cdot11$	23^2
53	$2\cdot5\cdot53$	$3^2\cdot59$	$2^2\cdot7\cdot19$	$13\cdot41$	$2\cdot3\cdot89$	$5\cdot107$	$2^3\cdot67$	$3\cdot179$	$2\cdot269$	$7^2\cdot11$
54	$2^2\cdot3^3\cdot5$		$2\cdot271$	$3\cdot181$	$2^5\cdot17$	$5\cdot109$	$2\cdot3\cdot7\cdot13$		$2^2\cdot137$	$3^2\cdot61$
55	$2\cdot5^2\cdot11$	$19\cdot29$	$2^3\cdot3\cdot23$	$7\cdot79$	$2\cdot277$	$3\cdot5\cdot37$	$2^2\cdot139$		$2\cdot3^2\cdot31$	$13\cdot43$
56	$2^4\cdot5\cdot7$	$3\cdot11\cdot17$	$2\cdot281$		$2^2\cdot3\cdot47$	$5\cdot113$	$2\cdot283$	$3^4\cdot7$	$2^3\cdot71$	
57	$2\cdot3\cdot5\cdot19$		$2^2\cdot11\cdot13$	$3\cdot191$	$2\cdot7\cdot41$	$5^2\cdot23$	$2^6\cdot3^2$		$2\cdot17^2$	$3\cdot193$
58	$2^2\cdot5\cdot29$	$7\cdot83$	$2\cdot3\cdot97$	$11\cdot53$	$2^3\cdot73$	$3^2\cdot5\cdot13$	$2\cdot293$		$2^2\cdot3\cdot7^2$	$19\cdot31$
59	$2\cdot5\cdot59$	$3\cdot197$	$2^4\cdot37$		$2\cdot3^3\cdot11$	$5\cdot7\cdot17$	$2^2\cdot149$	$3\cdot199$	$2\cdot13\cdot23$	
60	$2^3\cdot3\cdot5^2$		$2\cdot7\cdot43$	$3^2\cdot67$	$2^2\cdot151$	$5\cdot11^2$	$2\cdot3\cdot101$		$2^5\cdot19$	$3\cdot7\cdot29$
61	$2\cdot5\cdot61$	$13\cdot47$	$2^2\cdot3^2\cdot17$		$2\cdot307$	$3\cdot5\cdot41$	$2^3\cdot7\cdot11$		$2\cdot3\cdot103$	
62	$2^2\cdot5\cdot31$	$3^3\cdot23$	$2\cdot311$	$7\cdot89$	$2^4\cdot3\cdot13$	5^4	$2\cdot313$	$3\cdot11\cdot19$	$2^2\cdot157$	$17\cdot37$
63	$2\cdot3^2\cdot5\cdot7$		$2^3\cdot79$	$3\cdot211$	$2\cdot317$	$5\cdot127$	$2^2\cdot3\cdot53$	$7^2\cdot13$	$2\cdot11\cdot29$	$3^2\cdot71$

n	0	1	2	3	4	5	6	7	8	9
64	$2^7 \cdot 5$		$2 \cdot 3 \cdot 107$		$2^2 \cdot 7 \cdot 23$	$3 \cdot 5 \cdot 43$	$2 \cdot 17 \cdot 19$		$2^3 \cdot 3^4$	$11 \cdot 59$
65	$2 \cdot 5^2 \cdot 13$	$3 \cdot 7 \cdot 31$	$2^2 \cdot 163$		$2 \cdot 3 \cdot 109$	$5 \cdot 131$	$2^4 \cdot 41$	$3^2 \cdot 73$	$2 \cdot 7 \cdot 47$	
66	$2^2 \cdot 3 \cdot 5 \cdot 11$		$2 \cdot 331$	$3 \cdot 13 \cdot 17$	$2^3 \cdot 83$	$5 \cdot 7 \cdot 19$	$2 \cdot 3^2 \cdot 37$	$23 \cdot 29$	$2^2 \cdot 167$	$3 \cdot 223$
67	$2 \cdot 5 \cdot 67$	$11 \cdot 61$	$2^5 \cdot 3 \cdot 7$		$2 \cdot 337$	$3^3 \cdot 5^2$	$2^2 \cdot 13^2$		$2 \cdot 3 \cdot 113$	$7 \cdot 97$
68	$2^3 \cdot 5 \cdot 17$	$3 \cdot 227$	$2 \cdot 11 \cdot 31$		$2^2 \cdot 3^2 \cdot 19$	$5 \cdot 137$	$2 \cdot 7^3$	$3 \cdot 229$	$2^4 \cdot 43$	$13 \cdot 53$
69	$2 \cdot 3 \cdot 5 \cdot 23$		$2^2 \cdot 173$	$3^2 \cdot 7 \cdot 11$	$2 \cdot 347$	$5 \cdot 139$	$2^3 \cdot 3 \cdot 29$	$17 \cdot 41$	$2 \cdot 349$	$3 \cdot 233$
70	$2^2 \cdot 5^2 \cdot 7$		$2 \cdot 3^3 \cdot 13$	$19 \cdot 37$	$2^6 \cdot 11$	$3 \cdot 5 \cdot 47$	$2 \cdot 353$	$7 \cdot 101$	$2^2 \cdot 3 \cdot 59$	
71	$2 \cdot 5 \cdot 71$	$3^2 \cdot 79$	$2^3 \cdot 89$	$23 \cdot 31$	$2 \cdot 3 \cdot 7 \cdot 17$	$5 \cdot 11 \cdot 13$	$2^2 \cdot 179$	$3 \cdot 239$	$2 \cdot 359$	
72	$2^4 \cdot 3^2 \cdot 5$	$7 \cdot 103$	$2 \cdot 19^2$	$3 \cdot 241$	$2^2 \cdot 181$	$5^2 \cdot 29$	$2 \cdot 3 \cdot 11^2$		$2^3 \cdot 7 \cdot 13$	3^6
73	$2 \cdot 5 \cdot 73$	$17 \cdot 43$	$2^2 \cdot 3 \cdot 61$		$2 \cdot 367$	$3 \cdot 5 \cdot 7^2$	$2^5 \cdot 23$	$11 \cdot 67$	$2 \cdot 3^2 \cdot 41$	
74	$2^2 \cdot 5 \cdot 37$	$3 \cdot 13 \cdot 19$	$2 \cdot 7 \cdot 53$		$2^3 \cdot 3 \cdot 31$	$5 \cdot 149$	$2 \cdot 373$	$3^2 \cdot 83$	$2^2 \cdot 11 \cdot 17$	$7 \cdot 107$
75	$2 \cdot 3 \cdot 5^3$		$2^4 \cdot 47$	$3 \cdot 251$	$2 \cdot 13 \cdot 29$	$5 \cdot 151$	$2^2 \cdot 3^3 \cdot 7$		$2 \cdot 379$	$3 \cdot 11 \cdot 23$
76	$2^3 \cdot 5 \cdot 19$		$2 \cdot 3 \cdot 127$	$7 \cdot 109$	$2^2 \cdot 191$	$3^2 \cdot 5 \cdot 17$	$2 \cdot 383$	$13 \cdot 59$	$2^8 \cdot 3$	
77	$2 \cdot 5 \cdot 7 \cdot 11$	$3 \cdot 257$	$2^2 \cdot 193$		$2 \cdot 3^2 \cdot 43$	$5^2 \cdot 31$	$2^3 \cdot 97$	$3 \cdot 7 \cdot 37$	$2 \cdot 389$	$19 \cdot 41$
78	$2^2 \cdot 3 \cdot 5 \cdot 13$	$11 \cdot 71$	$2 \cdot 17 \cdot 23$	$3^3 \cdot 29$	$2^4 \cdot 7^2$	$5 \cdot 157$	$2 \cdot 3 \cdot 131$		$2^2 \cdot 197$	$3 \cdot 263$
79	$2 \cdot 5 \cdot 79$	$7 \cdot 113$	$2^3 \cdot 3^2 \cdot 11$	$13 \cdot 61$	$2 \cdot 397$	$3 \cdot 5 \cdot 53$	$2^2 \cdot 199$		$2 \cdot 3 \cdot 7 \cdot 19$	$17 \cdot 47$
80	$2^5 \cdot 5^2$	$3^2 \cdot 89$	$2 \cdot 401$	$11 \cdot 73$	$2^2 \cdot 3 \cdot 67$	$5 \cdot 7 \cdot 23$	$2 \cdot 13 \cdot 31$	$3 \cdot 269$	$2^3 \cdot 101$	
81	$2 \cdot 3^4 \cdot 5$		$2^2 \cdot 7 \cdot 29$	$3 \cdot 271$	$2 \cdot 11 \cdot 37$	$5 \cdot 163$	$2^4 \cdot 3 \cdot 17$	$19 \cdot 43$	$2 \cdot 409$	$3^2 \cdot 7 \cdot 13$
82	$2^2 \cdot 5 \cdot 41$		$2 \cdot 3 \cdot 137$		$2^3 \cdot 103$	$3 \cdot 5^2 \cdot 11$	$2 \cdot 7 \cdot 59$		$2^2 \cdot 3^2 \cdot 23$	
83	$2 \cdot 5 \cdot 83$	$3 \cdot 277$	$2^6 \cdot 13$	$7^2 \cdot 17$	$2 \cdot 3 \cdot 139$	$5 \cdot 167$	$2^2 \cdot 11 \cdot 19$	$3^3 \cdot 31$	$2 \cdot 419$	
84	$2^3 \cdot 3 \cdot 5 \cdot 7$	29^2	$2 \cdot 421$	$3 \cdot 281$	$2^2 \cdot 211$	$5 \cdot 13^2$	$2 \cdot 3^2 \cdot 47$	$7 \cdot 11^2$	$2^4 \cdot 53$	$3 \cdot 283$
85	$2 \cdot 5^2 \cdot 17$	$23 \cdot 37$	$2^2 \cdot 3 \cdot 71$		$2 \cdot 7 \cdot 61$	$3^2 \cdot 5 \cdot 19$	$2^3 \cdot 107$		$2 \cdot 3 \cdot 11 \cdot 13$	
86	$2^2 \cdot 5 \cdot 43$	$3 \cdot 7 \cdot 41$	$2 \cdot 431$		$2^5 \cdot 3^3$	$5 \cdot 173$	$2 \cdot 433$	$3 \cdot 17^2$	$2^2 \cdot 7 \cdot 31$	$11 \cdot 79$
87	$2 \cdot 3 \cdot 5 \cdot 29$	$13 \cdot 67$	$2^3 \cdot 109$	$3^2 \cdot 97$	$2 \cdot 19 \cdot 23$	$5^3 \cdot 7$	$2^2 \cdot 3 \cdot 73$		$2 \cdot 439$	$3 \cdot 293$
88	$2^4 \cdot 5 \cdot 11$		$2 \cdot 3^2 \cdot 7^2$		$2^2 \cdot 13 \cdot 17$	$3 \cdot 5 \cdot 59$	$2 \cdot 443$		$2^3 \cdot 3 \cdot 37$	$7 \cdot 127$
89	$2 \cdot 5 \cdot 89$	$3^4 \cdot 11$	$2^2 \cdot 223$	$19 \cdot 47$	$2 \cdot 3 \cdot 149$	$5 \cdot 179$	$2^7 \cdot 7$	$3 \cdot 13 \cdot 23$	$2 \cdot 449$	$29 \cdot 31$
90	$2^2 \cdot 3^2 \cdot 5^2$	$17 \cdot 53$	$2 \cdot 11 \cdot 41$	$3 \cdot 7 \cdot 43$	$2^3 \cdot 113$	$5 \cdot 181$	$2 \cdot 3 \cdot 151$		$2^2 \cdot 227$	$3^2 \cdot 101$
91	$2 \cdot 5 \cdot 7 \cdot 13$		$2^4 \cdot 3 \cdot 19$	$11 \cdot 83$	$2 \cdot 457$	$3 \cdot 5 \cdot 61$	$2^2 \cdot 229$	$7 \cdot 131$	$2 \cdot 3^3 \cdot 17$	
92	$2^3 \cdot 5 \cdot 23$	$3 \cdot 307$	$2 \cdot 461$	$13 \cdot 71$	$2^2 \cdot 3 \cdot 7 \cdot 11$	$5^2 \cdot 37$	$2 \cdot 463$	$3^2 \cdot 103$	$2^5 \cdot 29$	
93	$2 \cdot 3 \cdot 5 \cdot 31$	$7^2 \cdot 19$	$2^2 \cdot 233$	$3 \cdot 311$	$2 \cdot 467$	$5 \cdot 11 \cdot 17$	$2^3 \cdot 3^2 \cdot 13$		$2 \cdot 7 \cdot 67$	$3 \cdot 313$
94	$2^2 \cdot 5 \cdot 47$		$2 \cdot 3 \cdot 157$	$23 \cdot 41$	$2^4 \cdot 59$	$3^3 \cdot 5 \cdot 7$	$2 \cdot 11 \cdot 43$		$2^2 \cdot 3 \cdot 79$	$13 \cdot 73$
95	$2 \cdot 5^2 \cdot 19$	$3 \cdot 317$	$2^3 \cdot 7 \cdot 17$		$2 \cdot 3^2 \cdot 53$	$5 \cdot 191$	$2^2 \cdot 239$	$3 \cdot 11 \cdot 29$	$2 \cdot 479$	$7 \cdot 137$
96	$2^6 \cdot 3 \cdot 5$	31^2	$2 \cdot 13 \cdot 37$	$3^2 \cdot 107$	$2^2 \cdot 241$	$5 \cdot 193$	$2 \cdot 3 \cdot 7 \cdot 23$		$2^3 \cdot 11^2$	$3 \cdot 17 \cdot 19$
97	$2 \cdot 5 \cdot 97$		$2^2 \cdot 3^5$	$7 \cdot 139$	$2 \cdot 487$	$3 \cdot 5^2 \cdot 13$	$2^4 \cdot 61$		$2 \cdot 3 \cdot 163$	$11 \cdot 89$
98	$2^2 \cdot 5 \cdot 7^2$	$3^2 \cdot 109$	$2 \cdot 491$		$2^3 \cdot 3 \cdot 41$	$5 \cdot 197$	$2 \cdot 17 \cdot 29$	$3 \cdot 7 \cdot 47$	$2^2 \cdot 13 \cdot 19$	$23 \cdot 43$
99	$2 \cdot 3^2 \cdot 5 \cdot 11$		$2^5 \cdot 31$	$3 \cdot 331$	$2 \cdot 7 \cdot 71$	$5 \cdot 199$	$2^2 \cdot 3 \cdot 83$		$2 \cdot 499$	$3^3 \cdot 37$
100	$2^3 \cdot 5^3$	$7 \cdot 11 \cdot 13$	$2 \cdot 3 \cdot 167$	$17 \cdot 59$	$2^2 \cdot 251$	$3 \cdot 5 \cdot 67$	$2 \cdot 503$	$19 \cdot 53$	$2^4 \cdot 3^2 \cdot 7$	
101	$2 \cdot 5 \cdot 101$	$3 \cdot 337$	$2^2 \cdot 11 \cdot 23$		$2 \cdot 3 \cdot 13^2$	$5 \cdot 7 \cdot 29$	$2^3 \cdot 127$	$3^2 \cdot 113$	$2 \cdot 509$	
102	$2^2 \cdot 3 \cdot 5 \cdot 17$		$2 \cdot 7 \cdot 73$	$3 \cdot 11 \cdot 31$	2^{10}	$5^2 \cdot 41$	$2 \cdot 3^3 \cdot 19$	$13 \cdot 79$	$2^2 \cdot 257$	$3 \cdot 7^3$
103	$2 \cdot 5 \cdot 103$		$2^3 \cdot 3 \cdot 43$		$2 \cdot 11 \cdot 47$	$3^2 \cdot 5 \cdot 23$	$2^2 \cdot 7 \cdot 37$	$17 \cdot 61$	$2 \cdot 3 \cdot 173$	
104	$2^4 \cdot 5 \cdot 13$	$3 \cdot 347$	$2 \cdot 521$	$7 \cdot 149$	$2^2 \cdot 3^2 \cdot 29$	$5 \cdot 11 \cdot 19$	$2 \cdot 523$	$3 \cdot 349$	$2^3 \cdot 131$	
105	$2 \cdot 3 \cdot 5^2 \cdot 7$		$2^2 \cdot 263$	$3^4 \cdot 13$	$2 \cdot 17 \cdot 31$	$5 \cdot 211$	$2^5 \cdot 3 \cdot 11$	$7 \cdot 151$	$2 \cdot 23^2$	$3 \cdot 353$
106	$2^2 \cdot 5 \cdot 53$		$2 \cdot 3^2 \cdot 59$		$2^3 \cdot 7 \cdot 19$	$3 \cdot 5 \cdot 71$	$2 \cdot 13 \cdot 41$	$11 \cdot 97$	$2^2 \cdot 3 \cdot 89$	
107	$2 \cdot 5 \cdot 107$	$3^2 \cdot 7 \cdot 17$	$2^4 \cdot 67$	$29 \cdot 37$	$2 \cdot 3 \cdot 179$	$5^2 \cdot 43$	$2^2 \cdot 269$	$3 \cdot 359$	$2 \cdot 7^2 \cdot 11$	$13 \cdot 83$
108	$2^3 \cdot 3^3 \cdot 5$	$23 \cdot 47$	$2 \cdot 541$	$3 \cdot 19^2$	$2^2 \cdot 271$	$5 \cdot 7 \cdot 31$	$2 \cdot 3 \cdot 181$		$2^6 \cdot 17$	$3^2 \cdot 11^2$
109	$2 \cdot 5 \cdot 109$		$2^2 \cdot 3 \cdot 7 \cdot 13$		$2 \cdot 547$	$3 \cdot 5 \cdot 73$	$2^3 \cdot 137$		$2 \cdot 3^2 \cdot 61$	$7 \cdot 157$
110	$2^2 \cdot 5^2 \cdot 11$	$3 \cdot 367$	$2 \cdot 19 \cdot 29$		$2^4 \cdot 3 \cdot 23$	$5 \cdot 13 \cdot 17$	$2 \cdot 7 \cdot 79$	$3^3 \cdot 41$	$2^2 \cdot 277$	
111	$2 \cdot 3 \cdot 5 \cdot 37$	$11 \cdot 101$	$2^3 \cdot 139$	$3 \cdot 7 \cdot 53$	$2 \cdot 557$	$5 \cdot 223$	$2^2 \cdot 3^2 \cdot 31$		$2 \cdot 13 \cdot 43$	$3 \cdot 373$
112	$2^5 \cdot 5 \cdot 7$	$19 \cdot 59$	$2 \cdot 3 \cdot 11 \cdot 17$		$2^2 \cdot 281$	$3^2 \cdot 5^3$	$2 \cdot 563$	$7^2 \cdot 23$	$2^3 \cdot 3 \cdot 47$	
113	$2 \cdot 5 \cdot 113$	$3 \cdot 13 \cdot 29$	$2^2 \cdot 283$	$11 \cdot 103$	$2 \cdot 3^4 \cdot 7$	$5 \cdot 227$	$2^4 \cdot 71$	$3 \cdot 379$	$2 \cdot 569$	$17 \cdot 67$
114	$2^2 \cdot 3 \cdot 5 \cdot 19$	$7 \cdot 163$	$2 \cdot 571$	$3^2 \cdot 127$	$2^3 \cdot 11 \cdot 13$	$5 \cdot 229$	$2 \cdot 3 \cdot 191$	$31 \cdot 37$	$2^2 \cdot 7 \cdot 41$	$3 \cdot 383$
115	$2 \cdot 5^2 \cdot 23$		$2^7 \cdot 3^2$		$2 \cdot 577$	$3 \cdot 5 \cdot 7 \cdot 11$	$2^2 \cdot 17^2$	$13 \cdot 89$	$2 \cdot 3 \cdot 193$	$19 \cdot 61$
116	$2^3 \cdot 5 \cdot 29$	$3^3 \cdot 43$	$2 \cdot 7 \cdot 83$		$2^2 \cdot 3 \cdot 97$	$5 \cdot 233$	$2 \cdot 11 \cdot 53$	$3 \cdot 389$	$2^4 \cdot 73$	$7 \cdot 167$
117	$2 \cdot 3^2 \cdot 5 \cdot 13$		$2^2 \cdot 293$	$3 \cdot 17 \cdot 23$	$2 \cdot 587$	$5^2 \cdot 47$	$2^3 \cdot 3 \cdot 7^2$	$11 \cdot 107$	$2 \cdot 19 \cdot 31$	$3^2 \cdot 131$
118	$2^2 \cdot 5 \cdot 59$		$2 \cdot 3 \cdot 197$	$7 \cdot 13^2$	$2^5 \cdot 37$	$3 \cdot 5 \cdot 79$	$2 \cdot 593$		$2^2 \cdot 3^3 \cdot 11$	$29 \cdot 41$
119	$2 \cdot 5 \cdot 7 \cdot 17$	$3 \cdot 397$	$2^3 \cdot 149$		$2 \cdot 3 \cdot 199$	$5 \cdot 239$	$2^2 \cdot 13 \cdot 23$	$3^2 \cdot 7 \cdot 19$	$2 \cdot 599$	$11 \cdot 109$
120	$2^4 \cdot 3 \cdot 5^2$		$2 \cdot 601$	$3 \cdot 401$	$2^2 \cdot 7 \cdot 43$	$5 \cdot 241$	$2 \cdot 3^2 \cdot 67$	$17 \cdot 71$	$2^3 \cdot 151$	$3 \cdot 13 \cdot 31$
121	$2 \cdot 5 \cdot 11^2$	$7 \cdot 173$	$2^2 \cdot 3 \cdot 101$		$2 \cdot 607$	$3^5 \cdot 5$	$2^6 \cdot 19$		$2 \cdot 3 \cdot 7 \cdot 29$	$23 \cdot 53$
122	$2^2 \cdot 5 \cdot 61$	$3 \cdot 11 \cdot 37$	$2 \cdot 13 \cdot 47$		$2^3 \cdot 3^2 \cdot 17$	$5^2 \cdot 7^2$	$2 \cdot 613$	$3 \cdot 409$	$2^2 \cdot 307$	
123	$2 \cdot 3 \cdot 5 \cdot 41$		$2^4 \cdot 7 \cdot 11$	$3^2 \cdot 137$	$2 \cdot 617$	$5 \cdot 13 \cdot 19$	$2^2 \cdot 3 \cdot 103$		$2 \cdot 619$	$3 \cdot 7 \cdot 59$
124	$2^3 \cdot 5 \cdot 31$	$17 \cdot 73$	$2 \cdot 3^3 \cdot 23$	$11 \cdot 113$	$2^2 \cdot 311$	$3 \cdot 5 \cdot 83$	$2 \cdot 7 \cdot 89$	$29 \cdot 43$	$2^5 \cdot 3 \cdot 13$	
125	$2 \cdot 5^4$	$3^2 \cdot 139$	$2^2 \cdot 313$	$7 \cdot 179$	$2 \cdot 3 \cdot 11 \cdot 19$	$5 \cdot 251$	$2^3 \cdot 157$	$3 \cdot 419$	$2 \cdot 17 \cdot 37$	
126	$2^2 \cdot 3^2 \cdot 5 \cdot 7$	$13 \cdot 97$	$2 \cdot 631$	$3 \cdot 421$	$2^4 \cdot 79$	$5 \cdot 11 \cdot 23$	$2 \cdot 3 \cdot 211$	$7 \cdot 181$	$2^2 \cdot 317$	$3^3 \cdot 47$

n	0	1	2	3	4	5	6	7	8	9
127	$2\cdot5\cdot127$	$31\cdot41$	$2^3\cdot3\cdot53$	$19\cdot67$	$2\cdot7^2\cdot13$	$3\cdot5^2\cdot17$	$2^2\cdot11\cdot29$		$2\cdot3^2\cdot71$	
128	$2^8\cdot5$	$3\cdot7\cdot61$	$2\cdot641$		$2^2\cdot3\cdot107$	$5\cdot257$	$2\cdot643$	$3^2\cdot11\cdot13$	$2^3\cdot7\cdot23$	
129	$2\cdot3\cdot5\cdot43$		$2^2\cdot17\cdot19$	$3\cdot431$	$2\cdot647$	$5\cdot7\cdot37$	$2^4\cdot3^4$		$2\cdot11\cdot59$	$3\cdot433$
130	$2^2\cdot5^2\cdot13$		$2\cdot3\cdot7\cdot31$		$2^3\cdot163$	$3^2\cdot5\cdot29$	$2\cdot653$		$2^2\cdot3\cdot109$	$7\cdot11\cdot17$
131	$2\cdot5\cdot131$	$3\cdot19\cdot23$	$2^5\cdot41$	$13\cdot101$	$2\cdot3^2\cdot73$	$5\cdot263$	$2^2\cdot7\cdot47$	$3\cdot439$	$2\cdot659$	
132	$2^3\cdot3\cdot5\cdot11$		$2\cdot661$	$3^3\cdot7^2$	$2^2\cdot331$	$5^2\cdot53$	$2\cdot3\cdot13\cdot17$		$2^4\cdot83$	$3\cdot443$
133	$2\cdot5\cdot7\cdot19$	11^3	$2^2\cdot3^2\cdot37$	$31\cdot43$	$2\cdot23\cdot29$	$3\cdot5\cdot89$	$2^3\cdot167$	$7\cdot191$	$2\cdot3\cdot223$	$13\cdot103$
134	$2^2\cdot5\cdot67$	$3^2\cdot149$	$2\cdot11\cdot61$	$17\cdot79$	$2^6\cdot3\cdot7$	$5\cdot269$	$2\cdot673$	$3\cdot449$	$2^2\cdot337$	$19\cdot71$
135	$2\cdot3^3\cdot5^2$	$7\cdot193$	$2^3\cdot13^2$	$3\cdot11\cdot41$	$2\cdot677$	$5\cdot271$	$2^2\cdot3\cdot113$	$23\cdot59$	$2\cdot7\cdot97$	$3^2\cdot151$
136	$2^4\cdot5\cdot17$		$2\cdot3\cdot227$	$29\cdot47$	$2^2\cdot11\cdot31$	$3\cdot5\cdot7\cdot13$	$2\cdot683$		$2^3\cdot3^2\cdot19$	37^2
137	$2\cdot5\cdot137$	$3\cdot457$	$2^2\cdot7^3$		$2\cdot3\cdot229$	$5^3\cdot11$	$2^5\cdot43$	$3^4\cdot17$	$2\cdot13\cdot53$	$7\cdot197$
138	$2^2\cdot3\cdot5\cdot23$		$2\cdot691$	$3\cdot461$	$2^3\cdot173$	$5\cdot277$	$2\cdot3^2\cdot7\cdot11$	$19\cdot73$	$2^2\cdot347$	$3\cdot463$
139	$2\cdot5\cdot139$	$13\cdot107$	$2^4\cdot3\cdot29$	$7\cdot199$	$2\cdot17\cdot41$	$3^2\cdot5\cdot31$	$2^2\cdot349$	$11\cdot127$	$2\cdot3\cdot233$	
140	$2^3\cdot5^2\cdot7$	$3\cdot467$	$2\cdot701$	$23\cdot61$	$2^2\cdot3^3\cdot13$	$5\cdot281$	$2\cdot19\cdot37$	$3\cdot7\cdot67$	$2^7\cdot11$	
141	$2\cdot3\cdot5\cdot47$	$17\cdot83$	$2^2\cdot353$	$3^2\cdot157$	$2\cdot7\cdot101$	$5\cdot283$	$2^3\cdot3\cdot59$	$13\cdot109$	$2\cdot709$	$3\cdot11\cdot43$
142	$2^2\cdot5\cdot71$	$7^2\cdot29$	$2\cdot3^2\cdot79$		$2^4\cdot89$	$3\cdot5^2\cdot19$	$2\cdot23\cdot31$		$2^2\cdot3\cdot7\cdot17$	
143	$2\cdot5\cdot11\cdot13$	$3^3\cdot53$	$2^3\cdot179$		$2\cdot3\cdot239$	$5\cdot7\cdot41$	$2^2\cdot359$	$3\cdot479$	$2\cdot719$	
144	$2^5\cdot3^2\cdot5$	$11\cdot131$	$2\cdot7\cdot103$	$3\cdot13\cdot37$	$2^2\cdot19^2$	$5\cdot17^2$	$2\cdot3\cdot241$		$2^3\cdot181$	$3^2\cdot7\cdot23$
145	$2\cdot5^2\cdot29$		$2^2\cdot3\cdot11^2$		$2\cdot727$	$3\cdot5\cdot97$	$2^4\cdot7\cdot13$	$31\cdot47$	$2\cdot3^6$	
146	$2^2\cdot5\cdot73$	$3\cdot487$	$2\cdot17\cdot43$	$7\cdot11\cdot19$	$2^3\cdot3\cdot61$	$5\cdot293$	$2\cdot733$	$3^2\cdot163$	$2^2\cdot367$	$13\cdot113$
147	$2\cdot3\cdot5\cdot7^2$		$2^6\cdot23$	$3\cdot491$	$2\cdot11\cdot67$	$5^2\cdot59$	$2^2\cdot3^2\cdot41$	$7\cdot211$	$2\cdot739$	$3\cdot17\cdot29$
148	$2^3\cdot5\cdot37$		$2\cdot3\cdot13\cdot19$		$2^2\cdot7\cdot53$	$3^3\cdot5\cdot11$	$2\cdot743$		$2^4\cdot3\cdot31$	
149	$2\cdot5\cdot149$	$3\cdot7\cdot71$	$2^2\cdot373$		$2\cdot3^2\cdot83$	$5\cdot13\cdot23$	$2^2\cdot11\cdot17$	$3\cdot499$	$2\cdot7\cdot107$	
150	$2^2\cdot3\cdot5^3$	$19\cdot79$	$2\cdot751$	$3^2\cdot167$	$2^5\cdot47$	$5\cdot7\cdot43$	$2\cdot3\cdot251$	$11\cdot137$	$2^2\cdot13\cdot29$	$3\cdot503$
151	$2\cdot5\cdot151$		$2^3\cdot3^3\cdot7$	$17\cdot89$	$2\cdot757$	$3\cdot5\cdot101$	$2^2\cdot379$	$37\cdot41$	$2\cdot3\cdot11\cdot23$	$7^2\cdot31$
152	$2^4\cdot5\cdot19$	$3^2\cdot13^2$	$2\cdot761$		$2^2\cdot3\cdot127$	$5^2\cdot61$	$2\cdot7\cdot109$	$3\cdot509$	$2^3\cdot191$	$11\cdot139$
153	$2\cdot3^2\cdot5\cdot17$		$2^2\cdot383$	$3\cdot7\cdot73$	$2\cdot13\cdot59$	$5\cdot307$	$2^9\cdot3$	$29\cdot53$	$2\cdot769$	$3^4\cdot19$
154	$2^2\cdot5\cdot7\cdot11$	$3\cdot11\cdot47$	$2\cdot3\cdot257$		$2^3\cdot193$	$3\cdot5\cdot103$	$2\cdot773$	$7\cdot13\cdot17$	$2^2\cdot3^2\cdot43$	
155	$2\cdot5^2\cdot31$		$2^4\cdot97$		$2\cdot3\cdot7\cdot37$	$5\cdot311$	$2^2\cdot389$	$3^2\cdot173$	$2\cdot19\cdot41$	
156	$2^3\cdot3\cdot5\cdot13$	$7\cdot223$	$2\cdot11\cdot71$	$3\cdot521$	$2^2\cdot17\cdot23$	$5\cdot313$	$2\cdot3^3\cdot29$		$2^5\cdot7^2$	$3\cdot523$
157	$2\cdot5\cdot157$		$2^2\cdot3\cdot131$	$11^2\cdot13$	$2\cdot787$	$3^2\cdot5^2\cdot7$	$2^3\cdot197$	$19\cdot83$	$2\cdot3\cdot263$	
158	$2^2\cdot5\cdot79$	$3\cdot17\cdot31$	$2\cdot7\cdot113$		$2^4\cdot3^2\cdot11$	$5\cdot317$	$2\cdot13\cdot61$	$3\cdot23^2$	$2^2\cdot397$	$7\cdot227$
159	$2\cdot3\cdot5\cdot53$	$37\cdot43$	$2^3\cdot199$	$3^3\cdot59$	$2\cdot797$	$5\cdot11\cdot29$	$2^2\cdot3\cdot7\cdot19$		$2\cdot17\cdot47$	$3\cdot13\cdot41$
160	$2^6\cdot5^2$		$2\cdot3^2\cdot89$	$7\cdot229$	$2^2\cdot401$	$3\cdot5\cdot107$	$2\cdot11\cdot73$		$2^3\cdot3\cdot67$	
161	$2\cdot5\cdot7\cdot23$	$3^2\cdot179$	$2^3\cdot13\cdot31$		$2\cdot3\cdot269$	$5\cdot17\cdot19$	$2^4\cdot101$	$3\cdot7^2\cdot11$	$2\cdot809$	
162	$2^2\cdot3^4\cdot5$		$2\cdot811$	$3\cdot541$	$2^3\cdot7\cdot29$	$5^3\cdot13$	$2\cdot3\cdot271$		$2^2\cdot11\cdot37$	$3^2\cdot181$
163	$2\cdot5\cdot163$	$7\cdot233$	$2^5\cdot3\cdot17$	$23\cdot71$	$2\cdot19\cdot43$	$3\cdot5\cdot109$	$2^2\cdot409$		$2\cdot3^2\cdot7\cdot13$	$11\cdot149$
164	$2^3\cdot5\cdot41$	$3\cdot547$	$2\cdot821$	$31\cdot53$	$2^2\cdot3\cdot137$	$5\cdot7\cdot47$	$2\cdot823$	$3^3\cdot61$	$2^4\cdot103$	$17\cdot97$
165	$2\cdot3\cdot5^2\cdot11$	$13\cdot127$	$2^2\cdot7\cdot59$	$3\cdot19\cdot29$	$2\cdot827$	$5\cdot331$	$2^3\cdot3^2\cdot23$		$2\cdot829$	$3\cdot7\cdot79$
166	$2^2\cdot5\cdot83$	$11\cdot151$	$2\cdot3\cdot277$		$2^7\cdot13$	$3^2\cdot5\cdot37$	$2\cdot7^2\cdot17$		$2^2\cdot3\cdot139$	
167	$2\cdot5\cdot167$	$3\cdot557$	$2^3\cdot11\cdot19$	$7\cdot239$	$2\cdot3^3\cdot31$	$5^2\cdot67$	$2^2\cdot419$	$3\cdot13\cdot43$	$2\cdot839$	$23\cdot73$
168	$2^4\cdot3\cdot5\cdot7$	41^2	$2\cdot29^2$	$3^2\cdot11\cdot17$	$2^2\cdot421$	$5\cdot337$	$2\cdot3\cdot281$	$7\cdot241$	$2^3\cdot211$	$3\cdot563$
169	$2\cdot5\cdot13^2$	$19\cdot89$	$2^2\cdot3^2\cdot47$		$2\cdot7\cdot11^2$	$3\cdot5\cdot113$	$2^5\cdot53$		$2\cdot3\cdot283$	
170	$2^2\cdot5^2\cdot17$	$3^5\cdot7$	$2\cdot23\cdot37$	$13\cdot131$	$2^3\cdot3\cdot71$	$5\cdot11\cdot31$	$2\cdot853$	$3\cdot569$	$2^2\cdot7\cdot61$	
171	$2\cdot3^2\cdot5\cdot19$	$29\cdot59$	$2^4\cdot107$	$3\cdot571$	$2\cdot857$	$5\cdot7^3$	$2^2\cdot3\cdot11\cdot13$	$17\cdot101$	$2\cdot859$	$3^2\cdot191$
172	$2^3\cdot5\cdot43$		$2\cdot3\cdot7\cdot41$		$2^2\cdot431$	$3\cdot5^2\cdot23$	$2\cdot863$	$11\cdot157$	$2^6\cdot3^3$	$7\cdot13\cdot19$
173	$2\cdot5\cdot173$	$3\cdot577$	$2^2\cdot433$		$2\cdot3\cdot17^2$	$5\cdot347$	$2^3\cdot7\cdot31$	$3^2\cdot193$	$2\cdot11\cdot79$	$37\cdot47$
174	$2^2\cdot3\cdot5\cdot29$		$2\cdot13\cdot67$	$3\cdot7\cdot83$	$2^4\cdot109$	$5\cdot349$	$2\cdot3^2\cdot97$		$2^2\cdot19\cdot23$	$3\cdot11\cdot53$
175	$2\cdot5^3\cdot7$	$17\cdot103$	$2^3\cdot3\cdot73$		$2\cdot877$	$3^3\cdot5\cdot13$	$2^2\cdot439$	$7\cdot251$	$2\cdot3\cdot293$	
176	$2^5\cdot5\cdot11$	$3\cdot587$	$2\cdot881$	$41\cdot43$	$2^2\cdot3^2\cdot7^2$	$5\cdot353$	$2\cdot883$	$3\cdot19\cdot31$	$2^3\cdot13\cdot17$	$29\cdot61$
177	$2\cdot3\cdot5\cdot59$	$7\cdot11\cdot23$	$2^2\cdot443$	$3^2\cdot197$	$2\cdot887$	$5^2\cdot71$	$2^4\cdot3\cdot37$		$2\cdot7\cdot127$	$3\cdot593$
178	$2^2\cdot5\cdot89$	$13\cdot137$	$2\cdot3^4\cdot11$		$2^3\cdot223$	$3\cdot5\cdot7\cdot17$	$2\cdot19\cdot47$		$2^2\cdot3\cdot149$	
179	$2\cdot5\cdot179$	$3^2\cdot199$	$2^8\cdot7$	$11\cdot163$	$2\cdot3\cdot13\cdot23$	$5\cdot359$	$2^2\cdot449$	$3\cdot599$	$2\cdot29\cdot31$	$7\cdot257$
180	$2^3\cdot3^2\cdot5^2$		$2\cdot17\cdot53$	$3\cdot601$	$2^2\cdot11\cdot41$	$5\cdot19^2$	$2\cdot3\cdot7\cdot43$	$13\cdot139$	$2^4\cdot113$	$3^3\cdot67$
181	$2\cdot5\cdot181$		$2^2\cdot3\cdot151$	$7^2\cdot37$	$2\cdot907$	$3\cdot5\cdot11^2$	$2^3\cdot227$	$23\cdot79$	$2\cdot3^2\cdot101$	$17\cdot107$
182	$2^2\cdot5\cdot7\cdot13$	$3\cdot607$	$2\cdot911$		$2^5\cdot3\cdot19$	$5^2\cdot73$	$2\cdot11\cdot83$	$3^2\cdot7\cdot29$	$2^2\cdot457$	$31\cdot59$
183	$2\cdot3\cdot5\cdot61$		$2^3\cdot229$	$3\cdot13\cdot47$	$2\cdot7\cdot131$	$5\cdot367$	$2^2\cdot3^2\cdot17$	$11\cdot167$	$2\cdot919$	$3\cdot613$
184	$2^4\cdot5\cdot23$	$7\cdot263$	$2\cdot3\cdot307$	$19\cdot97$	$2^2\cdot461$	$3^2\cdot5\cdot41$	$2\cdot13\cdot71$		$2^3\cdot3\cdot7\cdot11$	43^2
185	$2\cdot5^2\cdot37$	$3\cdot617$	$2^2\cdot463$	$17\cdot109$	$2\cdot3^2\cdot103$	$5\cdot7\cdot53$	$2^6\cdot29$	$3\cdot619$	$2\cdot929$	$11\cdot13^2$
186	$2^2\cdot3\cdot5\cdot31$		$2\cdot7^2\cdot19$	$3^4\cdot23$	$2^3\cdot233$	$5\cdot373$	$2\cdot3\cdot311$		$2^2\cdot467$	$3\cdot7\cdot89$
187	$2\cdot5\cdot11\cdot17$		$2^4\cdot3^2\cdot13$		$2\cdot937$	$3\cdot5^4$	$2^2\cdot7\cdot67$		$2\cdot3\cdot313$	
188	$2^3\cdot5\cdot47$	$3^2\cdot11\cdot19$	$2\cdot941$	$7\cdot269$	$2^2\cdot3\cdot157$	$5\cdot13\cdot29$	$2\cdot23\cdot41$	$3\cdot17\cdot37$	$2^5\cdot59$	
189	$2\cdot3^3\cdot5\cdot7$	$31\cdot61$	$2^2\cdot11\cdot43$	$3\cdot631$	$2\cdot947$	$5\cdot379$	$2^3\cdot3\cdot79$	$7\cdot271$	$2\cdot13\cdot73$	$3^2\cdot211$

n	0	1	2	3	4	5	6	7	8	9
190	$2^2 \cdot 5^2 \cdot 19$		$2 \cdot 3 \cdot 317$	$11 \cdot 173$	$2^4 \cdot 7 \cdot 17$	$3 \cdot 5 \cdot 127$	$2 \cdot 953$		$2^2 \cdot 3^2 \cdot 53$	$23 \cdot 83$
191	$2 \cdot 5 \cdot 191$	$3 \cdot 7^2 \cdot 13$	$2^3 \cdot 239$		$2 \cdot 3 \cdot 11 \cdot 29$	$5 \cdot 383$	$2^2 \cdot 479$	$3^3 \cdot 71$	$2 \cdot 7 \cdot 137$	$19 \cdot 101$
192	$2^7 \cdot 3 \cdot 5$	$17 \cdot 113$	$2 \cdot 31^2$	$3 \cdot 641$	$2^2 \cdot 13 \cdot 37$	$5^2 \cdot 7 \cdot 11$	$2 \cdot 3^2 \cdot 107$	$41 \cdot 47$	$2^3 \cdot 241$	$3 \cdot 643$
193	$2 \cdot 5 \cdot 193$		$2^2 \cdot 3 \cdot 7 \cdot 23$		$2 \cdot 967$	$3^2 \cdot 5 \cdot 43$	$2^4 \cdot 11^2$	$13 \cdot 149$	$2 \cdot 3 \cdot 17 \cdot 19$	$7 \cdot 277$
194	$2^2 \cdot 5 \cdot 97$	$3 \cdot 647$	$2 \cdot 971$	$29 \cdot 67$	$2^3 \cdot 3^5$	$5 \cdot 389$	$2 \cdot 7 \cdot 139$	$3 \cdot 11 \cdot 59$	$2^2 \cdot 487$	
195	$2 \cdot 3 \cdot 5^2 \cdot 13$		$2^5 \cdot 61$	$3^2 \cdot 7 \cdot 31$	$2 \cdot 977$	$5 \cdot 17 \cdot 23$	$2^2 \cdot 3 \cdot 163$	$19 \cdot 103$	$2 \cdot 11 \cdot 89$	$3 \cdot 653$
196	$2^3 \cdot 5 \cdot 7^2$	$37 \cdot 53$	$2 \cdot 3^2 \cdot 109$	$13 \cdot 151$	$2^2 \cdot 491$	$3 \cdot 5 \cdot 131$	$2 \cdot 983$	$7 \cdot 281$	$2^4 \cdot 3 \cdot 41$	$11 \cdot 179$
197	$2 \cdot 5 \cdot 197$	$3^3 \cdot 73$	$2^2 \cdot 17 \cdot 29$		$2 \cdot 3 \cdot 7 \cdot 47$	$5^2 \cdot 79$	$2^3 \cdot 13 \cdot 19$	$3 \cdot 659$	$2 \cdot 23 \cdot 43$	
198	$2^2 \cdot 3^2 \cdot 5 \cdot 11$	$7 \cdot 283$	$2 \cdot 991$	$3 \cdot 661$	$2^6 \cdot 31$	$5 \cdot 397$	$2 \cdot 3 \cdot 331$		$2^2 \cdot 7 \cdot 71$	$3^2 \cdot 13 \cdot 17$
199	$2 \cdot 5 \cdot 199$	$11 \cdot 181$	$2^3 \cdot 3 \cdot 83$		$2 \cdot 997$	$3 \cdot 5 \cdot 7 \cdot 19$	$2^2 \cdot 499$		$2 \cdot 3^3 \cdot 37$	
200	$2^4 \cdot 5^3$	$3 \cdot 23 \cdot 29$	$2 \cdot 7 \cdot 11 \cdot 13$		$2^2 \cdot 3 \cdot 167$	$5 \cdot 401$	$2 \cdot 17 \cdot 59$	$3^2 \cdot 223$	$2^3 \cdot 251$	$7^2 \cdot 41$

C.13 DECIMAL EQUIVALENTS— INCHES AND MILLIMETERS

Inches			Inches						
Fraction	Decimals	Millimeters	Fraction	Decimals	Millimeters	mm	Inches	mm	Inches
1/64	0.015625	0.397	33/64	0.515625	13.097	.1	.0039	46	1.8110
1/32	0.03125	0.794	17/32	0.53125	13.494	.2	.0079	47	1.8504
3/64	0.046875	1.191	35/64	0.546875	13.891	.3	.0118	48	1.8898
1/16	0.0625	1.588	9/16	0.5625	14.288	.4	.0157	49	1.9291
5/64	0.078125	1.984	37/64	0.578125	14.684	.5	.0197	50	1.9685
3/32	0.09375	2.381	19/32	0.59375	15.081	.6	.0236	51	2.0079
7/64	0.109375	2.778	39/64	0.609375	15.478	.7	.0276	52	2.0472
1/8	0.1250	3.175	5/8	0.6250	15.875	.8	.0315	53	2.0866
9/64	0.140625	3.572	41/64	0.640625	16.272	.9	.0354	54	2.1260
5/32	0.15625	3.969	21/32	0.65625	16.669	1	.0394	55	2.1654
11/64	0.171875	4.366	43/64	0.671875	17.066	2	.0787	56	2.2047
3/16	0.1875	4.763	11/16	0.6875	17.463	3	.1181	57	2.2441
13/64	0.203125	5.159	45/64	0.703125	17.859	4	.1575	58	2.2835
7/32	0.21875	5.556	23/32	0.71875	18.256	5	.1969	59	2.3228
15/64	0.234375	5.953	47/64	0.734375	18.653	6	.2362	60	2.3622
1/4	0.2500	6.350	3/4	0.7500	19.050	7	.2756	61	2.4016
17/64	0.265625	6.747	49/64	0.765625	19.447	8	.3150	62	2.4409
9/32	0.28125	7.144	25/32	0.78125	19.844	9	.3543	63	2.4803
19/64	0.296875	7.541	51/64	0.796875	20.241	10	.3937	64	2.5197
5/16	0.3125	7.938	13/16	0.8125	20.638	11	.4331	65	2.5591
21/64	0.328125	8.334	53/64	0.828125	21.034	12	.4724	66	2.5984
11/32	0.34375	8.731	27/32	0.84375	21.431	13	.5118	67	2.6378
23/64	0.359375	9.128	55/64	0.859375	21.828	14	.5512	68	2.6772
3/8	0.3750	9.525	7/8	0.8750	22.225	15	.5906	69	2.7165
25/64	0.390625	9.922	57/64	0.890625	22.622	16	.6299	70	2.7559
13/32	0.40625	10.319	29/32	0.90625	23.019	17	.6693	71	2.7953
27/64	0.421875	10.716	59/64	0.921875	23.416	18	.7087	72	2.8346
7/16	0.4375	11.113	15/16	0.9375	23.813	19	.7480	73	2.8740
29/64	0.453125	11.509	61/64	0.953125	24.209	20	.7874	74	2.9134
15/32	0.46875	11.906	31/32	0.96875	24.606	21	.8268	75	2.9528
31/64	0.484375	12.303	63/64	0.984375	25.003	22	.8661	76	2.9921
1/2	0.5000	12.700	1	1.000	25.400	23	.9055	77	3.0315
						24	.9449	78	3.0709
						25	.9843	79	3.1102
						26	1.0236	80	3.1496
						27	1.0630	81	3.1890
						28	1.1024	82	3.2283
						29	1.1417	83	3.2677
						30	1.1811	84	3.3071
						31	1.2205	85	3.3465
						32	1.2598	86	3.3858
						33	1.2992	87	3.4252
						34	1.3386	88	3.4646
						35	1.3780	89	3.5039
						36	1.4173	90	3.5433
						37	1.4567	91	3.5827
						38	1.4961	92	3.6220
						39	1.5354	93	3.6614
						40	1.5748	94	3.7008
						41	1.6142	95	3.7402
						42	1.6535	96	3.7795
						43	1.6929	97	3.8189
						44	1.7323	98	3.8583
						45	1.7717	99	3.8976
								100	3.9370

1 mm = 0.03937 in; 0.001 in = 0.0254 mm.

Appendix D
SAMPLE PROBLEMS, MATHEMATICS PROGRAMS, AND MODERN HANDHELD CALCULATORS

Included in this appendix are samples of typical problems using the basic principles shown throughout this handbook. The samples shown cover the different sectors of mathematics including derivatives, maxima and minima, partial derivatives, differentiation, differentials, integration, and first- and second-order differential equations.

The sample problems are shown with all steps included, in order for the reader to follow the different procedures required to solve these problems. Also shown are the equations used in engineering economics, compound interest problems, percentages, vector analysis relationships, complex quantities, basic vector algebra, uses of the modern handheld calculators, samples from the PC program MathCad, and specialized programs used to solve highly intractable problems dealing with complex electrostatic field phenomena, such as corona discharge. Sources are given at the end of this section for obtaining these specialized computer programs.

Engineers and scientists who are required to keep accurate records of their calculations have available computer programs such as Mathcad 7, Mathematica, Mathtype 3.5, and a host of other highly effective programs developed for these purposes.

D.02 DERIVATIVES

The rate of change of a function is expressed by the derivative of the function.

Let $y = f(x)$ be a given function of one variable. Let x_1 be a chosen value of x, and let Δx be an increment of x to be added to x_1; then Δy will be the corresponding increment of y:

$$\Delta y = f(x_1 + \Delta x) - f(x_1)$$

Now form the increment ratio:

$$\frac{\Delta y}{\Delta x} = \frac{f(x_1 + \Delta x) - f(x_1)}{\Delta x}$$

Let $\Delta x \to 0$; then $\Delta y/\Delta x$ will usually approach a limit that is called the *derivative* of y with respect to x at the value $x = x_1$. Expressed in another manner, the derivative of a function $f(x)$ is the limit that is approached by the increment of $f(x)$ to the increment of x, when the increment of x approaches the limit 0.

The derivative of $y = f(x)$ with respect to x is symbolized in various ways:

$$\frac{dy}{dx} \quad \text{or} \quad D_x y \quad \text{or} \quad f'(x) \quad \text{or} \quad y' \quad \text{(1st derivative)}$$

$$\frac{d^2 y}{dx^2} \quad \text{or} \quad D_x^2 y \quad \text{or} \quad f''(x) \quad \text{or} \quad y'' \quad \text{(2d derivative)}$$

The derivative may be interpreted as

- *The slope of a curve:* If x and y are rectangular coordinates of a variable point, and if the function $y = f(x)$ represents a continuous curve, then the slope of the tangent line to the curve at a point $x = x_1$ is the value of the derivative of $f(x)$ at $x = x_1$.

- *The velocity of a moving particle:* If s represents the distance traveled by a particle moving in a straight line over time t, then a function $s = f(t)$ represents the law of motion, and the derivative of $f(t)$ at $t = t_1$ represents the velocity of the particle at the instant t_1.

For any function $y = f(x)$,

$$\text{Derivative} = D_x y' = \frac{dx}{dy} = y' \qquad y' = \text{slope of curve } f(x)$$

(1) Test for a Maximum

$y = f(x)$ is a maximum for $x = a$ if $f'(a) = 0$ and $f''(a) < 0$

(2) Test for a Minimum

$y = fx)$ is a minimum for $x = a$ if $f'(a) = 0$ and $f''(a) > 0$

(3) Test for a Point of Inflection

$y = f(x)$ has a point of inflection at $x = a$ if $f''(a) = 0$ and if $f''(x)$ changes sign as x increases through $x = a$

D.04 APPLICATIONS OF THE DERIVATIVE

(1) Maximum or Minimum

A container with planar sides and square ends is to be constructed from 50 square feet (ft^2) of material. What are the dimensions of the container (Fig. D.1) that will allow a maximum volume to be achieved?

Now if A_s = surface area and volume = $V = x^2 y$, then

$$A_s = x^2 + x^2 + 4xy$$

$$= 2x^2 + 4xy$$

$$50 = 2x^2 + 4xy$$

$$y = \frac{50 - 2x^2}{4x}$$

Restructuring the last equation, we get

$$y = \frac{50}{4x} - \frac{2x}{4}$$

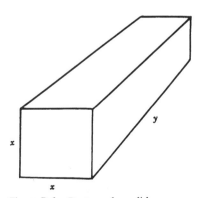

Figure D.1 Rectangular solid.

Next, we list the volume equation:

$$V = x^2 y$$

Substitute the expression for y back into the volume equation:

$$V = x^2 \left(\frac{50}{4x} - \frac{2x}{4} \right) = \frac{50}{4} - \frac{2x^3}{4} \qquad \frac{dv}{dx} = \frac{50}{4} - \frac{3x^2}{2}$$

Equate this expression to 0 and solve for x:

$$\frac{3x^2}{2} = \frac{50}{4} \qquad x = \sqrt{\frac{100}{12}} = 2.887$$

Substitute the value of x into the y equation above to solve for y:

$$y = \frac{50}{4x} - \frac{2x}{4} = \frac{50}{4(2.887)} - \frac{2(2.887)}{4} = 2.886$$

Substitute the calculated dimensions of x and y into the volume equation:

$$V = x^2 y = (2.887)^2 \times 2.886 = 24.054 \text{ ft}^3$$

It is interesting to note that this problem shows a cube to be the optimum six-sided solid for getting the most volume from a given amount of surface material.

(2) Rectilinear Motion

The motion of an object R along a straight line is described by an equation $S = f(t)$. If $t \geq 0$ is time and S is the distance of the object from the fixed point in its line of motion, then

$$\text{Velocity of } R \text{ at time } t = v = \frac{dS}{dt}$$

If $v < 0$, R is moving in the direction of decreasing distance S.

If $v > 0$, R is moving in the direction of increasing distance S.

If $v = 0$, R is instantaneously at rest.

The acceleration of R at time t is

$$a = \frac{dv}{dt} = \frac{d^2 S}{dt^2}$$

If $a > 0$, v is increasing.

If $a < 0$, v is decreasing.

If v and a have the same sign, the speed of R is increasing.

If v and a have opposite signs, the speed of R is decreasing.

Example: An object moves in a straight line according to the equation

$$S = \frac{1}{2} t^3 - 3.5t$$

Determine its velocity and acceleration at the end of 5 s.

Taking the first derivative, we obtain

$$\text{Velocity} = v = \frac{dS}{dt} = \frac{3}{2} t^2 - 3.5$$

When $t = 5$ s,

$$v = \frac{3}{2}(5)^2 - 3.5 = \frac{75}{2} - 3.5 = 34 \text{ ft/s}$$

Taking the second derivative, we have

$$\text{Acceleration} = a = \frac{dv}{dt} = 3t = 3(5) = 15 \text{ ft/s}^2$$

Note that the equation of motion in this example is for illustrative purposes only. In practice, the actual equation of motion must be determined or known before the problem can be solved.

The actual equation for motion of a body falling to earth is

$$S(t) = kt^2$$

where k is a constant due to earth's gravity and distance is a function of time. (We are neglecting air resistance.) In this case, $k = 16$ ft/s², or in SI units, 4.9 m/s².

Taking the first derivative of $S(t)$, we obtain

$$\frac{dS}{dt} = 2kt$$

Taking the second derivative, we have

$$\frac{dv}{dt} = 2k$$

With $k = 16$

$$\frac{dv}{dt} = 2k = 2(16) = 32 \text{ ft/s}^2$$

D.05 COMPUTING THE DERIVATIVE

Compute the average rate of change of $y = f(x) = x^2 - 2$ between $x = 3$ and $x = 4$.

Solution: The average rate of change is defined by

$$\frac{\Delta y}{\Delta x} \qquad \text{with} \qquad \Delta y = f(x + \Delta x) - f(x)$$

Given $x = 3$ and $\Delta x = 4 - 3 = 1$,

$$y = f(x) = f(3) = 3^2 - 2 = 7$$

For $x = 4$,

$$y + \Delta y = f(x + \Delta x) = 4^2 - 2 = 14$$

$$\Delta y = f(x + \Delta x) - f(x) = f(4) - f(3)$$

$$= (4^2 - 2) - (3^2 - 2) = 14 - 7 = 7$$

$$\frac{\Delta y}{\Delta x} = \frac{7}{1} = 7 \qquad \text{average rate of change}$$

Find $D_x^2 y = d^2/dx^2$ by implicit differentiation when $4x^2 + 9y^2 = 36$.

Solution: Differentiating implicitly with respect to x, we obtain

$$8x + 18y \cdot D_x y = 0$$

from which

$$D_x y = -\frac{4x}{9y}$$

To find $D_x^2 y$, differentiate $D_x y$ above by applying the quotient rule:

$$D_x^2 y = \frac{9y(-4) - (-4x)(9 D_x y)}{81 y^2}$$

Simplify the previous equation and substitute:

$$D_x y = -\frac{4x}{9y} \qquad D_x^2 y = \frac{-36y + (36x)[-4x/(9y)]}{81 y^2} = \frac{-36y^2 - 16x^2}{81 y^3}$$

D.07 MAXIMA AND MINIMA

Locate the maxima and minima of $y = 2x^2 - 3x$.

Solution: To obtain a maximum or minimum, we find dy/dx, set it equal to 0, and solve for x, obtaining the critical points. We then use the second-derivative test to determine whether the critical value is a maximum or a minimum. Thus, we have

$$\frac{dy}{dx} = 4x - 3 \qquad \frac{d^2 y}{dx^2} = 4 = + \text{ (positive)}$$

and this indicates a minimum. Setting dy/dx equal to 0 and solving to locate the minimum, we find

$$\frac{dy}{dx} = 4x - 3 = 0 \qquad x = \frac{3}{4}$$

and we substitute in the original equation to obtain

$$y = 2\left(\frac{3}{4}\right)^2 - 3\left(\frac{3}{4}\right) = \frac{18}{16} - \frac{9}{4} = -\frac{9}{8}$$

Therefore, the minimum is at $x = \frac{3}{4}$, $y = -\frac{9}{8}$.

D.08 PARTIAL DERIVATIVES

When a function is expressed in terms of several variables rather than only one variable, the concept of the *partial derivative* is normally applicable. For example, if z is a function of x and y [that is, $f(x, y)$], then $\partial z/\partial x$ is the derivative of z with respect to x, with y treated as a constant, and $\partial z/\partial y$ is the derivative of z with respect to y, with x treated as a constant.

Therefore, other than treating the variables one at a time, while the others are considered constant momentarily, there is no fundamental difference between the partial derivative and the total derivative.

When we consider the total differential, however, an additional concept is involved. Where for one variable we have

$$dy = \left(\frac{dy}{dx}\right) dx$$

for two variables, the differential must take into account both. Thus

$$dz = \frac{\partial z}{\partial x} dx + \frac{\partial z}{\partial y} dy$$

The total differential of z is the sum of the partial derivatives, each multiplied by its proper differential. This concept can be extended from two variables to any number of variables.

Example: Given $z = 4x^3 - 3y^2 + 4xy$, find

$$\frac{\partial z}{\partial x} \quad \text{and} \quad \frac{\partial z}{\partial y}$$

Solution: When solving for the partial derivative $\partial z/\partial x$, we differentiate with respect to x, treating y as a constant. We therefore obtain

$$\frac{\partial z}{\partial x} = 12x^2 + 4y$$

Solving for $\partial z/\partial y$, we differentiate with respect to y, treating x as a constant, to find

$$\frac{\partial z}{\partial y} = -6y + 4x \quad \text{or} \quad (4x - 6y)$$

Note: The first step in finding second partial derivatives is to find the first partial derivatives.

(1) Sample Problem Using Partial Derivatives

The length, depth, and width of a rectangular container are each increasing at a rate of 3 inches per minute (in/min). Find the rate at which the volume of the container is increasing at the instant when the length is 5 feet (ft), the width is 3 ft, and the depth is 2 ft.

Solution: Let x denote the depth, w the width, and l the length. The volume of the container is

$$v = xwl \quad \text{volume of a rectangular solid}$$

Its rate of increase can be found by partially differentiating the volume expression, and all the functions of time are

$$\frac{dv}{dt} = \frac{d}{dt}\left[\frac{\partial y}{\partial x} dx + \frac{\partial v}{\partial y} dy + \frac{\partial y}{\partial z} dz\right] = wl\frac{dx}{dt} + xl\frac{dw}{dt} + xw\frac{dl}{dt}$$

We know that $dx/dt = dw/dt = dl/dt = 3$ in/min $= \frac{1}{4}$ ft/min, and we wish to find dv/dt at the instant that $x = 2$ ft, $w = 3$ ft, and $l = 5$ ft. Thus

$$\frac{dv}{dt} = \frac{1}{4}(3 \cdot 5 + 2 \cdot 5 + 2 \cdot 3) = \frac{1}{4}(31)$$

$$= 0.25 \cdot 31 = 7.75 \text{ ft}^3/\text{min}$$

A differential is a change that the dependent variable (y) assumes when the independent variable (x) undergoes an infinitesimal change. The differential is also another way of interpreting the derivative. If

$$y = f(x) \qquad \text{function}$$

$$\frac{dy}{dx} = f'(x) \qquad \text{derivative}$$

$$dy = f'(x)\, dx \qquad \text{differential form}$$

This differential form is merely the derivative with the two sides multiplied by dx.

When we have more than two variables, the procedure is more complex. With three variables x, y, and z, the differential of z is

$$dz = \frac{\partial z}{\partial x}\, dx + \frac{\partial z}{\partial y}\, dy$$

Here, the symbol $\partial z/\partial x$ is the partial derivative of z with respect to x, that is, the derivative of z with respect to x while we consider y as a constant. The partial derivative of z with respect to y is defined similarly, and the concept can be extended to any number of variables.

If we wish to find the change in a dependent variable, where the change in the independent variable is small but not *infinitesimal,* we can apply the same equation as an approximation:

$$\Delta z = \frac{\partial z}{\partial x}\, \Delta x + \frac{\partial z}{\partial y}\, \Delta y$$

where the Δ values represent small but not infinitesimal changes. This method of approximation of dz by Δz is very useful in the numerical calculation of changes in quantities. Typical differential problems are presented in the following paragraphs.

Given $y = 2x^3$, find dy (differential y). By definition, for the differential $y = f'(x)\, dx = (dy/dx)\, dx$, we find $f'(x) = 6x^2$ (first derivative of $2x^3$), Therefore $dy = 6x^2\, dx$.

Find the differential of $(x^2 + 3)^3$:

$$d[(x^2 + 3)^3] = 3(x^2 + 3)^{3-1} \cdot d(x^2 + 3)$$

$$= 3(x^2 + 3)^2 \cdot d(x^2)$$

$$= 3(x^2 + 3)^2 \cdot 2x\, dx$$

$$= 6x(x^2 + 3)^2\, dx$$

If $y = 2x^{3/2}$, what is the approximate change in y when x changes from 9 to 9.02?

Solution: The differential dy is found from

$$\frac{dy}{dx} = \frac{3}{2}(2)x^{3/2-1} \qquad dy = 3x^{1/2}\, dx$$

The numerical value of dy may be found by allowing $x = 9.00$ and $dx = 0.02$ in this equation:

$$dy = 3(9.00)^{1/2}(0.02) = 0.18$$

The exact change in y may be found from

$$2(9.02)^{3/2} - 2(9.00)^{3/2} = 2(27.090\ 05) - 2(27) = 0.1801$$

However, this is a more laborious method of finding dy.

By using differentials, find an approximation for the value of $2x^3 - 2x^2 + 3x - 1$, when $x = 2.995$.

Solution: We will consider the value 2.995 as the result of applying an increment of -0.005 to an original value of 3. Then $x = 3$ and $\Delta x = -0.005 \approx dx$. Now, $dy/dx = 6x^2 - 4x + 3$ and $dy = (6x^2 - 4x + 3)\,dx$.

We now substitute $x = 3$ and $dx = -0.005$, obtaining

$$dy = [6(3)^2 - 4(3) + 3](-0.005) = -0.225$$

which is approximately the change in y caused by going from $x = 3$ to $x = 2.995$.

We must now find the value of y when $x = 3$ and add this value to dy. When $x = 3$ in the original equation, we obtain

$$dy = 2(3)^3 - 2(3)^2 + 3(3) - 1 = 44$$

Therefore, y at $x = 2.995 = y + dy = 44 + (-0.225) = 43.775$, which is the approximate value of the polynomial for $x = 2.995$.

To check, we substitute 2.995 in the original function $2x^3 - 2x^2 + 3x - 1$:

$$2(2.995)^3 - 2(2.995)^2 + 3(2.995) - 1 = 43.775$$

Using the differential in applications such as the preceding example provides very precise results.

D.10 DOUBLE OR ITERATED INTEGRALS

The double integral is often used in complex problems involving areas, volumes, and moments. The basic form of the double integral is

$$\int_a^b \int_{g_1(x)}^{g_2(x)} f(x, y)\, dy\, dx$$

where the variable of the first differential (y) is integrated first. While this is done, x is considered a constant. After integration with respect to y, the limits $g_2(x)$ and $g_1(x)$ are substituted. This results in an equation in x and dx alone, which is then integrated, and limits a and b are substituted for x.

The roles of the variables may be reversed, in which case the preceding integral can be written as

$$\int_c^d \int_{h_2(y)}^{h_1(y)} f(x, y)\, dx\, dy$$

In this case, integration is performed with respect to x first, and $h_1(y)$ and $h_2(y)$ are the limits. The constants c and d are the limits of y.

The end results are the same, and the choice of sequence depends on the complexity of the integration for either possibility. The choice of the order of integration is of no consequence, except that it may affect the amount of work required to solve the problem.

(1) Double-Integral References

Integrated areas in cartesian coordinates are

$$A = \int_R \int dy\, dx = \int_a^b dx \int_{f(x)}^{F(x)} dy$$

Integrated areas in polar coordinates are

$$A = \int \int p\, dp\, d\theta = \int_{\theta_1}^{\theta_2} d\theta \int_{f(\theta)}^{F(\theta)} p\, dp$$

(2) Fundamental Theorem of Calculus

When f is continuous on (a, b), let $F(x) = \int f(x)\,dx$. Then

$$\int_a^b f(x)\,dx = F(b) - F(a)$$

To evaluate a definite integral by use of the fundamental theorem given above, an antiderivative is required.

Example: Calculate the area projected from the parabolic curve (Fig. D.2) whose equation is $x = y^2 + 1$ to the Y axis between $y = -2$ and $y = 2$.

$$A = \int_{-2}^{2} f(y^2 + 1)\,dy$$

$$= \left[\frac{y^3}{3} + y\right]_{-2}^{2} = \left[\frac{2^3}{3} + 2\right] - \left[\frac{(-2)^3}{3} + (-2)\right]$$

$$= \left(\frac{8}{3} + 2\right) - \left(\frac{-8}{3} - 2\right) = \frac{28}{3} = 9.333 \text{ in}^2$$

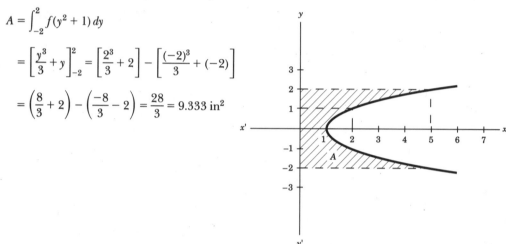

Figure D.2

D.11 INTEGRATION EXAMPLES

Example 1: Find the length of the curve $y = x^{3/2}$ from $x = 0$ to $x = 4$.

Solution: The equation for the arc length with respect to the x axis is

$$s = \int_a^b \sqrt{1 + \left(\frac{dy}{dx}\right)^2}\,dx$$

Since $y = x^{3/2}$, $y' = (3/2)x^{1/2}$ (first derivative),

$$s = \int_0^4 \sqrt{1 + \frac{9}{4}x}\,dx$$

$$= \frac{1}{2}\int_0^4 \sqrt{4 + 9x}\,dx = \left(\frac{1}{9}\right)\left(\frac{1}{2}\right)\left(\frac{2}{3}\right)(4 + 9x)^{3/2}\Big]_0^4$$

$$= \frac{1}{27}(40^{3/2} - 4^{3/2}) = \frac{8}{27}(10^{3/2} - 1) = 9.0734$$

Example 2: An extension spring is 6 inches (in) long and is stretched 1 in by a 5-pound (lb) weight. What is the work done when the spring is stretched from 8 to 11 in?

Solution: The force varies with the stretch of the spring. Therefore, $F = ks$ and 5 lb $= k \cdot 1$ in. (A 5-lb force stretches the spring 1 in; thus the spring has a rate of 5 lb/in.) Then $k = 5$ and $F = 5s$, $dW = F \cdot ds$.

A differential of work equals force times differential of stretch, or

$$dW = 5s \cdot ds$$

and

$$W = 5 \int_2^5 s \, ds$$

When the spring is 8 in long, the stretch is 2 in; and when it is 11 in long, the stretch is 5 in. Therefore, the limits of integration are 2 and 5.

$$W = \left| 5 \cdot \frac{s^2}{2} \right|_2^5 = 5 \cdot \frac{25}{2} - 5 \cdot \frac{4}{2} = 52.5 \text{ in} \cdot \text{lb}$$

In mechanical engineering practice, mathematicians have made the solution to a problem such as that shown above easier, by using the simple equation

$$P_e = \int_{s_1}^{s_2} Rs \, ds = \boxed{\frac{1}{2} R(s_2^2 - s_1^2)}$$

where R = spring rate, 5 lb/in
s_2 = final distance deflected, 5 in
s_1 = initial distance deflected, 2 in

From this we may calculate the energy required to stretch the spring as shown in the above integration problem, where 5 and 2 are the limits of stretch.
  Substituting values for R, s_2, and s_1 in the mechanical engineering equation gives

$$0.5(5)(5^2 - 2^2) = 2.5(25 - 4) = 52.5 \text{ in} \cdot \text{lb}$$

As can be seen, the answers to both problems are identical.

Example 3:   Find the equation, in polar coordinates, of the curve through the point $(a, \pi/4)$ from which the following relation is derived:

$$dr/d\theta = \frac{a^2}{r} \cos 2\theta$$

Integrating gives

$$\frac{r^2}{2} = \frac{a^2}{2} \sin 2\theta + C$$

where C is the arbitrary constant of integration. This expression represents a family of curves, any one of which can relate to the original given equation. But only one of these curves passes through the given point $(a, \pi/4)$. Therefore, substituting $(a, \pi/4)$ into the equation gives

$$\frac{a^2}{2} = \frac{a^2}{2} \sin \frac{\pi}{2} + C$$

from which $C = 0$. Therefore, the desired equation is

$$r^2 = a^2 \sin 2\theta$$

(1) Prismoidal Area Formula

For calculating approximate areas under curves (Fig. D.3), we have

$$A = \frac{h}{3}(y_1 + 4y_2 + y_3)$$

where A is exact only for third-degree or lower polynomials $y = f(x)$ and approximate for other-degree polynomial curves.

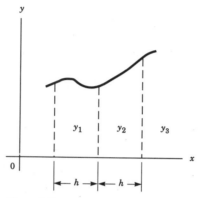

Figure D.3

(2) Simpson's Rule (for Areas)

Divide the required area (Fig. D.4) by an even number of vertical strips (the y ordinates). The more strips, the more accuracy. Then, if $y_1, y_2, y_3, \ldots, y_n$ = the values of the ordinates forming the boundaries of the strips, and Δx = distance between strips and ordinates,

$$A = \frac{\Delta x}{3}(y_1 + 4y_2 + 2y_3 + 4y_4 + 2y_5 + \cdots + 2y_{n-1} + 4y_n + y_{n+1})$$

The equation of the curve $y = f(x)$ does not have to be known if the y ordinates are known or measured.

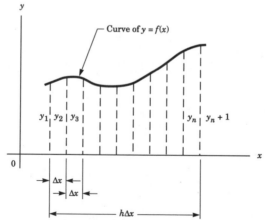

Figure D.4

(3) Approximate-Volume Calculations

Referring to Fig. D.5, if A_1, A_2, and A_3 equal areas of sections, then

$$A = \frac{h}{3}(A_1 + 4A_2 + A_3)$$

In practice, the area of the complex section is calculated with a mechanical integrating instrument, called a *polar planimeter*. This device is accurate enough for most work.

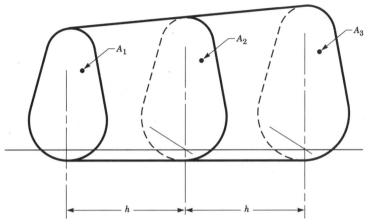

Figure D.5

D.13   FIRST-ORDER DIFFERENTIAL EQUATION

Find the general solution of

$$\frac{dy}{dx} = x^2 y^3$$

Solution: Separating the variables, we can write the given differential equation as

$$\frac{dy}{y^3} = x^2\, dx$$

The general solution may be obtained by integrating as shown:

$$\int \frac{dy}{y^3} = \int x^2\, dx + C \qquad \text{or} \qquad -\frac{1}{2y^2} = \frac{x^3}{3} + C$$

This may be rearranged, yielding

$$2y^2 x^3 + Cy^2 + 3 = 0$$

This equation satisfies the given differential equation, because differentiation will lead to

$$\frac{dy}{dx} = x^2 y^3$$

Solve the differential equation

$$y'' - 2y' - y = 0$$

Solution: This type of differential equation is called a *second-order, linear, homogeneous differential equation* with constant coefficients. We set $y = e^{mx}$ to see if it is a solution of the given equation, and since $y' = me^{mx}$ and $y'' = m^2 e^{mx}$, substitution yields

$$y'' - 2y' - y = e^{mx}(m^2 - 2m - 1) = 0 \tag{1}$$

In a finite range of x, e^{mx} is different from 0 unless m is carefully chosen. Thus, the only way for (1) to be valid is for

$$m^2 - 2m - 1 = 0$$

The roots are

$$m = 1 \pm \sqrt{2}$$

Thus, either

$$y = c_2 e^{(1+\sqrt{2})x} \qquad \text{or} \qquad y = c_2 e^{(1-\sqrt{2})x}$$

solves the equation. The general solution is the sum of the two terms:

$$y = c_1 e^{(1+\sqrt{2})x} + c_2 e^{(1-\sqrt{2})x}$$

D.15 ENGINEERING ECONOMICS

(1) Compound Interest

Where A_f = final value, P_s = amount saved, r = interest rate per year, $r/12$ = monthly rate, $r/365$ = daily rate, and n = term (years with r, months with $r/12$, and days with $r/365$),

$$A_f = P_s(1 + r)^n$$

Example: A person places \$5 000 in a savings account that pays $5\frac{1}{4}$ percent yearly interest. How much money will be in the account after 15 years?

$$A_f = 5\,000(1 + 0.055)^{15} = 5\,000(2.232\,48) = \$11\,162.40$$

(2) Amortization (Sinking Funds)

Where P_L = loan amount, n = number of months to pay the loan, r = yearly rate/12, and R = monthly payment,

$$P_L = R\frac{1 - (1 + r)^{-n}}{r}$$

Example: A person buys an automobile, and the total loan amount is \$7 500, to be paid in 48 monthly payments. The interest rate is $17\frac{1}{2}$ percent per year. What are the monthly payments?

$$7\,500 = R\frac{1 - (1 + 0.014\,583\,3)^{-48}}{0.014\,583\,3} \qquad R = \frac{7\,500}{34.347\,4} = \$218.36 \text{ per month}$$

Home loans are calculated using the above equation, as are other types of standard loans.

Factor Name	Converts	Symbol	Formula
Single payment compound amount	*P to F*	*(F/P, i%, n)*	$(1+i)^n$
Single payment present worth	*F to P*	*(P/F, i%, n)*	$(1+i)^{-n}$
Uniform series sinking fund	*F to A*	*(A/F, i%, n)*	$\dfrac{i}{(1+i)^n-1}$
Capital recovery	*P to A*	*(A/P, i%, n)*	$\dfrac{i(1+i)^n}{(1+i)^n-1}$
Uniform series compound amount	*A to F*	*(F/A, i%, n)*	$\dfrac{(1+i)^n-1}{i}$
Uniform series present worth	*A to P*	*(P/A, i%, n)*	$\dfrac{(1+i)^n-1}{i(1+i)^n}$
Uniform gradient ▲ present worth	*G to P*	*(P/G, i%, n)*	$\dfrac{(1+i)^n-1}{i^2(1+i)^n}-\dfrac{n}{i(1+i)^n}$
Uniform gradient ● future worth	*G to F*	*(F/G, i%, n)*	$\dfrac{(1+i)^n-1}{i^2}-\dfrac{n}{i}$
Uniform gradient ■ uniform series	*G to A*	*(A/G, i%, n)*	$\dfrac{1}{i}-\dfrac{n}{(1+i)^n-1}$

Definitions

A = Uniform amount per interest period
$B.V.$ = Book value
D_j = Depreciation in year j
F = Future worth, value or amount
G = Uniform gradient amount per interest period
i = Interest rate per interest period
i_e = Annual effective interest rate (see next column)
m = Number of compounding periods per year
n = Number of compounding periods
    in the expected life of the asset
P = Present worth, value or amount
r = Nominal annual interest rate

▲ $=P/G = (F/G)/(F/P) = (P/A) \cdot (A/G)$
● $=F/G = [(F/A) - n)]/i = (F/A) \cdot (A/G)$
■ $=A/G = [1-n(A/F)]/i$

Annual Effective Interest Rate

$$i_e = \left(1+\frac{r}{m}\right)^m - 1$$

Book Value

$$B.V. = initial\ cost - \Sigma D_j$$

(3) Annuities

A sum of $80 000 is in an account that pays $7\frac{1}{2}$ percent annual interest. What annuity can be paid out of this sum for 15 years at monthly intervals before the sum is fully depleted?

In this case,

$$A_m = \frac{P_s r (1 + r)^n}{(1 + r)^n - 1}$$

where A_m = monthly payout, P_s = $80 000, r expressed as a monthly rate = 0.075/12 = 0.006 25, and $n = 15 \times 12 = 180$ months.

$$A_m = \frac{80\,000(0.006\,25)(1.006\,25)^{180}}{(1.006\,25)^{180} - 1} = \$741.61 \text{ per month}$$

Example: If an annuity of A per month is to be paid for X years, what amount of money P_a must be put into the account if interest is R percent annually? Using $r = R/12$ and $n = X(12)$,

$$P_a = A \frac{(1 + r)^n - 1}{(1 + r)^n r}$$

(4) IRAs and Long-Term Accounts

Setting aside a sum A at the beginning of each year at r percent yearly interest, the total value P_a of the account in n years will be

$$P_a = A \frac{(1 + r)[(1 + r)^n - 1]}{r}$$

In this equation, eliminating the first $1 + r$ term will yield the total amount P_a when the sum A is set aside at the end of each year.

If an account with amount P is increased or decreased by a sum A at the end of each year, then the account will be valued at P_a after n years.

$$P_a = P(1 + r)^n \pm A \frac{(1 + r)^n - 1}{r}$$

The second term in this equation is added or subtracted according to whether the sum A is added to or subtracted from the account at the end of each year.

When you use the listed equations for compound interest, be careful to use the correct interest rate, whether it be yearly, monthly, or daily. Note also that when monthly rates are used, n must be the number of months; when yearly rates are used, n must be the number of years. The above equations are mathematically exact. With these equations, loan payments, annuities, and savings can be calculated to the penny.

(5) Percentages

Let us compare two arbitrary numbers, 22 and 37, as an illustration.

$$\frac{37 - 22}{22} = 0.681\,8$$

The number 37 is thus 68.18 percent larger than 22. We can also say that 22 increased by 68.18 percent is 37.

$$\frac{37 - 22}{37} = 0.405\,4$$

The number 37 minus 40.5 percent of itself is 22. We can also say that 22 is 40.54 percent less than 37.

$$\frac{22}{37} = 0.594\ 6$$

The number 22 is 59.46 percent of 37.

Example: A spring is compressed to 417 pounds pressure, or load, and later decompressed to 400 pounds load. The percentage pressure drop is $(417 - 400)/417 = 0.040\ 8$, or 4.08 percent.

The pressure is then increased to 515 pounds. The percentage increase over 400 pounds is therefore $(515 - 400)/400 = 0.287\ 5$, or 28.75 percent.

Percentage problem errors are quite common, even though the calculations are simple. If you remember that the divisor is the number of which you want the percentage, either increasing or decreasing, the simple errors can be avoided.

D.16 FUNDAMENTAL VECTOR ANALYSIS RELATIONSHIPS AND COORDINATE TRANSFORMATIONS

Refer to Fig. D.6.

(1) Vector Relationships

Vector $\mathbf{r} = r\underline{/\theta}$ polar form

$\mathbf{r} = (a + jb)$ complex form

$\mathbf{r} = (a, b)$ rectangular form

Length of $\mathbf{r} = \sqrt{a^2 + b^2}$

$a = r\cos\theta$ $b = r\sin\theta$

$$\tan\theta = \frac{b}{a}$$

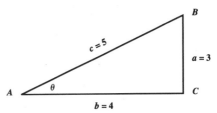

Figure D.6   Vector diagram.

(2) Coordinate Transformation

Polar ↔ rectangular: $r\underline{/\theta} = \sqrt{a^2 + b^2}\ \underline{\Big/\ \tan^{-1}\dfrac{b}{a}}$

Polar ↔ complex: $r\underline{/\theta} = a + jb = r\cos\theta + jr\sin\theta$

Example

$15\underline{/220°} = 15\cos 220° + 15j\sin 220°$ $\cos 220° = -\cos 40°$ $\sin 220° = -\sin 40°$

$15\underline{/220°} = -15\cos 40° - 15j\sin 40°$ $\cos 40° = 0.642\ 8$ $\sin 40° = 0.766\ 0$

$a = -11.5$ $b = -9.63$ $15\underline{/220°} = -11.5 - j9.63$

Answer

$15\underline{/220°} = (-11.5, -9.63) = -11.5 - j9.63$

Example

$$-3 + j^2 = \sqrt{-3^2 + 2^2} \bigg/ \tan^{-1}\left(\frac{2}{-3}\right) \qquad\qquad -3 + j^2 = 3.6/\tan^{-1}(-0.6667)$$

$$\tan^{-1}(-0.6667) = 33.7° \qquad 180° - 33.7° = 146.3° \qquad -3 + j^2 \text{ is in second quadrant}$$

Answer

$$-3 + j^2 = 3.6\underline{/146.3°}$$

D.17   COMPLEX QUANTITIES

(1) Imaginary Roots and Numbers

$$\sqrt{-1} = I \text{ or } j \qquad i^2 = -1 \qquad j^2 = -1$$

where j = active component in electric circuits and i = for use in nonelectrical work.

$$(a + jb) + (c + jd) = (a + c) + j(b + d) = (ac - bd) + j(bc + ad)$$

$$a + jb = \sqrt{a^2 + b^2} \cdot \epsilon^{j\theta} \qquad \text{where } \sqrt{a^2 + b^2} > 0$$

$$a + jb = \sqrt{a^2 + b^2}\,(\cos\theta + j\sin\theta) \qquad \sin\theta = \frac{b}{\sqrt{a^2 + b^2}}$$

$$\sin\theta = \frac{\epsilon^{j\theta} - \epsilon^{-j\theta}}{2j} = I_m[\epsilon^{j\theta}] \qquad \cos\theta = \frac{a}{\sqrt{a^2 + b^2}} \qquad \cos\theta = \frac{\epsilon^{j\theta} + \epsilon^{-j\theta}}{2} = \text{real}[\epsilon^{j\theta}]$$

$$(a + jb)(c + jd) = ac + j^2bd + j(ad + bc)$$

$$= ac - bd + j(ad + bc)$$

D.18   BASIC VECTOR ALGEBRA

For addition and subtraction, vectors must be in complex form.

Addition: $(a + jb) + (c + jd) = (a + c) + j(b + d)$

Subtraction: $(a + jb) - (c + jd) = (a - c) + j(b - d)$

Multiplication: $(a + jb)(c + jd) = ac + j(bc + ad) - bd \qquad (j^2 = -1)$

$$(r_1\underline{/\theta_1})(r_2\underline{/\theta_2}) = r_1 r_2\underline{/(\theta_1 + \theta_2)}$$

Division: $$\frac{(a + jb)}{(c + jd)} = \frac{(a + jb)}{(c + jd)} \cdot \frac{(c - jd)}{(c - jd)} = \frac{ac - j(bc + ad) + bd}{c^2 + d^2}$$

$$\frac{r_1\underline{/\theta_1}}{r_2\underline{/\theta_2}} = \frac{r_1}{r_2}\underline{/(\theta_1 - \theta_2)}$$

D.19   NEWTON'S METHOD FOR SOLVING INTRACTABLE EQUATIONS

This method of finding roots of difficult equations is accurate enough for most applications in engineering, in addition to being a powerful tool in mechanics.

The method is valid when there are no maxima and minima or inflection points between the intersection of the line $y = f(x)$ and the X axis and the x_1 point on the line.

To solve an equation such as

$$2.97x^3 - 2.075x^2 + 4.75x - 3.875 = 0$$

proceed as follows. Make a rough graph to find the approximate roots, or obtain the approximate roots using one of the handheld calculators, such as the Casio Fx-7000G or comparable Hewlett-Packard model. For the first approximation, we have

$$x_2 = x_1 - \frac{f(x_1)}{f'(x_1)}$$

where x_1 = first approximate root and x_2 = second approximate root. Solve for x_2 by substituting your first approximate root, say, 0.797 87, for x in the original equation. Substitute 0.797 87 for x in the first derivative of the original equation.

$$x_2 = x_1 - \frac{f(x)}{f'(x)} \qquad f(x) = 2.97(0.507\ 92) - 2.075(0.636\ 60) + 4.75(0.797\ 87) - 3.875$$

$$f'(x) = 8.91(0.636\ 60) - 4.15(0.797\ 87) + 4.75$$

$$x_2 = 0.797\ 87 - \frac{0.102\ 45}{7.110\ 95} = 0.783\ 46 \qquad \text{(2d approximation)}$$

Next, substitute 0.78346 back into the $f(x)$ and $f'(x)$ terms and solve for the third-root approximation x_3.

Continue the approximations until the desired accuracy is obtained. The general form of the Newton's method equation is

$$X_{n+1} = X_n - \frac{f(x_n)}{f'(x_n)}$$

For your information, the third-root approximation was 0.783 31. As a check, we will substitute this root back into the $f(x)$ function, $2.97x^3 - 2.075x^2 + 4.75x - 3.875 = 0$. If $x = 0.783\ 31$,

$$1.427\ 44 - 1.273\ 17 + 3.720\ 72 - 3.875 = 0 \qquad 0.000\ 010\ 00 \approx 0$$

This root is accurate to four decimal places.

D.20   FOUR-BAR LINKAGE

Refer to Sec. 2.04(e) for the standard Freudenstein equation which is the basis for the general solution and its variations. Below are the relational equations for solving the very important four-bar linkage used in many engineering mechanics and geometry problems.

Following the equations are examples of how these equations are entered into the latest generation of handheld calculators such as the Texas Instruments TI-85 and the Hewlett-Packard HP-48G. Both of these new-generation calculators operate much as small computers do, and both have enormous capabilities for solving general and very difficult engineering mathematics problems.

The general four-bar linkage equation

$$L_1 \cos \alpha - L_2 \cos \beta + L_3 = \cos(\alpha - \beta)$$

where

$$L_1 = \frac{a}{d} \qquad L_2 = \frac{a}{b} \qquad L_3 = \frac{b^2 - c^2 + d^2 + a^2}{2bd}$$

is the short form and is shown below in correct working form:

$$\frac{a}{d} \cos \alpha - \frac{a}{b} \cos \beta + \frac{b^2 - c^2 + d^2 + a^2}{2bd} = \cos(\alpha - \beta)$$

Transposing the above equation to solve for c, we obtain

$$c = \left\{ \left[-\cos(\alpha - \beta) + \frac{a}{d}\cos\alpha - \frac{a}{b}\cos\beta \right] 2bd + d^2 + b^2 + a^2 \right\}^{0.5}$$

The above equation must be entered into the modern handheld calculators, such as the TI-85 and HP-48G as shown, except that the brackets and braces shown here must be substituted by parentheses in the calculators. The calculator will give an error message unless the equation is correctly separated with parentheses according to the proper algebraic order of operations. On the TI-85 the above equation must appear as shown below:

$$C = (((-\cos(A - B) + (R/T)\cos A - (R/S)\cos B)(2ST)) + s^2 + T^2 + R^2)^{0.5}$$

Note: $A = \alpha$, $B = \beta$, $R = a$, $T = d$, and $S = b$. The TI-85 cannot show α, β, a, b, c, d; it can show only capital letters.

When we know angle α, angle β may be solved by [where $h^2 = a^2 + b^2 + 2ab \cos \alpha$ and $h = (a^2 + b^2 + 2ab \cos \alpha)^{0.5}$]

$$\beta = \cos^{-1}\frac{h^2 + a^2 - b^2}{2ha} + \cos^{-1}\frac{h^2 + d^2 - c^2}{2hd}$$

The transmission angle θ is therefore

$$\theta = \cos^{-1}\frac{c^2 + d^2 - a^2 - b^2 - 2ab \cos \alpha}{2cd}$$

In Fig. D.7, the driver link is b and the driven link is d. When driver link b moves through a different angle α, we may compute the final follower angle β and the transmission angle θ.

The equation for the follower angles β, shown previously, must be entered into the calculator as shown here:

$$\beta = \cos^{-1}((H^2 + R^2 - S^2)/(2HR)) + \cos^{-1}((H^2 + T^2 - K^2)/(2HT))$$

and the transmission angles θ must be entered as shown here:

$$\theta = \cos^{-1}((K^2 + T^2 - R^2 - S^2 - 2RS \cos A/(2KT))$$

As before, the capital letters must be substituted for actual equation letters as codes.

Note: In the preceding calculator entry form equations, the calculator exponent symbols ^ have been omitted for clarity, i.e.,

$$\cos^{-1}((K^{\wedge}2 + T^{\wedge}2 - R^{\wedge}2 - S^{\wedge}2 - 2RS \cos A)/(2KT))$$

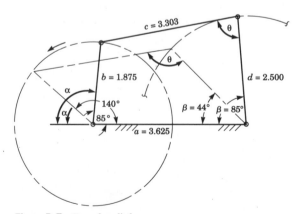

Figure D.7   Four-bar linkage.

It is therefore of great importance to learn the proper entry and bracketing form for equations used on the modern calculators, as illustrated in the preceding explanations.

When you work with trigonometry to solve plane triangles (see Chap. 3), solving the Mollweide equations on the TI-85 or HP-48G saves a substantial amount of time. The Mollweide equations (see Sec. 3.08) may be entered on these calculators in the following manner:

$$\frac{a - b}{c} = \frac{\sin[(A - B)/2]}{\cos(C/2)} \qquad \text{or} \qquad \frac{a + b}{c} = \frac{\cos[(A - B)/2]}{\sin(C/2)}$$

The Mollweide equation on the left above may be rearranged as

$$\frac{\sin[(A - B)/2]}{\cos(C/2)} - \frac{a - b}{c} = 0$$

and entered into the calculator as

$$(\sin((A - B)/2)/\cos(C/2)) - ((X - Y)/Z)$$

The TI-85 and the HP-48G cannot display a, b, and c, so we must substitute symbols as shown:

$$a = X$$

Sides: $\quad b = Y \qquad A, B,$ and C are angles

$$c = Z$$

Enter the three sides and three angles of the calculated triangle into the memory positions A, B, C, X, Y, and Z, and press Enter. The closer the answer is to zero, the more accurate are the calculated parts of the triangle we wish to check.

D.21 TYPICAL MATHCAD PRINTOUT

(1) Load on Compression Spring

$$G := 11500000 \qquad d := 0.125 \qquad D := 0.875 \qquad N := 17 \qquad c := \frac{D}{d} = 7$$

$$\frac{4 \cdot c - 1}{4 \cdot c - 4} + \frac{0.615}{c} = 1.213 \quad \text{(Wahl factor)} \qquad K := 1.213$$

$$\frac{G \cdot d^4}{8 \cdot N \cdot D^3} = 30.816 = (\text{RATE} = R) \qquad K = \text{Wahl factor} \qquad P := 35, 40 .. 110$$

P

$$\frac{8 \cdot K \cdot P \cdot D}{\pi \cdot d^3} = \text{stress, psi}$$

P	$\frac{8 \cdot K \cdot P \cdot D}{\pi \cdot d^3}$ = stress, psi	Load, lbf
35	$4.843 \cdot 10^4$	35
40	$5.535 \cdot 10^4$	40
45	$6.227 \cdot 10^4$	45
50	$6.919 \cdot 10^4$	50
55	$7.611 \cdot 10^4$	55
60	$8.303 \cdot 10^4$	60
65	$8.995 \cdot 10^4$	65
70	$9.687 \cdot 10^4$	70
75	$1.038 \cdot 10^5$	75
80	$1.107 \cdot 10^5$	80
85	$1.176 \cdot 10^5$	85**
90	$1.245 \cdot 10^5$	90
95	$1.315 \cdot 10^5$	95
100	$1.384 \cdot 10^5$	100
105	$1.453 \cdot 10^5$	105
110	$1.522 \cdot 10^5$	110

P = load on spring in lbf in increments to show the stress in psi in the chart on the right, as the spring is loaded. 30.816 lb/in was the spring rate shown above, and the Wahl stress factor was 1.213. The allowable stress on 0.125 dia. wire or music wire, ASTM A228, is .45 × 261,000 = 117,450. You may select the maximum load that can be placed on this spring by analyzing the chart on the right, for max. stress under different loads. 85 lbf load is the maximum that can be placed on this spring. Note that you may change any of the variables shown above to arrive at different spring rates and stress levels under varying loads, using this basic setup for many applications. (See Fig. D.8.)

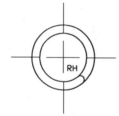

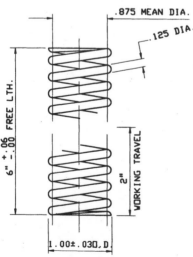

SPRING DATA

TYPE : COMPRESSION

O.D.-1.00±.03"

WIRE DIA.-.125"
NO. OF COILS -
19 TOTAL
17 ACTIVE
MATERIAL -MUSIC WIRE
    (ASTM-A228)

FREE LTH.6" +.06
          -.00
RATE-31LBS/IN ± 10% & 2" DEFL.
ENDS-CLOSED & GROUND
WIND-RIGHT HAND HELIX

Figure D.8   A typical compression spring showing engineering design data.

(2) Surface Plot

$$f(x, y) := \sin[x^2 + y^2] \qquad N := 20 \qquad i := 0 .. N$$
$$j := 0 .. N$$

$$x_i := -1.5 + .15 \cdot i \qquad y_j := -1.5 + .15 \cdot j \qquad M_{i,j} := f[x_i, y_j]$$

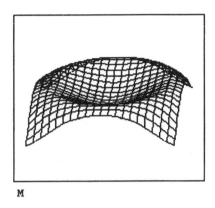

M

(3) Anharmonic Oscillations

Over a mesh of N points ...

$$N := 100 \qquad i := 1 .. N$$

... spanning the range from $-\alpha$ to α, where $\alpha := 3.1$

$$t_0 := 0 \qquad x_0 := \alpha \qquad x_i := \alpha \cdot \left[2 \cdot \frac{i}{N+1} - 1 \right]$$

We compute the times at which these points are reached by an oscillator having a cosine potential energy, and we plot their trajectory.

$$t_i := t_{i-1} + \frac{\alpha}{\sqrt{|\cos[x_i] - \cos(\alpha)|}}$$

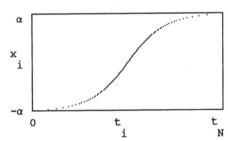

(4) Simulate Diffusion

Based on the diffusion equation ...

$$\frac{d}{dt^2} \cdot f(x, t) := \frac{d^2}{dx^2} \cdot f(x, t)$$

Approximate this differential equation with a difference equation

$$t := 0 .. 9 \qquad \text{... ranges over time increments}$$

$$x := 1 .. 49 \qquad \text{... ranges over space increments}$$

$$\alpha := .25 \qquad \text{... } \alpha \text{ diffuses to left and right in each time increment}$$

Initial conditions: time t = 0

$$f_{0,x} := 0 \qquad f_{0,0} := 0 \qquad f_{0,50} := 0 \qquad \ldots 0 \text{ everywhere but} \ldots$$

$$f_{0,25} := 1 \qquad \ldots 1 \text{ in the middle.}$$

Difference equation for diffusion:

$$f_{t+1,x} := f_{t,x} + \alpha \cdot [f_{t,x-1} - 2 \cdot f_{t,x} + f_{t,x+1}]$$

Now plot concentration at t = 0, t = 4, and t = 9.

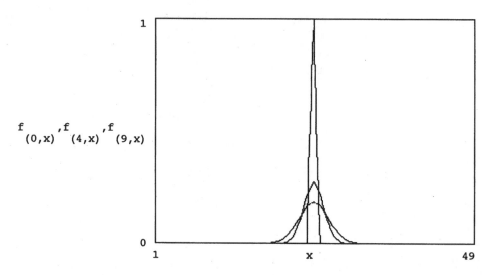

$$f_{(0,x)}, f_{(4,x)}, f_{(9,x)}$$

(5) Radioactive Decay

This document shows the results of radioactive decay. Element A decays with a half-life of thalfA into element B. Element B decays with a half-life of thalfB.

gm ≡ 1M   yr ≡ 1T

Initial conditions:

Initial amount of element A: A0 := 100 · gm

Half-lives: thalfA := 10 · yr
                                      thalfB := 100 · yr

Ratio: mass of 1 atom of A to
1 atom of B: α := 1.1

Compute decay constants:

$$kA := \frac{\ln(2)}{\text{thalfA}} \qquad kB := \frac{\ln(2)}{\text{thalfB}}$$

Decay functions:

$$A(t) := A0 \cdot e^{-(kA \cdot t)}$$

$$B(t) := -\alpha \cdot A0 \cdot e^{-(kB \cdot t)} \cdot [e^{-(kA \cdot t)} - 1]$$

Graph the mass of A and B over time:

t := 0 · yr, 2 · yr .. 200 · yr   t2 := 0 · yr, 10 · yr .. 200 · yr

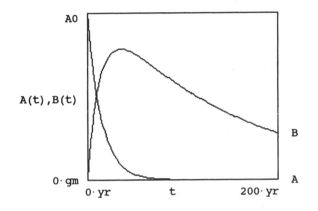

t2 yr	A(t2) gm	B(t2) gm
0	100	0
10	50	51.3
20	25	71.8
30	12.5	78.2
40	6.3	78.2
50	3.1	75.4
60	1.6	71.4
70	0.8	67.2
80	0.4	62.9
90	0.2	58.8
100	0.1	54.9
110	0	51.3
120	0	47.9
130	0	44.7
140	0	41.7
150	0	38.9
160	0	36.3
170	0	33.9
180	0	31.6
190	0	29.5
200	0	27.5

(6) Dwight's Break-Jaw Force Equation Application—Switching Devices, Electrical

$$c := 1 \qquad D := 1.62 \qquad r := 0.81 \qquad A := 1.81 \qquad B := 12 \qquad I := 40\,000, 50\,000 \;..\; 100\,000$$

$$\frac{I^2}{4.45 \times 10^7} \cdot 2.30 \cdot \log\left[\frac{2 \cdot A}{r}\right] - \frac{2}{3} - \frac{A}{2 \cdot B} - \frac{c^2}{6 \cdot A^2} + \frac{3 \cdot r^2}{20 \cdot A^2} + \frac{A^3}{24 \cdot B^3} + \frac{A \cdot c^2}{24 \cdot B^3} + 0$$

53.009 lbf	40 000 amp
83.255	50 000
120.223	60 000
163.912	70 000
214.322	80 000
271.454	90 000
335.308	100 000

Above table gives break-jaw force in pounds-force for peak currents from 40 000 to 100 000 A. The above equation is known as Dwight's break-jaw force equation for switching devices. Although the geometry of the switch pole for this equation is slightly different than Powercon's PIF switches, the equation's prediction of the break-jaw forces should be adequate for good results.

The closing energy available in a standard 15-kV PIF switch is 98 ft·lb. That is equal to 32.7 ft·lb/pole. If the break-jaw force is at its maximum when the main switch blades are 2 in from the jaw contact, and the peak current is 70 000 A, we will need 81 × 2 = 162 in·lb of energy to overcome the magnetic force and close the switch. The PIF has available $32.7 \times 12 = 392$ in·lb of energy per pole, so there should be no problem since we have available more than 2 times the required energy. The PIF switch closes so rapidly that each pole may never see the maximum magnetic force until after the switch is closed and latched, when the magnetic forces cannot open the blades since they are mechanically latched and locked in the closed position.

Due to the high-speed closing action of Powercon-type PIF switches, minimal damage is sustained on the main switch blades during high fault current close and latch operations. At the moment of arc-over during a fault-closing operation, the PIF 15-kV switch series blades are traveling at a velocity of 9.2 m/s or 30.2 rad/s angular velocity. If the arc is established when the blades are 2 in from the jaw contact, the arc time is only 0.005 sec. or ¼ cycle (Hertz). Peak current of the power system is never attained within the arc because the contacts close before the current wave can attain its maximum peak value during the arc time. Thus the arc power is maintained at a low value during the fault close action.

(7) Filtering a Noisy Signal with FFT (Fast Fourier Transform)

Define the signal:

$$i := 0 .. 127$$

$$q_i := \sin\left[\frac{i}{128} \cdot 14 \cdot \pi\right] + \cos\left[\frac{i}{128} \cdot 19 \cdot \pi\right]$$

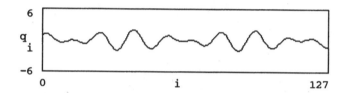

Add some noise:

$$s_i := q_i + \text{rnd}(2) - 1$$

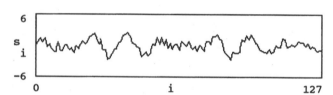

Take its discrete fourier transform:

$$f := \text{fft}(s) \qquad j := 0 .. 64$$

$$\alpha := 2.5 \qquad \text{... define threshold for spectral noise rejection.}$$

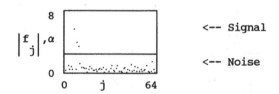

Filter, and take the inverse transform:

$$g_j := f_j \cdot \Phi[|f_j| - \alpha] \qquad h := ifft(g)$$

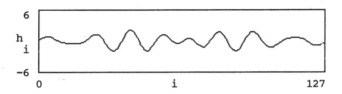

(8) Corona Test

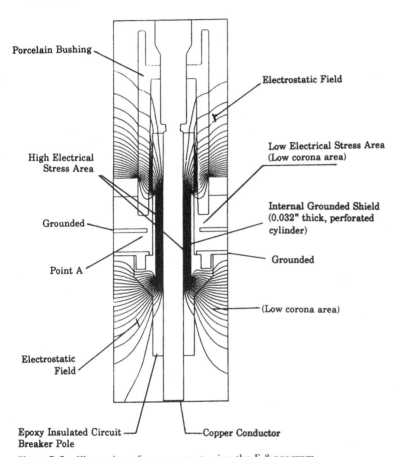

Figure D.9 Illustration of a corona test using the E-3 program.

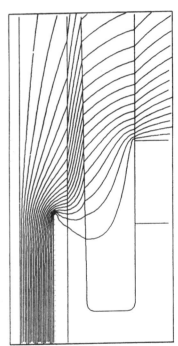

Figure D.10 Detail (enlarged view) of high electrical stress area of Fig. D.9.

Appendix E
REFERENCES AND BIBLIOGRAPHY

E.1 Algebra

1.01 Aitken, A. C.: "Determinants and Matrices," 8th ed., Interscience, New York, 1956.
1.02 Ayres, F., Jr.: "Matrices," McGraw-Hill, New York, 1962.
1.03 Birkhoff, G., and S. MacLane: "A Survey of Modern Algebra," Macmillan, New York, 1941.
1.04 Ferrar, W. L.: "Algebra," Oxford, London, 1941.
1.05 Frazer, R. A., W. J. Duncan, and A. R. Collar: "Elementary Matrices," Cambridge, London, 1938.
1.06 Lipschutz, S.: "Linear Algebra," McGraw-Hill, New York, 1968.
1.07 Middlemiss, R. R.: "College Algebra," McGraw-Hill, New York, 1952.
1.08 Smirnov, V. I.: "Linear Algebra and Group Theory," McGraw-Hill, New York, 1961.
1.09 Upensky, J. V.: "Theory of Equations," McGraw-Hill, New York, 1948.
1.10 Zurmühl, R.: "Matrizen," 3d ed., Springer, Berlin, 1961.

E.2 Geometry

2.01 Bronstein, I. N., and K. A. Semendjajew: "Taschenbuch der Mathermatik," 7th ed., Deutsch, Zurich, 1967.
2.02 Court, N. A.: "College Geometry," 2d ed., Barnes & Noble, New York, 1952.
2.03 Coxeter, H. S. M.: "Introduction to Geometry," Wiley, New York, 1961.
2.04 Klein, F.: "Famous Problems of Elementary Geometry," 2d ed., Dover, New York, 1956.

E.3 Trigonometry

3.01 Kells, L. M., W. F. Kern, and J. R. Bland: "Plane and Spherical Trigonometry," 3d ed., McGraw-Hill, New York, 1951.
3.02 Loney, S. L.: "Plane Trigonometry," Cambridge, London, 1900.

E.4 Plane Analytic Geometry

4.01 Carmichael, R. D., and E. R. Smith: "Mathematical Tables and Formulas," Dover, New York, 1962.
4.02 Cell, J. W.: "Analytic Geometry," 3d ed., Wiley, New York, 1954.
4.03 Middlemiss, R. R.: "Analytic Geometry," 2d ed., McGraw-Hill, New York, 1955.
4.04 Oakley, C. O.: "Analytic Geometry," Barnes & Noble, New York, 1961.

E.5 Space Analytic Geometry

5.01 Albert, A.: "Solid Analytic Geometry," McGraw-Hill, New York, 1949.
5.02 Bartsch, H. J.: "Mathematische Formeln," 5th ed., Veb Fachbuchverlag, Leipzig, 1966.
5.03 Láska, W.: "Sammlung von Formeln der Mathematik," Vieweg, Braunschweig, 1894.
5.04 Tideström, S. M.: "Manuel de Base de L'ingénieur," Dunod, Paris, 1959.

E.6 Elementary Functions

6.01 Abramowitz, M., and I. A. Stegun: "Handbook of Mathematical Functions," National Bureau of Standards, Washington, D.C., 1964.
6.02 Hayashi, K.: "Fünfstellige Tafeln der Kreis- und Hyperbelfunktionen," De Gruyter, Berlin, 1955.

E.7 Differential Calculus

7.01 Courant, R.: "Differential and Integral Calculus," vol. 1., Interscience, New York, 1936.
7.02 Goursat, E.: "A Course in Mathematical Analysis," vol. 1., Dover, New York, 1959.
7.03 Guggenheim, H. W.: "Differential Geometry," McGraw-Hill, New York, 1963.
7.04 Vojtěch, J.: "Základy Matematiky," vol. 1., Jednota ČMF, Prague, 1946.
7.05 Yakolev, K. P.: "Handbook for Engineers," vol. 1., Pergamon, New York, 1965.

E.8 Infinite Series

8.01   Fort, T.: "Infinite Series," Oxford, New York, 1930.
8.02   Jolley, L. B. W.: "Summation of Series," 2d ed., Dover, New York, 1961.
8.03   Knopp, K.: "Theory and Application of Infinite Series," Hafner, New York, 1948.

E.9 Integral Calculus

9.01   Apostol, T. M.: "Mathematical Analysis," Addison-Wesley, Reading, Mass., 1957.
9.02   Bartle, R. G.: "The Elements of Real Analysis," Wiley, New York, 1964.
9.03   Buck, R. C.: "Advanced Calculus," 2d ed., McGraw-Hill, New York, 1965.
9.04   Courant, R.: "Differential and Integral Calculus," Interscience, New York, 1936.
9.05   Ernyei, W. J., Private communication, 1982.
9.06   Sokolnikoff, I. S.: "Advanced Calculus," McGraw-Hill, New York, 1939.
9.07   Taylor, A.: "Advanced Calculus," Ginn., Boston, 1955.
9.08   Thomas, G. B.: "Calculus and Analytic Geometry," Addison-Wesley, Reading, Mass., 1953.
9.09   Vojtěch, J.: "Základy Matematiky," vol. 2., Jednota ČMF, Prague, 1946.

E.10 Vector Analysis

10.01   Coffin, J. G.: "Vector Analysis," Wiley, New York, 1938.
10.02   Hawkins, G. A.: "Multilinear Analysis. . . ," Wiley, New York, 1963.
10.03   Margenau, H., and G. M. Murphy: "Mathematics of Physics and Chemistry," Van Nostrand, Princeton, N.J., 1943.
10.04   Whittaker, E. T., and G. N. Watson: "A Course of Modern Analysis," Macmillan, New York, 1944.

E.11 Functions of Complex Variable

11.01   Churchill, R. V.: "Complex Variables and Applications," 2d ed., McGraw-Hill, New York, 1960.
11.02   Goursat, E.: "A Course in Mathematical Analysis," vol. 2., part 1, Dover, New York, 1959.
11.03   Souders, M.: "Engineer's Companion," Wiley, New York, 1966.
11.04   Spiegel, M. R.: "Complex Variables," McGraw-Hill, New York, 1964.

E.12 Fourier Series

12.01   Churchill, R. V.: "Fourier Series and Boundary Value Problems," McGraw-Hill, New York, 1941.
12.02   Miller, K. S.: "Partial Differential Equations in Engineering Problems," Prentice-Hall, Englewood Cliffs, N.J., 1953.
12.03   Salvadori, M. G., and R. J. Schwartz: "Differential Equations in Engineering Problems," Prentice-Hall, Englewood Cliffs, N.J., 1954.
12.04   Sokolnikoff, I. S., and R. M. Redheffer: "Mathematics of Physics and Modern Engineering," 2d ed., McGraw-Hill, New York, 1966.
12.05   Tolstov, G. P.: "Fourier Series," Dover, New York, 1976.
12.06   Zygmund, A.: "Trigonometric Series," Dover, New York, 1955.

E.13 Higher Transcendent Functions

13.01   Andrews, L., C.: "Special Functions for Engineers and Applied Mathematicians," Macmillan, New York, 1985.
13.02   Davis, H.: "Tables of Higher Mathematical Functions," Principia, Bloomington, Ind., 1935.
13.03   Erdelyi, A.: "Higher Transcendental Functions," 3 vols., McGraw-Hill, New York, 1953–1954.
13.04   Janke, E., and F. Emde: "Tables of Functions," Dover, New York, 1945.
13.05   Lebedev, N. N.: "Special Functions and Their Applications," Dover, New York, 1972.
13.06   Magnus, W. F., Oberherringer, and R. P. Soni: "Formulas and Theorems for the Special Functions of Mathematical Physics," Springer, New York, 1966.
13.07   Sneddon, I. N.: "Special Functions of Mathematical Physics and Chemistry," Oliver and Boyd, Edinburgh, 1961.

E.14 Ordinary Differential Equations

14.01   Abramowitz, M., and I. A. Stegun: "Handbook of Mathematical Functions," National Bureau of Standards, Washington, D.C., 1964.

14.02   Dwight, H. B.: Tables of Integrals and Other Mathematical Data," 4th ed., Macmillan, New York, 1967.

14.03   Frank, R., and R. V. Mises: "Die Differential und Integral Gleichungen Der Mechanik und Physik," 2 vols., Dover, New York, 1961.

14.04   Ince, E. L.: "Ordinary Differential Equations," Dover, New York, 1956.

14.05   Kamke, E.: "Differential Gleichungen, Lösungsmethoden und Lösungen," 3d ed., Chelsea, New York, 1948.

14.06   MacLachlan, N. W.: "Bessel Functions for Engineers," Oxford, London, 1946.

14.07   Madelung, E.: "Die Mathematischen Hilfsmittel des Physikers," 7th ed., Springer, Berlin, 1964.

14.08   Sauer, R., and I. Szabo: "Mathematische Hilfsmittel des Ingenieurs," vol. I., Springer, Berlin, 1967.

14.09   Szabo, I.: "Hütte Mathematische Formeln und Tafeln," Ernst, Berlin, 1959.

14.10   Szegö, G.: "Orthogonal Polynomials," American Mathematics Society, New York, 1939.

E.15 Partial Differential Equations

15.01   Bender, L., and S. Orzag: "Advanced Mathematics for Engineers," McGraw-Hill, New York, 1978.

15.02   Carrier, G. F., and C. E. Pearson: "Partial Differential Equations," Academic, New York, 1976.

15.03   Korn, G. A., and T. M. Korn: "Mathematical Handbook for Scientists and Engineers," 2d ed., McGraw-Hill, New York, 1968.

15.04   Sommerfeld, A.: "Partial Differential Equations of Physics," Academic, New York, 1949.

15.05   Vladimirov, V. S.: "Equations of Mathematical Physics," Dekker, New York, 1971.

15.06   Zauderer, E.: Partial Differential Equations for Applied Mathematics," Wiley, New York, 1983.

E.16 Laplace Transforms

16.01   Churchill, R. V.: "Operational Mathematics," 3d ed., McGraw-Hill, New York, 1971.

16.02   Doetsch, G.: "Handbuch der Laplace-Transformation," 3 vols., Birkhäuser, Basel, 1950–1956.

16.03   Nixon, F.: "Handbook of Laplace Transforms," Prentice-Hall, Englewood Cliffs, N.J., 1960.

16.04   Tuma, J. J.: "Handbook of Structural and Mechanical Matrices," McGraw-Hill, New York, 1987.

E.17 Numerical Methods

17.01   Fröberg, C. E.: "Numerical Mathematics," Benjamin/Cumming, Reading, Mass., 1985.

17.02   Hamming, R.: "Numerical Methods for Scientists and Engineers," 2d ed., McGraw-Hill, New York, 1973.

17.03   Hildebrand, F. B.: "Introduction to Numerical Analysis," McGraw-Hill, New York, 1956.

17.04   Jain, M. K., S. R. K. Iyengar, and R. K. Jain: "Numerical Methods for Scientific and Engineering Computation," Halstead, New York, 1985.

17.05   Johnson, L. W., and R. D. Riess: "Numerical Analysis," 2d ed., Addison-Wesley, New York, 1982.

17.06   Salvadori, M. G., and M. L. Baron: "Numerical Methods in Engineering," 2d ed., Prentice-Hall, Englewood Cliffs, N.J., 1961.

17.07   Sanden, V.: "Praktische Mathematik," Teubner, Leipzig, 1953.

17.08   Scarborough, J. B.: "Numerical Mathematical Analysis," 3d ed., Johns Hopkins Press, Baltimore, 1955.

17.09   Yakolev, K. P.: "Handbook for Engineers," vol. 1., Pergamon, New York, 1965.

E.18 Probability and Statistics

18.01   Burington, R. S., and D. C. May: "Handbook of Probability and Statistics with Tables," Handbook Publishers, Sandusky, Ohio, 1953.

18.02   Cramer, H.: "Mathematical Methods of Statistics," Princeton University Press, Princeton, N.J., 1951.

18.03   Hald, A.: "Statistical Theory and Engineering Applications," Wiley, New York, 1952.

18.04   Korn, G. A., and T. M. Korn: "Mathematical Handbook for Scientists and Engineers," 2d ed., McGraw-Hill, New York, 1968.

18.05   Spiegel, M. R.: "Statistics," McGraw-Hill, New York, 1961.

E.19 Tables of Indefinite Integrals

19.01   Bois, G. P.: "Tables of Indefinite Integrals," Dover, New York, 1964.

19.02   Bronstein, I., and K. Semendjajev: "Pocketbook of Mathematics," 6th ed., Soviet Government Press, Moscow, 1956.

19.03   Burington, R. S.: "Handbook of Mathematical Tables and Formulas," 4th ed., McGraw-Hill, New York, 1965.

19.04   Dwight, H. B.: "Tables of Integrals and Other Mathematical Data," 4th ed., Macmillan, New York, 1961.

19.05   Gröbner, W. and N. Hofreiter: "Integraltafeln," vol. 1, 4th ed., Springer, Vienna, 1965.

19.06   Hirsch, M.: "Integral Tables," Baynes, London, 1823.

19.07   Meyer Zur Capellen, W.: "Integraltafeln," Springer, Berlin, 1950.

19.08   Spiegel, M. R.: "Mathematical Handbook of Formulas and Tables," McGraw-Hill, New York, 1968.

E.20 Tables of Definite Integrals

20.01   de Haan, B.: "Nouvelles Tables d'Intégrales Définies," Hafner, New York, 1957.

20.02   Gradshteyn, I. S., and I. M. Ryzhik: "Table of Integrals, Series and Products," 4th ed., Academic, New York, 1965.

20.03   Gröbner, W., and N. Hofreiter: "Integraltafeln," vol. 2, 4th ed., Springer, Vienna, 1965.

E.21 Plane Curves and Areas

21.01   Bruce, J. W., and P. J. Giblin: "Curves and Singularities," Cambridge University Press, Cambridge, 1984.

21.02   Coolidge, J. L.: "Treatise on Plane Curves," Dover, New York, 1959.

21.03   Lawrence, J. D.: "Catalog of Special Plane Curves," Dover, New York, 1972.

21.04   Tuma, J. J.: "Dynamics," Quantum, New York, 1974.

21.05   Tuma, J. J.: "Technology Mathematics Handbook," McGraw-Hill, New York, 1975.

21.06   Walker, R. J.: "Algebraic Curves," Dover, New York, 1950.

21.07   Zwikker, C.: "The Advanced Geometry of Plane Curves and Their Applications," Dover, New York, 1963.

E.22 Space Curves and Surfaces

22.01   Blaschke, W., and H. Reichard: "Vorlesungen über Differentialgeometrie," 2d ed., Springer, Berlin, 1960.

22.02   DoCarmo, M.: "Differential Geometry of Curves and Surfaces," Prentice-Hall, Englewood Cliffs, N.J., 1976.

22.03   Eisenhart, L. P.: "A Treatise on the Differential Geometry of Curves and Surfaces," Ginn, Boston, 1909.

22.04   Guggenheimer, H. W.: "Differential Geometry," McGraw-Hill, New York, 1963.

22.05   Kommerell, V., and K. Kommerell: "Vorlesungen über the Analytische Geometrie des Raumes," 3d ed., Teubner, Leipzig, 1953.

22.06   Struik, D. J.: "Lectures on Classical Differential Geometry," 2d ed., Addison-Wesley, Reading, Mass., 1961.

22.07   Tuma, J. J.: "Statics," Quantum, New York, 1974.

E.23 Computer Programs

23.01   "EMSS," Ansoft Corporation, Pittsburgh, Pa. 15219, (412) 261-3200. (For design and analysis of electromechanical systems, using finite element analysis and engineering equations.)

23.02   "E3," Field Precision, Albuquerque, NM, (505) 296-6689. (For solutions to electrostatic field problems. Also, other scientific programs for professionals.)

23.03   "Mathcad 7," MathSoft, Inc., 101 Main St., Cambridge, MA 02142, (617) 577-1017. (For scientific and engineering calculations, covering a large segment of mathematics. WYSIWYG printouts of the monitor screen for record purposes. Samples of the MathCad program are shown in App. D.)

23.04   Mathematica, Wolfram Research, Champaign, IL 61820, (217) 398-0700.

23.05   "Mathtype 3.5," Design Sciences (800) 827-0685. (A full-feature equation-writing program with plain TeX output.)

23.06   "Microsoft Word for Windows 95," Microsoft Corporation, Edmonds, Wash. (A word processor with excellent equation-writing features.)

23.07   TechWriter, a program for writing any types of equations and text. (Obsolete, for older DOS PCs.)

23.08   "WordPerfect 5.1" to "WordPerfect 7 for Windows 95," Corel Corporation, Ottawa, Canada (800) 772-6735. (A writing program with equation-writing capability which is similar to a desktop publishing program.)

E.24 Addenda

24.01   Ayres, F., and E. Mendelson, "Schaum's Outline of Differential and Integral Calculus," 3d ed., McGraw-Hill, New York, 1990.

24.02   Calter, P., "Schaum's Outline of Technical Mathematics," McGraw-Hill, New York, 1979.

24.03   Downing, D., "Calculus the Easy Way," Barron's Educational Series, New York, 1982.

24.04   Hicks, T., "Standard Handbook of Engineering Calculations," 3d ed. McGraw-Hill, New York, 1994.

24.05   Mendelson, E., "Schaum's Outline of Beginning Calculus," 2d ed., McGraw-Hill, New York, 1996.

24.06   Research and Education Association, "Calculus Problem Solver," Piscataway, N.J., 1994. (This is an extremely well done and valuable work on all branches of the calculus. It is excellent for university students as well as professionals who need an excellent reference of solved problems in advanced mathematics.)

Index

References are made to section numbers for the text material. Numbers preceded by the letters A and B refer to the Appendixes. In the designation of units and systems of units the following abbreviations are used:

avd	= avoirdupois	thm	= thermochemical
liq	= liquid	try	= troy(apothecary)
nat	= nautical	FPS	= English system
phs	=physical	IST	= international steam tables
stu	= statute	MKS	= metric system
tec	= technical	SI	= international system